Introduction to Plant Automation and Controls

Introduction to Plant Automation and Controls

Raymond F. Gardner

CRC Press is an imprint of the
Taylor & Francis Group, an **informa** business

First edition published 2021
by CRC Press
6000 Broken Sound Parkway NW, Suite 300
Boca Raton, FL 33487-2742

and by CRC Press
2 Park Square, Milton Park, Abingdon, Oxon OX14 4RN

ISBN 13: 978-0-367-49420-9 (hbk)

Library of Congress Cataloging-in-Publication Data
Names: Gardner, Raymond F., author.
Title: Introduction to plant automation and controls / Raymond F. Gardner.
Description: First edition. | Boca Raton : CRC Press, 2021. | Includes index.
Identifiers: LCCN 2020020498 (print) | LCCN 2020020499 (ebook) | ISBN 9780367494209 (hbk) |
ISBN 9780367560928 (paperback) | ISBN 9781003091134 (ebk)
Subjects: LCSH: Production control – Automation.
Classification: LCC TS155.8 .G37 2021 (print) | LCC TS155.8 (ebook) | DDC 670.42/75 – dc23
LC record available at https://lccn.loc.gov/2020020498
LC ebook record available at https://lccn.loc.gov/2020020499

Visit the Taylor & Francis Web site at
http://www.taylorandfrancis.com

and the CRC Press Web site at
http://www.crcpress.com

Contents

Preface

In the early days of power generation, steam engines were driven by hand-stoked coal-fired boilers, whose controls were mechanical in nature and rudimentary by today's standards. Power plants required significant manning, continuous monitoring, and human intervention to maintain safe and economical operation. Many early control systems used simple mechanical feedback loops that were functional, but the response left much to be desired with respect to speed, accuracy, autonomous adjustments, and reliability, all while trying to keep stable pace with fast-changing loads.

Over time, the mechanical controls were refined to incorporate sophisticated features that could anticipate the proper response during large load swings. These improvements reduced setpoint errors with quicker action. Eventually, proportional, integral, and derivative continuous-control schemes were combined with electric, pneumatic, and hydraulic devices to obtain high precision and stability. The electric and pneumatic controllers evolved over many decades and are still used in today's plant systems; although modern microprocessor-based programmable-logic controllers (PLCs) are now commonly used to accomplish the controlling functions, coupled with the electrical and pneumatic devices that provide the actuation.

Today, most plant automation systems use microprocessor-based PLCs to control individual devices, complete sub-systems, and even entire plants. These PLCs maintain extremely high levels of performance and reliability with minimal manning. In addition to the high degree of automation and the superior precision afforded by these systems, modern controllers acquire large amounts of data that are used for real-time performance calculations, historical trending, predictive maintenance, local and off-site monitoring of plant status, and the safe orderly shutdown of equipment during abnormal conditions. In some cases, engineers do not need to be present to operate, adjust, and troubleshoot plant equipment.

Chapters 1–4 begin with instrumentation, which is the starting point for essentially all automated systems. The chapters cover the theory and background for electronics, continuous-control fundamentals, digital control, and logic.

Chapters 5–10 provide an overview of equipment and associated control devices for plants that provide central heating, cooling, and power, as opposed to the types of control systems used for manufacturing applications. The goal is to provide the integration of instrumentation and controls with the automation of boilers, pumps, fans, heat exchangers, control valves, governors, and their related systems. The motor-controller chapter introduces relay logic, which is the basis for the ladder diagram (LD) programming language used for many industrial PLCs.

Chapters 11–19 cover PLC construction, operation, programming, distributed control systems (DCS), supervisory control and data acquisition systems (SCADA), and the networking that brings plant devices and systems together. An overview of Allen-Bradley RS Logix PLC ladder-diagram language is provided, for both digital and analog input and output signals, including the 4–20mA analog signal that is common with plants. An overview of the five IEC 61131–3 PLC languages, and an introduction to Industrial Control System (ICS) security is included. PLC and communications-related terms and definitions are included in Chapter 20.

Disclaimer

The information provided herein does not necessarily represent the view of the U.S. Merchant Marine Academy and Department of Transportation.

Although reasonable care has been taken in preparing this book, the author or publisher assumes no responsibility for any consequences of using the information. The text, diagrams, figures, technical data, and trade names presented herein are for illustration and education purposes only, and may be covered under patents. For design and analysis, the equipment manufacturers should be consulted for their exact data and operating recommendations. Manufacturer instructions and documentation should be used for all system designs and installations.

This book contains information obtained from authentic and highly regarded sources. Reasonable efforts have been made to publish reliable data and information, but the author and publisher cannot assume responsibility for the validity of all materials or the consequences of their use. The author and publisher have attempted to trace the copyright holders of all material reproduced in this publication and apologize to copyright holders if permission to publish in this form has not been obtained. If any copyright material has not been acknowledged, please write and let us know so we may rectify in any future reprint.

Acknowledgments

I would like to thank my friend Jim Smits of Green Bay, Wisconsin. His proactiveness, self-sacrifice, dedication, and assistance brought the instruction of PLC-based controls at the United States Merchant Marine Academy into the twenty-first century. Jim's proactive involvement with the Kings Point engineering program has made this book possible. Go Pack Go! I would also like to thank his colleague Zac Everard and the Professional Control Solutions company for their development of the PLC-based laboratory equipment at Kings Point, and Mr. Daniel Meehan, KP Class of 1952, for his equipment donations to the engineering PLC lab.

I thank Capt. Ann Sanborn for her support in obtaining a sabbatical to work on this project, and especially, Capt. William J. Sembler for making the sabbatical happen despite being shorthanded of staff.

I thank Cornelius J. Breen and Kevin M. Duffy who are two of the most upstanding individuals I have had the pleasure of practicing engineering with. Their loyalty, support, and tolerance during this undertaking has been extraordinary. They are more than partners, they are great engineers and friends, whom I have come to count on.

I thank my parents, Joan and Leon, for a lifetime of love and encouragement, and my lovely wife, née Marianne Giovanniello, for her unconditional love and support. I could not have asked for a better, more loving life partner. Her kind encouragement and personal sacrifice has helped me through this process.

I am very appreciative of the contributions within this book by ABB Marine, AMOT Control Valves, Cadillac Meter, CIRCOR | Leslie, Eaton Corporation, E Craftsmen, Emerson, MTE Corporation, Flowserve Corporation, Fluke, Mitsubishi Electric Corporation, Texas Instruments, VAF Instruments, the United States Merchant Marine Academy, and Woodward Inc.

Thank you for the support provided by Ramboll, formerly O'Brien and Gere, and Schuyler Engineering.

It is my hope that this book provides a useful source of information that ties together the very broad topics of instrumentation, controls, and automation with their application to the safe, reliable, and efficient operation of various types of plants and equipment.

Raymond F. Gardner
Kings Point, New York, August 2020

Author

Raymond F. Gardner, P.E., is a professor of engineering at the U.S. Merchant Marine Academy at Kings Point, New York. He has more than forty years of design and hands-on engineering experience, including shipboard power plant and auxiliary system operations, design of large dc power supplies, and the design of mechanical, electrical, and control systems for central heating, cooling, and power plants. Mr. Gardner obtained his M.S. in Mechanical Engineering from Polytechnic University in Brooklyn, his B.S. in Marine Engineering from the U.S. Merchant Marine Academy, and he holds a U.S. Coast Guard Chief Engineer license.

Author

[illegible]

Chapter 1

Measurement fundamentals and instrumentation

Instrumentation is used to measure equipment performance, indicate plant operating parameters, and record trends, and instrumentation forms the starting point for automatic control systems. Since many plant processes cannot be directly observed, instrumentation provides the means for visualizing what the system is doing and how the system is behaving. Instrumentation is used for quantifying operating levels and performance, verifying efficiency, troubleshooting, and for comparing manufacturer's data or plant heat-balance parameters for making informed operational decisions. Instrumentation is key to proper plant operation. Typical parameters that are measured in plants include:

- Pressure
- Temperature
- Liquid level
- Flow rates
- RPM
- Viscosity
- Acceleration
- Smoke opacity
- Stack-gas oxygen content
- Electrical parameters, etc.

INSTRUMENTATION DEFINITIONS

Several terms are used to define instrumentation performance, including sensitivity, accuracy, precision/repeatability, and resolution. They are defined as follows:

- *Sensitivity*: is the change in output reading versus the change in input measurement. For instance, if an electronic pressure gage reads between 0-100psi and uses 4-20mA as the signal, the sensitivity is (20-4mA)/(100-0psi) = 0.16mA/psi.
- *Accuracy*: is the comparison of the measured value versus the true value. For instance, if many people step on a scale and their weights always read 5 pounds too low, the scale is accurate to within 5 pounds.
- *Precision*: is somewhat synonymous with "repeatability."

  If the same parameter is measured over and over, the output reading should be the same every time, but it is probable that there will be some variation in the reading. For instance, if a person were to weigh himself on a scale many times, and each scale reading was within ± 1 lb of all other readings, then the scale is precise to within 1 lb.

It is possible that an instrument is precise, but not accurate. If the true weight is 150 lb, but every reading exactly indicates 145 lb, it is interpreted that the scale is very precise, but also, that it is inaccurate. In other words, the scale is precisely wrong by 5 lb every time.

- *Resolution*: is the smallest change in input that yields an interpretable change in output. As an example, an analog pressure-gage scale may have index marks for each psi, implying that the resolution is 1 psi, assuming that the pointer is frictionless and moves smoothly as pressure changes. However, if during operation, gradual pressure changes cause no movement until the pointer suddenly jumps two index marks to overcome friction, its resolution would be 2 psi.
- *Turndown*: is often expressed as a ratio. Turndown is the lowest reading before the gage becomes unacceptably inaccurate.

  For instance, the full-scale reading of a flow meter may be 1,000 gpm, but the lowest reading on the meter scale might be 100 gpm. This meter has a 10:1 turndown ratio.

GAGE ACCURACY

When procuring instruments, the manufacturer presents accuracy in terms relative to its "% Full Scale" value, i.e., the worst accuracy is compared to the gage's highest reading. For example, if a pressure gage has a full-scale reading of 1,000 psig and a rated accuracy of ±1% F.S., it is expected to be accurate to within $1{,}000\,psig \times 0.01 = \pm 10$ psig, high or low (see Figure 1.1). Now assume that this gage is installed on a system whose normal pressure is 100 psig. In this case, it is possible that the true pressure can fall anywhere between 90 psig to 110 psig and still be within specifications. Therefore, even though this gage has ±1% full-scale accuracy, in this application, its accuracy is really within ±10% when used in a 100-psig system:

$$\frac{\pm 10\,psig\,deviation}{100\,psig\,system\,pressure} \times 100 = \pm 10\%\,accuracy$$

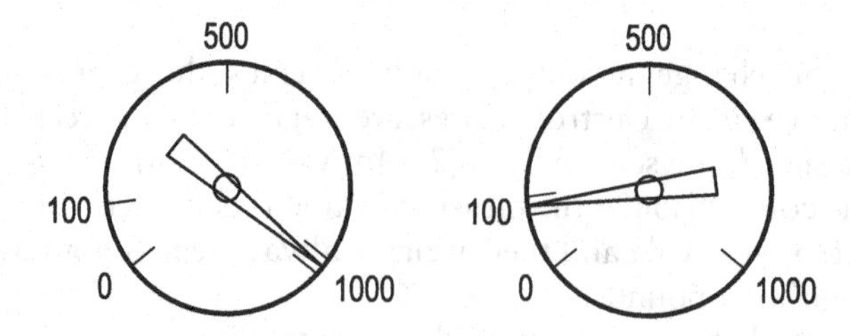

Figure 1.1 A 1,000-psig pressure gage having an advertised full-scale accuracy of 1% would be accurate to within ±10 psig. When installed in a 100-psig system, its true accuracy is ±10%.

GAGE SELECTION

Gages are normally selected to read near the middle or upper range[1] for general-service applications. For most applications, the gage should be selected so that the normal operating point is within 45-80% of the full scale, with consideration for the system maximum value, such as when a relief valve lifts. General practice is to calibrate gages to be most accurate at the mid-scale, so that errors are mitigated when the pointer swings either high or low.

- The engineer selects gage accuracy based on the importance of the service, and the accuracy is generally related to the gage physical size.
  - *Cursory operation*: ±3% F.S. (e.g., Compressed-air receiver) 2–6" dial
  - *System operation*: ±1% F.S. or ±1.5% F.S. (e.g., Main steam pressure) 4–10" dial
  - *Calibration gages*: ±0.25% F.S. 4–8" dial / ±0.10% F.S. 6–12" dial

CALIBRATION STANDARDS

Calibration should be "traceable" to the National Institute of Standards and Technology (NIST)—formerly the National Bureau of Standards. Every generation of calibration equipment removed from the NIST can introduce its own inaccuracy into the calibration procedure. Oftentimes, calibration must be done under strictly controlled environments, such as close control of temperature and humidity. Some companies specialize in calibrating instrumentation to very high accuracy levels, and calibration may be mandated when the instruments are used to prove that systems or products meet the contractual performance levels. However, in-situ calibration of plant equipment using comparison gages is generally good enough for plant systems.

CALIBRATION OF INSTRUMENTATION

All instrumentation works on the principle that a sensed phenomenon is related to a measured parameter, and then that measurement is converted to a signal that is interpreted as a meaningful value. Sometimes the relationship between the measured parameter, the sensing mechanism, and the signal conversion is linear, but more often it is not. The output mechanism needs to convert the signal into an accurate measurement reading, despite any non-linearity.

In the case of the manometer, the *pressure* input signal is directly and linearly related to the *column-height* output in the manometer. Calibration consists of marking a graduated scale, like a ruler, located adjacent to the column. Calibration procedures may include using a bubble-level to correctly set the gage angle, and the indexed scale adjacent to the column can be repositioned relative to the column to compensate for variations in the amount of liquid in the gage.

In the case of the bourdon-tube pressure gage, the flexing motion produced as a semicircular bourdon tube unwinds is not linear. In this case, the manufacturer characterizes the

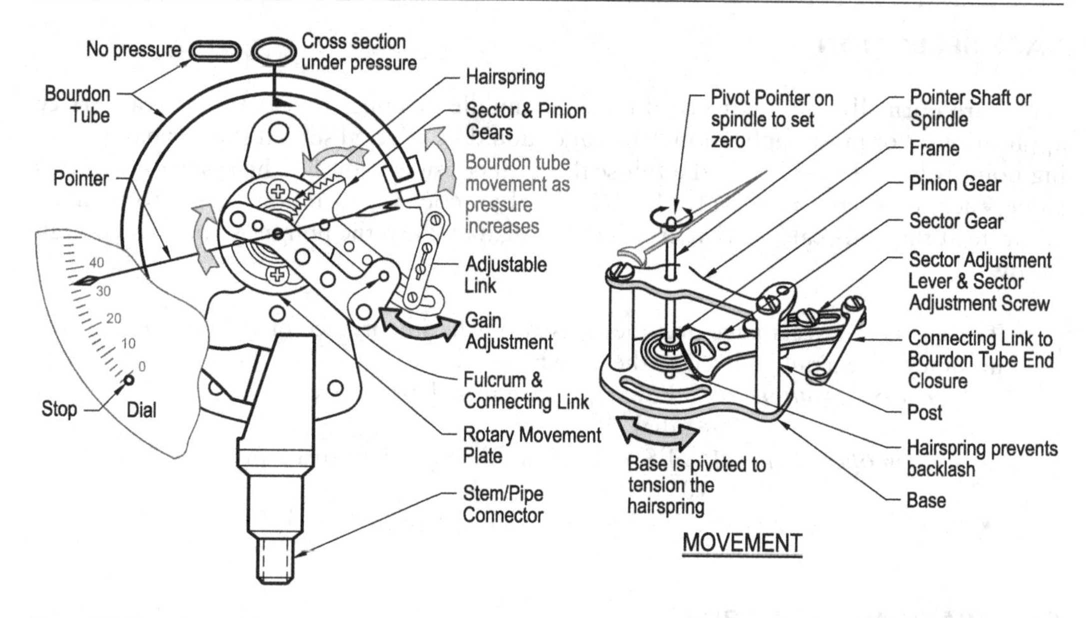

Figure 1.2 Bourdon-tube pressure gage and movement. The zero is adjusted by pivoting the pointer on its spindle, and the span is adjusted by swinging the adjustable link to change the gain. Gain is increased by moving the adjustable link closer to the pointer, so that less bourdon tube movement causes more pointer deflection.

output reading onto the scale located behind the pointer by adjusting the spacing between the printed index marks. Unequally spaced marks compensate for the non-linearity, so as to provide an accurate measurement reading at all pressures. The gage movement typically has adjustable links as shown in Figure 1.2 to change the ratio of the input pressure signal to the output measurement reading.[2] By swinging the adjustable linkage outward, the bourdon-tube motion produces less rotation of the pointer, while swinging it inward produces more rotation. These linkage adjustments provide a method of adjusting the span of movement that accounts for minor manufacturing tolerances. The pointer itself is pivoted on the spindle to set the zero reading.

Three variables are required in calibrating all instrumentation and control devices: zero, span, and the characterization between input signal and output reading. Oftentimes, the characterization between input and output is fixed by the manufacturer, as is the case of a printed index on a pressure gage. In some cases, the engineer can re-characterize the output. An example is a type of calibration gage, where the index-mark positions may be repositioned so that the gage is extremely accurate at multiple settings. Another example of characterization includes modern electronic measurement and control devices, where mathematical equations can provide an accurate relationship between an input signal and output reading.

The other adjustments are the zero and span. In calibrating a gage, the instrument is observed while there is no signal, and the pointer is set to zero. Then the instrument is loaded, often to half-scale. At this point, if the reading is off, the span is tweaked using a gain adjustment until the mid-scale reading is correct. In the case of the pressure gage a deadweight tester may apply the very accurate half-scale pressure, and the adjusting links are repositioned to make the half-scale reading accurate. The load is then removed and the zero is rechecked and readjusted if required, as there may be some interaction between the zero and span adjustments. This process is iterated until both the zero and span are within acceptable limits.

At this point, measurements at intermediate pressures from zero to full scale would indicate whether the device remains within the full-scale error specifications.

When calibration is complete, a calibration sticker should be initialed by the technician, dated, and applied onto the device.

PRESSURE CONVERSION

Pressure conversions can be determined from equalities relative to 1 Atmosphere of pressure and setting up ratios:

$$1\text{Atm} = 14.7 \text{ psi} = 29.92 \text{ in Hg} = 101.3 \text{ kPa} \approx 34 \text{ ft w.c.} \approx 1 \text{ bar} = 100 \text{ kPa} = 1.0197 \text{ kg/cm}^2$$

Sample calculation: A pressure gage reads 10 bar. Find the approximate pressure in psi and ft w.c. (use ratios relative to 1 atmosphere)

$$10bar\frac{14.7psi}{1bar} \approx \underline{147psi} \quad 147psi\frac{34ft\,w.c.}{14.7psi} \approx \underline{340\,ft\,w.c.}$$

PRESSURE GAGES

Some methods of measuring pressure include bourdon tubes, manometers, and diaphragms or bellows that flex under pressure. Different gages types are shown in Figure 1.3, including a generic connection installation shown in "a."

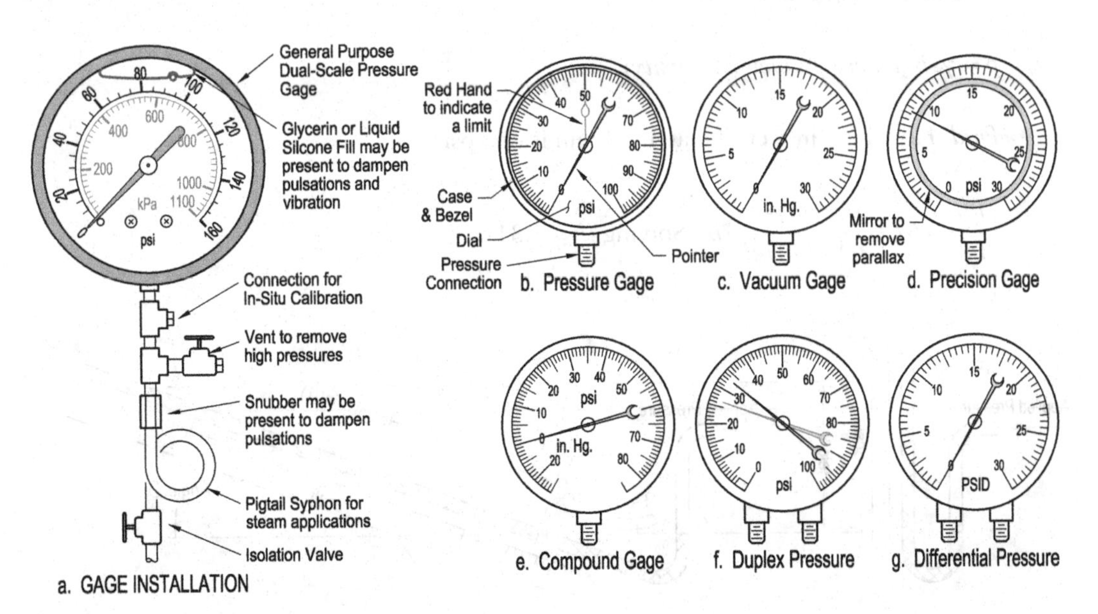

Figure 1.3 Various pressure-gage types are shown where (a) shows a typical gage installation, (b) pressure gage, (c) vacuum gage, (d) precision gage, with a parallax mirror, (e) compound gage, measuring pressure and vacuum, (f) duplex gage, measuring two pressures, and (g) differential pressure gage.

- A *mirror* on the faceplate of calibration gages is used to remove *PARALLAX error*. Parallax occurs because the pointer is located above the faceplate surface, so when looking at the gage other than straight on, the reading appears to shift relative to the faceplate. When the mirrored image of the pointer is aligned and hidden behind the pointer, parallax is removed, and the reading is accurately taken. Mirrored faceplates are used on "instrument quality" gages.
- A connection may be provided for **"In-situ" testing**. Sometimes a calibration comparison gage is paralleled with the system gage for recalibration in the field.
- A *snubber* is a small-orifice restriction in the line that delays and dampens pulsations that can cause the pointer to bounce during operation. Some operators throttle the isolation valve to act as a snubber, but care is required to avoid shutting the sensing line completely.
- A *pigtail siphon* is used in steam lines to avoid direct contact of hot steam against the bourdon-tube measuring element. Steam condenses in the loop, which is part of a static line.[3] The room-temperature condensate protects the gage's bourdon tube from direct contact with hot steam, which could distort the bourdon tube or alter its material properties.

Pressure measurement

Manometer (U-tube)

Fluid pressure—usually air—is applied against a liquid in a U-tube as shown in Figure 1.4. The pressure forces the liquid to rise in the U-tube until the liquid weight balances the pressure force. The measurement is the height of the liquid in the manometer. For small pressure measurements, such as fan-discharge pressures, water is often used, and the pressure is given in inches of water column. For larger pressures or vacuum, mercury is used, and the units are given in inches of mercury. Standard atmospheric pressure is slightly less than 30 inches of mercury. To increase the sensitivity, manometers are often inclined at an angle.

- To convert height of liquid to psi

$$p = \gamma \cdot h \ (where\ \gamma\ is\ weight\ density)$$

- *Find*: How high in feet of water column is one psi?

$$1\frac{lb_f}{in^2} = 62.4\frac{1b}{ft^3}\cdot\frac{ft^2}{144\,in^2}\cdot h \quad \text{Solving}: h = 2.31\,ft$$

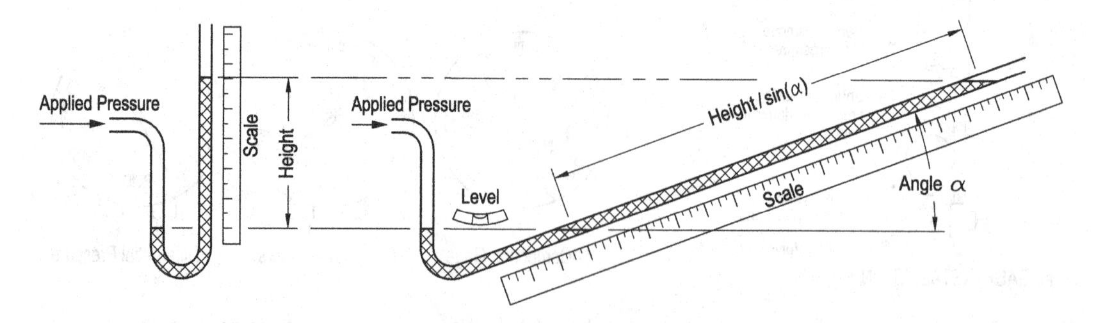

Figure 1.4 Simple manometer (left). To increase sensitivity and resolution, the manometer is inclined so that small changes in height result in large linear movements of fluid in the tube (right).

Bourdon tube pressure gage

Fluid pressure is applied into a flattened metal tube, which is coiled into a "C," spiral, or helical shape. Internal pressure inflates the flattened shape, straightening the tube. The straightening motion is linked to a geared mechanism that is connected to a pointer in front of a gage dial, which reads pressure.

Calibrating a pressure gage:

- *Zero adjust*: The pointer is pivoted on the spindle, either by a screw adjustment on the pointer or by spindle friction until zero is indicated when there is no pressure.
- *Span adjust*: Good-quality gages will have an adjustable linkage as part of the gage movement. By pivoting the "adjustable link" arm, the ratio of pointer motion to bourdon-tube motion is changed (gain), thus changing the span.

The zero and span are rechecked until both are correct.

Pressure transducer

This uses solid-state electronics to measure pressure. Pressure is applied against a metallic diaphragm, with integral strain gages to produce an electrical effect that is converted into an electronic signal. Figure 1.5 shows general purpose and heavy industrial styles. Some pressure-transducer options include the strain-gage, capacitive, and reluctance types.

The strain-gage pressure transducer has four *strain gage* resistors mounted on a diaphragm and wired into a Wheatstone bridge circuit, as shown in Figure 1.6. As the pressure flexes the diaphragm, each strain gage stretches or compresses, depending on the side it is installed. The stretching or compression causes the strain gage electrical resistance to change. For instance, when the strain gage is stretched, its resistance increases, and when it is compressed, its resistance decreases. The four strain gages are wired into a Wheatstone bridge circuit so that two resistances on one side will increase and the other two will decrease, producing four sensing elements and very high sensitivity. The Wheatstone voltage imbalance measured between the opposite nodes in the bridge becomes the signal. The signal is conditioned, amplified, and often converted to a 4-20mA signal, to ultimately become an analog signal into a meter or into a programmable logic controller.

Later generation pressure transducers use silicon piezoresistive materials that are manufactured using micromachining techniques, like those used to fabricate integrated circuit chips. The resistors are deposited directly onto the diaphragm surface at the molecular level, so that the resistors are free from creep to avoid long-term drift, less hysteresis for improved repeatability, and no thermal expansion mismatches for improved accuracy.

A capacitive pressure sensor is shown in Figure 1.7. Either a metal or a conductive-silicon pressure-sensing diaphragm is used at the heart of the sensor, depending on the maximum operating pressure. System pressure is applied to isolating diaphragms, which are made from materials that are compatible with the system fluid. The system pressure is hydraulically transmitted from the isolating diaphragm to the sensing diaphragm through a dielectric fluid, often silicone oil. The silicone oil separates the system fluids from direct contact with the capacitor plates, while its dielectric properties contribute to its capacitance.

As the pressure changes, the sensing diaphragm deflects causing the capacitance on one side to increase, while simultaneously decreasing capacitance on the other side. Sensing two capacitances essentially doubles the signal to improve sensitivity, while canceling undesirable common-mode noise. The output signal originates from a high-frequency AC excitation signal that is used to produce the capacitive reactance ($X_C = \frac{1}{2\pi fC}$). That signal, which represents

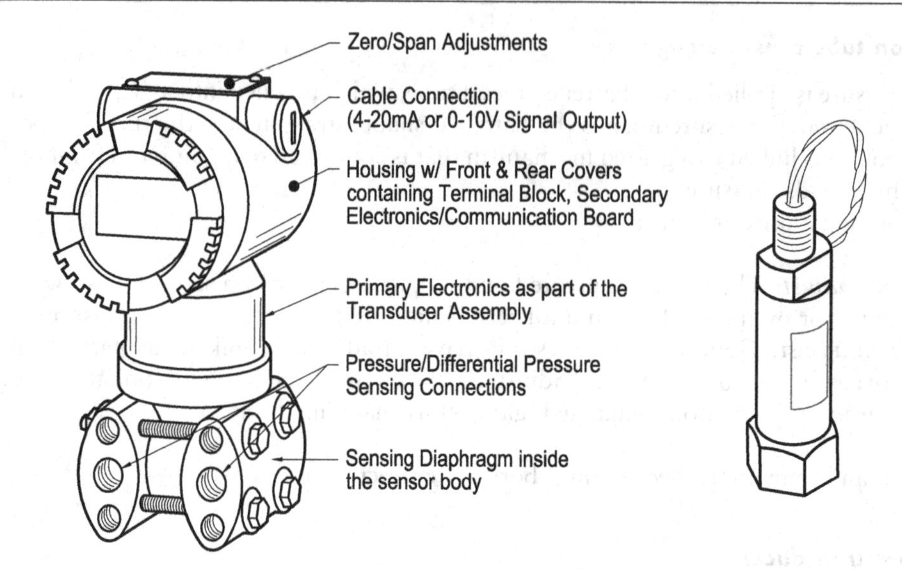

Figure 1.5 Left: Heavy-duty industrial differential-pressure transducer. If one side is unconnected relative to the atmosphere, the transducer directly measures gage pressure. Right: A general-purpose pressure transducer.

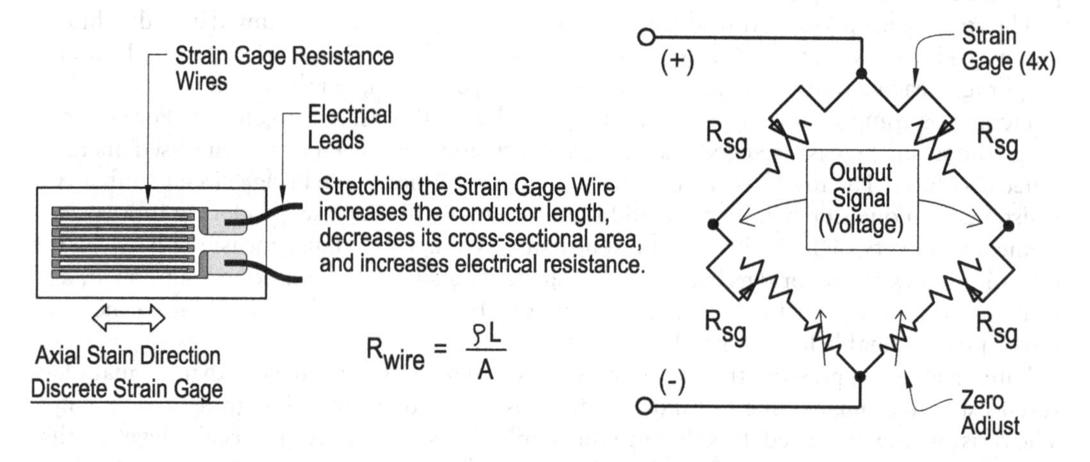

Figure 1.6 Strain gages (left) are wired into a Wheatstone bridge (right), where two of the four strain gages are mounted on each side of a diaphragm to form the sensing element of a pressure transducer. The signal is produced from the voltage imbalance that occurs between the two intermediate nodes, as the strain-gage resistances vary with diaphragm deflection.

pressure or differential pressure, is ultimately converted to a 4-20mA signal for transmission as a PLC analog input.

Capacitive pressure sensors are very accurate, up to 0.1% full scale and can be configured to measure very high, very low, or vacuum ranges. They are very robust and good for harsh service environments. The isolating fluid dampens pressure fluctuations, and the capacitor plates form mechanical stops to avoid damage from excessive deflection of the diaphragm during over-pressurization.

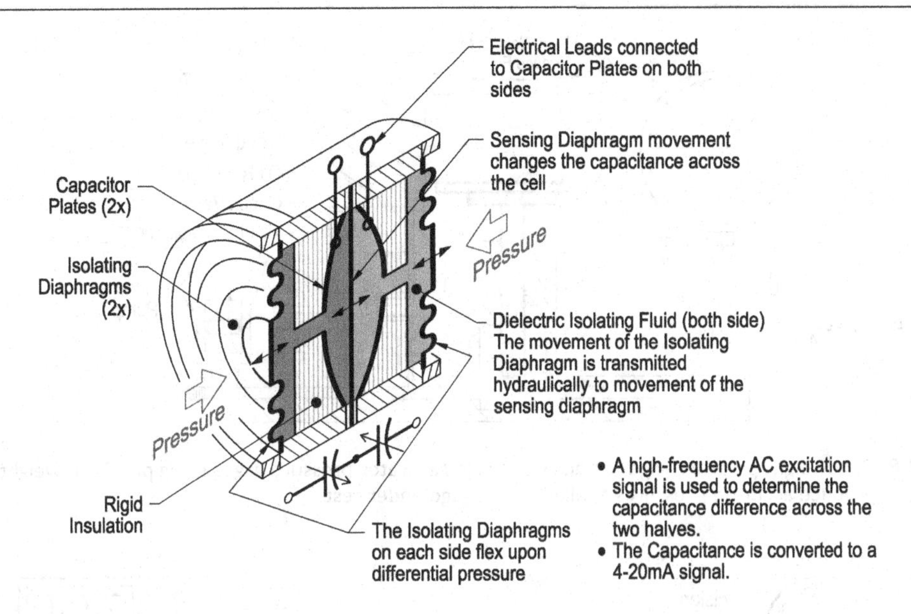

Figure 1.7 Capacitive pressure or differential-pressure sensor. Higher capacitance occurs to one capacitor when the sensing diaphragm approaches its plate, and lower capacitance simultaneously occurs to the other plate, as the sensing diaphragm moves away.

Reluctance pressure transducers are another variation of the diaphragm pressure-sensing transmitter; however, they use coils surrounding permeable materials to change the sensor inductance.

Calibrating pressure gages

Pressure gages are calibrated against accurate standards traceable to the NIST, where the gage adjustments are described earlier in this chapter. Two methods used for pressure-gage calibration are the deadweight tester and calibrated comparison gages, illustrated in Figure 1.8 and Figure 1.9.

- *Deadweight tester*: the gage is mounted on the deadweight tester and weights are applied onto a piston having an accurately machined diameter. Hydraulic fluid is then pumped up to the point where the piston and weights are "floating" on the pumped fluid, with fine adjustments made using a volume displacer. To reduce friction, the weights are slowly spun by hand. The weights are stamped with units of pressure, corresponding to the piston diameter. Figure 1.8 shows a deadweight tester being used to test a gage.
- *Comparison gages*: precision test gages are piped in parallel with the gage under test. Pressure is then simultaneously applied to both gages. The pressure source can be from a hand pump or from a nitrogen bottle for high pressures. Figure 1.9 shows a comparison gage being used to calibrate a 4-20mA electronic pressure transducer, and Figure 1.10 shows in-situ testing of an in-service gage. For the electronic gage, the 4-20mA transmission signal is first calibrated at the zero and span, and later the meter is also calibrated to produce the correct zero and span readings.

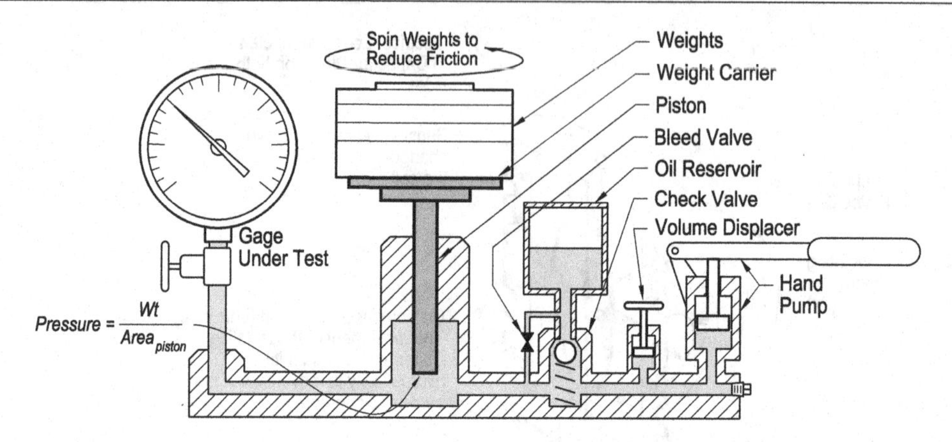

Figure 1.8 Deadweight pressure gage tester. This device calibrates pressure gages using precision weights that produce accurate pressures applied to the gage under test.

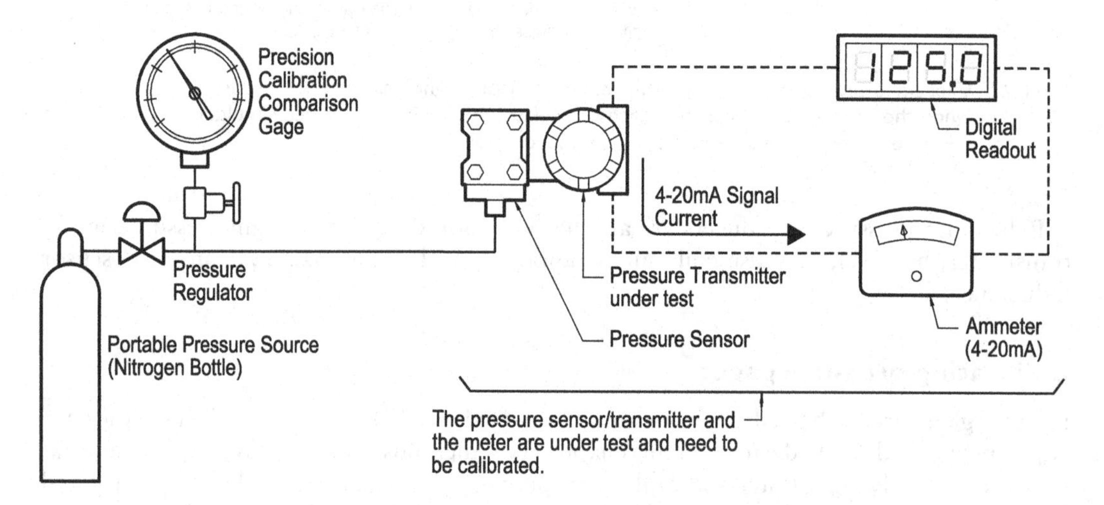

Figure 1.9 A comparison calibration gage is placed in parallel with a pressure transducer undergoing calibration. The 4-20mA transmission signal is calibrated first, and the meter is then calibrated from the corrected 4-20mA signal.

- *Shunt-resistor calibration*: is a quick way to recalibrate strain-gage pressure transducers formed using a Wheatstone bridge circuit. After a pressure transducer is calibrated to meet specifications, the pressure is removed to obtain a 0-psig measurement. At that point, a shunt resistor is temporarily inserted in parallel with one resistor in the Wheatstone bridge, knocking the bridge out of balance. This resistor simulates a reading, which is recorded for future recalibration.

During recalibration, the pressure is removed, and the transducer should indicate 0 psig. If not, the zero adjustment is tweaked until it reads zero. While the transducer is at 0 psig, the calibration resistor is then replaced into the circuit, providing a simulated reading (see Figure 1.11). If the simulated reading differs from the recorded value obtained during the initial calibration, then an error exists. The span adjustment is tweaked to reconcile the error back to the recorded value. To ensure there is no interaction between the span and zero, the

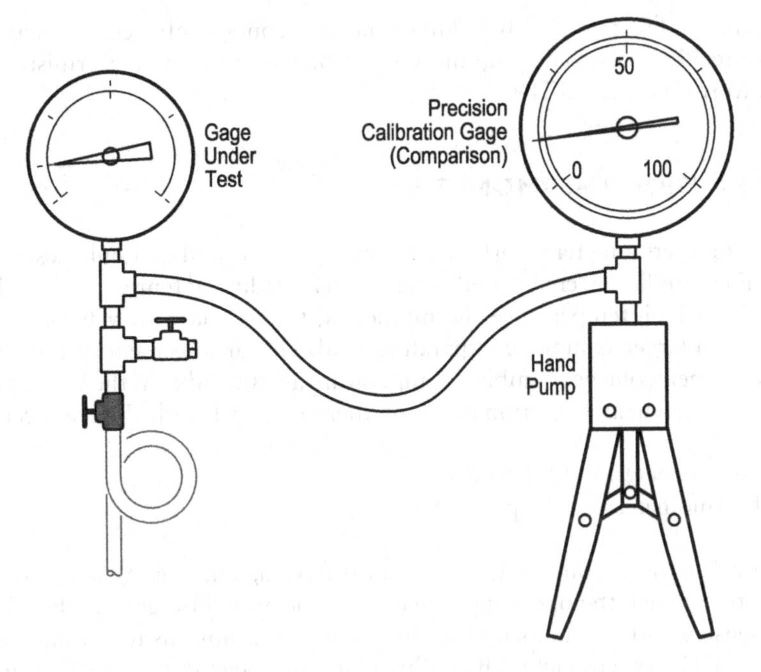

Figure 1.10 A handheld comparison calibration gage is used for in-situ testing.

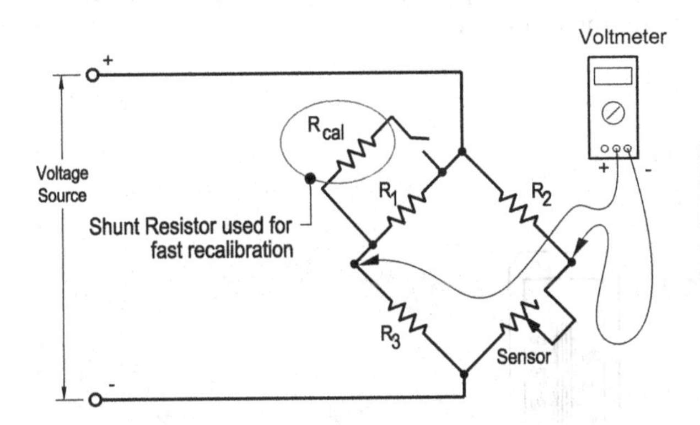

A Shunt (or Parallel) Resistor is used to intentionally change the Wheatstone Bridge Resistance and the meter reading. Quick meter recalibration can be done later by switching-in the shunt resistor and adjusting the meter to read the value obtained during the original calibration.

Figure 1.11 A shunt resistor can be used for quick in-situ recalibration of a strain-gage pressure transducer.

pressure is removed and the 0 psig value is rechecked. The process may require several iterations until both the zero and span values are correct.

It should be noted that this method is not a complete calibration, since no actual NIST-referenced pressure sources were used to set the span, but this method removes drift errors associated with the diaphragm, strain gages, and electrical circuit, and it usually replicates accuracies that are close to the original calibration.

Temperature measurement

Temperature measurement can be done using local indicators or remote electronic devices that are suitable for both indication and control input signals. Local indicating types consist of liquid expansion, filled system, bimetallic, and non-contact infrared. The electronic types are conducive to PLC-based control, and consist of thermocouples, thermistors, and resistance temperature detectors (RTDs).

LIQUID-EXPANSION THERMOMETER

Liquid-expansion thermometers work on the principle that fluid in a bulb reservoir expands and rises within a capillary stem located adjacent to a graduated temperature scale, as shown in Figure 1.12. In high-temperature thermometers, the capillary may have a bulge in the middle to accept a larger volume of expanding fluid. This arrangement creates two zones of accurate measurement, one near ambient temperature and the other at the high operating temperature, while avoiding an objectionably long thermometer length. Typical accuracies are:

- 300°F Thermometer: ± 0.5°F to F.S.
- 600°F Thermometer: ± 2°F up to ± 5°F

Thermometer calibration is often referenced to the freezing and boiling points of water, or to other temperatures when the operating range is significantly different. In plant installations, temperature sensors and thermometers are installed in thermowells to maintain the pressure boundary, to avoid direct contact with the fluid, and to protect the device. Figure 1.13 shows the proper installation of thermowells in a piping system. The interstitial space in the thermowell is typically packed with a thermally conductive paste, and the sensor probe should be installed where the flowrate is highest and most turbulent.

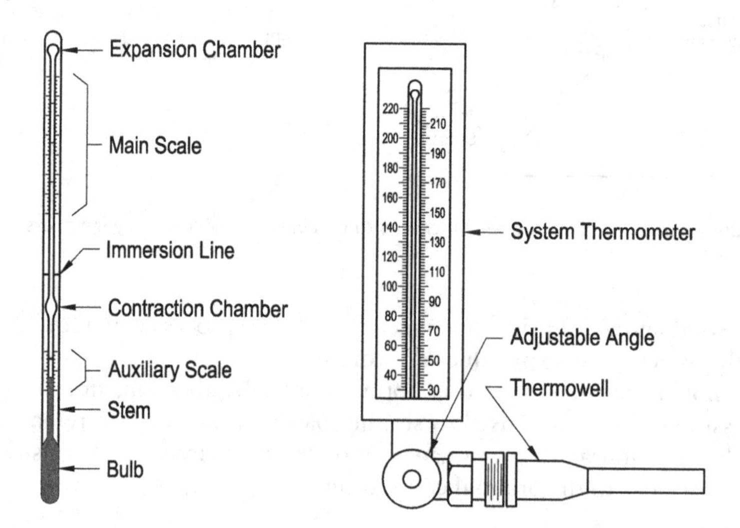

Figure 1.12 Typical liquid expansion thermometer (alcohol or mercury column). The main and auxiliary scales, as well as the correction chamber, on the left-hand thermometer, indicates it is designed for a high-temperature application.

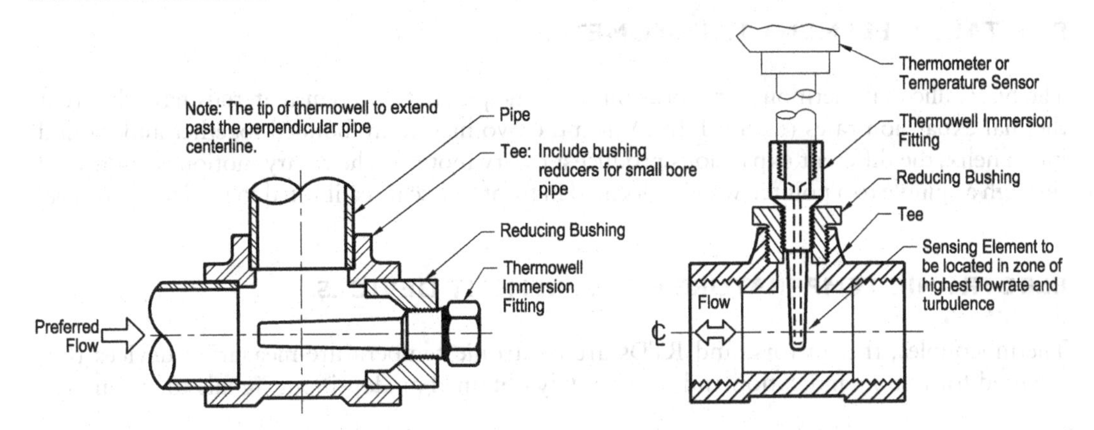

Figure 1.13 Non-contact thermowells are installed in high-velocity turbulent locations for good temperature measurement. The thermowell is filled with a thermally conductive compound to provide fast response.

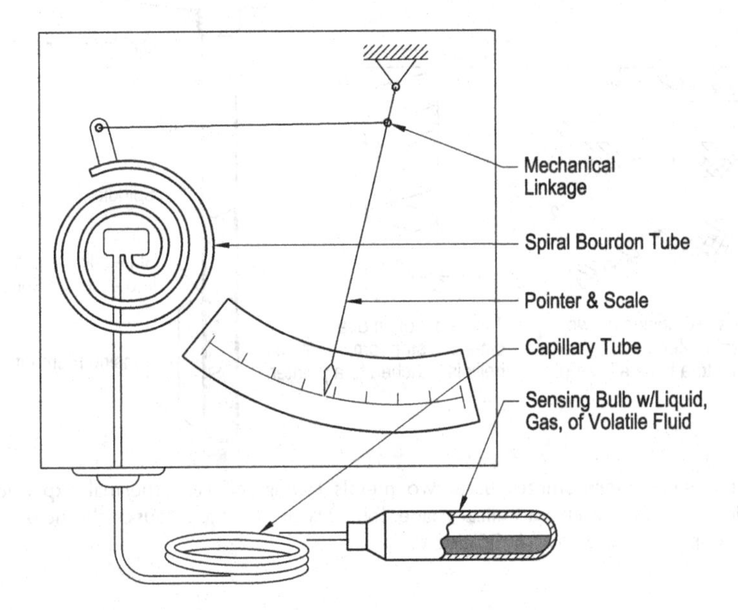

Figure 1.14 Filled-system thermometer. The sensed temperature is converted to pressure, which becomes the signal that is transmitted to the measuring mechanism.

FILLED-SYSTEM THERMOMETER

The filled-system thermometer uses a sensing bulb containing a volatile fluid whose saturation pressure changes with temperature (see Figure 1.14). The change in fluid pressure is transmitted via a capillary tube to a Bourdon tube. The Bourdon tube is linked to a pointer to create a motion in response to the pressure change, but the scale is indexed in degrees of temperature. The full-scale accuracy is often ±1°F (±1°C).

BIMETALLIC-ELEMENT THERMOMETER

The bimetallic dial thermometer works on the principle that different materials have different thermal expansion rates (Figure 1.15). When the two metals are bonded together and wound into a helix, the different expansions produce a rotary motion. The rotary motion is connected through a spindle to a pointer, which is located in front of a scale calibrated to read temperature.

ELECTRONIC TEMPERATURE MEASUREMENT DEVICES

Thermocouples, thermistors, and RTDs are electronic temperature-measuring devices that are used for remote indication and for remotely obtained analog-input signals to automated

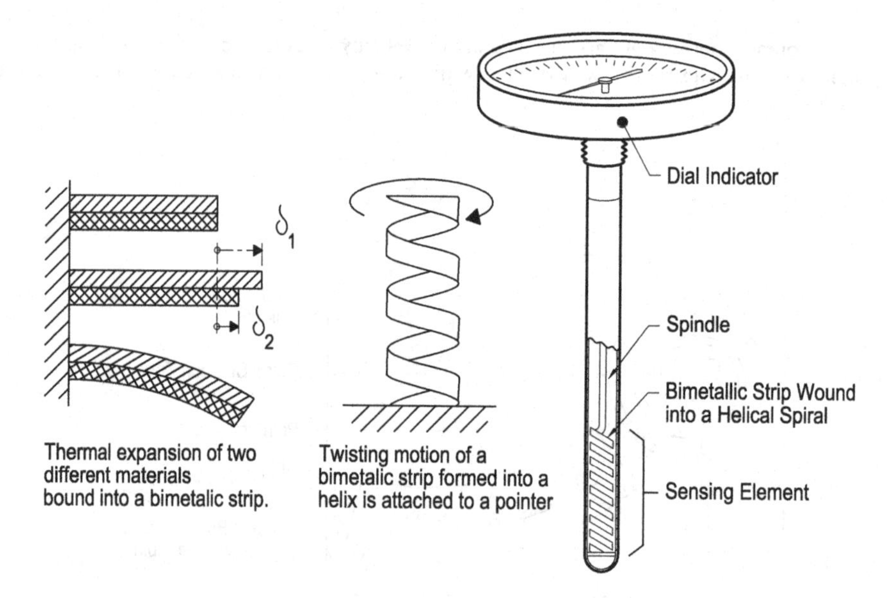

Figure 1.15 The bimetallic thermometer uses two metals having different thermal expansion coefficients, bonded and formed into a helical shape. Temperature changes causes the helix to twist, and its pointer spins relative to an indicating scale.

Table 1.1 Characteristics of electronic temperature indicators used for automated-control systems

	Thermocouple	*Thermistor*	*RTD*
Temperature Range (typical)	–450 to 2,300°F –270 to 1,250°C	–150 to 600°F –100 to 325°C	–325 to 1,100°F –200 to 650°C
Accuracy (typical)	0.9 to 9°F 0.5 to 5°C	0.09 to 2.7°F 0.05 to 1.5°C	0.18 to 1.8°F 0.1 to 1°C
Long-term stability @ 100°C	Variable	0.2°C/year	0.05°C/year
Linearity	Very non-linear	Exponential-like	Fairly linear
Response time	Fast: 0.10 to 10s	Fast: 0.12 to 10s	Slow: 1 to 50s
Susceptibility to electrical noise	Susceptible	Rarely susceptible	Rarely susceptible
Cost	Low	Generally Low	High

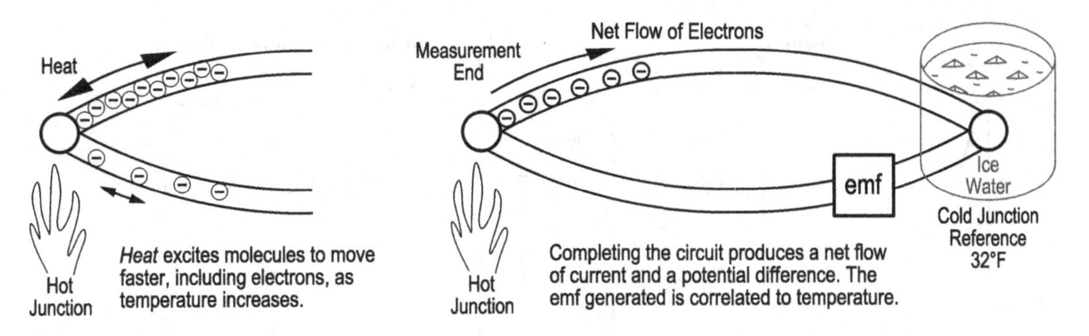

Figure 1.16 Thermocouple theory of operation. Heat and electrons move faster through a better conductor, creating an electromotive force.

control systems. The characteristics of three types of remote, electronic temperature sensors are summarized in Table 1.1 and discussions of their operation follow.

Thermocouple

Thermocouples use two conductors having dissimilar metals. One material has higher thermal conductivity and conducts heat faster than the other. Thermal conductivity and electrical conductivity tend to be related, so the flow of heat causes faster movement of electrons in the better conductor. By completing the circuit, the overly-excited electrons in the better conductor flow from its junction faster than the less-excited electrons in the poorer conductor, producing an electromotive force (emf or voltage), as illustrated in Figure 1.16. The emf is the signal and is a function of the two conductor materials and the temperatures between the two junctions. The emf is correlated to read temperature. For the system to work accurately, a precise temperature at the "cold" or "reference" junction is required, typically referenced to the freezing point of water. Since ice-water is not normally available at the meter, electronic circuits are used to simulate the 32°F reference temperature and to compensate for variations in ambient temperature.

Type J thermocouples use iron and constantan, where constantan is a copper-nickel alloy. They function up to 800°F. Type K thermocouples use chromel and alumel, both being corrosion-resistant nickel alloys, and they are used in high-temperature applications reaching 2300°F. Thermocouples are known for rapid response, high-temperature capability, durability, accuracy, relatively low cost, remote indication, and are easily adapted to electronic control. The disadvantage of thermocouples is that noise or stray voltages can be picked up and its output voltage is very small, requiring amplification.

The hot junction is often formed by fusing the lead ends, but the conductors can be simply twisted together. The hot junction may be placed in direct contact with the fluid or material whose temperature is being measured, or it is often enclosed in a protective metal sheath, which is inserted into a thermowell. Protective sheathing is required when degradation from oxidation is possible, and the junction may be bonded to the case for very faster response, or it may be electrically insulated from the case to avoid interaction with other circuits.

Thermistor

A thermistor is a type of resistor that exhibits large but predictable changes in resistance with respect to temperature. Typically, the thermistor is made from a semiconductor material such

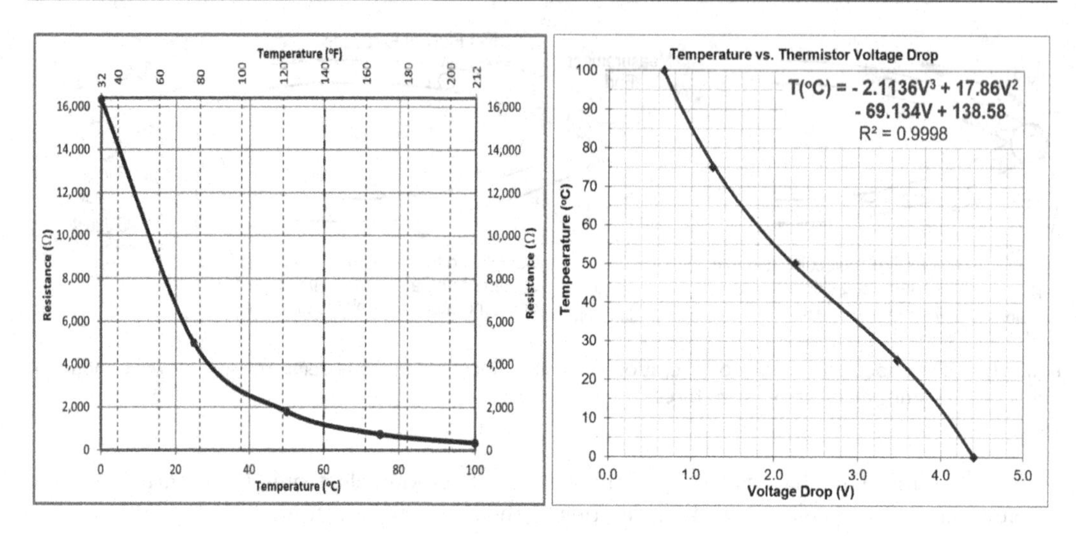

Figure 1.17 A 5,000-Ω thermistor is installed in a circuit having a 2,200-Ω fixed resistor. The resistance vs temperature (left) is a non-linear relationship. The graph (right) shows a plot of the calculated thermistor voltage drop along with its third-order polynomial curve-fit equation. The trendline equation yields a temperature measurement from the thermistor voltage drop.

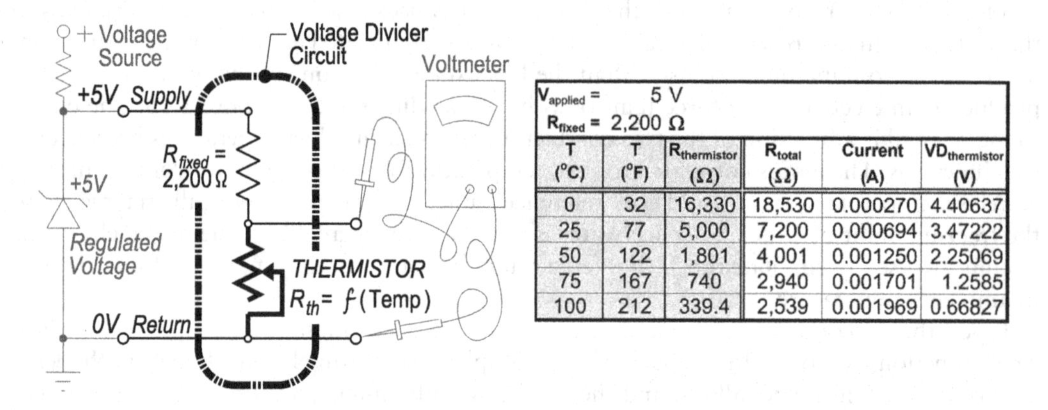

$V_{applied}$ = 5 V
R_{fixed} = 2,200 Ω

T (°C)	T (°F)	$R_{thermistor}$ (Ω)	R_{total} (Ω)	Current (A)	$VD_{thermistor}$ (V)
0	32	16,330	18,530	0.000270	4.40637
25	77	5,000	7,200	0.000694	3.47222
50	122	1,801	4,001	0.001250	2.25069
75	167	740	2,940	0.001701	1.2585
100	212	339.4	2,539	0.001969	0.66827

Figure 1.18 A 5,000-Ω thermistor is installed in a +5Vdc voltage divider circuit. The thermistor signal is formed by its voltage drop. Resistances for several temperatures are provided by the manufacturer in the shaded half of the table, while the thermistor's corresponding calculated voltage-drop signals are in the unshaded half.

as a metallic oxide, ceramic, or a polymer, and can be obtained in positive (PTC) or negative (NTC) temperature coefficient versions. The resistance of PTC thermistors increases with temperature, like most metallic conductors, and they are not used for temperature measurement, but rather for current-limiting applications. NTC thermistors are used for temperature measurement, and NTC means that resistance decreases as the temperature increases, which is contrary to most conductors. Unlike the RTD, the thermistor temperature-and-resistance relationship is very non-linear, as shown in Figure 1.17 (left), and thus, they are somewhat limited to a measurement range near the knee of the curve. Thermistors are very accurate, typically within a half a degree Fahrenheit, and inexpensive, but they are limited to narrow temperature ranges, such as between the freezing and boiling points of water, as an example.

Since resistance is a passive characteristic, it cannot be measured directly, so therefore, its value must be converted into either a voltage or current for signal transmission. Figure 1.18 (left) shows a thermistor connected as part of a voltage-divider circuit and driven by a constant-voltage source. The thermistor voltage drop becomes the output signal. The voltage drop is calculated by finding the total circuit current passing through the divider circuit using Ohm's Law, and then the total circuit current is used to calculate the voltage drop across the thermistor alone, again using Ohm's Law. The calculation is iterated for several data points obtained from the thermistor manufacturer to produce enough temperature-vs-voltage-drop data points to characterize a temperature reading from a voltage signal, as summarized in the table in Figure 1.18.

This example uses a 5,000-Ω thermistor[4] in a +5VDC voltage-divider circuit having a fixed-resistor value of 2,200 Ω. The table in Figure 1.18 shows several thermistor resistances and the calculated voltage-output signals. The manufacturer-furnished temperature versus resistance relationship is graphed in Figure 1.17 (left) and the calculated temperature-versus-voltage drop is graphed on the right. The third-order-polyline trendline equation within the graph is derived using curve-fit techniques. The trendline equation directly relates the analog-input voltage signal to its output temperature readout as part of automatic control in a PLC.

The supply voltage and fixed-resistor values are judiciously selected to produce a voltage-drop signal that falls within an analog-input-device signal range. In the 5,000-Ω thermistor example above, the signal is well-suited as a 0-5V analog-input signal corresponding to an operating temperature range of 0°C to 100°C. The trendline equation could be used to translate a 0-5Vdc PLC analog-input voltage signal into a temperature measurement for indication or control in a PLC.

Resistance temperature detector (RTD)

The resistance temperature detector works on the principle that the resistivity (ρ) of metals increases with temperature. The RTD sensor probe is placed on equipment in the machinery space and its signal is sent to a conveniently located meter or controller. The most common industrial RTD is the 100-ohm platinum resistor, although nickel is sometimes used for less-expensive, lower-importance installations. The resistor is created by forming the platinum into a thin, hair-like strand[5] having 100 ohms at 0°C (32°F). The sensor element is connected in a Wheatstone bridge arrangement, as shown in Figure 1.19.

Figure 1.21 and Figure 1.22 show two- and three-wire Wheatstone-bridge alternatives for remote sensor locations. The Wheatstone-bridge circuit works by measuring the voltage imbalance between the mid-point nodes of the two parallel branches. The signal voltage is then

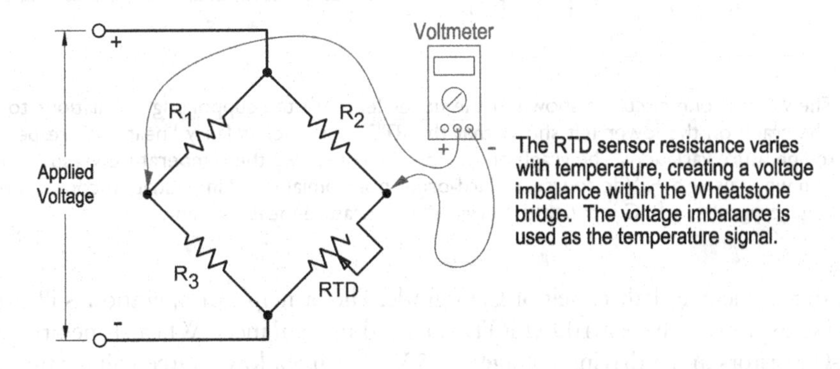

Figure 1.19 An RTD is installed in a Wheatstone-bridge circuit. The output signal is produced by the voltage imbalance between the midpoint nodes.

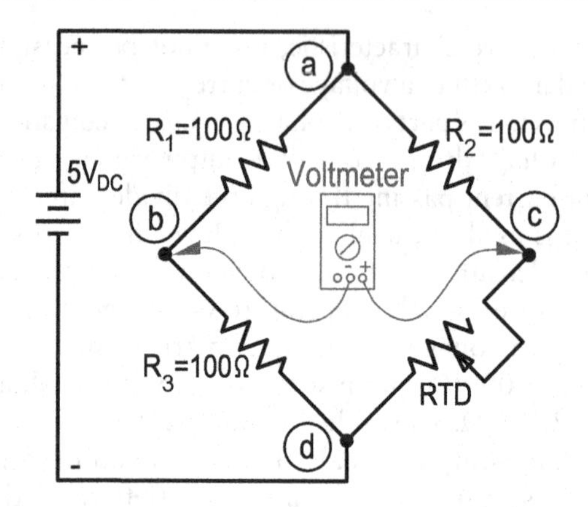

CALCULATIONS

Left side:

1. $V_b = +5.000V/2 = +2.500V$ *(equal voltage drop across equal series resistances)*

Right Side: *ex. at* $T = 100^0 C$ — $R_{RTD} = 138.5\,\Omega$

2. *Right-side resistance:*	$R_{acd} = R_2 + RTD = 100\Omega + RTD$	$= 100\Omega + 138.5\Omega$	$= \mathbf{238.5\Omega}$
3. *Right-side current:*	$I_{acd} = \frac{V_{Source}}{R_{acd}} = \frac{5V}{R_{acd}}$	$= \frac{5V}{238.5\Omega}$	$= \mathbf{0.02096A}$
4. *V.D. across* R_2:	$V_2 = V_{ac} = I_{acd} \times R_2$	$= 0.02096A \times 100\Omega$	$= \mathbf{2.096V}$
5. *Voltage level at node c:*	$V_c = V_{Source} - V_{ac} = 5V - V_{ac}$	$= 5V - 2.096V$	$= \mathbf{2.904V}$
6. *Voltmeter (nodes b to c):*	$V_{bc} = V_c - V_b$	$= 2.904V - 2.5V$	$= \mathbf{0.4036V}$

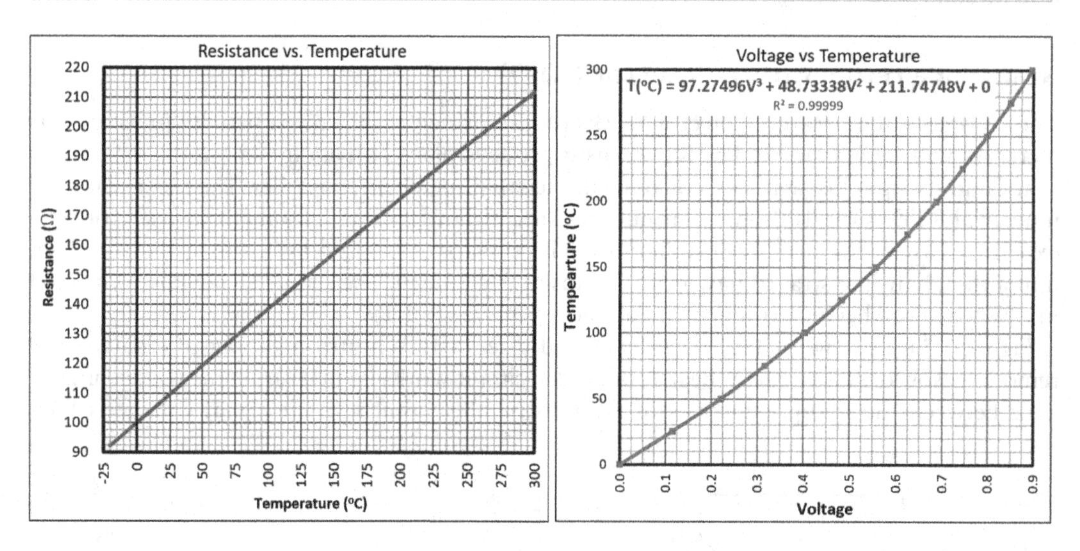

Figure 1.20 The Wheatstone circuit is shown in the upper left with the supporting calculations to the right. The graph on the lower left shows that the RTD resistance is fairly linear with respect to small temperature variations. The graph on the lower right shows the temperature-versus-voltage solution when using a +5Vdc source. A third-order polynomial trendline equation converts the analog voltage signal as a PLC input to the original temperature measurement.

calibrated to read degrees Fahrenheit or Centigrade. The principle of operation is illustrated by the following example, where a 100-Ω RTD is placed in a balanced Wheatstone bridge having three 100-Ω resistors and a driving voltage of +5 Vdc. Using a low source voltage mitigates the I^2R self-heating effects through the platinum sensing element. The calculation shown is for a measurement at 100°C (212°F), where the RTD resistance is at 138.5 Ω, as obtained from

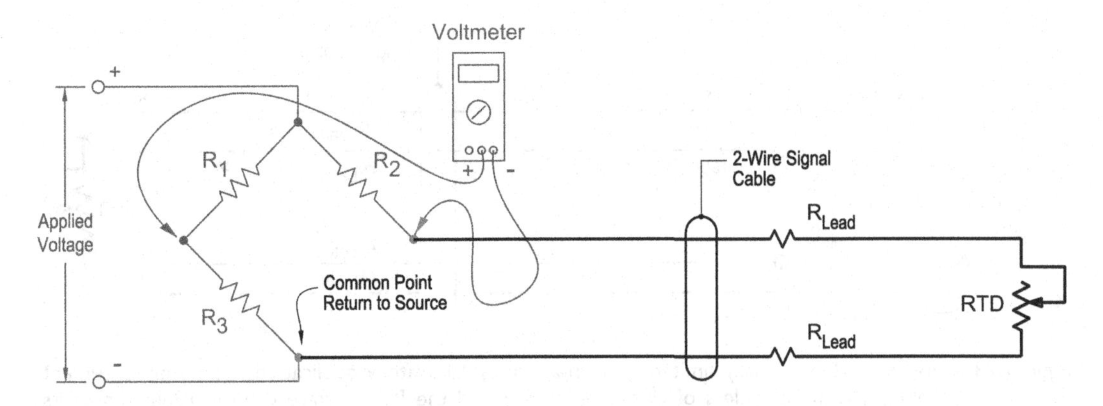

Figure 1.21 If a two-wire RTD is placed remotely, the resistances of both RTD leads add to the sensor resistance, inducing a measurement error during ambient-temperature variations.

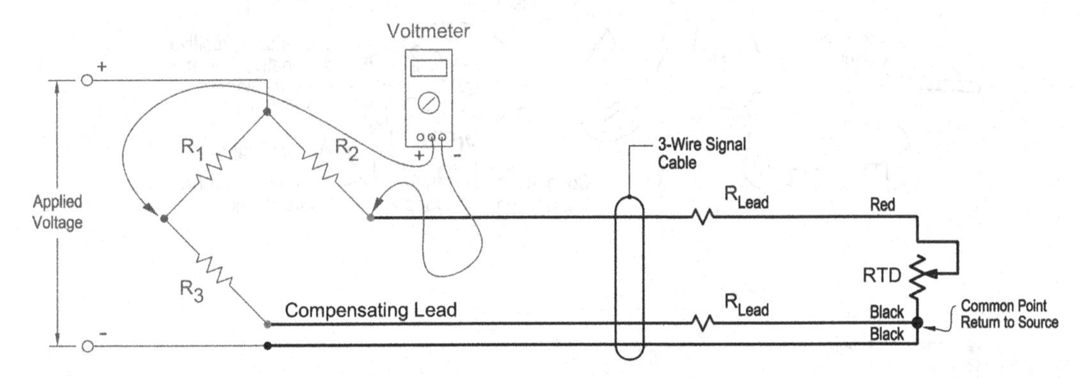

Figure 1.22 The three-wire remote RTD uses a compensating lead. The compensating lead shifts the left-to-right-junction node from its location near the Wheatstone bridge to be near the remote RTD. The third lead causes one lead's resistance to be added to the left side of the bridge and the other lead's resistance to be added to the right side of the bridge, which balances the small resistance variations between both sides and reduces lead errors.

Table 1.2. The resistance-versus-temperature and corresponding Wheatstone voltage-versus-temperature graphs are shown in Figure 1.20 for a 5-Vdc source.

When the two-wire RTD is located remotely from the Wheatstone-bridge circuit, as shown in Figure 1.21, a lead-resistance measurement error occurs. The problem is that the two RTD leads add their resistances only to the right-side branch of the Wheatstone bridge. Therefore, ambient temperature variations cause the lead resistance to change, but only to the right side of the bridge, imposing a non-measurement-induced temperature-based resistance imbalance to the bridge, and an associated error.

To compensate for the imbalanced lead resistances seen with a two-wire RTD, a third lead is added. Effectively, the third compensating lead functions to shift the common junction point of the Wheatstone return circuit to add an equal amount of lead resistance to both sides of the bridge, as shown in Figure 1.22. As the ambient-space temperature varies, lead resistances change equally to both sides of the bridge, balancing the effect between the two halves, thus mitigating lead-resistance errors.

A 4-wire RTD is shown in Figure 1.23. The 4-wire RTD is more complex and more expensive than the 3-wire version, but it is also more accurate, as the leads and connector resistances do not influence the voltage-measurement signal at all. The 4-wire RTD provides power from a constant

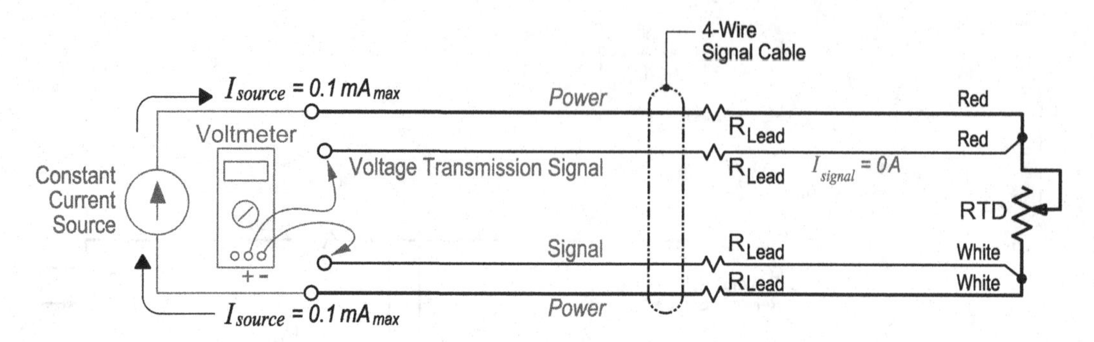

Figure 1.23 A 4-wire RTD is usually limited to high-accuracy laboratory applications. The source current is held constant regardless of circuit resistance, and the RTD voltage drop is a function of its temperature only.

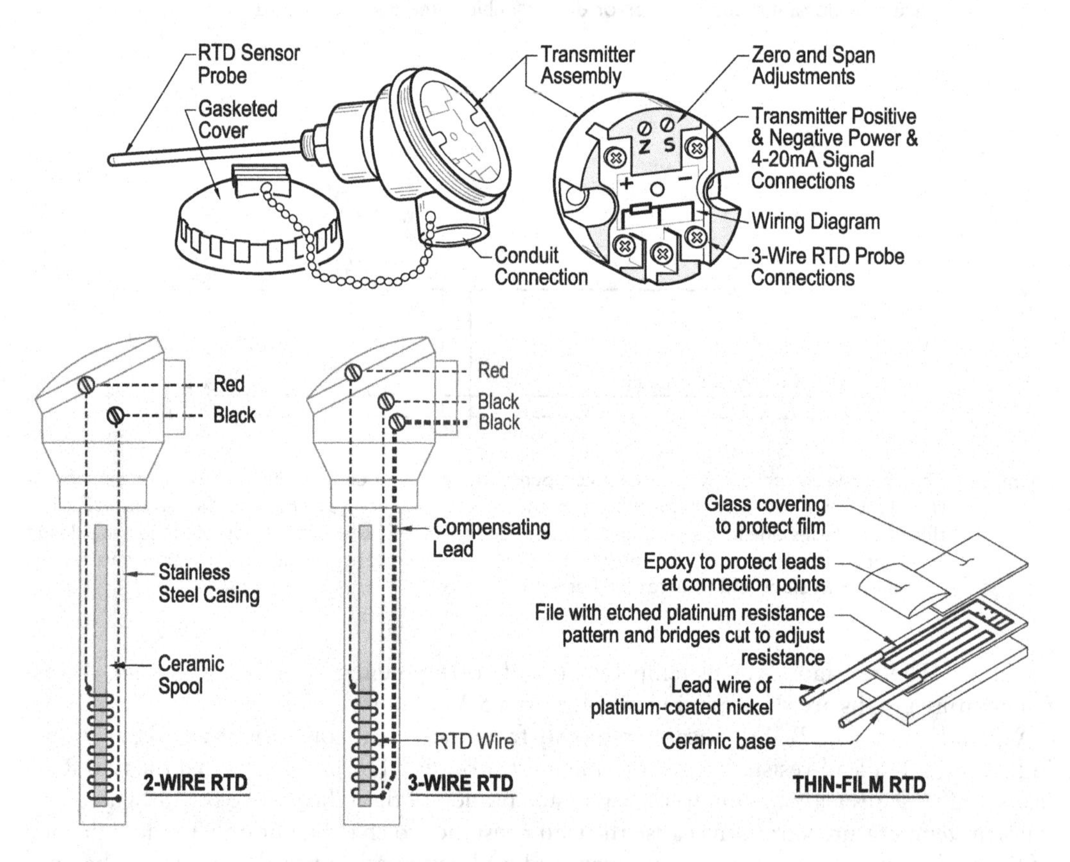

Figure 1.24 The construction of two- and three-wire RTD sensors and the transmitter assembly are shown on the left, and a thin-film RTD alternative is shown on the right.

current source through two leads, and the RTD voltage-drop signal is statically measured across two independent leads, which do not have any current, nor the associated voltage drops that would corrupt the signal. The RTD resistance, and its corresponding temperature, is related to the voltage signal using the constant value of applied current and Ohm's Law. The circuit current is limited to 0.1 mA to limit I^2R heating of the platinum sensor element. Generally, 3-wire RTDs

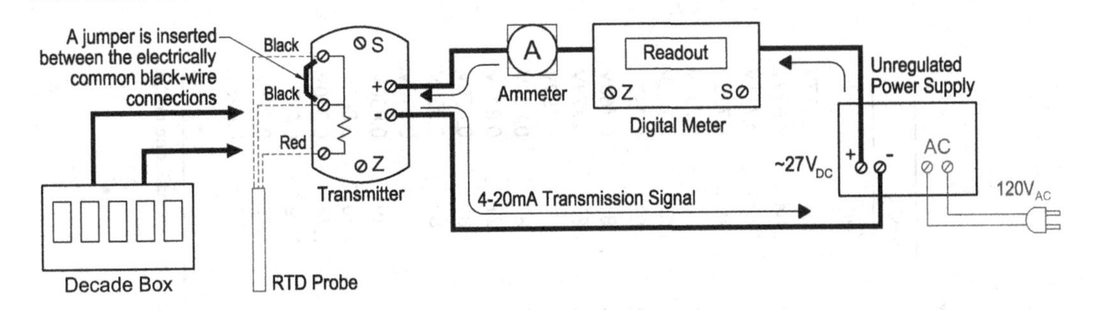

Figure 1.25 Calibration of an RTD. The RTD sensor is go/no-go tested for accurate functionality by measuring its resistance at a known temperature and comparing it to the table values. The 4-20 mA transmission signal is calibrated at both the zero and span values, followed by similar zero and span calibrations at the meter.

provide truly excellent accuracy for plant applications, while the 4-wire systems are reserved for extreme-accuracy laboratory measurements. Figure 1.24 shows RTD construction techniques.

Calibration of a 4-20mA sensor

Figure 1.25 shows a 4-20mA RTD digital meter circuit. The voltage imbalance in the Wheatstone bridge is converted into a 4-20mA signal by electronics in the transmitter subassembly and the digital readout. 4-20mA analog signals are covered in Chapter 14. The 4-20mA signal is transmitted to the digital meter where the amperage is converted to a temperature readout. After the sensor is go/no-go tested for proper function, the circuit is calibrated in two steps, where at the zero and span points are adjusted for both the transmitter circuit, and for the signal-receiving device. A decade box is used to simulate the sensor resistance at any temperature.

Verification of the RTD sensor

The RTD sensor is disconnected and its resistance is measured using an ohmmeter. The ohmmeter value is compared to the resistance it should have at its temperature. Resistance-versus-temperature tables are available for the RTD material. Ice-water or boiling water is often conveniently used as a source of calibration temperatures, although any source of precisely known temperature will work. If the sensor resistance is wrong, then the sensor is defective and needs to be replaced. The transmitter and receiver circuits are calibrated when it is determined that the sensor is functioning properly.

Calibration of the 4-20mA transmission signal

- The RTD is disconnected and replaced by a "decade box" of precision resistors. The resistances are dialed in to simulate the RTD sensor temperature. Table 1.2 shows an abridged list of resistances versus temperatures for a platinum RTD, up to 300°F (572°F).
- The two black-wire connections are connected with a jumper to serve as the compensation leads in the bridge circuit, and the precision resistors emulate the RTD value.
- Using a 0-200°C (32-396°F) meter as an example, the zero reference would be 0°C (32°F), and its midspan calibration reading would be 100°C (212°F). First the transmission

Table 1.2 Abridged resistance versus temperature for a 100-ohm platinum RTD at 5°C increments, stopping at 300°C for brevity

			Resist.	*Volt*				*Resist.*	*Volt*				*Resist.*	*Volt*
°C	*°F*	*Ohm*	*Diff.*	*Diff.*	*°C*	*°F*	*Ohm*	*Diff.*	*Diff.*	*°C*	*°F*	*Ohm*	*Diff.*	*Diff.*
−20	−4	92.16	1.97	−0.102V	90	194	134.70	1.90	0.370V	200	392	175.84	1.84	0.687V
−15	5	94.12	1.96	−0.076V	95	203	136.60	1.90	0.387V	205	401	177.68	1.84	0.699V
−10	14	96.09	1.97	−0.050V	100	212	138.50	1.90	0.404V	210	410	179.51	1.83	0.711V
−5	23	98.04	1.95	−0.025V	105	221	140.39	1.89	0.420V	215	419	181.34	1.83	0.723V
0	32	100.00	1.96	0.000V	110	230	142.29	1.90	0.436V	220	428	183.17	1.83	0.734V
5	41	101.95	1.95	0.024V	115	239	144.17	1.88	0.452V	225	437	184.99	1.82	0.746V
10	50	103.90	1.95	0.048V	120	248	146.06	1.89	0.468V	230	446	186.82	1.83	0.757V
15	59	105.85	1.95	0.071V	125	257	147.94	1.88	0.483V	235	455	188.63	1.81	0.768V
20	68	107.79	1.94	0.094V	130	266	149.82	1.88	0.499V	240	464	190.45	1.82	0.779V
25	77	109.73	1.94	0.116V	135	275	151.70	1.88	0.514V	245	473	192.26	1.81	0.789V
30	86	111.67	1.94	0.138V	140	284	153.58	1.88	0.528V	250	482	194.07	1.81	0.800V
35	95	113.61	1.94	0.159V	145	293	155.45	1.87	0.543V	255	491	195.88	1.81	0.810V
40	104	115.54	1.93	0.180V	150	302	157.31	1.86	0.557V	260	500	197.69	1.81	0.820V
45	113	117.47	1.93	0.201V	155	311	159.18	1.87	0.571V	265	509	199.49	1.80	0.830V
50	122	119.40	1.93	0.221V	160	320	161.04	1.86	0.585V	270	518	201.29	1.80	0.840V
55	131	121.32	1.92	0.241V	165	329	162.90	1.86	0.598V	275	527	203.08	1.79	0.850V
60	140	123.24	1.92	0.260V	170	338	164.75	1.85	0.611V	280	536	204.88	1.80	0.860V
65	149	125.16	1.92	0.279V	175	347	166.61	1.86	0.625V	285	545	206.67	1.79	0.870V
70	158	127.07	1.91	0.298V	180	356	168.46	1.85	0.638V	290	554	208.45	1.78	0.879V
75	167	128.98	1.91	0.316V	185	365	170.31	1.85	0.650V	295	563	210.24	1.79	0.888V
80	176	130.89	1.91	0.334V	190	374	172.16	1.85	0.663V	300	572	212.02	1.78	0.898V
85	185	132.80	1.91	0.352V	195	383	174.00	1.84	0.675V					

Note: The resistance variation between each 5°C increment shows the non-linearity over large temperature spans. The signal voltage is produced from a Wheatstone bridge source having +5 Vdc source, as in the prior example problem.

signal is calibrated, to be correct, and later the meter is calibrated to indicate the correct temperature.

- The transmission circuit is opened and a milliammeter is placed in the circuit.
- The decade box is set to 100 ohms to simulate a measurement of 0°C. Then, the "ZERO" (Z) trimming potentiometer, also called a trim pot, is adjusted to produce a 4mA transmitter signal, where 4 mA corresponds to the lowest temperature in the range.
- The decade box is then readjusted to the temperature where the highest accuracy is desired, which is the mid-span reading for general purpose applications. For this example, the mid-span reading of 100°C (212°F) corresponds to 138.5 ohms.
- At the transmitter, the "SPAN" (S) potentiometer is then adjusted to provide the correct mid-span amperage. For a 4-20mA signal, the mid-span transmission signal is $(20+4mA)/2=12mA$.
- After calibrating the "SPAN," the "ZERO" is then rechecked to see if some interaction was introduced between the two adjustments. If the "ZERO" still reads correctly, then the circuit calibration is complete. If not, the "ZERO" and "SPAN" adjustments are repeated until both are correct.
- If desired, the RTD can be tested at several setpoints over the range of the meter. The worst error would be used to calculate the full-scale accuracy of the meter.

Calibration of the digital meter

After the 4-20mA signal-transmission circuit is calibrated, the meter is then adjusted in a similar manner.

The decade box is set to emulate the 100-ohms "ZERO" resistance, which produces a 4mA transmission signal. The meter "ZERO" potentiometer within the meter is then adjusted to indicate 0°C (or 32°F).

The decade box is then set to emulate the mid-span value of 138.5 ohms, which produces a 12mA transmission signal. The meter "SPAN" potentiometer is then adjusted to 100°C (or 212°F) mid-span temperature.

The "ZERO" and "SPAN" values are rechecked and readjusted until they are both correct.

TANK LIQUID-LEVEL MEASUREMENT

Manual and automatic methods are used for measuring tank liquid levels. Manual methods include gage glasses to see the level, sounding tubes used to measure levels or ullages, and sampling petcocks. Methods that are useful for automated systems include proximity switches, magnetic, capacitance, ultrasonic or radar systems that measure tank ullages, or the pressures at the bottom of a tank, such as pneumercators or pressure transmitters. If the tank is pressurized, differential pressure is used to create a static reference, which is compared to a variable liquid level.

Gage glasses, petcocks, and sounding tube

The simplest level measurements are gage glasses, sampling petcocks, and sounding tubes that use measuring tapes, as shown in Figure 1.26. With gage glasses, the level is directly observed. Sounding tubes use bobs attached to a measuring tape. A plumb bob directly measures the liquid height, where chalk may be applied to the tape to highlight wetting by clear

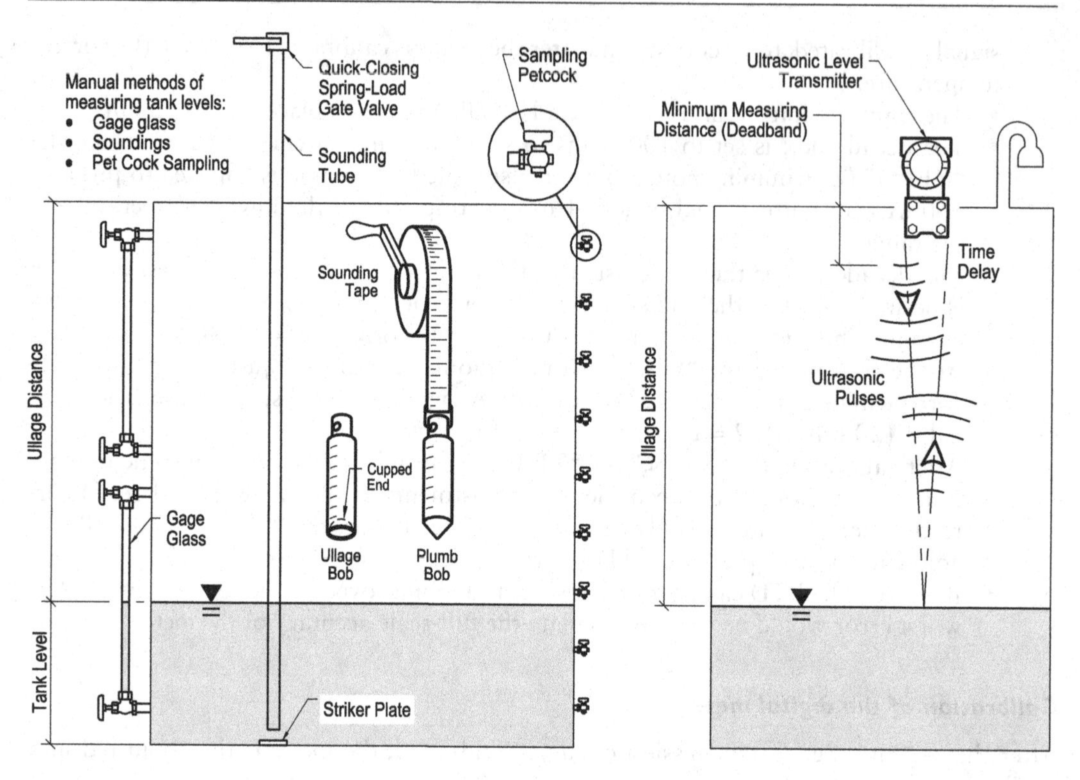

Figure 1.26 Several methods of tank level measurements are shown; manual methods include gage glasses, sounding tubes and tapes, and sampling petcocks (left), and one automatic method includes an electronic ultrasonic level transmitter (right).

fluids. The ullage bob is lowered using jerking motions, so that its cupped end makes a popping sound on the liquid surface, measuring the ullage distance, which is the clearance above the surface to the top of the sounding tube. Measuring ullages is good for messy fluids, such as oil. Petcocks or try cocks are used for harmless fluids, like water, by opening the valves until the fluid issues.

Ultrasonic or radar level transmitters

These are non-contact instruments that transmit pulses of sound or radio-frequency waves that reflect off the liquid surface and back to a receiver in the device. The ullage distance is determined from the wave velocity and the half-time it takes for the pulse to travel. This method is especially good for dangerous fluids, such as liquid cargo aboard tankers, where the sensor is relatively immune to condensation and hostile environments. Transmitter frequencies range from 15-200 kHz, where lower frequencies are used for more difficult fluids, solids, or deep tanks, while higher frequencies are used for tanks with small ullage. Some disadvantages are:

- the speed of sound varies with the temperature of the air or vapor density above the level, so temperature compensation may be required,
- foamy surfaces can absorb sound energy and degrade the signal,
- turbulence or agitation on the surface can lead to fluctuating readings.

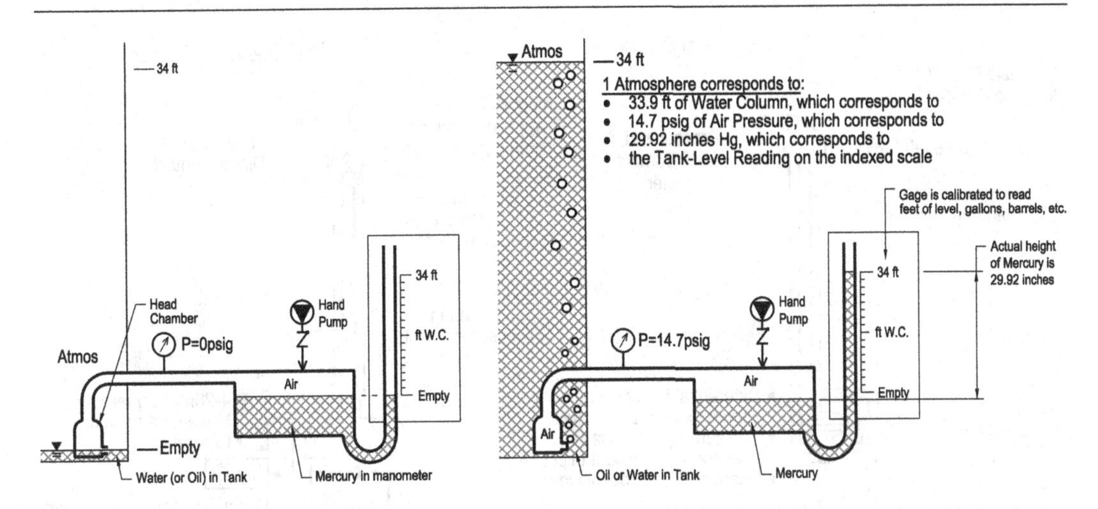

Figure 1.27 Pneumercators use hydrostatic principles to obtain tank-level measurements.

Pneumercators

These use the hydrostatic principles of the manometer to measure tank levels, as shown in Figure 1.27. Liquid mercury, which is a hazardous heavy metal, is used in the pneumercator because of its very high density relative to the fluid being measured. Air is used to transmit the signal. Air is pumped, sometimes by hand, into the head chamber to displace any tank liquid that may have entered the sensing line. As the air bubbles out the head chamber, it reaches an internal pressure that balances the height of the fluid applied to it. The transmitted air pressure is applied to the mercury manometer. As an example, a 30-inches tall mercury pneumercator could measure water level to a height of about 34 feet. The pneumercator scale would be calibrated in tank volume or in liquid depth, or both.

Pressure transducers and liquid level measurement

Liquid levels create predictable pressures at the bottom of a tank when the tank is vented to atmosphere and the liquid density is known. Technically, it is differential pressure that is measured between the bottom of the tank and atmospheric pressure, where the pressure measurement is calibrated to indicate level. Very sensitive differential-pressure transducers are used to create very precise electronic liquid-level measurements, sometimes in fractions of an inch of water column, for applications such as boiler steam-drum water-level control, which are maintained very precisely, as shown in Figure 1.28.

Often, the high connection is piped to a constant head chamber, which helps dampen tank pressure and level fluctuations. The constant head chamber contains an overflow which is connected to the bottom of the tank and which forms the level transmitter's variable leg. The variable leg freely tracks the tank level by gravity. The constant-head chamber creates a reference leg at the overflow, which is called the static leg. The liquid in the static leg is trapped, so that the pressure caused by its height is constant. Differential pressure is obtained by "hydraulically subtracting" the lower variable-leg pressure from the upper static-leg reference, using a sensitive intervening diaphragm between the two. The static leg is directed to one side of the sensing diaphragm and the variable leg to the other, so that the tank's internal

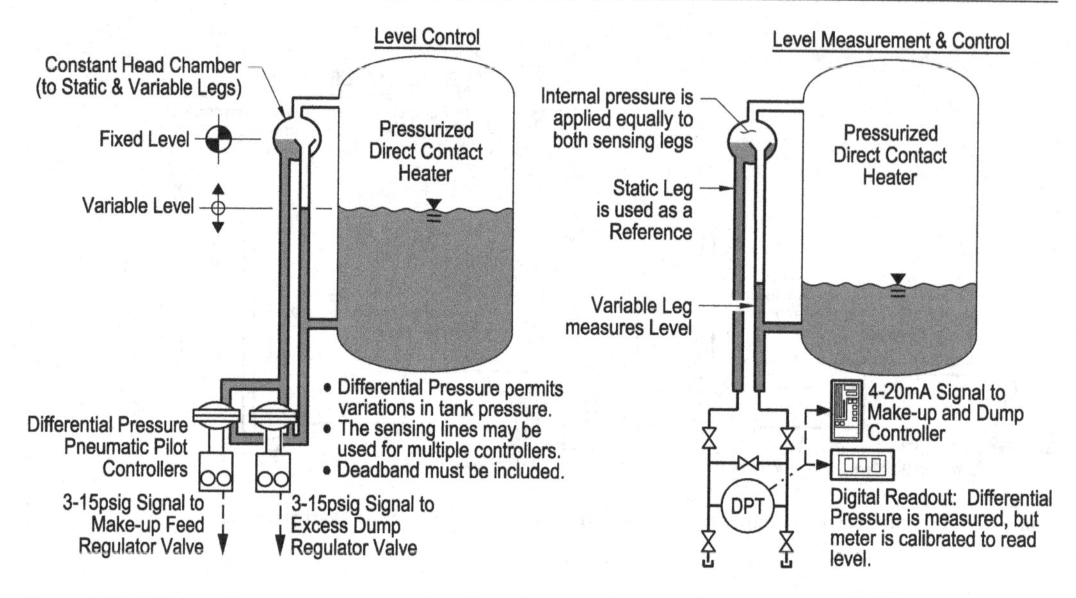

Figure 1.28 Differential pressure is applied across diaphragms in two pneumatic pilot controllers, shown on the left, used to automatically actuate low-level make-up and high-level dump valves. On the right, an electronic differential-pressure transmitter is used to measure level which is displayed by a digital readout. Commonly, this differential-pressure signal is also used in electronic controllers to do makeup and dump.

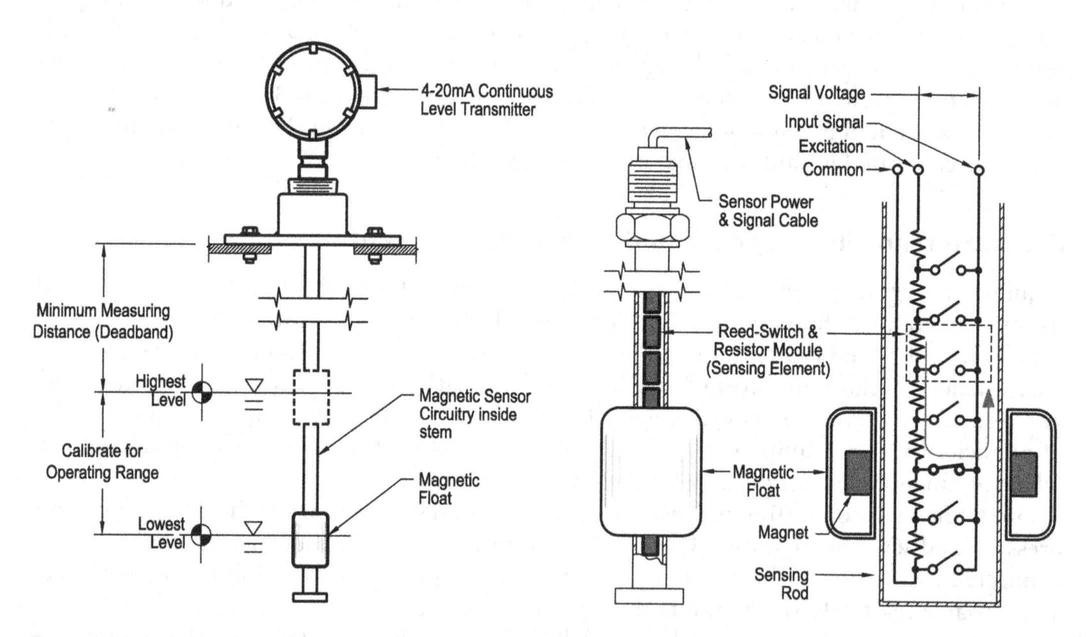

Figure 1.29 Magnetic float-level transmitters used magnets in a float to toggle adjacent switches.

pressure is applied equally to both sides, thus canceling. Consequently, the diaphragm senses only the difference in pressure between the static reference leg height and the variable-leg height. In this arrangement, the level transmitter is naturally self-correcting with any variations in tank pressure.

Magnetic float level transmitters

These use a magnet housed in a float and guided vertically by a tube containing a chain of resistors and reed switches, as shown in Figure 1.29. Excitation voltage is applied through all resistors in series. As the level rises, the float moves the magnet upward, closing the nearest reed switch contained inside the sensing rod. The voltage drop through all resistors leading to the closed switch becomes the sensor input signal, which may be converted to 4-20mA for transmission. The voltage corresponding to the position of the float is calibrated as the level signal.

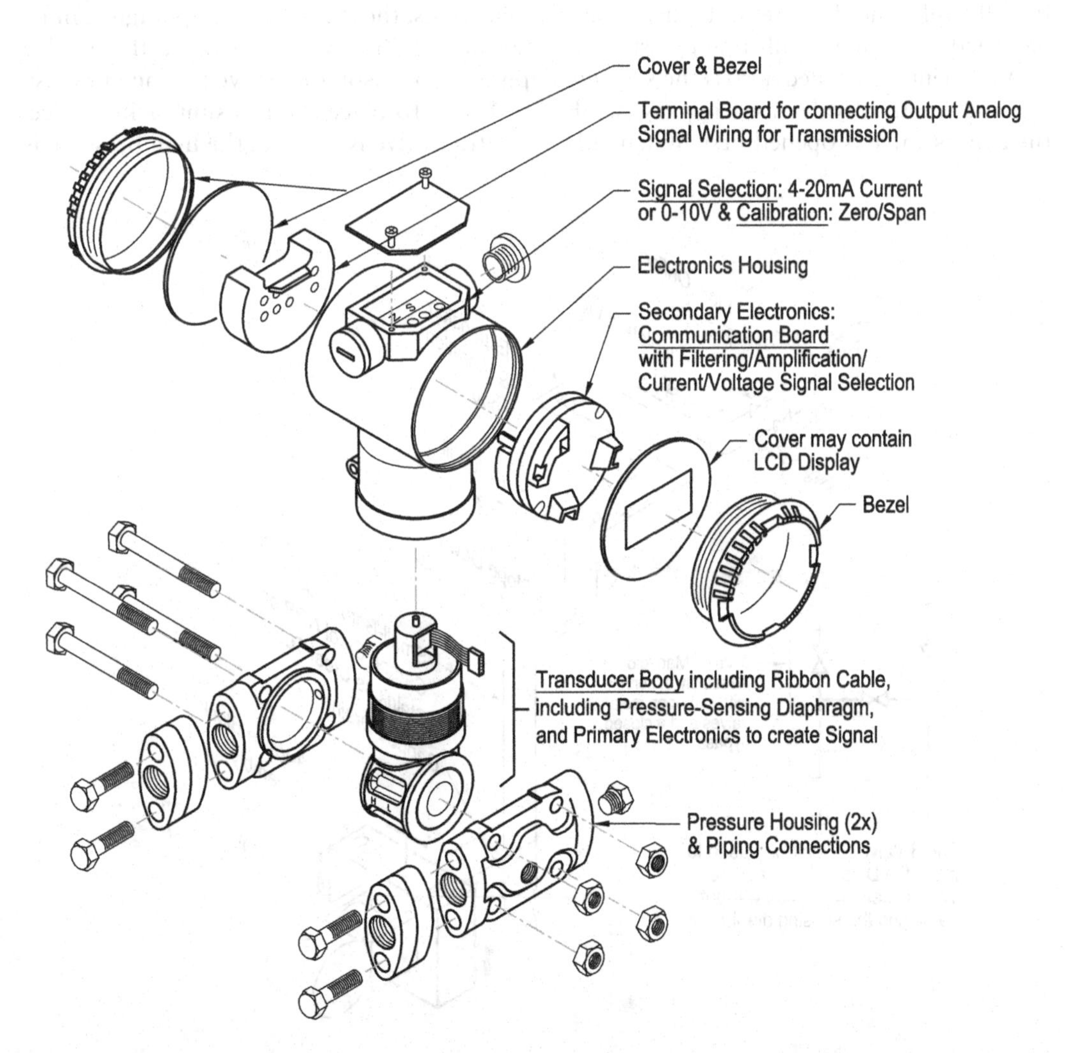

Figure 1.30 Exploded view of a differential-pressure transmitter. The lower transducer body contains the sensing diaphragm and primary electronics that create the analog-input measurement. The upper electronics converts the measurement into a voltage or current signal for transmission, and contains zero and span calibration adjustments. Some models have LCD displays built into the cover for local indication, in addition to the intended remote signal transmission. (Adapted from ABB.)

DIFFERENTIAL PRESSURE AND DIFFERENTIAL-PRESSURE TRANSMITTERS

In addition to measuring tank level as described above, differential-pressure transmitters can be used to measure flow rates, fluid viscosity, or pressure losses across strainers and heat exchangers for determining cleanliness. An exploded view of a differential-pressure transmitter is shown in Figure 1.30. The construction of a differential-pressure transmitter is identical to a pressure transmitter, except there are two impulse or sensing connections, and the pressure measurement range is often very small, in the range of inches of water column, requiring a very flexible diaphragm.

In measuring flowrates or level, the system pressure may be very high, such as for a 1,000 psig high-pressure boiler, while the differential-pressure, itself, used for measurement may well be under 1psid. To obtain the required sensitivity, the sensing diaphragm is very thin, flexible, and delicate. In high-pressure applications, the transducer diaphragm can be damaged by applying full line pressure on one side only. To avoid damage, a three-valve arrangement is installed across the sensing diaphragm, one isolating valve for each impulse leg, and the third valve as a bypass (see Figure 1.31). To place the transmitter in service, the bypass valve is opened first. When either isolation valve is opened, the high pressure is

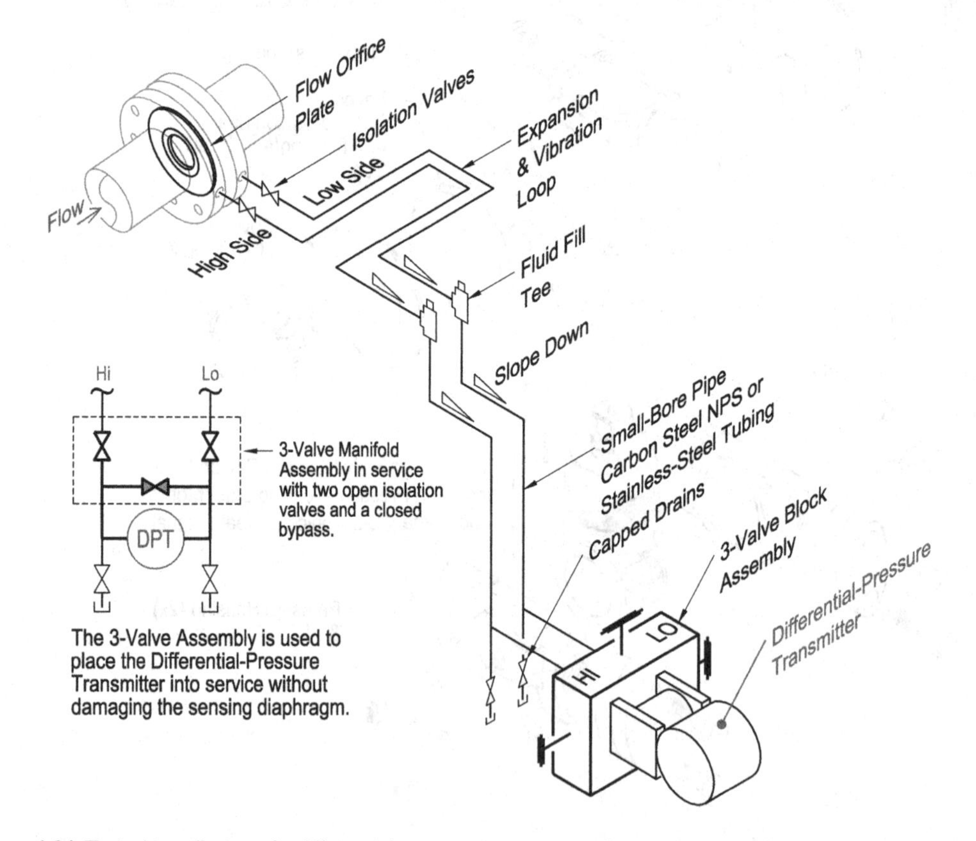

Figure 1.31 Typical installation of a differential-pressure transmitter. Sensing lines need to be pitched to vent air when used for liquid service, or care must be taken to ensure that condensate fills the sensing lines when measuring steam. A double block and equalizing valve is required for placing the transducer in service without damaging the sensitive diaphragm.

applied equally to both sides of the sensing diaphragm, avoiding damage. The bypass valve is then closed after the second isolation valve is opened, thus placing the transmitter into service.

Flow measurement

Positive-displacement totalizing meters

These consist of several types of flow meters, including positive-displacement, turbine, variable-area, vortex-shedding, ultrasonic, velocity, and differential-pressure types, as well as other types that use other effects.

Positive-displacement flow meters are shown in Figure 1.32, including the nutating-disk, oval-gear, and lobe positive-displacement-types. Using the nutating-disk meter as an example, fluid movement pushes the volume trapped between the nutating disk and chamber, causing both rotary and wobbling motions, which is called nutation. The disk movement is linked to a gear set having a rotating pointer and mechanical register that totalizes the flow. The gear-type and oval-gear-type-flow meters work on similar positive-displacement principles. Positive-displacement meters can also be fitted with analog-to-digital devices that produce electronic signals for PLC-based systems.

Differential-pressure flow meters

These are used very often in electronic flow-measurement systems and include flow nozzles, orifice plates, or Venturi elements, as shown in Figure 1.33. The differential-pressure device provides a restriction to flow, which induces a related pressure drop. Generally, the relationship between flow and pressure-drop is quadratic, where $Q \propto \sqrt{\Delta p}$. In the case of compressible fluid, the relationships are more complex and need to account for density, in terms of temperature and pressure. The pressure drop across the flow-element is characterized over the flowrate range. Although these devices may be used with analog differential-pressure gages, they are particularly well-suited for electronic instrumentation and control systems. Some have limited turndown and require a number of straight lengths of pipe before and after the flow-sensing element.

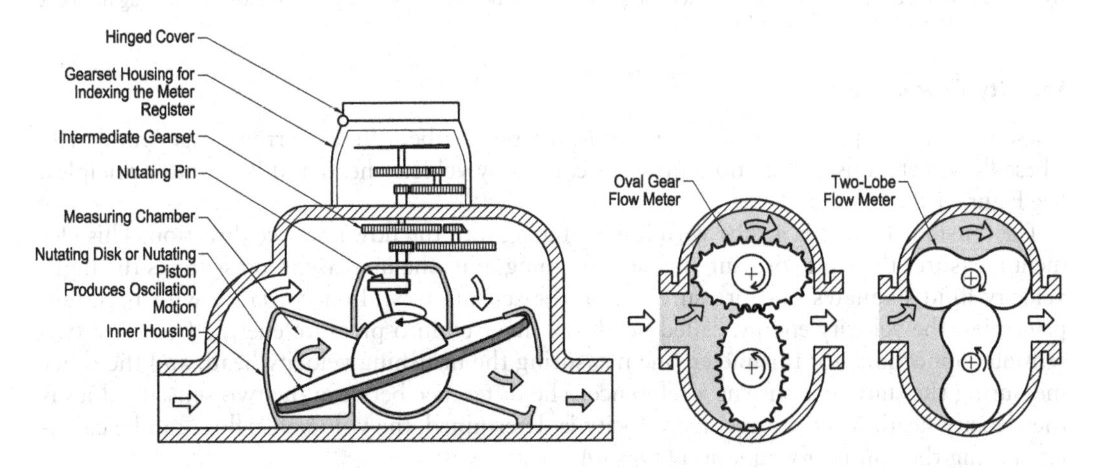

Figure 1.32 Positive displacement, totalizing flow meters, including the nutating-disk, oval-gear, and lobe-types.

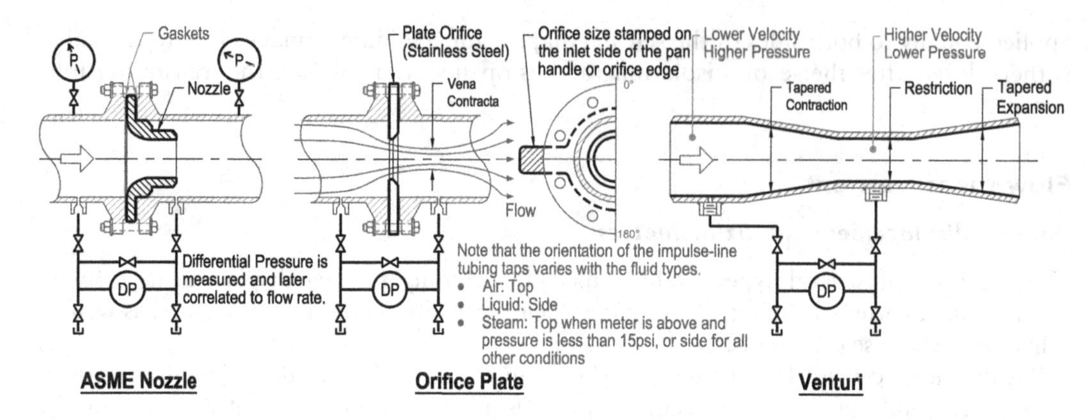

Figure 1.33 Differential-pressure type flow meters, including nozzle, orifice plate, and Venturi elements.

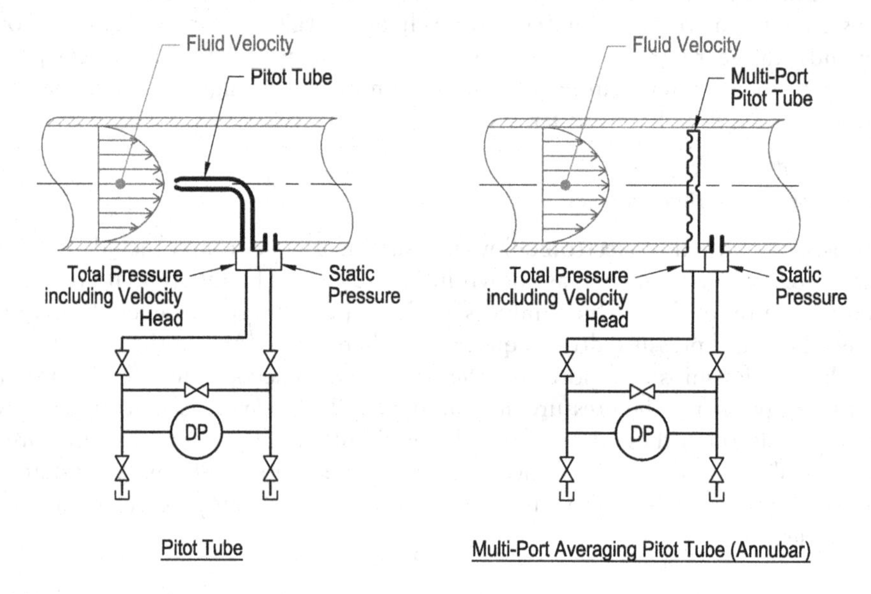

Figure 1.34 Velocity-type flow meters, including the pitot tube on the left and the multi-port averaging pitot tube (Annubar®) on the right.

Velocity flow meters

These include the pitot tube and the multi-port pitot tube, often referred to as Annubar®. These flow meters use differential pressure caused by velocity head and Bernoulli principles. See Figure 1.34.

The sensing element is a tube with hole(s) facing into the flow's source direction. This element measures the static pressure inside the piping, plus the increase in pressure as the high-velocity fluid stagnates into pressure within the sensing tube. In accordance with Bernoulli principles, the velocity energy, called head, is converted into pressure energy. There are two sensing connections for this meter, one measuring the inrushing velocity head, and the other measuring the static pressure as a reference. The difference between the two sensed values is the velocity head. After the average velocity is determined, the volumetric flow can be calculated using the continuity equation ($Q = vA$).

The multiport averaging pitot tube is a refinement to the simple pitot tube, and is often referred to as an Annubar®, which is a registered trademark of Emmerson/Rosemount. It uses multiple holes directed into the flow designed to measure the effective stagnation pressure.

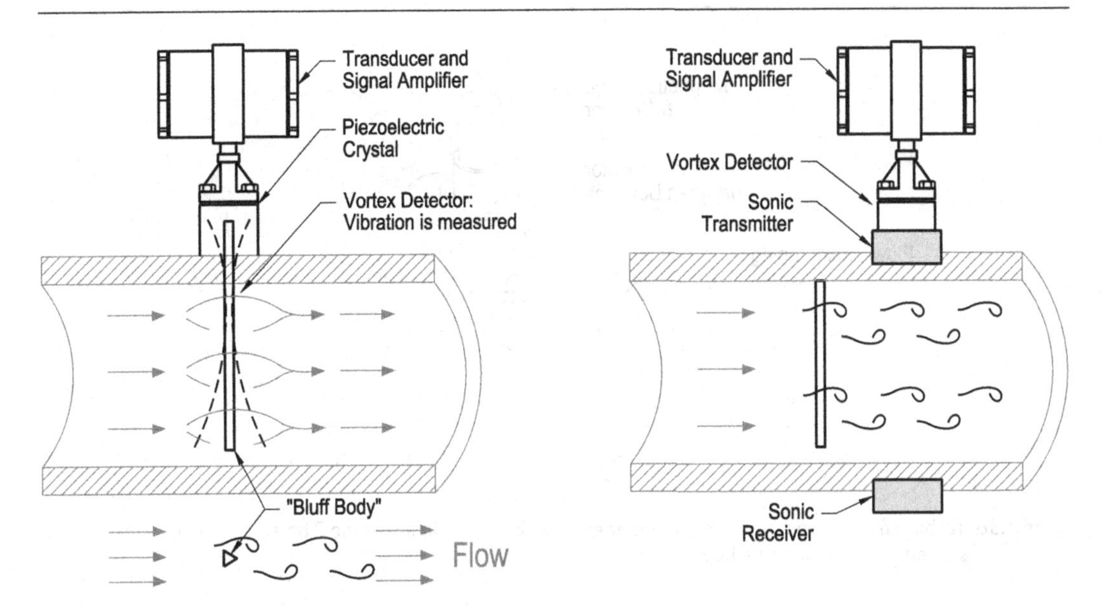

Figure 1.35 Two designs of the vortex-shedding flow meter. The flowmeter on the left measures vibrational frequency, while the flowmeter on the right uses an acoustic sensor to pick up eddies.

This arrangement results in more turndown. The device may have a small bleed port on the trailing edge to provide faster response. One manufacturer uses a T-shaped element with a thin slot down its length in lieu of holes, where the claim is that it produces more stagnation pressure in the "T" and better averaging of velocity head for improved accuracy.

Vortex-shedding flow meter

These work on the principle that when a flow stream encounters an obstruction, eddy currents and vortices are produced downstream. In addition to the turbulence, which can be measured acoustically, the vortices produce alternating high- and low-pressure forces on the obstruction or "bluff body," resulting in vibration. Figure 1.35 shows two vortex-shedding flow meters. The frequency of vibration is directly proportional to the flow velocity. The meter is calibrated to read volumetric flow using the continuity equation ($Q = vA$), average velocity, and pipe flow area.

Compared to the orifice flow meter, vortex-shedding flow meters have good linearity, are good for measuring flows of both liquids and compressible fluids, and maintain better accuracy over a larger turndown range, especially when they are installed in a reducing tube to increase velocity. Some models can operate with low flowrates having Reynolds numbers as low as 50. The meter does not use external impulse lines and they are well suited for measuring superheated, saturated, or wet steam flow.

Turbine flow meter

This is a kinetic device. It works on the principle that as flow passes over inclined impeller blades, rotary motion is created. The greater the flow rate, the faster the turbine spins. A magnetic pickup counts each passage of the blade tips, and the rotational speed is correlated to the average fluid-flow velocity and flow rate. The turbine flow meter is shown in Figure 1.36.

Flow measurements are most often made on a volumetric basis; however, sometimes mass flow measurements are preferred. Mass is a more accurate representation of flow when energy

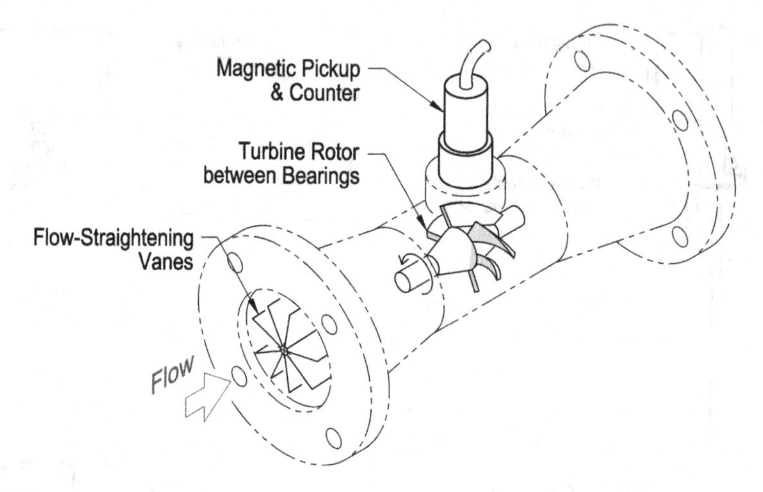

Figure 1.36 Turbine flow meter uses the kinetic energy of flow to spin a turbine. The signal comes from pulses sensed by a magnetic pickup.

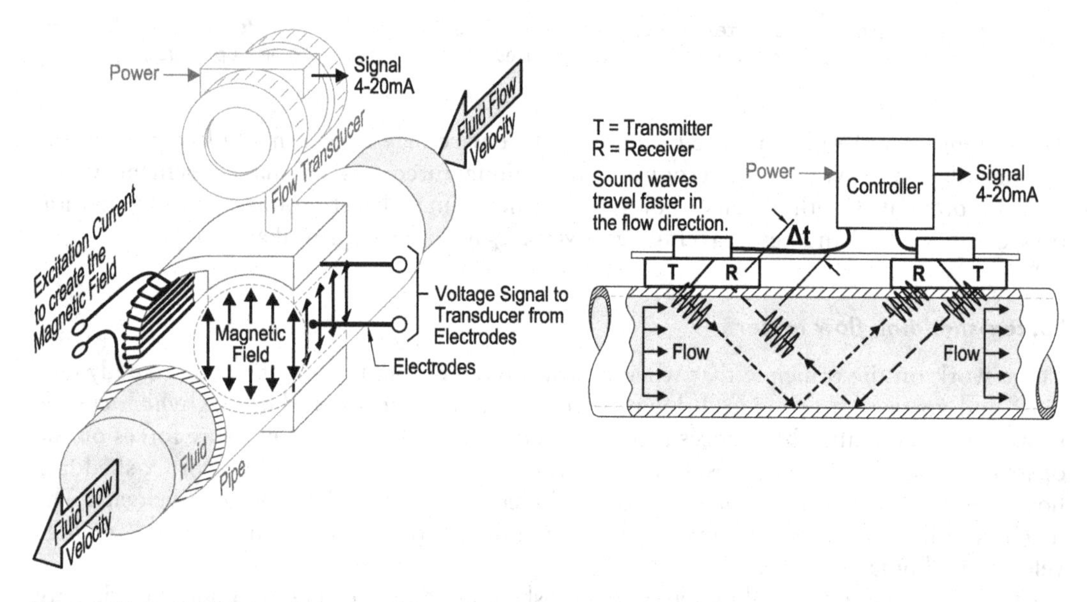

Figure 1.37 (Left) Magnetic flow meter (Cadillac Meter EMAG). (Right) Ultrasonic flow meter and transmitter.

measurement or utility billing is involved. Fluid density is used to convert from volumetric measurement to mass flow. To determine the fluid density, temperature is required and with compressible fluids such as steam or natural gas, the pressure is also required. Some instrumentation devices are available with multi-variable sensors, which means that in addition to measuring the flow, the same device has sensors to measure temperature and maybe pressure as well. The signal-processing electronics or the PLC program would then be characterized to accurately convert the three inputs into mass flow.

Non-intrusive flow meters/transmitters

Two types are shown in Figure 1.37, where the left side shows a magnetic flow transmitter and the right side shows an ultrasonic flow transmitter. The measurement relies on the fluid having a minimum amount of electrical conductivity. The magnetic type has high turndown and can measure all types of liquids, as long as they have an electrical conductivity above 3 micro-Siemens/cm^2 and a velocity exceeds 1 ft/sec. This transmitter is piped into the system with all metering components located outside the flow. The operation uses Faraday's Law, where voltage is produced in a conductor moving perpendicular to a magnetic field ($v(t) = -N\frac{d\varnothing}{dt}$). This flow meter creates a magnetic field from a coil having an excitation current, and the flowing fluid induces a voltage that is picked up by two electrodes and sent as a signal to the controller.

A transit-time type of ultrasonic flow meter/transmitter is shown in the right side of Figure 1.37. Although there are variations of ultrasonic meters, this meter uses two transmitters and receivers. Two ultrasonic sound waves are simultaneously transmitted, one into the direction of flow and the other opposite. In a similar manner to the way a boat moves slower when moving into the current, the sound waves transmitted upstream arrive after a longer delay than those transmitted downstream. The time difference between the two received signals is correlated to flow. This flowmeter is available for both liquids and gases and has very accurate measurements to extremely low flowrates.

Viscosity measurement

Viscosity measurement is important for some systems that use intermediate or heavy fuel oil for combustion. For example, precise viscosity is required for adequate injection and atomization of heavy fuel oil in large slow-speed diesel engines, and to a lesser extent, for heavy fuel-oil atomization in boiler burners. The heart of the control system is the viscometer, such as the Viscomater shown in Figure 1.38.

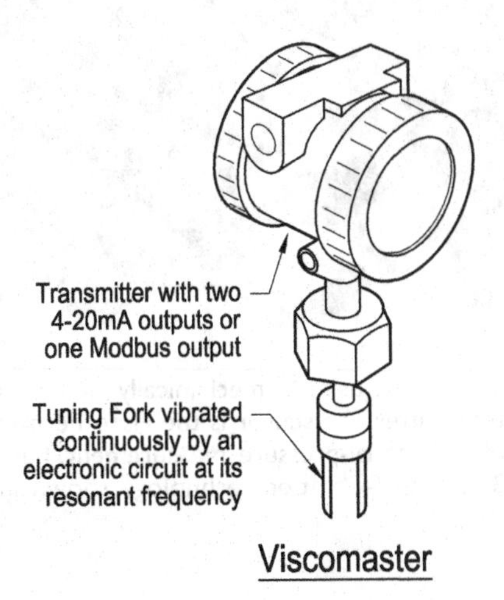

Figure 1.38 Fuel-oil viscosity measurement using a Viscomaster multivariable sensor and a tuning fork mechanism. (Courtesy of Emerson.)

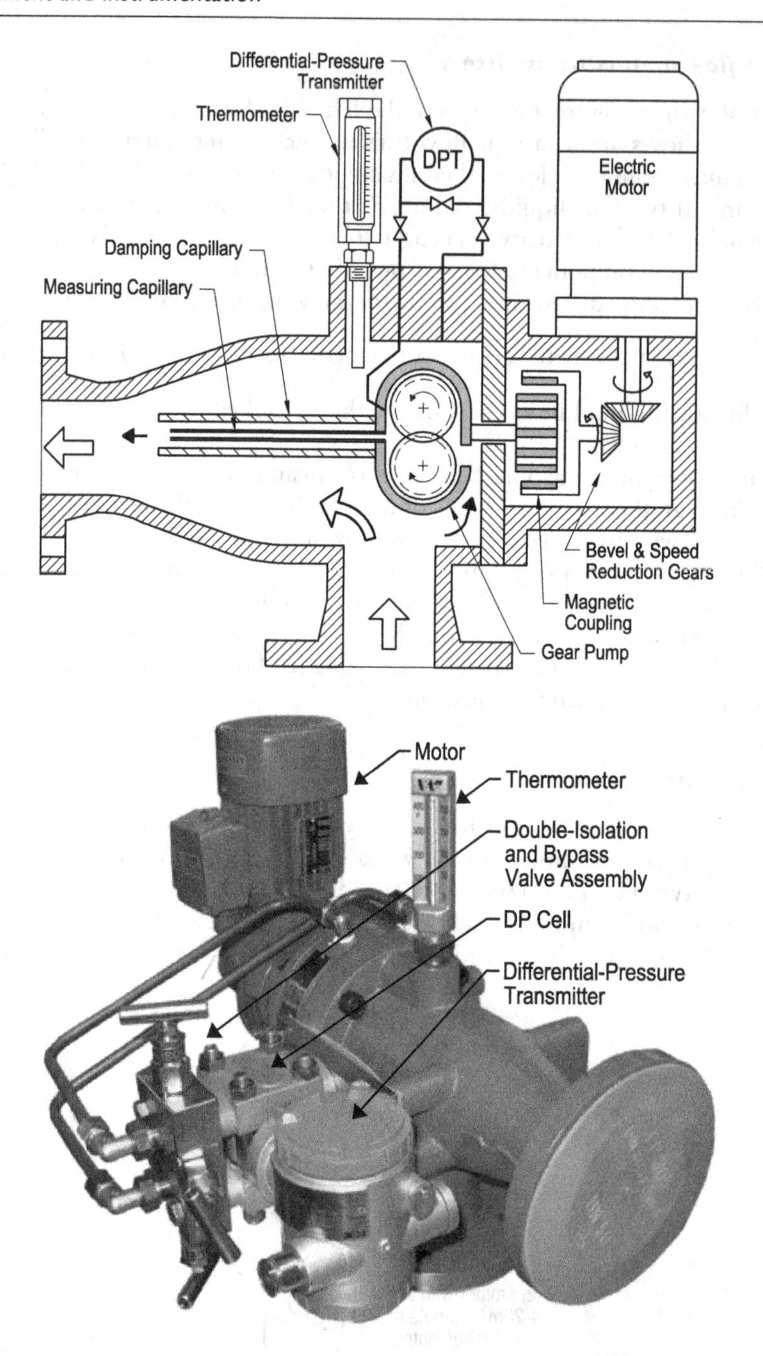

Figure 1.39 Fuel-oil viscosity measurement using a mechanically driven capillary tube and an electronic differential-pressure transmitter. The sensor is the Viscotherm® Series V92. This product shows another application of differential-pressure measurement, but has been discontinued in favor of the Viscosense® 3 torsional-vibration technology shown in Figure 1.5. (Courtesy of VAF Instruments B.V.)

The capillary-tube *Viscotherm* (VAF Instruments) in Figure 1.39 works by drawing a small sample of the flowing fuel oil through a slow-rotating gear pump and then through a measuring capillary tube. A differential-pressure sensor measures the back pressure between the gear pump discharge, whose flow is restricted by the measuring capillary, and the surrounding

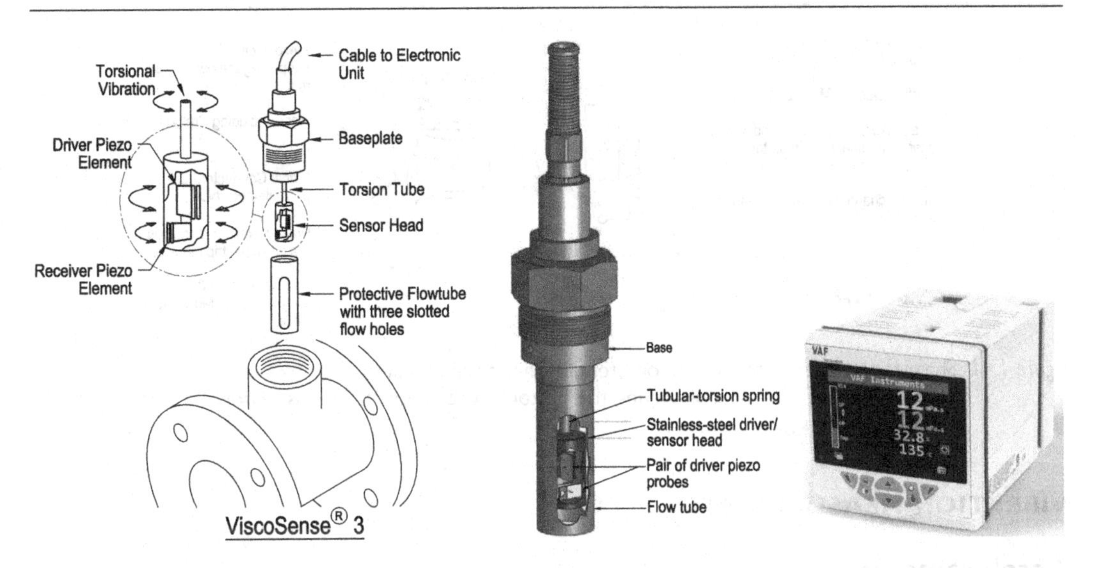

Figure 1.40 Fuel-oil viscosity measurement using vibration technology having two piezo elements in a torsional-pendulum arrangement, where resonant vibration is dampened by the fluid viscosity. An exploded view is shown on the left, the Viscosense®3 sensor in the center, and electronic controller on the right. (Courtesy of VAF Instruments B.V.)

line pressure. Higher viscosity causes more back pressure and more differential pressure. The device uses an electric motor to turn the gear pump at roughly 40 rpm through a reduction gear set to create low-velocity laminar flow. The differential pressure for laminar flow is directly proportional to the absolute viscosity.

A thermometer is usually installed, as temperature is an excellent predictor of viscosity, and its measurement is useful in adjusting the associated controller. A magnetic coupling on the motor drive allows slip and prevents over torqueing the motor/shaft when the viscosity is too high. A damping capillary tube smooths pressure fluctuations for more consistent readings, and the differential-pressure cell is mounted on the viscometer housing to eliminate the need for heating the impulse lines.

The *ViscoSense® 3* (VAF Instruments) in Figure 1.40 (left) is a newer technology intended to replace the Viscotherm, having no moving parts. The sensor consists of a torsional-pendulum mass mounted to a torsion tube and connected to a supporting baseplate. The torsional pendulum is excited by an alternating electronic signal to a set of piezoelectric elements causing it to vibrate at its resonant frequency. The frequency of the vibrating mass is affected by the amount of system damping, caused by the fluid viscosity. The thicker the oil, the greater the damping and the lower the oscillating frequency. A second set of piezoelectric sensors picks up the vibration, and an electronic processor determines the phase shift between the excitation signal and the measured feedback signal. The viscosity is directly proportional to the square of the phase shift.

Viscomaster (Emerson), shown in Figure 1.38, in is a multivariable device that simultaneously measures absolute viscosity and temperature. The Viscomaster differs from the other vibration-type viscometers in that it uses an excited tuning fork as a vibration-sensing element. Because the sensor also measures fork displacement, both density and kinematic viscosity can also be derived from the measurements.

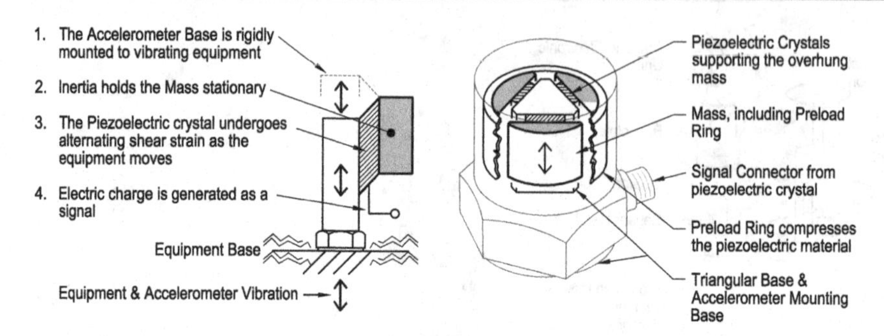

Figure 1.41 Shear-type piezoelectric accelerometer. The mass remains stationary by inertia as the base moves with equipment vibration, distorting the piezoelectric support. The distortion induces voltage as an acceleration signal.

VIBRATION MEASUREMENT

Accelerometer

Vibration measurements are used in plant operations to perform condition assessment of rotating machinery, trending of vibration levels, and predictive maintenance. Accelerometers are also installed on some high-speed machinery, such as large turbines or high-speed centrifugal lube-oil purifiers, to safely shut down the equipment before catastrophic failure occurs.

The accelerometer works on the principle that voltage is produced when a changing strain is applied to a piezoelectric material. Figure 1.41 shows the construction of one type of accelerometer and its operation. A mass, which tends to remain stationary by inertia, is mounted to an accelerometer base through a piezoelectric support. When the base moves by vibration, the piezoelectric material is distorted as the mass remains stationary by inertia. The strain caused by the deflection produces a measurable charge that is directly proportional to the acceleration. The piezoelectric material can be from naturally occurring quartz crystal, or synthetically produced doped ceramic.

DEFINE:

1. Sensitivity
2. Accuracy
3. Precision
4. Resolution

QUESTIONS

1. Where and why are snubbers and pigtail siphons used with pressure gages?
2. What is a strain gage, and how does it function?
3. List and describe six types of pressure gages, and where they would be used.
4. Describe the general procedure for calibrating a pressure gage. How is a precision shunt resistor used for fast "recalibration" of an electronic meter?

5. How are comparison gages used in calibrating gages? Describe how a deadweight tester is used in calibrating a pressure gage.
6. List three *mechanical* types of temperature gages and describe how they work.
7. List three types of *electronic* temperature indicators and describe how they work.
8. Why is a compensating lead needed in a three-wire RTD circuit? What does the compensating lead do?
9. Describe the construction of the 100-ohm RTD temperature probe and how it works.
10. List and describe three types of level-measuring devices.
11. What is the difference between a totalizing meter and a flow meter? How can a totalizing meter be programmed in a PLC to yield the total flow?
12. List and describe five methods of measuring flowrate.

PROBLEMS

1. A pressure gage has a maximum reading of 500 psig. During calibration, the highest error was observed at 200 psig, where the gage read 215psig. What is the true error in percen t at 200 psig? What is the manufacturer's gage accuracy based on the full-scale reading? Based on a 3% full-scale error, what could be the lowest and highest readings when the true pressure is 50 psig for this gage while still in compliance with its specifications?
2. A pressure gage measures 15 bar. Using references to one atmosphere, estimate the pressure reading in psig, kPa and feet of water column.
3. A compound gage measures 5-inches of mercury vacuum. What is the gage reading in lb/in^2 absolute (psia)? What is the gage reading in millimeters of mercury vacuum, knowing that 1-inch = 25.4 mm?
4. A water manometer having an angle of 15 degrees is used to measure fan discharge pressure, which operates between 1 to 5 psig. How far does the water column move in the manometer tube between its lowest and highest readings?
5. A 100-ohm platinum resistor, whose temperature-versus-resistance values are provided in Table 1.2, is installed in a balanced Wheatstone-bridge circuit with 6 volts applied into the bridge. The sensor is measuring exactly 212°F. What is the voltage signal measured across the bridge?
6. Using a spreadsheet, calculate the signal voltage for each temperature from -20°C up to 300 °C as obtained across the Wheatstone bridge. The circuit uses a 100-ohm platinum resistor in a balanced bridge circuit having an applied voltage of 10 Vdc across its midpoint junctions. Plot the relationship of voltage versus temperature, and using the trendline function, determine the best polynomial-fit equation for use in a PLC program.
7. A pressurized tank has a 4–20mA level sensor designed to indicate level from empty at 0-10 feet. The 0-foot level is to be calibrated to produce a signal of 4mA, and the 10-foot level is calibrated to produce 20mA. The fluid is water having a specific gravity of 1.0 and a density of 62.38 lb/ft^3. The normal tank level is 6-feet. What is the electronic signal in milliamps when operating at the normal water level? What is the differential pressure in inches of water column across the transmitter's level-sensing diaphragm when operating at the normal water level? What is the sensor's sensitivity in mA/inch? What is the milliamp signal for a low-level alarm set at 4.5 feet?

NOTES

1. Vacuum and boiler pressure gages are exceptions to this generalization. Since perfect vacuum at STP is 29.92 inches of Hg Vacuum, the vacuum gage has a full scale of 30 inches of Hg, and in high-vacuum systems, the pointer normally operates near the full-scale reading. Consequently, the gage would be calibrated for highest accuracy near this level. Likewise, if boiler pressure is constant at 850 psig, the gage would be calibrated to be most accurate near this pressure. Readings that deviate from normal generally indicate transient conditions where accuracy is less important, such as start up or abnormal conditions that need attention.
2. The relationship between input signal and output indication is called "gain" and is discussed in more detail in the Chapter 2 on control theory.
3. The sensing or "impulse" line lacks continuous fluid flow and is static. Except for thermal conduction or natural drain-back of condensate when the gage is located above the steam line, the static impulse line fills with condensate and reaches room temperature. The pigtail is only required when the gage is located above the steam line.
4. The thermistor rated ohmic value is its resistance at 25°C (77°F). Thermistors can be obtained for various temperature ranges. In this example, a 5,000-Ω thermistor was used, where its nominal resistance represents its value at 25°C (77°F).
5. The resistance of the wire is given by the equation $R = \rho\frac{L}{A}$. Resistivity (**r**) is a material property representing "specific resistance," or the resistance per unit length and area. Platinum is an expensive, high-conductivity, noble metal, but because it is such a good conductor, it takes only a very small amount of material to form the sensing element. The 100-ohm nomenclature represents resistance at the freezing temperature of water.

Chapter 2

Control terminology, theory, and tuning

Safe and efficient plant operation depends on well-designed control systems to keep operating parameters within acceptable limits, and to do so automatically and with good stability. The control systems can vary from crude but straightforward mechanical systems using diaphragms, springs, and levers that balance forces, or pneumatic systems whose signals create large actuation forces using air pressure, or very sophisticated electromechanical systems that use state-of-the-art electronic instrumentation with microprocessor controls. The choice of system is a trade-off between the control-system precision and response requirements compared against the economics. Regardless of the type of control system employed, the basic theory and terminology is the same.

In plant continuous control, some important terms include process, controlled quantity, setpoint, manipulated quantity, and system disturbances, which are shown in the block diagram of Figure 2.1.

The *process* is the plant function that is being controlled. To control tank level, the process is the mass balance of liquid inflow to outflow, whereas for boiler control, the primary process is the modulation of burner firing rate so as to match load, and the secondary process is the adjustment of the air-fuel ratio, so as to maintain efficient combustion.

The *controlled quantity* is the plant parameter that is to be regulated, maintained, or adjusted for the plant to operate properly. Controlled quantity examples are flow rates, pressure, temperature, tank levels, load, etc. For a tank, the controlled quantity is the level itself.

The *setpoint* is the desired value or goal. For a tank containing liquid, the setpoint might be to maintain a tank to be half or two-thirds full.

The *manipulated quantities* include the parameters that are being controlled. For instance, if tank level is the controlled quantity, the manipulated variable could be the inflow of fluid flow as make-up, or it could be the outflow of excess level as dump.

System disturbances are changes from outside the process that tend to knock the system out of steady-state equilibrium. For the tank, the disturbance may be consumers demanding more fluid.

OPEN- AND CLOSED-LOOP CONTROL

Control may be described as open-loop control or closed-loop control. *Open-loop control* does not use feedback to the controller and is much less common in industrial control. Instead, control is based on time. A good example of open-loop control is a washing machine, where the wash, spin, and rinse cycle times are preprogrammed, based on experience, and there is no feedback to indicate cleanliness has been achieved. For plant operations, automatic backwashing filters may use open-loop control for a cleaning cycle followed by

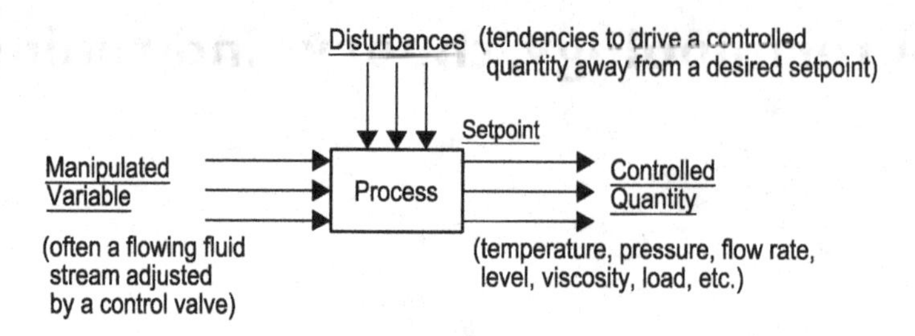

Figure 2.1 Block diagram showing basic concepts and important control terms.

differential-pressure measurement across the filter to determine whether fouling was resolved. A more sophisticated backwashing filter might be programmed to make several open-loop cleaning attempts if the pressure differential does not come within acceptable limits, and the controller may initiate an alarm after several unsuccessful attempts.

Closed-loop control is industrial control. Closed-loop control uses feedback and feedforward techniques to monitor the system and make fine-tuned continuous-control adjustments. Measurements of the controlled quantity that represent the process are sensed, and an output signal is produced for adjusting the manipulated variable, so as to keep the process within an acceptable range.

MODES OF CONTROL

Closed-loop control having feedback consists of the following options:

- On-Off Control
- Proportional Control
- Integral Control, also called Reset Control
- Derivative Control, also called Rate Control

On-off control

On-off control is the simplest and least expensive type of feedback control and is widely used. The manipulated variable is quickly toggled between maximum and minimum values, where 100% output represents ON, and 0% output represents OFF. On-off control is appropriate when precise control of the process is not required. An example is a steam-plant condensate-return system, where a pump may remain off until enough condensate has been collected to warrant starting the pump. Another example is a heating boiler controlled by a room thermostat, where the thermostat calls for heat and starts the system, and when satisfied, it stops the system. For this control scheme to work, the tank accepting condensate must have enough surge capacity to allow for unobjectionable level variations that do not adversely affect the system operation.

Figure 2.2 shows an atmospheric drain collecting system with on-off pump control and its response curves. When the level rises, a single float-actuated switch starts a pump and water is pumped away. Eventually the level drops enough for the float switch to stop the pump. In this system, the level rises and falls several inches or more, and if this level variation can be tolerated and motor short cycling is not a concern, then this control system might be a good inexpensive option. To obtain a longer duty cycle, two floats may be used, one for high-level cut-in and the other for low-level cut-out.

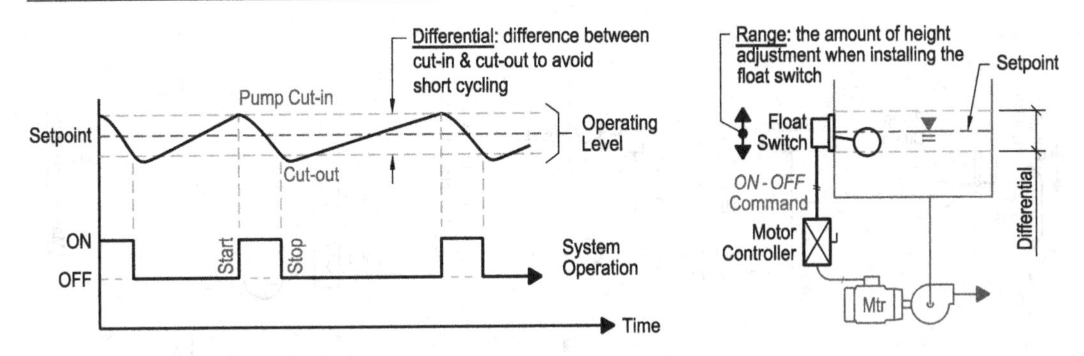

Figure 2.2 On-off control. This mode is the simplest and least expensive control type, but it does result in sudden level swings. If the system can tolerate the swings, then the control method is generally economical.

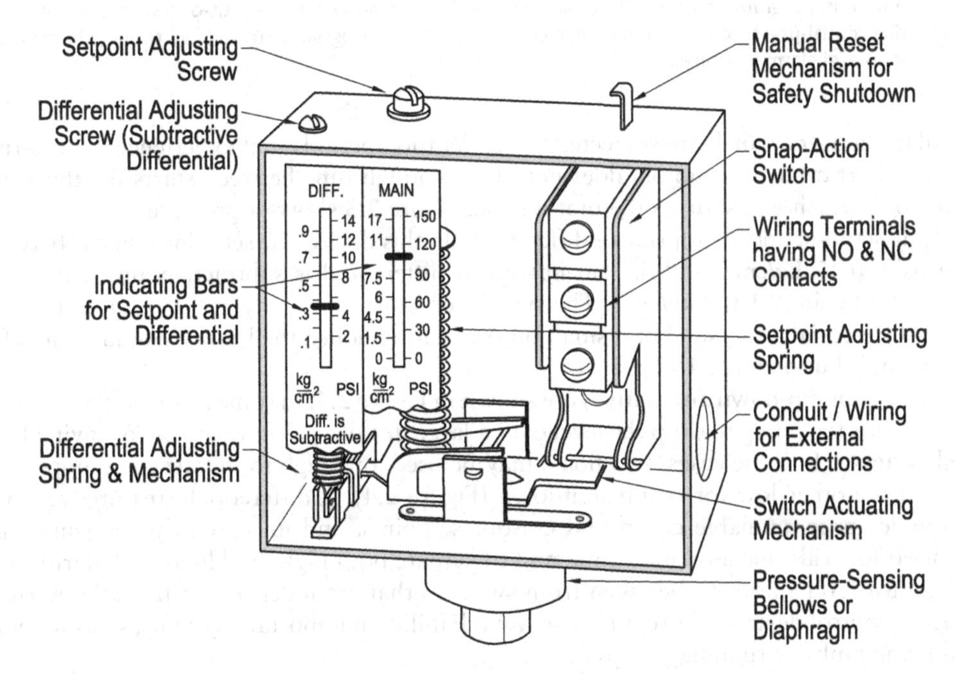

Figure 2.3 Pressure-actuated switch. Note the setpoint range on the left is 0-150 psig (0-17 kg/cm^2) with adjustable cut-in / cut-out differential between 2-15 psid (0.1-1.0 kg/cm^2). The 5-psig differential is subtractive from the setpoint for this device, which is used to determine the cut-in setting of 145 psig. As an alternative, the switch can be procured with additive differential to determine the cut-out setting above the setpoint.

Terms associated with on-off control and associated switches are range and differential. The *range* is the span of possible setpoints. For example, a home-thermostat may have a setpoint range between 45-80°F (7-27°C), where 45°F may be set for energy conservation when away and 80°F is higher than most people would want. *Differential* is the delta between the cut-in and cut-out toggle points that straddle the setpoint. For the home thermostat, the differential might be 1°F, so that if the setpoint is 68°F, the heating system would start at 67.5°F and cut out at 68.5°F. Differential is important for avoiding short-cycling of equipment,

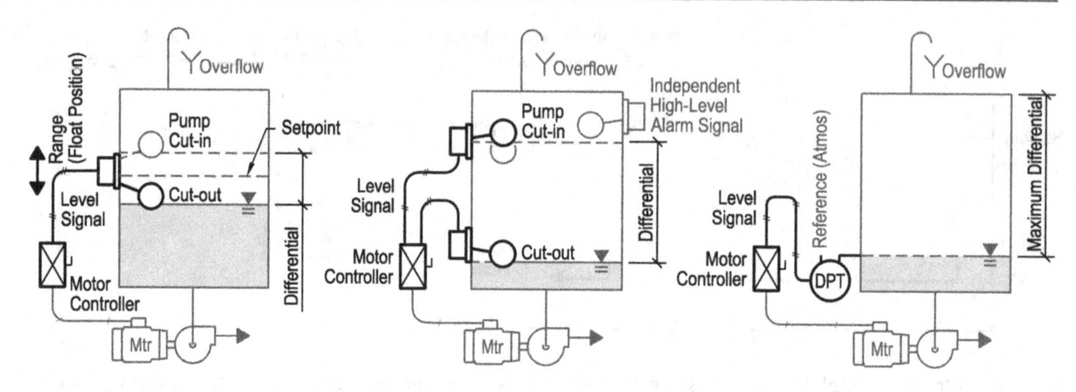

Figure 2.4 Three alternative methods of tank-level regulation using on-off pump-down operation. The single float has relatively tight differential (left). Two-float control increases the pump-down range (center) and extends the duty cycle to eliminate short cycling. An electronic differential-pressure transmitter is programmed for adjustable high- and low-level cut-in and cut-out settings (right). The differential-pressure-based controller can include more programming, such as to include high- and low-level alarm warnings.

particularly motors, which are subjected to acceleration periods that experience high in-rush currents. Short cycling of motors does not allow enough time between starts for the windings to cool, which will shorten the motor's life. Figure 2.3 shows a pressure switch with an adjustable setpoint and an adjustable differential. In this device, the setpoint range is between 0-150 psig (0-17 kg/cm^2) and the cut-in/cut-out differential is subtractive and may be set between 2-14 psid (0.1-0.9 kg/cm^2). The switch setpoint is presently at 105 psig with a subtractive differential of 5 psid. For a small on-off heating boiler, the burner would light-off at 100 psig and shut down at 105 psig.

Three tank pump-down alternatives are shown in Figure 2.4. In some instances, it is desired to arrange for the pump to drop the level over a large distance, as in waste tanks having large stored volumes. In those cases, two floats may be used, one high to initiate the pump-down cycle and the other low for pump shutdown (Figure 2.4). A differential-pressure transmitter provides programmable cut-in and cut-out setpoints, and its signal can be simultaneously used for real-time level indication and to actuate both high- and low-level alarms. It is common for level alarms to be taken from switches that are independent from the normal-operation control devices, in order to increase reliability in important systems, so as to avoid overflowing tanks or running pumps dry.

Proportional control with feedback

Unlike the system interruptions that occur from using on-off control, proportional control provides continuous modulation of the process. With proportional control, the corrective output signal is proportional to the error signal, where the error is the amount that the level deviates from its setpoint. Small deviations from a setpoint will result in small corrective actions, while large deviations will initiate large corrections.

As an analogy to automatic continuous control, Figure 2.5 shows the equivalent steps involved with manual control. The operator observes a parameter that represents the system behavior, decides whether the reading is acceptable compared to a desired value, and then actuates a corrective adjustment to change the system. To automate the system, those manual functions are replicated by a controller, as shown in the block diagram in Figure 2.6.

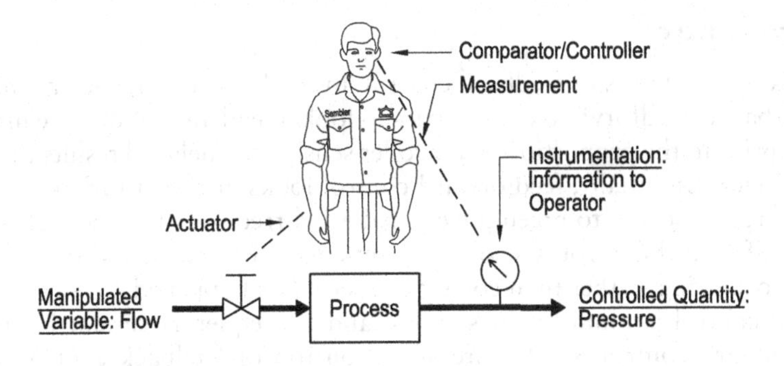

Figure 2.5 Manual control system. The operator observes the system, processes a decision, and makes an adjustment. An automatic-control system replaces the operator's actions.

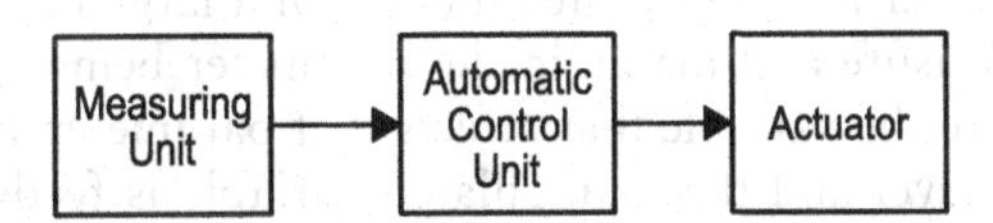

Figure 2.6 The plant operator is replaced by an automatic system that emulates his actions. Control is accomplished by using instrumentation to measure an appropriate system parameter, an automatic control unit to compare the measurement against a setpoint so as to produce an output signal, and an actuator to make the adjustment. The three control tasks are measure, compare, and adjust.

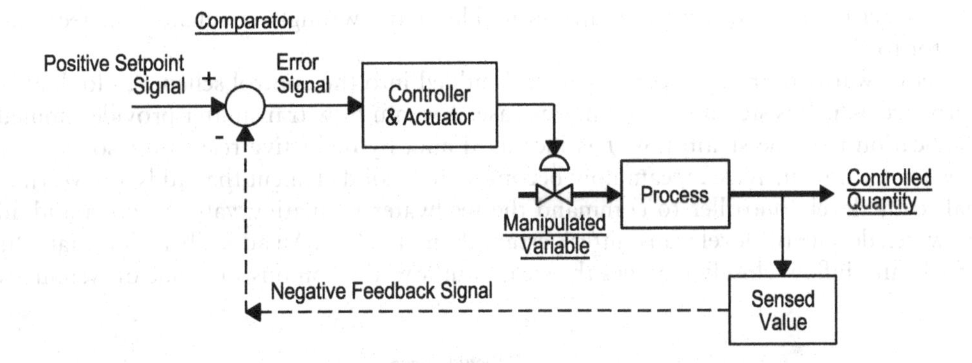

Figure 2.7 Feedback control. The simplest type of continuous control is the negative-feedback loop controller. The controller compares a sensed value against the desired setpoint, determines the error between the two, and then outputs an adjusting signal, whose goal is to eliminate the error.

The basis for continuous automatic control is from a closed-loop measurement-and-actuation system that uses negative feedback, which is shown schematically in Figure 2.7. In automatic control, a parameter that represents the process state is measured and fed back as a negative signal into a comparator device, where it is summed into a positive setpoint signal. Any non-zero output from the comparator is forwarded as an error signal into the controller for action. Often the output signals from a controller are very small and must be boosted to create adequate actuation forces. The actuator manipulates the process, while the feedback control-loop functions to remove error.

Feedforward control

Feedforward is the antithesis of feedback control. Instead of looking back at the result from a system disturbance, feedforward control tries to look ahead and predict the impending corrective action, prior to the proportional controller sensing the delayed results that occur from the system changes. Essentially, feedforward control looks at the disturbances to anticipate an appropriate response and to preemptively make a corrective adjustment (Figure 2.8). For this reason, feedforward control is also called anticipatory control, and it strives to prevent the error from occurring, rather than correct it after it has happened.

Feedforward control is blind to the setpoint and consequently cannot be used by itself. Instead, feedforward control is a feature added on top of feedback control to produce a faster, more accurate control response. For feedforward control to work, the nature of the disturbance's effect on the system must be understood. Feedforward control is often applied as second or third elements of control.

A good example of feedforward control is a two-element feedwater regulator designed to maintain a very precise water level in the steam drum of a large industrial boiler. In control systems, the goal is to measure and maintain the parameter being controlled as directly as possible and to manipulate the variable that affects that parameter. In this case, the primary control element is drum level and the manipulated variable is feedwater flow. However, a problem occurs when measuring boiler water level during load changes, where the level initially moves opposite to the intuitive direction from phenomena called shrink and swell. For example, when the demand increases, steam flow increases immediately, and a lower level is expected to follow. But, as steam is extracted from the drum, the pressure begins to drop, causing entrained steam bubbles to expand, leading to a rise in water level. Consequently, the feedwater flow should have increased along with the steam flow increase, but instead, the feedwater level controller senses an elevated level and wrongly commands the feedwater regulator to close.

In feedforward control, a second element is added into the control scheme to look at the disturbance, which is steam demand in this case. A steam-flow transmitter provides immediate indication that the steam flow has increased and, by deductive reasoning, so too must the feedwater flow increase. Steam flow becomes the second element that adds an overriding signal to the level controller to command the feedwater regulating valve to open and add more water, despite the level transmitter's errant desire to close. An added benefit is that when the feedwater inflow closely matches the steam outflow, the combustion-control system fires

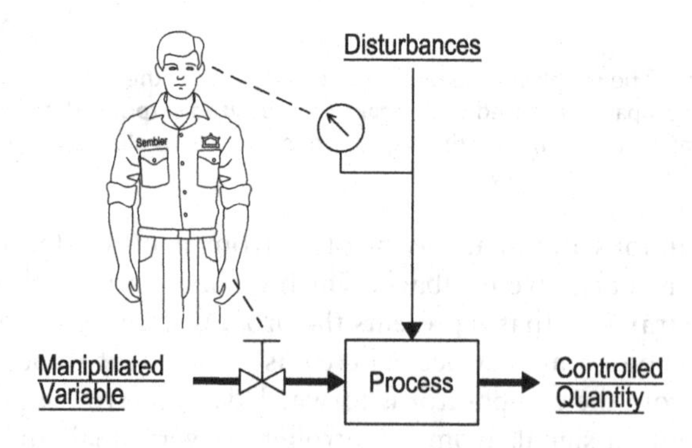

Figure 2.8 In feedforward control, the system disturbances are observed, and a preemptive adjustment is made in anticipation of the impending system changes. Since there is no relationship to the setpoint, feedforward action cannot be used by itself, but it can produce faster, more accurate response when it is combined with negative-feedback control.

the boiler at a steadier rate and the boiler pressure remains more constant, as sudden surges of cold water are avoided. A third benefit is that the same steam-flow transmitter used for feedwater anticipatory control can also be used by the combustion control system to preemptively command the heat input to increase, so as to further avoid a dip in boiler pressure.

The second element, steam flow, anticipates that more water is required when it observes the steam-flow increase, and the system is tuned to react appropriately. The feedforward element only works during transients, and after steady-state flowrates are reached, the second element does not contribute to the level control, but instead waits in abeyance for the next transient. The extra expense of the second element is worthwhile on large industrial boilers supplying steam turbines but is often not justified for small heating boilers.

To obtain even better control, another feedforward function can be added as a third control element. The third element is feedwater flow, and it forms a check-and-balance to the second element. Upon a steam-flow increase, the second element should be tuned to override the level controller to rightly command the feedwater regulator valve to open. The third element then checks whether the feedwater flow did indeed increase, and if so, it moderates the feedwater flow rate to avoid overreaction.

Proportional control

Proportional control provides continuous modulation of a process and is the basis for simple feedback control and for more sophisticated schemes. With proportional control, a sensed value representing the controlled quantity is compared against a setpoint, to generate an error signal, as shown in Figure 2.9. The error is used to produce a proportional corrective signal, which is transmitted to an actuator to make the change. Proportional action implies that small deviations from a setpoint will produce small corrective actions, while larger deviations will produce larger corrections. Proportional control looks at the present value of error.

A good way to understand proportional control is through a conceptual example, which is done in Figure 2.10. In this fictitious tank-level-control system, a float is directly linked to a makeup-water valve so that as level drops, the valve will proportionately open to replenish the extra usage. The load or demand is the rate at which water flows from the tank. In the left sketch in Figure 2.10, the demand is at half of its maximum value, the tank level drops to be half full, and the linkage serendipitously positions the valve to be half open, reaching steady-state equilibrium. When the demand rises to 75%, the level drops to 25% full and the

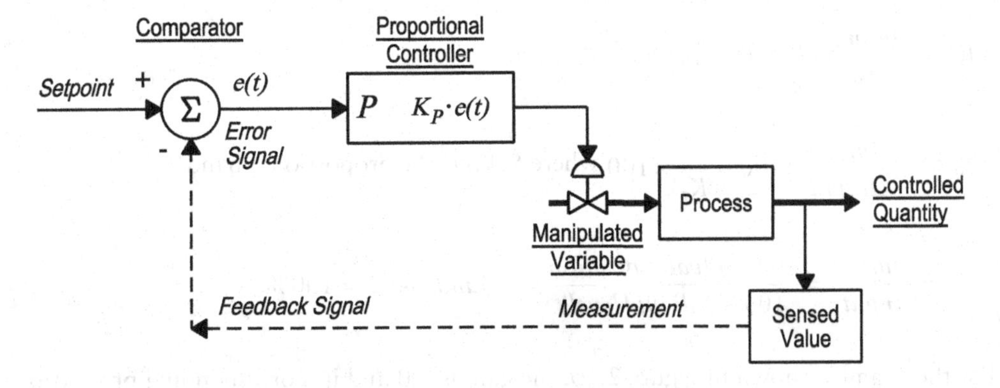

Figure 2.9 Block diagram showing proportional-only control using negative feedback. Proportional-only control looks at the present value of error and produces a corresponding corrective action, where the greater the error, the larger the corrective action will be.

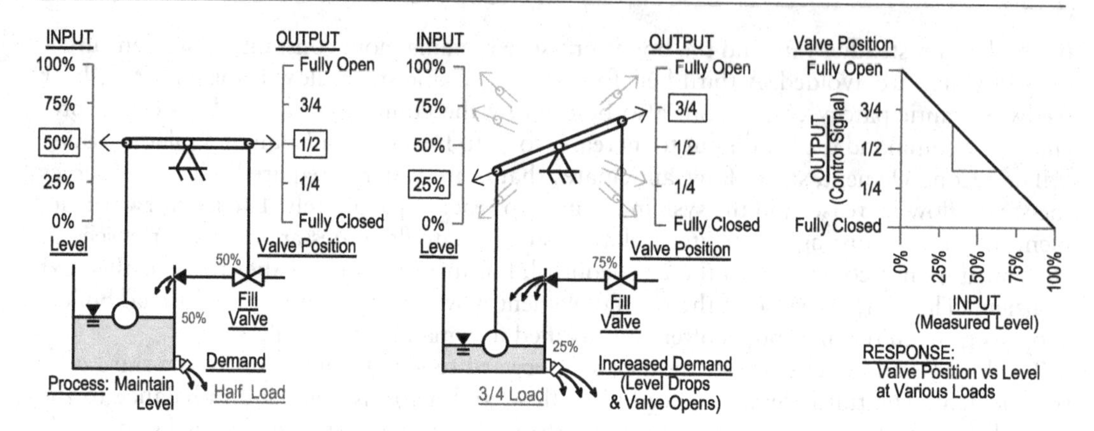

Figure 2.10 Proportional control provides a method to make continuous adjustments that avoid large, sudden swings.

valve opens 3/4 of the way, establishing a new equilibrium at the new level. In this manner, the valve position is directly proportional to the demand and inversely proportional to the tank level, while the valve varies from being fully closed when the tank is topped off to fully open when the tank reaches empty.

The output is represented by the valve position, and it can be plotted against the input, which is represented by tank level. The plot between output and input is illustrated at the right of Figure 2.10. Mathematically, the controller output is algebraically proportional to the input-error signal, and the slope of the line describes the response for the valve's full excursion between fully closed and fully open.

Gain and proportional band

Several important concepts can be observed from this example. First, the slope represents the relationship between signal output to its input and is called *gain*. In controls mathematics, gain is represented by the variable "K." *Proportional band* (PB) is another way to look at gain, where PB is given in percent. Mathematically, proportional band is the inverse of gain, so it is the ratio of the input to the output. Physically, PB defines the tightness of control. The equations for gain and proportional band are:

$$K_P = \frac{Output}{Input} \, where\, K_P \, is\, proportional\, gain$$

$$\%PB = \frac{Input}{Output} \times 100 = \frac{1}{K_P} \times 100 \text{ where } \%PB \text{ is the proportional band}$$

$$K_P = \frac{Output}{Input} = \frac{100 - 0\ (valve\, position)}{100 - 0\ (liquid\, level)} = 1.0\, and\, the\, PB = 100\%$$

For the example shown in Figure 2.10, the gain is 1.0 and its corresponding proportional band is 100%. The proportional band of 100% means that the valve's full range of modulation extends over its entire possible physical span of level variation. This variation of level would be considered loose and poor, as the tank level can be anywhere from empty to full, depending on load.

Offset

This example illustrates an important characteristic of proportional control called *offset*. Before the corrective action can take place, the level needs to deviate from its setpoint. In this example, the desired setpoint is to maintain the 50% level, which occurred at the 50% load with the valve half-open. When the water usage increased, the valve did not open until the level dropped. In other words, the objectionable characteristic of offset occurs during demand changes when using proportional control, where the level will deviate from the setpoint at different loads. In effect, each new demand establishes a new level, with a different deviation from the setpoint, illustrated by the offset in Figure 2.11. Offset is the difference between the desired setpoint and the actual level after the new steady state is reached. For a 75% demand in the example above, the level dropped to be one-quarter full before the valve correctly opened three-quarters of the way, to ultimately balance the load.

For speed-control governors, the total offset measured between no-load and full-load operation is referred to as droop. A governor's sensitivity is defined by its speed droop. The speed droop characterizes an electrical generator's ideal synchronous-speed frequency relative to its actual frequency at various loads, and is used to predict the load-sharing characteristics between paralleled generators.

Since 100% proportional band in this example is unacceptable, the controller performance needs improvement. To tighten control, the system gain can be increased, so that more valve-position output signal is obtained from less liquid-level input movement. In this example, the gain is increased simply by shifting the fulcrum to the left, as shown in Figure 2.12, producing the better controller response shown in Figure 2.13. With higher gain, the valve excursion still goes from 0 to 100% open, but the water level is tightened to remain between 25% and 75%. The new gain and proportional band are calculated as:

$$Gain: K_p = \frac{Output}{Input} = \frac{100-0}{75-25} = 2.0 \text{ and the } PB = \frac{1}{K} \times 100 = 50\%$$

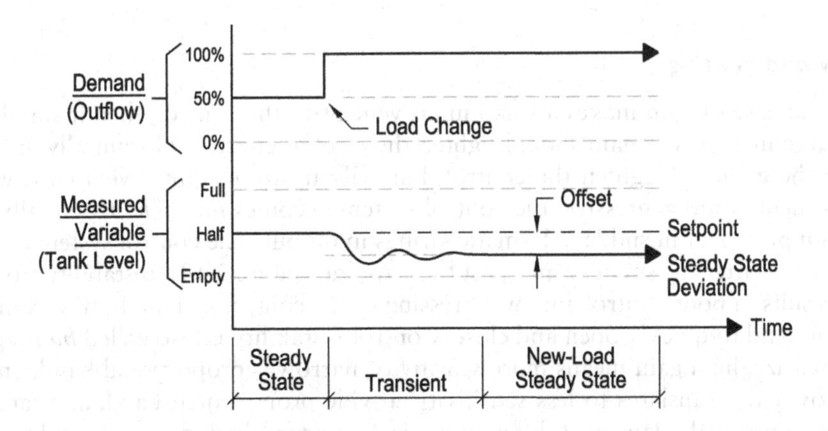

Figure 2.11 The system response curve (lower) is shown after undergoing a load change in a proportional control system. The response shows some settling time and residual offset from the setpoint after steady state is reached.

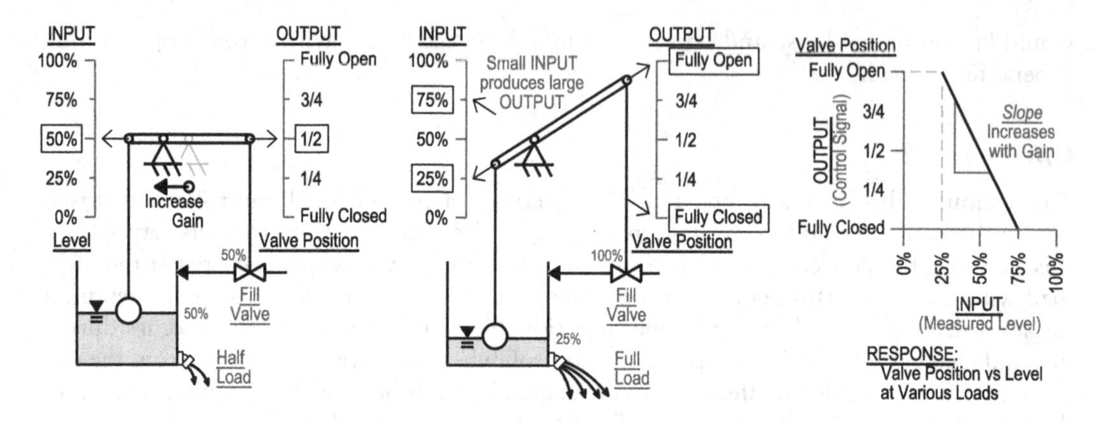

Figure 2.12 The disadvantage of proportional control is that it produces different amounts of offset or deviations from the setpoint at various loads. The gain is the ratio of the output signal compared to the input signal. Increased gain results in more sensitivity, faster response, and better accuracy with less offset.

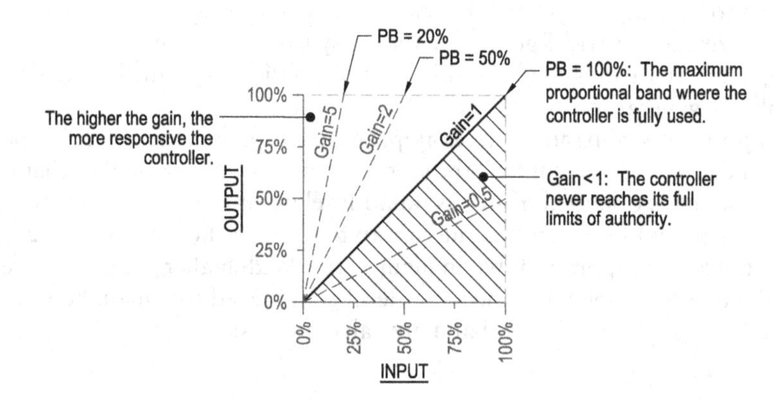

Figure 2.13 Controller response. The gain is plotted as the ratio of output versus input signals. The higher the gain, the more responsive the controller will be. Gains less than 1.0 do not use the controller's full authority and are to be avoided.

Instability and hunting

If a slight increase of gain makes a small improvement to the control, then it stands to reason that large increases of gain should tighten the level of control substantially. In practice, increasing the gain will tighten the control, but only up to a point. Eventually, when the gain adjustment is too aggressive, the control system becomes *unstable*. Essentially, minute changes in input result in sudden, dramatic swings in output. The control system overshoots and then chases after the small swings past the setpoint (Figure 2.14). Instability from excessive gain results in poor control and overstressing of the control system. In this example, the valve would tend to quickly open and close. Control instability is also called *hunting*.

To summarize, high gain means more sensitivity, narrower proportional band, and better control. Low gain translates to less sensitivity, a wide proportional band, and large variations of the controlled quantity at different loads. In a properly designed control system, the proportional band should never be adjusted beyond 100%, as this value prevents the full extents of controller output from being reached. In the tank example above, a proportional band exceeding 100%, or a gain less than 1.0, means that when the tank is completely full or

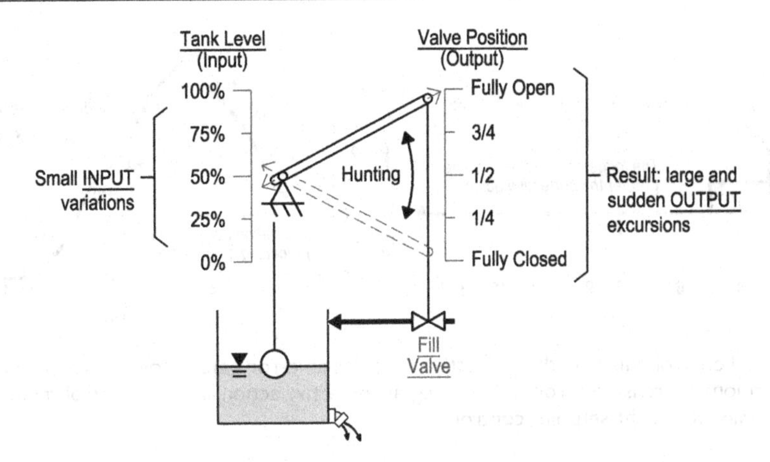

Figure 2.14 Increasing the gain reduces offset by reacting faster and more aggressively. Too much gain results in instability, called hunting, with wild swings in the output response and poor control.

empty, the valve has not reached its full extent of travel, and the controller has unused capability. Excessively small gain may also be indicative of oversizing of the controlling devices, which is generally a bad practice.

Integral or reset control

When the amount of offset produced by a well-tuned proportional-only controller produces adequate, reliable control, then proportional control is justified and economical. If the offset is objectionable, then additional control features are needed, including integral and/or possibly derivative control.

Integral control is designed to remove offset by shifting the setpoint back to the desired value after a load change has occurred. Essentially, the desired setpoint is "reset" to its original value after the change, so consequently, integral control is also called *reset* control. Integral control cannot be used by itself and is always coupled with proportional control. Together they are referred to as "Proportional plus Integral" control, "Proportional plus Reset" control, or simply as "PI" control.

Integral action produces a correction signal based on the mathematical integration of the error signal with respect to time, constrained within a limited period. It is the summation of each error multiplied by the duration of time that the error has existed. As in calculus, integral-control signal becomes the area underneath the error-versus-time curve. The integral action modifies the output control signal and its manipulated variable at a rate proportional to both the duration of the error's existence while looking at the *past*, and the magnitude of the error, while looking at the *present*. *Integral* control looks at the *past* for making a real-time setpoint adjustment. The longer the operating point has strayed from the desired setpoint, the greater the corrective action that will be applied. The result of integral control is the eventual removal of offset, after steady-state conditions have been reached. Integral control takes the summation of discrete values of past errors multiplied by their times away from the setpoint, to produce its output signal.

Figure 2.15 shows an illustrative example of integral control over the preceding 4 time periods for a system having a gain of +10. At time equals 3, a total error of 1 unit has existed, and an integral corrective signal of +10 is produced. At time equals 4, the error increased to 2 to bring the total area under the error curve in the preceding 4 time periods to 3 units,

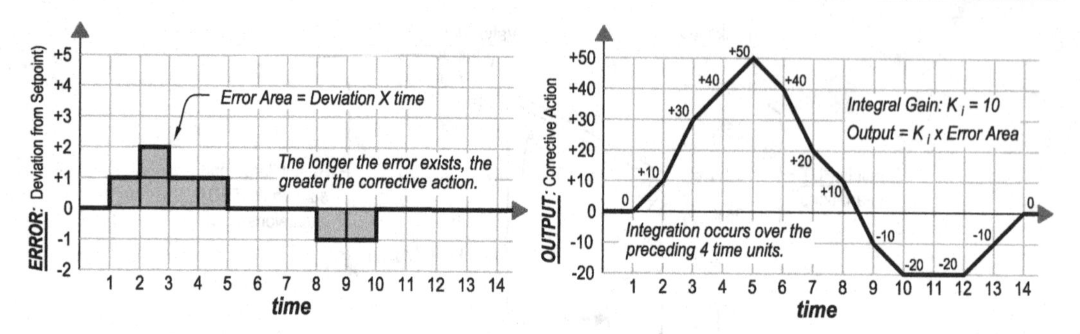

Figure 2.15 Integral control builds up the corrective action as an error persists over time. Under steady-state conditions, there is no error and no integral corrective action. Integral control removes offset and maintains very tight setpoint control.

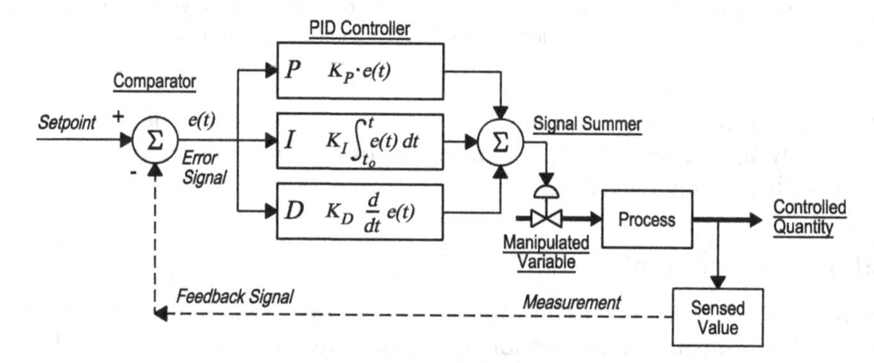

Figure 2.16 General block diagram for PID control. To obtain P-only, PI, or PD control, the integral and derivative functions can be omitted from the control-system design, or more simply, just not used by the control function.

producing a corrective signal that is now increased to +30. This action continues ad infinitum. The illustration shows how the corrective action continues to increase, until it is strong enough to reverse the error trend. Since the error accumulates over time, the sampling time is important. In plant control, most systems are considered to be slow acting, especially when compared to some manufacturing systems. The units for reset control are typically given in repeats per minute, where each repeat is its error-measurement action. Although integral action eliminates offset and mitigates overshoot, it tends to slow the control process.

Derivative or rate control

When faster response is required, derivative control can be added, producing the PID control shown in Figure 2.16. From calculus, the derivative is the rate of change or slope of the curve, which in this case is represented by the error with respect to time. Like integral control, derivative control remains dormant during steady-state conditions and is initiated only when there is a system transient, and like integral control, derivative control cannot be used by itself. The amount of corrective action is determined by tuning the rate function in the controller, and is given in units of time, such as minutes.

Figure 2.17 illustrates how the rate function works. From the time period between 1 to 2, the error increased 3 units, so the slope representing the corrective action becomes +3.

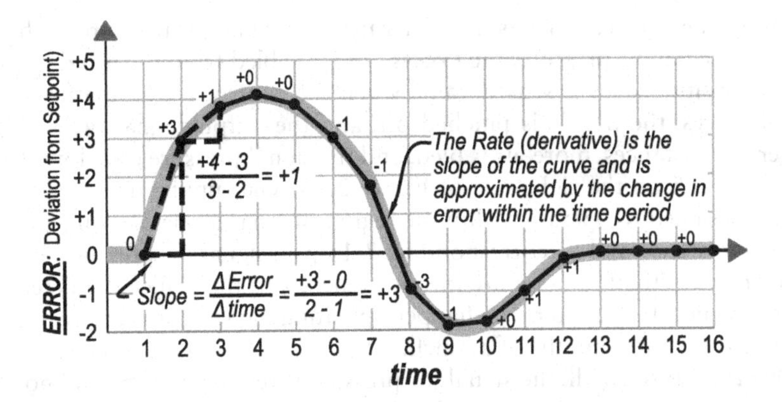

Figure 2.17 Derivative or rate control uses discrete measurements to examine the change in error with respect to a given time period. The rate control looks at how fast the error is changing and whether it is departing from or approaching the setpoint. Essentially, the rate corrective action is backed off when the setpoint approaches very fast, and conversely, the rate corrective action is enhanced when the setpoint is approaching slowly.

From the time between 2 to 3, the error slope increased only 1 unit, indicating that the rate of setpoint deviation is still increasing, but not as fast as before, meaning that the amount of rate-control action should be mitigated.

To summarize, rate control is a function that determines how fast the error is changing, and its direction. In operation, derivative control looks at the present value of the process variable compared to its previous value to determine the most recent rate of error change, which is then used to anticipate where the controlled quantity is heading. The present value of error-change rate is compared to its previous value to determine the trend. The rate controller then tries to "predict" what the future error will be, and to preemptively adjust the controller output based on the expected future value. For instance, if the level had physically dropped one inch in one minute during the last evolution and is presently observed to be dropping at a rate of two inches per minute, the rate-controller corrective action might attempt to fill at three inches per minute.

Pneumatic control: nozzle-flapper, pneumatic relays, and PID control

Pneumatic control systems are characterized by relative simplicity and ease of installation, operation, and maintenance. Although PLCs have replaced many pneumatic-control applications today, they were the mainstay for plant control systems for many decades. Presently, they are used for more specialized applications, such as for explosion-proof environments, but many final-device actuators continue to use pneumatics through current-to-pneumatic or I/P converters. Pneumatic controllers provide a straightforward way to visualize control concepts in a practical application, while illustrating examples of P-I-D control. Further, nozzle-flapper control and feedback mechanisms continue to be used in the closed-loop pneumatic positioners used for control-valve actuators, as discussed in Chapter 9.

Nozzle-flapper systems

The heart of the pneumatic system is the nozzle-flapper mechanism, as shown in the left side of Figure 2.18. In this arrangement, compressed air is supplied at 20 psig to the nozzle-flapper controller through an orifice, which is then forwarded as signal-air pressure to the controlled device. To set the signal-air pressure, the excess output is bled from the nozzle by the flapper valve, which is positioned by a sensed value when the control is automatic. At very small flapper-gap distances, the nozzle is pinched off, and the output-signal pressure goes high, while at larger gap distances, more air is bled, and the signal pressure goes low. The response is shown in the graph on the right side of Figure 2.18. The supply air pressure and device dimensions are designed to operate within the region that has a relatively linear relationship between the gap distance and its corresponding 3-15 psig signal-air pressure. In general, the nozzle diameter is less than one-hundredth of an inch, and the orifice diameter is less than one-quarter of the nozzle diameter. The flapper operating movement is between a fraction of a thousandth-to-three-thousandths of an inch.

Because the orifice is so small, the signal-air pressure is relatively static and not suitable by itself for pneumatic devices having large air requirements. In most cases, a pneumatic relay is incorporated as part of the controller to act as a volume amplifier or volume booster, as shown in Figure 2.19. This device uses the static pressure applied from the controller nozzle-flapper against a booster diaphragm to create a force to move a double-acting valve, which is opposed by the force of a leaf spring. The valve seats in both directions, so when the signal-air force applied on the left exceeds the leaf-spring force, the valve on the right opens directing a large volume of actuation air to the controlled device. But when the signal pressure is

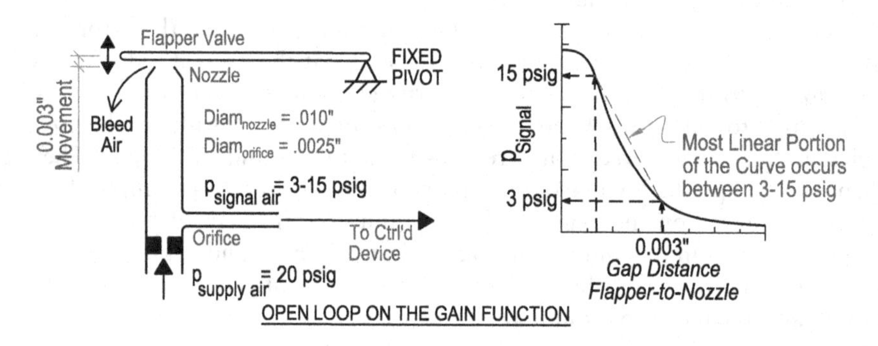

Figure 2.18 Nozzle-flapper mechanism for setting the signal-air pressure. The signal air is ultimately directed to actuate the pneumatically controlled device. The fixed pivot produces open loop on the gain, meaning the gain function lacks feedback. The dimensions are designed to produce a relatively linear 3-15 psig output over a small range of gap distance.

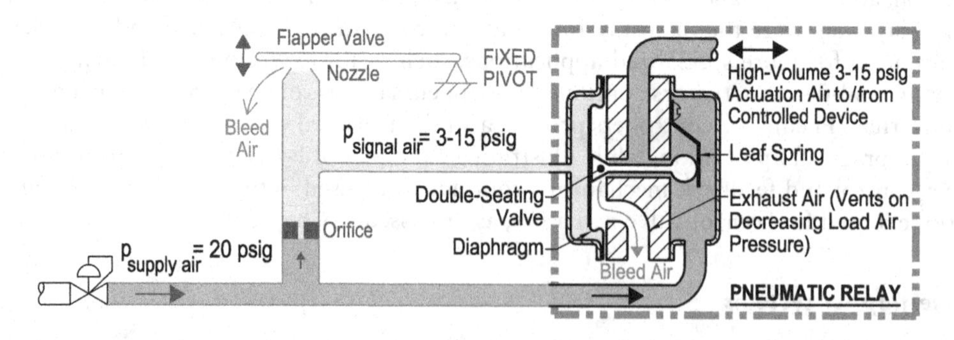

Figure 2.19 The pneumatic relay is used as a signal amplifier to boost a relatively static signal pressure into a high-volume actuation-air flow.

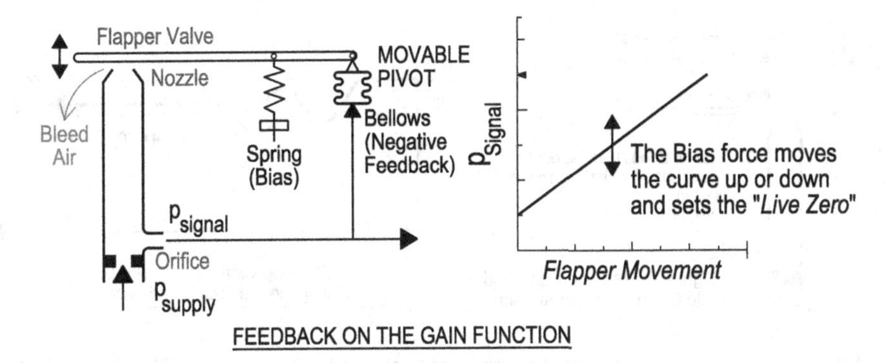

Figure 2.20 Negative feedback for the gain is achieved by replacing the fixed pivot with a bellows, which senses the signal-air pressure. Feedback added to the gain mechanism produces better linearity, faster action, and less offset. A spring force can be superimposed onto the flapper mechanism to provide "bias," which causes the gain curve to shift up or down. The bias can be used to adjust a setpoint or to tune the controller's "live-zero" setpoint.

less than the spring force, the actuation valve on the right closes and the exhaust valve on the left opens, bleeding actuating air pressure until a balance is reached.

To improve the control, the open-loop gain from the fixed pivot in Figure 2.18 is replaced with a movable-pivot mechanism for closing the loop on the gain. Closed-loop gain is obtained from the signal-pressure-sensing bellows that repositions the pivot in response to the output-signal-air pressure, as shown in the left side of Figure 2.20. For instance, when the flapper moves down, the nozzle exhaust is pinched off, slowing the nozzle bleed rate, and the output signal-air pressure increases. The bellows then expand under the increasing air pressure, moving the pivot upward, which mitigates some of the flapper's initial downward movement. The result is adjustable gain, less offset, and a more linear output signal. Various combinations of P-I-D control are illustrated using a pneumatic system for easier visualization in Figure 2.21 through Figure 2.24.

Proportional-only control with open-loop gain

Figure 2.21 shows a bourdon tube added to supply a *sensed value* for automatic positioning of the flapper. This controller could be part of a boiler combustion-control system, as an example, where the *process* is to balance the burner heat input against the steam *demand*. The heat input is modulated by varying the fuel flow in response to the steam pressure, as the *sensed value* is used for controller *negative feedback*. In this process, the burner firing rate is increased in response to greater steam demand. A second independent combustion-control *process* is to maintain the correct ratio of air to fuel for smokeless operation, which uses its own devices.

The right side of Figure 2.21 shows the control response just as its steady-state operation is upset by a *disturbance* consisting of a sudden increase in steam demand. As the steam load increases, the steam pressure drops, curling the bourdon tube downward. In this case, the steam pressure is both the *sensed value* and the *controlled quantity*, providing the *negative-feedback signal*. The initial position of the bourdon tube is the *setpoint*. The deviation of the flapper from its initial position is the *error signal*, where the goal of the *corrective action* is to put the bourdon tube back to its original position. If the initial position were attained, the original setpoint pressure would be completely reestablished without any offset. As the bourdon tube curls downward, the flapper pinches down toward the nozzle vent and the

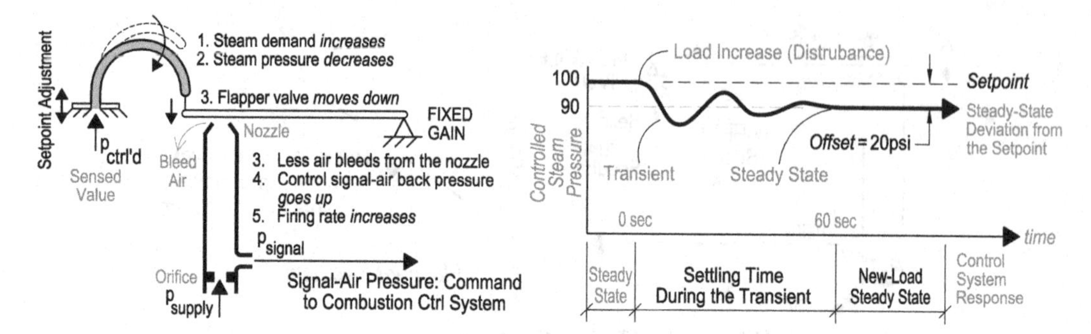

Figure 2.21 Pneumatic feedback-control system for increasing a boiler's firing rate. The flapper makes control adjustments after the bourdon tube senses changing steam pressure. This controller uses proportional-only control with *fixed gain*. In this example, the new steam-flow demand caused a 20-psi *offset* from the *desired setpoint*. Note that if signal air pressure is lost, the direct-acting nature of this controller causes the boiler to conservatively go to its low-fire position. Pressure-setpoint adjustment is obtained by shifting the bourdon tube up or down (far left).

signal-air pressure increases, serving as the *controller output signal*. In most cases, the small signal from the controller output is directed to a *pneumatic relay* to create a combination of *amplified signal and actuating force*.

The output from this controller is *direct-acting*, so an increasing air signal to the combustion-control system would increase the firing rate. In this arrangement, the combustion-control system would go to the minimum firing rate on loss of control air, which is desirable as the conservative *fail-safe* direction. Because this arrangement is proportional only, residual deviations from the desired steam pressure occur at different loads, which is the proportional-control *offset*.

The gain is non-adjustable, so the transient overshooting, undershooting, and resulting offset are constrained only by the mechanism design. The mechanism serendipitously balances the bourdon-tube deflection and its linked flapper-to-nozzle position to produce a new burner firing rate and a new post-adjustment steam pressure. With non-adjustable gain, the controller would need to be designed to have a *large proportional band* or *low gain* to avoid *instability*. In this example, the low gain causes significant undershoot and overshoot and a relatively long delay before a new steady state is reached, along with significant offset. The steam-pressure *setpoint* could be readjusted by up-down *biasing* motion at the bourdon-tube linkage, if the offset is objectionable, as shown at the far left of Figure 2.21.

Proportional-only with adjustable closed-loop gain

To improve the system response, the controller would be designed to have a very tight proportional band before any gain adjustment is incorporated, so that the gain adjustment could serve to loosen the proportional band. In the nozzle-flapper controller, adjustable gain is obtained by replacing the fixed pivot with a movable pivot from a bellows. The bellows receive a delayed feedback signal from the controller output-air pressure as shown in Figure 2.22. The bellows close the loop to provide negative feedback to the gain function. When the sensed value causes the flapper to close, the signal-air pressure rises, which in turn is transmitted to the bellows. The bellows mitigate the flapper-closing movement by lifting the flapper with its follow-up motion, causing the output signal to decrease a little. The bellows mechanically confirm that correction has indeed taken place, so that the initial high

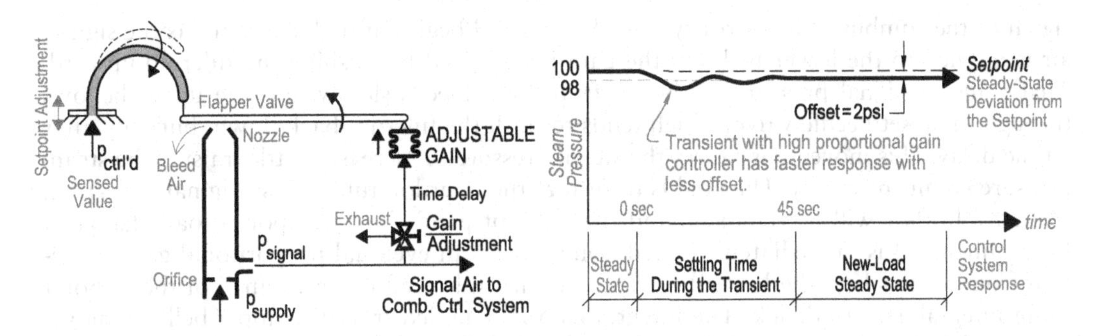

Figure 2.22 Pneumatic-control system. *Adjustable gain* is added to the proportional feedback control to reduce offset.

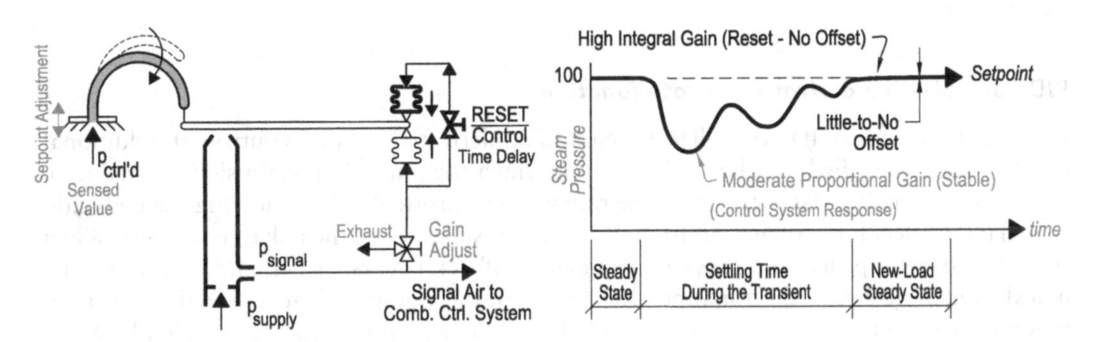

Figure 2.23 Pneumatic control system having proportional with integral or reset action for PI control. The reset function only works during transients. Reset removes offset to maintain very tight setpoint levels.

gain giving good response is replaced with lower steady-state gain for stability. The initial high gain speeds the response, mitigates the overshoot, and reduces the offset, while the final lower gain avoids hunting.

Gain adjustment is obtained by delaying its feedback signal to the bellows using a needle valve, as shown in the Figure 2.22 controller. The highest gain occurs with the valve closed where the pivot becomes fixed, and the lowest gain occurs when the needle valve is wide open, creating no delay to the bellows' follow-up motion. When the gain is adjustable, the highest gain setting might allow instability to occur under large load changes, however, in practice, the gain would be adjusted during tuning for the highest value that avoids hunting. In this manner, the gain-adjusted controller would be significantly better than a conservatively designed controller having low fixed gain.

PI proportional with integral or reset control

Figure 2.23 shows a pneumatic PI controller. A second set of bellows is added to oppose the proportional-gain bellows to produce the reset function. At first glance, it appears that the two bellows negate each other, however, a needle valve and sometimes a volume chamber is installed between the upper and lower bellows, adding a time delay to the air-signal transmission to the upper bellows. If the steam pressure increases, the bourdon tube curls down and reduces the flapper bleed rate, and the controller sends an increasing air-pressure signal to the combustion-control system. The control begins at high gain, but as the signal-air is applied to the lower bellows, the gain is mitigated by pushing the fulcrum upward. That same air-signal pressure to the lower bellows bleeds slowly into the upper bellows through the reset needle valve, which tends to push the fulcrum back down, but only after a time delay. This movement causes the signal pressure to increase a little, raising the steam pressure commensurately. The goal is to restore the bourdon tube to its original deflection, so that all offset will be removed, and the setpoint precisely met. Upon a load change, a high-gain signal is immediately created, followed by an eventual proportional-gain reduction from feedback to the lower bellows, and then followed by "resetting" of the setpoint using integral-gain feedback. The integral gain is obtained from the upper-bellows movement of the flapper fulcrum. Consequently, high gain is inserted during the transient and is later reduced to lower gain during steady-state conditions, but only after the integral action of the upper bellows has repositioned the fulcrum. The bottom line is that the integral action "resets" the setpoint to remove offset, and that integral control only works during the transient.

PID control—The derivative or rate function

Figure 2.24 shows a PID controller. When adding derivative or rate control, an additional needle valve is installed to adjust the speed at which the initial high-gain signal is reduced. For a system needing faster response, the rate valve is throttled more, allowing the high gain to remain in effect for a longer duration during a transient, but at the risk of instability. When the rate valve is opened, the proportional-gain feedback occurs earlier, and control occurs in a slower, more circumspect manner. Regardless of the amount of rate correction, the reset function eventually removes offset and provides adherence to the setpoint at all loads. When tuned properly, the PID controller provides faster response with fewer tendencies to overreact, overshoot, and oscillate. Table 2.1 summarizes the advantages and disadvantages of the different control modes.

Figure 2.25 shows the internal components of a Bailey AR80 computing relay, used as part of a combustion-control system. This device is capable of PID control with adjustable gain. Presently, this unit is configured for PI control, as it has a micrometer needle valve with a Vernier dial installed adjacent to the reset bellows, and a rate needle valve is not used. This unit can be configured for P, PI, PD, or PID control and can be connected to be direct or reverse acting. A separate pneumatic steam-pressure transmitter would provide the

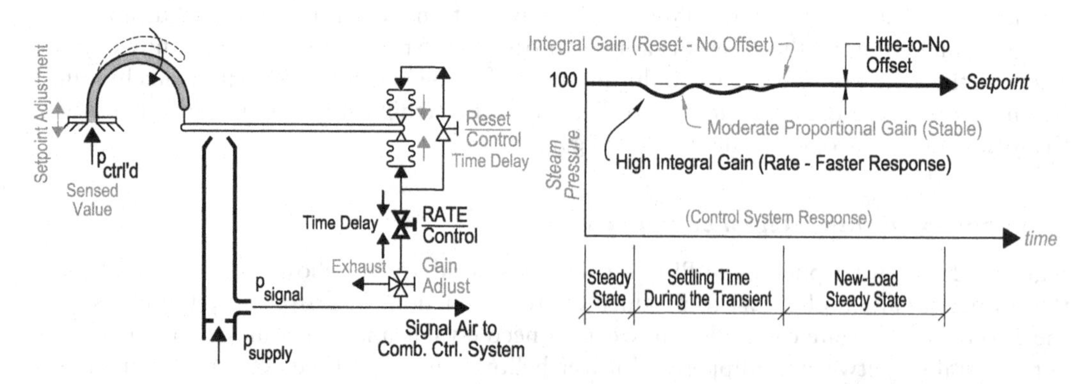

Figure 2.24 Pneumatic control system with proportional-plus-integral-plus-derivative feedback (PID). This controller provides the fastest response with tight setpoint adherence when properly tuned.

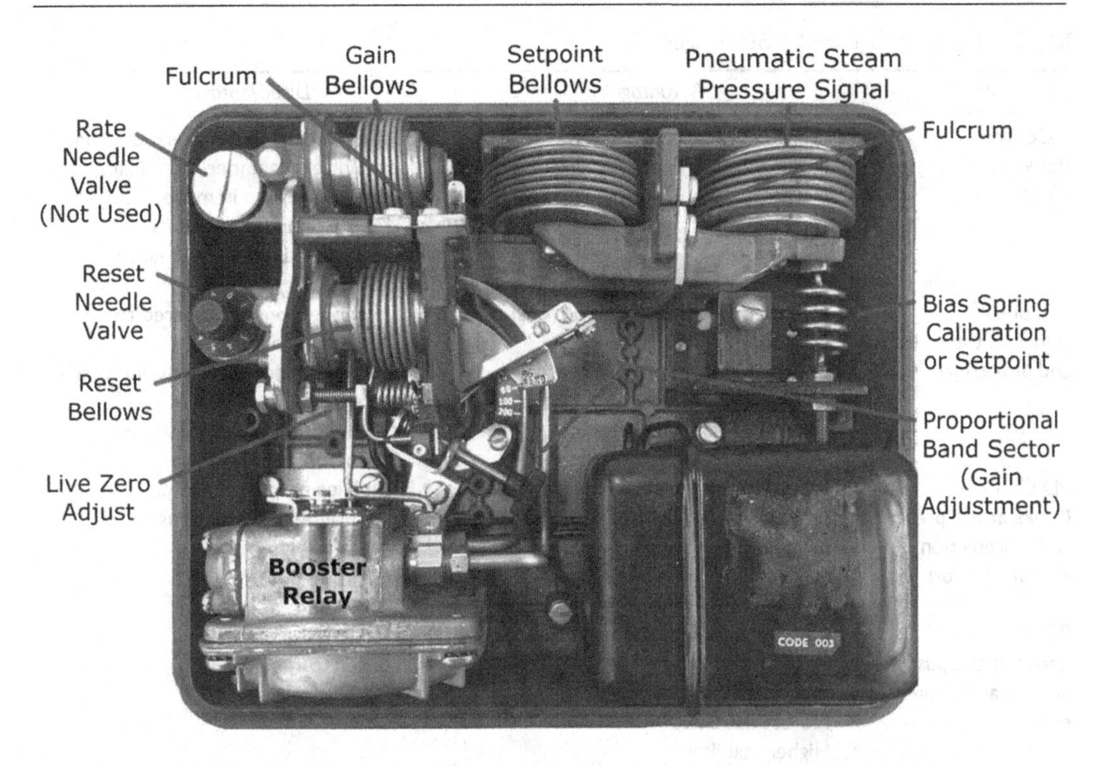

Figure 2.25 Bailey AR80 computing relay used for industrial PI boiler combustion control. This unit is capable of direct or reverse action using P, PI, PD, or PID control, and can be configured to multiply, divide, subtract, average, or totalize. (Courtesy of U.S. Merchant Marine Academy.)

measurement input signal for this unit to function in its control system, and a pneumatic pressure-reducing valve would provide the setpoint signal.

Pneumatic PD control

PD control can be obtained from the PID controller in Figure 2.24 by eliminating the integral function. The integral function is removed by completely closing the reset needle valve and venting its bellows to the atmosphere, thus eliminating all effects of the reset components. PD control might be needed when fast response without overshoot is required.

Pneumatic P, PI, PD, and PID control

Using all three of the PID-control functions results in very fast response, high accuracy, and small overshoot, but the additional features beyond proportional-only control add cost. Adding integral control removes offset and adding derivative control improves response time. If variations of setpoint can be tolerated without adversely affecting the process, then integral control is not needed. If slow responses can be tolerated without adversely affecting the process, then derivative control is not needed. Consequently, proportional-only control can be applied relatively inexpensively, PI control can be added for more precise setpoint adherence, PD control for faster response, or PID control for very accurate and very fast response. In addition to requiring more devices for mechanical- or pneumatic-control systems, tuning becomes more labor intensive for each added feature, and if controllers are not

Table 2.1 PID advantages and disadvantages

Control Mode	*Advantages*	*Disadvantages*
P Control (looks at the *present error*)	• Immediate reaction • Increasing gain improves reaction speed • Increasing gain produces improved steady-state accuracy • Inexpensive	• Offset—steady state deviation from setpoint exists under most loads • High gain results in more overshoot and oscillations • Too much gain produces hunting and instability
PI Control (looks at the present and *past error*)	• Eliminate steady-state error (offset) for best accuracy • Looks at past errors and the compensates future output based on time away from the setpoint	• Slower response compared to P-only control • Too much reset is likely to increase overshoot • Will not eliminate oscillations • Can be unstable
PD Control (looks at the present and a prediction of the *future error*)	• Predicts future based on trending the past error rate corrective signals • Increases stability by reducing settling time • Decreases overshoot	• Amplifies process noise or measurement fluctuations
PID Control Looks at the *past, present*, and *future error*)	• Optimum control can be achieved • Zero steady-state error • Fast response • No oscillations • Higher stability • D reduces or eliminates overshoot	• More expensive • Longer to tune

tuned properly, instability can result from any of the three PID functions. In mechanical or pneumatic systems, reset and rate functions are made by adjusting mechanisms. In analog-electronic controllers, reset and rate functions are made using circuit adjustments, while in digital-electronic-control systems, reset and rate settings are made through mathematical instructions in the PLC programming language.

In addition to PID control to improve performance, feedforward control can be added to look at the disturbances and begin the corrections before the PID controller determines a need to act. The feedforward control needs to look for the immediately occurring parameters that affect the process, such as steam-flowrate variations always precede the drum-pressure and water-level changes. The feedforward signal would be summed into the primary signal produced by the measured system parameter to act as a second element of anticipatory control. The feedforward signal would sum a value of zero during steady-state conditions, but its signal contribution would be fast-acting and strong during transients.

TUNING THE PID CONTROLLER

Good control depends on proper tuning of the PID functions. The goal of tuning is to obtain stable output under all conditions, fast response for modulating the process, strict adherence to the setpoint, and little-to-no overshoot or oscillations during the transients. Additionally, good process control should be obtained without excessive controller overreactions, even during large disturbances. A controller meeting those goals is considered robust. Tuning involves strategies that are based on observations and measurements, then applying quantified or empirical adjustments, and finalized using some trial and error. Many tuning methods are heuristic in nature, that is, they use experience-based procedures that involve observation and discovery, educated guesses, judgments, rules of thumb, etc. The optimal tuning solution is usually somewhat subjective, and generally, "close enough is perfect" for most tuning activities. Observation of the system after being placed in service, particularly during large transients, is particularly useful for verifying that the control system operation is stable and satisfactory at all loads, or whether additional adjustments are needed. Various process responses are shown in Figure 2.26.

There are several formal methods of tuning controllers as well as experience-based methods. Tuning methods may use open-loop or closed-loop techniques. In all closed-loop methods, the controller is placed into service and the system behavior is observed after a new setpoint is entered; almost like displacing a pendulum and observing the decay in oscillations. Generally, the tuning begins with the integral and derivative functions removed, so that the controller is operating in proportional-only mode and set for low gain, which is the same as large proportional band. The gain of the system is incrementally increased, typically doubled, while the system response is observed under large transient disturbances, such as going from low load to high load. The gain is repeatedly increased until the control system becomes unstable during the transient, and hunting ensues. At this point, the gain is substantially backed off, placing the control system into a stable operating mode, even under large transients. Integral and derivative adjustments are then made.

The classic form of the PID equation follows. In its general form, each of the P-I-D coefficients are simple gains, and this form of the PID equation is general and flexible for analysis.

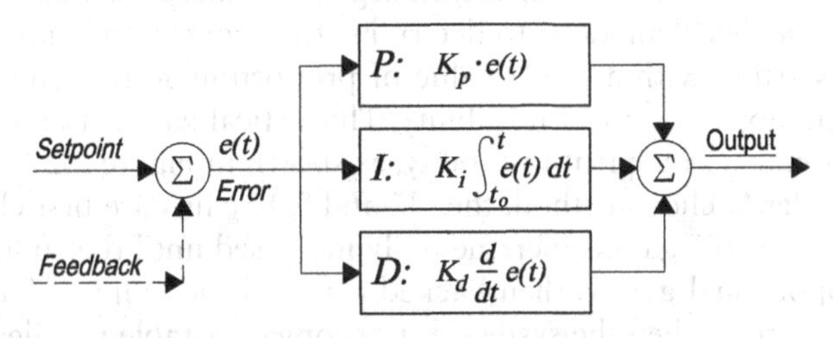

$$\text{Output: } u(t) = K_p e(t) + K_i \int_0^t e(\tau)\,d\tau + K_d \frac{d}{dt} e(t)$$

where K_p: Proportional gain, a tuning parameter
K_i: Integral gain, a tuning parameter
K_d: Derivative gain, a tuning parameter
e: Error = SP - PV
t: Time or instantaneous time (the present)
τ: Integration variable from time t=0 to the t=present

Today, an alternative to the classic PID equation is commonly used with microprocessor controllers. The alternative equation adds physical meaning by using reset and rate time constants, so that all three PID functions produce a single controller output that are all simultaneously adjusted by one overall controller-gain coefficient (K_c). The single controller gain simultaneously adjusts the present (P), past (I), and future (D) errors.

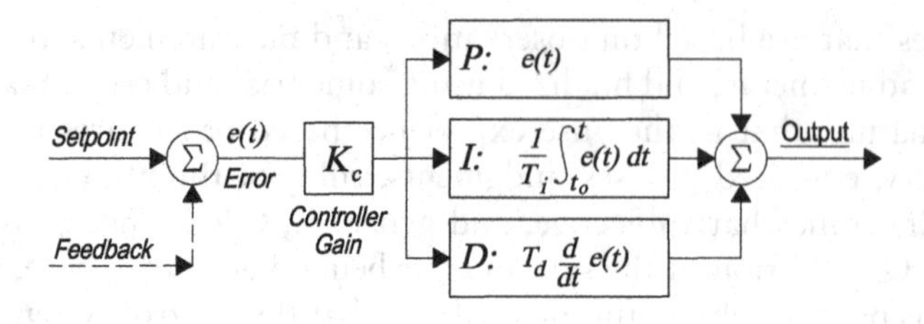

$$\text{Output}: u(t) = K_c\left[e(t) + \frac{1}{T_i}\int_0^t e(t)dt + T_d\frac{d}{dt}e(t)\right]$$

where
T_i: Controller Integral Time Constant
T_d: Derivative Time Constant,
K_c: Controller Overall Gain is applied to both the integral and derivative gain outputs
$K_i = \frac{K_c}{T_i}$ and $K_d = K_c T_d$

Ziegler-Nichols closed-loop method

One common tuning method is known as the Ziegler-Nichols closed-loop oscillation method. This method uses simple calculations to determine the proportional, integral, and derivative gain coefficients, starting with a single value of proportional-only gain found during trial-and-error determinations of critical instability. The critical gain is the highest proportional-only gain that just reaches incipient instability, as shown in Figure 2.27.

In using the Ziegler-Nichols method, the "I" and "D" gains are first eliminated by setting $T_i = \infty$ and $T_D = 0$. The "P" gain is incrementally increased until the control system becomes unstable. The proportional gain is then backed off until the point of incipient instability is reached. The gain setting when the system just becomes unstable is called the "critical gain" or K_C. At the critical gain, the loop-controller output begins steady sinusoidal oscillations. The critical gain value K_C and the period of oscillation P_C are noted, and they are used to obtain the integral gain (K_i) and derivative gain (K_d) from the Ziegler-Nichols predetermined relationships included in Table 2.2.

The Ziegler-Nichols tuning constants are based on delivering quarter-amplitude decay in the control response. By design, each cycle of overshoot has one-quarter of the amplitude of the previous cycle. For this method, the second cycle has one-quarter of the overshoot compared to the first, the third cycle has one-sixteenth, and the fourth one-sixty-fourth, which is essentially steady state.

Calculations yielding the proportional, integral, and derivative coefficients for the Ziegler-Nichols and Microstar closed-loop tuning methods for the classic PID control equation are listed in Table 2.2. Similarly, the Ziegler-Nichols and Microstar closed-loop coefficients in terms of integral and derivative time constants combined with a single controller gain are listed in Table 2.3.

As a tuning example, the PID-controller gain for a cooling-water temperature-regulating valve would be adjusted so that it hunts excessively. The gain is then reduced until the point of critical instability, K_c, is found to be 6 degrees per minute on the controller scale and the

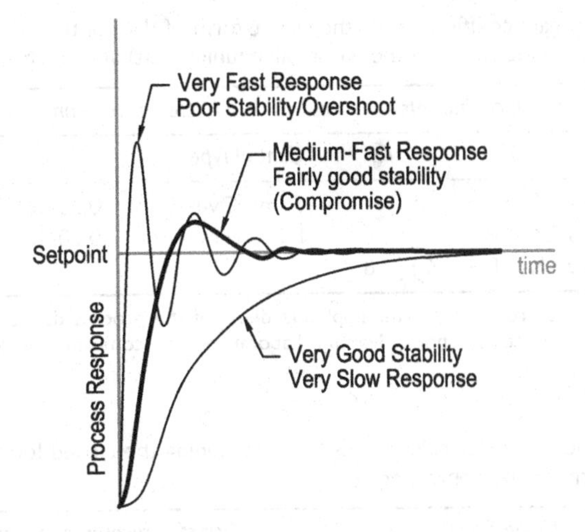

Figure 2.26 The process responses are shown for three controller gain settings. In general, the final tuning will be a compromise between stability, overshoot, and speed to reach setpoint. In special cases very fast response may be required, and in other cases no overshoot is permitted for safety or product protection reasons. A medium-fast response shows a good trade off between speed, without excessive overshoot, which is a good goal for most control systems.

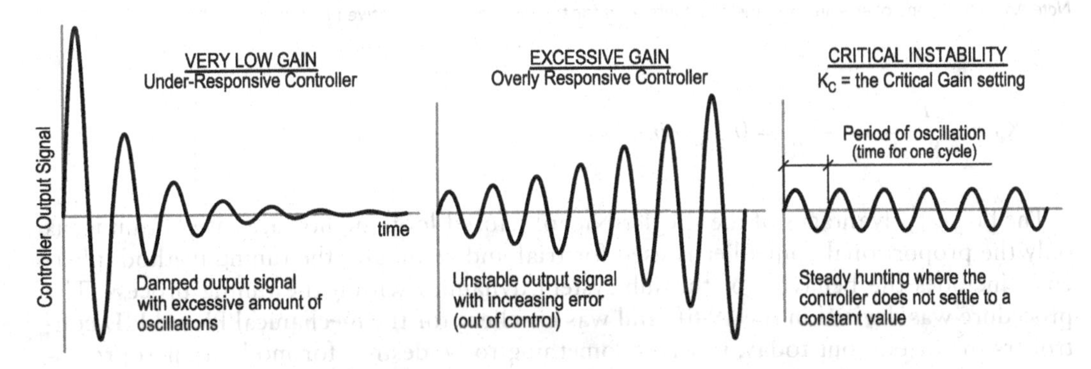

Figure 2.27 Output responses for a proportional-only controller set for very low, excessive, and critical gains, where integral and derivative gains have been disabled. Low proportional-only gain produces large overshoot with many oscillations before reaching steady state conditions (left), while excessive gain produces instability with increasing overshoot (center). The point of critical instability produces steady oscillations (right). The gain setting and period at critical instability are used in the Ziegler-Nichols closed-loop PID tuning method.

period of steady oscillation is found to be 1.5 minutes. The Ziegler-Nichols proportional PID coefficients are calculated as follows, and then programmed into the controller.

$$K_p = 0.60K_c = 0.60 \times 6 = 3.6 \textit{ degrees per minute}$$

$$K_i = \frac{2.0K_p}{P_c} = 2 \times \frac{3.6}{1.5} = 4.7 \textit{ repeats per minute}$$

Table 2.2 Calculations of the gain coefficients for the classic form of the control equation using the Ziegler-Nichols closed-loop method of controller tuning (left) and the Microstar coefficients (right)

Ziegler–Nichols proportional coefficients				*Microstar recommendations, for example*			
Control Type	K_p	K_i	K_d	*Control Type*	K_p	K_i	K_d
P	$0.50 \bullet K_c$	-	-	Some Overshoot	$0.33 \bullet K_c$	$2.0K_p / P_c$	$K_pP_c / 3$
PI	$0.45 \bullet K_c$	$1.2K_p / P_c$	-	No Overshoot	$0.20 \bullet K_c$	$2.0K_p / P_c$	$K_pP_c / 3$
PID	$0.60 \bullet K_c$	$2.0K_p / P_c$	$K_pP_c / 8$				

Note: The Ziegler-Nichols method produces quarter-amplitude decay of the process deviation from setpoint, while the Microstar goal is to reduce or eliminate overshoot. Microstar Laboratories is a company that manufactures data acquisition and controller products.

Table 2.3 Ziegler-Nichols and Microstar calculations for determining the closed-loop tuning coefficients for the alternative form of the control equation

Ziegler–Nichols proportional coefficients				*Microstar recommendations, for example*			
Control Type	K_c	T_i	T_d	*Control Type*	K_c	T_i	T_d
P	$0.50 \bullet K_u$	-	-	Some Overshoot	$0.33 \bullet K_u$	$P_u / 1.2$	$P_u / 3$
PI	$0.45 \bullet K_u$	$P_u / 1.2$	-	No Overshoot	$0.20 \bullet K_u$	$P_u / 2.0$	$P_u / 3$
PID	$0.60 \bullet K_u$	$P_u / 2.0$	$P_u / 8$				

Note: An overall controller gain and time constants used for the integral and derivative functions.

$$K_d = \frac{K_p P_c}{8} = 3.6 \times \frac{1.5}{8} = 0.67 \ minutes$$

The biggest advantages of the Ziegler-Nichols closed-loop method are ease of tuning, as only the proportional controller is used for trial and error, and the tuning method inherently includes the behavior of the sub-system dynamics within the tuning process. The procedure was devised in the 1940s and was excellent for the mechanical PI and PID controllers of the era, but today, it leaves something to be desired for modern microprocessor controllers. Its disadvantages include its time-consuming nature for slow-responding systems, and the system can become unstable and go out-of-control during tuning, or later during an unexpectedly large transient. Quarter-amplitude overshoot will always occur, so that the controller will always exhibit more reaction than is necessary during each transient.

The biggest problem with the Ziegler-Nichols method is its use of quarter-amplitude damping as its tuning objective, guaranteeing its less-than-optimum controller behavior. Figure 2.28 shows quarter-amplitude oscillatory decay. The purpose of using quarter-amplitude damping is to obtain fast elimination of the disturbance-based error that occurs between the setpoint and the process-variable feedback. Quarter-amplitude damping produces a controller output that is halfway between a dead controller and a critically unstable controller. The result of the Ziegler-Nichols tuning method is a controller and actuator that naturally overshoots and undershoots several times during each transient before settling out, producing more controller and actuator reaction than is necessary. Some processes are not tolerant of overshooting beyond the setpoint or control loop oscillations, in which case this method is problematic.

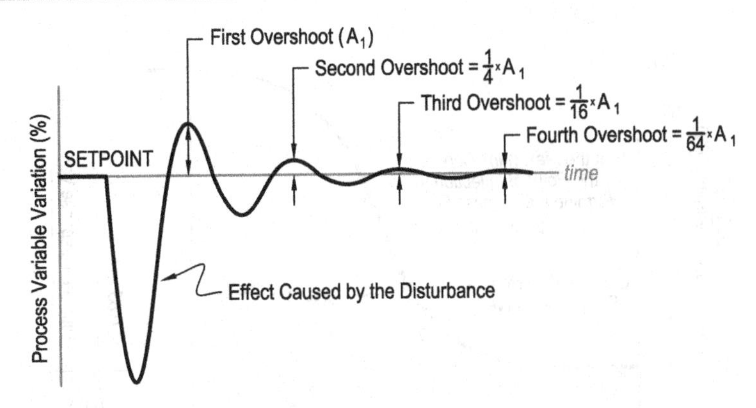

Figure 2.28 Quarter amplitude damping is the design control response using the Ziegler-Nichols method. Each overshoot is one-fourth of the previous overshoot.

Table 2.4 Effects of independently increasing the proportional, integral, and derivative gains

Gain mode	*Reaction time*	*Overshoot*	*Settling time*	*Steady-state error*	*Stability*
K_P	Decrease	Increase	Small change	Decrease	Decrease
K_I	Decrease	Increase	Increase	Eliminate	Decrease
K_D	Small change	Decrease	Decrease	No real effect	Improve if K_d small Amplifies noise if K_d is large

Note: These behaviors are useful in fine-tune adjustments for improving desirable control characteristics.

The solution for mitigating quarter-amplitude problems is to detune the controller from the Ziegler-Nichols settings. Recommended tuning improvements are available from many published resources, based on experience with various processes and operating conditions, judgment, and even some trial and error. One such method is recommended by Microstar Laboratories and is included in Table 2.2 and Table 2.3 for closed and open loop methods, respectively. Depending on the process being controlled, it should be expected that some fine-tuning will be necessary in any systems needing robust control. Table 2.4 shows the general effects that each control mode has for fine tuning.

Open loop tuning

In addition to closed-loop tuning methods, there are also open-loop methods. The open-loop methods are sometimes called "process-reaction methods" because they test the process without any control for the purpose of observing and quantifying how it reacts. To perform this tuning method, the controller is bypassed, or the system is placed in manual operation. After steady state is reached in hand-bypass operation, either the disturbance or the load is manually changed for the purpose of observing the system reaction during the entire transient period. The time-dependent behavior is plotted as a "process-reaction curve," from which system-dynamics measurements are obtained, as shown in Figure 2.29. The measurements include the lag or dead time that occurs before the controller observes and begins changing the output (t_{dead}), the total amount of time to reach 63% of the response, where 63% represents one time constant (τ) and five time constants would be the time to reach steady state, the input step-load change (X_o), and ultimate final load value (M_u). The maximum slope of

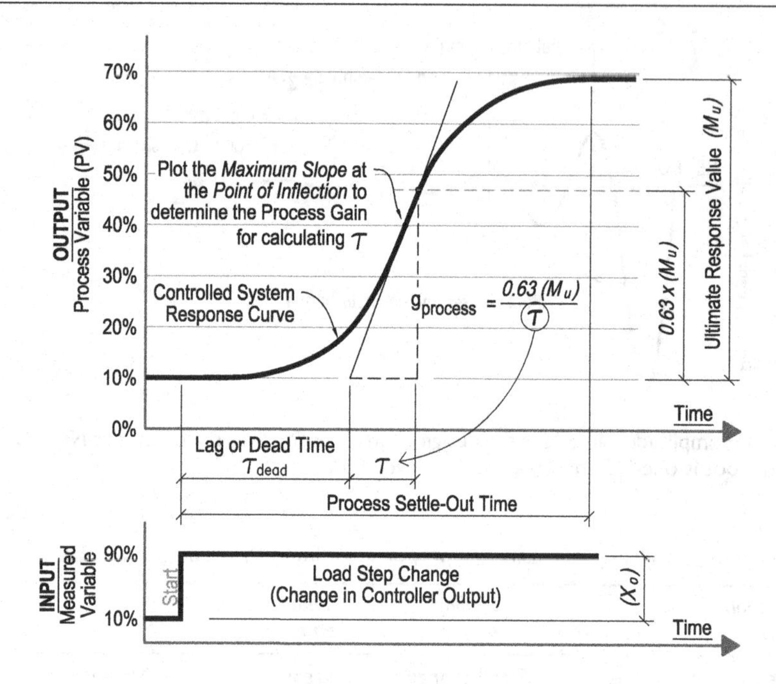

Figure 2.29 The Ziegler-Nichols open-loop reaction curve showing the process variable versus time. The lag time, time constant, and slope are used in the tuning process.

Table 2.5 Ziegler-Nichols proportional coefficients and time constants for open-loop control

	Ziegler–Nichols proportional coefficients		
Control Type	K_c	T_i	T_d
P	K_0	-	-
PI	$0.9 \bullet K_0$	$3.3t_d$	-
PID	$1.2 \bullet K_0$	$2.0t_d$	$t_d / 2$

the output process-value reaction curve is determined at the point of inflection, where the process ramp-up begins its transition from a progressively increasing value to a progressively decreasing value. The values obtained from this curve are used in straightforward algebraic equations to obtain the PID gain coefficients.

Ziegler-Nichols and the Cohen-Coon are two popular open-loop methods, although there are others. The Ziegler-Nichols method works for processes having small time delays and more complicated higher-order systems, while the Cohen-Coon method is preferred for systems having longer dead times and is limited to simpler first-order systems.

For the Ziegler-Nichols open-loop tuning method, the overall adjustment coefficient (K_c) and the time constants (T_i and T_d) are determined from Table 2.5, using $K_0 = \dfrac{X_o}{M_u} \cdot \dfrac{\tau}{\tau_{dead}}$.

For the Cohen-Coons open-loop method, the equations for determining the loop tuning constants are given in Table 2.6. The procedure is similar to the Ziegler-Nichols open-loop method, but the Cohen-Coons table also provides coefficient equations for PD control. In

Table 2.6 Cohen-Coon open-loop equations for determining P, PI, PD, and PID controller tuning coefficients

	Cohen-Coon tuning coefficients		
Control type	*Controller gain K_c*	*Integral time constant T_i*	*Derivative time constant T_D*
P	$K_c = \frac{1.03}{g_p}\left(\frac{\tau}{t_d}+0.34\right)$	-	-
PI	$K_c = \frac{0.9}{g_p}\left(\frac{\tau}{t_d}+0.092\right)$	$T_i = 3.33t_d\left(\frac{\tau+0.092t_d}{\tau+2.22t_d}\right)$	-
PD	$K_c = \frac{1.24}{g_p}\left(\frac{\tau}{t_d}+0.129\right)$	-	$T_D = .027t_d\left(\frac{\tau-0.324t_d}{\tau+129t_d}\right)$
PID	$K_c = \frac{1.35}{g_p}\left(\frac{\tau}{t_d}+0.185\right)$	$T_i = 2.5t_d\left(\frac{\tau+0.185t_d}{t_d+0.611t_d}\right)$	$T_D = .037t_d\left(\frac{\tau}{\tau+185t_d}\right)$

any event, these formulaic methods can be useful in obtaining a good first guess that can be fine-tuned into better, more robust control.

STABILITY

No matter the controller scheme and tuning method, stability is always of utmost importance. Hunting results in processes that can swing out of control, exceed allowable limits, set off nuisance alarms, trigger unscheduled safety shutdowns, and create surges that add to the wear and tear on equipment and machinery. Where plant systems interact, hunting in one system can cause erratic behavior to affect other systems. Valves that hunt are more prone to wiredrawing, seal leakage, and actuator failure. Motors that start and stop often are prone to degradation and damage from short-cycling. Generators that hunt produce synchronous-frequency fluctuations that result in surging of motors, shaft fatigue loads, and so on.

Excessive gain in a proportional-only controller causes severe hunting. However, excessive proportional gain is not the only source of instability. With proportional gain, the tuning is easy; increase the gain until instability appears, then decrease the gain by a large enough amount to reliably eliminate hunting based on observations during transients. If hunting reappears during a large transient, then detune the gain a little more.

In operation, controllers ultimately reposition real-world devices, such as valves. The controlled devices have their own travel limits, such as a valve can only vary between fully closed and fully open. Even if the controller sends a corrective-action signal that exceeds the device operating limits, the actuator cannot oblige past its travel extents, and the actuator saturates. When the corrective signal exceeds the actuator limits, feedback is lost, and the controller becomes ineffective. Worse, when integral control is used under these extreme-output conditions, the controller-reset function continues to integrate the error in the direction of the actuator saturation. Recalling that integral action is the magnitude of the error multiplied by the time away from the setpoint, the corrective signal mistakenly continues to build, although the actuator has reached its limit. The behavior where the error continues to be integrated by the controller when the controlled device is beyond its limits is called "windup," and the result is unrecoverable instability. In addition, if the integral controller was adequately tuned to handle moderate transients, a substantial change in setpoint or a large disturbance can have the potential to produce unexpected windup and complete loss of control.

One method of avoiding windup is to limit the amount of setpoint adjustment that can be suddenly introduced, so that an actuator does not need to reach its extents of travel. A disadvantage of limiting the amount of setpoint change includes sacrificing some controller robustness. Also, this method is not guaranteed to avoid windup if the cause is from an extreme disturbance.

Another method of avoiding windup is to physically track the actuator position and then feed its position back to the controller. The true actuator position is compared to the controller command signal to form its own error signal. When the actuator is within its normal operating range, the error signal sent back to the controller is zero and it has no effect on control. However, when the actuator approaches saturation, an error is generated, which is programmed to prevent the integrator from winding up.

A third method of avoiding windup is to use conditional integral control, so that integration is switched off when the operating level is far from its steady-state setpoint. Simply stated, when the error is large, the integration function is ignored. Removal of integral control can be initiated by observing when the controller-output signal has reached its limit, such as 4mA or 20mA, while also observing that the error is progressing in a direction that would increase the error.

Although windup problems can be addressed by complicated control schemes, they form an excellent reason for installing hand-auto stations for important control applications in manned plants. Temporarily switching to hand or manually bypassing a control device during a large transient can help an operator to ride through unusual disturbances without initiating safety or shutdown sequences.

DEFINE:

1. Negative Feedback
2. Feedforward Control
3. Open-Loop Control
4. Closed-Loop Control
5. Setpoint, Range, and Differential
6. Gain and Proportional Band
7. Offset
8. Integral or Reset Control
9. Derivative or Rate Control
10. PID Control

QUESTIONS

1. List the four modes of control.
2. What is the disadvantage of proportional control?
3. What is gain? What is the disadvantage of too much gain?
4. Describe when reset and rate functions become activated? Can reset and rate control be used by themselves?
5. Describe the difference between closed-loop tuning and open-loop tuning.
6. Describe the Ziegler-Nichols method of tuning a controller. What is its tuning goal?
7. What is windup? Explain how windup happens and how it can be avoided.

Chapter 3

Basic electronics

Electronic devices and microprocessors form the heart of modern plant-control systems today. Although a black-box understanding of control systems is enough to successfully work with controls, having a fundamental understanding of electronics is helpful in relating how these systems function, how to configure devices, component behavior, device limitations, and troubleshooting.

Electronic semiconductors are also called solid-state devices. Semiconductors are formed from crystals made of silicon or germanium, which are two periodic-table elements having four valance electrons. Semiconductors are produced by "doping" the pure crystalline-lattice structure with impurities that replace one of the crystalline atoms during the lattice formation. Some doping elements, such as phosphorous or arsenic, produce donor regions at what is called the P-junction. Other doping elements, such as boron or aluminum, produce acceptor regions, which is at the N- junction. The donor region has a shortage of electrons, where the missing valence electrons are called holes, and the acceptor region has a surplus of electrons. The free holes and free electrons are known as charge carriers. In electronics, conventional current is defined as the flow of "holes" or positive charges from the positive voltage-source terminal to the negative or return terminal, which is often the circuit ground. The formation of donor or acceptor patterns tends to allow current to pass more easily in one direction than the other. When germanium is used for the crystalline material, the good-conduction forward-bias voltage drop across the adjacent regions is 0.3V, and when silicon is used, the forward-bias voltage drop is 0.7V. Reverse bias results in very high resistance, essentially becoming a non-conductor.

DIODE

The simplest solid-state device is the diode. Figure 3.1 shows different styles of diodes, and their relative polarities. The operation of a diode is analogous to a check valve in a fluid-flow system, where the check valve permits flow in one direction when the inlet pressure exceeds the outlet pressure, but it prevents reverse flow when the inlet pressure drops below the outlet pressure (see Figure 3.1). The diode is formed into two junctions, one doped to have a surplus of holes and the other doped to have a surplus of electrons (see Figure 3.2(a) and (b)). The hole side has a net positive charge and is called the P-junction, which is the diode's anode. The electron side has a negative charge and is called the N-junction, which is the cathode.[1] The diode has the characteristic that when positive voltage is applied to the P-junction in a closed circuit, the diode is *forward-biased*, and the diode freely conducts current. When negative voltage is applied to the P-junction, the diode is *reverse-biased*, and current does not flow (see Figure 3.3(a) and (b)). Figure 3.3(c) shows a simplified representation of a diode circuit,

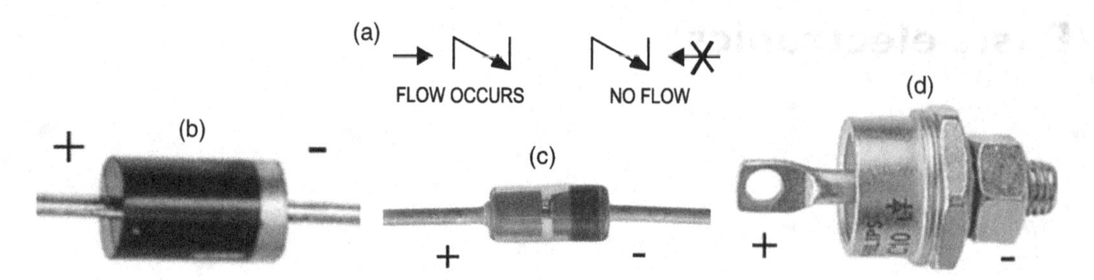

Figure 3.1 Diodes. (a) The unidirectional flow of check valves can be used as an analogy to visualize diode operation. Current flows through the diode when it is forward biased, and flow is blocked when it is reverse biased. (b) Power diode: The band indicates the cathode or negative connection for forward biasing. (c) Glass signal diode: The band indicates the cathode. (d) Larger power diode: the casing is the cathode.

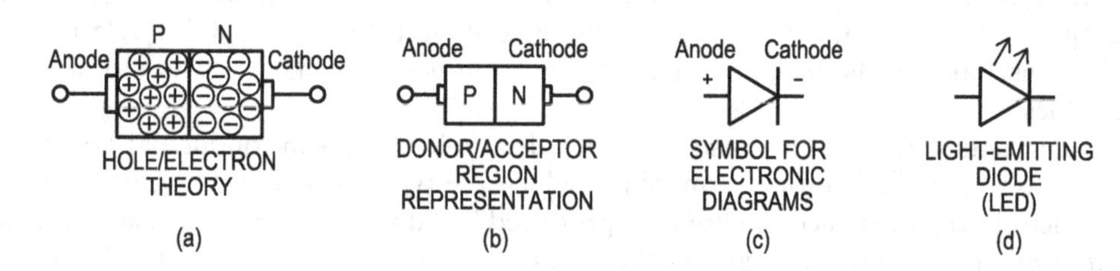

Figure 3.2 (a) and (b) show conceptual representations of the doped donor/acceptor regions of a diode's crystalline microstructure. (c) shows the diode symbol used on schematic drawings. In the symbol, the anode is on the left and the cathode on the right. Conventional current is the flow of holes, where the diode conducts when positive voltage is applied to the anode. Using this convention, the symbol has the appearance of being an arrow indicating current flowing from left to right when the diode is forward biased. (d) shows a light-emitting diode (LED).

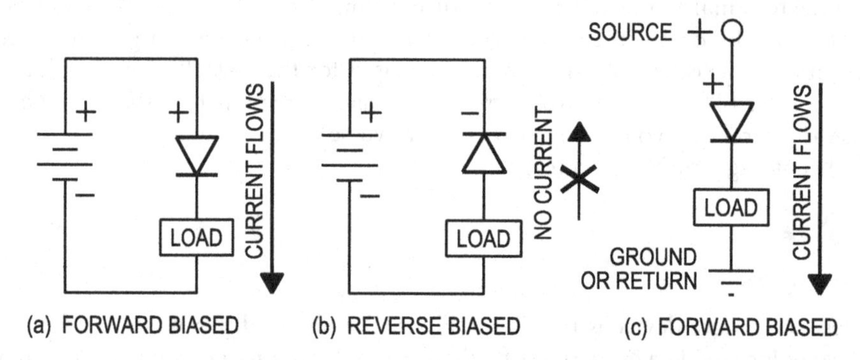

Figure 3.3 (a) and (b) Simple diode circuit showing forward and reverse biasing. The diode has a small voltage drop (0.3V for silicon and 0.7V for germanium) and the load provides most of the current-limiting impedance. (c) Simplified circuit representation, where the complete power-supply circuit is replaced with a positive source voltage and a common ground or "return." By omitting some of the complete-circuit line work, the diagram becomes easier to read.

where the closed-loop electrical connections are replaced as a positive voltage source and the completion of the circuit is shown by a common return or ground.

Figure 3.4 shows a diode characteristic curve. With forward-bias polarity, the diode conducts well, having very little voltage drop. The forward-biased voltage drop depends on the crystalline base material used. When the diode is made from germanium, the diode

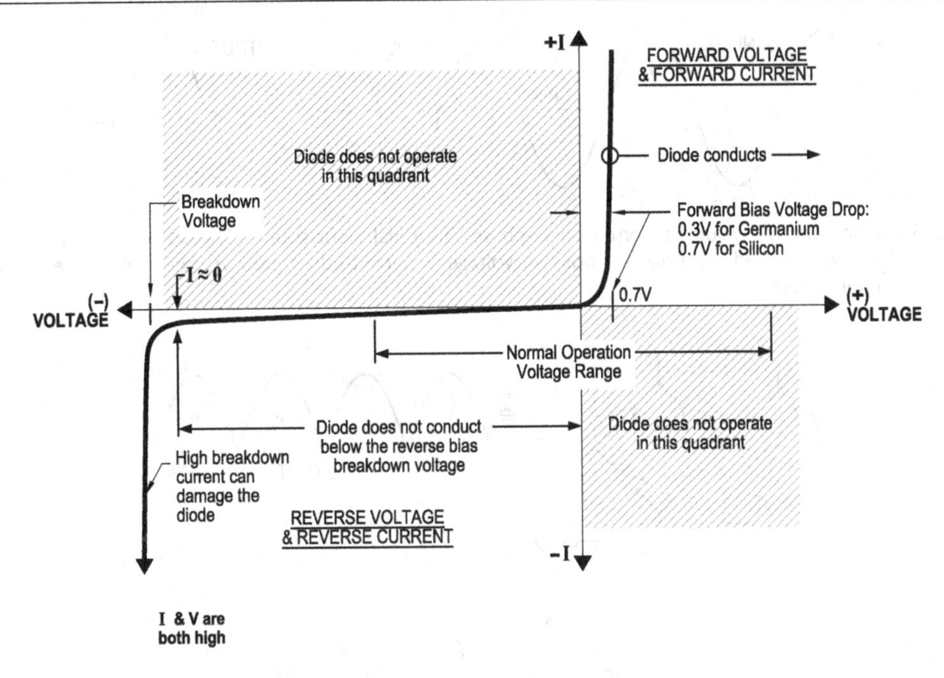

Figure 3.4 Diode characteristic curve.

forward-bias voltage drop is 0.3V and when it is made from silicon, the voltage drop is 0.7V. In the reverse-bias direction, the diode has the characteristic where it does not conduct, except for very small amounts of leakage current, until the applied voltage becomes large. The point where the diode conducts in the reverse-bias direction is the breakdown voltage. A diode is selected so that the applied reverse-bias voltage is below the breakdown voltage, and it is intended to conduct only in the forward-bias direction. Exceeding the reverse voltage breakdown point will damage conventional diodes.

Often, diodes are installed in alternating current circuits to rectify ac into dc voltage. During the positive-voltage portion of the ac waveform, the diode is forward biased[2] and it conducts. During the negative-voltage portion of the ac waveform when current would flow in the opposite direction, the diode is reverse biased, and the diode does not conduct. Figure 3.5 shows the ac input voltage (left) and resulting pulsating dc output voltage (right) for a single diode. In many cases, multiple diodes are connected into bridge arrangements for rectifying ac into better-quality dc. The bridge essentially reroutes amperage to flow from the terminal having the most-positive voltage to the terminal having the most-negative voltage, as time progresses and sinusoidal voltages vary. Figure 3.6 shows a single-phase bridge rectifier, which produces better-quality pulsating dc than a single diode. If the ac frequency is 60Hz, the pulsations occur at 120Hz, as the waveform inverts. Figure 3.7 shows a three-phase bridge rectifier, which produces pure dc with ripples. If the ac frequency is 60Hz, the three-phase ripples occur at 360Hz and the voltage variation would be less than 15% of the peak voltage. Three-phase bridge rectifiers are the basis for producing the rotating field current in large brushless ac alternators, as one application.

Some diodes are designed to emit light and are called light-emitting diodes or LEDs. Like all diodes, they conduct in one direction, but LEDs use very little power, and are excellent for use as indicator lamps. In digital control circuits used in PLCs, they form part of the optical isolators that can significantly limit fault damage without disabling an entire control system. The symbol for an LED is shown in Figure 3.2(d).

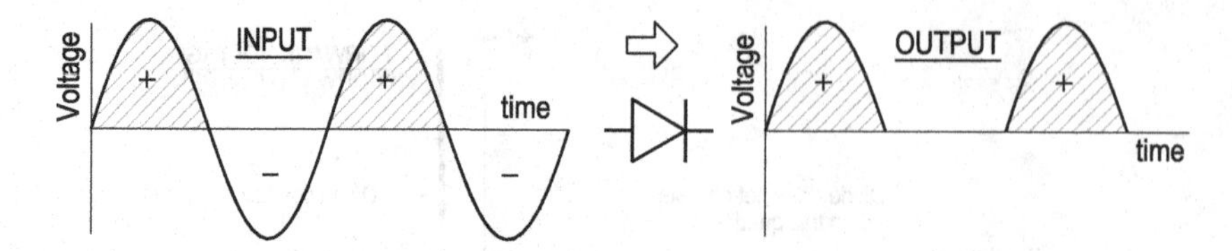

Figure 3.5 Diode behavior resulting from a single-phase sinusoidal input source. The output is very pulsating direct current, where only the positive voltage is permitted to pass at times when the diode is forward biased.

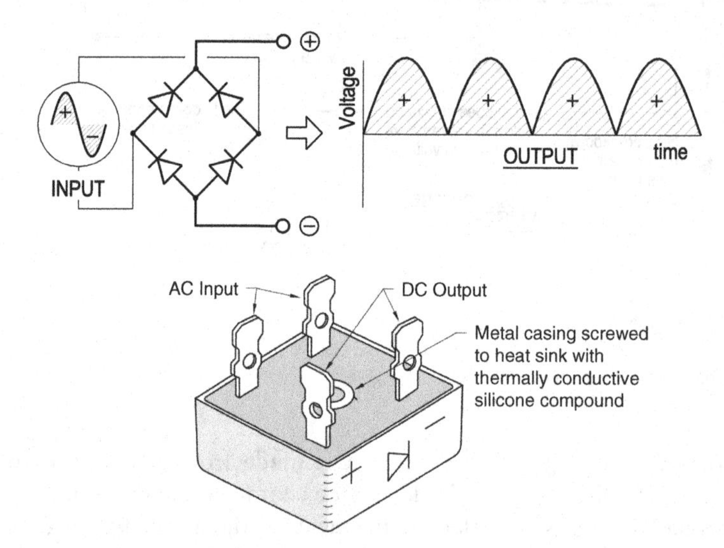

Figure 3.6 Full-wave rectification of single-phase ac voltage using a four-diode bridge (upper). The output voltage is pulsating dc, but the dc is better quality and of higher RMS value than the voltage produced by a single diode. The image (lower) shows a potted single-phase, full-wave rectifier, having two ac inputs and two dc outputs with the wiring diagram printed on the body.

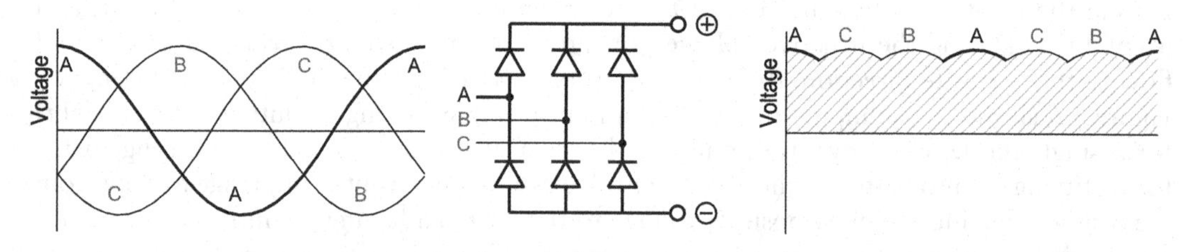

Figure 3.7 Full-wave rectification of three-phase ac voltage using a three-phase/six-pulse bridge rectifier. The output voltage is pure dc with some ripple. This arrangement is used in three-phase electrical generators to produce its dc excitation voltage, or in the rectifier section of some variable-frequency drives. This arrangement is also called a six-pulse rectifier, having six diodes and producing six ripples per cycle.

FILTERS

Filters are energy-storage devices that smooth out pulsations or ripples in dc-power systems, or filters can be used to block the transmission of particular signal frequencies. Filters are classified as either active or passive. A passive filter uses combinations of resistors, capacitors, and inductors to store or delay the transmission of energy in an R-L-C circuit. Active filters use passive filters in conjunction with amplifying components, such as transistors, operational amplifiers (OPAMPS), or digital signal processors. Active filters require a source of power to function.

Figure 3.8 shows a simplified schematic of a capacitor used to absorb energy during the sinusoidal rise of current, and later the capacitor's stored current continues to supply the circuit as the sinusoidal source drops. This filter converts larger pulsations of current into smoother dc with small amounts of ripple. Filters can be used with single diodes, single-phase or three-phase bridge rectifiers, SCR bridges, etc. It should be noted that the attached loads typically contain some impedance, which also has filtering effects that oppose the pulsation changes.

Figure 3.9 shows simplified RC low-pass and high-pass filters used for signal processing. In a low-pass filter, low frequencies are permitted to pass, but high-frequency noise is attenuated or essentially blocked. The frequency where the attenuation is effective is called the cut-off frequency. The product of RC is called the time constant. The RC filter cut-off frequency is calculated from the equation:

$$f_c = \frac{1}{2\pi RC}$$

VOLTAGE DIVIDER

Voltage dividers can be used to drop circuit voltage levels to match device requirements or to create a reference voltage. Voltage dividers are commonly used for signal attenuation to obtain low-voltage signals that are proportional to the higher voltage source signals. Reducing the voltage may be required when signals are excessive enough to "peg-out" the sensing-device to its limits. A voltage divider is essentially two resistors in series, where the output is obtained between the resistor junction and ground, as shown in Figure 3.10. The output voltage can be calculated using Ohm's Law.

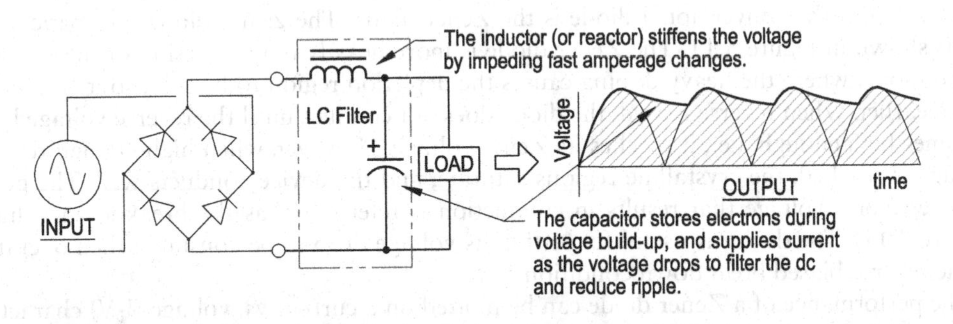

Figure 3.8 An inductor and capacitor is added to the circuit shown in Figure 5 to form an LC filter. The capacitor stores current during its increase in charge, and it supplies the current to the circuit as the sine wave drops. The capacitor smooths large pulsations into smooth dc with some ripple. The inductor inhibits fast voltage variations and squashes voltage spikes.

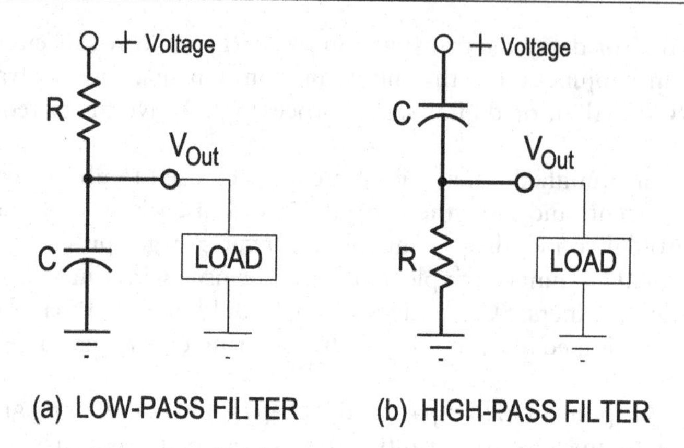

Figure 3.9 (a) Low-pass filter and (b) high-pass filter used for mitigating EMI/RFI noise within a signal. The low-pass filter permits low-frequency signals to pass, while filtering high frequency noise, and conversely for the high-pass filter.

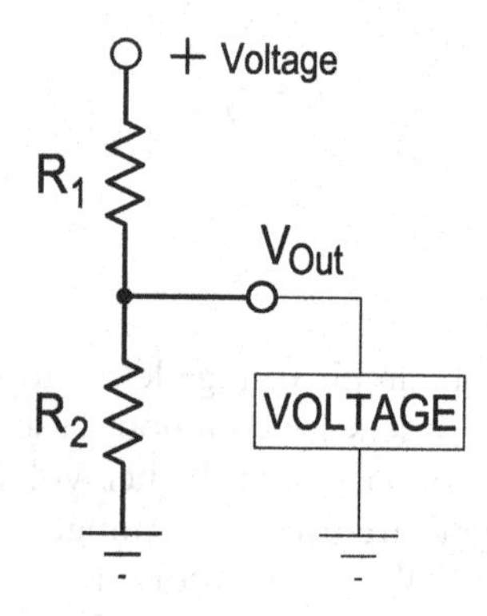

Figure 3.10 A voltage divider circuit is used to provide signal attenuation, i.e., to drop the signal when its upper range exceeds the analog-input limit. Essentially, the voltage divider makes the signal fit within the range of the measuring equipment.

ZENER DIODE

A variation of the conventional diode is the Zener diode. The Zener diode schematic symbol is shown in Figure 3.11. The Zener diode is more heavily doped version of the conventional diode, where the heavy doping causes the depletion region to be thin, so it can break down earlier. When reverse biased, the diode does not conduct until the reverse voltage level becomes high enough, i.e., it reaches its Zener voltage. However, when high-enough reverse voltage is reached, the crystalline regions saturate, and the device conducts well. The point of reverse-bias voltage that results in conduction is referred to as the breakdown voltage (Figure 3.11). The device is useful for setting its voltage drop to be constant when operated in the reverse-biased breakdown condition.

The performance of a Zener diode can be plotted on a current vs. voltage (I-V) characteristic curve. When the diode is forward biased, it conducts quickly with very little resistance, like a regular diode. The forward-biased direction is of little interest in Zener-diode circuits, and its usefulness comes when wired in the reverse-biased direction, where is acts as a simple voltage regulator.

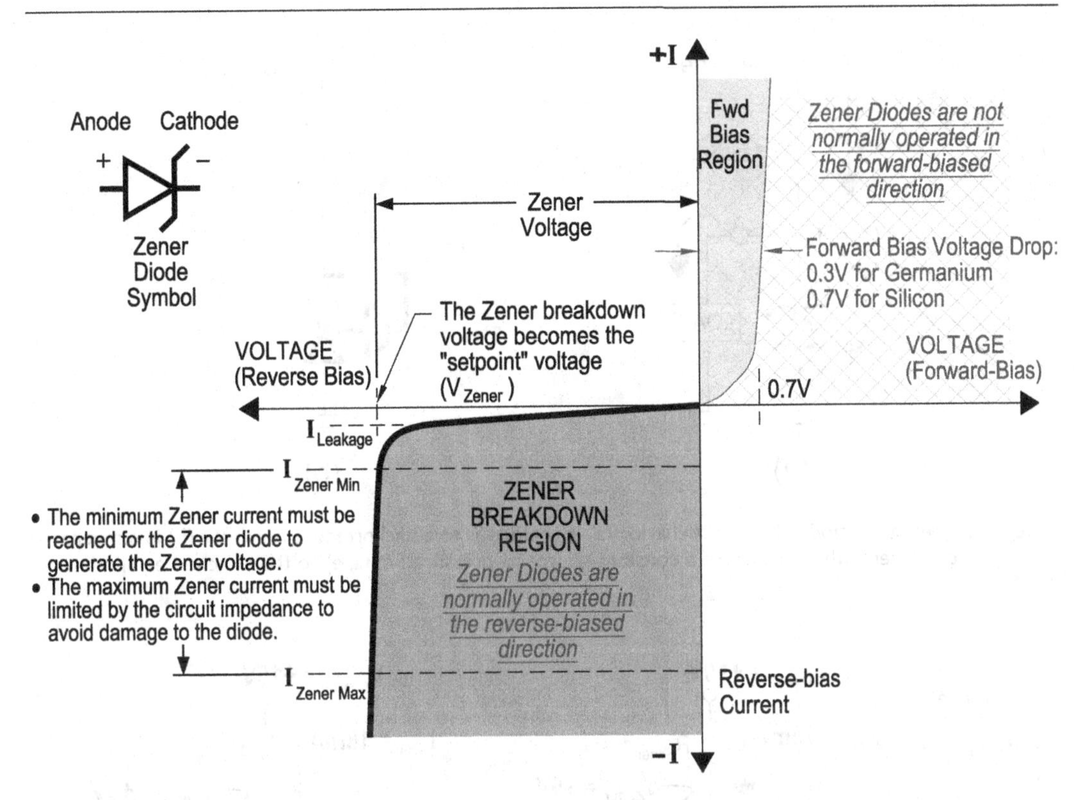

Figure 3.11 Zener diode symbol and characteristic curve. The Zener diode is normally operated in the reverse-bias direction (third quadrant), which sets the downstream circuit voltage to be the breakdown- or reverse-bias voltage level.

In a mechanical system, the Zener diode is comparable to a backpressure-regulating valve whose purpose is to maintain constant inlet/system pressure[3] to the device (Figure 3.12).

Normal diodes can be damaged when reverse-biased breakdown occurs; however, the Zener diodes are carefully doped with just enough impurities to make the breakdown voltage both predictable and constant. Zener diodes are designed to operate continuously in the reverse-bias polarity and the resulting voltage drop across the Zener diode is always the breakdown voltage. Because these diodes are designed to produce a fixed voltage level, they can be used to set or "regulate" voltage to other parts of the electronic sub-circuit, behaving like an inexpensive voltage regulator. Another function of a Zener diode is to prevent voltage surges from reaching sensitive parts in a circuit. Figure 3.13 shows a Zener diode properly installed in the reverse-bias direction in a dc circuit.

For a Zener diode to function as a dc voltage regulator it must have a current-limiting resistor placed in series (R_S), the diode must be connected in the reverse-bias direction, and the applied voltage reaching the diode must exceed the breakdown voltage after the resistor. Using Figure 3.13 as an example, if a 5V Zener diode is installed in a circuit powered by a +10V source, the voltage at the common diode-to-load junction and consequently voltage across the load is always set at +5V by virtue of the Zener breakdown voltage. If the total source current was 10mA and the load current happened to be 4 mA, then the current flowing through the Zener diode would be the 6mA difference. If the source voltage was raised to 18V causing the circuit to draw a total of 18mA, the junction voltage would remain at +5V,

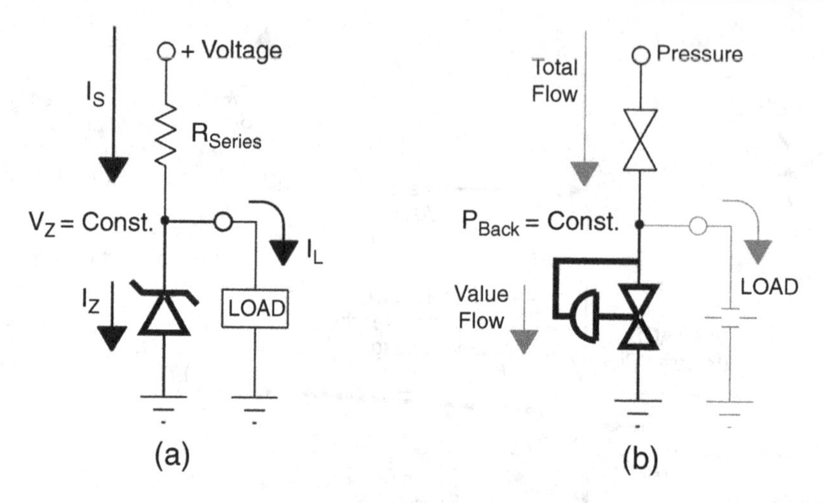

Figure 3.12 The Zener diode electrical behavior (a) is analogous to a backpressure regulator (b) in a mechanical system, which maintains a constant source of pressure (voltage) to the attached equipment.

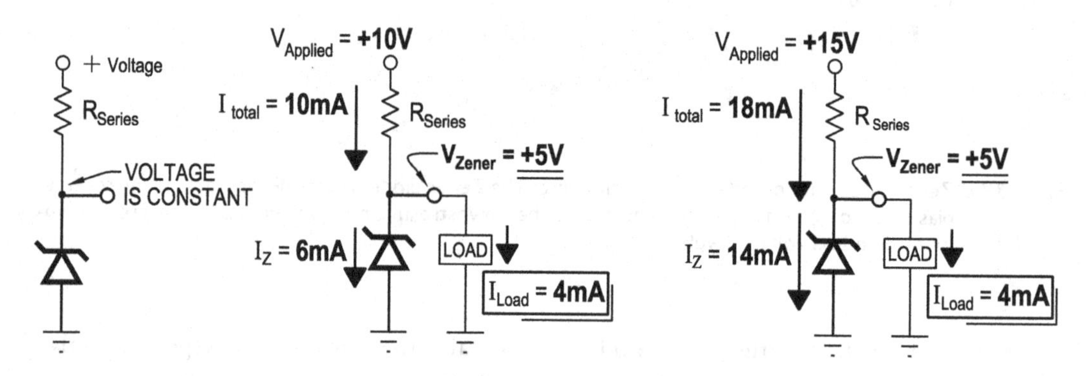

Figure 3.13 Behavior of a Zener diode installed as a voltage regulator. The current through the Zener diode varies as the applied voltage changes. The resulting output voltage is constant regardless of the input voltage to the Zener diode.

and the load current would remain unchanged at 4 mA. In the higher-voltage case, the Zener diode flows the additional current for a total diode amperage of 14mA.

BIPOLAR JUNCTION TRANSISTOR (BJT)

A common electronic component is the bipolar junction transistor (BJT). The nomenclature "bipolar junction" is derived from its appearance of having two diodes connected back-to-back or front-to-front, in either an NPN or PNP arrangement (Figure 3.14). However, unlike diodes, the transistor has three connections. While the diode is a passive device, the transistor is an active device. The two extremes of the transistor are the collector and emitter, and the center connection is the base. The symbol has an arrow on emitter leg for identifying polarity. By default, the leg without the arrow is the collector. Like the diode symbol, the emitter arrow distinguishes the P- and N-regions, where the arrow points toward the N-region and in the direction of conventional current flow when the

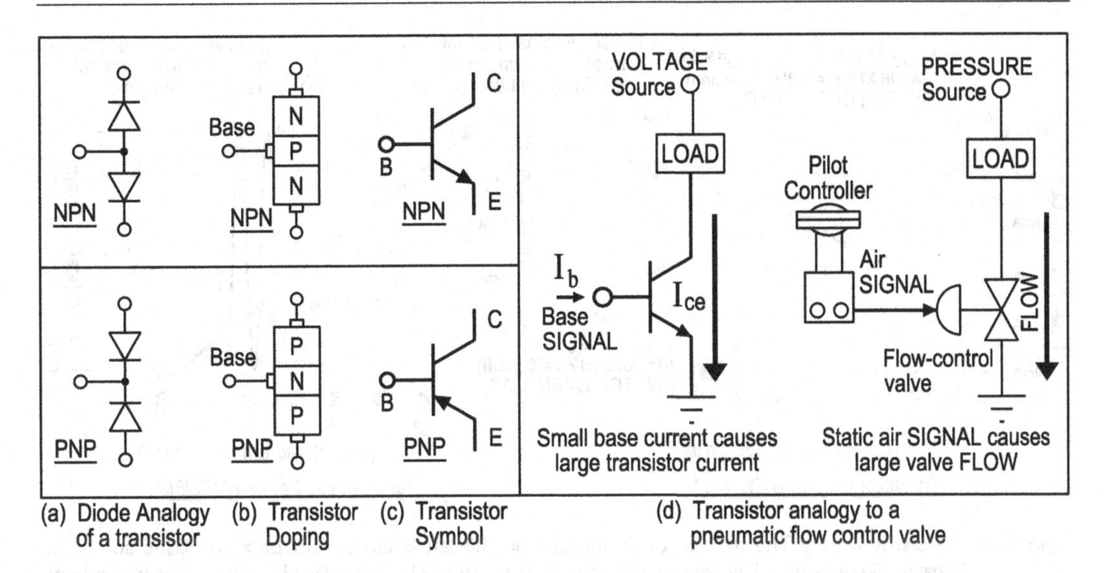

Figure 3.14 (a) The transistor construction is like two back to back center-tapped diodes, (b) NPN and PNP transistor doping, (c) transistor symbols, (d) transistor operating analogy to a mechanical pilot-actuated control valve, where small pilot signals modulate large system flowrates.

transistor is forward biased, in the same manner as the arrow pointed to the N-junction in the diode.

BIASING

Biasing in electronics is the application of correct voltage or current levels and polarity within a circuit to permit component operation. In the case of a diode, conduction occurs when polarity is applied in the forward-biased direction so that the diode behaves like a closed switch, while conduction does not occur in the reverse-biased direction, where the diode behaves like an open switch.

In the case of a transistor, forward-biasing voltage is applied from the emitter to the base to permit current flow. However, transistor conduction will not occur until base current is applied. When the base is properly biased, the transistor will conduct large amounts of current between the collector and emitter. If large enough biasing current enters the base, saturation is reached, and the transistor conducts like a closed switch (Figure 3.15 left). When only a small biasing current is input into the base, the transistor only partially conducts. Maintaining a small fixed amount of biasing base current causes the transistor to operate part way within the active region at its quiescent point[4] (Figure 3.15 right). For amplifiers, the active region is between the cut-off, where there is no collector current, and saturation, where there is maximum collector current, and the quiescent or no-signal operating point is found using a load line. For an amplifying transistor, proper biasing produces partial conduction between the collector and emitter, so that the resulting large range of "amplified" collector-to-emitter current is not clipped. A moderate amount of biasing current into the base allows larger variations of dc collector-to-emitter current defined by the transistor's *quiescent point*, or q-point, which is the transistor's operating point based on the collector-to-emitter voltage and the biasing base current. For bipolar-junction transistors, biasing is directly related to the base current, for field-effect transistors, biasing is directly related to the

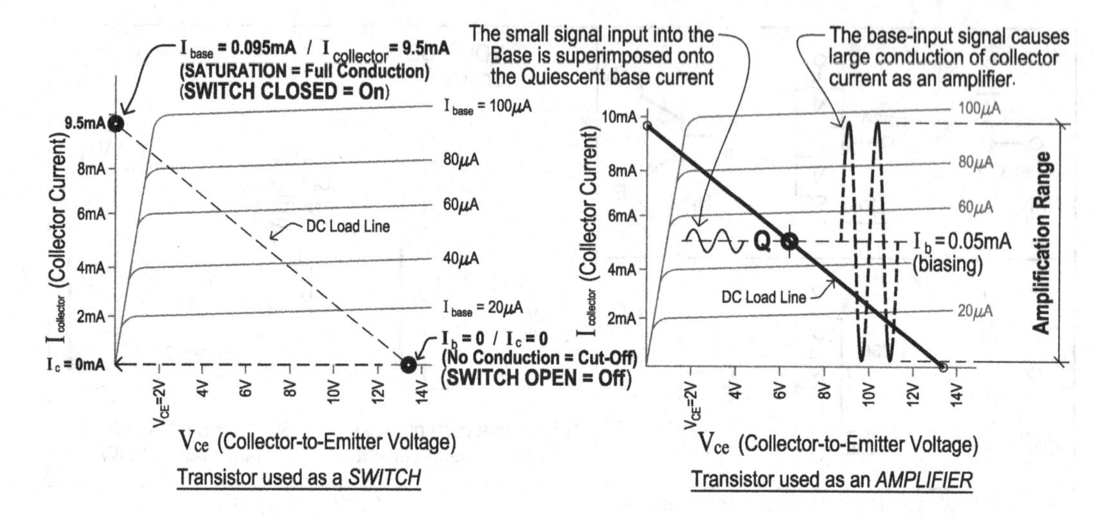

Figure 3.15 Transistor biasing. The base of the transistor on the left is either fully or zero biased so that it behaves like a switch. The base of the transistor on the right is partially biased so that it conducts halfway. The output is amplified when a small input signal is superimposed onto the biasing current, so that small input signals become large output amperages.

gate voltage. When operated at the saturation or cut-off bias levels, the transistor behaves as a base-controlled switch.

To function, the transistor collector to emitter must be wired in the forward bias direction, that is, conventional current flows in the direction of the arrow. However, even with the emitter-to-collector forward biased, the transistor still will not conduct until correct-polarity voltage and its resulting current is applied to the base. As the base current flows, the holes or electrons bridge the charge space between the collector to the emitter, allowing the transistor to conduct. If a small dc voltage is applied to the base, the transistor collector-to-emitter conducts just a small amount. As the base voltage increases, the transistor conducts more and more. When enough voltage is applied, the transistor becomes saturated and it fully conducts, like a closed switch.

The transistor is typically configured for one of two functions; either as a switch or as an amplifier. As a switch, the transistor output is connected in series with a dc load, such as a LED light or relay coil. To turn the switch on, the full biasing current via its voltage is applied to the base, to produce saturation and to cause it to conduct. To turn the transistor off, the biasing current via its voltage is removed.[5] In this manner, electronic signals applied to the base perform transistor-switching operations. The switching operations can be ganged together to produce control logic, and in fact, microprocessor integrated chips contain millions of embedded transistors that perform logic functions.

To behave as an amplifier, the transistor collector and emitter must be installed in a forward-biased arrangement within a dc circuit. A limited amount of biasing current via voltage control is applied to the base of the transistor, causing partial transistor collector-to-emitter current. In the partially biased state, the transistor does not reach saturation, which allows the transistor current to be increased or decreased through the load. When the biasing current is fixed in the active region, the transistor operates at its "quiescent point," or point of partial conduction.

Amplification is obtained by superposing small signal amperages into the base on top of the biasing current. When the signal voltage into the base increases, the transistor collector-to-emitter current follows with a significant increase in output current. When the biasing signal decreases, the transistor current decreases significantly. The result is the amplification of very small base input signals that produce significantly high output currents that can be used to power devices, such as audio speakers as an example. When the transistor's quiescent operating point is located judiciously, and the signal is not over-amplified beyond the active limits of the transistor, the signal amplification will be true and without distortion.

A transistor with a modulating output can be likened to a 3–15 psi pilot-actuated pneumatic control valve in a piping system (Figure 3.14d). The relatively static, small 3–15 psi input signal is like the small biasing current applied at the transistor base. At 3 psi, the control valve is closed, analogous to a switched-off transistor with no base current. Stepping the pressure to 15 psi would be like fully biasing a switched-on transistor, and the valve would open fully. By setting the pilot pressure to 9 psi, the valve would be half open (analogous to the ideal Q-point), and the valve could be modulated to open or close as the pilot pressure varies. The large control-valve flow caused by the small pilot signal would be analogous to the transistor large amplified collector-to-emitter output current in proportion to the small base signal.

THYRISTOR OR SILICON-CONTROLLED RECTIFIER (SCR)

A variation of the diode is the thyristor or silicon-controlled rectifier (SCR). The SCR is a three-connection device, where power connections are the anode and cathode, the third signal connection is the gate (Figure 3.16). The SCR is a device that conducts in one direction, like a diode, but only when it has forward-biased polarity and only after its gate is triggered by a signal. After the trigger signal is stopped, the SCR continues to conduct indefinitely in the forward-biased direction until the voltage goes to zero, called latch-up. When the gate voltage goes to zero, the SCR stops conducting, until the gate is triggered again *while* the SCR is in forward-biased state. Like a reverse-biased diode, a reverse-biased SCR will never conduct, even if triggered.

As an example, if a continuous gate signal is applied to an SCR in an ac circuit, the SCR would behave the same as a diode; it would conduct during the positive-voltage portion of the waveform during the time when the SCR is forward biased. Like the diode, the SCR will not conduct when the voltage is negative or in a reverse-biased state, regardless whether the gate voltage is applied. Under application of continuous gate voltage triggering, the output would be identical to the output shown in Figure 3.5 and the latched-up SCR "ON" regions of Figure 3.17. When the gate voltage is toggled off, the output will always be "off" without application of a gate voltage, until the gate is triggered again. Figure 3.18 shows the output from a triggered SCR.

Like diodes, SCRs can be connected to form single-phase or three-phase bridges, with the added advantage that the SCR output voltage can be varied. The outputs would correspond to those illustrated in Figure 3.5, Figure 3.6, and Figure 3.7, except that the output

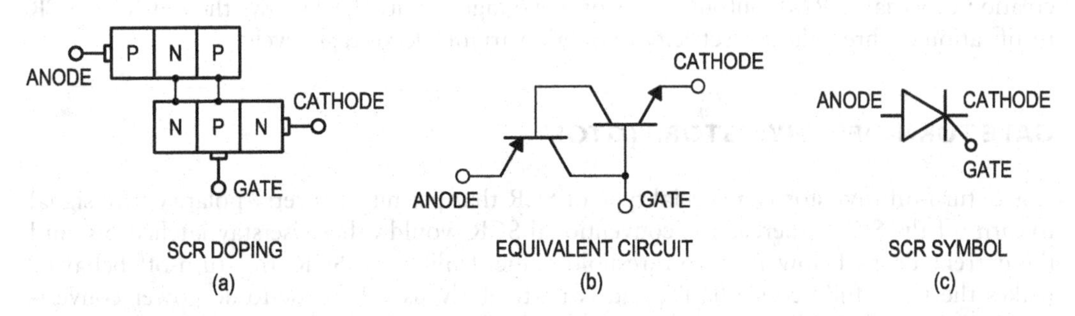

Figure 3.16 Thyristor or SCR. (a) shows the physical diagram of the doping, (b) shows the equivalent circuit diagram, which can be explained using transistor theory, and (c) is the schematic diagram symbol.

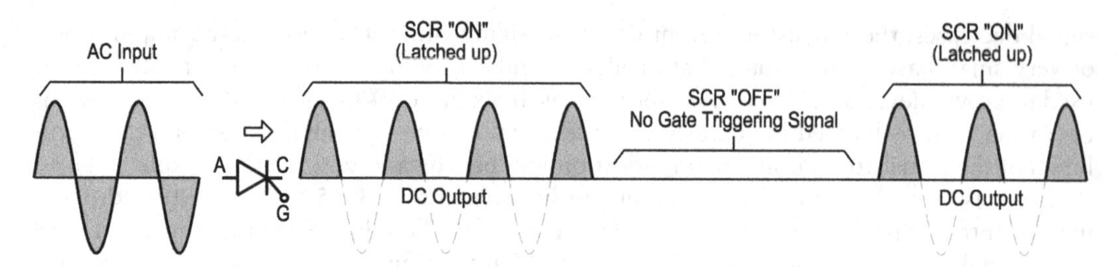

Figure 3.17 The SCR output is shown with the gate toggled on (latched up) and toggled off (no gate signal). The SCR is essentially a gate-triggered diode.

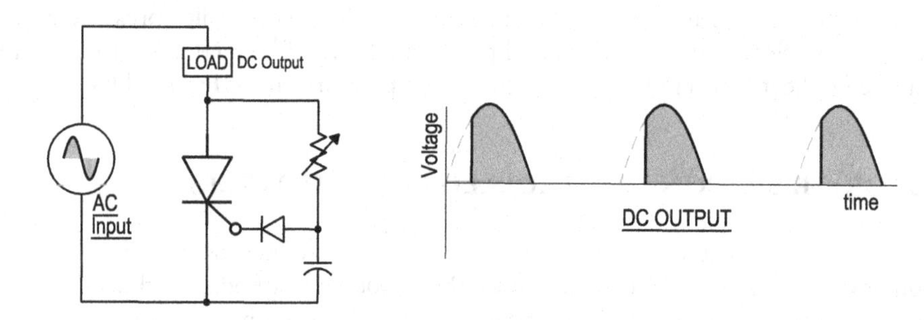

Figure 3.18 SCR circuit with a simplified time-delayed gate trigger. When the capacitor charge reaches the triggering voltage, the SCR "fires" and conducts through the remaining positive waveform. Once the anode-to-cathode ac voltage passes through 0V, the SCR becomes reverse-biased and turns off until the gate is both triggered *and* the ac waveform is positive.

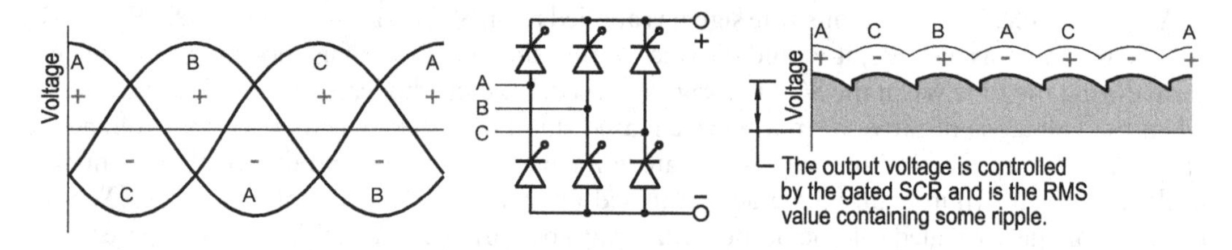

Figure 3.19 SCRs connected into a three-phase bridge-rectifier circuit. The result is rectified dc electricity; however, the RMS dc-voltage output can be modulated by varying the SCR gate triggering.

waveforms would be truncated based on the amount of gate triggering. The result is the creation of variable-RMS output values of dc voltage. Figure 3.19 shows the full-wave SCR rectification of three-phase electricity to obtain variable dc-voltage levels.

GATE TURN-OFF THYRISTORS (GTOS)

A gate turn-off thyristor is a special type of SCR that permits a reverse-polarity gate signal to turn off the SCR, whereas the conventional SCR would otherwise stay latched on until the current drops below its zero-threshold value. Unlike the SCR, this turn-off behavior makes the GTO fully controllable, and is particularly useful for dc-to-dc power converters, dc drives, and VFD ac inverters, where it can be used in both the rectifier and inverter sections. The GTO symbol and turn-off behavior are shown in Figure 3.20. Essentially, the

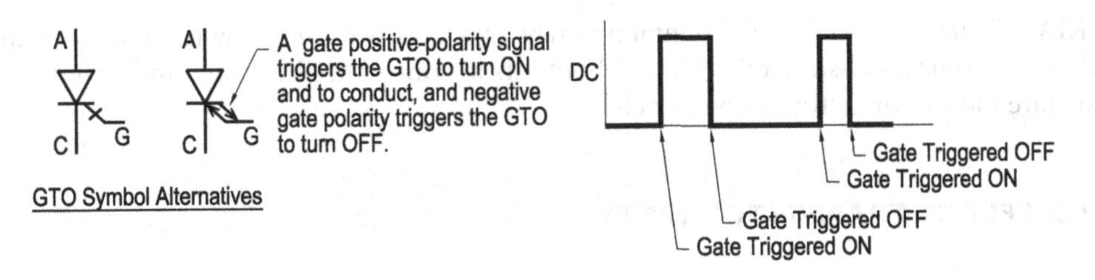

Figure 3.20 GTO symbol alternatives on the left, and GTO gate-triggering behavior on the right.

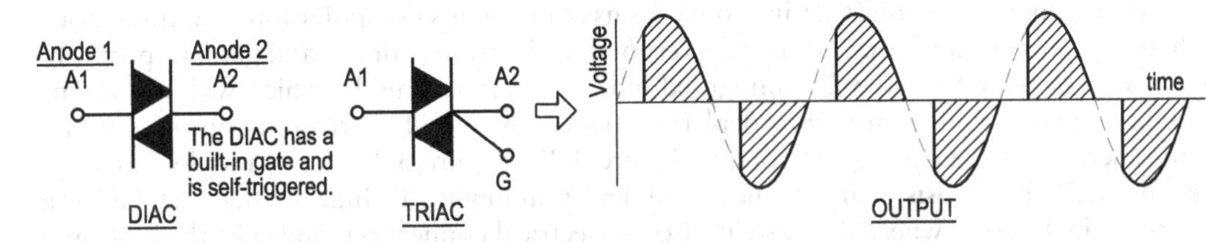

Figure 3.21 DIAC and TRIAC symbols, with accompanying output. The DIAC is a two-terminal device, triggered by a built-in gate circuit, while the TRIACs output is adjustable via an externally triggered gate circuit.

GTO behavior is nearly identical to the SCR, in that it permits forward-bias conduction only, requires a positive-current gate signal to turn on, and it will remain on until the current goes to zero; however, the GTO can be forced off at any time by applying a reversed-polarity gate signal. Like the SCR, the GTO is a high-power, three-terminal device, having an anode, cathode, and gate. It too exhibits low conduction losses and is limited to relatively low switching frequencies, around 1 kHZ, however, the GTO turn-off time is about ten times faster than the SCR, providing less commutation losses, less heat, and higher efficiency in a bridge circuit.

GTOs come in two types, the symmetrical version and the more common asymmetrical version. The symmetrical version can block the same amount of forward-bias voltage when it is reverse-biased, while the asymmetrical GTO reaches a breakdown voltage, which is often in the range of tens of thousands of volts. Symmetrical GTOs are used with current-source inverters (CSI), and asymmetrical GTOs are used with voltage-source inverters (VSI). Each asymmetrical GTO is often paralleled with a reverse-biased "snubber" diode to protect the GTO against voltage-spike damage that could exceed breakdown.

DIACS AND TRIACS

DIAC (Diode for Alternating Current) and TRIAC (Triode for Alternating Current) are constructed as two back-to-back SCRs, where the devices conduct in both alternating-current polarities and therefore do not have any specific terminal labels beyond anode. The DIAC has two connections and the TRIAC has three connections, where the third connection is the gate. The gate is used to trigger the device, so it starts conducting, and like the SCR, the device stops conducting when the voltage reaches zero, until it is triggered again (Figure 3.21). The DIAC has its gate built into the device, so that the device self-triggers to begin conduction when the gate breakdown voltage is reached, essentially becoming a non-controllable TRIAC. Sometimes a small DIAC is used as a gate input into a larger TRIAC to trigger its conduction. The bidirectionality of the TRIAC makes it useful for switching ac circuits.

TRIACs behave like relays and contactors when the gate is triggered without delay, and are used as electronic solid-state relays in PLC-outputs in control applications, and as non-arcing solid-state motor contactors in power circuits.

FIELD-EFFECT TRANSISTOR (FET)

Field-effect transistors are electronic devices that use an electric field, instead of current, to control the transistor output through a physical channel of doped material. The FETs are voltage-controlled at the gate in contrast to BJTs, which are current-controlled at the base. The terminology for the FETs wiring connections is analogous to bipolar-junction transistors, where the emitter, collector, and base correspond to the source, drain, and gate, respectively. However, the MOSFET may be configured into enhancement mode, which will increase its output current, or configured into depletion mode, which will decrease its output current, upon increase of the gate-signal voltage (Figure 3.22). Figure 3.23 shows the flow analogy for the MOSFET transistor in enhanced and depletion modes. Technically the MOSFET is a four-terminal device, where the body is also an electrical connection; however, the body and its attached semiconductor substrate is typically connected to the source terminal for power applications, so the symbol is simplified to three terminals.

The operation of the FET is different and somewhat opposite from the transistor. In a fluid-flow analogy, the FET would be likened to a reverse-acting pilot-actuated control valve, where

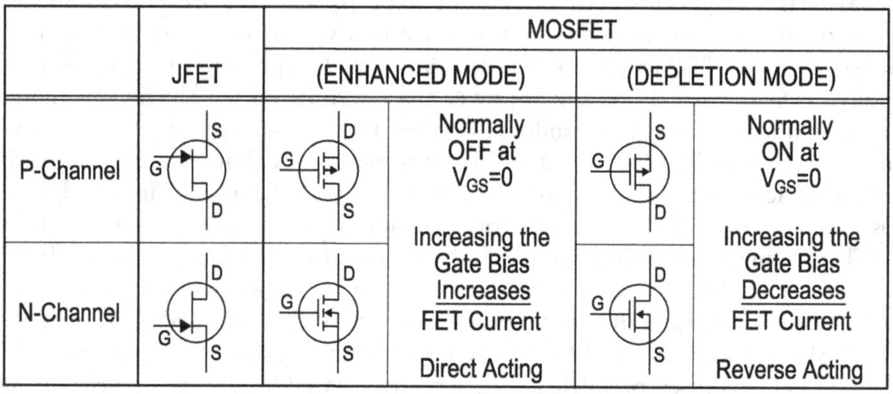

		MOSFET			
	JFET	(ENHANCED MODE)		(DEPLETION MODE)	
P-Channel	G, S, D	G, D, S	Normally OFF at $V_{GS}=0$ Increasing the Gate Bias Increases FET Current Direct Acting	G, S, D	Normally ON at $V_{GS}=0$ Increasing the Gate Bias Decreases FET Current Reverse Acting
N-Channel	G, D, S	G, D, S		G, D, S	

Figure 3.22 JFET and MOSFET symbols and MOSFET behavior in enhanced and depletion modes.

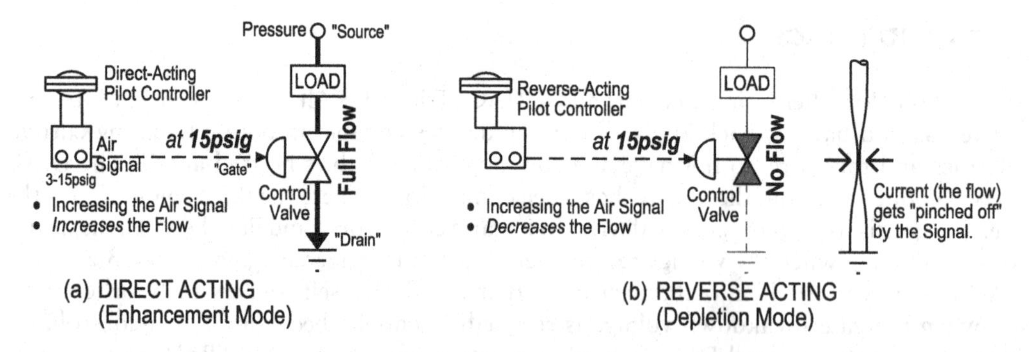

Figure 3.23 MOSFET behavior is analogous to a "direct-acting" pilot-actuated control valve when it is in enhancement mode, and a "reverse-acting" control valve in depletion mode. In depletion mode, the current (flowrate) essentially gets pinched off by the gate signal (pilot signal).

an increasing signal produces a decreasing output. Another analogy is like water flowing through a hose from source to drain, where applying gate voltage is analogous to pinching the hose to reduce the flow rate. In the case of the FET, the device naturally conducts from the source to the drain without any gate signal. Application of voltage to the gate causes the source-to-drain resistance to increase, and the FET conduction decreases. When the gate voltage reaches its "pinch-off" value, the FET essentially turns off like a switch. Full source-to-drain current flow is obtained by removing voltage to the gate. The source-to-drain current can be smoothly modulated all the way from full output to shutoff by increasing the gate voltage levels.

The two types of FETs are the Junction Field-Effect Transistor (JFET) and Metallic-Oxide Semi-Conductor Field-Effect Transistor (MOSFET). The big difference is that the JFET operates in the depletion mode by controlling the flow of holes, and the MOSFET operates in the enhancement mode by controlling the flow of electrons. The JFETs' P-N junction provides high input impedance, while MOSFETs use an isolated gate, which results in lower gate-leakage current and better control. The JFET gate must never be forward biased, while reversing polarity on the MOSFET has no deleterious effects.

MOSFETs are more widely used because they have better operating characteristics and are easier to manufacture. MOSFETs are very sensitive to overvoltage at the gate, electrostatic discharge (ESD), device overvoltage, overcurrent, and heat. Often the MOSFET is supplied with a Zener-diode voltage clamp on the input to avoid problems associated with its high sensitivity to ESD. Despite its sensitivities, careful handling and good circuit designs avoid these problems.

MOSFETs are used on switching power supplies. In these applications, it is important that the gate's driving voltage is high enough to completely turn on the MOSFET. When the MOSFET is completely turned on, its internal resistance is very low, and the power supply is very efficient. However, when the MOSFET is less-than-fully turned on, its I^2R losses can produce enough heat to damage the device. MOSFETs can be used for high-voltage, high-current applications, such as in large dc switching power supplies, but they are also used in ultra-small applications, and they are often the transistor type used in logic gates and integrated-circuit chips. Their near-perfect switching characteristics at very high switching speeds, very low resistive losses, and very low gate current make them very good for switching applications. With no bias, they are normally non-conducting, and the high gate input resistance means that very-little-to-no standby current is needed.

INSULATED-GATE BIPOLAR TRANSISTOR (IGBT)

The insulated-gate bipolar transistor (IGBT) is a three-terminal power semiconductor primarily used as an electronic switch (Figure 3.24). It combines the low forward-biased conduction losses of a bipolar-junction transistor with the fast switching speeds seen with MOSFETs. Also, like the MOSFET, the gate drive circuitry is relatively simple. The doped crystalline structure of the device makes it behave as if it has a high-speed, low-loss MOSFET gate input coupled into a high-current, low-saturation-voltage, low-resistance, fast-acting BJT. In plants, the IGBT is becoming the standard for use in larger switched-mode variable-frequency drives. It uses a collector and emitter, like a BJT, and a gate, like a MOSFET.

The IGBT has significantly lower forward-bias voltage drops than the MOSFET, and it can also withstand higher reverse-bias blocking voltages than the MOSFET. Unlike MOSFETs, IGBTs cannot conduct in the reverse direction, and where reverse current is needed, a free-wheeling diode is required to mitigate voltage spikes when powering inductive loads from the MOSFET-body terminal. Although the doped crystalline structure is somewhat similar to an SCR, it does not have latching characteristics. The relative characteristics between BJTs, MOSFETs, and IGBTs are summarized in Table 3.1.

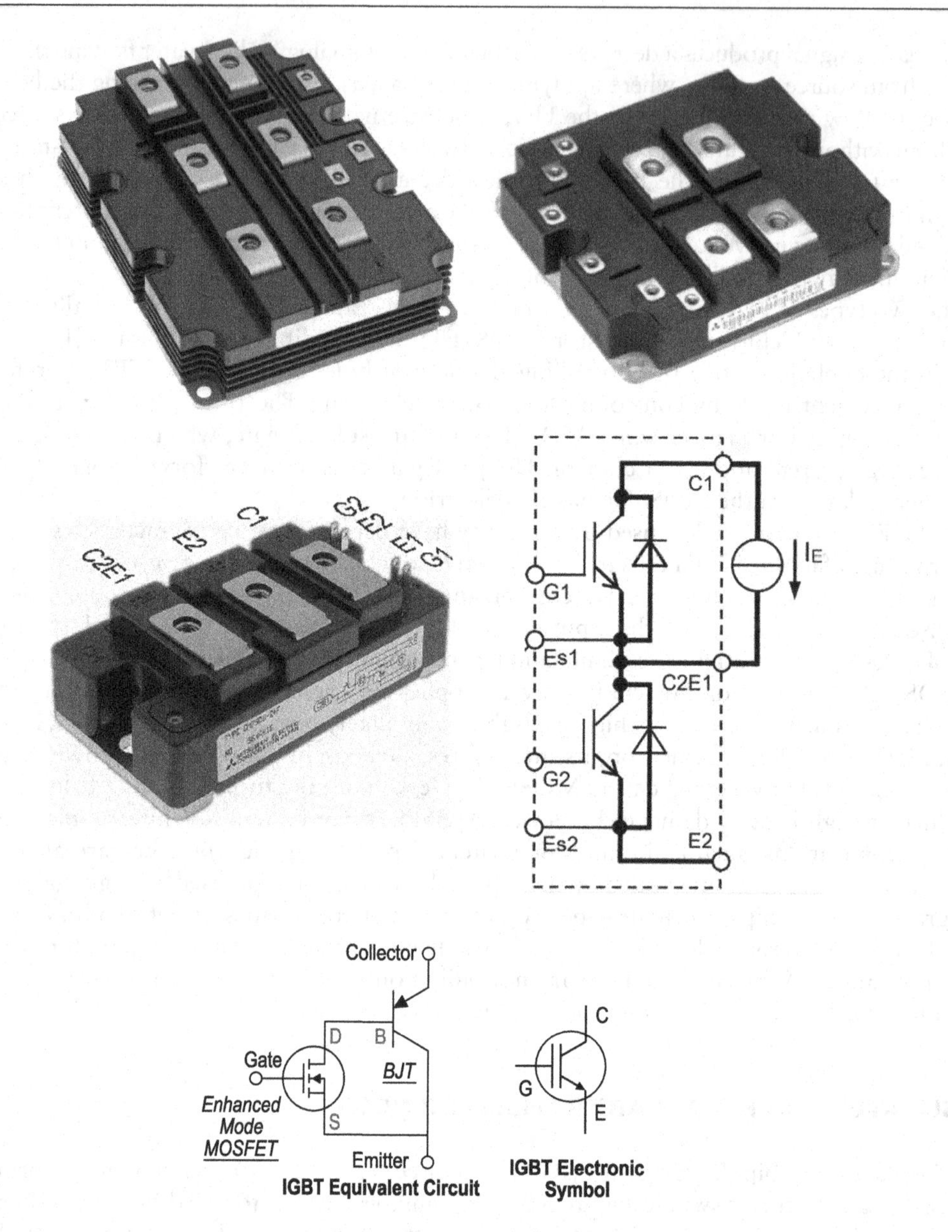

Figure 3.24 Insulated-gate bipolar transistor (IGBT) used in high-power switching applications for variable-frequency drives. The baseplate would be mounted to a heat sink. The connections are shown on the schematic diagram. (Courtesy of Mitsubishi Electric Corporation.)

Table 3.1 Relative characteristics between bipolar-junction transistors, MOSFETs, and IGBTs used in power applications

Device Characteristic	*Power BJT*	*Power MOSFET*	*IGBT*
Voltage Rating	High (< 1,000V)	High (<1.000V)	Very High (> 1,000V)
Current Rating	High (< 500A)	Low (< 200A)	High (<>500A)
Input Drive	Current: h_{FE} = 20–200μA	Voltage: V_{GS} = 3-10V	Voltage: V_{GES} = 4-8V
Input Impedance	Low	High	High
Output Impedance	Low	Medium	Low
Switching Speed	Slow	Fast (nanoseconds)	Medium

DESCRIBE AND SKETCH

For each of the following devices, label all device terminals and indicate whether they are ac or dc devices. Briefly describe how they work and list applications where they are used. Sketch the flow valve arrangement whose behavior is analogous to the device.

1. Diode
2. Zener diode
3. Bipolar-junction transistor (BJT)
4. Diode bridge rectifier
5. Thyristor or silicon-controlled rectifier (SCR)
6. Low-pass filter and high-pass filter
7. Voltage-divider circuit
8. Gate turn-off thyristors (GTOs)
9. DIACs and TRIACs
10. Metallic-oxide field-effect transistor (MOSFET)
11. Insulated-gate bipolar transistor (IGBT)

QUESTIONS

1. How are solid-state electronic devices formed? What is the doping process? What materials are used in the doing process?
2. What are the two crystalline elements used in the doping process? What is the forward-bias voltage drop of the two elements?
3. What is the difference between electrons and holes? What is the difference between electron current and conventional current?
4. What do the terms forward bias and reverse bias mean with diodes and transistors? What happens to transistor collector-to-emitter current when a transistor is partially forward biased?
5. What are the two applications for BJTs? What is meant by quiescent point, and when is it important in transistor-circuit design? What is meant by a fully biased transistor? What automation and control applications use fully biased transistors?
6. Describe the operation of thyristors (SCRs) and provide sketches of a single-SCR rectifier with its resulting waveform? How is the newer gate turn-off thyristor different from the conventional thyristor? Where are thyristors (SCRs) used?
7. Describe the operation of DIACs and TRIACs. Where are TRIACs used in power electricity and PLCs?
8. Where are metallic-oxide semiconductor field-effect transistors (MOSFETs) commonly used? Describe MOSFET operation in enhanced mode and depletion mode.
9. Compare insulated-gated bipolar transistors (IGBTs) to MOSFETs. List the IGBT characteristics that make it attractive for large variable-frequency drives.

PROBLEMS

1. Using Figure 3.9a as a reference, 24Vdc is applied to a 6V Zener diode through a 100-ohm resistor to a load. Neglecting any load impedance, find the maximum current and power that can be provided to the load.
2. A 5,000-ohm thyristor is used to measure temperature as a 0-5Vdc voltage signal into

a PLC using the circuit shown in Figure 3.9a, having a fixed resistor of 2,200-ohms. The design temperature range is 0°C to 100°C. Thyristor manufacturer data indicates 16,330-ohms at 0°C, 5,000-ohms at 25 °C, 1801-ohms at 50 °C, 740-ohms at 75 °C, and 212-ohms at 100 °C. Using a spreadsheet, provide a table showing the total circuit resistance, the thyristor current, and the voltage signal produced across the thyristor. Plot the data points and trendline describing the relationship of signal voltage versus temperature, and using the plotted trendline, obtain a polynomial equation whosen R^2 regression calculation is better than 0.9990, which would be programmed into the PLC to indicate temperature.

NOTES

1. The anode is defined as the connection where conventional current or holes flow into a polarized device. Conversely, the cathode is the connection where conventional current, or holes, leaves the polarized device. These terms can be confusing in some instances, so the terminology will be avoided in preference for p-region or positive terminal, etc.
2. Biasing is analogous to check-valve behavior. When a device is forward biased, it conducts and current flows. When a device is reverse biased, the check valve is closed, and the device does not conduct. In the way that forward-flow pressure difference actuates the check valve, properly forward biasing the voltage polarity actuates the diode.
3. Pressure in a mechanical system is the potential for fluid to flow and is analogous to voltage in an electrical system, which is the potential for current to flow. Zener diodes, like backpressure regulators, create a source of constant potential.
4. The quiescent point is also referred to as the Q point or biasing point. The Q point falls on the DC load line, or along the line that connects between zero base current and saturation base current. The DC load line is a function of the transistor characteristics and the circuit voltage levels between the emitter and collector. For amplifiers, the Q-point design goal is ideally halfway into the active region to permit maximized amplification without clipping the output signal.
5. Technically, although the biasing current into the base permits the transistor to conduct, it is the correct polarity base-to-emitter voltage (V_{BE}) that produces the base current.

Chapter 4

Digital theory, logic, and two-state control

One of the most important aspects of a control system is its ability to receive conditional, two-state signal-input information and then make a "decision" for setting a two-state output to produce a desired outcome. As simple examples, if system air pressure is too low, then a compressor should start, or if water-tank level is too high, then a dump valve should open. The two-state decision-making process is part of a logic scheme that can be complex for some systems. This chapter gives an overview of digital theory and logic, to provide conceptual information used with switching theory, microprocessor-based control, and PLC programming.

In general, digital control implies two-state switching, represented by binary 1 or 0, "true" or "false," "yes" or "no," "on" or "off," "open" or "closed," "running" or "stopped," and so on. For control, *digital input* means a two-position switch, whose control signal comes from either being open or closed. Likewise, *digital output* implies two-state devices that are to be either turned on or off, such as status-indicator lights, motor-starter contactors, solenoid-actuated valves, commands to initiate start or stop actions, audible and visual alarms, etc. Digital signals are contrasted against analog signals, which are used to modulate the control. Analog implies continual variations in input measurements followed by modulations in the output actions that affect the "controlled quantity." Analog-input signals originate from transmitters, while analog-output signals modulate device positions, such as speed settings in VFDs, readings for electronic meters or indicators, regulating-valve positions, damper control, etc. This chapter is limited to digital-input and digital-output signal processing.

With digital control, there are two types of logic, combinational and sequential. Combinational logic depends on the present state of input signals, where the input originates from switches that are either open or closed, and which in turn sets the output to be either *on* or *off*. In contrast, sequential logic uses the same digital-input signals combined with the output signals produced from the previous state to create an updated output signal. Often, sequential logic uses edge-triggered pulses to enable the sequence. The pulses may be synchronously generated in a repetitive manner from an electronic oscillator, called a "clock," or the pulses may be asynchronously initiated by external switches at random times, such as from manual actuation. Sequential logic needs a built-in form of memory to produce an output that depends on the prior state.

There are several ways to describe logic strategies, including logic gates, flow diagrams having if-then conditional statements, truth tables, and relay logic. Logic schemes can be complicated, but often they are broken into simpler, more manageable modules, procedures, or blocks of code.

BOOLEAN LOGIC

Boolean logic is named after George Boole, and uses two-state inputs and two-state outputs in a *logic-gate* decision-making process. The two states are often described as "true" or "false," and are commonly represented as "1" or "0." Conditional states are passed into logic gates where they are manipulated into a single true/false or 1/0 output. In Boolean logic, there are seven types of simple logic gates. The three primary gates are AND, OR, and NOT gates and are shown in Figure 4.1. The three primary gates form building blocks for creating the remaining four gates. The NOT gate is simply an inverter that toggles a single input to its opposite value as its output. Input/output maps and truth tables are two ways of showing every combination of input states and their resulting outputs, and they are useful for visualizing the output produced by logic functions. The input/output maps, truth tables, and Boolean representation symbols are shown in Figure 4.1 for the primary gates.

Secondary logic gates are produced by combining the NOT with the AND gate to produce the NAND, and combining the NOT with the OR gate produces the NOR. The NAND gate is the inverted AND, which reverses the AND output. Likewise, the NOR gate reverses the OR output. The remaining two gates are the "exclusive" versions of the OR and NOR gates, represented as XOR—"exclusive OR"—and XNOR—"exclusive NOR." For the exclusive XOR to produce a true output, only one or the other input must be true, which also means both inputs cannot be true or both inputs cannot be false at the same time. The exclusive XNOR is exactly the opposite of the XOR gate. For the XNOR output to be true, both inputs must be true or both must be false. Figure 4.2 shows the Boolean symbols and truth tables for the secondary logic gates, and Figure 4.3 shows Venn diagrams to help visualize the gate inputs and outputs.

Other Boolean logic-gate symbols are sometimes used. Figure 4.4 shows the symbols used in recent editions of the IEEE and the IEC standards and Figure 4.5 shows the logic symbols used in the German DIN standards.

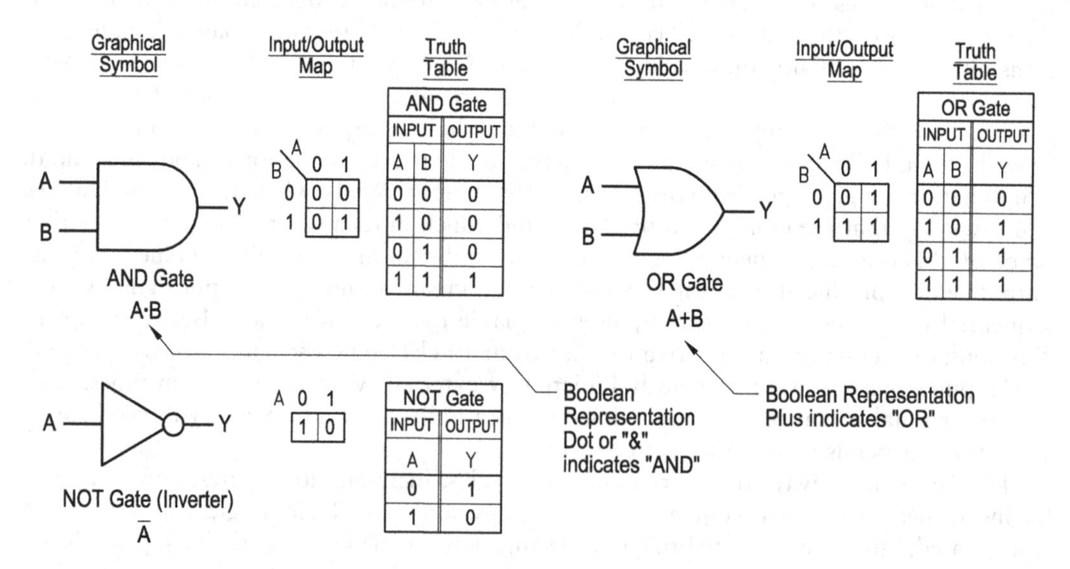

Figure 4.1 U.S. customary logic symbols with associated truth tables and input-output maps for the three primary Boolean gates; AND, OR, and NOT functions. The gate symbols shown are taken from ANSI/IEEE standards.

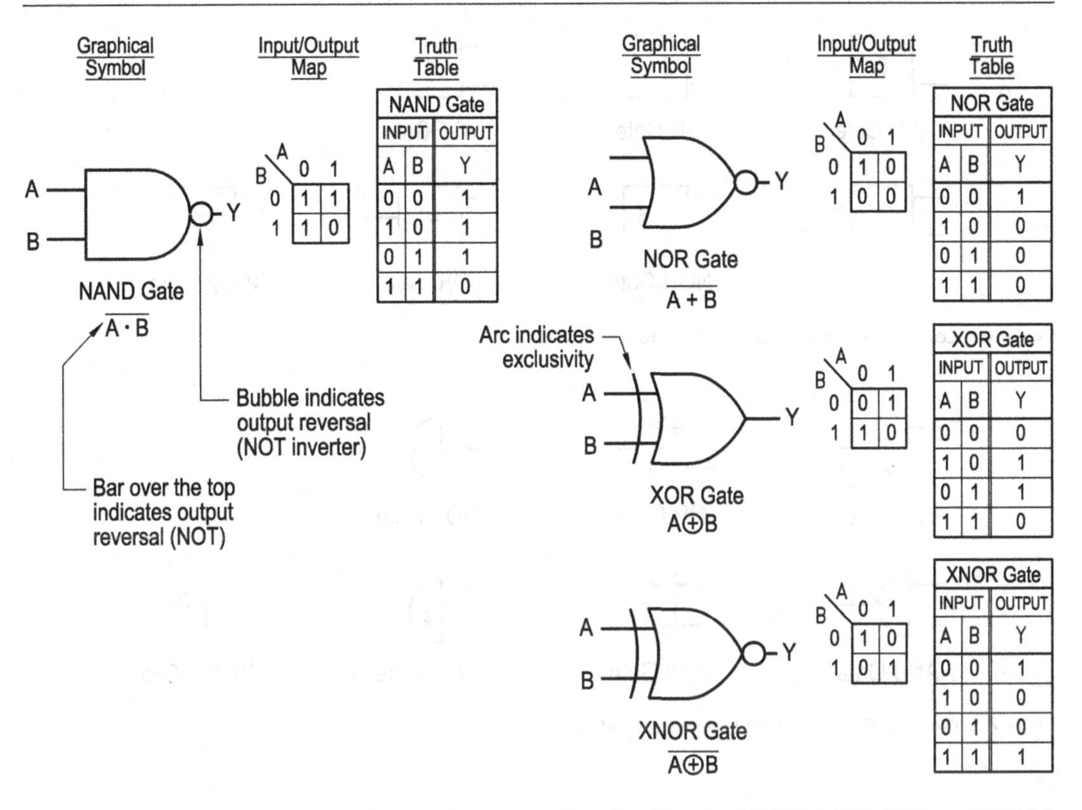

Figure 4.2 U.S. customary logic symbols with associated truth tables for NAND, NOR, XOR, and XNOR logic gates.

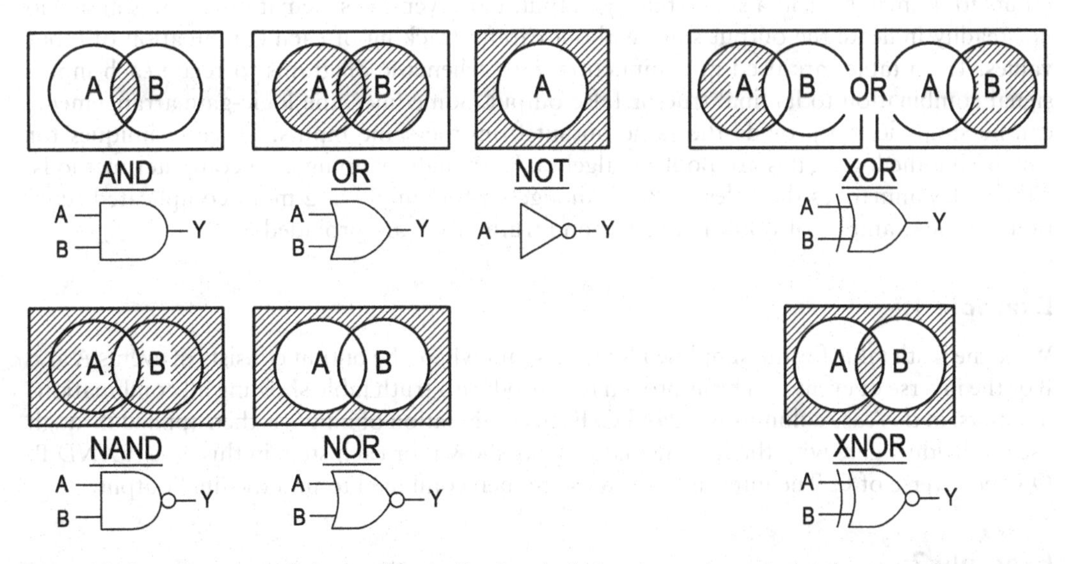

Figure 4.3 Venn diagrams to help visualize the Boolean-logic functions, where the hatch pattern shows output logic state 1.

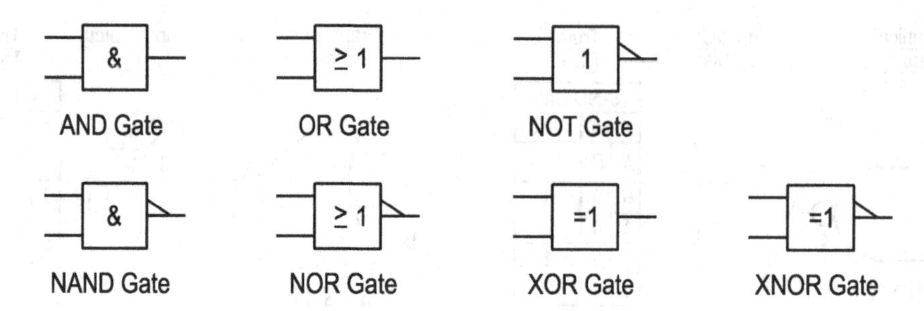

Figure 4.4 Logic gate symbols using IEEE and IEC standards.

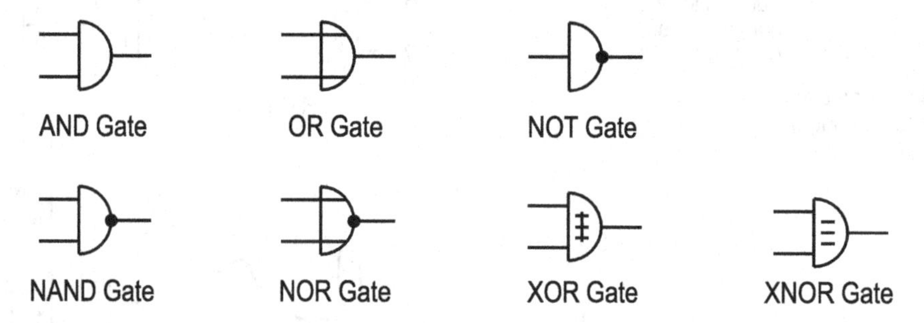

Figure 4.5 German DIN-standard logic gate symbols.

COMBINATIONAL LOGIC AND BOOLEAN ALGEBRA

Schematic diagrams are useful when ganging several Boolean-logic operations having many inputs to ultimately yield a single binary output. However, the schematic diagrams alone do not readily indicate the output state without careful tracking of each combination of input values. Truth tables are used in conjunction with schematic diagrams to relate each input-signal combination to its final true or false output. Sometimes, the logic-gate arrangements can be simplified to provide the same output from the same inputs. Three techniques for simplifying the logic gates are Boolean algebra, Karnaugh mapping, and computer methods. Table 4.1 summarizes the rules of Boolean algebra for simplifying more complicated functions. Some examples of Boolean algebra and truth tables are provided.

Example 1:

Write the truth table for the simplified logic diagram where the output consists of events A and B or the inverse of event C. The approach is to produce a truth table showing all combinations of inputs, in this case columns A, B, and C. Each combination of inputs is then applied through each individual gate with the intermediate results shown for each step, in this case A *AND* B, *OR* the inverse of C. The intermediate steps are then combined to find the final output.

Example 2:

Use Boolean algebra to simplify the logic function: $Y = (A \cdot B) + B + C$. Create a truth table for the original logic and a second truth table for the simplified logic to show that the results after simplification are identical.

Solution to Example 2 using Boolean algebra: note that the Boolean operations are cross-referenced to the rules contained in Table 4.1.

Table 4.1 Boolean algebra rules

Commutative Laws		*Associate Laws*		*Distributive Laws*	
$A+B=B+A$	(1)	$(A+B)+C=A+(B+C)$	(3)	$A\cdot(B+C)=A\cdot B+A\cdot C$	(5)
$A\cdot B=B\cdot A$	(2)	$(A\cdot B)\cdot C=A\cdot(B\cdot C)$	(4)	$A+(B\cdot C)=(A+B)\cdot(A+C)$	(6)
Identity Laws		*Redundancy Laws*		*DeMorgan's Theorem*	
$A+A=A$	(7)	$A+(A\cdot B)=A$	(11)	$\left(\overline{A+B}\right)=\overline{A}\cdot\overline{B}$	(21)
$A\cdot A=A$	(8)	$A\cdot(A+B)=A$	(12)	$\left(\overline{A+B}\right)=\overline{A}+\overline{B}$	(22)
$A\cdot B+A\cdot\overline{B}=A$	(9)	$0+A=A$	(13)		
$(A+B)\cdot\left(A+\overline{B}\right)=A$	(10)	$0\cdot A=0$	(14)		
		$1+A=1$	(15)		
		$1\cdot A=A$	(16)		
		$\overline{A}+A=1$	(17)		
		$\overline{A}\cdot A=0$	(18)		
		$A+\overline{A}\cdot B=A+B$	(19)		
		$A\cdot\left(\overline{A}+B\right)=A\cdot B$	(20)		

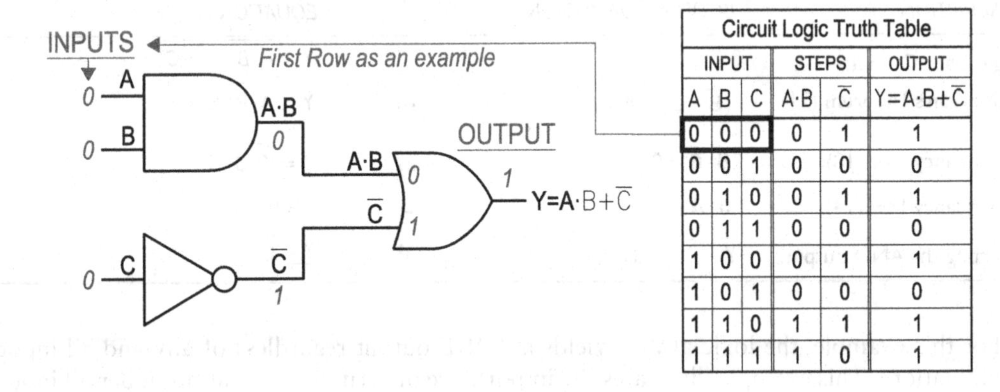

INPUT			STEPS		OUTPUT
A	B	C	A·B	C̄	Y=A·B+C̄
0	0	0	0	1	1
0	0	1	0	0	0
0	1	0	0	1	1
0	1	1	0	0	0
1	0	0	0	1	1
1	0	1	0	0	0
1	1	0	1	1	1
1	1	1	1	0	1

Example 1 The truth table showing all input combinations and each corresponding output for the adjacent Boolean logic diagram.

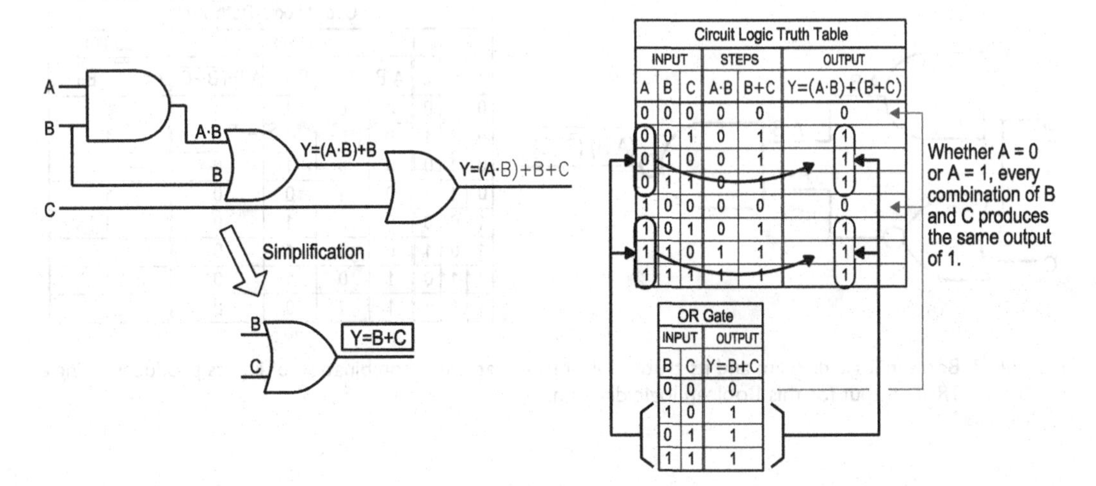

INPUT			STEPS		OUTPUT
A	B	C	A·B	B+C	Y=(A·B)+(B+C)
0	0	0	0	0	0
0	0	1	0	1	1
0	1	0	0	1	1
0	1	1	0	1	1
1	0	0	0	0	0
1	0	1	0	1	1
1	1	0	1	1	1
1	1	1	1	1	1

INPUT		OUTPUT
B	C	Y=B+C
0	0	0
1	0	1
0	1	1
1	1	1

Example 2 The initial Boolean logic diagram and its equivalent logic diagram after simplification.

DESCRIPTION	*BOOLEAN OPERATION*		*EQUATION*
Logic Gate Output:			$Y = (A \cdot B) + B = C$
Factoring (15):	$Y = B \cdot (A + 1) + C$	→	$A + 1 = A$
Redundancy Law (16):	$Y = (B \cdot 1 + C$	→	$B \cdot 1 = B$
Simplifying:	$\boxed{Y = B + C}$		

Example 2 shows that it is sometimes possible to simplify a logic schematic by using Boolean algebra, so as to reduce the number of gates. Note that in this example, the truth table shows that the output depends only on combinations of B and C, which indicates that A is not needed, and the logic can be simplified. Reducing the number of logic gates results in simpler, less expensive, smaller circuit boards that consume less power while producing the same results, or more efficient programs.

Example 3:

Using Boolean algebra, simplify the Example 3 schematic and show the truth table output.

DESCRIPTION	*BOOLEAN OPERATION*		*EQUATION*
Logic Gate Output:			$Y = \overline{A \cdot B \cdot (\overline{\mathbf{B} + \mathbf{C}})}$
DeMorgan's Theorem (21)	$(\overline{\mathbf{B} + \mathbf{C}}) = \overline{B} \cdot \overline{C}$	→	$Y = \overline{A \cdot \mathbf{B} \cdot \overline{\mathbf{B}} \cdot \overline{C}}$
Redundancy Law (18):	$\mathbf{B} \cdot \overline{\mathbf{B}} = 0$	→	$Y = \overline{\mathbf{A} \cdot \mathbf{0} \cdot \overline{C}}$
Redundancy Law (14):	$\mathbf{0} \cdot \mathbf{A} = 0$	→	$Y = \overline{0}$
Inverting the AND output:	$Y = \overline{0}$ yields	→	$\boxed{Y = 1}$

For this example, the logic always yields a *TRUE* output regardless of any and all input combinations. This example illustrates the importance of verifying the output under all input combinations to avoid mistakes.

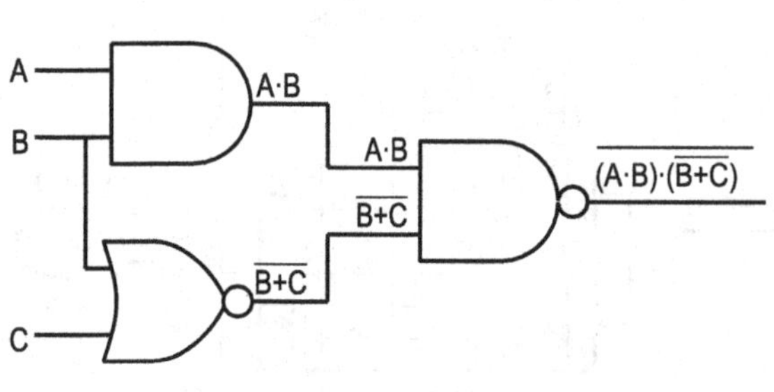

Circuit Logic Truth Table							
INPUT			STEPS				OUTPUT
A	B	C	A·B	B+C	$\overline{B+C}$	$A \cdot B \cdot (\overline{B+C})$	$Y = \overline{A \cdot B \cdot (\overline{B+C})}$
0	0	0	0	0	1	0	1
0	0	1	0	0	1	0	1
0	1	0	0	0	1	0	1
0	1	1	0	1	0	0	1
1	0	0	0	0	1	0	1
1	0	1	0	0	1	0	1
1	1	0	1	0	1	0	1
1	1	1	1	1	0	0	1

Example 3 Boolean logic diagram and its truth table. Note that every combination of inputs produces a single *TRUE* output for this Boolean logic diagram.

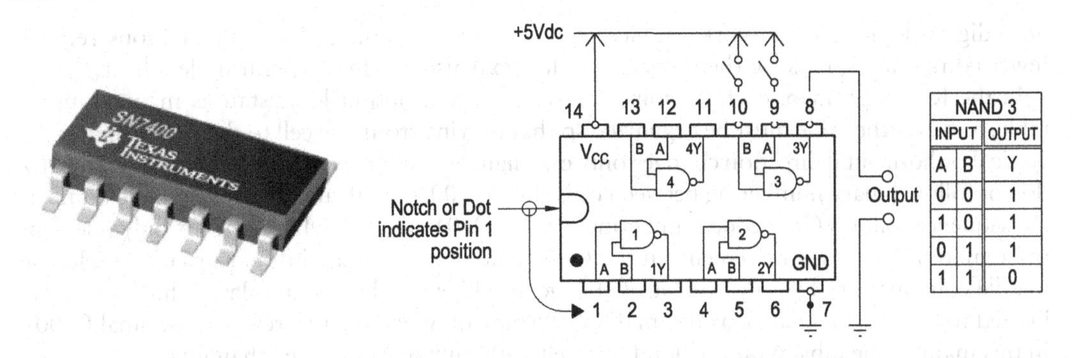

NAND 3		
INPUT		OUTPUT
A	B	Y
0	0	1
1	0	1
0	1	1
1	1	0

Figure 4.6 The TI 7400 J-Package surface-mount quadruple-NAND-gate dual-inline-pin IC chip. The chip has two additional pins for power (+5Vdc) and return (ground). From the truth table at the right, a 0V input at either pin A, B, or both produces a NAND output of 1, indicating *TRUE*, which is seen as +5V at the Y pin. If both A and B switches are closed (*TRUE*), then the output at Y goes *FALSE*, and 0V shows at the Y pin. (Courtesy of Texas Instruments.)

The universal NAND gate

NAND and the NOR gates are considered universal gates, because all other gates can be constructed from combinations of either of them. Although either would work, the NAND gate is preferred for constructing electronics-based gates. The electronics for the NAND circuit is faster and more efficient than the NOR circuits for several reasons. First, it can handle more than two inputs into a single transistor, so fewer transistors may be required. Additionally, the NAND gate uses a different crystalline structure that contributes to faster speed while also having off-states that have less gate leakage current. The better NAND-gate characteristics come from its NMOS construction, which relies on electron movement, whereas the NOR gate has a PMOS structure that relies on the movement of much-slower holes. These features result in faster speed, higher efficiency, and less heat generation. NAND gates are more advantageous for integrated circuits in terms of size, speed, and thermal reasons.

Figure 4.6 shows a surface-mount version of the TI-7400 integrated-circuit chip, which has four built-in NAND gates. A positive-voltage source is applied to V_{CC} at pin 14 for chip operation and to produce voltage outputs, and the internal circuitry is completed through the ground at pin 7. With each switch input designated by A and B pins and the corresponding output designated as Y, the NAND-3 input-signal switches are connected to pins 9 and 10 with the corresponding logic output at pin 8. Whenever either switch is open, meaning *FALSE*, no voltage is applied to least one gate and the output at pin 8 is set to *TRUE*, having a value of +5V. When both switches into pins 9 and 10 are closed, having a +5V *TRUE* input signal applied to both the A and B inputs, the output from the NAND-3 gate at pin 8 is set to *FALSE*, and it goes to 0V.

INTRODUCTION TO KARNAUGH MAPS

A Karnaugh map is an alternative method to graphically simplify logic-gate functions without using Boolean algebra. A Karnaugh map is an "expanded-table" representation of the truth tables. Karnaugh maps are practical for simplifying relatively small Boolean functions of up to six variables, but for larger logic gates, more complex computer algorithms are available. Karnaugh maps can reduce the number of gates, and they can provide some insight into how digital logic circuits work. In electronic circuits, simplified Boolean functions require fewer components, making them faster and less expensive, while generating less heat.

In the Karnaugh map, each combination of input and output logic state is mapped into a table, however the table must be organized so that moving from one cell to the next in either the vertical or horizontal direction changes only one digit. For that reason, the columns in Figure 4.7 do not follow binary numbering between cells (Binary: 00, 01, 10, and 11). Instead, the numbering sequence follows Gray code numbering (Gray: 00, 01, 11, 10), which permits only one digit to be modified for each increment. In the Gray-code sequence used in the Karnaugh table, the middle columns vary by only one bit of change, but likewise the first and last columns are also limited to the one-bit change as decimal 3 (10) shifts only one digit to revert to decimal 0 (00). In this manner, the table wraps around on itself with only one digit ever changing.

The two Karnaugh map methods are the product of sums, which looks at the 1s within the table, and the sum of products, which looks at 0s. The rules for Karnaugh mapping for the sum-of-products method require the largest collections of adjacent horizontal or vertical cells containing 1s, so that the collections for grouping multiples have 2^n items (2, 4, 8, 16), and that the number of groupings is minimized so that all 1s are selected at least once. The grouping includes overlapping cells, as well as groups formed by wrapping the table around so that the right-and-left columns and the top-and-bottom columns close on themselves. In some "don't care" cells that do not affect the desired outcome, the Karnaugh result can be further simplified.

Example 4:

Consider an application where at least two out of three pieces of equipment must be available for operation, or if not, an alarm shall be sounded when the Y output becomes true to indicate inadequate redundancy. For this example, adequate redundancy occurs when $Y = (A \cdot B) + (B \cdot C) + (A \cdot C) + (A \cdot B \cdot C)$ is *TRUE*. These conditions are summarized in the truth table shown in Figure 4.7, where the last term was found to be unnecessary.

In the Karnaugh map in Figure 4.7, three groups are found, where one variable in the group does not affect the output. For example, in the upper grouped cells, B can be 0 or 1, but A and C are always 1, yielding A · C. Similarly, procedures are followed for the other two groups to result in A · B or B · C. This example shows that the last A · B · C function is not necessary and is the basis for two-out-of-three (2oo3) voting when arranging three controllers able to tolerate a single failure, when very high reliability is required.

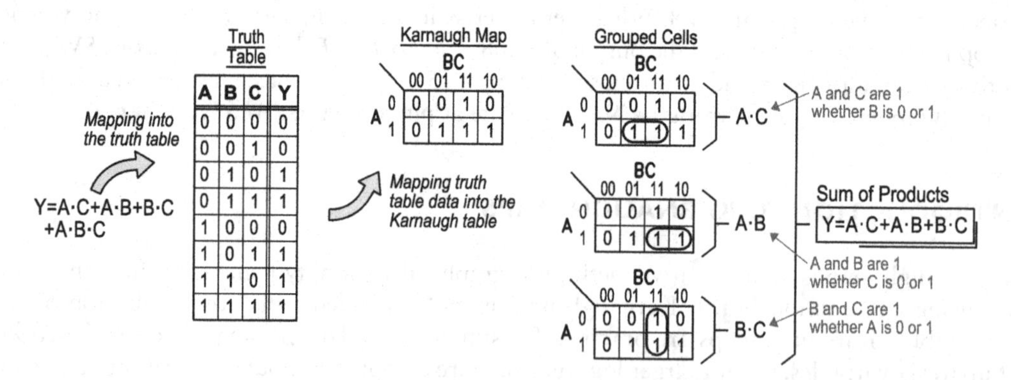

Figure 4.7 Example problem showing simplification using Karnaugh maps, applying the sum of products method.

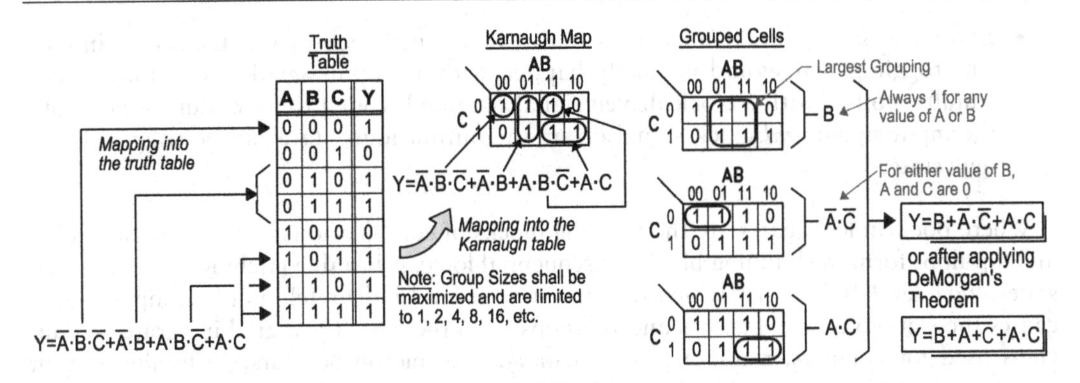

Figure 4.8 Karnaugh map example using the sum of products method and applying DeMorgan's Theorem on the middle term. The lower result exchanges an AND function for an OR function and reduces the result by one inverter.

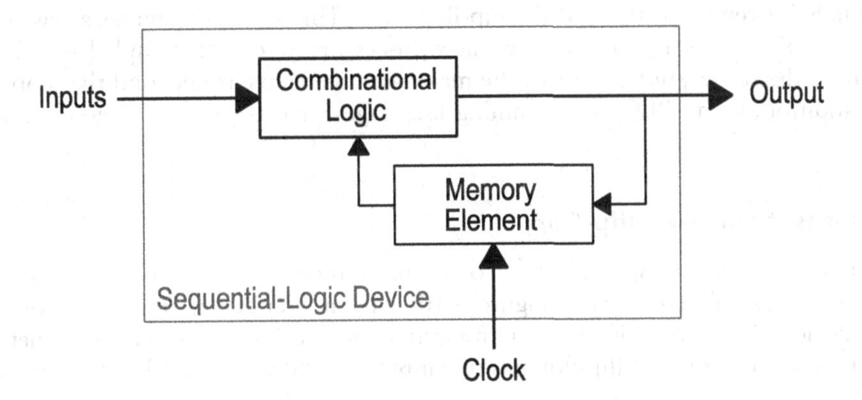

Figure 4.9 Functional block diagram showing sequential logic.

Example 5:

Simplify the three-variable Boolean expression $Y = \overline{A \cdot B \cdot C} + \overline{A} \cdot B + A \cdot B \cdot \overline{C} + A \cdot C \cdot$ using Karnaugh mapping, with the solution shown in Figure 4.8:

SEQUENTIAL LOGIC AND FLIP-FLOPS

Whereas combinational logic uses discrete condition-based input signals, sequential logic follows a series of events, and to function, it requires memory of its previous output. Sequential logic can be event driven, clock driven, or pulse driven. Figure 4.9 shows a general functional block diagram of sequential logic.

- *Event-driven* sequential logic uses an external signal, such as a hand- or pressure-actuated switch, to change the output logic state when toggled, and is called asynchronous since it is independent of a clock.
- *Clock-driven* sequential logic is synchronized to a clock, which obtains its triggering function from a timer circuit.

- *Pulse-driven* sequential logic uses both event- and clock-driven triggering inputs. The toggling that would normally happen with an input-signal switch must occur simultaneously with a clock-driven signal obtained from a timer circuit. When both the input signal *and* clock signal occur, the output logic state can be triggered to a new state.

Where Boolean logic gates are the building blocks of combinational logic, bistable latches and flip-flops form the building blocks of sequential logic. A bistable latching device has two states, logic level "0" or "1," which remains at a set level indefinitely until an input trigger causes the state to toggle. Latches and flip-flops form the basis for digital integrated-circuit chips used for counters, random-access memory, and microprocessors. A flip-flop output stores one bit of information as either "0" or "1."

There are several variations of flip-flops, including the toggle (T), delay (D), set-reset (S-R), and the J-K flip-flops; and there are many ways to configure flip-flops to obtain many desirable functions. A flip-flop can have none, one, or two inputs, and it often uses a clock-triggering signal to initiate the logic action. Some flip-flops have "preset" and "clear" input signals for reinitialization of the flip-flop state. This section provides a few overview examples to obtain a sense as to how microprocessor systems accomplish control functions behind the scenes, but details of the many arrangements are beyond the scope of this text. In addition, many PLC programming languages provide ability to emulate flip-flop behavior.

The set-reset (or S-R) flip-flop

As a relatively simple example, flip-flop logic can be formed using two NAND gates combined to form a single latch, as in Figure 4.10. The latch-circuit is important for creating memory devices. Its purpose is to catch an input value and hold it until a subsequent signal changes it. As a latch, the S-R flip-flop has two inputs, set "S" and reset "R," and two outputs, Q and Q.

Consider the low-latch S-R flip-flop shown in Figure 4.10, where low-latch means that the input latching signal is *TRUE* when the input is 0. When either input to a NAND gate is zero, the NAND output is 1. At the start, the gates are set at S=1, R=1, and Q=1. When the SET "$\overline{S}$" is triggered-low to 0, the NAND1 output toggles ON or $Q = 1$. When $Q = 1$ and

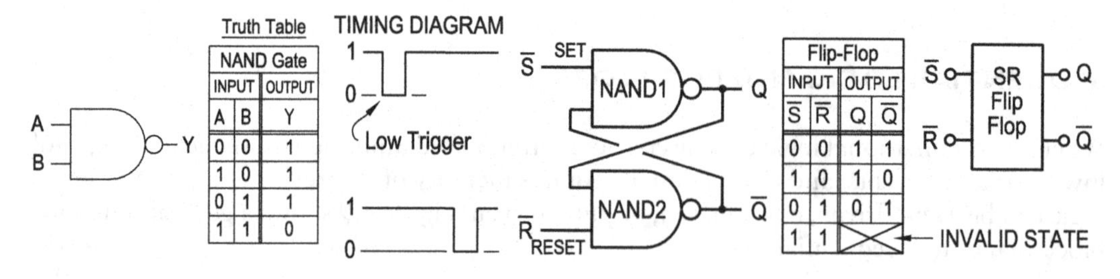

Figure 4.10 Set-Reset flip-flops are formed by cross-connecting two NAND gates. A low-trigger S-R flip-flop latching circuit is shown. The low trigger means that the signal originates from the 0 value as the toggling input into the NAND gate, the TRUE output is indicated by 1. The low-signal input is designated by a bar over the S and R input connections. Simultaneous set and reset triggering signals are not permitted, which is a condition that causes an unpredictable output.

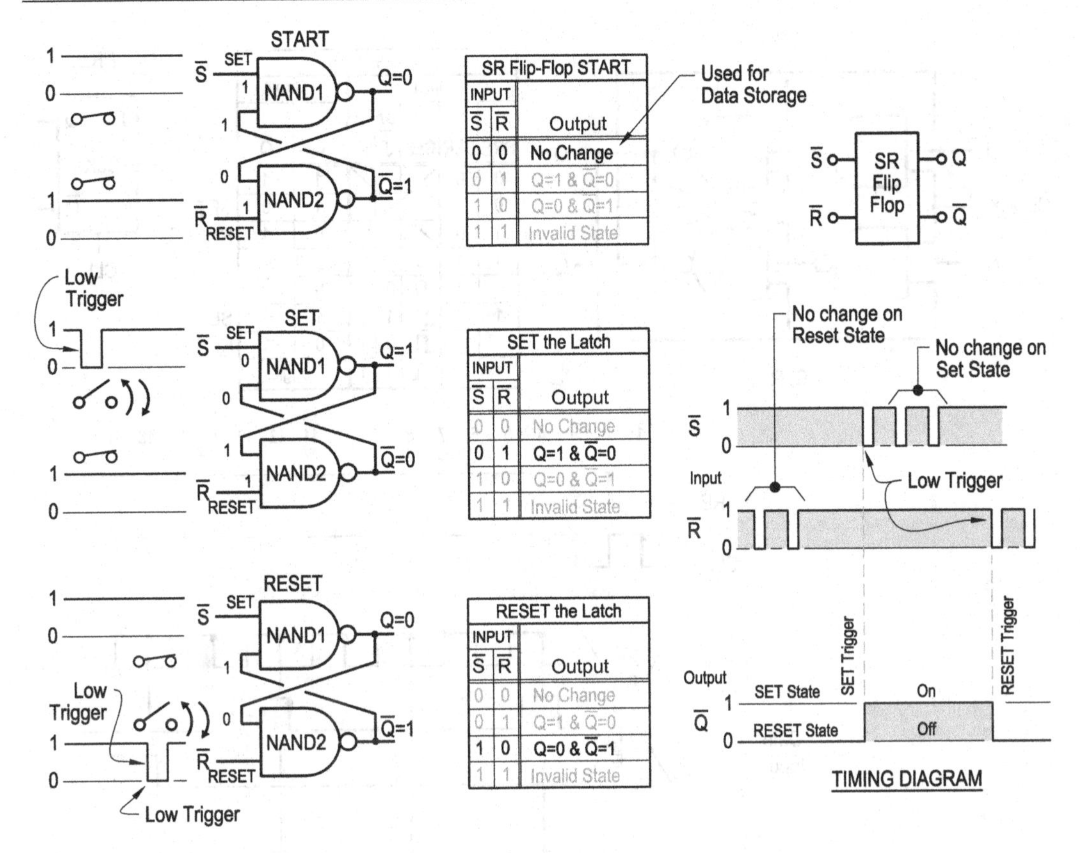

Figure 4.11 S-R flip-flop behavior is shown for various input signals. Note that simultaneous *TRUE* input signals into both the S and R inputs results in an invalid output state.

$\overline{R} = 1$ are input into NAND2, $\overline{Q}$ toggles OFF. The flip-flop stays in this state for any S input value. Later, when the RESET is triggered OFF, the NAND2 output toggles ON or $\overline{Q} = 1$. The NAND2 *TRUE* output subsequently inputs into NAND1 along with $\overline{S} = 1$, which causes the NAND1 output to reset back to $Q = 0$. Figure 4.10 shows the progression of outputs using the sequencing tables as the SET/$\overline{R}$ESET inputs are toggled, along with a diagram showing the input/output ON-OFF states. In other words, the SR latch becomes active when set low.

When dealing with sequential logic circuits, timing diagrams are very important for observing the behavior of the logic device for all input combinations. A timing diagram for the S-R flip-flop is provided as an example in Figure 4.11.

The J-K flip-flop

The J-K flip-flop is a five-input device with two outputs and is shown in Figure 4.12. The J and K inputs are for data and are analogous to the SET and RESET inputs of the S-R flip-flop. The sequential operation of the J-K flip-flop is similar to the gated S-R flip-flop; however, the J-K flip-flop includes a clock requirement (CLK) as a third input to avoid the S-R invalid state when simultaneous set-and-reset *TRUE* inputs are applied. The result is four valid outputs for any combination of data inputs, where the fourth output, which was invalid for the S-R flip-flop, toggles the $Q$and $\overline{Q}$ outputs, but only when both J and

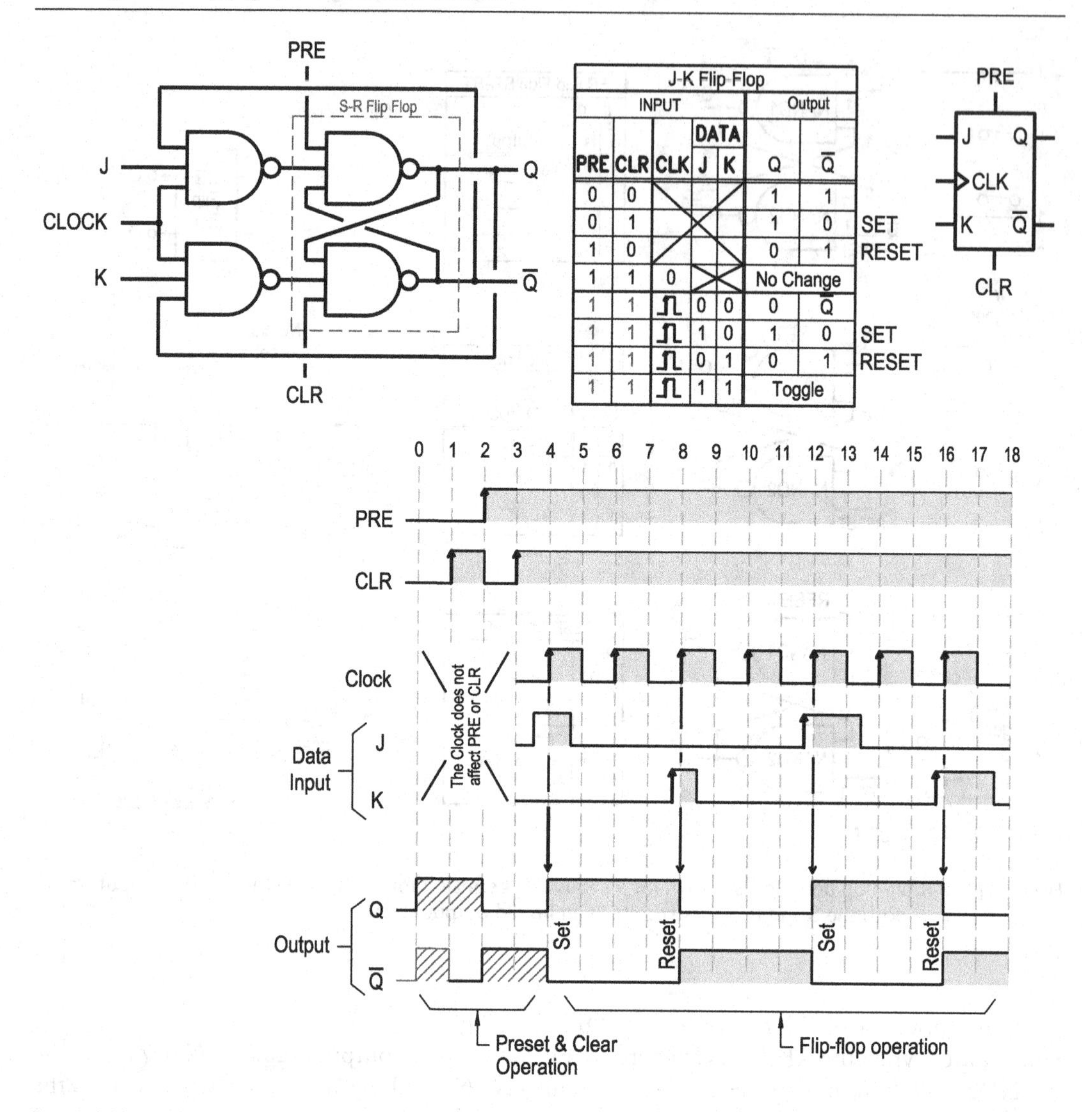

J-K Flip Flop							
INPUT					Output		
			DATA				
PRE	CLR	CLK	J	K	Q	$\overline{Q}$	
0	0				1	1	
0	1				1	0	SET
1	0				0	1	RESET
1	1	0			No Change		
1	1	⎍	0	0	0	$\overline{Q}$	
1	1	⎍	1	0	1	0	SET
1	1	⎍	0	1	0	1	RESET
1	1	⎍	1	1	Toggle		

Figure 4.12 J-K flip-flop logic, truth table, and sequence of operation for a positive-triggered device. Note that a negative-triggered device would be represented by a NOT symbol into the clock, and toggling would occur on the falling-direction of the set or reset trailing edge.

K inputs are *TRUE and* a clock signal is received. The fourth and fifth inputs in the J-K flip-flop are preset (PRE) and clear (CLR) functions. The PRE and CLR signals override the other gate signals, forcing the outputs to their steady-state levels. PRE and CLR are asynchronous signals in that they directly control the Q and $\overline{Q}$ outputs independently from the clock status. The preset acts to drive the flip-flop to the SET state, while the CLR acts to drive the flip-flop to the RESET state. PRE and CLR signals can occur at any time, but because those signals override the internal feedback, they do have the potential to drive the flip-flop to an invalid state.

Figure 4.13 shows two usages of J-K flip-flops. Figure 4.13(a) shows a single J-K flip-flop being used in a D latch. The output (Q) is latched to match the Data (D) input *when* the clock signal triggers the device. When the Data input changes, the output will toggle at the next

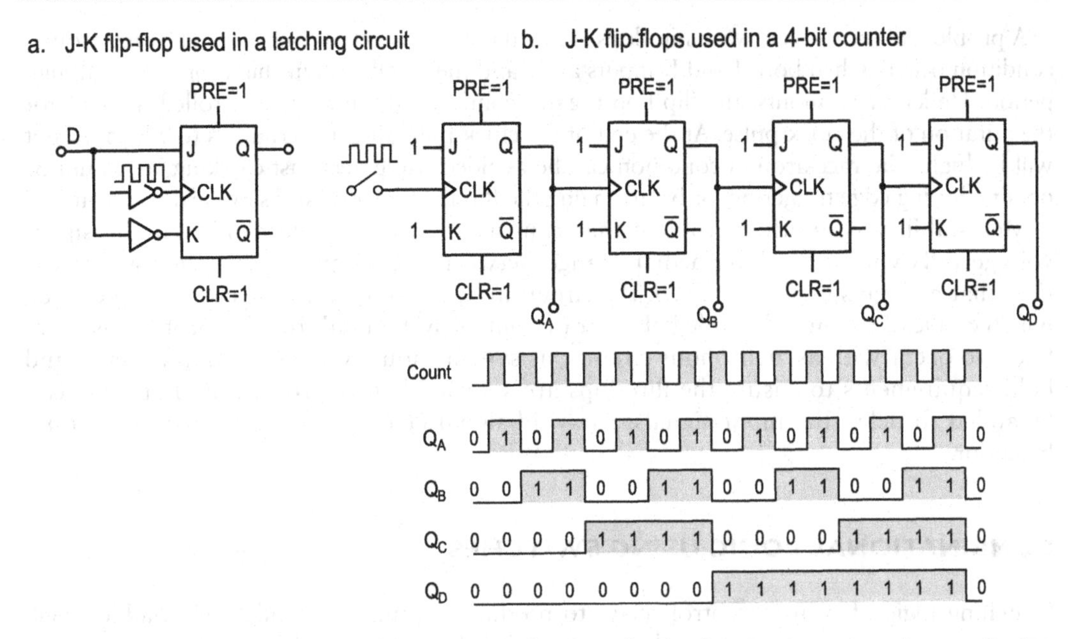

Figure 4.13 A J-K flip-flop application where outputs from one flip-flop become inputs into the next flip-flop to produce a counting function.

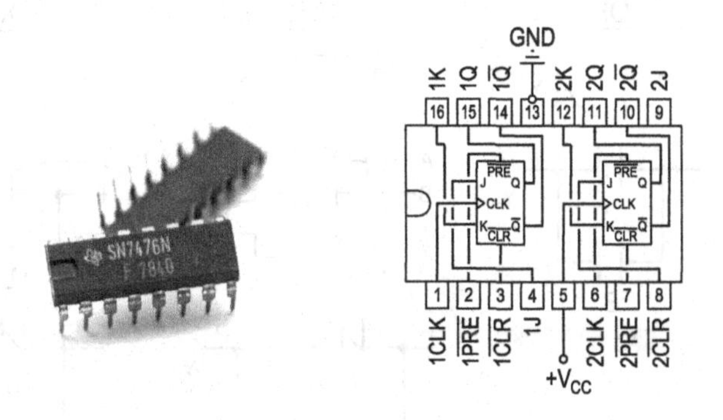

Figure 4.14 TI 7476 dual J-K flip-flops in a single chip. Each flip-flop contains preset and clear functions and uses positive edge triggering. A plug-in dual in-line package (DIP chip) is shown on the left, and a schematic representation of the chip connections is shown on the right. Power is provided between pins 5 and 13. (Courtesy of Texas Instruments.)

clock signal. Latches are important for designing central processor units, microprocessors, and computer peripherals.

Figure 4.13(b) shows four J-K flip-flops arranged to create a four-bit counter. An alternating 0 and 1 counting signal is input into the leftmost flip-flop. The counter senses each triggering edge as the signal to toggle its output. The four-bit number increments by one, counting from 0 to 15, at which point the 16th trigger cycles the count back to zero to start over. For counting in the decimal system, logic needs to be added to clear the counter and restart at zero when nine is exceeded. The clock can be synchronous for timing purposes, or it can be replaced by a discrete switch that counts the switching actions, for asynchronous operation. Flip-flops are obtained from IC chips, such as the TI 7476 chip shown in Figure 4.14 which includes two flip-flop circuits in one package.

A problem that can occur with flip-flops is called the race-around condition. The race-around condition occurs when both J and K inputs are 1 and the clock remains high for a long enough period. Under those inputs, the flip-flop toggles continuously in an uncontrolled manner for the duration of the clock pulse. At the end of the clock pulse, it is uncertain as to which output will be high. The race-around condition can be avoided using very fast clocking, as would be obtained using edge triggering, or by arranging the flip-flops in a master-slave arrangement.

Metastability is another problem that can occur in digital electronics. If the signal voltage falls within the intermediate range between the 0- and 1-state voltages at the time that a clock signal is received, the circuit may not settle into the correct logic level, which can lead to unpredictable behavior or control system failure. This problem is more likely to occur with asynchronous circuits, as synchronous systems often have setup and hold requirements to ensure the flip flops are satisfied before proceeding. Buffering can be added to delay the input signals, to avoid signal-error propagation, causing erratic behavior.

COMBINATIONAL LOGIC USING SWITCHES

Switching logic often uses control relays to produce output commands. Relay/ladder logic predates microprocessors, but to a large extent, it is how integrated-circuit microprocessors

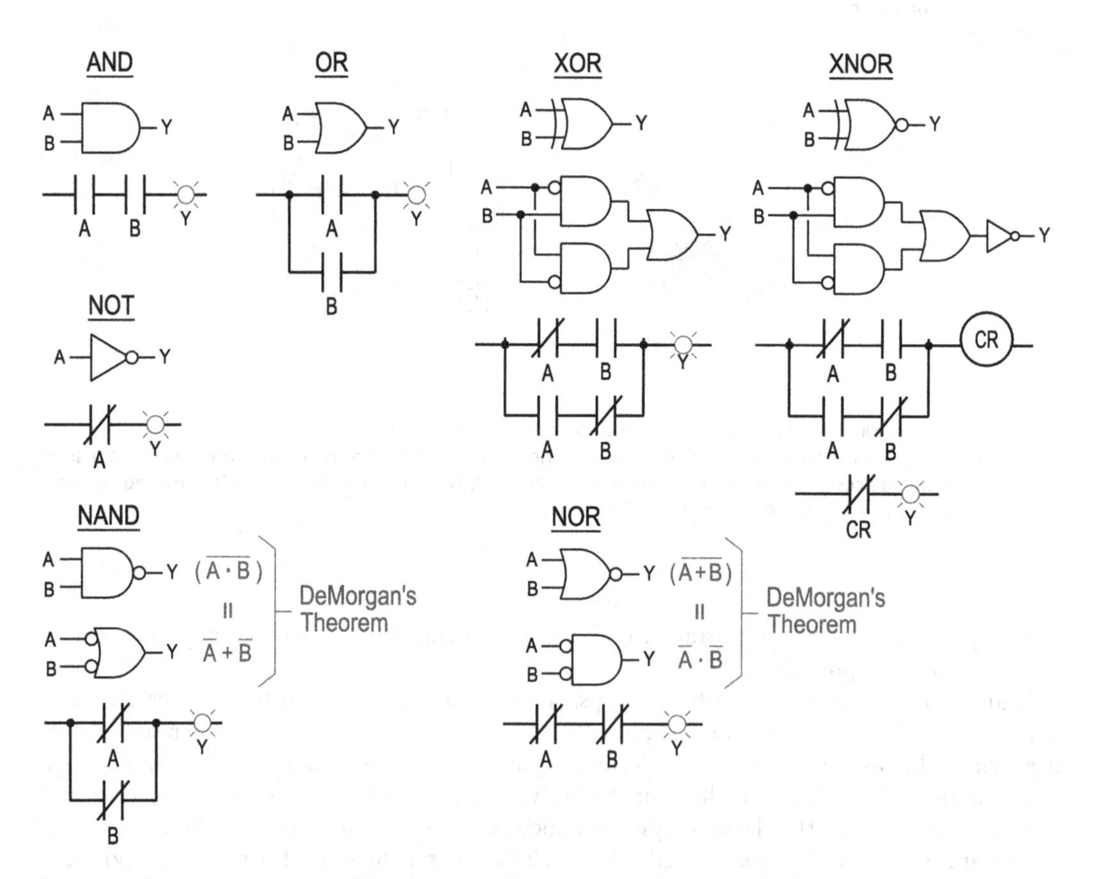

Figure 4.15 Relay-logic switch configurations are used to produce Boolean logic functions in ladder diagrams.

function, except by using transistor switches instead of electromechanical relays. Chapter 5 on motor controllers contains wiring diagrams as practical examples for the design and operation of switching-logic circuits.

Figure 4.15 shows how switching arrangements can be configured to produce Boolean operations. In general, the functions can be summarized as follows:

- Series open switches produce an AND gate
- Parallel open switches produce an OR gate
- A normally closed switch produces a NOT gate or signal inverter
- Parallel closed (NOT) switches produce a NAND gate
- Series closed (NOT) switches produce a NOR gate
- A relay coil may be required to invert the output of complicated gates

SEQUENTIAL LOGIC USING CONTROL RELAYS

In addition to combinational logic, relays can be arranged to produce sequential logic. Figure 4.16 shows relays configured to produce sequential logic to emulate the behavior of an S-R latch. The ladder-logic diagram shows the conflict that arises when the two mutually exclusive outputs (CR1 and CR2) cause a race-around condition, and worse, the uncontrolled toggling that would follow. The diagram on the right shows the addition of a small time delay added to prevent racing that would occur when the circuit is first energized. The invalid condition occurs when both the SET and RESET are *TRUE* simultaneously. Identical behavior to the S-R flip flop is shown by the truth tables.

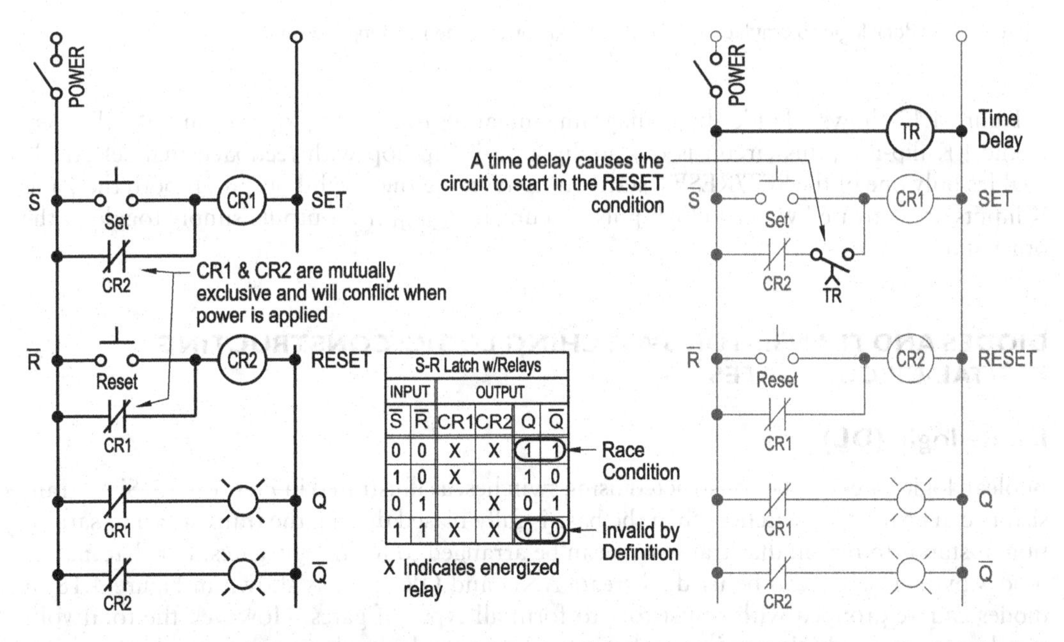

Figure 4.16 The relays are arranged to produce S-R latching behavior, shown by the same truth table. The ladder on the left illustrates the race-around condition, and the right shows race-around avoidance via a time delay added to the SET function.

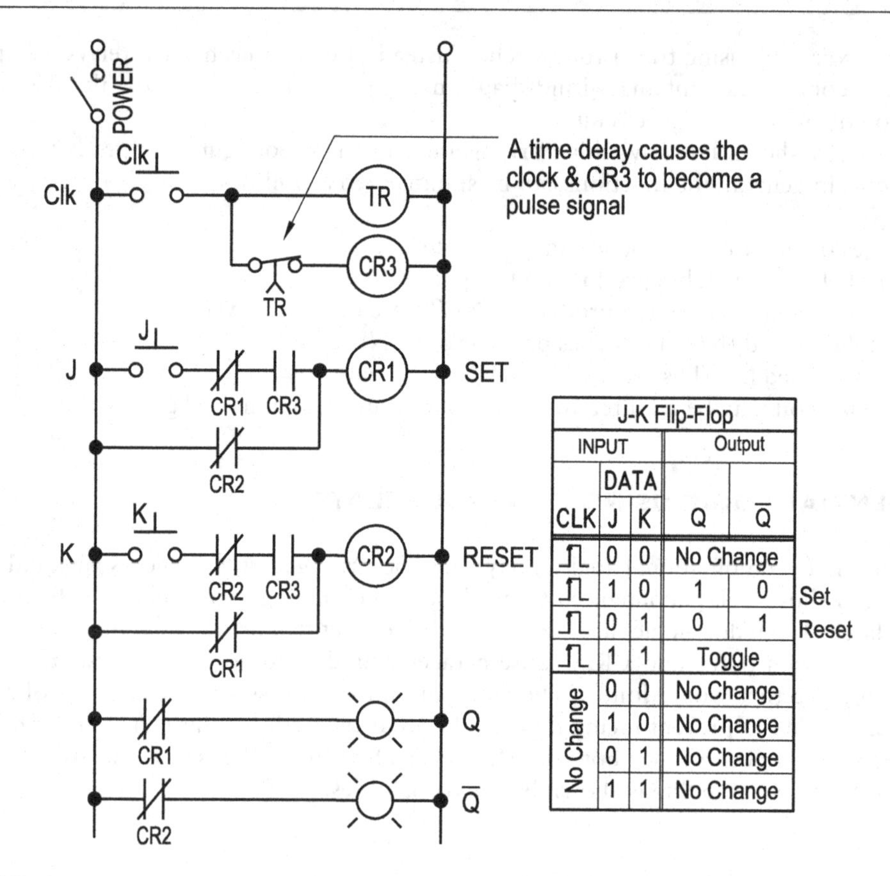

J-K Flip-Flop				
INPUT			Output	
CLK	DATA J	DATA K	Q	$\overline{Q}$
⎍	0	0	No Change	
⎍	1	0	1	0
⎍	0	1	0	1
⎍	1	1	Toggle	
No Change	0	0	No Change	
No Change	1	0	No Change	
No Change	0	1	No Change	
No Change	1	1	No Change	

Figure 4.17 Ladder logic to emulate J-K flip-flop behavior, and the resulting truth table.

Figure 4.17 shows a ladder-logic diagram emulating J-K flip-flop operation. Like the electronic J-K flip-flop, this circuit is essentially an S-R flip-flop with feedback that selectively enables only one of the SET/RESET states, thus avoiding the invalid output. If both the J and K inputs are activated when a clock pulse occurs, the $Q(\text{and}\overline{Q})$ outputs simply toggle to the other state.

DIODES AND TRANSISTOR SWITCHING LOGIC: CONSTRUCTING DIGITAL-CIRCUIT GATES

Diode logic (DL)

Boolean logic gates can be constructed using switches, as illustrated in Figure 4.15. Since transistors can be used as switches when the base is fully biased driving the transistor into saturation, it stands to reason that transistors can be arranged to form logic gates. For that matter, diodes by themselves can be used to create AND and OR gates as shown in Figure 4.18 or diodes can be grouped with transistors to form all types of gates. However, the total voltage drop across multiple passive diodes accumulates, and complex gates formed by multiple diodes quickly become impractical. In general, voltages between 0 to +1.5V_{DC} are logic state 0 (*FALSE*), and voltages between 3.5 to 5.0V_{DC} are logic state 1 (*TRUE*). The metastable condition occurs when voltages fall between 1.5 to 3.5V, which might be interpreted as either true or false, so this range is not permitted.

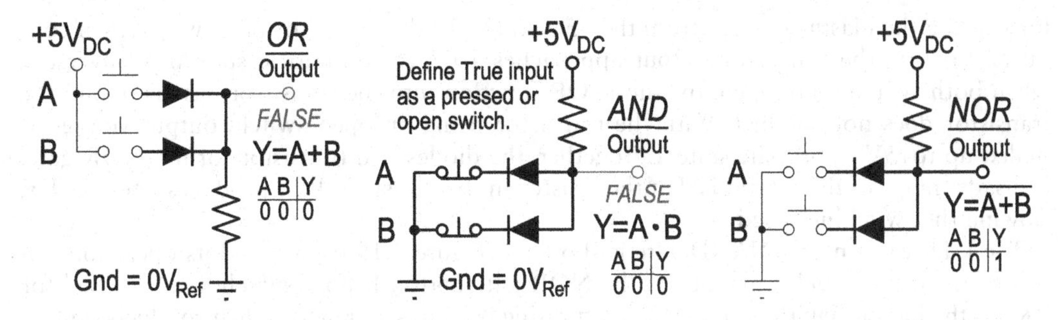

Figure 4.18 Diode logic forming an OR gate on the left, and an AND gate on the right.

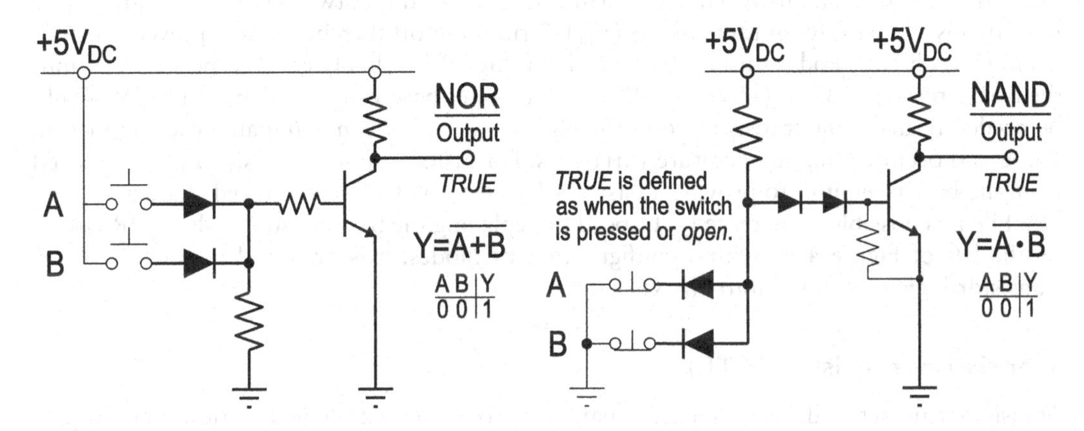

Figure 4.19 Diode-transistor logic forming a NOR gate on the left, and a NAND gate on the right.

Figure 4.18 shows diode circuits to create OR and AND gates. In the left circuit, if both switches are open, then the output is at $0V_{Ref}$ and is logic state 0. If either or both switches are closed, the relatively high resistance causes a voltage divider, so that the voltage after the diode is approximately 4.3Vdc = 5.0 - 0.7 for a silicon diode, which corresponds to logic state 1, and thus the circuit is a logic OR gate.

The right circuit in Figure 4.18 shows an AND gate. When either or both switches are not pressed, logic-input state 0, the diodes conduct bringing the output voltage at the junction to 0.7Vdc = $0V_{Ref}$ + 0.7V, corresponding to logic state 0. However, when both switches are pressed, the infinite resistance to ground pulls the output voltage at the junction to $+5V_{DC}$ or logic state 1.

Diode-transistor logic (DTL)

The problem with diode logic is that the output signal deteriorates with each stage of diodes when gates are ganged together. To compensate for the decreasing voltage, bipolar junction transistors (BJTs) can be used to re-amplify the signal to avoid the 1.5 to 3.5V metastable range of unreliability.

Figure 4.19 shows diode-transistor-logic (DTL) circuits for NOR and NAND gates. The OR logic on the left is formed by paralleling diodes to supply the signal, as in Figure 4.18; however, the output is used to bias the base of a transistor that both amplifies the voltage and inverts the signal. When either switch is closed, the OR output logic goes to 1, forwarding the biasing current from the +5V source to the transistor base, which then conducts. In turn, the transistor output approaches +0.7 volts, or logic state 0. Conversely, when both switches are open, the diode OR-function into the transistor base is 0, and the transistor does not conduct. With the transistor being an open switch, output voltage is pulled up to+5V_{DC}, or logic state 1. Together, the diodes and transistor form a NOR gate. A disadvantage of this circuit is that the resistor in the transistor base produces a time delay, slowing the switching speed.

The DTL version of a NAND gate is shown in Figure 4.19 (right), and its operation can be described in a similar manner as the NOR gate above. If both switches are pressed for *TRUE*, the base-to-emitter sees +5V_{DC}, saturating the transistor, which then conducts pulling the output low. If one or both switches are closed, the base current is diverted, and the base voltage is pulled down to +0.7V, which marginally should cause the transistor to cut off and the output to go high. However, it is worth noting that adding two diodes in the transistor base avoids any base-to-emitter voltage (V_{BE}) during shut off thereby reliably preventing conduction. With two diodes in the base, transistor cut-off is reliably ensured below 1.4V into the base circuit and 3.6 V (+5V_{DC} - 1.4$V_{2\ diodes}$) a higher base voltage, greater than 2V, would be needed to make the transistor conduct. This arrangement can eliminate false conduction that could occur during temperature variations. For faster response, a resistor can be placed from the base to emitter to provide a base discharge path, dissipating stored charges.

While it is possible to increase DTL speed by replacing the base resistor in the NOR circuit on the left of Figure 4.19 with a configuration of diodes, it is preferred to use transistor-transistor logic configured into NAND gates.

Transistor-transistor (TTL)

Transistors are active devices that are capable of producing both logic functions and output-signal amplification, eliminating the signal degradation associated with logic circuits that use passive diodes. TTL circuits tend to be faster and easier to construct, and they are consequently preferred for integrated circuits (IC chips) used in computers, instrumentation, and controls.

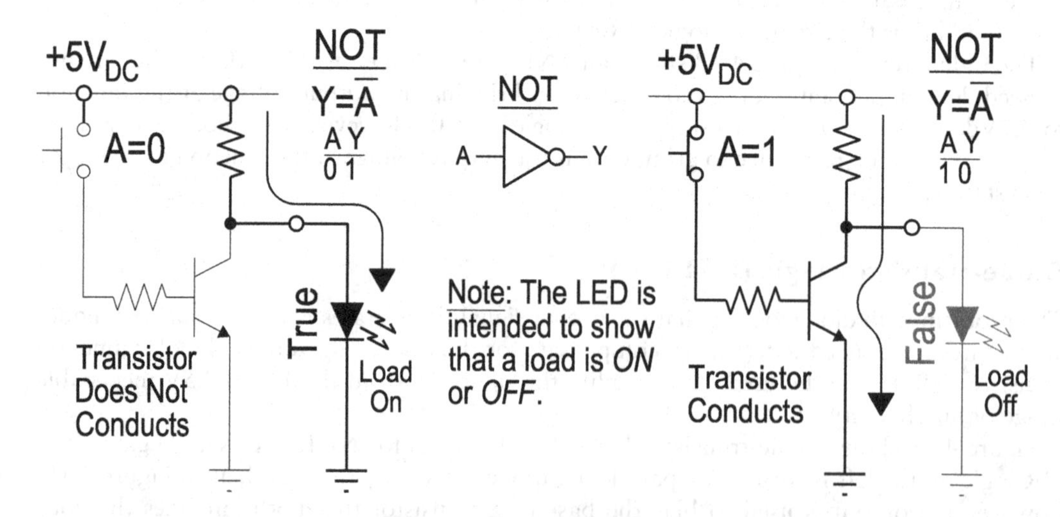

Figure 4.20 This transistor logic is creating a NOT gate or signal inverter. The left figure shows an unpressed switch or null input signal that produces a TRUE output at the LED. When the switch is pressed making the input TRUE (right), the LED output goes FALSE and turns off.

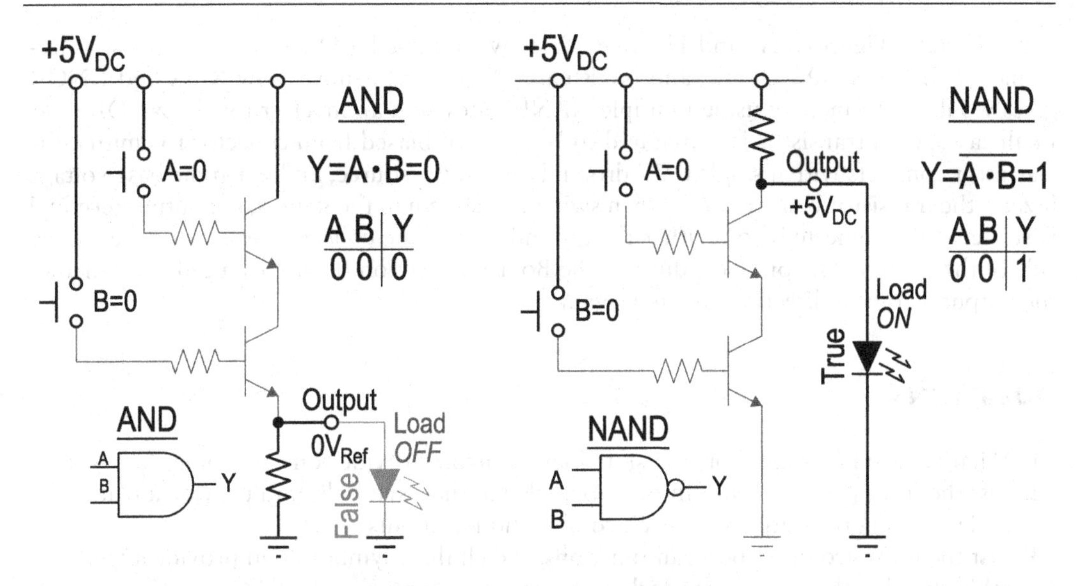

Figure 4.21 This TTL logic is forming an AND gate (left) and a NAND gate (right), which are shown with both inputs at logic 0. For these inputs, the AND gate produces a *FALSE* output and the NAND gate a *TRUE* output, shown by the LEDs.

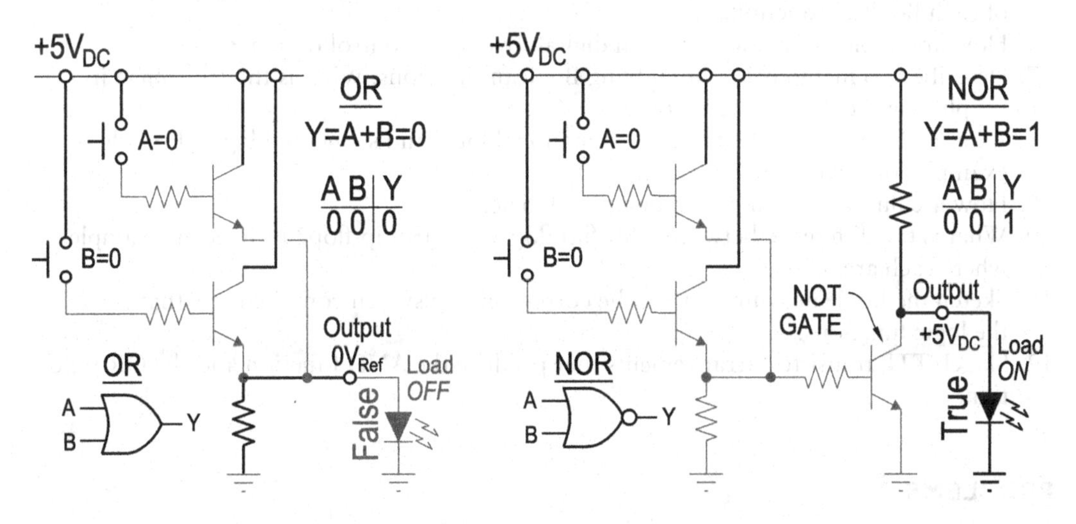

Figure 4.22 TTL to produce OR and NOR gates. The NOR function is obtained by adding a NOT function to the OR output.

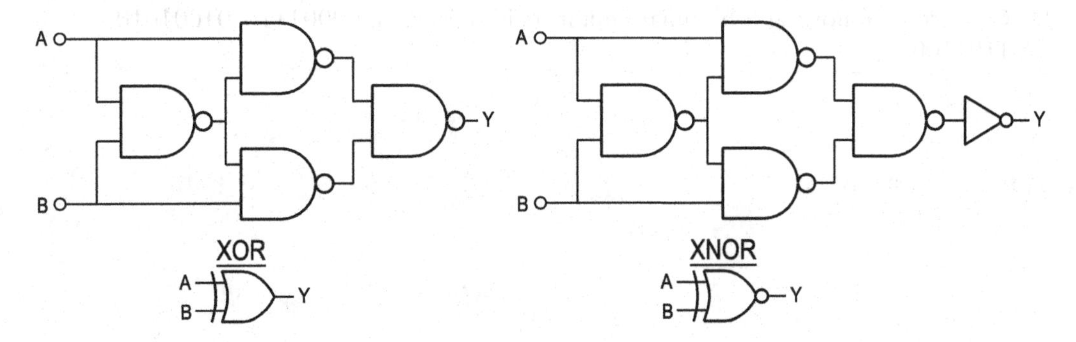

Figure 4.23 XOR gates are produced using TTL NAND gates that are ganged together. The XNOR gate simply inverts the XOR output.

Figure 4.20, Figure 4.21, and Figure 4.22 show simplified TTL circuits to produce electronic NOT, AND, NAND, OR, and NOR gates. Figure 4.23 shows how XOR and XNOR gates would be formed by using multiple NAND gates, and an inverter for the XNOR gate. In all cases, each transistor is configured to be forward biased from collector to emitter, the transistor conducts with a single 0.7V drop relative to the source; and when the base voltage is zero, the transistor behaves as an open switch. As shown in the figures, the output terminal is strategically connected from either the ground side of the collector resistor or the supply side of the emitter resistor, depending on the Boolean function being produced and whether the output is to be a direct or inverted signal.

QUESTIONS

1. What is meant by digital or two-state logic control? Provide some examples.
2. List the three primary Boolean functions, sketch their symbols, and provide a truth table showing the outputs for each combination of inputs.
3. List the four secondary Boolean functions, sketch their symbols, and provide a truth table showing the outputs for each combination of inputs.
4. Sketch the Boolean functions using the DIN and IEC symbols.
5. Provide Venn diagrams associated with each Boolean symbol, to illustrate the behavior of each Boolean function.
6. How are Boolean functions used in digital, two-state-control designs?
7. Describe two methods for simplifying Boolean functions. What is the advantage in simplifying the Boolean functions?
8. What is the difference between combinational logic and sequential logic? Provide two examples of each.
9. How are flip-flops used in combinational logic?
10. What is the difference between a SR flip-flop and a JK flip-flop? Provide an example where each are used.
11. Sketch the Boolean symbols with the corresponding switch combinations that achieve the logic functions.
12. Sketch TTL transistor arrangements that produce the AND function and the OR function.

PROBLEMS

1. Using an 8-bit word, convert the following numbers into binary: 10, 50, 100, 200, and 255.
2. Convert the following 8-bit binary numbers into decimal: 00001110, 01001010, 11011001.

Chapter 5

Motor controllers

A motor controller is a group of electromechanical devices used to manage the proper operation of motors, often under the direction of automated controls. Motor controllers start and stop motors and provide other functions that make systems work properly. Control can be initiated by pressure, temperature, flow, level, limit-sensing, or other types of switches. Motors can be wired using several techniques to operate at a single speed, several discreet speeds, or at variable speeds, and they may be wired to be unidirectional or reversing. Controllers may start small motors across-the-line or they may use one of several reduced-voltage options to bring large motors up to speed without excessive electrical stresses and large voltage dips. Motor controllers are an integral part of the plant operation.

Understanding motor controller components and operation is important not only for plant design, operation, and maintenance, but it also forms a basis for the ladder-diagram programming language for PLCs. While motor controller devices consist of real-world hardware, PLC ladder-logic instructions now replace control relays with virtual devices that exist only in PLC memory. Virtual relays are really programming instructions that produce the same functions as their real-world counterparts. Further, each motor-control device has an equivalent PLC instruction that emulates the same function, but the PLC can provide more sophisticated control in a smaller package.

SWITCHES AND ELECTRICAL DEVICES

Switches come in many forms and variations. Switches may be manually operated, process-actuated using pressure, thermostats, limit-sensors, etc., or electrically actuated using relays and contactors.

Switches are classified by the number of switching devices that are toggled by a single action, which are called *poles*. Thus, flipping a two-pole switch would be the same as simultaneously toggling two switches with the same motion. Switches are also classified by the number of electrically conductive positions that can be "made," which are called *throws*. Thus, a single-throw switch has only one position that conducts and is limited to strictly on-off functions, such as lighting. A double-throw switch has two electrically conducting positions, one normally open (NO) or deenergized in its untoggled state, and the other normally closed (NC) or conducting, in its untoggled state. It is possible for a single switch to have many poles (switches) and many throws (positions), such as a rotary selector switch that may has several decks or poles of switches, with a dozen or more selector throws or positions.

Switches are also classified in the way they changeover their positions. Switch action that closes and energizes a circuit is called *make*, while the action that opens and deenergizes a circuit is called *break*. Most multi-throw switches and relays have the characteristic of *break-before-make*, where one set of contacts opens and kills its circuit before the next set of contacts closes to energize its circuit. Conversely, make-before-break switches use a wiping action to make electrical contact with the next set of contacts, before breaking the previous set. Some specialty applications use *make-before break* switches, such as tap changers to control the secondary voltage for utility-transformers while under load. This type of switch keeps the transformer windings continuously energized during changes, eliminating the arcing that occurs with inductive loads. Another example is a "closed-transition" automatic-transfer switch using a make-before-break sequence that avoids interruptions to critical services, such as when restoring normal power from an emergency source. The normal-power switch would close when the voltages are in-sync, followed by opening of the emergency switch immediately afterward.

Switches are classified by their actuation mechanism. Some types include toggle, pushbutton, rocker, and rotary. They are also classified by their final position after actuation, where switches are either maintained contact or momentary. A maintained-contact switch stays toggled into its new position by a snap-action mechanism or detent to hold its new position after the actuating force is removed. Momentary switches are spring-loaded to return them to the starting position, after the they are released.

Switches often have two amperage ratings, one for resistive loads and a lower value for inductive loads, which are prone to arcing. Its maximum amperage is limited by its construction, including contact size and clamping force, and there would be different amperage ratings for ac and dc applications. Strong spring force is important to obtain quick action, low contact resistance, and to avoid chatter or contact bounce, which can cause problems with high-speed electronic controls. Its maximum voltage rating is limited by creepage and clearance dimensions needed to avoid surface conduction or air ionization. Various switch symbols used for motor-control wiring diagrams are shown in Figure 5.1.

Switches can be ganged together to perform simple or complex logic functions. Some examples of simple switching circuits are provided in Figure 5.2, which shows loads toggled from one, two, or three locations. Since each switch has two states, the number of switch-position

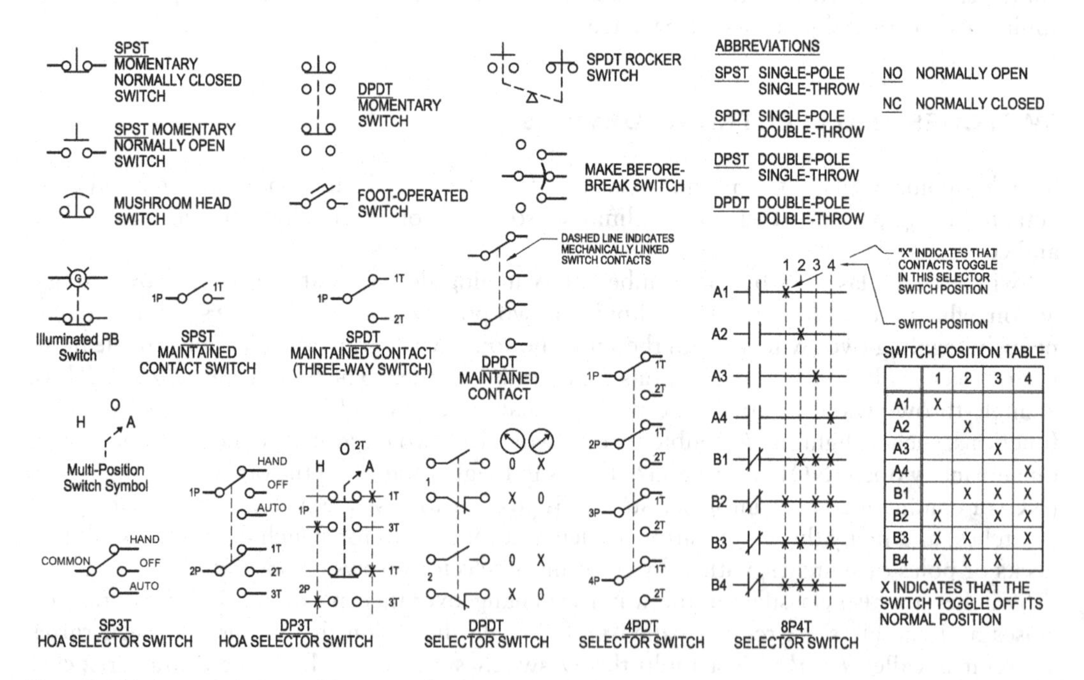

	1	2	3	4
A1	X			
A2		X		
A3			X	
A4				X
B1		X	X	X
B2	X		X	X
B3	X	X		X
B4	X	X	X	

Figure 5.1 Various symbols used for manually operated switches.

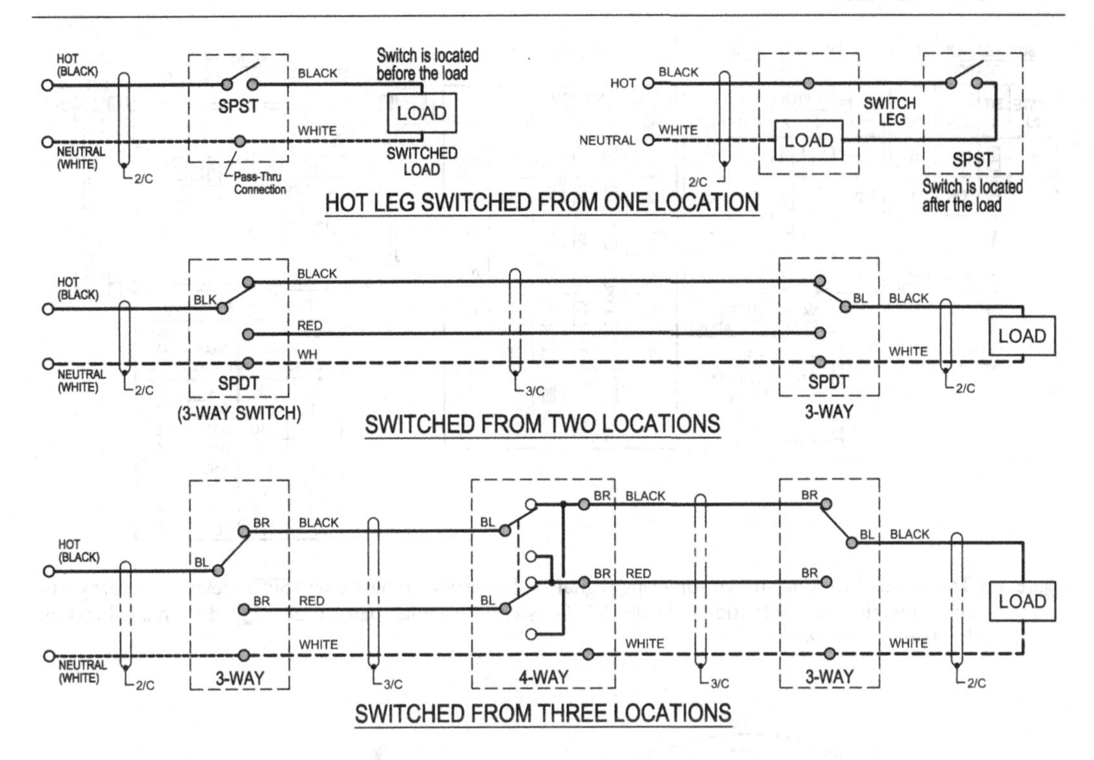

Figure 5.2 Load switching from one or more locations, using specialty switches to obtain control.

permutations is 2^{SW}. A single switch has binary $2^1 = 2$ switch permutations, either on or off. Switching from two locations has potentially $2^2 = 4$ switch permutations and switching from three locations has $2^3 = 8$. Tracing circuit continuity in Figure 5.2 is helpful for seeing the effect of different switch-state combinations in toggling the load.

Figure 5.3 shows the bit logic and permutations needed when switching is done using simple SPST switches at different locations for achieving the same multi-location control in the Figure 5.2 circuits. This type of SPST switching might be used in PLC control systems, and this example shows the significant difference between wiring control circuits and digital control circuits. The tables illustrate the concept that, although individual SPST switches are on-off binary devices, toggling the switches does not follow the binary numbering system. In some cases, multiple switches need to be simultaneously toggled to obtain the next binary number. For example, a total of three bits need to change state when counting from binary 3 (011) to binary 4 (100).

Incremental toggling of SPST switches is shown in the left three tables of Figure 5.3. The rightmost column in those tables shows how the Gray numbering system tracks the switch positions, where only one switch or one bit is toggled at a time. The rightmost table shows the conversion between decimal, binary, and Gray numbers for a three-bit numbering system. Gray code is important because it represents real-world sequential switching actions, where only one Gray digit differs during each number increment. Gray coding forms the basis of optical position encoders and glitch-free counters so that all starting positions are unambiguously interpreted. A three-bit absolute-position rotary Gray encoder is shown as an example in Figure 5.4.

The "state machine diagrams" for the two-switch and three-switch designs are also shown in Figure 5.3. The state machine diagrams show the complexity that occurs when using SPST switches to be actuated from multiple locations. State machine diagrams are directional

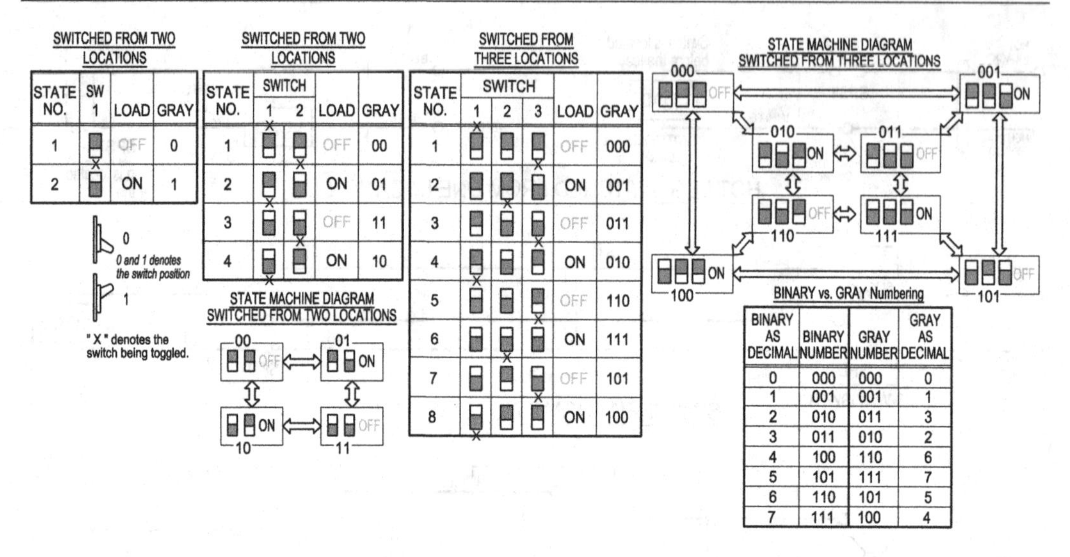

STATE NO.	SW 1	LOAD	GRAY
1		OFF	0
2		ON	1

STATE NO.	SWITCH 1	SWITCH 2	LOAD	GRAY
1			OFF	00
2			ON	01
3			OFF	11
4			ON	10

STATE NO.	SWITCH 1	SWITCH 2	SWITCH 3	LOAD	GRAY
1				OFF	000
2				ON	001
3				OFF	011
4				ON	010
5				OFF	110
6				ON	111
7				OFF	101
8				ON	100

BINARY AS DECIMAL	BINARY NUMBER	GRAY NUMBER	GRAY AS DECIMAL
0	000	000	0
1	001	001	1
2	010	011	3
3	011	010	2
4	100	110	6
5	101	111	7
6	110	101	5
7	111	100	4

Figure 5.3 The control logic is shown for using digital or bitwise switching using SPST devices to carry out multi-location load actuation in Figure 5.2. Gray code permits loads to be toggled from any location by flipping one switch.

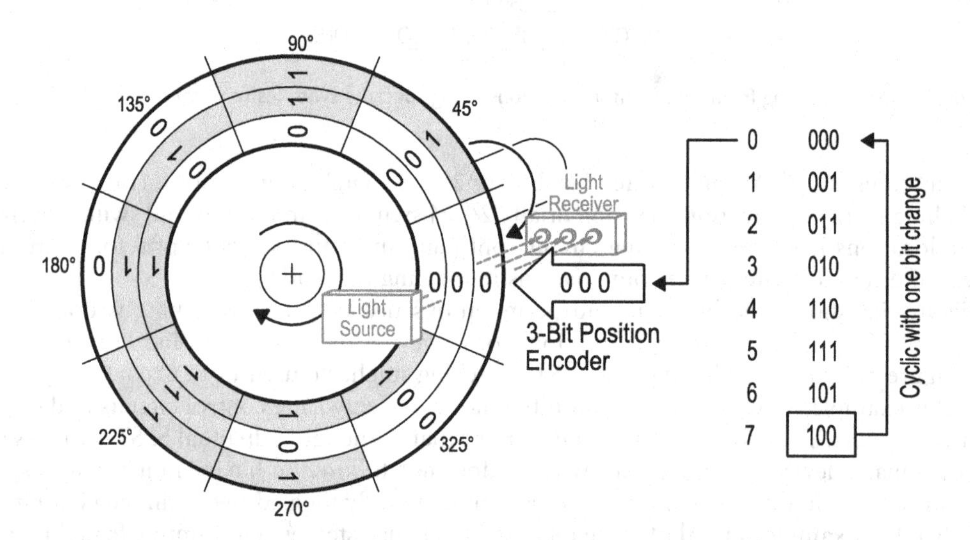

Figure 5.4 A three-bit Gray rotary absolute position encoder, as an example, is used to identify the rotated shaft angle at each 45° increment. Increasing the number of bits increases the angular resolution. The Gray encoder is "cyclic" so that only one bit ever changes, even when the count resets back from 325° (Gray 100) back to 0° (Gray 000).

graphs that show the initial and final states for each switching option, and they are useful for organizing one's thought processes and designing logic functions.

MOTOR CONTROLLER /COMBINATION STARTER CONSTRUCTION

A typical motor controller consists of an enclosure, a disconnect switch, a motor contactor, overload protection, control-voltage transformer, control relays, switches, indicator lights,

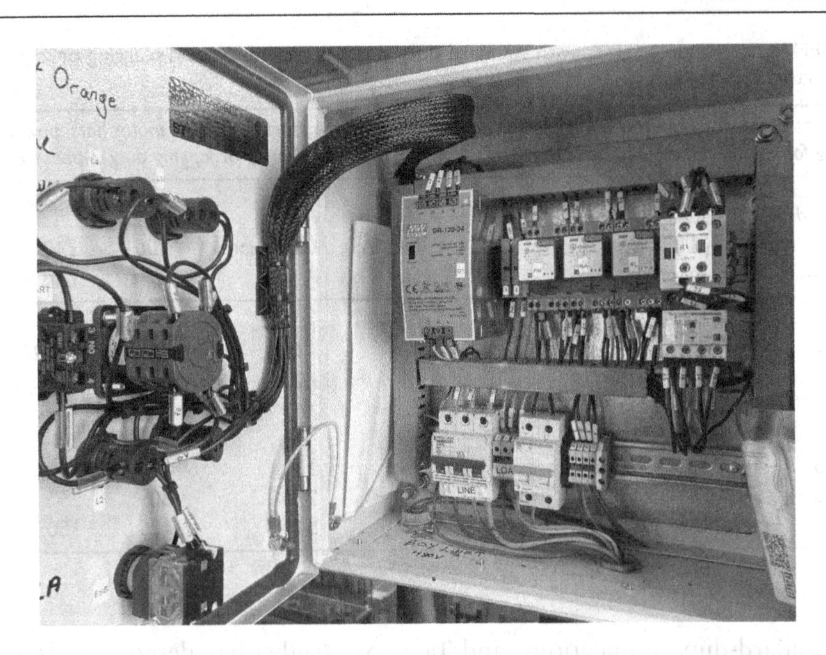

Figure 5.5 Three-phase motor controller with a DIN rail-mounted combination circuit breaker and service disconnect switch, fuse block, dc power supply, control relays, and motor contactor with overload device, connected via a plastic wireway. Switches and indicator lights are mounted on the hinged door of a NEMA 4 gasketed enclosure. (Courtesy of United States Merchant Marine Academy.)

and terminal boards for interfacing external devices with the controller. The assembly is referred to as a combination starter and is shown in Figure 5.5.

Contactors and relays are electromechanical switches, using small electrical signals applied to an electromagnetic coil to remotely actuate the associated switch contacts. For control relays, low-amperage switch contacts may be normally open (NO) or normally closed (NC), where normal refers to the deenergized position. When the coil is energized during operation, the contacts are toggled away from their normal, deenergized position, so that normally open contacts are closed to energize a circuit, and the normally closed contacts are opened to deenergize a circuit.

MOTOR CONTACTORS

Motor contactors are larger versions of relays, capable of conducting very large amperages and having the ability to accommodate the arcing associated with making and breaking inductive loads. Generally, the term motor starter refers to the motor contactor and its integrally installed overload-protection device having thermal current-sensing elements and their associated normally closed tripping contacts, as well as auxiliary contacts that can be wired to provide control-logic functions. A contactor's main contacts are physically large normally open switches, called poles, that connect the power source when the coil is energized. The main contacts are closed by electromagnetic force, and the contacts come together with a sliding or wiping action designed to minimize contact resistance. The contactor size is selected to match the motor horsepower and voltage using a NEMA starter size numbering system, to pass the in-rush current and continuously carry the running current. Table 5.1 (left) summarizes the minimum NEMA starter

Table 5.1 NEMA motor-starter sizes, with standard-duty starters on the left, and plugging or jogging starters on the right

NEMA starter sizes vs. motor horsepower rating for standard duty, 3-phase 60Hz					*NEMA starter sizes vs. motor horsepower rating for plugging or jogging duty, 3-phase 60Hz*				
		Horsepower					*Horsepower*		
NEMA Size	*Rated Continuous Amps*	*200 Volts*	*230 Volts*	*460 Volts*	*NEMA Size*	*Rated Continuous Amps*	*200 Volts*	*230 Volts*	*460 Volts*
00	9	1-1/2	1-1/2	2	00				
0	18	3	3	5	0	18	1-1/2	1-1/2	2
1	27	7-1/2	7-1/2	10	1	27	3	3	5
2	45	10	15	25	2	45	7-1/2	10	15
3	90	25	30	50	3	90	15	20	30
4	135	40	50	100	4	135	25	30	60
5	270	75	100		5	270	60	75	150
6	540	150	300	400	6	540	125	150	300

size for standard-duty applications, and Table 5.1 (right) has derated NEMA sizes for jogging or plugging applications.[1] Contactors usually come in three-pole arrangements, but are also available in two-, four-, and five-pole versions. The main poles are wired in series with each motor leg, so a three-phase motor would require a three-pole contactor with three sets of large contacts.

Motor starters are matched with overload (OL) devices consisting of two main parts;

- thermally actuated overload-heater elements that are wired in series with each motor phase and which pass the entire motor amperage,
- and a small set of normally closed contacts that are thermally actuated by the overload heaters and which are wired in series with the contactor's electromagnetic coil.

The OL device initiates the safe shutdown of equipment that might otherwise become damaged by excessive amperage. The device uses the I^2R heat generated by the motor current to flex a bimetallic element and initiate the tripping action. The thermally actuated overload device has an inverse-time characteristic, meaning the greater the overload current, the shorter the time delay for tripping. When an overload device trips, a built-in mechanism holds its contacts open, preventing the motor contactor from being reenergized, until after the heater element has cooled and the OL device is manually reset. Solid-state-electronic and eutectic/melting variations of overload protection are also available.

For most motors, the small normally closed OL auxiliary contacts associated with the heaters are wired to deenergize the contactor's low-voltage electromagnetic coil, stopping the motor. However, for applications involving life-safety, such as a fire pump or a ship's steering gear, the OL contacts are wired to set off alarms instead, without interrupting the vital service. Alternatively, fractional horsepower motors that do not use contactors often have overload protection built directly into the motor.

Contactors are usually fitted with at least one set of auxiliary contacts, usually both NO and NC, although many contactors can accept a stack of several sets of auxiliary contacts. The auxiliary contacts are small, low-voltage, low-amperage switches that are used in the control logic, such as to turn on a "motor-running" indicator light. An auxiliary contact on

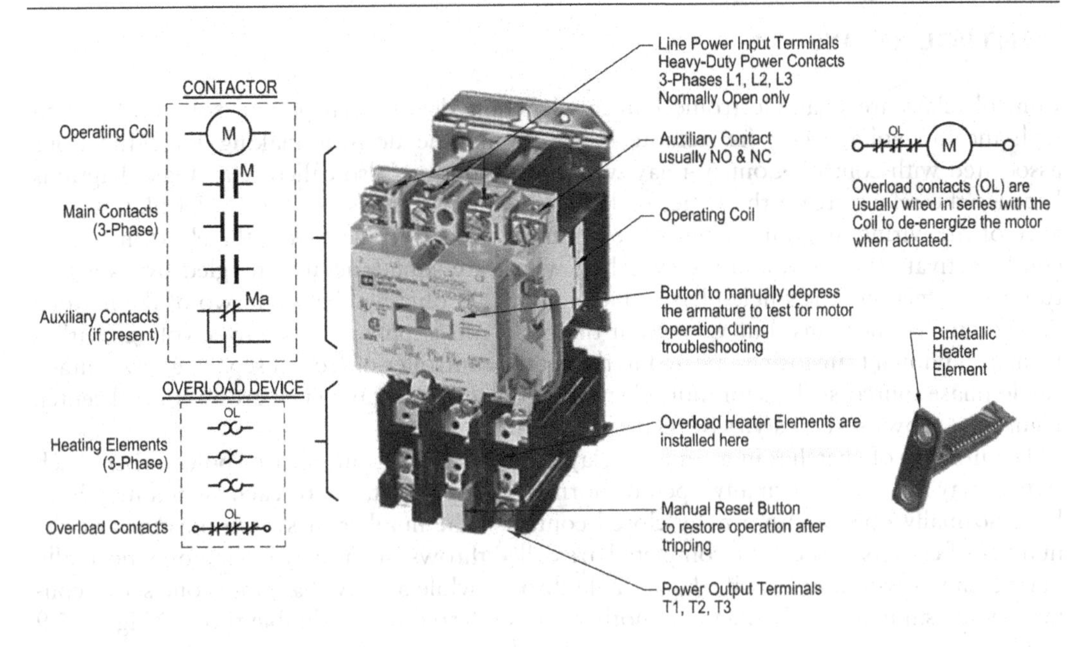

Figure 5.6 Motor Contactor with overload device and heater element. (Courtesy of Eaton.)

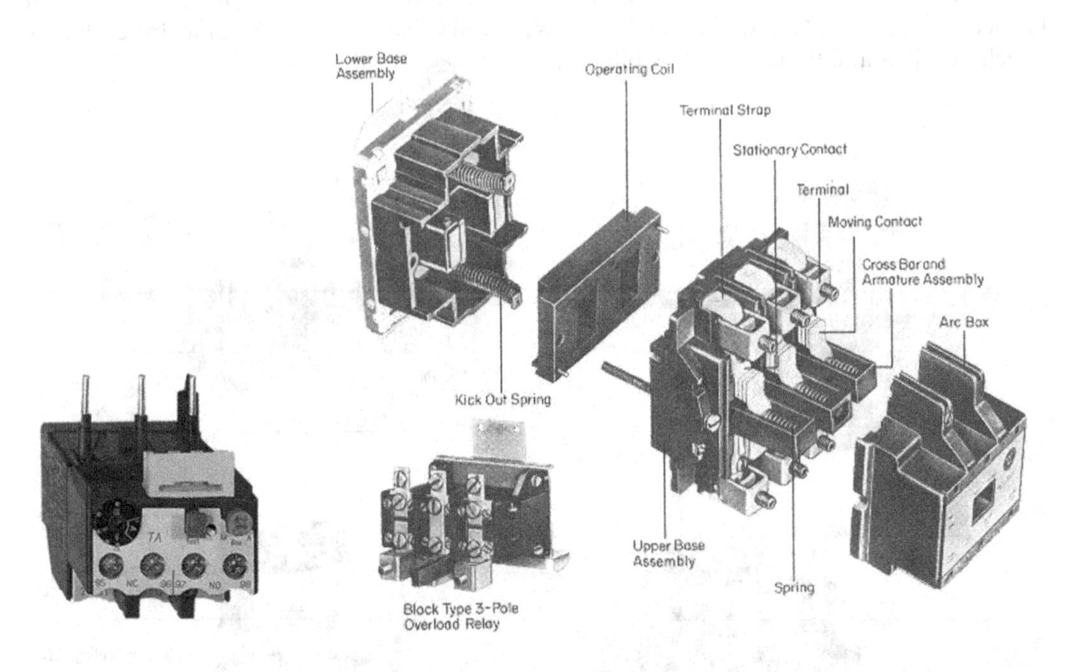

Figure 5.7 Exploded view of a motor contactor. (Courtesy of Eaton.)

a motor starter can provide a reliable proof-of-actuation signal, indicating that the contactor armature has indeed pulled-in after being energized.

Figure 5.6 shows a typical motor-starter assembly, consisting of a motor contactor and its overload device, shown with its bimetallic heater element removed. Figure 5.7 shows an exploded view of a contactor.

CONTROL RELAYS

Control relays are small electromechanical switching devices that permit isolation between high-and-low-voltage circuits, remote starting, and the decision-making logic functions associated with control. Control-relay wiring diagrams are also called relay-logic diagrams because the relays create the logic, or ladder diagrams because of the ladder-like appearance of the wiring diagrams. Control relays have two functional parts, the electromagnetic coil to activate the relay, and the switches, which are simultaneously toggled by its armature bar. Sometimes the single-phase control voltage is obtained between two of the motor's line-voltage connections, but very often the relay coil is powered by lower voltage, either from a stepdown transformer located in the motor controller or from a separate low-voltage single-phase source, such as for multiple motor-controllers installed in a group control center. Figure 5.8 shows several styles of relays.

The number of switches in a control relay are referred to as number of poles, where each switch may have only normally open or normally closed contacts, or each switch may have both normally open and normally closed contacts. The number of switch positions or the number of circuits that can be completed are called throws; so, a relay having only normally open contacts would be classified as "single throw," while a relay that closes one set of contacts while simultaneously opening another set is referred to as "double throw." Figure 5.9 shows different symbols used for single-pole relays having one set of switches, including single-pole single-throw (SPST) and single-pole double-throw (SPDT) relays. The figure also shows the construction variations, where Form A, B, and C share a common connection between the NO and NC contacts and Form X, Y, and Z have two break points for each pole, which provides additional logic flexibility.

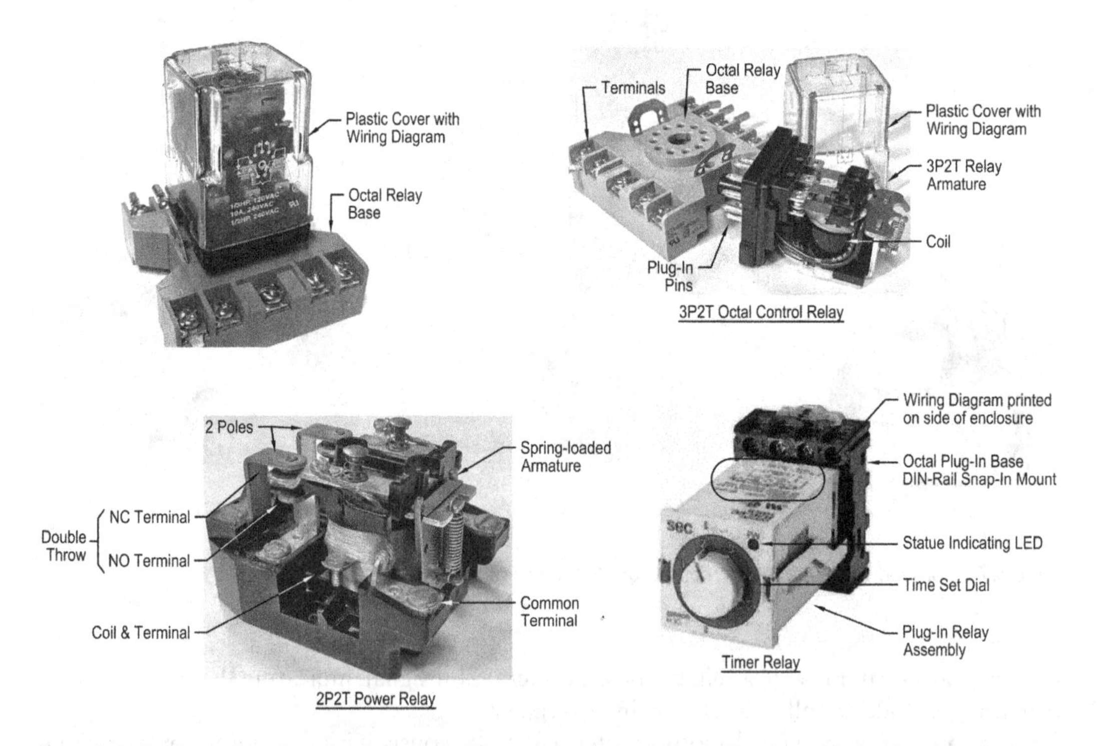

Figure 5.8 Various styles of mechanical relays. (Courtesy of United States Merchant Marine Academy.)

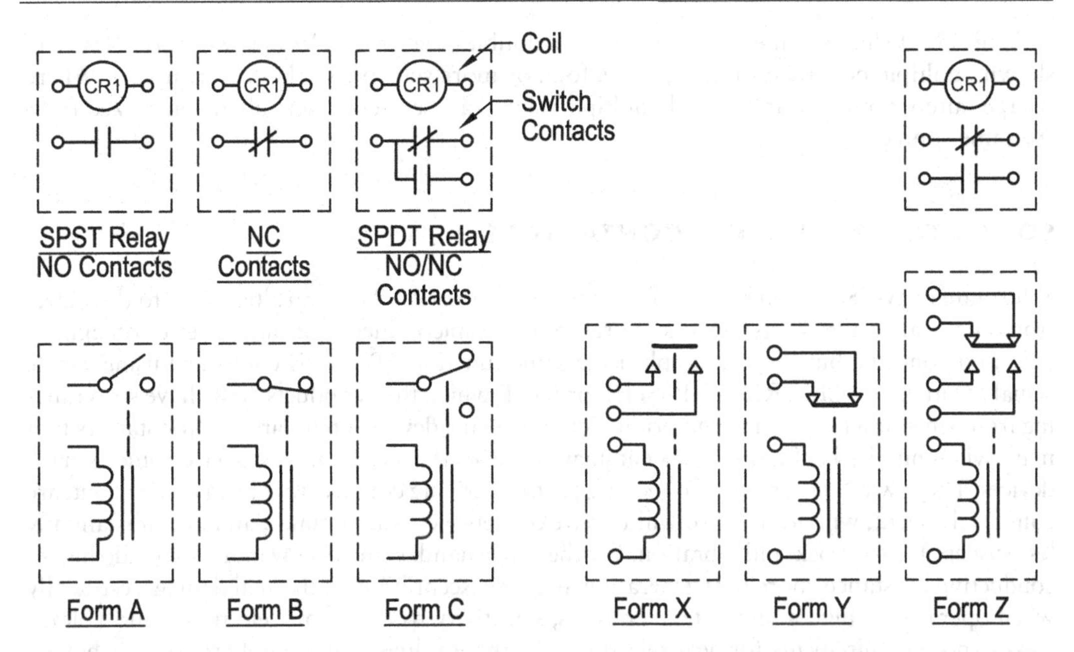

Figure 5.9 Relay symbols for *single-pole relays*. The Form A, B, and C relays will make-and-break only one contact, while the Form X, Y, and Z relays break two contacts. Note that the Form A double-throw contact shares a common connection, thus limiting this relay to one switch having two positions, whereas a jumper can be added to the Form Z relay to do the same thing, or the Form Z relay can be wired for two independent functions, one for NO logic and the other for NC logic.

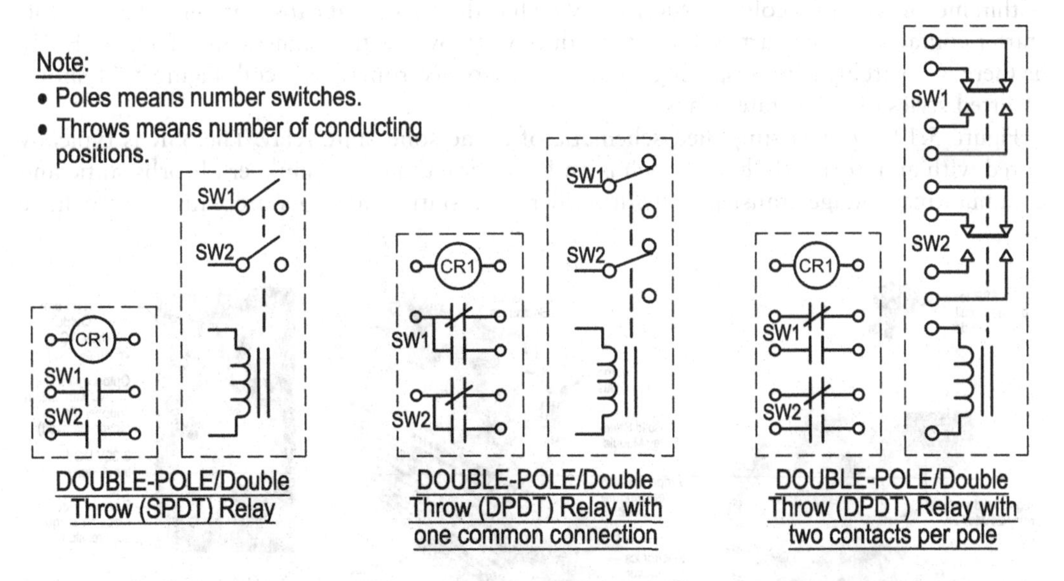

Figure 5.10 Relay symbols for *double-pole relays*. Note that the double-throw relay can be obtained with a common connection between the NO and NC make and break contacts, or separate make and two break contacts can be used.

Figure 5.10 shows some symbols used for double-pole relays. While two-pole relays are shown, multiple pole relays having three, four, or more sets of switches are available. Where a large number of poles are needed, multiple relays can be used, where all are energized from the same source.

SOLID-STATE RELAYS AND CONTACTORS

Solid-state relays (SSR) and power solid-state contactors (PSSR) are an alternative to their electromechanical counterparts. Solid-state relays use semiconductors in lieu of electromagnetic coils and contacts, but they accomplish the same functions. Power is controlled using a gate signal to either an SCR, TRIAC, MOSFET, or IGBT, which then conducts. SSRs have sizes ranging from small fraction-of-an-amp printed-circuit-board devices up to large motor starters that may switch hundreds of amps. SSRs eliminate some of the shortcomings of electromechanical devices. They switch faster, have no moving parts, no arcing contacts that burn or wear out, are completely silent, will not chatter, will never experience contact bounce, and are substantially less susceptible to shock and vibration. On the other hand, some disadvantages include higher conducting resistance, more heat generation, higher susceptibility to thermal damage, especially when operated at high ambient temperatures, sensitivity to both static electricity and voltage spikes, more requirements for heat rejection, slightly non-linear, non-purely resistive behavior, and they can experience unexpected, spurious switching during voltage spikes.

Many SCRs use optical coupling between the low-voltage control signals and the high-power circuit, to protect the electronic controller. SCRs and TRIACs used in AC circuits only turn off when the sine wave reaches zero amps, which is called zero-crossover switching. Consequently, the circuit will never be interrupted in the middle of the sine wave. Zero-crossover switching nicely avoids the large transient voltages that occur as magnetic fields within motor windings collapse suddenly, which is the effect that causes arcing across motor-starter contacts as they part. SSRs also produce very low electromagnetic interference (EMI), as there is no arcing across opening contacts and no electromagnetic coil. Figure 5.11 shows assorted styles of solid-state relays.

Figure 5.12 shows a simplified schematic of an ac solid-state relay. The SSR is typically wired with an integral snubber, which is an RC filter circuit. The snubber absorbs static and commutating voltage transients (dv/dt), where the static transients are caused by voltage

Figure 5.11 Various styles of solid-state relays (SSR). Miniature SSR for printed circuit board (upper), 2-amp DC-switch slim SSR (left), 30-amp AC single-phase SSR with heat sink (center), 25-amp SSR (right). The SSRs are turned on by a DC signal in the range of 3V to 32V. (Courtesy of United States Merchant Marine Academy.)

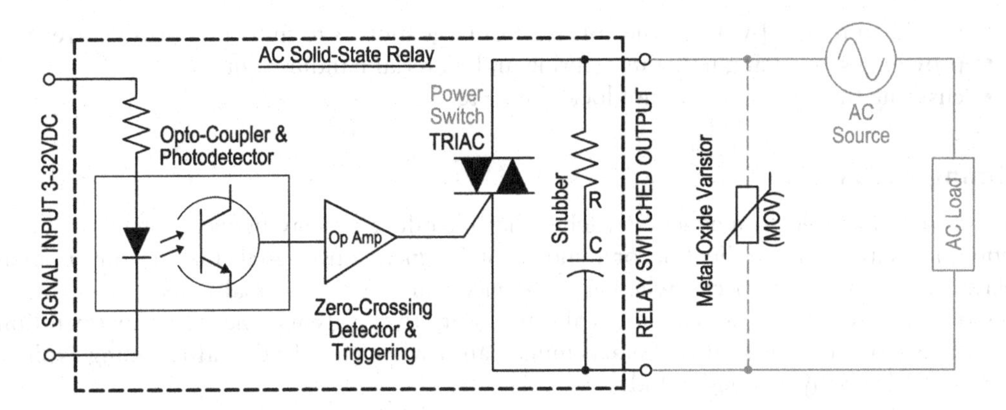

Figure 5.12 Simplified schematic of an AC solid-state relay. Snubbers or metal-oxide varistors (MOVs) are used to protect the electronics against voltage spikes.

spikes as the SSR is turned on, and commutating transients are caused by voltage spikes that occur at low power-factor operation, when the current lags the voltage by quite a bit at the time of switching. Although snubbers are often built into the SSR electronic package, there are times when the snubbers are not sufficient by themselves at absorbing transient voltage spikes, so back-to-back Zener diodes or transient-voltage suppression diodes (TVS) are used, or metal-oxide varistors (MOV) are added in the power circuit beyond the SSR to protect against large voltage spikes that may occur during shutoff transients or utility power surges. Back-to-back Zener diodes simply clamp the voltage at the reverse-bias breakdown level and are used for smaller SSRs, while the MOVs are voltage-dependent resistors with similar behavior to Zener diodes, but are used for high-power SSRs.

SPECIALTY RELAYS

Simple relays operate by applying voltage to a coil, which causes switch contacts to toggle. When voltage is removed from the coil, springs return the contacts back to their deenergized or normal position. However, other types of specialty relays are available to permit more sophisticated control, beyond simple Boolean on-off logic. These relays consist of timing relays, counting relays, and latching relays. Latching relays retain their last state upon loss of power, in the same way that a maintained-contact switch remains in its last position. Timing and counting relays are very useful in sequential-logic functions, such as:

- sequencing lights, as for traffic control, or for repeated flashing to indicate an active alarm.
- sequencing the starting of many LVR motors for spreading out their in-rush currents when transferring to an emergency generator during an outage.
- permitting large motors to start first while delaying smaller motors when shifting back from emergency power to normal power, ensuring coordinated and reliable power restoration.
- starting auxiliary equipment at the proper time during automatic startup sequence of engines or systems.
- providing pre-purge, post-purge, and loss-of-flame sequencing in boiler burner management systems.
- providing alarming after a number of attempts have failed to produce an effect.

- providing reduced-voltage and soft-start of large motors to mitigate in-rush current.
- providing sequencing in manufacturing and material handling automation.
- ensuring that systems have not locked up, etc.

Timing relays

Timing relays are electromagnetic switches that introduce a delay between switch-toggling action and either the application or removal of its energizing signal. Like a conventional relay, it is necessary to specify whether the device contacts are normally open or normally closed in the deenergized state, but it is also necessary to specify whether the relay operation should open or close the switch contacts immediately when energized or after timing expires. The time-delay relay options include:

- normally open, timed closed (NOTC),
- normally open, timed-open (NOTO),
- normally closed, timed open (NCTO),
- normally closed, timed closed (NCTC),
- one-shot and watchdog functions.

The original mechanical timing relays used pneumatic or oil-filled dashpots, needle valves, and springs to create the delaying action before the switch toggled. Newer designs use potentiometers and capacitors in electronic circuits to create the time delay, a design that is much more robust and less prone to failure (Figure 5.13), while also avoiding occasional readjustments to compensate for drift in the time-delay mechanism. Figure 5.14 shows the sequencing behavior of the relay-contacts after energizing the timer-relay coil for the four basic timer configurations. Table 5.2 summarizes alternative nomenclature, which is helpful for understanding the relay behavior.

Although understanding timing-relay action can be a little confusing at first, the timing diagrams in Figure 5.14 are helpful in distinguishing behavior. "Normal" describes the contacts starting position in the deenergized state, and it is the first descriptor in the naming convention, so that a normally open switch starts with its contacts open and a normally closed switch starts with its contacts made. The key to understanding the sequencing behavior lies in the direction of the arrow in the symbol. Referring to Figure 5.15, the two "on-delay" contacts, NOTC and the NCTO, are on the left side and are easier to understand. These symbols indicate that the switch starts in its "normal" position, and then shifts in the direction of the arrow to the other position when timeout is reached. For example, the NOTC contacts begin open, and at timeout, they close. In a similar manner, the NCTO contacts begin closed and they open at timeout. The toggled position at timeout is then maintained until the signal is removed from the timer coil, at which point the contacts revert to their normal positions.

The NOTO and NCTC symbols are less intuitive at first, but again, the arrows describe the behavior. It should be recognized that the contacts for these two symbols need to finish in the same pre-energized starting position after timeout, as they were before timing began. Using the NOTO contact symbol in Figure 5.15 as an example, the contacts start open (NO) and must finish in the open (TO). Therefore, when the timer coil is first energized, the contacts immediately shift to close, but timing does not yet start. The contacts remain in the shifted position for the entire duration that the signal is applied to the timer coil, however, once the timer signal is removed, timing begins. It is not until timeout that the contacts return to their starting position, and then the timer is reset for its next cycle. The NCTC behaves in a similar manner, but in the opposite direction.

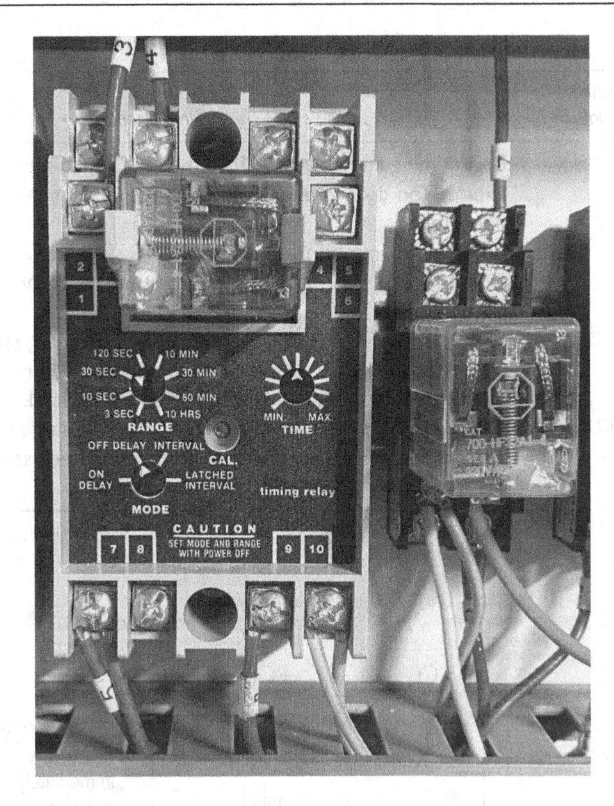

Figure 5.13 Universal timing relay (left) alongside a control relay. The timing relay has on-delay, off-delay, interval, and latched interval capability with adjustable time settings from 3 seconds to 10 hours. A replaceable DPDT plug-in relay does the switching. (Courtesy of United States Merchant Marine Academy.)

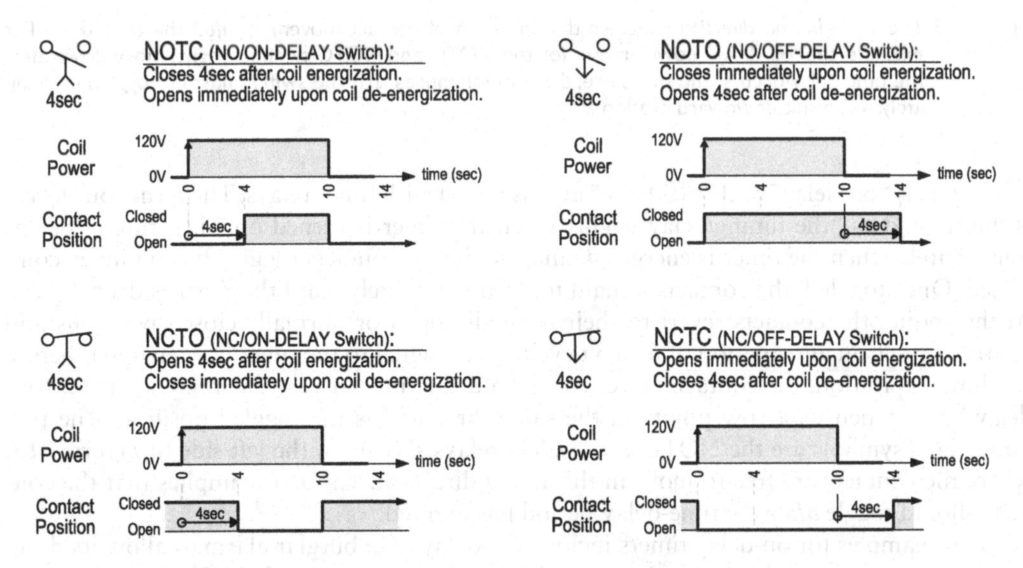

Figure 5.14 Timing relay switch behavior in response to timer energization. The arrow in the timer symbol indicates the direction that the contact shifts at timeout, and is key to understanding the behavior exemplified in the timing diagrams.

Table 5.2 Nomenclature and timer action for the different timing-relay types

Time-delay contact type	*Alternative nomenclature*	*Timer action*
NOTC	NO, *On-Delay*	Begin timing immediately. Close the NO switch after the timing period has expired (on delay). Reopen the switch and reset the timer when the coil is deenergized.
NCTO	NC, *On-Delay*	Begin timing immediately. Open the NC switch after the timing period has expired (on delay). Reopen the switch and reset the timer when the coil is deenergized.
NOTO	NO, *Off-Delay*	Close the NO switch immediately upon coil energization. Delay the timing until after the coil is deenergized (off delay). Then reopen the switch and reset the timer after the timing period has elapsed.
NCTC	NC, *Off-Delay*	Open the NC switch immediately upon coil energization. Delay the timing until after the coil is deenergized (off delay). Reclose the switch and reset the timer after the timing period has elapsed.

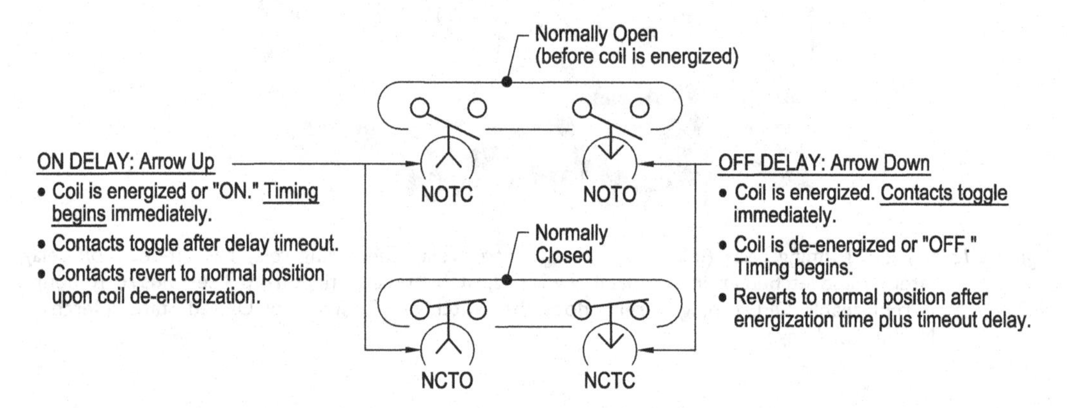

Figure 5.15 The *symbol arrow direction* indicates the direction of contact movement *after* the time delay. For example, the "off-delay" down arrows for the NOTO and NCTC symbols (right) move down after the time delay occurs, and therefore, the symbol implies that the switch must be toggled immediately, to permit downward motion later.

The terms "on-delay" and "off-delay" are also used with timer relays. The term "on-delay" is interpreted as "the timing/delay begins when the timer is turned on," i.e., timing begins immediately when the timer is energized, and the timer's contacts toggle after timing is completed. Once toggled, the contacts remain toggled indefinitely, until the timer is deenergized. At that point, the contacts revert to their normally open or normally closed positions and the timer resets in anticipation of its next cycle. If the signal to the timer is interrupted before reaching expiration, the contacts never toggle and the timer resets. The relay is of the "on-delay" type when its arrow points in the same direction as the toggled position. The two "on-delay" symbols are the NOTC and NCTO relays shown on the left side of Figure 5.15, where the contacts are free to move in the arrow direction. The arrow implies that the contacts should toggle *after* the time-delay period has expired.

Some examples for on-delay timers include the delay on a burglar alarm to allow its deactivation by a homeowner, the staging of multiple fans for an air-cooled HVAC condenser to mitigate in-rush current, or the delay before attempting a boiler-burner light-off sequence until an adequate pre-purge period has elapsed.

The term "off-delay" is interpreted as "the timer/delay begins when the timer is *turned off*." Because the arrows showing the initial position of the off-delay symbol before timer activation are in the same position as after timeout, it is necessary for the contacts shift away from the final position. The shifting away happens immediately when the timer is energized. Once the timer is energized, the "off-delay" contacts remain shifted during the entire time that the timer is energized, and beyond. Timing begins only after the timer is turned off. At timeout, the contacts shift back to their "normal" starting position, and since the signal has already been removed, the timer resets. If the timer signal is reapplied before timeout, the contacts remain toggled and timing will start over only when the timer signal is removed. The two "off-delay" symbols are the NOTO and NCTC timers shown on the right side of Figure 5.15, where the contacts move away immediately, and they will later

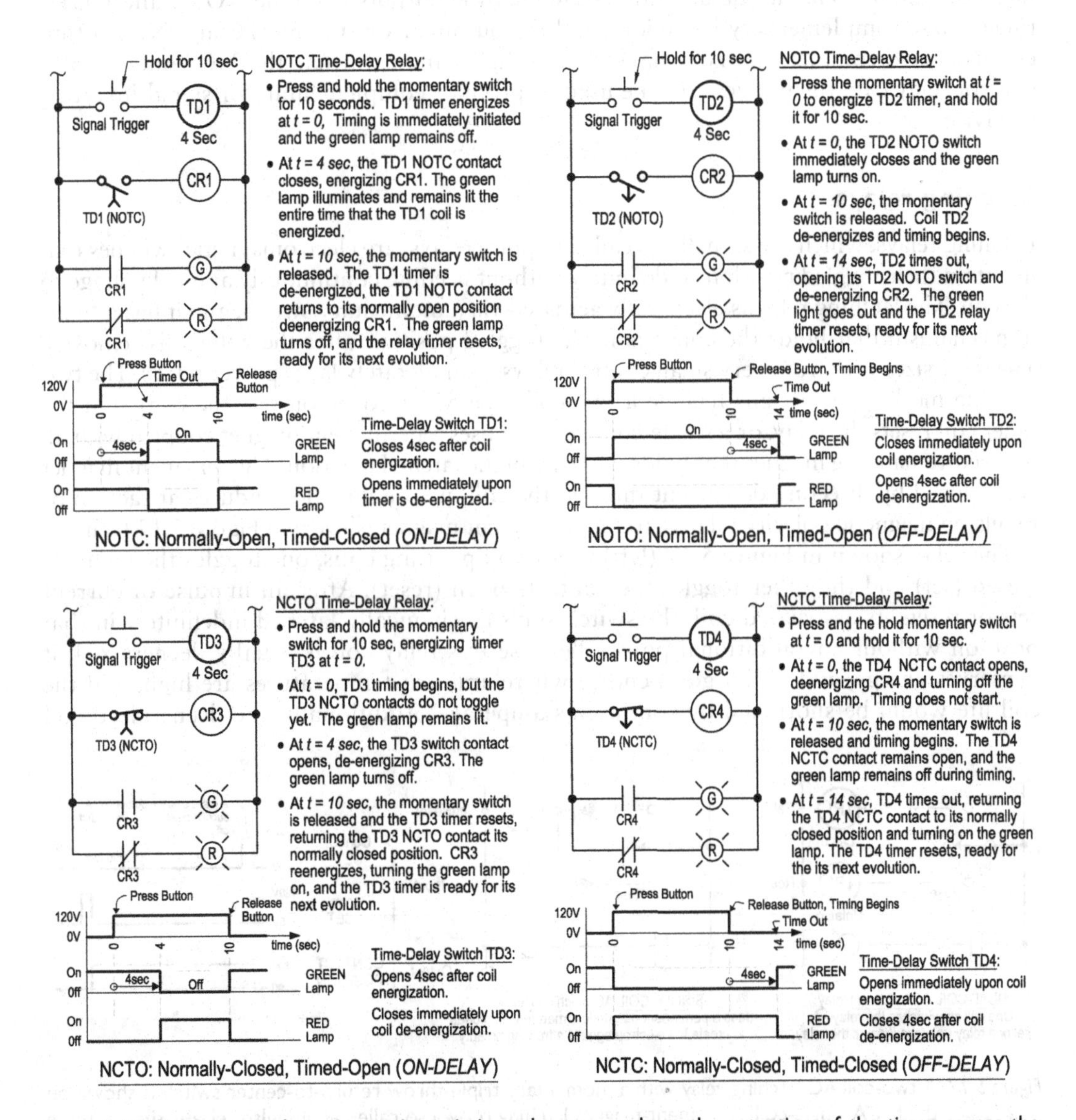

Figure 5.16 Circuit examples using various timing-relay types, where the actuation of the timer causes the behavior that is specific to the relay type. Often, the timer contacts energize a control relay, whose normally closed contacts can reverse the timing-relay output logic.

toggle back in the arrow direction. Essentially, the "off-delay" contacts toggle immediately and remain toggled until the total duration consisting of the coil-energized period *plus* the elapsed time is reached.

Some examples of off-delay timers include, a boiler combustion-air fan that is kept running after burner fuel is secured as part of a post-purge process to ensure no explosive gases are retained in the furnace, or an air-conditioning system where a fan may operate for a short period after a heater or compressor is stopped.

Figure 5.16 includes simplified circuits, along with a concise description of the circuit response, to show the behavior of each of the four types of timing relays. The illustrations include the timing-relay coil (TD) and its associated time-delay contact (with the arrow), which in turn actuates a control relay (CR) whose NO and NC contacts toggle indicator lights on and off. The timing diagrams show that the outputs from the NOTO and NCTC timers have complementary behaviors, and the outputs from the NOTC and NCTO timers also have complementary behaviors. The lights indicate that the NC contacts of a control relay can, themselves, also be used to produce complementary or signal-inverting behavior.

Latching relays

Latching relays, which are sometimes called impulse relays, are electromagnetic switches that maintain their toggled positions indefinitely without needing continuously applied voltage to the coil. The latching relay is electrically actuated via a momentary switch, and it has a built-in mechanism that holds the contacts in the toggled position after the voltage is removed. The relay sizes vary from very small control relays to moderately large power relays. The two latching-mechanism designs include a two-coil type, where one coil sets the relay and the other coil resets the relay, or a single-coil type that uses a permanent magnet coupled with an induced-remanence magnet that holds the relay in the latched position. The remanent magnet is produced by directing dc current through the coil, whose direction induces attraction or repulsion magnetic polarity relative to the permanent magnet, shown in Figure 5.17 (center).

The relay shown in Figure 5.17 (left) uses two operating coils, one toggles the contacts closed (set) and the other toggles the contacts open (reset). After an impulse of current sets or resets its associated coil, the switch stays mechanically latched indefinitely in that position without any additional power. Because this relay coil is usually deenergized, it conserves energy, and it is a good choice where ambient temperatures are high, and the coil life would be shortened by continuous amperage. Additionally, the deenergized coil

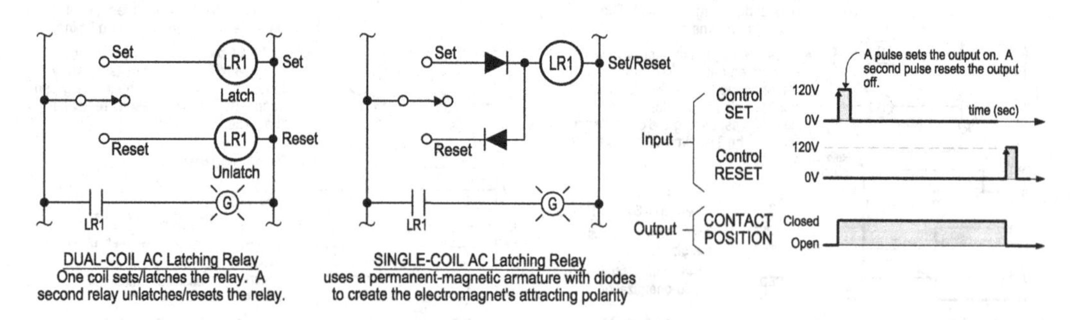

Figure 5.17 A two-coil AC latching relay with a momentary triple-throw return-to-center switch is shown on the left. A single-coil permanent-magnet latching relay, also called an impulse relay, is shown in the center. Input signals and output responses are shown on the right.

avoids 60-cycle hum, which is particularly good for turning on a large lighting system in quiet occupied spaces. For plant equipment, the latching relay has the advantage of retentive memory, i.e., it leaves the switch in its last operating position after an unscheduled interruption of power occurs. Upon restoration of power, the equipment continues where it left off.

Alternating relays

Alternating relays are used in pumping or compressor applications where one machine operates, while the other machine is kept in ready standby. The alternating relay automatically swaps operation after each cycle, equalizing run times and wear and tear. Alternate starting of the motors has the additional benefit of extending the duty-cycle off-time, allowing the motor windings more time to dissipate the heat produced by in-rush current, and extending motor life.

In an alternator circuit, every time a motor operates, the alternator realigns the circuit to start the other motor at the next cycle. Figure 5.18 shows two methods for producing alternating action, where conventional relays and logic are used in the diagram on the left, and a dedicated alternator relay is used on right. The diagram on the left has no memory associated with it, so upon restoration of lost power, motor 1 always starts first. Motor 1 starts through normally closed contacts 2Ma and the deenergized CR2. Once the 1M relay energizes, it pulls in a 1Ma holding contact, energizing the alternator control relay (CR2). CR2 remains energized throughout the motor-1 operation and its holding contact keeps it energized throughout the subsequent off period. The energized CR2 becomes the permission for the next cycle to start motor 2, while inhibiting motor 1 from starting. The operation alternates at each subsequent start cycle.

Figure 5.18 (right) uses a latching type relay (ALT) to alternate motor operation. When the high-level switch (LSH) calls for operation, motor 1 starts and alternator coil ALT energizes. When LSH opens, the ALT coil deenergizes and the ALT armature toggles its SPDT contacts to enable motor 2 for the next cycle. This design has retentive memory and will alternate to

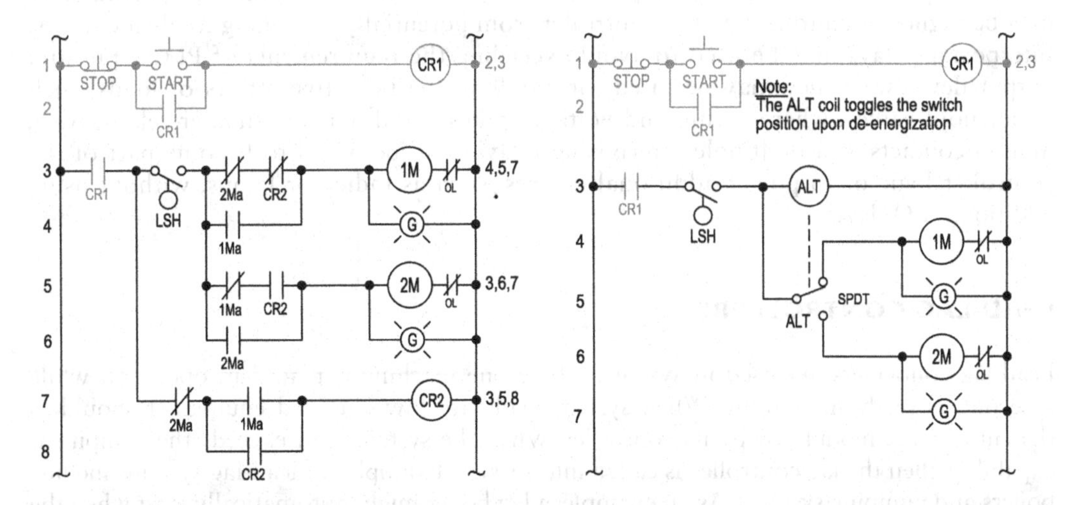

Figure 5.18 Two methods for alternating device operation. The left diagram uses conventional relay logic, but upon restoration of lost power, motor 1 always starts first. The right diagram uses a latching alternator having memory, where upon restoration of lost power, the next motor in the sequence will start first.

Figure 5.19 Solid-state motor-alternator relay. In the normal position, the relay toggles to the next load after the next off cycle, which is indicated by the LED lights, or the switch can be set to select only one load.

the next motor even after a power interruption. Figure 5.19 shows a commercially available alternator latching relay.

INTERPOSING RELAYS

Interposing relays are conventional relays used to segregate and control another part of the circuit, where "interposing" describes its purpose, rather than its construction. The other part of the circuit may operate at a different voltage or the interposing relay may be higher rated to isolate the controller from potentially damaging load-side faults. Interposing relays may be used to avoid exceeding the requirements of PLC input and output devices, or they may be rated for handling the inductive effects of loads, such as solenoid valves, where arcing and voltage spikes could damage smaller relays. Also, unused contacts in a multipole interposing relay could be wired to become part of the controller logic or to power additional devices, such as indicator lamps, without using additional I/O devices.

LEAD-LAG CONTROLLERS

Lead-lag controllers are used in systems where one machine is placed in operation while a second is ready in standby. When system loads are low, the lead equipment shoulders the entire load, modulated by its controller. When the system load exceeds the equipment capability, then the lag controller is called into service. Examples of lead-lag systems include boilers and pumping systems. As an example, a lag boiler might automatically start when the lead boiler reaches 80-90% of its rated output, and it might shut down when the lead boiler drops to between 25-35% of its rating. As another example, pumps, which come online much faster than a boiler, may incorporate lead-lag functions coupled with alternator functions.

ONE-SHOT, WATCHDOG, AND OTHER TIMERS

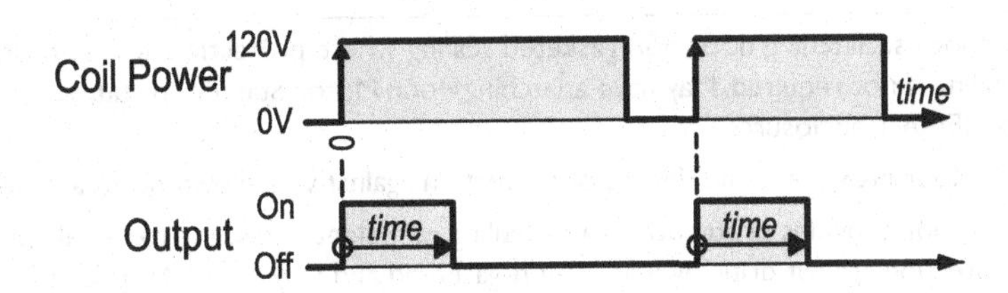

One-shot timers are also called "interval-on" timers. Immediately upon energizing the relay coil, the one-shot timer contacts toggle and timing begins. Upon timeout, the contacts return to their normal deenergized state, however the relay coil does not reset. At this point, the one-shot timer is inhibited from being reactivated until the coil voltage is removed, resetting the timer. A one-shot timer might be used for shutting a welding machine when its duty cycle is reached, automobile seat-belt warnings, or dispensing equipment.

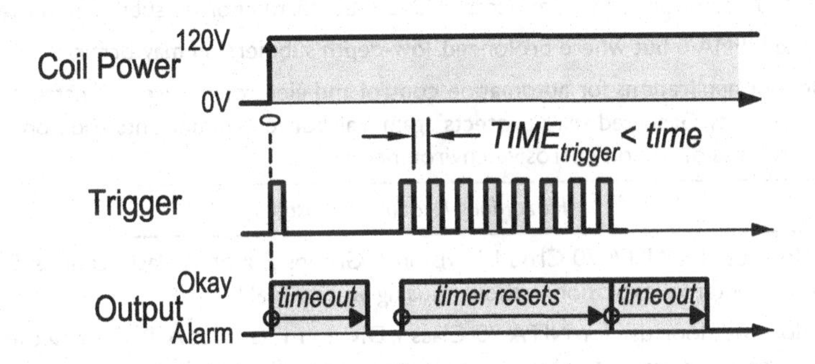

A watchdog relay is a retriggerable one-shot relay. The watchdog behaves like the one-shot timer in that the contacts toggle immediately upon application of coil voltage and a trigger signal, and it will toggle back to the normal deenergized position at timeout, while the coil voltage is maintained. However, where the one-shot cannot be retriggered until the relay is de-energized, the watchdog can be. If the retriggering occurs in the middle of the timing cycle, the timing period is extended. Eventually, if the trigger is lost for a period reaching timeout, the relay goes to its normal deenergized state. The watchdog is useful for actuating an alarm in the event of a system lockup. As long as a repeating function triggers the relay, the assumption is that all is in apparent good order. When the repeating function fails to trigger the watchdog beyond the timeout period, an alarm is actuated.

Other timers consist of various types of flashers, repeat cycles, delayed cycles, and other specialty relays.

NEMA RATINGS

Motor controls are installed in enclosures that may be located near the motor or remotely in a motor-control or group-control center. The electrical enclosures are rated by the National Electrical Manufacturers Association (NEMA) based on the location and environment using NEMA designations. Some of the NEMA enclosure types are listed in Table 5.3 for general-purpose applications and for use in hazardous locations.

Table 5.3 NEMA ratings for electrical and electronic enclosures

	General purpose enclosures
NEMA 1	Indoor use, latching door, non-gasketed sealing where protection against dust, oil, and water sealing is not required. May have a latching door. Motor Start/Stop stations are often housed in NEMA 1 enclosures.
NEMA 3	Outdoor, provides some degree of protection against windblown dust, rain, sleet, external ice.
NEMA 3R	Outdoor, provides protection against falling rain, sleet, snow, and external ice. Indoor protection against dripping water. Non-gasketed and may have a hasp for padlocking.
NEMA 3S	Outdoor, provides some degree of protection against windblown dust, rain, sleet, and external ice while providing operation of external mechanisms when laden with ice.
NEMA 4	Indoor/outdoor use where occasional washdown and splashing may occur. Uses gasketed doors with compression clamping for sealing, continuous hinges, mounting feet, and padlock hasps. Available in small wall-mounted units up to large two-door floor mount.
NEMA 4X	Indoor/outdoor use, like NEMA 4 but constructed of corrosion-resistant materials for use where contact with caustic cleaners may occur. Typically constructed with stainless steel or plastic.
NEMA 6	Indoor/outdoor use where occasional low-depth, temporary submersion may occur.
NEMA 6P	Like NEMA 6 but where prolonged, low-depth submersion may occur.
NEMA 12	Indoor applications for automation control and electronic drives. Constructed without knockouts. Gasketed seals protects against airborne contaminants and non-pressurized water and oil in non-corrosive environments.
	Hazardous locations enclosures
NEMA 7	Indoor use for NFPA 70 Class I, Division 1, Groups E, F, or G classifications. Designed to contain internal explosion without causing an external hazard.
NEMA 8	Indoor/outdoor use for NFPA 70 Class I, Division 1, Groups A, B, C, or D classifications. Designed to prevent combustion by using oil-immersed equipment.
NEMA 9	Indoor use for NFPA 70 Class II, Division 1, Groups E, F, or G classifications. Designed to prevent the ignition of combustible dust.
NEMA 10	Mine Safety and Health Administration, 30 CFR Part 18 use. Designed to contain internal explosion without causing an external hazard.

MOTOR CONTROLLER DIAGRAMS

Motor controller diagrams are used to show the device wiring connections and often the control logic behind the motor operation. Electrical diagrams are either schematic diagrams or point-to-point. The relay-logic diagram, also called a ladder diagram or ladder-logic diagram, is schematic in nature and arranged in an organized, progressive manner, so that the circuit logic can be easily followed. The schematic diagram is not constrained by the physical locations of devices within the panel, and oftentimes device sub-parts are split up and located on the schematic where the logic is easy to follow. Ladder diagrams are especially conducive to both circuit design and troubleshooting. These diagrams show all wiring within the controller, plus externally connected field devices, where the field wiring is often shown connected via a terminal-board interface using dashed lines. "Ladder Diagram" (LD) is a popular PLC programming language, which is similar in philosophy to relay logic, and is discussed in Chapter 16. Figure 5.20 shows schematic-diagram electrical symbols used for the power circuit and Figure 5.21 shows electrical symbols used for control devices. Figure 5.22 shows the breakdown of a relay and contactor so its sub-parts fit logically into a ladder diagram.

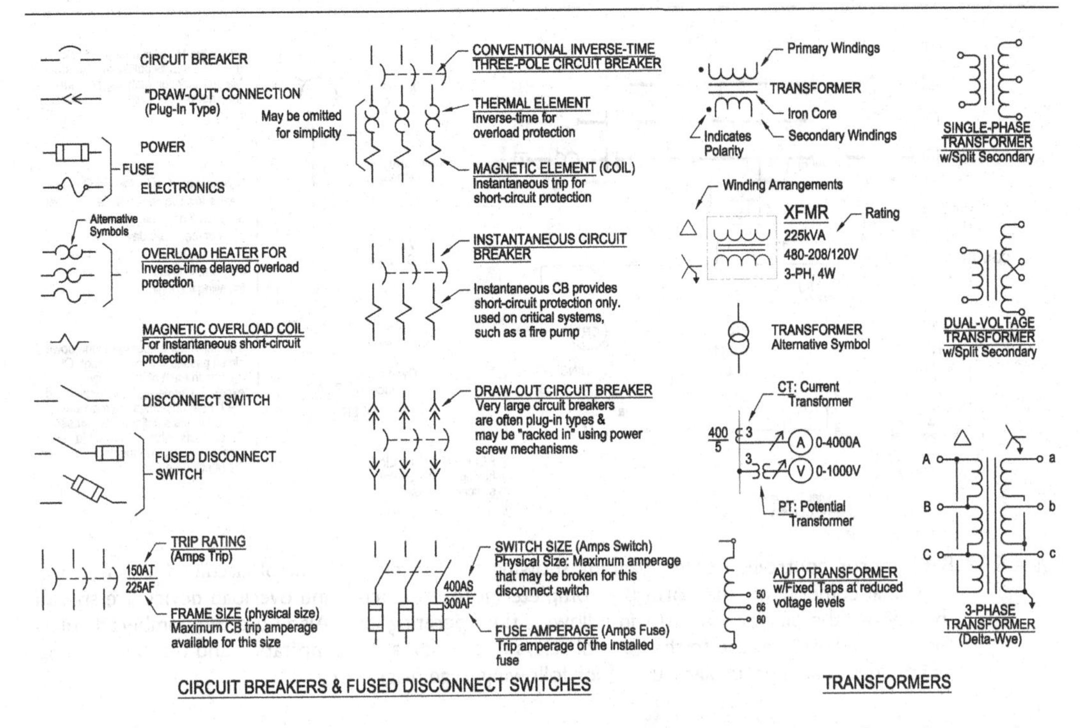

Figure 5.20 Power electrical symbols for circuit breakers, fuses, disconnect switches, and transformers, showing various symbol combinations and some symbol alternatives.

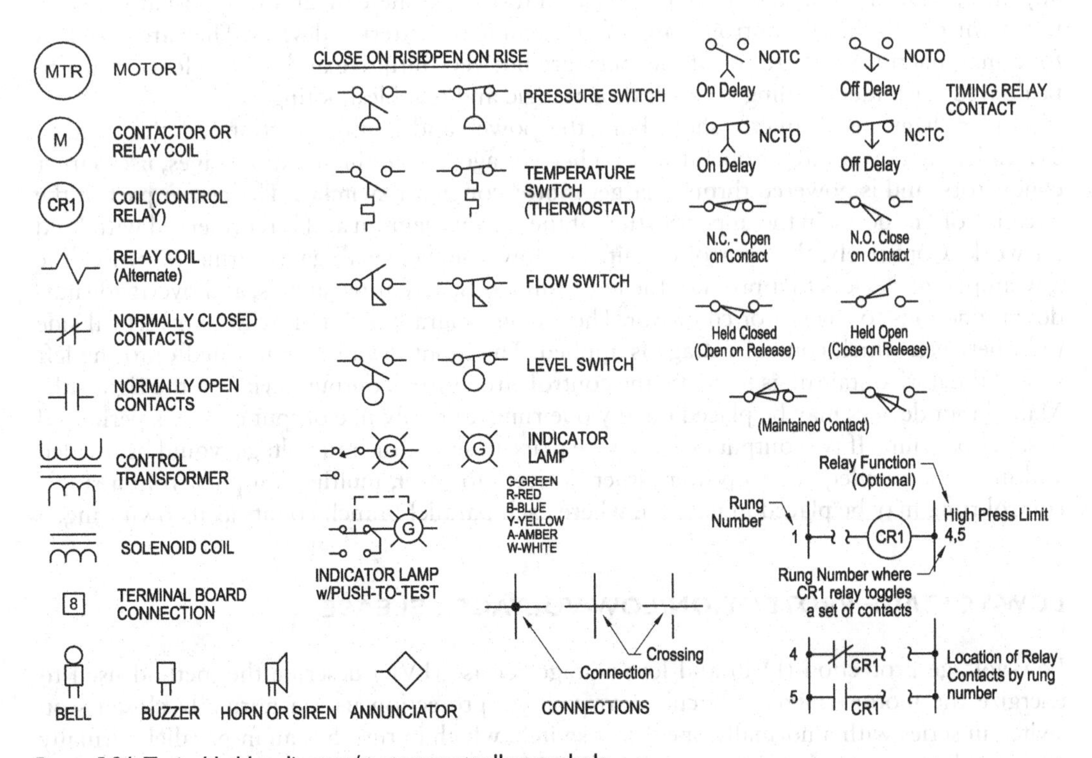

Figure 5.21 Typical ladder diagram/motor controller symbols.

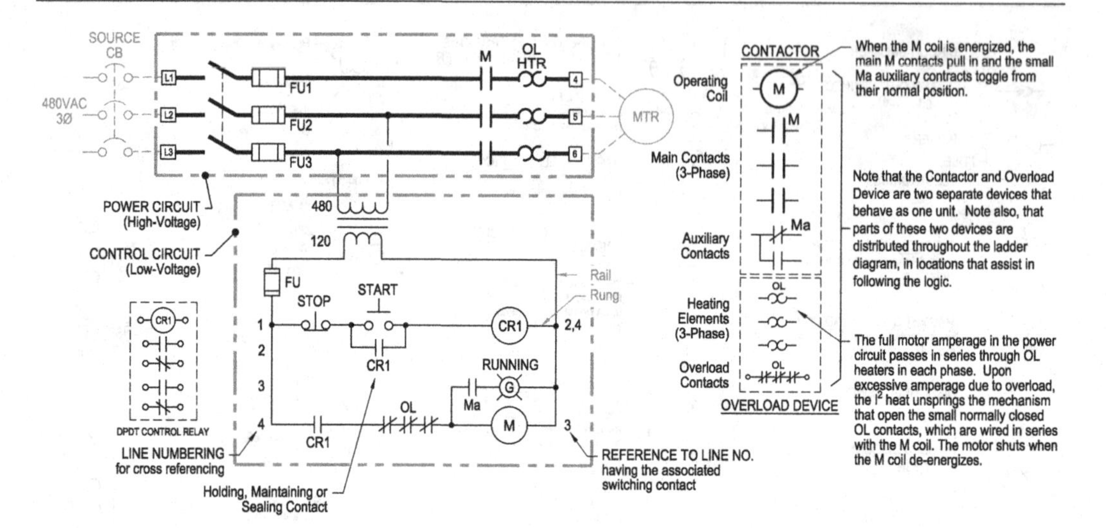

Figure 5.22 Basic motor controller ladder diagram, showing the power and control circuits. Note that the individual components that form the complete motor contactor and overload device are spread throughout the diagram to assist in following the operating logic. Also note the numbered rungs that are cross-referenced to the control relays to aid diagram interpretation and troubleshooting, a technique that is particularly useful for following large diagrams.

Wiring diagrams show all the controller electrical components in their entirety and in their locations within a panel. A wiring diagram indicates all point-to-point connections, without any interpretation or variations required by a fabricator. These diagrams are usually limited to the wiring within the controller and often exclude the external devices. They are excellent for a manufacturing environment, but they are not very intuitive and tend to leave much to be desired for understanding the operational logic and troubleshooting.

Ladder diagrams distinctly show both the power and control portions of the circuits. Typically, the power circuit operates at higher voltage, carries higher amperages, uses larger conductors, and is powered through larger motor-contactor switches. The power part of the circuit is often shown in the upper portion of the circuit diagram and is represented with bold linework. Conversely, the control circuit uses low-voltage, small-gage wiring to power the low-amperage devices that produce the logic functions, actuation signals, and overload shut-down functions to the motor contactor. The ladder diagram is drawn with two vertical side rails, between which control voltage is applied. The input devices are inserted onto the left side of the horizontal rungs to create the control circuit to an output device on the right side. Many input devices may be placed on any one rung, but only one output device is permitted in series per rung. If two output devices were place in series, the rail voltage would be divided and may not be adequate to power either device. However, multiple outputs from a single control rung may be placed in parallel, where each parallel branch counts as its own rung.

LOW-VOLTAGE PROTECTION/LOW-VOLTAGE RELEASE

Low-voltage protection (LVP) and low-voltage release (LVR) describe the method used to energize the motor-contactor circuit. Low-voltage protection uses a normally closed stop switch in series with a normally open start switch, which in turn has an in-parallel normally open holding contact from its relay. Figure 5.23 (left) shows that when the start switch is pressed, the control relay coil (CR1) energizes, pulling in its holding contact. When the start switch is released, the holding contact maintains or seals the relay coil in an energized state. If

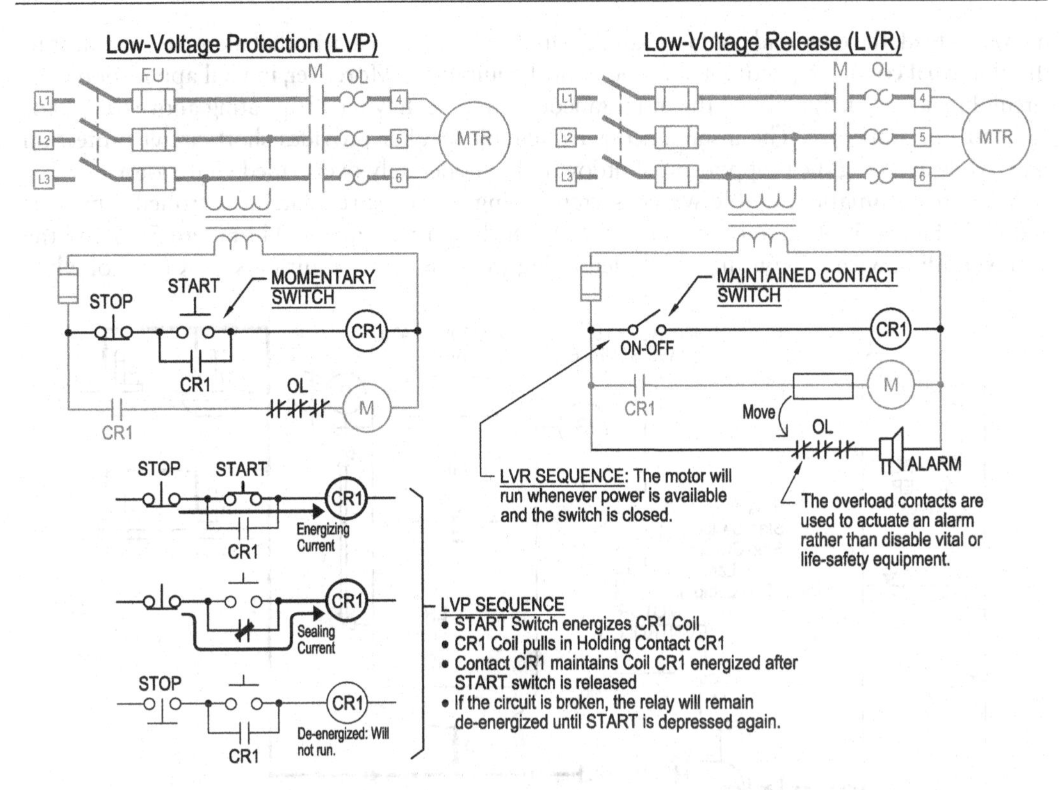

Figure 5.23 Low-voltage protection (left) requires restarting after power interruption, and low-voltage release (right) will automatically restart the motor, as would be desirable with vital or life-safety systems.

there is an interruption of power to CR1 for any reason, the coil deenergizes and the holding contact drops out. The motor will remain off indefinitely, until the plant operator re-presses the start switch. With low-voltage protection, it is personnel that are protected against an unexpected motor start, since the normally open start switch and its associated normally open control-relay holding contact inhibit motor operation.

Low-voltage release (LVR) uses a maintained contact switch to energize the control relay coil, as shown in Figure 5.23 (right). Whenever the switch is closed, the relay coil CR1 is energized, pulling in the motor starter. Upon a power outage, or even a low-voltage transient event, the CR1 coil demagnetizes and the relay drops out, killing power to the motor contactor. However, upon power restoration, coil CR1 re-energizes right away and the motor starts without warning. The immediate restart action occurs on every LVR-controlled motor whose start switch was left in the switch-ON position. LVR circuits are used in vital services, such as applications involving life-safety or the avoidance of machine damage.

LVP is used for most motors in a plant. The relay drops out upon a low-voltage condition, requiring manually restarting to resume operation. Since ac motor in-rush current during starting is roughly five to seven times its normal running current, it would be very possible to overwhelm an electrical source, such as a generator or utility transformer, by the sudden application of many loads. It is possible to cause power-source circuit breakers to trip and blacken the plant, but source damage is also possible. By manually restarting the motors, the in-rush currents are spread out, and plant operation is recovered in a controlled manner.

On the other hand, LVR is used on vital or life-safety equipment, such as fire pumps, engine lube-oil pumps, or shipboard steering-gear motors, etc., where a prolonged interruption of service cannot be tolerated. Often, the motor overload (OL) contacts for LVR applications may not be wired to protect the motor by tripping it, as the safety consequences far outweigh the value of the motor. Instead, the overload contacts can be wired to actuate an alarm that alerts the operator to the abnormal condition, without shutting critical equipment. Moreover, in vital applications, the circuit breaker is likely to be of the instantaneous-trip type, having a trip rating above the motor locked-rotor amperage. The instantaneous-trip circuit breaker provides short-circuit protection only, while avoiding the inopportune shutdown of a moderately overloaded vital motor.

A hydropneumatic potable-water system using a pressure-switch-controlled pump is shown in Figure 5.24. Its motor-controller ladder diagram is included in Figure 5.25 and the corresponding wiring diagram is included in Figure 5.26. The circuit uses LVR control when

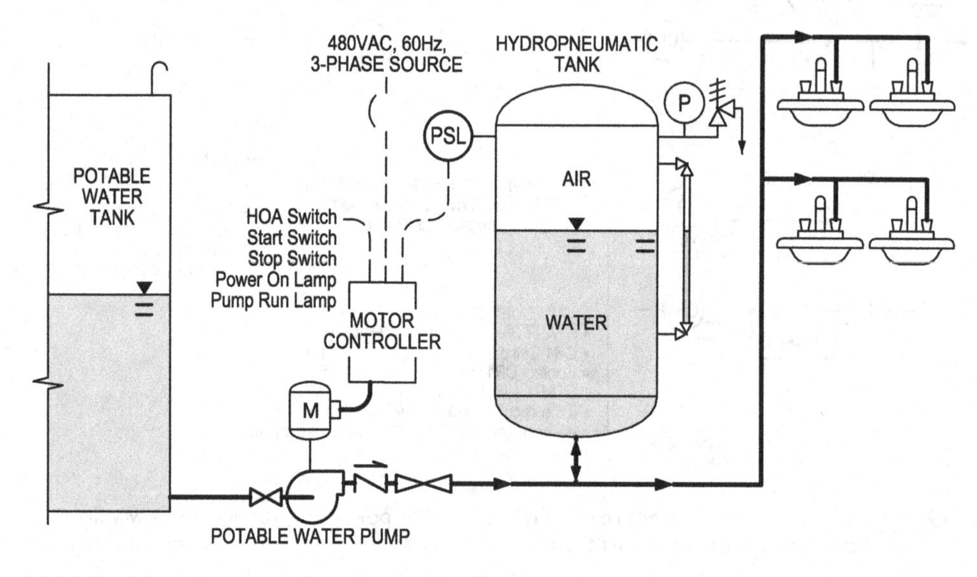

Figure 5.24 Potable-water system flow diagram. System operation can be better understood when the flow diagram is combined with the instrumentation and control devices.

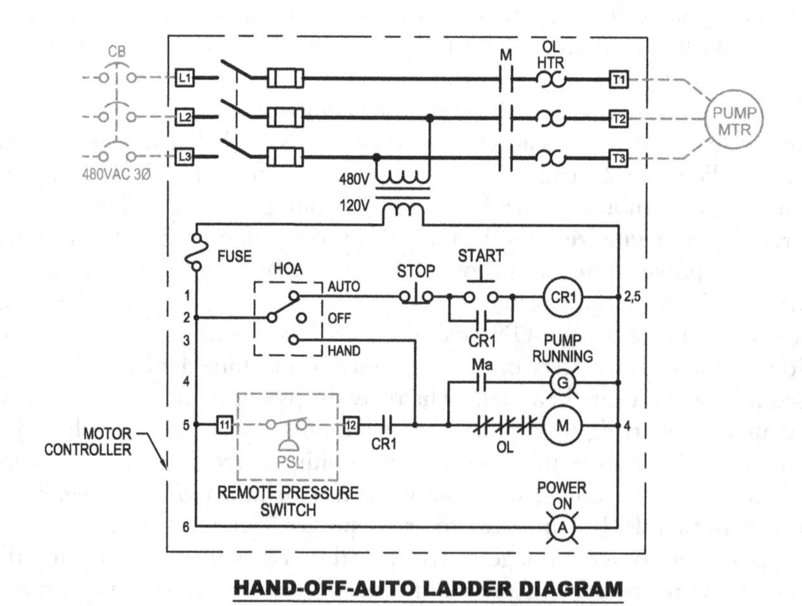

Figure 5.25 Motor starter wiring diagram for the potable-water system shown in Figure 5.24, producing continuous pump operation in "Hand," and low-voltage-protection with pressure-actuated operation in "Auto." This diagram is useful for following the logic that causes operation.

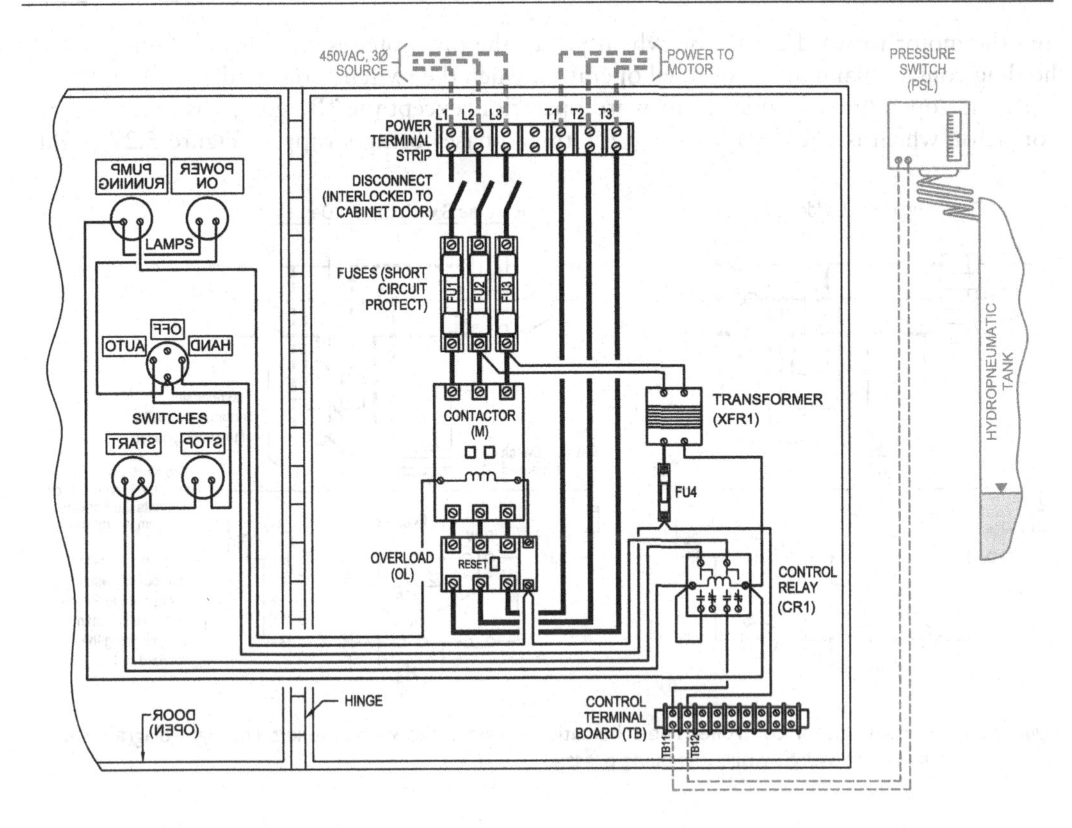

Figure 5.26 Point-to-point wiring diagram for the potable water system shown in Figure 5.24, having continuous pump operation in "Hand" and low-voltage-protection with pressure-actuated operation in "Auto." This diagram is useful in the factory, during controller fabrication.

the HOA maintained-contact switch is placed in "hand," so that the M coil is directly energized, and the pump runs continuously. In hand mode, the motor will automatically restart after lost power is restored. "Hand" is useful for testing motor operation or for attended operation if the control system is malfunctioning.

LVP control is obtained when the HOA maintained-contact switch is set to "auto," aligning the series-connected stop, start, and parallel holding-contact switches into the circuit. In automatic mode, the CR1 parallel holding contact becomes the permissive that allows pressure-switch control after the "start" button is pressed. Upon loss of power, the start switch must be repressed to restart the automatic operation.

REVERSING MOTORS

Three-phase motors are reversed by swapping any two of the line connections leading to the motor. Reversing motors have two motor contactors in the controller, one for forward (F) and the other for reverse (R). The reversing contactor swaps two of the three phases in the power circuit. Figure 5.27 shows a reversing arrangement having LVP control. This control circuit has two double-pole momentary switches, each with a normally closed contact wired to break the circuit in one direction before its normally open contact makes the circuit in the other direction. When the forward button is pressed, the reverse normally closed contact breaks, ensuring that the reverse contactor is deenergized, followed by the forward normally open contact closing and energizing the "F" coil. The "F" coil pulls in the "F" sealing contact and the motor forward contactor, which passes the three phases straight through. The "F" holding contact maintains continued operation when the switch is released.

Reverse operation is similar to forward operation, except the "R" coil pulls in the reverse contactor, which is wired with two of the three phases being swapped. Figure 5.27 (right)

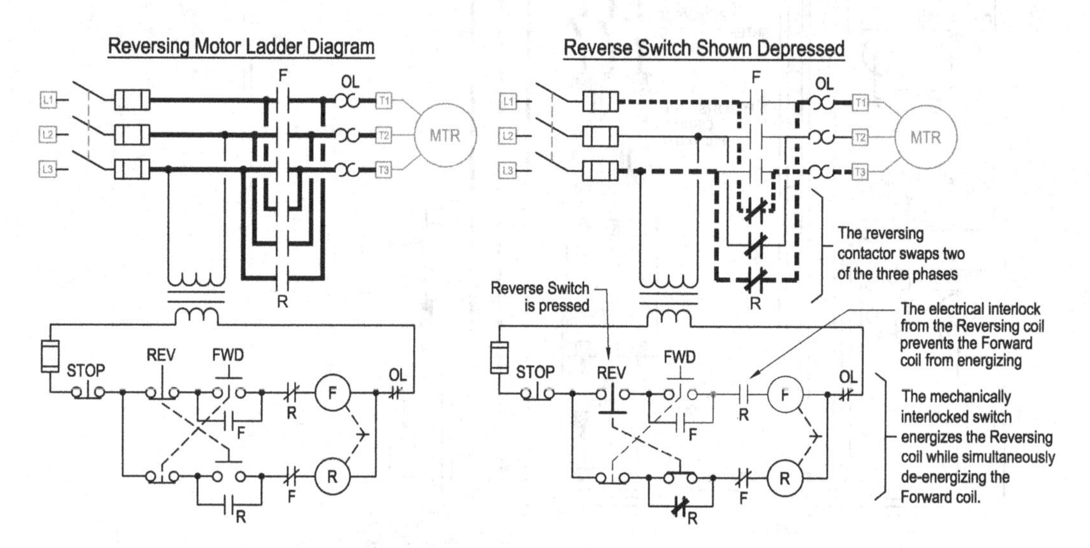

Figure 5.27 Reversing motor controller. The schematic diagram is shown on the left. The right diagram shows the reverse switch while it is being pressed.

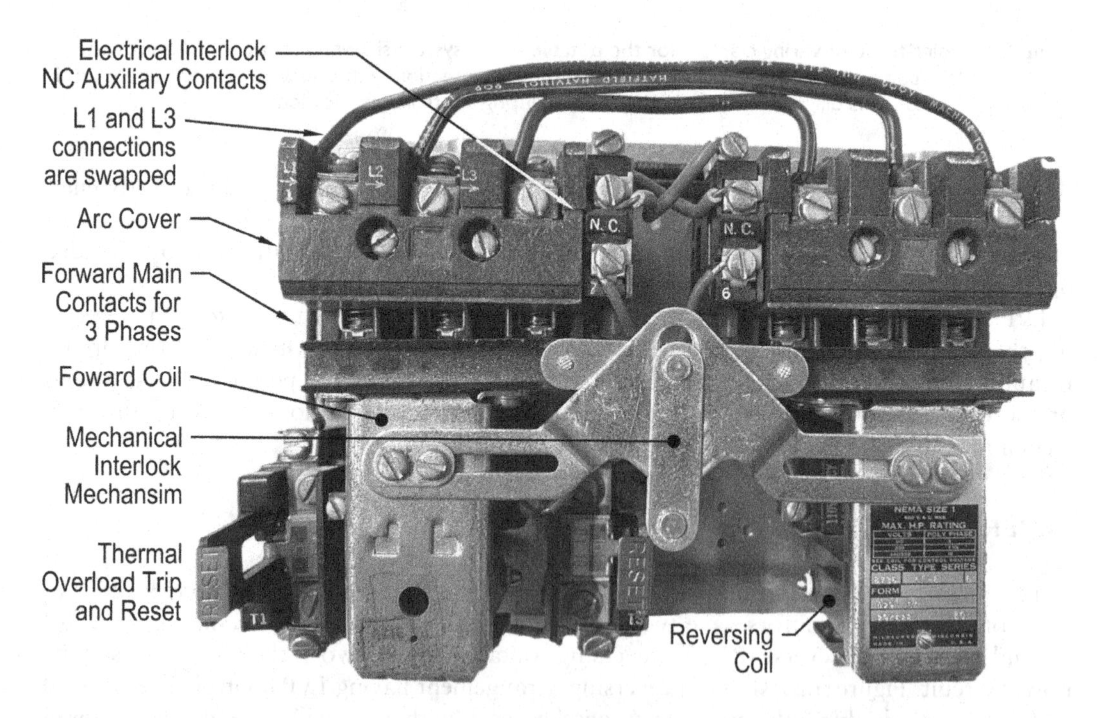

Figure 5.28 Reversing motor starter having two side-by-side main contactors and a mechanical interlock mechanism that prevents the armatures of both contactors from actuating simultaneously. The contactors swap the L1 and L3 phases to reverse the motor rotation and are wired to share the overload-protection device. (Courtesy of United States Merchant Marine Academy.)

shows the incoming L1 and L3 line connections cross over to the motor T3 and T1 connections, while the L2 phase continues straight through to T2. The dashed lines between the F and R contactor coils in the ladder diagram show mechanical interlocks are installed between the "F" and "R" contactors to make it physically impossible to close both contactors simultaneously, as shown in Figure 5.28. As an additional failsafe against simultaneous closure, normally closed "F" and "R" auxiliary contacts are wired in series with the opposite contactor coil. These contacts serve as electrical interlocks to prevent the simultaneous energization of both contactors, if the mechanical interlock was inadvertently removed. In this wiring arrangement, the opposite-direction button can be pushed, and although the original-direction contactor will deenergize, the reversing contactor could energize before the motor stops spinning. If the stop button is not pressed before motor reversal, the duration of in-rush current can double as it must decelerate the machine before reversing it. The extra in-rush current increases heating of the motor windings and may cause damage if it is repeated often. Moreover, extra in-rush current can cause nuisance tripping of the overload device, interrupting service. Good operating practice is to press the stop button and allow the machine to coast before attempting to start it in the other direction, or as an automatic alternative, to include a centrifugal zero-speed switch or a timer relay to delay the motor reversal.

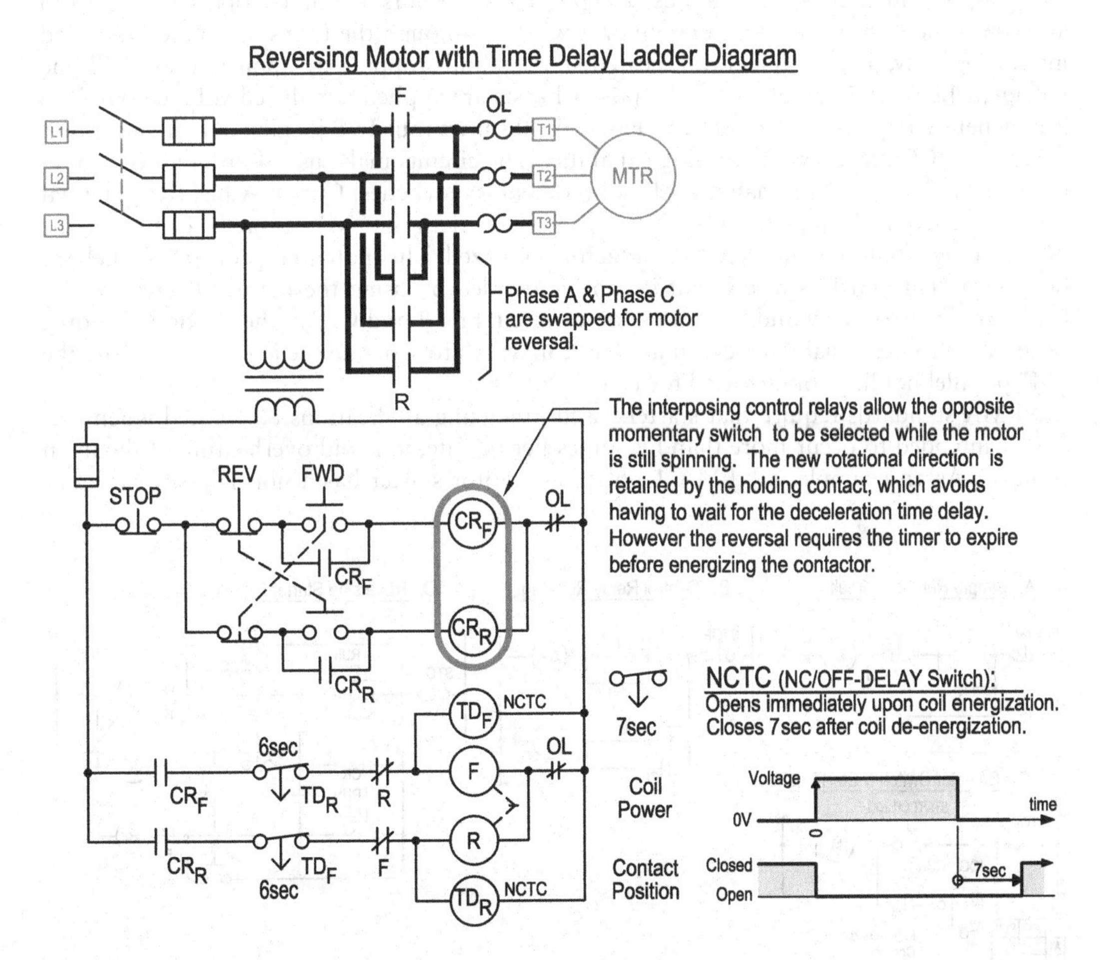

Figure 5.29 Reversing motor controller with deceleration time delay. The interposing control relays allow the reversing switch to be pressed without waiting for the motor to stop, which allows the motor to slow without using phase-swapping plugging action. The time delay reduces the duration of in-rush current.

Figure 5.29 shows a schematic diagram for a reversing controller having a deceleration time delay. The controller uses two normally closed, off-delay timing relays that open immediately when energized and remain open until the off-delay time has elapsed. At that point, the timer energizes its motor contactor. The forward contactor has normally closed auxiliary contacts wired in series with the reverse contactor, and vice versa, serving as electrical interlocks against simultaneous operation of the "F" and "R" coils. The contactors may be mechanically interlocked, as well, for the same purpose, as shown here. The two control relays (CR_F and CR_R) allow the operator to press the reversal-direction momentary switch while the motor is running, and the control-relay holding contacts will "remember" that the reversal command was issued after the momentary switch is released. Without the two control relays, the operator would need to wait until the time delay has expired before the momentary switch would have an effect. The control relay initiates the timer, providing the necessary deceleration time before the motor is energized in the other direction.

JOGGING MOTORS

"Jogging" or "inching" circuits are used to provide short bursts of motor operation, often to advance a mechanism, such as a crane or a winch. Although the terms are sometimes used interchangeably, jogging refers to repeated starting and stopping by application of full line voltage, while inching implies that the power bursts are applied at reduced voltage to reduce heat generated by the in-rush-current. Figure 5.30 shows four LVP jogging circuit variations. Schematic "C" has reversible run-jog capability. The circuits make use of different combinations of SPST and DPDT maintained and momentary switches. Circuit A has the potential problem where if the jog-switch spring returns faster than the "M" contactor can react, the "M" sealing circuit could keep the contactor energized. This potential problem is likely to be intermittent and frustrating, but it can be avoided by using the natural lagging action between a control relay and the contactor, as in circuits B and C. Another option is shown in circuit D to use a maintained-contact selector switch to separate the jog function from the "M" parallel holding contact used for run.

NEMA standards require that starters used in jogging applications be derated when jogging is intended to occur more than five times per minute, to avoid overheating of the main contacts. As an example, a NEMA 1 polyphase-motor starter has a normal-duty rating of

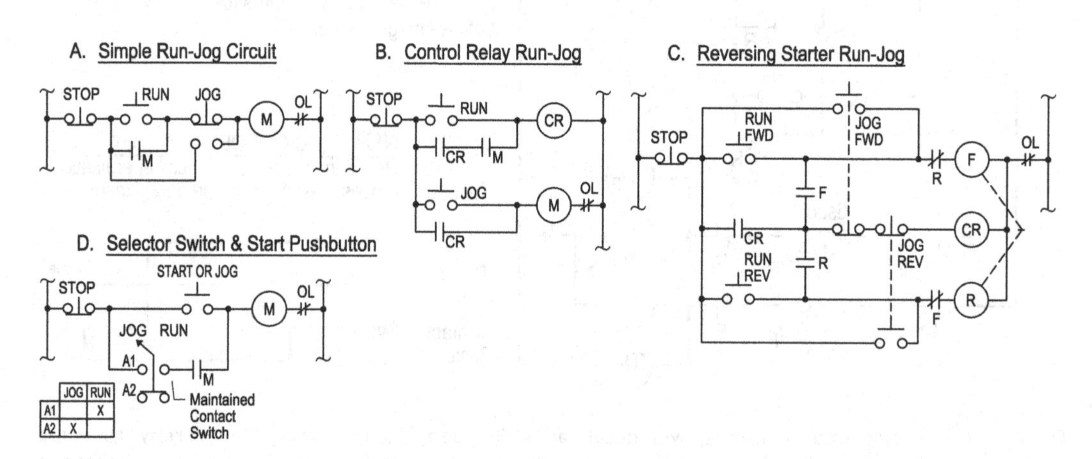

Figure 5.30 Four start-stop methods for jogging control.

10HP at 460V, but the starter would be derated to maximum of 5HP for a jogging application (see Table 5.1).

REDUCED-VOLTAGE STARTING

Starting ac induction motors across-the-line draws in-rush current that is approximately 5-7 times[2] greater than the normal running current for several seconds or longer. Starting large motors causes transient slowdown of generators and significant voltage sags, until the governor and voltage regulator catch up. If the voltage dip is severe, the generator can trip on its under-voltage protection. Further, the plant distribution voltage could dip enough to cause other motor controller contactors to drop out, randomly shutting some equipment. Reduced-voltage starting techniques are used to mitigate very high in-rush amperages and electrical stress to the system.

The reduced-voltage starting techniques include:

- Multi-Speed Motors
- Wye-Delta Starting
- Series Resistance/Series Reactance/Autotransformers
- Electronic Soft-Starting/Variable-Frequency Drives

Several general techniques are used in reduced-voltage starting, including:

- rewiring the motor stator-windings though external contactors,
- dropping the stator voltage to increase motor slip
- changing the source frequency.

Some of the techniques are applications of the relationship $f = \frac{pn}{120}$, where speed (n) in *rpm* can be changed by adjusting the frequency (f) in *Hz* or the number of stator poles (p), where changing the poles changes the speed incrementally. Table 5.4 summarizes various reduced-voltage starting techniques along with a comparison of the voltage, current, and starting torque.

Table 5.4 Comparison of voltage, current, and starting torque values for reduced-voltage starting options relative to across-the-line starting

Staring method	*% of line voltage*	*% of line amps*	*% of DOL starting torque*
Across-the-Line	100	100	100
Series Resistance (Typical)	80	80	64
Autotransformer 80% Tap	80	64	64
65% Tap	65	42	42
50% Tap	50	25	25
Wye-Delta	$V_{line}/\sqrt{3}$	33	33
Part Winding Low Speed	100	50	50
High Speed	100	70	50
Solid-State Soft Start	Variable	Variable	Variable
VFD Soft Start	100	100	Up to 100

Acceleration time is directly related to the load plus motor-rotor inertia and inversely related to the motor developed torque. DOL = Direct-On-Line.

Note that torque is proportional to the voltage squared ($T \propto V^2$).

Multi-speed motors

Although multiple-speed operation is designed to match a motor's speed to its instantaneous loads, especially where the load varies substantially, multi-speed motors also draw less in-rush current when starting at the low speed, and later shifted to high speed. Technically, although low-speed starting of an across-the-line multi-speed motor is not reduced-voltage starting, it provides reduced-current starting, which is the desired effect. Accelerating at low speed overcomes much of the rotor and load inertia at lower amperage, before transitioning to full speed.

Multi-speed motors are common in fuel pumps, ventilation fans, boiler forced-draft blowers, circulator pumps, and any application where there are large variations in load. Examples include a ship propulsion-plant, which has very small loads in port versus very large loads when underway, or HVAC systems that undergo significant daily or seasonal variations based on the weather.

Motor-nameplate wiring diagrams are illustrated in Figure 5.31 A-D. Three diagrams are for two-speed, single-voltage motors, which could be applied to multi-speed motor starting. The fourth diagram is for a star-wye single-speed, dual-voltage motor. Figure 5.32 shows two power and control circuit options for single-voltage, two-speed-motors. Six conductors are required between the low- and high-speed contactors in the motor controller and motor junction box, where the two contactors essentially rewire the motor stator windings within the controller to obtain the multiple speeds.

Wound-rotor induction motor—Multiple speed and reduced-current starting

The wound rotor induction motor substitutes the robust solid-aluminum squirrel-cage rotor bars with insulated-copper windings, but the induction-motor operating principle is the same. The open rotor windings are, however, connected to the stator through slip rings and brushes, where the rotor-winding circuit is externally completed. Current is induced in the wound-rotor windings, just as it is in a squirrel-cage rotor, however, the induced rotor

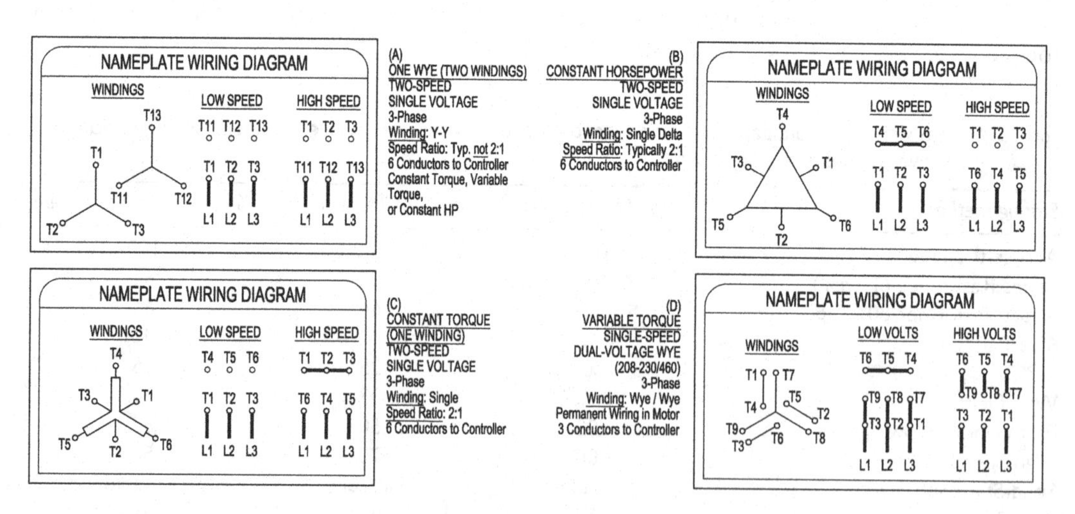

Figure 5.31 Motor nameplate wiring diagrams are mounted on motor housings to ensure correct wiring for multiple speed, multiple voltage, and wye-delta starting. Nameplates A, B, and C show various two-speed wiring diagrams, while nameplate D is for a dual-voltage motor. The wiring diagram on the motor housing should always be used for making field connections.

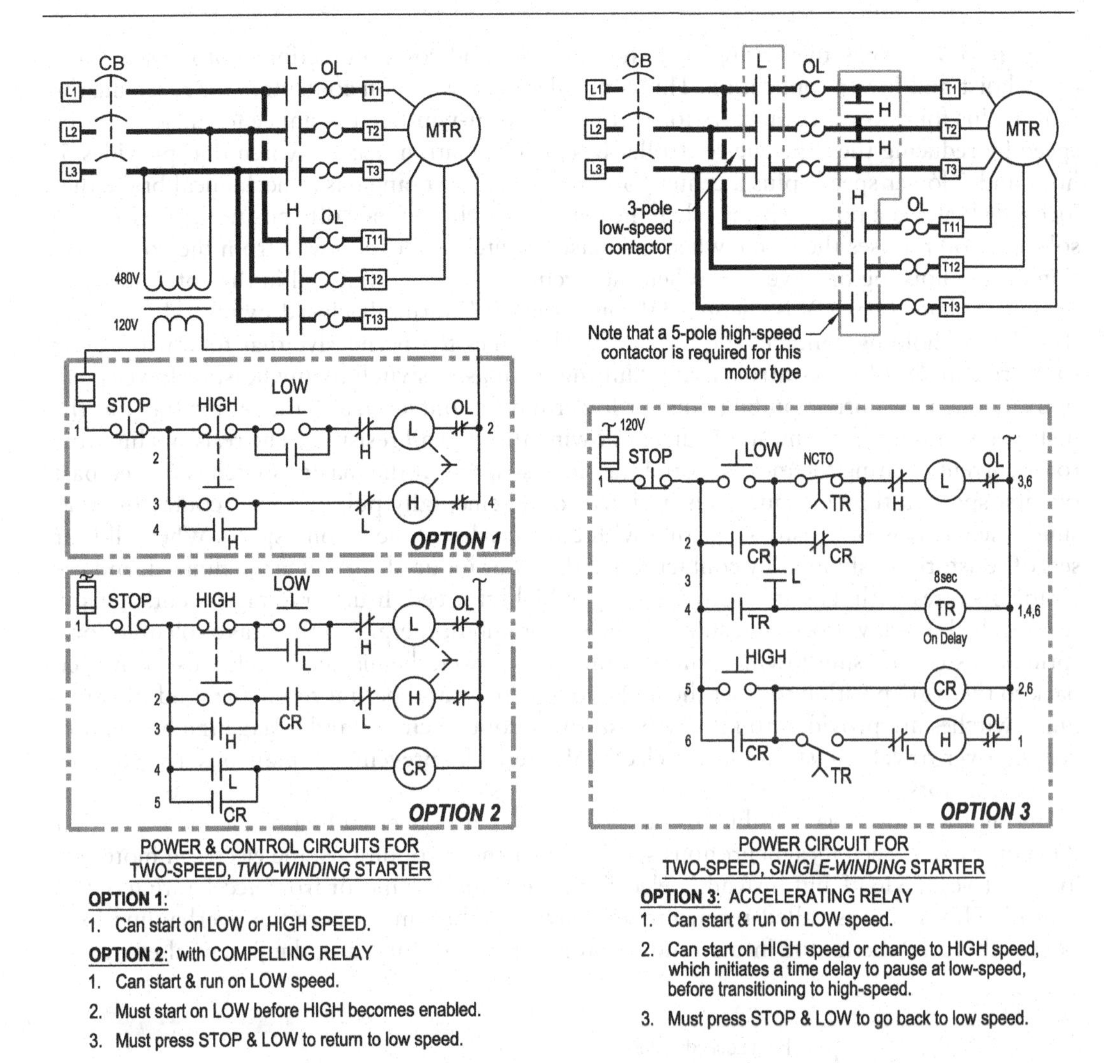

Figure 5.32 Three control-circuit options to obtain two-speed motor operation. Options 1 and 2 can be used with motors having nameplate A in Figure 5.31. The option 1 motor can be started or shifted into high-speed operation at any time, while the option 2 motor requires low-speed starting before transitioning to high speed. Option 3 uses a single-winding motor and incorporates a time delay to force low-speed starting before permitting high-speed operation.

current can be passed through stator-mounted resistors to reduce the rotor's induced current. Less rotor current leads to a weaker rotor magnetic field, more slip, and slower speed. The external resistors not only provide the means for reducing operating speed, but they reduce the in-rush current when started at low speed, while nicely increasing the locked-rotor starting torque when started with the maximum rotor-winding resistance, as shown in Figure 5.33. This motor is suitable for starting high-inertia loads. Upon reaching intermediate speeds, the rotor series resistors can be shunted to increase the speed, and when fully shunted, the wound-rotor induction motor replicates the behavior of the squirrel-cage induction motor. In practice, this method can be used to control speed in a range of 50-100% of its rated speed.

Figure 5.34 shows the wiring diagram for a wound-rotor induction motor used for a crane hoist-and-lower application. This controller uses two main contactors for reversing the motor, plus four power contactors for cutting out rotor-winding resistance for increasing the speed by reducing the slip. The controller has an 11-position master switch that provides 5-hoist and 5-lower speeds plus a center "off" position. The crane has a mechanical brake that locks the load via a power spring when the master switch is in the "off" position. An electrical solenoid coil releases the brake when the master switch is moved away from the "off" position during hoisting or lowering. When the circuit is energized but still in the "off" position, an undervoltage relay UV is energized through the MS1 normally closed switch, which closes the UV NO holding contact. The UV contact becomes the permissive that forwards power to the remainder of the control circuit. Shifting the master switch to any hoist or lower position energizes the corresponding hoist or lower motor contactor, and the motor starts at the first speed with the rotor induced current flowing through all resistors. The resistors limit the rotor current creating the most slip and the slowest speed. If the master switch is moved past the first speed, a timing sequence is initiated to incrementally pull in power contactors that shunt the last row of resistors, starting with contacts 1A at the second speed. When all four sets of resistors are shunted by contactor 4A, the rotor induced current is maximized, and the motor operates with its minimum slip and the highest speed. If undervoltage occurs for any reason, the UV relay drops out, shutting the motor and deenergizing the brake solenoid for a spring-assisted safe shutdown. Upon restoration of power, the master switch must be moved back to the "off" position to reset the undervoltage relay and permit restarting. End-of-travel limit switches are provided to kill the hoist-lower control circuit and engage the brake, preventing over travel. Hoist and lower electrical interlocks prevent simultaneous actuation of both contactors.

During lowering, it is possible for heavy loads to cause overhauling of the motor, so that the motor can exceed its synchronous speed. Under those conditions, the electrical motor no longer drives the load, but instead, it electrically restrains the motor from accelerating out of control. This action is called regenerative or dynamic braking, where the overhauling load essentially turns the motor into an induction generator. In this case, the potential energy of

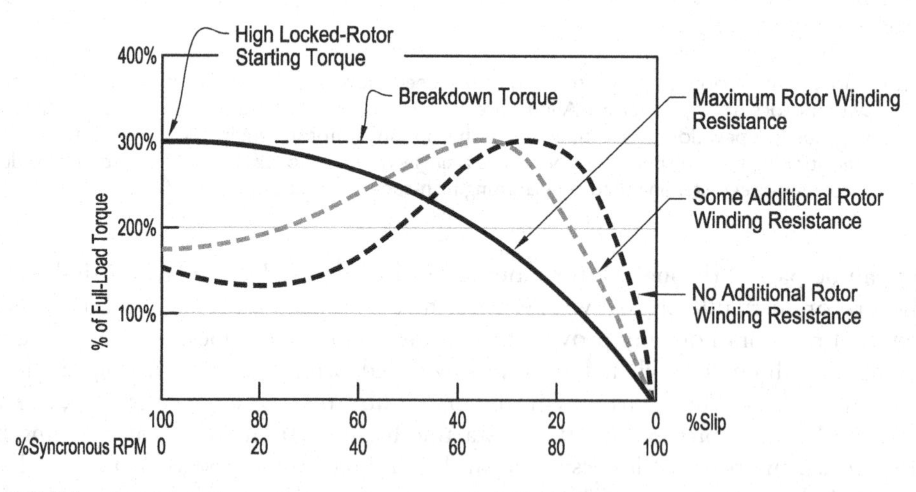

Figure 5.33 Wound-rotor induction motor torque-speed curves for different rotor-winding resistances. Note that the maximum wound-rotor motor starting torque, also called locked-rotor torque, occurs at the lowest speed setting just as the motor is starting and when it is needed most.

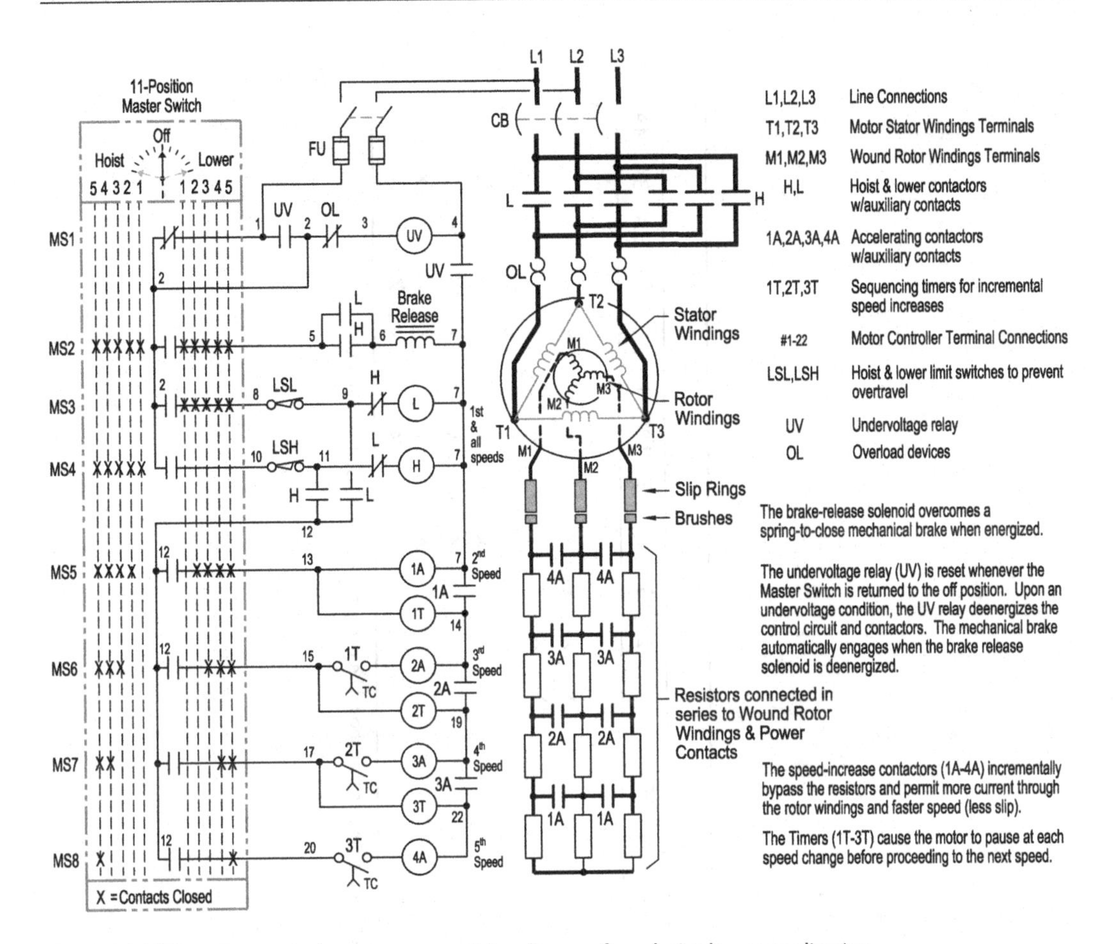

Figure 5.34 Wound-rotor induction motor wiring diagram for a hoist-lower application.

the lifted load is converted into kinetic energy by the dropping load, and the energy is recovered back into the electrical system. Table 5.5 shows a sequence table for the motor controller in Figure 5.34. The sequence table is helpful in troubleshooting, as it shows the energization state of each relay for every master-switch position.

An interesting thing to note about the control circuit is that the control power is taken from the line side of the motor circuit breaker and is entirely protected and isolated by its own two-pole fused-disconnect switch. Powering the control circuit separately permits the controller to be placed in service without the motor, allowing troubleshooting without concern of hoist movement. The contactors can be safely observed for their proper actuation as the master-switch position is moved, as summarized by Table 5.5.

Wye-delta reduced-voltage starting

This works like the multi-speed-motor arrangement, in that the motor stator windings are externally connected by contactors in the controller. However, in wye-delta reduced-voltage starting, the motor stator coils are first connected across-the-line in a wye arrangement, and later, the windings are shifted into a delta configuration after adequate time has been allowed for the motor to accelerate. This starter works on the principle that the wye-configuration

Table 5.5 Sequence table for the circuit shown in Figure 5.33

	Hoist													Lower											
	5	4	3	2	1	UV	H	L	1A	2A	3A	4A	OFF	1	2	3	4	5	UV	H	L	1A	2A	3A	4A
MS1						X							X						X						
MS2	x	x	x	x	x	x	x							x	x	x	x	x	x		x				
MS3						x	x							x	x	x	x	x	x		x				
MS4	x	x	x	x	x	x	x												x		x				
MS5	x	x	x	x		x	x		x						x	x	x	x	x		x	X			
MS6	x	x	x			x	x		x	x						x	x	x	x		x	X	x		
MS7	x	x				x	x		x	x	x						x	x	x		x	X	x	x	
MS8	x					x	x		x	x	x	x						x	x		x	X	x	x	x

Note: This table assists during controller troubleshooting by quickly identifying the devices that should be energized for any master-switch position.

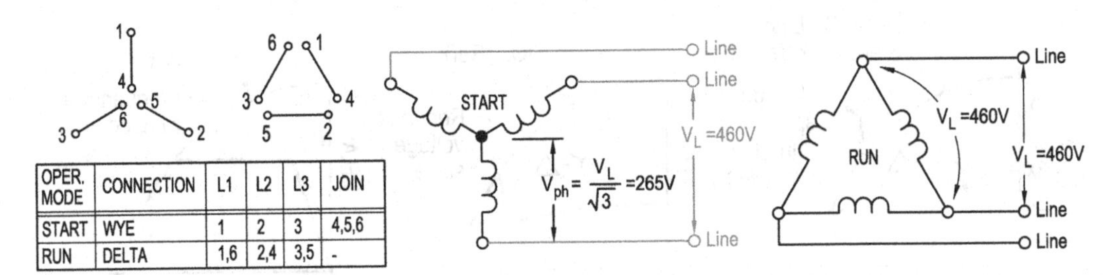

OPER. MODE	CONNECTION	L1	L2	L3	JOIN
START	WYE	1	2	3	4,5,6
RUN	DELTA	1,6	2,4	3,5	-

Figure 5.35 The wiring connections and the resulting voltage levels are shown during the start and run sequences for wye-delta reduced-voltage starting. The motor windings experience 265V during acceleration and 460V during operation.

line-to-neutral voltage (phase voltage) is less than the line-to-line applied voltage by factor of $1/\sqrt{3}$, so that the phase voltage is 57.7% of the applied voltage. The wye reduced-voltage level is not adjustable, and consequently, the starting torque is low. For a 460V motor, the reduced starting voltage is 265V, as shown in Figure 5.35, and the starting torque is $(265V/460V)^2 \times 100 = 33\%$ of the across-the-line torque. It is necessary that the starting torque exceeds the torque required to overcome the system friction and load inertia, or this method will not work. This reduced-voltage starting technique is very effective for fan, centrifugal pump, and other relatively low-inertia applications.

Just as the reduced-voltage starting torque is one-third of the across-the-line torque, the starting current will be about a third of the across-the-line in-rush current. The wye in-rush current works out to be roughly twice the full-speed amperage, based on one-third of a motor's estimated 5-7 times across-the-line in-rush current relative to its running current. As an example, if a motor draws *100 A* in operation, it would draw about *600 A* if it was started direct across-the-line, but only about *200 A* (600A × 1/3) in the wye starting arrangement. In the delta configuration, the full line-to-line voltage is applied directly across each motor phase and the motor operates as intended.

The wye delta transition can occur either with an open transition for smaller motors or a closed transition for larger motors. For the open transition, the contactors forming the wye connection deenergize followed by another contactor quickly making the delta connection. The inductive effect of interrupted current in the motor windings during an open transition introduces transient voltage spikes during the changeover of the contactors. Figure 5.36 shows a comparison between across-the-line starting, which is also called direct-on-line (DOL) and where the in-rush current will be about 5-7 times the normal running current, open-transition wye-delta reduced voltage starting, which creates a voltage spike, and closed-transition wye-delta reduced voltage starting, which avoids the spike. Electronic soft starting is shown for comparison.

A traditional open-transition wye-delta starter has a timer relay, three contactors, where two are full-amperage contactors and one is slightly smaller, and a single shared overload device. Its schematic diagram is shown in Figure 5.37. Closed-transition wye-delta starters use four contactors (Figure 5.38). Like the six-lead two-speed motors that use external contactors to make the motor-winding connections, the wye-delta starter externally connects the six motor-winding terminals to form a single set of either wye or delta windings.

The open-transition controller shown in Figure 5.37 runs when the start switch is pressed, energizing time-delay relay TD1 and beginning the timing sequence. TD1 immediately pulls in its untimed NO "emerging" contact, energizing start contactor 1S, which ties together the wye common terminals. Contactor 1S locks out delta-contactor 2M and immediately energizes the main contactor 1M, which closes the wye across-the-line connections to begin the

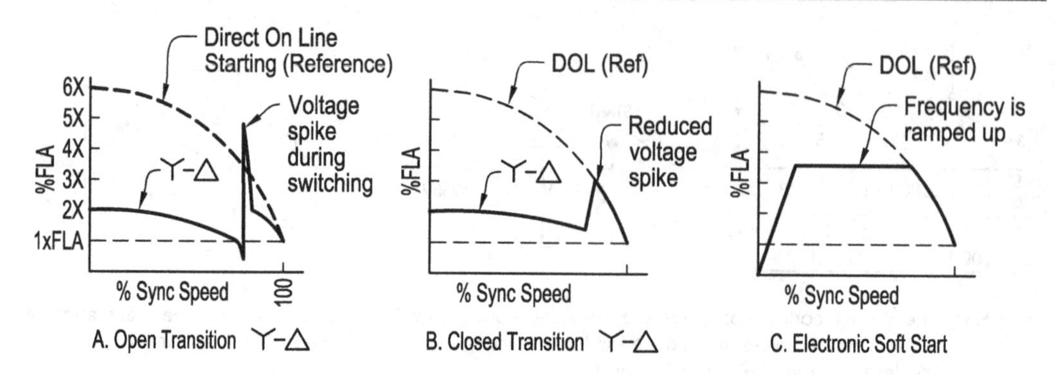

Figure 5.36 Comparison of starting current between across-the line, wye-delta open transition, wye-delta closed transition, and electronic soft start controllers. The dashed line shows the across-the-line starting current vs. speed as a reference.

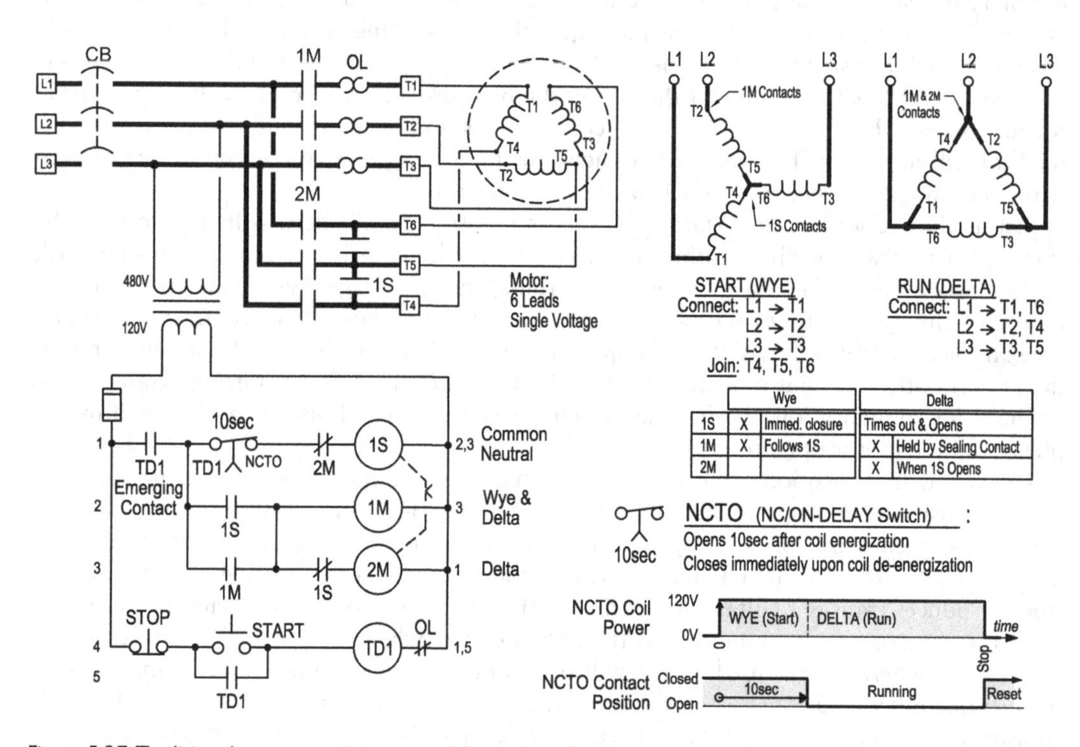

		Wye		Delta
1S	X	Immed. closure	Times out & Opens	
1M	X	Follows 1S	X	Held by Sealing Contact
2M			X	When 1S Opens

Figure 5.37 Traditional open-transition wye-delta reduced voltage starter. Contactor 1S closes immediately, creating the common wye junction, followed shortly by wye contactor 1M. After TD1 times out, 1S deenergizes, allowing contactor 2M to make the full-voltage delta circuit in conjunction with the energized 1M contactor.

reduced voltage starting. Contactor 1M pulls in its 1M sealing contact, and 1M remains continuously energized during both wye and delta operation. Contactor 2M completes the delta connection along with contactor 1M, which remained energized At timeout, the timer TD1 NCTO contact opens, deenergizing contactor 1S and breaking the wye common connection, which, with the time it takes for the 1S contacts to swing its position, allows delta-contactor 2M to energize, putting the motor on line. throughout the process. A mechanical interlock between contactors 1S and 2M prevents shorting if the wye and delta contactors were to

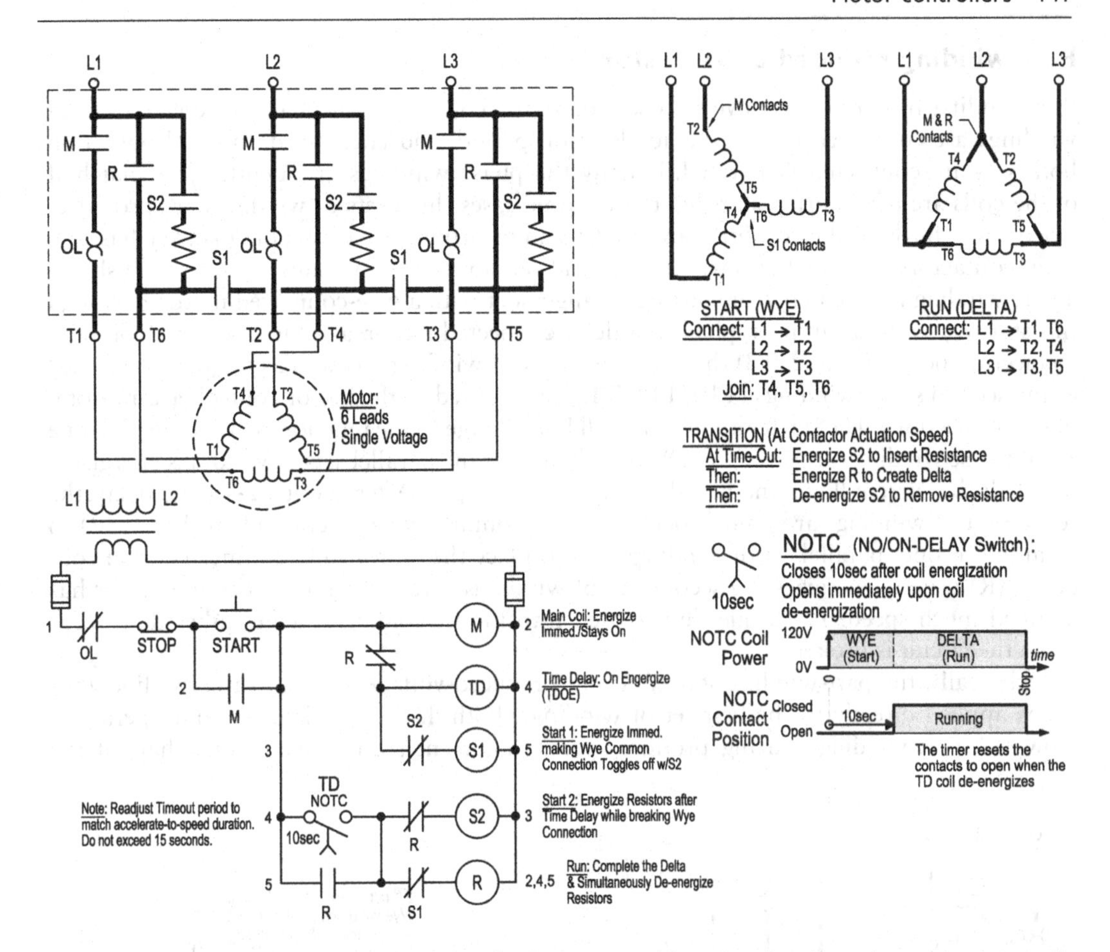

Figure 5.38 Closed transition wye-delta reduced-voltage starter. Note that shunting resistance is temporarily inserted in the wye circuit during the transition so that current continues to flow as the S1 contacts open and just prior to the R contactor closing. The severity of the voltage spikes is mitigated by maintaining current during the wye-to-delta transition.

close simultaneously, and an electrical interlock is added by the 1S and 2M NC auxiliary contacts does the same thing to enhance reliability.

A closed-transition wye-delta starter is shown in Figure 5.38. When the start switch is pressed, the main contactor M is energized, pulling in its holding contact, along with the time-delay relay TD, and the start contactor 1S, which ties the wye common terminals together. The motor starts under reduced voltage. The timer TD contact is of the normally open on-delay type (NOTC), and at timeout its contact closes to energize the transition contactor 2S. Contactor 2S inserts resistors between the delta terminals and creates a current-limited high-resistance short circuit between the wye and delta windings. As soon as the 2S contactor makes, its normally closed auxiliary contact opens and it deenergizes contactor 1S, breaking the wye common connection and removing the short circuit. When the 1S contactor deenergizes, its normally closed auxiliary contact energizes the run contactor R, which closes its own holding contact. The run contactor deenergizes S2, completely isolating the resistors, and the motor is placed direct across the line. The resistors are necessary, as there is an instant during the closed transition when the line connections are shorted, so the resistors limit the amperage from being excessive.

Part-winding reduced current starters

Part-winding motors often have two sets of stator coils for each phase, where one set of windings are connected during the acceleration period, and later during normal operation both sets are connected in parallel. Usually the phase windings are identical, so that half of the coils are used during starting, but in some cases three sets of windings are provided, where two-thirds of the windings are used during starting. A part-winding starter has two main contactors, one for the start winding and one for the run winding. Figure 5.39 shows a part-winding reduced-current starting arrangement with a wye-connected motor, although this starting method can be applied to a delta-connected motor, as well. The three common connection points (T1, T2, T3) that join each pair of windings, as well as the three free ends of the second set of windings (T10, T11, T12), are routed to the motor controller for a total of six connections. The starter applies the full line voltage to the junction, so that only the one set of phase windings are energized. With only half of the parallel sets of windings energized, only half the current flows and half the torque is developed. After about a 2-second delay, the second set of windings are joined together at the common wye junction (T10, T11, T12) to form the second, parallel set of windings and to place the motor fully on-line. The run contactor (R) usually energizes the second set of windings across the line before the motor has attained much speed. The sequence is closed transition, avoiding voltage spikes that occur when the circuit is broken.

Technically, the part-winding starter is not a reduced-voltage starter, as the full line voltage is applied directly to the first set of windings, both during acceleration and operation. However, part-winding starting produces reduced-current starting, since only half of the

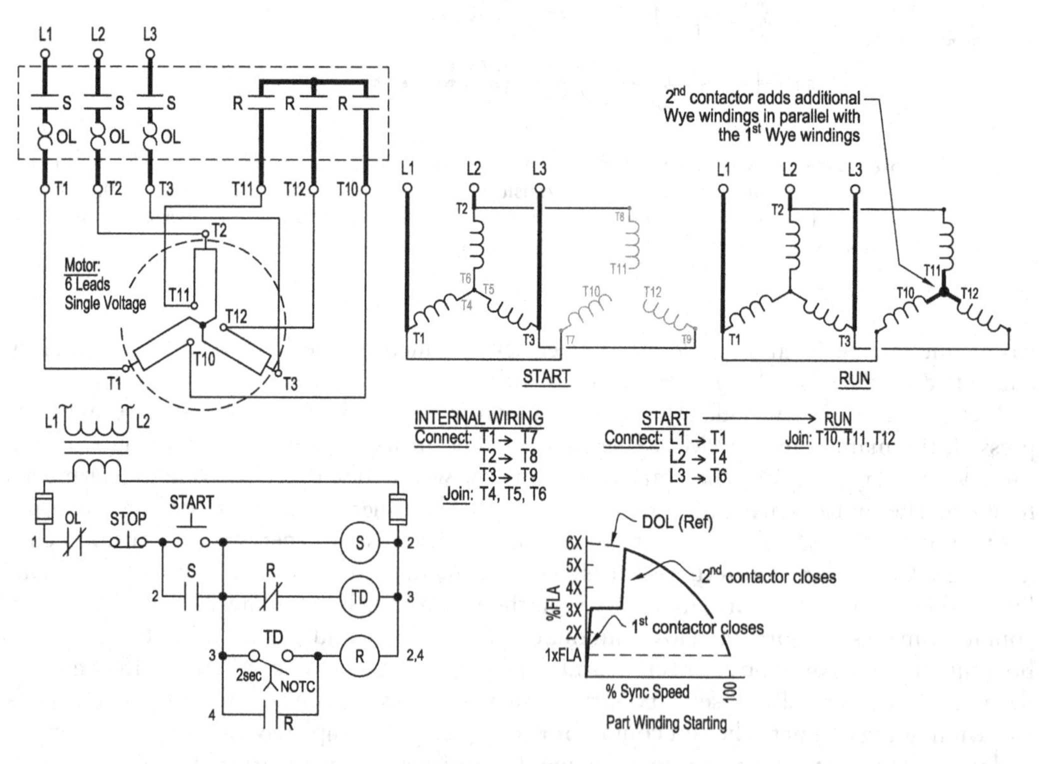

Figure 5.39 Part winding starting. Note that a wye-connected motor is shown, but part-winding motors may also use delta windings. Part-winding starting limits stress to the electrical source by limiting current.

paralleled windings draw current initially. Since one set of windings does not have the thermal capacity to operate alone at the full in-rush current for more than a few seconds, the transition to two-winding parallel operation needs to take place quickly, typically within two or three seconds, as the start windings tend to heat up rapidly. After its transition to two-winding operation, both coils draw the remaining in-rush current, but the first winding has already drawn more than its share prior to the transition while overcoming inertia, causing it to be hotter. The real benefit of this starting technique is that amperage is increased in equal increments, so that "soft" electrical distribution systems have time to respond to a stepped demand, providing a manageable voltage dip. In effect, the generator governor and voltage regulator are given a chance to respond gradually, mitigating excessive voltage sag from a suddenly applied large load. In some cases, the reduction of starting-current is enough to avoid shedding of non-vital loads before attempting a start.

Some wye-connected 230/460V dual-voltage motors are constructed with stator windings that can be isolated, in which case it is possible to use part-winding starting. However, it is prudent to contact the motor manufacturer to determine the motor capability before attempting. Delta-connected dual-voltage motors cannot be used for part-winding starting, as delta-connected motors must be specifically designed to be part-winding capable.

Primary resistor/primary reactor/autotransformer

Reduced-voltage starting may be obtained by placing artificial loads, such as resistors or inductors, in series with the motor, so as to force a voltage drop to the motor before transitioning to across-the-line operation. Figure 5.40 shows simplified arrangements for a series resistor, series reactor, and an autotransformer for reduced-voltage starting. Primary resistors are sometimes used for smaller motors where infrequent starts are expected, as they are cheaper than autotransformers. An added benefit of primary resistor starting is very smooth acceleration, which is very desirable for torque-sensitive applications. Series reactors consist of either air-core or iron-core inductors and are a better choice for slightly larger motors. For large motors, the autotransformer is more economical; taking less space, generating less heat, producing more torque per starting amp, along with having a much shorter duty cycle.

Primary-resistor starting

Figure 5.41 shows an elementary diagram for a primary resistor starter. When energized, timer TR has two instantaneous NO "emerging" contacts one that acts as an LVP holding contact and the other that energizes the start contactor S. Contactor S directs line voltage through the resistors in series with the motor, creating a voltage drop. The motor accelerates under reduced voltage and reduced torque. As the motor speed increases, the in-rush current naturally decreases, and since the resistor amperage follows the same decreasing motor amperage, its voltage also drops ($V=IR$), nicely resulting in a proportionally higher voltage to the motor. The result is very smooth acceleration. As the motor approaches its full speed after enough time delay, the timer energizes run contactor R, shorting the resistors and placing the motor direct on line. Sometimes contactors are added to completely isolate the resistors, in which case a make-before-break circuit must be used to for an uninterrupted smooth transition without voltage spikes. Where loads are torque sensitive, multiple series resistors may be installed in each phase and sequenced from the circuit to further smooth the acceleration.

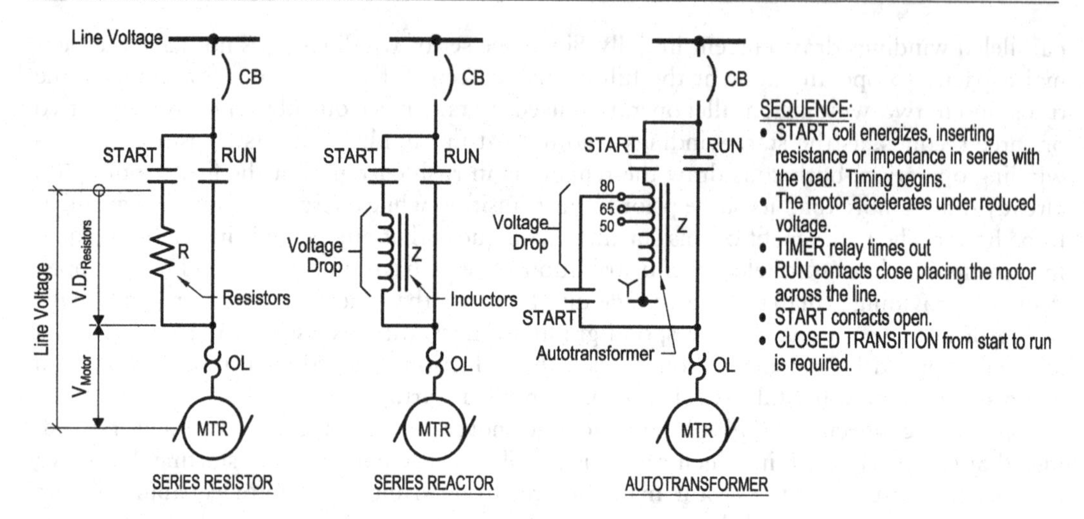

Figure 5.40 Simplified one-line diagram showing reduced-voltage starting using resistors in series, reactors in series, or an autotransformer to drop voltage to the motor during acceleration. The motor accelerates at reduced voltage and is placed direct-on-line after a time delay allows the motor to approach its full speed.

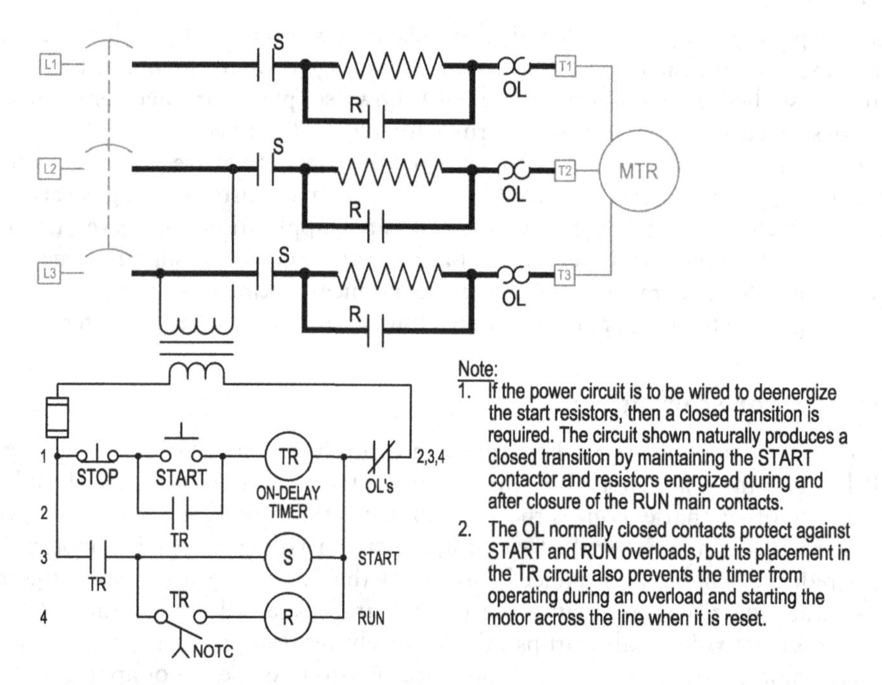

Figure 5.41 Primary-resistor reduced-voltage starter. The resistors drop voltage during acceleration, which reduces the in-rush current. The resistors are later bypassed when the time-delay relay energizes the direct-on-line run contactor, which occurs as the motor approaches its full speed. Substitute inductors instead of the resistors for primary reactance starting.

Primary-reactor starting

Primary-reactor starters are like primary-resistor starters except inductors are used to drop voltage, with one reactor for each phase. The operation is essentially the same as with primary resistors, except the reactors take advantage of inductive reactance to drop voltage, so they tend to be smaller, generate less heat, and are more likely to have taps for fine-tuning the reduced-voltage level to match motor-starting requirements. Primary-reactor starters are used on medium-voltage motors above 1,000V and on low-voltage motors in applications needing low starting torque.

Autotransformers

Autotransformer[3] starters use single-winding stepdown transformers to supply each motor phase with reduced voltage and reduced starting current during acceleration. Conventional two-winding transformers having separate primary and secondary windings could be used, but the relatively close turns ratio, and the significantly smaller size and weight makes a single-winding autotransformer far more practical. The primary and secondary windings of the autotransformer share some windings and are linked both electrically and magnetically. The autotransformer works on the principle that when line voltage is applied to a coil, a fraction of the line voltage appears at an intermediate tap, whose value is based on the turns ratio to the tap. Taps are usually made at 5:4, 3:2, and 2:1 ratios to produce a choice of 80%, 65%, and 50% reduced-voltage levels for motor starting.

Example 1

It is desired to start a full-load 100HP, 460V motor at the reduced-voltage obtained from the 65% tap of an autotransformer having 200 turns, as shown in Figure 5.42. Its full-load amperage is 120A at 460V, and the direct-on-line starting current is 6 times the normal running current. Find the following:

a. DOL in-rush current without reduced voltage: $I_{in\ rush} = 6 \times 120A = 720A$

b. Autotransformer Voltage at the 65% Tap: $V_{sec} = 0.65 \times 460V = 300V$

c. Number of Turns at the 65% Tap: $N_{sec} = 0.65 \times 200T = 130\ Turns$

d. Secondary Amperage at Reduced-Voltage Starting: $I_{sec} = 720A \times \dfrac{300V}{460V}$
Solving: $I_{sec} = \mathbf{470}\,\boldsymbol{Amps}$

e. Motor kVA at Reduced-Voltage Starting: $kVA = \sqrt{3} \cdot 0.300kV \cdot 470A = 245kVA$

f. Source Amperage at Reduced-Voltage Starting: $245kVA = \sqrt{3} \cdot 0.460kV \cdot I_{pr}$
Solving: $I_{pr} = \mathbf{305}\ \boldsymbol{Amps}$

g. Ratio of the Source $I_{Reduced\ Voltage\ Starting}$ to I_{DOL}: $\dfrac{I_{Reduced\ Volt}}{I_{DOL}} = \dfrac{305A}{720A} = 0.42 = 42\%$

h. Current in the Autotransformer "Common Windings": $I_{common} = I_{sec} - I_{pr} = 470A - 305A$
Solving: $I_{com} = \mathbf{165}\ \boldsymbol{Amps}$

During starting, the reduced-voltage at the autotransformer secondary and the corresponding amperage to the motor follows the same ratio as the transformer turns ratio, so

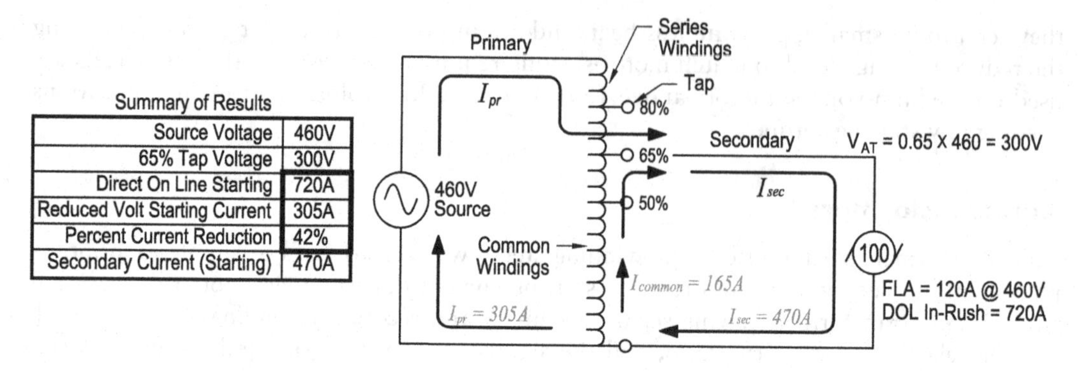

Figure 5.42 Autotransformer example calculations. The autotransformer leverages the motor starting current, where although most current originates from the electrical source, a large amount of current is provided by the autotransformer.

that a motor started from the 65% tap will see 65% of the line voltage and 65% of the in-rush current at the motor. For a100-HP, 460V motor having an across-the-line in-rush current of 720A, the motor would see 300V and 470A from the autotransformer secondary. At the reduced-voltage and reduced-current conditions, the motor apparent power is calculated to be 245kVA, which is the power that needs to be supplied by the plant's 460VAC source. Using this kVA value at the 460-V source, the transformer-primary current is found to be 305A during in-rush, rather than the 720 across-the-line amps or the 470A from the transformer secondary. This calculation demonstrates that the autotransformer-secondary amperage is further reduced by the transformer ratio, so the electrical distribution system and the autotransformer primary experiences only 42% (*0.65 × 0.65=0.42*) of the motor's full across-the-line amperage (*305A/720A=0.42*). Consequently, the autotransformer produces the most starting torque per ampere of source current compared to the other reduced-voltage methods; i.e., the motor starting torque is based on 65% of the across-the-line voltage and current, but the source only sees 42% of the across-the-line amperage.[4] Consequently, autotransformers are significantly smaller and more reliable than primary resistor or primary reactor starting. For NEMA motor-starter size 5 or smaller, open transition controllers are generally used, and closed transition is used for larger starters. The autotransformer starting torque-versus-speed curves are similar to the wye-delta starting shown in Figure 5.36; however, the magnitude of starting torque and in-rush current can be set by the tap connection.

Figure 5.43 shows the wiring diagram for an open-transition autotransformer motor starter. When started, timer TR energizes first, and timing begins. Its two "emerging" contacts immediately close, one as an LVP-holding contact and the other as a permissive to the start-run circuit. The 1S starting contactor is energized through the R normally closed run contact and the TR normally closed NCTO contact, tying the three autotransformer windings into a wye common connection. The 2S contactor pulls in as soon as the 1S normally open contacts toggle, connecting the autotransformer primary across-the-line to complete the wye circuit. The motor then accelerates up to speed at the tap reduced-voltage level. When the timer expires, its NOTC normally closed on-delay contact causes contactor 1S to deenergize, breaking the wye common connection, followed shortly by run contactor R energizing as both the NOTC normally open on-delay contact and the 1S normally closed contact close. The run contactor energizes after the several electrical cycles that it takes for the TR and 1S contacts to toggle, placing the motor direct-on-line in an open transition. The R normally closed run contact opens, deenergizing the 2S contactor, which isolates the

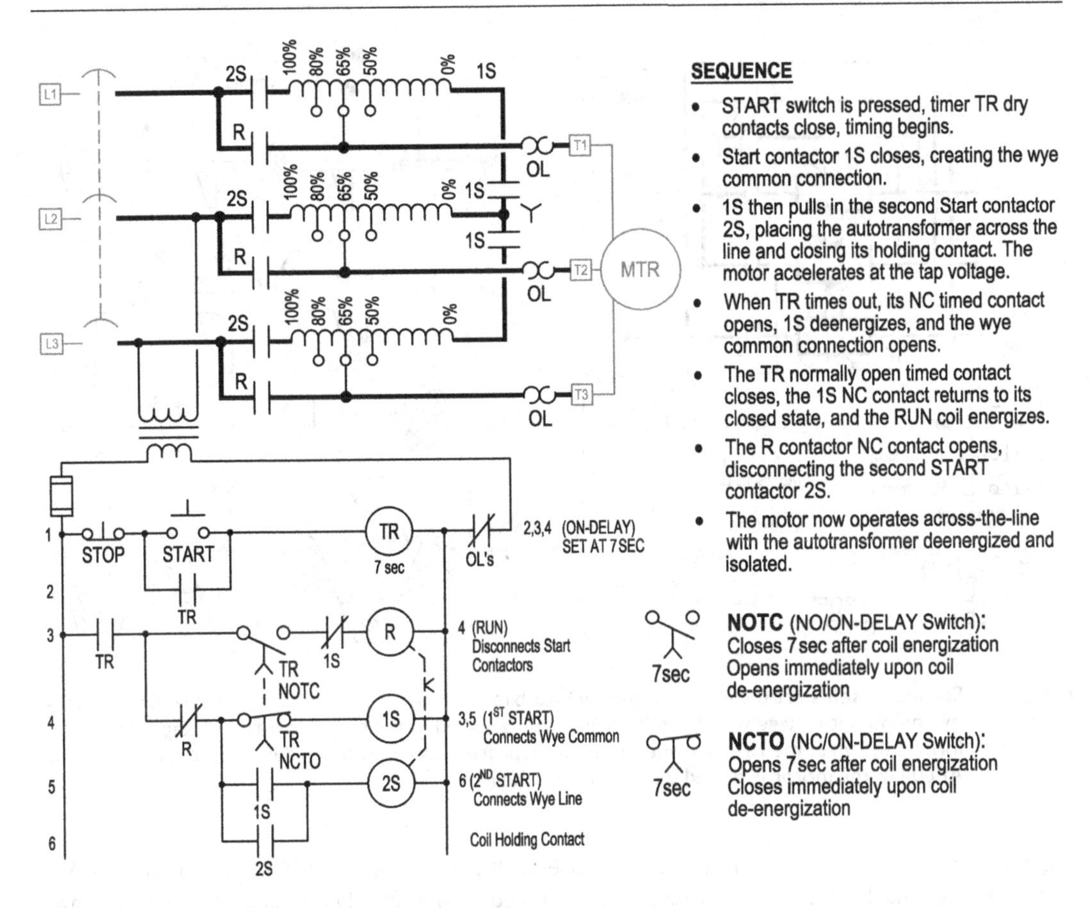

Figure 5.43 Autotransformer reduced-voltage starting.

autotransformer and acts as an interlock to prevent the start circuit from being energized while the motor is running.

Electronic solid-state soft start

Electronic solid-state soft-start is a newer technology compared to other reduced-voltage starting techniques. Electronic soft-start can be obtained using dedicated soft-starters or as an inherent part of variable-frequency drives. Some soft starters work by varying the voltage while maintaining 60Hz, and others work by ramping up the driving frequency until it reaches 60Hz. In both cases, the motor is placed across-the-line after reaching full speed using a bypass contactor. The most common dedicated soft-starters use thyristors, which are also known as silicon-controlled rectifiers or SCRs, and are shown in Figure 5.44. SCRs produce variable-output-voltage while maintaining 60Hz, where the AC sinusoidal current is truncated at the beginning of the waveform as the SCR is "triggered on." The RMS voltage is reduced while maintaining 60 cycle electricity causing a weaker field and more slip. Although this approach is not good for continuous operation at reduced speeds, it is less expensive than other electronic technologies and is only in service during the short acceleration period.

VFDs using MOSFETs (metallic-oxide semiconductor field-effect transistors) or IGBTs (integrated-gate bipolar transistors) have inherent soft-start capability. MOSFETs and IGBTs

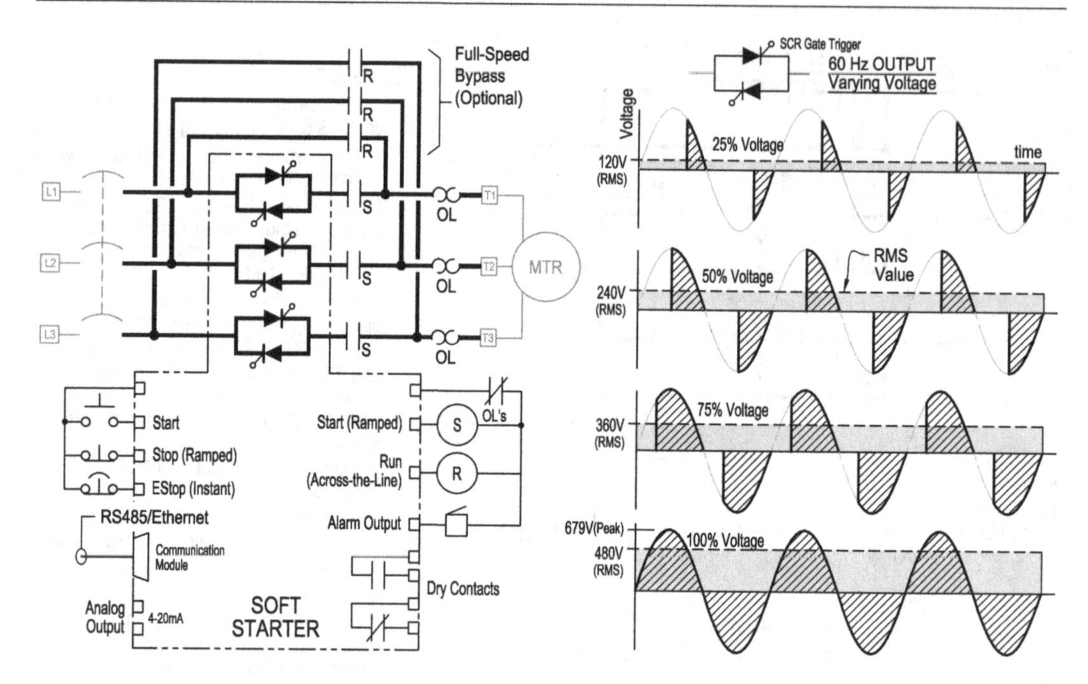

Figure 5.44 Solid-state soft start technology using back-to-back SCRs to create variable voltage 60Hz electricity. The output voltage waveform is an alternating spiked shape having low-rms voltage. The waveform is shown at the right for one of the three phases at various times during motor acceleration. At 60 Hz, the motor is placed directly across the line.

use switching technologies coupled with pulse-width modulation (PWM) to create an AC waveform having both reduced voltage and reduced frequency. These electronic devices are discussed in Chapter 3, "Basic Electronics," and VFDs are discussed in Chapter 6, "Variable-Frequency Drives and Harmonics." MOSFETs and IGBTs with PWM control are used in the latest- technologies for better-quality variable-frequency drives, where they create less harmonics problems compared to earlier SCR-based variable-frequency drives, technologies which are still used with smaller motors.

In addition to soft starting, most electronic starters, including VFDs, have a programmable interface for adjusting the starting parameters. Programmable features include adjustable ramped acceleration and deceleration, speed, voltage, amperage, and frequency status indication, at-speed signal indication, manual or automatic programmable resetting of overload protection, programmable time-trip characteristics for improved coordination with the electrical distribution-system protection devices, ability to temporarily bypass the overload-protection to avoid nuisance tripping, ground-fault protection, shunt-trip actuation, phase-rotation trip, dry contacts output for additional control, SCADA communication links, self-diagnosis and fault reporting for shorted SCRs, loss of a phase, overload, over temperature, short-cycling, lockout, and other features. Soft starters are mostly configured for bypass operation, where the motor power is transferred to an across-the-line electromechanical contactor after full-speed is reached, providing higher reliability and redundancy.

Solid-state starters and VFDs provide the smoothest operation, having the ability to adjust current/torque/speed during starting and coast-down stopping. The SCR generates heat that needs to be rejected while they are conducting, but the heat is eliminated when using bypass contactors, and the harmonics problems only occur during the ramp-up/ramp-down transients.

Variable-frequency drives

Variable-frequency drives are used to adjust motor operating speed. These drives use the principle that the motor speed (n) is directly related to the line frequency (f) by the following relationship: $f = \frac{pn}{120}$ when the number of motor poles (p) is constant, as would be the case with permanently connected motor windings. As an example, if a four-pole motor rotates at 1750RPM at 60 Hz, it would run at 50Hz/60Hz = 1460RPM when operated at 50Hz.

Variable-frequency drives use solid-state electronic technology to convert 60-cycle electricity to other frequencies. In very basic terms, VFDs use three main sections, an ac-to-dc converter, a dc-energy-storage bus, and a dc-to-ac inverter. Fundamentally, these drives apply ac electricity through a solid-state rectifier to produce dc electricity. Later, the dc electricity is pulsed through an inverter to approximate sinusoidal ac electricity at a desired frequency. Typically, filtering is installed to improve the quality of the output sinusoidal shape and to mitigate harmonic disturbances caused by the switching technology. The inverter is electronically controlled to create the desired frequency. In addition to modulating frequency, some VFD drives also adjust the voltage. Figure 5.45 shows a VFD functional block diagram and its power connection to a motor circuit. The VFD soft-start acceleration and adjustable operating speed is produced through a built-in microprocessor-based programmable-logic controller having its own digital and analog inputs and outputs.

The types of drives available are

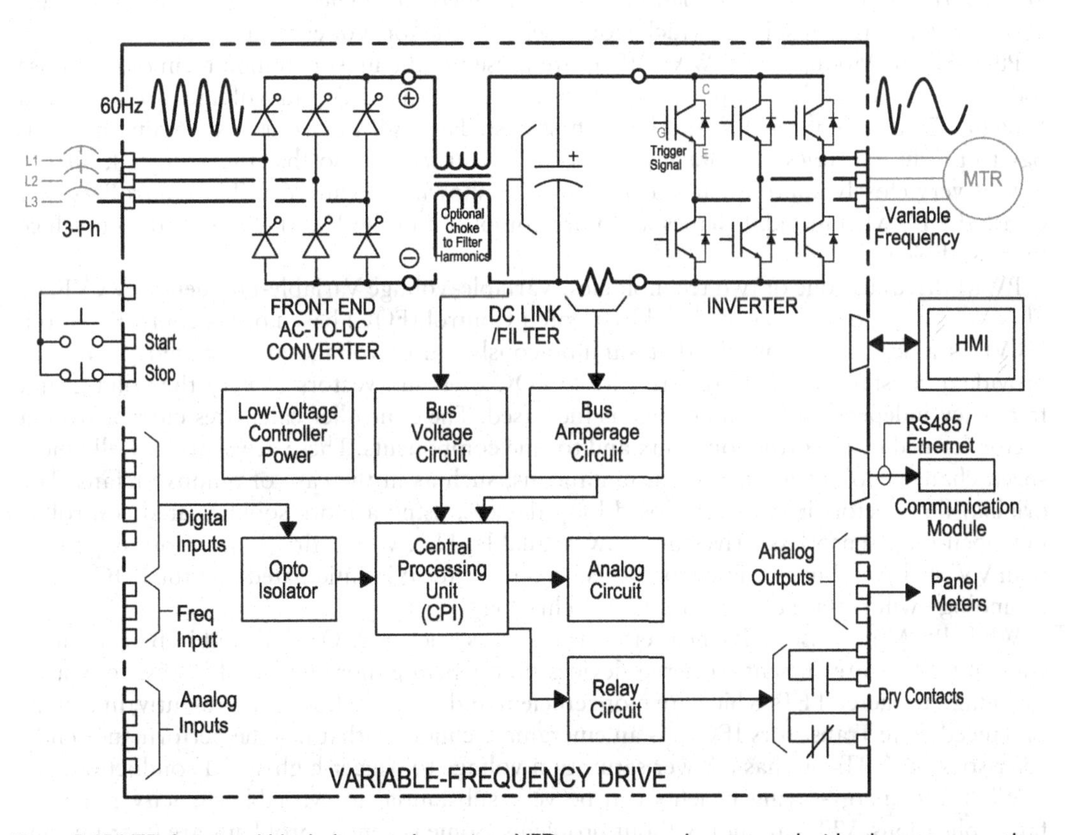

Figure 5.45 VFD functional block diagram. As options, VFD starters can be procured with a bypass across-the-line contactor for redundancy, and/or braking modules to accommodate overhauling loads for non-regenerative drives.

- Variable-Voltage Inversion (VVI)
- Current-Source Inversion (CSI)
- Pulse-Width Modulation (PWM)

Variable-Voltage Input (VVI) is the oldest of the acceptable solid-state technologies for speed control. It is also called a six-step design (Figure 6.4), as it used stepped pulses to approximate a sine wave, with filtering to smooth the edges. VVI technology provides good speed range and is a relatively simple technology, but at the same time it has poor power factor,[5] poor "ride-through" characteristics during power-source disturbances, significant harmonics production, "motor cogging" or torque pulses at low speeds, they are non-regenerative if driven by overhauling loads, and may require an isolation transformer to mitigate harmonics from creating distortion of the sinusoidal distribution system. Although these legacy VVI devices exist and are still used for soft-start controllers, the VVI technology has been superseded by better technologies for VFD applications.

Current-Source Inversion (CSI) is the next generation of VFD. This technology provides an output voltage level that is closer to the motor rating than the VVI drives. Typically, the drives were provided as a matched set with the motor. It is a high-efficiency device and is the only drive technology capable of regenerating overhauling energy back to the electrical supply. The technology provides a low harmonics amperage waveform but requires large inductors. However, the CSI technology exhibits similar disadvantages as the VVI drives. The biggest difference between the VSI and CSI drives is the energy-storage section between the rectifier and inverter. VSI drives use inductors to store magnetic-field energy in the form of voltage, and CSI drives use capacitors, which store energy in the form of current.

Pulse-Width Modulation (PWM) VFDs are presently the most common technology in use today. The have excellent input power factor due to a fixed dc bus voltage, no low-speed "cogging," higher efficiencies, and lower first cost. The PWM drive uses switching technology to produce a series of voltage pulses of different lengths, so that the averaged, filtered output very closely approximates a sine wave as shown in Figure 6.8. The controllers that create the PWM sinusoidal shapes are more complex and the PWM VFDs tend to produce more audible noise.

PWM drives use one of two technologies: Variable-Voltage Variable-Frequency (VVVF) or Flux Vector Control, also called Field-Oriented Control (FOC) based on its control method. VVVF is a scalar relationship that simultaneously varies voltage and frequency so as to provide a constant Volts/Hertz ratio, while FOC uses flux vectors to vary the voltage and frequency independently as amperage is increased. The controller calculates current from a vector formed by the orthogonal flux and torque components. These drives work well where speed changes do not need to be instantaneous, such as in the case of pumps or fans. The original flux vector drives were closed-loop devices, using a more sophisticated controller, but open-loop sensorless drives are now available. Flux-vector drives are more expensive than VVVF drives, but as microprocessors become faster, it is anticipated that the FOC vector technology will replace the scalar VVVF technology.

While PWM may use SCRs, field-effect transistors, such as MOSFETs and IGBTs are more common. SCRs are current triggered devices that generate more heat and losses, compared to voltage-triggered FETs which are more efficient and generate less heat. Presently, Injection-Enhanced Gate Transistors IEGTs is an emerging technology that has the performance characteristics of IGBTs but has a lower saturation voltage to trigger high-speed conduction.

While the energy-saving benefits can be very substantial for variable-capacity pump or fan applications, VFDs are not without problems. Some potential problems are distribution-system harmonics, motor heating, premature bearing failures, and insulation degradation.

In balanced sinusoidal systems, zero sequencing is defined as when the current returning from all three phases into the neutral of a wye-connected transformer all sum to zero amps. Harmonics are created by non-linear loads, i.e., loads that generate non-purely sinusoidal currents. Harmonic currents produce harmonic voltages, following Ohm's Law. Examples of non-linear loads are ballasted fluorescent lighting, diode bridge rectifiers, switching power supplies, VFDs, etc. Harmonics are currents and corresponding voltages in a distribution system that are integer multiples of the first or fundamental frequency, which is 60Hz in the United States, such that the second harmonic is 120Hz, the third harmonic is 180Hz, and so on. Switching power supplies are notorious for creating high harmonic levels at odd multiples of the third, e.g., 3rd, 9th, 15th, 21st, etc. When the harmonics are injected back into the distribution system, it can create problems with other electronic devices, transformers, and motors. In the case of dual-voltage grounded-neutral wye-connected transformers, harmonics can produce additive, rather than subtractive amperages that overload and overheat the neutral connections.

In the case of motors, VFDs can cause stator-windings to overheat at low speeds as the rotor cooling blades may not create enough cooling-air flow. Bearing problems can result from residual shaft voltages caused by zero-sequence harmonics that break through the lubrication film, causing pitting or fluting damage to the contact surfaces of rolling contact bearings. Voltage spikes from the harmonics may eventually break down motor insulation causing partial short circuits or ground faults. These voltage spikes can be exacerbated by long cable lengths that reflect the spikes.

Techniques to mitigate VFD problems include, tuned filtering to reduce the harmonics, isolation transformers to avoid backfeeding harmonics into the distribution system, shaft grounding systems to prevent voltage buildup between the rotor bearing inner race and the housing bearing outer race, and the avoidance of low motor speeds that do not produce adequate cooling-air flow. Where possible, drives should be located near motors. K-type transformers with upsized neutrals may be used. Some of motor-related problems are mitigated to acceptable levels by using inverter-duty-rated motors, that cool better at slower speeds and use insulation that withstands higher voltage spikes without failing. Variable-frequency drives are covered in more detail in

QUESTIONS

1. Sketch the following symbols: momentary normally open and normally closed switches, maintained-contact normally open and normally closed switches, an HOA switch, a 3P2T selector switch along with its switch-position table.
2. Sketch a conventional electrical circuit for switching from two locations using two SPDT maintained-contact switches.
3. List the components commonly found in a motor-starter panel.
4. List and describe the components that form a motor contactor.
5. Describe jogging and plugging. What is the NEMA size contactor required for a 460-VAC 100-hp motor used for standard duty? What is the NEMA size contactor required for a 460-VAC 100-hp motor used for jogging duty? Why is there a difference?
6. What is the purpose of the overload device (OL) used with a motor contactor? How does the OL device function?
7. What are control relays? How are control relays used in motor controllers?
8. What is a solid-state relay or contactor?
9. What are timing relays? What are the four types of timing relays? Sketch the four types of timing contacts. What does the arrow indicate on the timer symbol?

10. What is a latching relay? What is its advantages, and list two applications where they might be used?
11. What is an alternating relay? Where are alternating relays used?
12. Describe low-voltage-protection (LVP) and low-voltage release (LVR) circuits. Sketch the control-switch arrangement to produce LVP and LVR. Provide examples where each would be used and explain why.
13. What is the difference between a motor-controller point-to-point wiring diagram and a ladder diagram? Where would each be used?
14. How is three-phase motor reversal achieved in a motor controller?
15. What is reduced-voltage starting? When is reduced-voltage starting needed? List five methods of obtaining reduced-voltage/reduced-current starting.
16. Describe how variable-speed is obtained from a wound-rotor induction motor. Describe how variable speed is obtained from an electronic variable-frequency drive.
17. Using the *Example 1* parameters for a 100 HP motor, calculate the reduced-voltage starting current drawn from the source when using the 50% and the 80% taps. Assuming a four-pole motor having 50 RPM slip and a starting torque equal to 180% of the full-load torque, calculate starting torque for the DOL and the 80%, 65%, and 50% reduced starting voltages.

NOTES

1. Jogging is quick, intermittent operation, such as might be required when taking up cable with a winch. Jogging results in many pulses of in-rush current. Plugging is the reversal of the phase sequence while the motor is spinning to create a dynamic break followed by motor-direction reversal.
2. Some dc motors can draw up to 200 times the normal running current during across-the-line starting, and therefore require reduced-voltage starting in a stepped-speed start sequencing and having specially designed blowout contactors that draw the inductive sparking into an arc-quenching chute.
3. An autotransformer is a single-winding transformer, where line voltage is applied to all turns and reduced voltage is obtained by tapping between a portion of the turns. The reduced-voltage level is directly proportional to the number of turns in service. For three-phase systems, three autotransformers are formed onto a single iron-core frame and wired into a wye arrangement.
4. Transformer power is $V_p I_p = V_s I_s$. As the secondary voltage goes down, the secondary current goes up. It is the current through motor windings that creates the magnetic-field strength and rotor torque.
5. Poor power factor causes additional amperage to be carried over the distribution system to power the reactive portion of loads. Although the reactive power technically is not consumed, the extra amperage on the line limits the amount of real power that can be transmitted and does contribute to I^2R losses.

Chapter 6

Variable-frequency drives and harmonics

Variable-frequency drives (VFDs) are electronically controlled devices for modulating motor speeds. VFDs reduce unnecessary electrical energy consumption by better matching motor speeds to the application's instantaneous need. As a "non-linear load," VFDs tend to draw current intermittently from an ac distribution system, in a non-sinusoidal manner. Non-sinusoidal current induces voltage distortions in the source impedances, where the distortions are imposed into the distribution system and transmitted to other loads. Most distortions to the ac sine wave are induced by harmonics produced by VFDs, where harmonics are multiples of the 60Hz fundamental frequency, and the cumulative effects from all harmonic sources can create potentially severe problems with other attached equipment. Options exist for avoiding or mitigating harmonics, and for keeping the electrical system within acceptable limits and trouble free. This chapter covers VFDs and harmonics, and is organized as follows:

- Variable-Frequency Drives
- VFD Design
- Harmonics
- Causes of Harmonics
- Classification of Harmonic Types
- Harmonics Mitigation Techniques
- Harmonics and Power Quality Analyzers

VARIABLE-FREQUENCY DRIVES (VFDS)

A variable-frequency drive is an electronic device used to soft-start a motor, smoothly accelerate it, and to finely set the operating speed, with two models shown in Figure 6.1. In addition to mitigating large in-rush motor currents during starting, VFDs can produce significant energy savings, by drawing only the power needed to accomplish the task at hand when loads such as pumps or fans can meet their functional requirements at lower speeds.

Historically, variable speed was obtained using dc motors or through variable-speed mechanical transmission systems. Later, motor-generator sets created variable speed using an ac motor to drive a dc generator, which in turn powered a variable-speed dc motor. For alternating current, multiple-speed motors used contactors to rewire the number of stator poles, but true variable speed from ac motors did not exist until the wound-rotor induction motor incorporated rheostats through slip rings to vary the rotor-induced current. Decreasing the wound-rotor current caused increased motor slip and lower speed. Nowadays, variable-frequency drives are a fully developed technology that finely varies motor speed by changing its delivered frequency, and often the voltage as well.

Figure 6.1 Three-phase low-voltage variable-frequency drives with the covers removed. The microprocessor controllers and keypad displays are shown, with the power electronics located behind the circuit boards. The VFDs are fitted with digital and analog input and output connections for self-control or remote command from a system PLCs or SCADA systems. (Courtesy of United States Merchant Marine Academy.)

Neglecting rotor slip, ac-motor synchronous speed is related to frequency by the relationship $n = \frac{120f}{p}$. Variable-frequency drives modulate the motor speed by changing the VFD-output frequency. Typically, VFDs are used to slow motors to exactly match the equipment speed requirements, but in fact, some VFDs can also increase motor speeds above synchronous by producing output above 60Hz. VFDs not only control speed, but often they also control voltage, and consequently torque. Speed is dropped by decreasing the frequency, but torque is reduced by dropping voltage, an action that reduces electrical stress on the motor conductors. Electrically, motor torque is proportional to the voltage squared ($T_{motor} \propto V^2$), but mechanically, power is directly related to the motor torque and speed

$$\left(P = T\omega \;\; or \;\; HP = \frac{Tn}{63{,}025} \right)$$

At lower operating speeds, less power and torque are required by the attached load; so the voltage is reduced commensurately for many VFDs. Most VFDs having variable-voltage output maintain the relationship where the voltage divided by frequency is held constant. For 460V motors at 60Hz, the voltage reduction with respect to speed is $\frac{V}{f} = \frac{460V}{60Hz} = \frac{7.67V}{Hz}$. As an example, if the 60Hz line frequency is dropped to 50Hz, the original 460V would be dropped to 383V. Reducing the voltage along with the frequency is important for limiting the current drawn by the motor at lower frequencies.[1] Conversely, some VFDs are programmed to intentionally increase the voltage-to-frequency ratio during acceleration, boosting the starting torque. When operating above 60Hz, most VFDs maintain the motor at its rated voltage so as to stay within its insulation limits. When operating above 60Hz, it is important to consider whether the fall off in motor torque occurring from the lower volts-to-frequency ratio may shorten motor life.

Early generation VFDs were installed on conventional induction motors, which led to unforeseen motor problems. The problems were primarily caused by high-voltage spikes during inverter switching, which created electrical potential differences between the rotor and housing. These voltage spikes often discharged from the rotor to the stator housing through the bearings and lubricant, breaking down the grease and causing pitting and fluting of the bearing running surfaces. This pitting process is similar to electrical discharge machining. Nowadays, "inverter-duty" motors are specified for better performance and reliability. The inverter-duty motors use stator insulation materials designed to withstand the steep voltage spikes that occur during switching, and vacuum-impregnation of the windings using high-dielectric-strength varnish increases winding robustness. Also, the rotor squirrel-cage bars are shaped to have more surface area, enabling them to run smoother, cooler, and with less slip. Since voltage buildup between the rotor and housing along with the resulting voltage spikes degrades insulation and shortens bearing life, the electrical potential difference between the rotor and motor frame can be avoided using a grounding ring and brush between the rotor and stator. Conductive greases also help to mitigate the damaging voltage build up between the rotor and stator.

Induction motors are constructed with integral fan blades installed both inside and outside the stator housing to provide forced-convection cooling of the windings. When VFD-driven motors operate at slow speeds, there is less forced convection and motors tend to run hotter, reducing the insulation life. Therefore, manufacturers specify minimum permissible speeds, which for many applications, is roughly one-quarter to one-third of the rated speed. In addition, the operating characteristics of the driven load should be considered in determining the minimum speed. For example, a high-turndown boiler-burner forced-draft fan will reach a low-enough speed where the air flow becomes "lazy," hard-to-control, and unsatisfactory. In that case, the fan minimum speed should be electronically limited and low combustion-air flow obtained using dampers.

High-speed switching in the VFD PWM inverters also tends to create high-voltage spikes that are sent down the line to the motor. If the motor is located far from the VFD, the waves can be reflected back through the cabling. In some instances, the reflected waves can approach double the voltage, leading to premature failure of insulation. Reflected waves can be mitigated using line reactors installed close to the motor in its inverted variable-frequency ac supply. The line reactor is usually a series-connected inductor whose iron core does not go into saturation, or it can use an air core. Line reactors are a commonly used method to limit extreme short-circuit fault currents, but in the case of avoiding reflected waves, line reactors mitigate the voltage spikes. As an alternative, a 1:1 isolation transformer is sometimes used just before the VFD to achieve the same result.

VFD DESIGN

A variable frequency drive consists of three fundamental sections; the rectifier assembly, which is also called the "front end," the dc link, and the inverter. Variable-frequency drives work on the general principle that ac electricity is converted into dc using a front-end rectifier assembly, and later the dc is converted back into ac using an SCR-bridge, or a switching-type inverter after it passes through a dc link. The dc link is designed to smooth and "stiffen" the voltage. Schematic representations are illustrated in Figure 6.2 and Figure 6.3. The inverted ac is produced electronically by one of several technologies at a frequency dictated by the VFD microprocessor-based controller. During the dc-to-ac conversion, the inverter electronically approximates a sinusoidal shape at the desired frequency, often while simultaneously adjusting the RMS voltage output.

The VFD controller is essentially a specialized microprocessor-based programmable-logic controller that contains a user interface for starting, stopping, and setting motor speed and voltage/torque. An alphanumeric display, keypad, and indicator lights provide local status indication, and it is typically possible to scroll through the display to show operating parameters, such as rpm, frequency, voltage, current, power factor, and other useful information. The controller typically has signal connections for analog and digital inputs and outputs, and communication ports that can link to a distributed-control system. Speed-control setpoints can be set manually at the VFD panel, automatically from attached feedback instrumentation, or remotely from a distributed-control system.

Older technology inverters, which are still available, use stepped output from an incrementally triggered SCR bridge in a three-phase, six-pulse arrangement, as shown in Figure 6.4. The stepped arrangement is intended to grossly approximate a sine wave after the load's inductive reactance helps to smooth the steps. This technology is still used for small motors and for inexpensive "soft-starters" as a reduced-voltage motor-starting technique. For

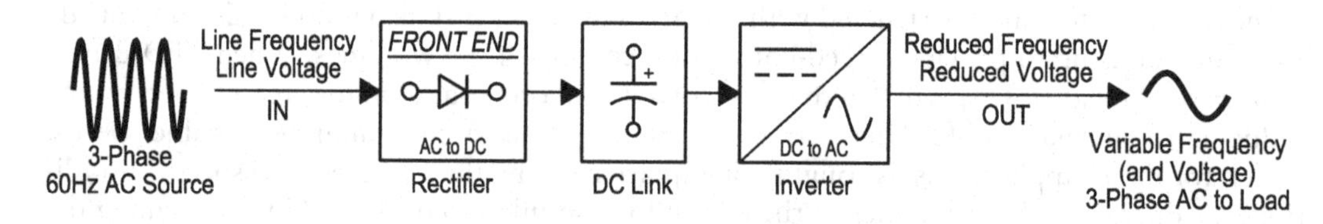

Figure 6.2 Block diagram showing the general operating principle of a variable-frequency drive (VFD). Electricity passes through the ac-to-dc rectifier assembly, is stiffened in the dc link, and is inverted back into ac at a lower frequency. Often the output voltage is dropped to better match the reduced torque requirement at the lower speed.

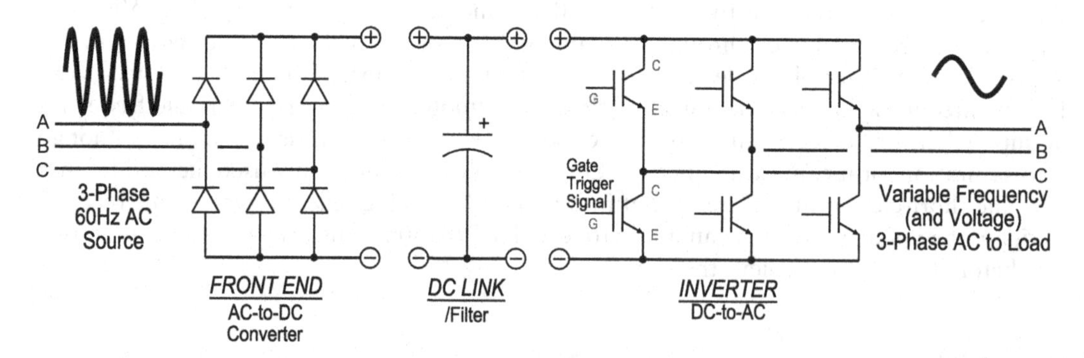

Figure 6.3 A diode rectifier is shown for the front-end dc converter, although SCRs or IGBTs are also used for the front end. IGBTs are shown in the inverter section, which would use pulse-width-modulation; although MOSFETs or SCRs are sometimes used in low-voltage applications.

soft-starters, the pulses occur only during the acceleration transient, after which, the motor is subsequently placed direct-on-line using a bypass contactor. Newer VFD technologies are tending toward pulse-width modulation to better approximate the sine wave, which is described later.

Several inverter technologies exist to create variable speeds. Figure 6.5 shows ac motor-drive technologies compared to system voltage and motor power requirements. Variable-speed technologies include:

- Voltage-Source Inverters (VSI),
- Phase-Controlled Current-Source Inverters (CSI),
- Load-Commuted Inverters (LCI),
- Cycloconverters (CCV), and
- Pulse-Width Modulation (PWM).

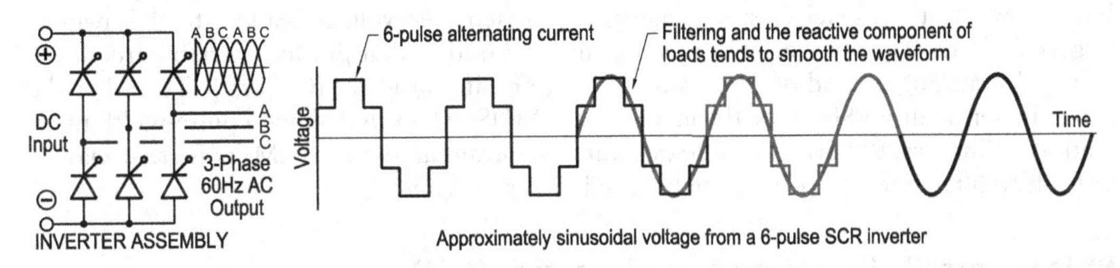

Figure 6.4 Six-pulse inverter using older six-step SCR technology. This technology is still used for electronic soft start.

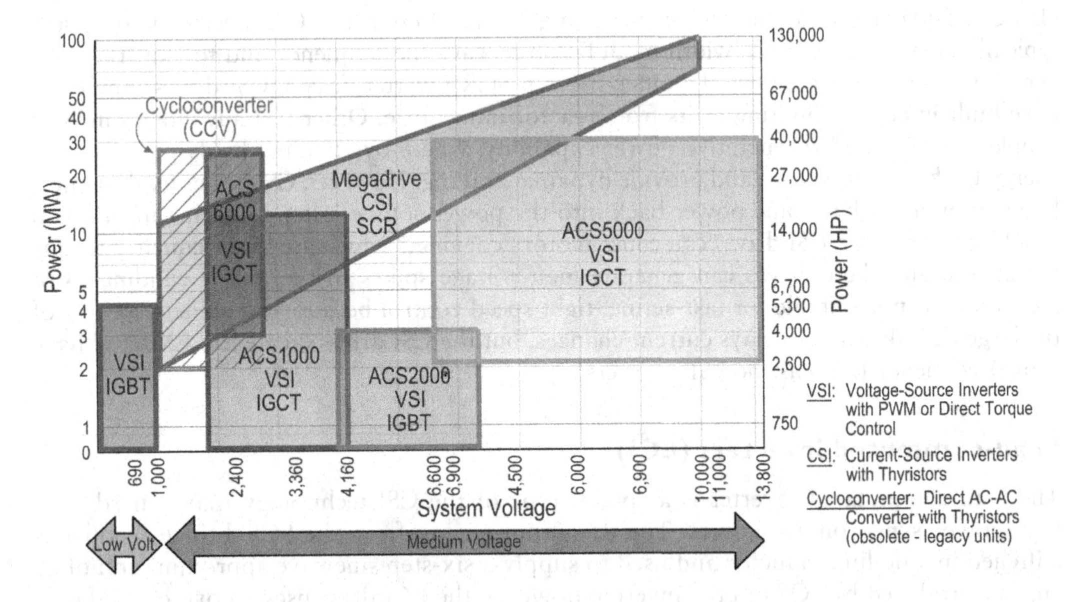

Figure 6.5 Large-power AC motor-drive technologies. (Courtesy of ABB Marine.)

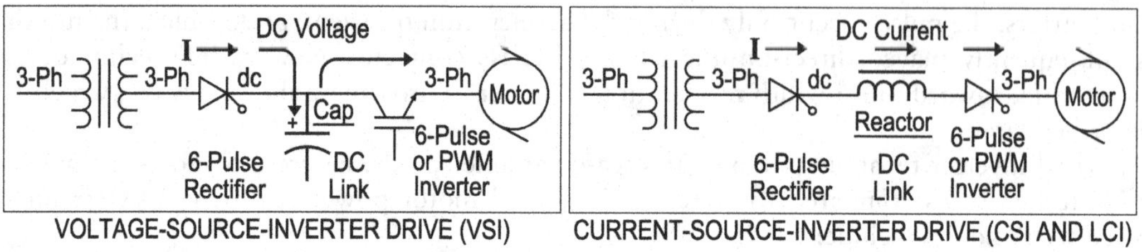

Figure 6.6 A capacitor in the dc-link distinguishes the VSI drive (left) from the CSI/LCI Drive (right), which uses a reactor. The capacitor stores energy to smooth dc variations, while the reactor inhibits fast transient voltages, such as spikes.

Voltage-source inverters (VSI)

Voltage-source inverters use a diode or SCR bridge rectifier to create a constant-voltage dc source applied to the dc link. A capacitor in dc-link defines the unit as being a voltage-source inverter, where the capacitor stores energy to "stiffen" the voltage leading to the inverter (Figure 6.6 left). The inverter produces six pulses of output that produce variable-frequency voltage by turning on and off transistors, SCRs, or gate turn-off thyristors (GTOs). For the newer better-quality VSI drives, the inverter uses MOSFETs or IGBTs with pulse-width modulation techniques. VSI drives tend to generate significant amounts of EMI noise that may be difficult to filter, but they produce tight speed or torque control.

Phase-controlled current-source inverters (CSI)

Current-source inverters use an SCR bridge rectifier, but it stores energy in a series reactor, which is an inductor installed in the dc link (Figure 6.6 right). As an alternative to capacitors, the reactor's inductive effect creates a counter-emf in the coil to impede fast amperage changes. The reactor stiffens the current going to the inverter. The CSI inverter ac output is typically created using pulse-width-modulation to vary the frequency and to vary the voltage, but some drives may use the older six-step SCR inverter design. CSI drives inherently have built-in current-limiting traits from reactor inductance. Other CSI advantages include simple circuitry, and regenerative power capability, where overhauling loads put electrical energy back into the source and provide dynamic braking. However, CSI drives tend to inject large amounts of harmonic power back into the power source, often requiring installation of a large ac reactor. CSI drives can cause motor "cogging," which are low-frequency torque pulsations, and these drives can generate high-voltage spikes to the motor windings. CSI technology is not suitable for fast-acting, tight speed control because the inductive effect of the large dc link reactor delays current changes, but the CSI drives create less EMI and have similar efficiency levels as the VSI inverters.

Load-commuted inverters (LCI)

The load-commutated inverter is a special type of the CSI technology that is used with high-power synchronous motors. The dc output energy from the SCR bridge rectifier is stiffened in a dc-link inductor and used to supply a six-step sinewave approximation of ac current produced by SCRs in the inverter, however, the LCI drive uses an overexcited synchronous motor to help mitigate harmonic currents and reduce reactive power for improving power factor. Table 6.1 compares the VSI and CSI/LCI technologies.

Table 6.1 Comparison of high-power VSI and CSI inverter characteristics

	VSI system	*CSI / LCI system*
Max. Motor Power	100 MW	120MW
Motor Types	Induction or Synchronous	Synchronous only
Max. Drive Voltage	7.2 kV	11 kV
Motor Design	Independent from converter	Special design, affected by converter and harmonic heating
Converter Size	Typical	More space for inverter, harmonic filter, pf correction, and line reactor
Efficiency	Approx 1% more efficient	
AC Power Disturbance	Stable	Sensitive
Line Current THD	2% THD	Up to 12% THD
Ripple induced torque cogging	Less than 1%	Up to 7%
Input power factor	0.95	0.5 – 0.92 capacitive power factor correction required

Cycloconverters (CCV)

The cycloconverter is an ac-to-ac drive that eliminates the dc-conversion circuit and the intermediate dc link, thus eliminating the associated dc-conversion losses. Cycloconverters are high-voltage, high-current devices used with large synchronous motors, and nicely, they can supply large amounts of power at low speeds. The cycloconverter takes a constant voltage and constant frequency source of ac electricity and modulates it into lower frequency and lower voltage by strategically combining portions of the input ac waveform in a manner that directly creates the low-frequency output ac waveform, by using many properly timed SCRs. There are two types of cycloconverters; blocking mode, which is more common, and circulating-current.

At a minimum, cycloconverters use six sets of six-pulse SCR bridges for a total of 36 SCRs (see Figure 6.7). However, if phase-shifting transformers are used to produce more pulses, the total becomes 72 SCRs. CCVs are relegated to high-power, low-speed applications, such as a ship's main-propulsion motor; however, at this time, the CCV technology has pretty much been superseded by PWM technologies for new installations, although future improvements in the electronics could bring these drives back.

Pulse-width modulation (PWM)

This is presently the state-of-the-art VFD inverter technology, and its development continues to be refined. PWM is a switching technology used in the inverter section of a drive to strategically pulse dc electricity so as to create a sinusoidal waveform of current. By alternating the dc polarity coming from the inverter, and by letting the dc current through for varying lengths of time, the electricity well approximates a sine wave (Figure 6.8) and it uses much of the source ac current, mitigating harmonics. Typically, the inverter section of the PWM is accomplished using MOSFETs for lower-power applications and IGBTs for higher-power outputs. The gate signal originates from the microprocessor controller. The IGBT/FET gates are rapidly triggered from a digital signal to create the pulses emulating the sine wave.

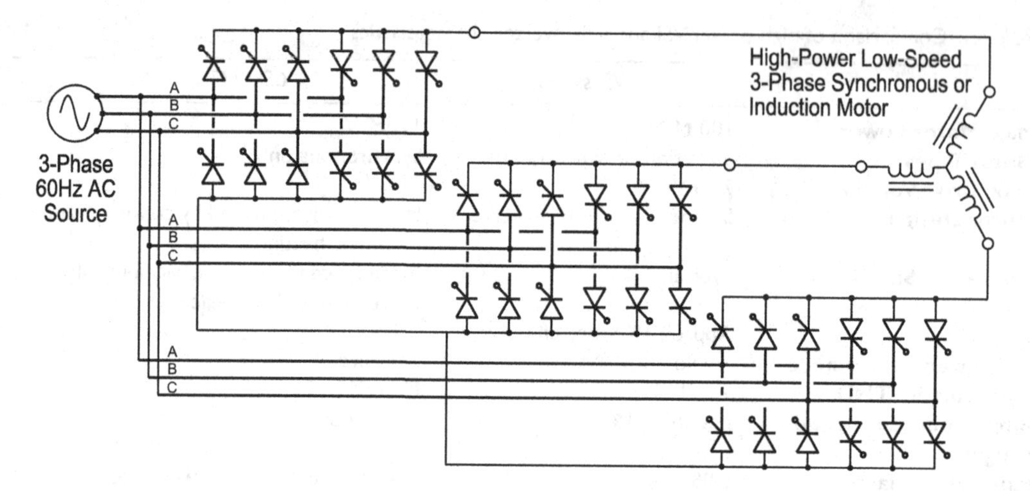

Figure 6.7 SCR arrangements used with a three-phase-to-three-phase cycloconverter. The dc rectifier and link are eliminated. This legacy drive is presently superseded by other technologies, but could be redeveloped using the commutation improvements obtained with newer power-electronic devices and microprocessor controllers.

As a flow analogy, pulse-width modulation is like a piping-system whose outlet flow rate and pressure are controlled by fully opening and fully closing a quick-acting valve in bursts of different durations. When low flow and low pressure is required, such as at the beginning of the sine wave, the valve is fully opened for a very short duration and closed for a longer period. When high flow and high pressure is required, such as near the peak of the sine wave, the valve is opened for a longer burst. If the valve is properly timed and pulsed very fast, the water pressure will fluctuate smoothly in a semi-sinusoidal manner. For the VFD, the pulsing frequency is called the carrier frequency, and is typically fixed to one value within a range of 2 kHz up to 20 kHz. The tradeoff is that while higher frequencies provide a smoother sinusoidal shape, the higher frequencies also produce more switching losses, motor bearing problems, and higher voltage spikes from reflected current. Carrier frequencies below 5 kHz are less prone to creating the bearing problems that emerged with the advent of VFD technologies; however, the lower frequencies also tend to produce more audible noise.

Flux-vector drives

Flux-vector drives are sometimes called field-oriented control (FOC) drives. Flux-vector drives are a specialty application of PWM technology that are designed to modulate both output torque and speed. The flux-vector drive is a non-scalar but more-precise alternative to the more common scalar Volts-per-Hertz method of setting voltage. This control method considers the motor-stator current to be the resultant of two mutually orthogonal vectors, where one vector represents the radial magnetic flux and the other vector represents the tangential mechanical-torque. The idea is that some of the stator current is used to create the changing magnetic fields that interact through the air gap between the stator and rotor, and some of the stator current creates the rotational torque. The output current from the flux-vector drive is adjusted in proportion to the motor speed, with the purpose of optimizing the amperage required for the most efficient production of both the flux and torque. Flux-vector ac drives

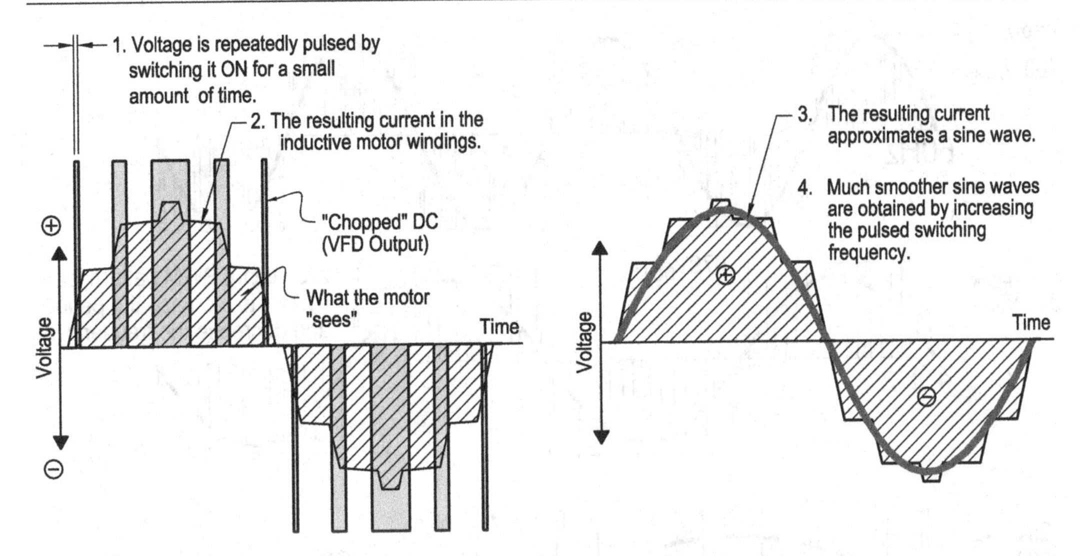

Figure 6.8 Pulse-width modulation is used to create nearly sinusoidal ac electricity using a switching power supply. It works by fully switching on a constant-amplitude voltage pulse for varying durations as shown on the left. The pulses on the left figure are intentionally very coarse for illustrative purposes, but the figure shows how low current is let through in bursts.

are particularly good for powering high-torque loads that undergo large speed variations; which are the types of applications where dc motors excelled in the past. Flux-vector-drive outputs are pulse-width modulated, where the driving signal can come directly from a closed-loop feedback signal originating from a shaft-mounted digital encoder to accurately control torque; or indirectly from a mathematically preprogrammed VFD microprocessor whose torque-versus-flux profile does the torque control. The flux-vector calculations are mathematically intensive, requiring a very fast microprocessor; but this computing technology is readily available today. While flux-vector drives are particularly beneficial in high-torque, low-speed applications, the extra cost is generally not justifiable for general purpose motor applications.

VFD voltage control

In VFDs, the maximum dc voltage is set by the rectifier-section, and the ac frequency is produced by the inverter section. For 480 VAC_{RMS}, the dc-link voltage can approach the peak voltage of $\sqrt{2} \cdot V_{RMS}$, or 679V. The dc-link voltage can be reduced by strategically triggering the rectifier SCRs or GTO thyristors, or the dc-link voltage can be reduced by using MOSFET or IGBT switching technologies in a dc-to-dc converter, sometimes called a chopper. The variable-frequency sinusoidal output is later produced in the inverter from the voltage-adjusted dc using a fixed-pattern of PWM signals from the VFD microprocessor by triggering the inverter gates with various durations of on-and-off pulsations. Moreover, the relative amounts of the on-to-off durations provides the added advantage of simultaneously reducing the output rms voltage. Examples of PWM signals being used to produce variable frequencies and variable voltages are illustrated in Figure 6.9. Ultimately, PWM electronically produces very high-quality sinusoidal output to the attached motor with the ability to set the voltage. Reducing voltage more efficiently controls torque, while avoiding extra amperage that could overheat a motor. A high-power, medium-voltage variable-speed motor drive is shown in Figure 6.10.

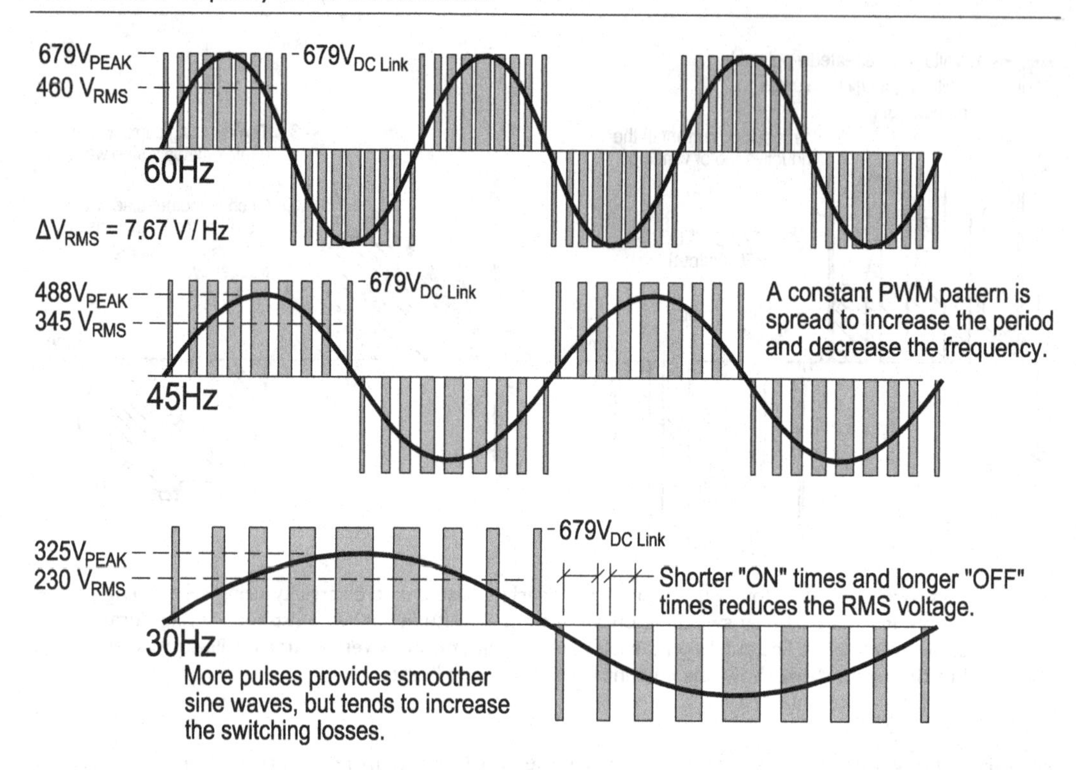

Figure 6.9 The PWM pattern produces the sine wave. Increasing the width of the PWM pattern increases the period, which decreases the frequency. By decreasing the relative amount of "ON-to-OFF" time, the RMS voltage is also reduced. In this manner, the voltage can be controlled to better match the required motor torque.

Soft starters

VFD technologies are also applied to the "soft-starting" of motors by gradually ramping up frequency, in lieu of more conventional reduced-voltage starting techniques, such as autotransformers or wye-delta starting. Soft-start technologies essentially work by converting 60Hz ac electricity into dc using a three-phase, six-pulse bridge rectifier having six diodes, SCRs, or IGBTs. Like a VFD, the rectified current is passed through a dc link and the soft starter uses a six-pulse inverter to create ac. Either the voltage is gradually ramped up to line voltage in the front end, or the output ac frequency is gradually ramped from 0 to 60 Hz by triggering the SCRs. Since soft starting is only needed during the acceleration transient, short-term harmonics are generally not a concern, as they disappear when 60Hz is reached, and a "bypass" contactor transitions the motor to be direct-on-line. The bypass contactor removes the "non-linear load" from the electrical system after the motor comes up to speed. Soft starters are discussed in Chapter 5, "Motor Controllers."

HARMONICS

Harmonics are distortions in the sinusoidal ac-current waveform caused by non-linear loads, where the current distortions lead to voltage perturbations in the source electrical supply

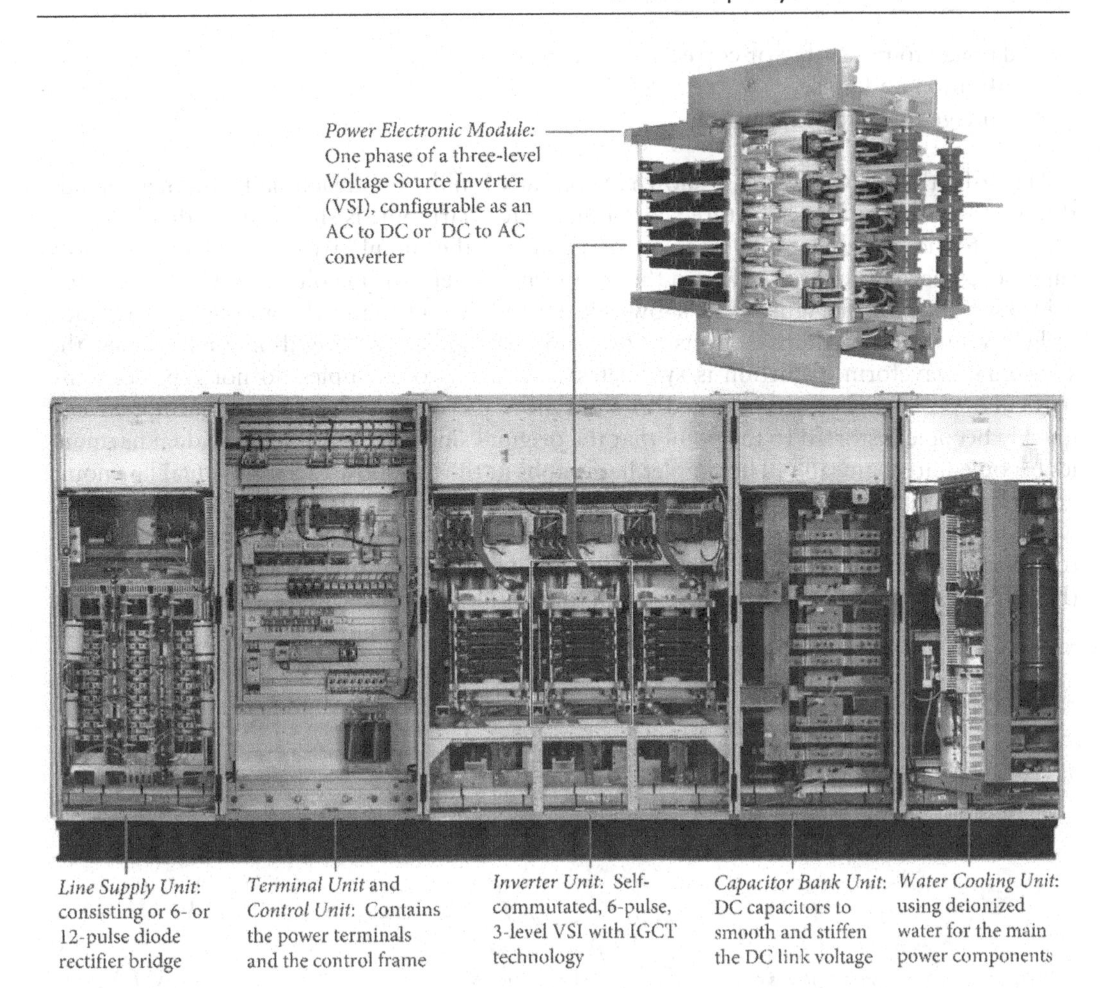

Figure 6.10 High-power, medium-voltage ac motor-drive technology in a 7-foot tall cabinet. This unit is used in a shipboard ac electric-drive propulsion system. (Courtesy of ABB Marine.)

system. The voltage distortions are spread through the distribution system to create plant-wide power-quality problems, such as:

- increased copper and iron losses in motors and transformers, causing overheating,
- overloading of transformer neutrals,
- excessive skin effect and overheating of conductors,
- deterioration of insulation,
- low voltages at the loads,
- misfiring of VFDs and switching power supplies,
- notching[2]
- amperage-measurement problems,
- voltage-regulation problems in standby generators,
- false tripping of circuit breakers,
- motor-torque pulsations (cogging),

- damage to power-factor-correction capacitors,
- vibration and noise,
- damage to motor bearings.

The 3rd, 6th, 9th, and 12th are shown in columns 1 and 2 on the left half, and they include the even and odd multiples-of-three harmonics. Each harmonic is shown individually superposed onto the fundamental 60Hz frequency, and then the cumulative effects of all previously superposed harmonics are shown on the composite waveforms in column 2. Columns 3 and 4 in the right half of Figure 6.11 show odd harmonics only, and the graphs are provided to help visualize the additive effects of odd-only harmonics. With odd-only harmonics, the sinusoidal waveform distortion is symmetrical. These two examples do not represent any particular problem but are intended to show how the composite of many harmonics can quickly become distorted to the point that the original sine wave and the individual harmonics become unrecognizable. The simpler bar graphs at the top show each individual harmonic and its amplitude in the frequency domain, corresponding to the complicated composite waveforms in the time domain. The frequency domain is produced by deconstructing the distorted time-domain waveform into its individual frequencies using Fast-Fourier Transform (FFT) techniques, and in practice, is essential for troubleshooting power-quality problems and

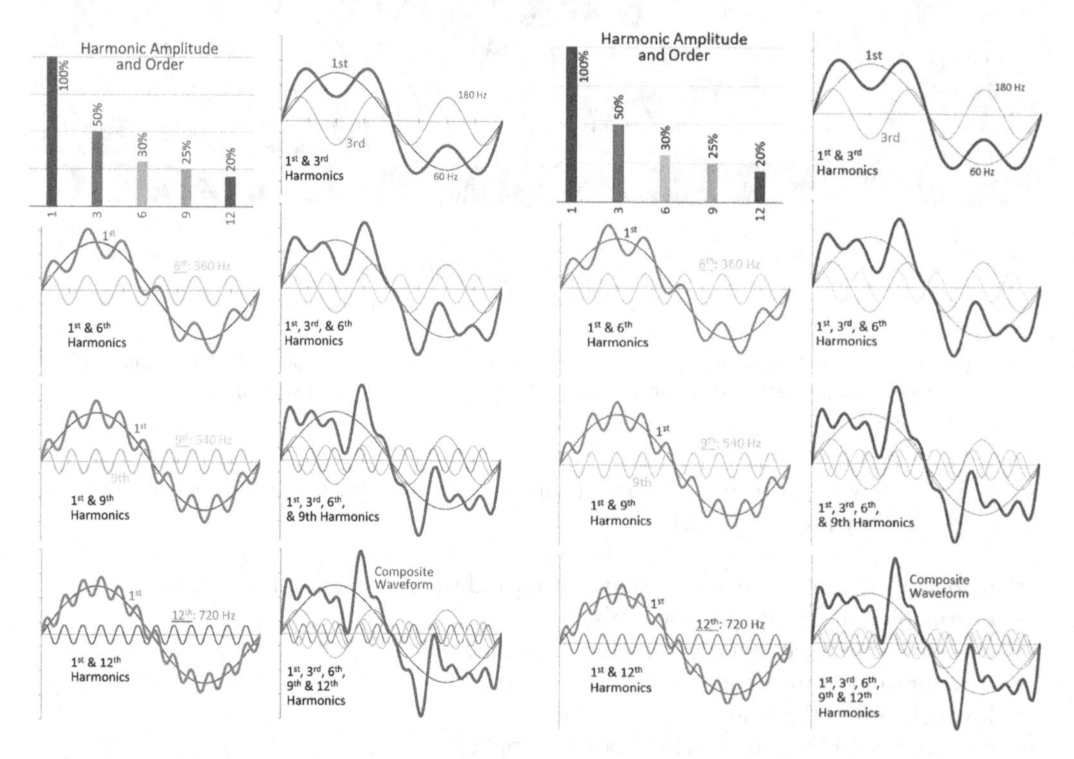

Figure 6.11 Superposition of 3rd, 6th, 9th, and 12th current harmonics onto a 60Hz sine wave are shown in the time domain in columns 1 and 2. Each individual harmonic is superposed onto the fundamental frequency in column 1, and the composite with all previous harmonics are shown in column 2, where it can be seen that the even harmonics distort the waveform symmetry. The frequency-domain bar graph is shown at the top of column 1, where each individual harmonic is easily distinguished. The curves in columns 3 and 4 are similar, but include only the 3rd, 5th, 7th, and 9th odd harmonics, where interestingly, symmetry of the distorted waveforms remains.

for developing mitigation strategies. While the figure represents equipment amperage draw, the distorted current produces voltage distortions as electricity is pulled from the distribution system. Quantifying the voltage distortion is more complicated than current distortion, as it is strongly influenced by the system impedance, system stiffness and the relative amount of harmonic amperage to the plantwide amperage. For example, small current distortions on a very large system may have no observable effect, but large current distortions on a small soft system would be objectionable.

Harmonics in the time and frequency domains

Figure 6.12 (left) shows the composite of a 60Hz fundamental frequency superimposed with a distorting 3rd harmonic sine wave. High 3rd harmonics are inherently present in systems having a substantial number of single-phase dc power supplies, or three-phase systems having amperage imbalances between the three phases, typically from unequal distribution of single-phase loads between phases. Referring to Figure 6.12 (left), the 1st and 3rd composite ac waveform takes on a distorted shape, whose superimposed shapes may be somewhat intuitive in appearance, where the 60Hz sine wave has two additive humps for the positive half waves and a center dip for when the negative half wave is added.

The right side of Figure 6.12 shows both 3rd and 5th harmonic distortions superimposed on the fundamental frequency. In this case, individual frequencies in the composite become less intuitive. In real situations, the composite waveform becomes so distorted that it is virtually impossible to visually dissect out the individual harmonic waveforms. In that case, a spectrum analysis using Fast Fourier Transforms deconstructs the individual harmonic frequencies from the distorted composite. A spectrum analysis uses the frequency

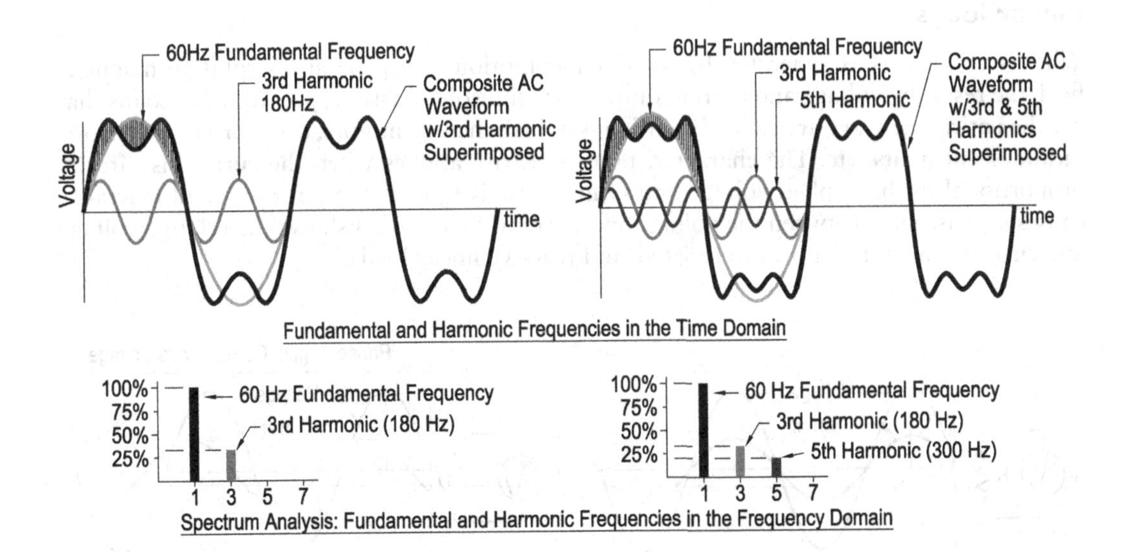

Figure 6.12 **3rd harmonic distortion occurs from imbalanced three-phase systems (left). 5th harmonics are the first objectional frequency with three-phase full-wave bridge rectifiers. As a generalization, lower-order harmonics tend to produce higher amplitudes of distortion and increase the total harmonic distortion (THD). The two distorted sine-wave curves in the upper part of the figure show the superposition of 3rd and 5th harmonics in the time domain, while the two lower bar graphs show the same information in the easier-to-visualize frequency domain.**

domain, instead of time, and each amplitude is plotted in a bar-graph compared to the first harmonic. The signature of the harmonic spectrum analysis is useful in identifying the predominant causes of the harmonics. The spectrum analysis allows engineers to easily discern which harmonics are the most problematic, and to aid in designing a solution to mitigate them.

The cumulative effect of all the individual harmonics represents the severity and is quantified by calculating the total harmonic distortion (THD) for either current or voltage. THD represents the overall severity caused by a single VFD drive, and the equation is

$$\mathrm{THD}_{\mathrm{Current}} = \frac{\sqrt{\sum_{n=2}^{50} I_n^2}}{I_{1st}} \qquad \mathrm{THD}_{\mathrm{Voltage}} = \frac{\sqrt{\sum_{n=2}^{50} V_n^2}}{V_{1st}}$$

Troublesome harmonics originate from the third harmonic and odd multiples of the third harmonic, i.e., the 3rd, 9th, 15th, 21st, 27th, and so on. These harmonic multiples are called triplens, and the triplens tend to overheat the neutral of a three-phase, four-wire, wye-connected transformer. For linear loads, the neutral carries differences of current between phases, which would be zero amps for a balanced three-phase system. However, the triplens, being odd zero-sequence harmonics, cause additive current on the neutral. This problem is addressed by using "K-factor" transformers having oversized neutral buses and other transformer-construction techniques.

CAUSES OF HARMONICS

Linear loads

When ac electricity is generated by the circular motion of a generator's rotating magnetic field, the resulting voltage and current outputs are sinusoidal with respect to time. Loads that use the entire sine wave are called linear loads, which include motors, transformers, resistors, incandescent lights, etc. The characteristic of linearity applies when the current is directly proportional to the applied voltage and Ohm's Law is followed, even if the load reactance causes a phase shift between the voltage and current. Figure 6.13 shows the relative voltage and current waveforms for both resistive and reactive linear loads.

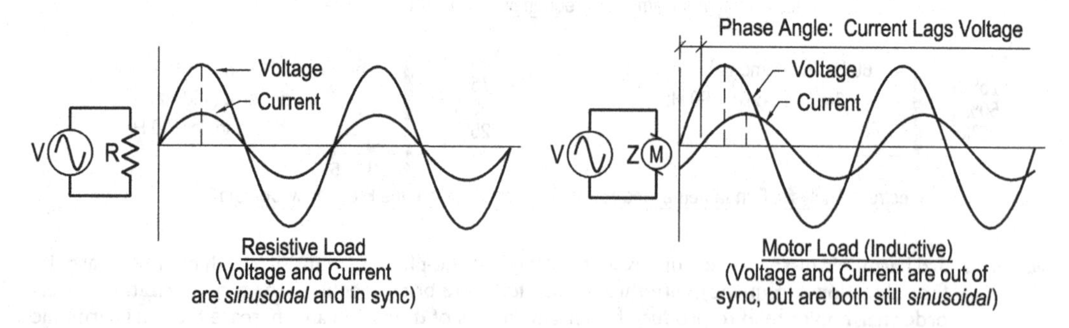

Figure 6.13 A resistor forms a linear load (left) where both sinusoidal current and voltage are in phase with each other and at unity power factor. A motor/inductive load is also a linear load. Both voltage and current are still sinusoidal (right), even though there is a lagging phase shift and a non-unity power factor. Linearity is met because Ohm's Law applies ($V=IZ$).

Non-linear loads

Some types of loads use only pieces of the available sinusoidal current, and they are called non-linear loads. An easy-to-visualize example of a non-linear load is a single diode inserted into an otherwise linear sinusoidal circuit. In the forward-biased direction, the current and voltage track together when supplying a resistive load, but in the reverse-biased direction, the diode current is zero even though the voltage is still non-zero and semi-sinusoidal. Figure 6.14 illustrates a linear load on the left next to a single-diode non-linear load on the right, with calculations showing its non-linearity.[3] 20A of sinusoidal current is not equal to 9A of semi-sinusoidal current draw, so Ohm's Law does not work with the original ac voltage source into the half-wave rectifier.

When a full-wave diode bridge is used with capacitive filtering, the non-linearity becomes even more pronounced. In this case, current is drawn only when the voltage approaches the peak of the sine wave and exceeds the voltage stored in the capacitor (see Figure 6.15). The inductive components in the electrical distribution system experience the spiked non-linear current draw, where the non-sinusoidal current induces voltage distortions in the source,

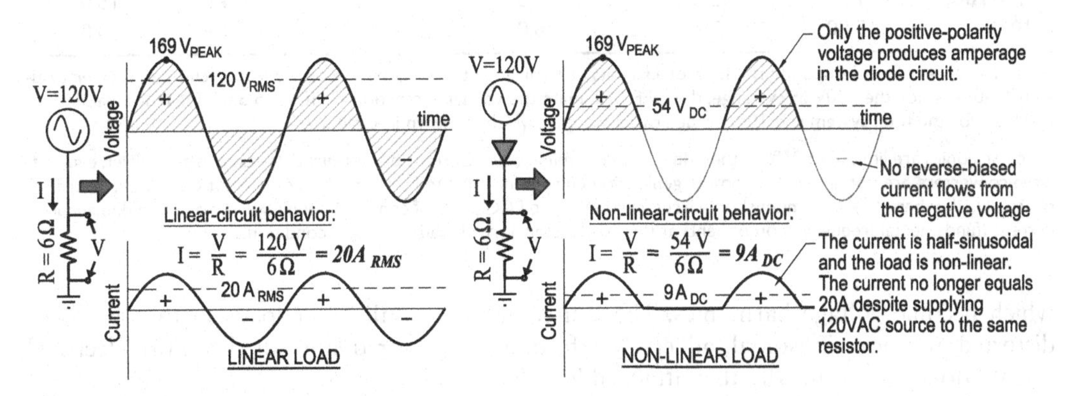

Figure 6.14 A single-diode circuit produces non-linear current draw as exemplified by the semi-sinusoidal curve in the lower right of the figure, even though fully sinusoidal electricity is input into the circuit. Nonintuitively, the diode semi-sinusoidal amperage is less than half of the non-diode 20A at 9A, quantifying the non-linearity.

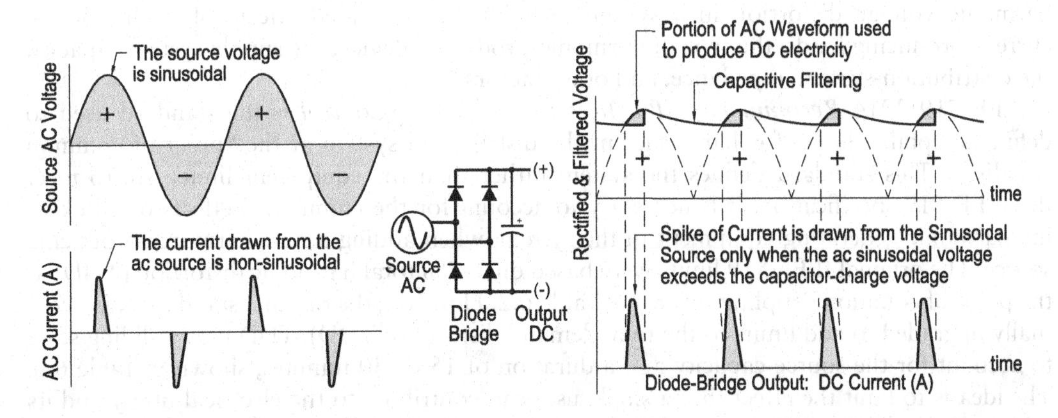

Figure 6.15 Diode rectification with filtering draws current from only the peaks of the sine wave. The spiked amperage drawn by the load distorts the current flow on the ac source. Total harmonic distortion is the cumulative effect of all distortions being superimposed, where each harmonic occurs at an integer multiple of the 60Hz fundamental frequency. Each amplitude and its frequency are a harmonic.

Table 6.2 IEEE 519 Low-voltage system classification and THD distortion limits

	Special applications	*General system*	*Dedicated system*
THD (Voltage)	3.0%	5.0%	10%

Note: Special applications include hospitals and airports. A dedicated system is exclusively used by the non-linear loads.

Table 6.3 Current distortion limits for systems rated 120V through 69kV from Table 2 of IEEE 519–2014

Maximum harmonic amperage distortion in % of I_L						
	Individual harmonic drder (odd harmonics)					
I_{SC}/I_L	*h < 11*	*11 ≤ h <17 16*	*17 ≤ h <23 23*	*23 ≤ h <35 34*	*h >35*	*%TDD*
<20	4.0	2.0	1.5	0.6	0.3	5.0
20 < 50	7.0	3.5	2.5	1.0	0.5	8.0
50 < 100	10.0	4.5	4.0	1.5	0.7	12.0
100 < 1000	12.0	5.5	5.0	2.0	1.0	15.0
> 1000	15.00	7.0	6.0	2.5	1.4	20.0

Note: This table defines ranges for the harmonics limits compared to the system's available short-circuit current per load-amp. Note that for the 120V-69kV range, the IEEE 519 limits the voltage distortion (THDv) to 3.0% for individual devices and 5.0% for the total system and 10% for dedicated systems, which is shown in Table 6.2.

Even harmonics are limited to 25% of the odd harmonic limits above. Current distortions that result in a dc offset, e.g., half-wave converters, are not allowed. All power generation equipment is limited to the values of current distortion, regardless of actual I_{SC}/I_L where I_{SC} = maximum short-circuit current at the PCC point of common coupling, I_L = maximum demand load current (fundamental frequency component) at the PCC under normal load operating conditions.

which become voltage harmonics. The source voltage to all other loads in turn becomes distorted and non-sinusoidal, adversely affecting the power quality of the entire electrical distribution system and all other attached loads.

Non-linear loads cause sinusoidal distortions in the form of harmonics. The harmonics occur at integer multiples of the fundamental frequency. Harmonics are produced by many types of electronic loads, including variable-frequency drives, fluorescent-lighting electronic ballasts, switching-power supplies, arc-discharge devices, iron cores that reach magnetic saturation, resonant conditions in power-factor correction capacitors, etc. The amount of harmonic voltage distortion in a system includes the combined effects of *all* harmonic-current-producing loads, the type of harmonic-producing device, the supply system capacity, the distribution-system impedance, and other factors.

IEEE 519 2014 *Recommended Practices for Harmonic Control* is the standard used to define acceptable limits for harmonics in the distribution system at the "point of common coupling." This standard defines the system rather than the equipment limits; so, to meet the IEEE 519 specification, it is necessary to account for the cumulative effects of all existing harmonic-generating equipment in the system, when adding a new harmonic-producing source. The original IEEE 519 limits were based on 3–5% total harmonic distortion (THD) at the point of common coupling, shown in Table 6.2. However, the original standard was eventually upgraded to add limits to the total demand distortion (TDD). TDD uses a sliding scale to account for the source capacity and a duration of 15 or 30 minutes, shown in Table 6.3. The idea is to limit the effect that a single user can contribute to the electrical utility and its other customers. The TDD is based on the ratio (I_{SC}/I_L), which represents the utility's short-circuit capability (I_{SC}) relative to the user load (I_L). Europe, on the other hand, uses the IEC 61800–3 *Adjustable Speed Electrical Power Drive Systems—EMC Requirements and Specific*

Test Methods standard to specify equipment limits. The European standard seeks to limit the harmonics contribution of any individual piece of equipment instead of addressing the system.

CLASSIFICATION OF HARMONIC TYPES

As described earlier, harmonics are caused by non-linear loads. The lower-left diagram in Figure 6.15 exemplifies an extreme amount of current distortion imposed by a load onto the ac source, however, the total source distortion also depends on the size of the electrical source, the source impedance, and other factors, such as phase imbalances. For example, harmonics from a single-phase SCADA-computer would be easily absorbed in a large plant electrical system, but if the same plant has many sizeable VFDs, serious harmonic problems would not be surprising. Harmonics recur regularly at multiples of the fundamental frequency, such as 60Hz, as opposed to other power-quality problems such as voltage dips, spikes from lightning strikes, load surges causing armature reaction, and other transient problems, which are non-harmonic and random in nature. Types and sources of power-quality issues are:

- Positive-, Negative-, and Zero-Sequence Harmonics
- Triplens
- Intraharmonics and Subharmonics
- Third Harmonic
- Harmonics from the Front-End Rectifier Section
- Reflected Waves
- Notching

Positive-, negative-, and zero-sequence harmonics

Technically, harmonics include all integer multiples of the fundamental frequency superposed onto that frequency to create one distorted waveform. Ignoring the non-repeating, transient spikes, and noise, the distorted wave shape is the compilation of all harmonic amplitudes and frequencies onto the fundamental frequency. Using 60Hz as an example, the 2nd harmonic is 120Hz, the 3rd is 180Hz, the 4th is 240Hz, the 5th is 300Hz, the 6th is 360Hz, and so on. In practice, troublesome harmonics occur at predictable frequencies, which are specific to the cause. The harmonics are classified by their name, frequency, and phasor-rotational direction relative to the fundamental frequency. In addition, even-number harmonics typically cancel themselves, and in most cases are non-problematic.

Positive-sequence harmonics consist of each integer *multiple of three, plus 1* for three-phase electricity. The positive-sequence harmonics therefore consists of the 4th, 7th, 10th, 13th, 16th, and so on. A characteristic of positive-sequence harmonics is that voltage and current phasors for these harmonics rotate faster than the first harmonic and are therefore in the same or positive direction as the fundamental frequency.

Negative sequence harmonics consist of *multiples of 3, minus 1*. The negative-sequence harmonics are therefore the 2nd, 5th, 8th, 11th, 14th, 17th, and so on. The phasors for these harmonics rotate slower than the fundamental frequency, and therefore in the opposite or negative direction. For wye-sourced circuits, all positive- and negative-sequence harmonics self-cancel at the common neutral.

Zero sequence harmonics consist of multiples of 3, and are the 3rd, 6th, 9th, 12th, 15th, 18th, 21st harmonics, and so on. Since the even harmonics tend to be non-problematic, then only the 3rd, 9th, 15th, 21st, etc., tend to be the troublesome zero-sequence harmonics. This

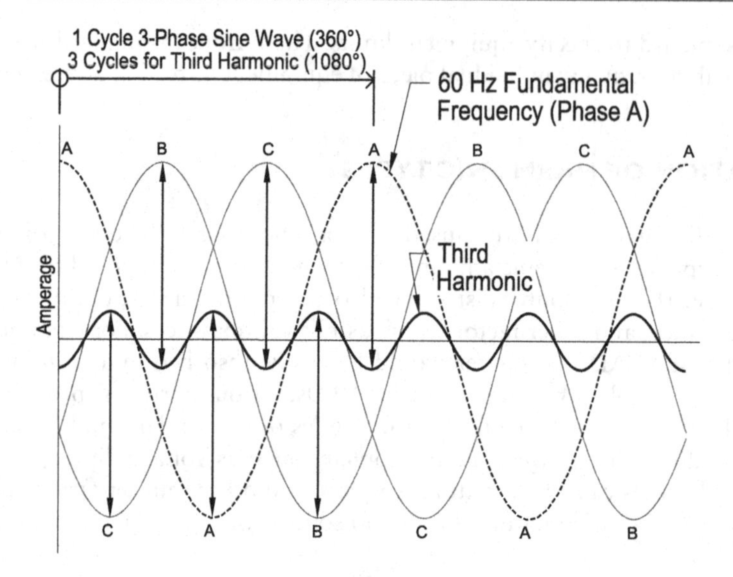

Figure 6.16 The 3rd harmonic is the lowest-order zero-sequence harmonic. Each 3rd-harmonic peak corresponds with an opposite peak for each of the fundamental-frequency phases. The result is that the 3rd-harmonic amperages are superimposed onto each phase, creating current distortions, and worse, each 3rd harmonic completes its circuit through the common neutral in an additive manner, rather than being balanced and cancelling each other. 3rd harmonics have a tendency to overload and overheat transformer neutrals in a dual-voltage, three-phase wye secondary.

special case of odd-only zero-sequence harmonics is called the triplen harmonics. The current and voltage phasors for zero sequence harmonics rotate at the same rate as the fundamental frequency. For grounded-wye circuits, each zero-sequence-harmonic frequency causes circulating amperages between the phase and the neutral. So instead of canceling, as they do in positive and negative sequence harmonics, zero-sequence harmonic currents *add* arithmetically to the neutral, leading to neutral problems.

Figure 6.16 shows a third harmonic current being superposed onto the three-phase electricity. so that three third-harmonic cycles occur for each fundamental-frequency cycle. The diagram shows that each third-harmonic peaks at the same time but in the opposite polarity of each phase. The alignment of third-harmonic peaks causes circulating phase currents, which add onto the neutral when completing the circuit.

Triplens

Another set of harmonics are referred to as the triplens, and they consist of offset multiples of the third harmonic. Triplens follow the relationship of $h = 3 + 6n\,(n = 0,1,2,3,etc.)$. The triplens are the 3rd, 9th, 15th, 21st, 27th harmonics, and so on. The triplens consist of all odd-only zero-sequence harmonics, excluding all even zero-sequence harmonics and are shown in Table 6.4. Triplen phasors rotate at the same speed and in sync with the fundamental frequency, so there is no phasor-angle difference. Linear three-phase circuits are designed so the amperages on each phase are balanced and sum to zero at the common neutral point in a three-phase, four-wire grounded-neutral wye circuit. Unfortunately, the triplen currents align with each other and are additive. In a theoretically worst-case scenario, the triplen harmonic currents could reach three times the phase current. In practice, the mix of linear and non-linear loads reduces the harmonics below the theoretical possibility.

Table 6.4 Frequencies and harmonic sequencing. The triplens are the odd zero-sequence harmonics in bold

Harmonic:	*Fund.*	*2nd*	*3rd*	*4th*	*5th*	*6th*	*7th*	*8th*	*9th*	*10th*	*11th*	*12th*
Frequency. (Hz):	**60**	120	**180**	240	300	360	420	480	**540**	600	660	720
Sequence:		Neg	**Zero**	Pos	Neg	Zero	Pos	Neg	**Zero**	Pos	Neg	Zero

In general, positive-sequence harmonics cause overheating of conductors, transformers, generator windings, etc., because the waveforms are additive, and do not cancel. Negative-sequence harmonics cause circulating currents between phases. In a motor, the positive-sequence harmonics try to make the motor run faster, and the negative-sequence harmonics try to slow the motor, but in both cases, they cause circulating currents within the motor that weaken the induced magnetic fields causing less torque, reduced efficiency, and overheating. Triplen harmonics add directly into the common neutral of transformers, causing neutrals to overheat. Fortunately, even-number harmonics, whether they originated from the positive, negative, or zero sequence, are typically small and non-problematic. In general, the even-number harmonics only tend to exist when the alternating current has unequal half-sine waves, a condition that is not common.

Intraharmonics and subharmonics

Intraharmonics have frequencies that fall between the integer multiples of the fundamental frequency. Although they can exist, they are not common. Subharmonics are a specific case of intraharmonics, where the subharmonic frequency falls below the fundamental frequency, so it would be less than 60Hz. Although subharmonics can exist, they too, are rare in power systems. It should be recognized that many power-quality meters use Fast Fourier Transform (FFT) techniques that are designed to ignore intraharmonics to improve the data-acquisition and calculation-speed capability. Although intraharmonics are rare, subharmonics, however, can be caused by resonance in power-factor-correction capacitors or passive filters having very high inductances. Subharmonics produce slow, undamped oscillations that exhibit themselves as voltage sags or flickering of lights, even if they do not show up on some power quality analyzers.

Third harmonic

The third harmonic is caused by imbalanced amperages on the three phases and it shows up as extra neutral current. Third harmonics are also the first of the triplen harmonics, if they exist. In an ideal three-phase, four-wire electrical distribution system having three "hot" wires and a common wye-connected grounded neutral, single-phase amperages should be evenly distributed between each phase. In that case, the common neutral current is zero. When phase imbalances exist, the neutral returns the differences in a manner that causes circulating currents between the phases. Circulating currents from this cause are third harmonics. Another cause of third-harmonic currents comes from single-phase switching power supplies used by computers, televisions, uninterruptible power supplies, induction heating elements, etc. These devices cause triplens, of which the third harmonic is the strongest.

The third harmonic currents are notorious for causing overheating and damage of transformer neutrals. In installations where the third harmonics are not filtered, a solution is to use "K-factor" transformers in preference to oversized transformers. A K-factor rating is essentially a survival rating for the additional heat caused by non-linear loads and does nothing

Table 6.5 K-Factor transformer designations and applicability

K-factor	Applications	% Non-linear loads
K-1	Conventional transformer. Motors, incandescent lighting, resistance heating, non-solid-state drives	Linear loads only
K-4	HID lighting, induction heaters, welders, PLC and solid-state controls. Supplies the rated kVA without overheating to 100% of the normal 60Hz current plus 16% of fundamental current at the 3rd harmonic current, 7% at the 5th, 7% at the 7th, 5.5% of the 9th, etc., up to the 25th harmonic.	35% of attached loads are non-linear
K-9	163% of the K-4 loading	75% of loads are non-linear
K-13	200% of the K-4 loading. Telecommunications, unfiltered UPS, health care multi-wire receptacle circuits	75%
K-20	Solid-state variable-speed motor drives, multi-wire receptacles in hospital critical-care areas	100%
K-30	Multi-wire receptacles	
K-40	Other loads identified as producing very high amounts of harmonics, especially of higher orders	
K-50	Used for the harshest harmonic environment possible	

to address system power quality. The K-factor number represents the transformers ability to carry 100% of the transformer rating, including a defined percentage of the loads being non-linear. K-factor transformers use an oversized neutral bus to accommodate the extra third-harmonic amperage. They also have a larger high-grade silicon-steel iron core to reduce flux density and avoid saturation, their windings may use parallel conductors to reduce high-frequency skin effect heating, and they are constructed with higher-temperature-rated insulation, more air-ducts for cooling, and electrostatic shielding to reduce transient noise. In procuring K-factor transformers, K-Factor 1 corresponds to a conventional transformer carrying linear loads only, and the sizes increment up to K-factor 50 as shown in Table 6.5. The K-factor is calculated from $K\,Factor = \frac{\sum_1^{34} (i_h \times h)^2}{\sum_1^{34} i_h^2} = \sum_1^{34} i_h (pu)^2 h^2$ as specified in the ANSI/IEEE C57.110–2018 *Recommended Practice for Establishing Liquid Immersed and Dry-Type Transformer Capability When Supplying Nonsinusoidal Load Currents.*

Harmonics from the front-end rectifier section

The front end of the variable-frequency-drive consists of the rectifier assembly where three-phase electricity is applied to bridges formed by either diodes, SCRs, GTO thyristors, MOSFETs, or IGBTs. In the case of a six-diode, six-pulse bridge rectifier assembly shown in Figure 6.3, the bridge produces the rippled dc current like that shown in Figure 3.7. The non-linear current draw that produces this output causes distortions on the 5th and 7th harmonic, 11th and 13th, and so on. Harmonics from the six-pulse rectifier follow the pattern of $h = kq \pm 1$, where h is the harmonic order produced by the drive, k is every integer from 1 up to a practical limit, which is the 40th harmonic when following IEC 61800–3 or the 50th harmonic when following the IEEE 519 standard, and q is the number of rectifier pulses that in the front end, which shows as the number of humps forming the dc ripple.

Table 6.6 The most problematic harmonics caused by VFD multi-pulse rectifier front-end assemblies follow the pattern of $h = kq \pm 1$

6-Pulses rectifer			*12 Pulses rectifer*			*18-Pulses rectifer*			*24-Pulses rectifer*			*36-Pulses rectifer*		
	Harmonic no.			*Harmonic no.*			*Harmonic no.*			*Harmonic no.*			*Harmonic no.*	
K(Integer)	*(-1)*	*(+1)*	*K(Integer)*	*(-1)*	*(+1)*	*K(Integer)*	*(-1)*	*(+1)*	*K(Integer)*	*(-1)*	*(+1)*	*K(Integer)*	*(-1)*	*(+1)*
1	5	7												
2	11	13	1	11	13									
3	17	19				1	17	19						
4	23	25	2	23	25				1	23	25			
5	29	31												
6	35	37	3	35	37	2	35	37				1	35	37
7	41	43												
8	47	49	4	47	49				2	47	49			
9	53	55				3	53	55	3					

Note: The harmonic multiples most likely to be problematic are highlighted. In general, the higher the harmonic number, the less likely it is to be problematic, and the smoother and more sinusoidal the ac output. The table shows that multipulse transformers provide a natural advantage in avoiding problematic harmonics. The magnitude of all harmonics is described by their percent of total harmonic distortion.

Percent total harmonic distortion

%THD is a measure of the severity of harmonics by a single device. %THD can be measured by the manufacturer, however, predicting %THD requires an individual system analysis, and includes quantifying the system capacity, the system impedance and stiffness, the motor load, and other parameters. Table 6.6 show s that more rectifier pulses will inject less harmonics back into the distribution system, e.g., a six-pulse rectifier injects 16 harmonics below the 50th, and a 36-pulse rectifier injects only two. Table 6.7 shows the general effect that electrical system factors have on harmonics. Figure 6.17 illustrates that as a trend, front-end rectifiers having more pulses produce fewer individual harmonics and less THD. More pulses are obtained from specialty multi-pulse transformers.

Reflected waves

Although not a sinusoidal harmonic, reflected waves are another source of system problems particular to PWM inverters. Reflected waves become problematic when distances between

Table 6.7 Electrical systems factors and their effects on harmonics when using VFDs

	Change	*Effect on harmonics*
Motor Size:	Increase	Higher
Motor Load:	Increase	Higher
Inductance (ac source or dc link):	Increase	Lower
Rectifier Pulses	Increase	Lower
Generator or Transformer Rating:	Increase	Lower
Generator or Transformer Internal Impendance:	Increase	Lower
Supply Short-Circuit Capacity:	Increase	Lower

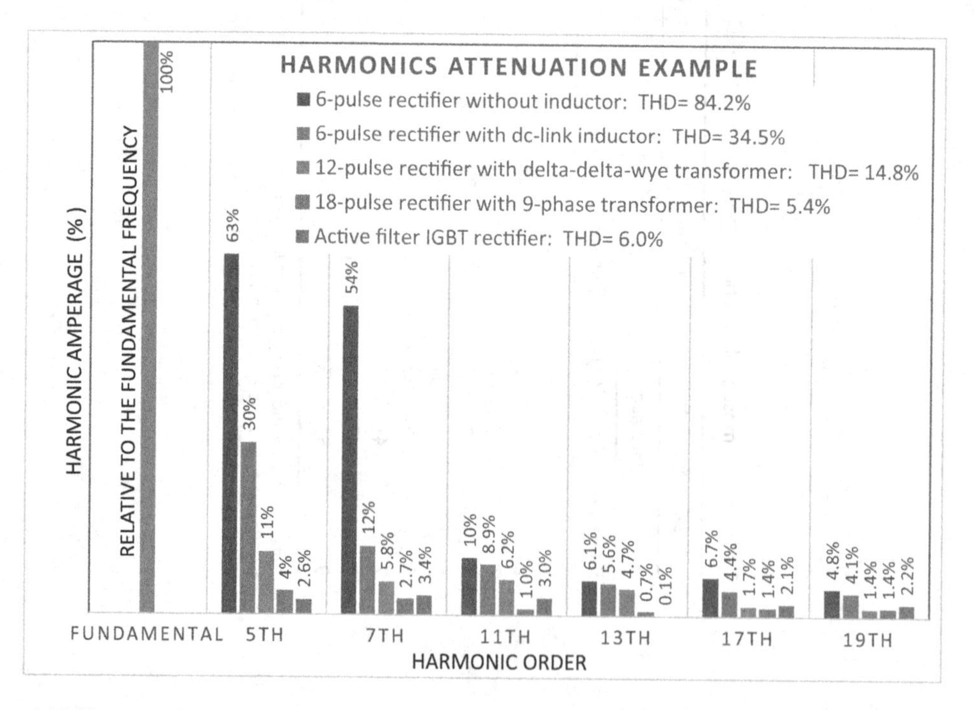

Figure 6.17 The trend for various front-end types to attenuate harmonic production.

the VFD and motor are long, and the carrier frequency is high. Reflected waves should be expected when a motor is more than 60 ft from the VFD for a motor less than 20 HP and 100 ft for a motor larger than 25 HP. In any event, manufacturer instructions should be consulted, as permissible lengths vary, especially when switching frequencies are very high. The other factor that exacerbates reflected waves is the PWM carrier frequency, which is set for a value between 3,000–20,000Hz based on the make and model. For carrier frequencies between 3–12 kHz for VFDs powering 460V motors, the voltage spikes on "short" cables typically fall within the insulation ratings of inverter-duty motors. Figure 6.18 (right) shows the carrier high frequency superposed onto the inverted ac waveform to the motor. Typically, the reflected-wave amperage is very low, and the harmonic order is very high, so harmonics themselves are not the problem. However, high rates of very high-voltage spikes can occur, and in severe cases, motor life can be shortened to a matter of hours.

Reflected waves are a phenomenon that appeared after PWM high-speed switching technologies were developed using MOSFETs and IBGTs in the inverter sections of VFDs. The reflected-wave phenomenon is explained by transmission-line theory, and the overvoltage it produces is sometimes referred to as transmission-line effect or a standing wave. The reflected waves produce voltage spikes, which can break down motor insulation, even for inverter-duty rated motors. When damage occurs, it is typically in the turns nearest to the winding terminals. As a rule, the voltage spikes can reach twice the dc-link voltage, which for a 460V motor would be $V_{spike} = 2 \times V_{rms} = 2 \times \left(\sqrt{2} \times 460V\right) = 1{,}300V$.

Notching

Although notching is not a harmonic phenomenon, it is a steady-state, repeating event caused by six-pulse rectifier assemblies in the VFD front end, particularly those using SCRs. Notching is a condition where the sinusoidal voltage drops very suddenly for a brief period, typically nanoseconds, and then the voltage resumes (Figure 6.19). Objectionable levels of notching are usually limited to six-pulse SCR-based rectifiers, where the voltage notching coincides with current switching associated with the phase commutation that occurs twice per cycle per phase. The commutation time it takes for the current to toggle from one SCR to another is the cause of abrupt voltage drop.

If the notch voltage drop is severe enough to reach zero, equipment that relies on zero crossing may experience operational problems, such as VFDs misfiring or digital clocks running

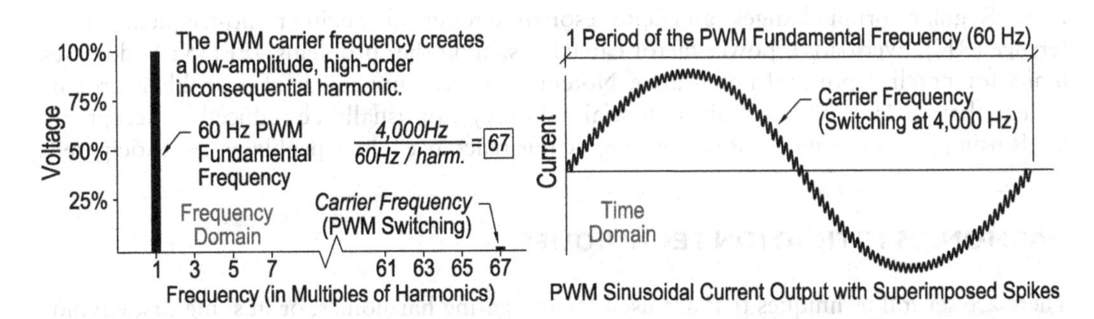

Figure 6.18 Frequency spectrum (left) showing the fundamental frequency compared to the carrier high switching frequency. The right illustration shows the carrier high-fequency spikes of inverter current that are superimposed onto the PWM sine wave during switching. The high-frequency current spikes coupled with motor inductance produces high circuit impedance, leading to heating ($\mathbf{X_L} = 2\pi f\mathbf{L}$).

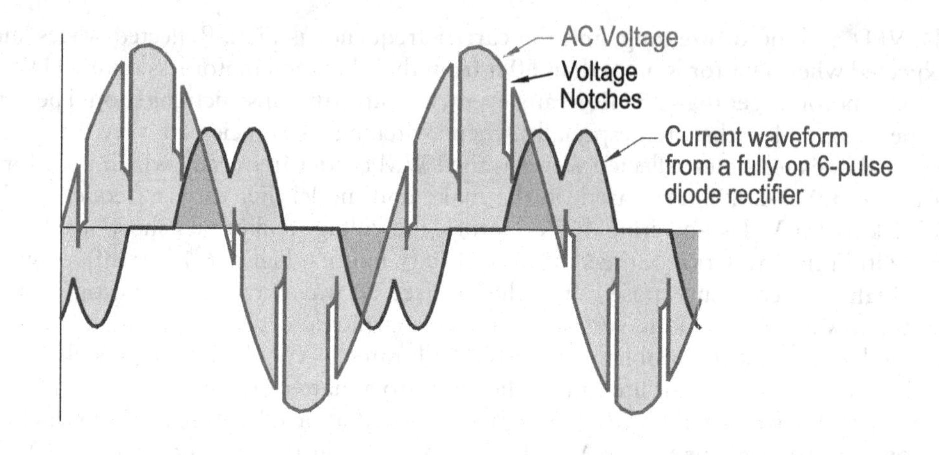

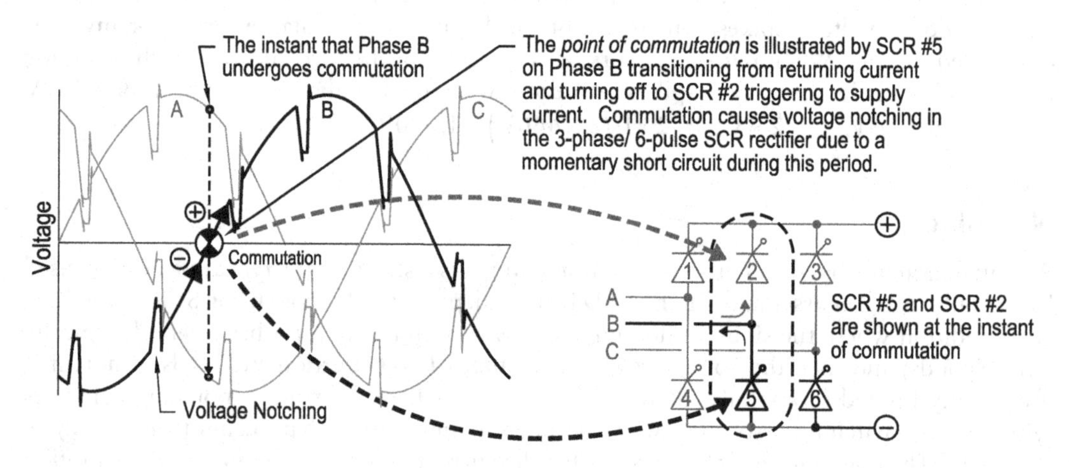

Figure 6.19 Voltage notching occurs during commutation. Commutation is the switching of polarity that occurs at the instant that two SCRs undergo phase transition from returning current to supplying current, and vice versa in the next half cycle.

faster. Regular abrupt changes can excite resonant circuits, producing radio-frequency interference (RFI), overload of power-factor capacitors, and other problems. IEEE 519 addresses limits for notch depth and notch area. Notching is often not much of a problem on stiff systems having large short-circuit potential, and they can usually be reduced to acceptable levels using a small-to-moderate amount of ac line reactance where problems are anticipated.

HARMONICS MITIGATION TECHNIQUES

There are several techniques that are used for mitigating harmonics, or in some cases avoiding harmonics. They consist of:

- DC Link Choke
- AC Line Reactor
- Tuned Harmonic Filters

- Multi-Pulse Transformers
- Active Front Ends
- Dv/dt Filters

DC link choke

The dc link choke, also called a dc link reactor, is a passive inductor that is placed in the VFD circuit between the rectifier assembly and the inverter section (Figure 6.20 upper left). The dc link choke stiffens against surges, filters the ripple, mitigates ac-source voltage spikes and sags (dv/dt) and the current spikes and sags (di/dt) produced by the front-end rectifier section.

Typically, the dc link choke is constructed around an iron core, which should be designed to avoid reaching saturation. Link chokes help to reduce the harmonic current (THDi) distortion of the ac supply, which leads to mitigation of voltage distortion, and in some installations, a dc link choke may be enough to meet the IEEE 519 THD requirements. The natural behavior of an inductor is to delay changes in current, so chokes have historically been used as current-limiting devices to avoid damaging levels of short-circuit amperage. A single, large choke may be installed in the positive leg before the inverter section, or preferably, two smaller chokes are used, one in the positive leg and the other in the negative leg. The double-inductor choke arrangement is more effective at filtering EMI than a single choke. Depending on the harmonic order being removed, the dc link choke can possibly be more or less effective than an ac line reactor. The dc link choke attenuates 5th and 7th harmonics fairly well, where the 5th and 7th are the largest harmonic contributors from VFDs.

The dc link choke is located after the rectifier assembly, so it does not afford any buffering of the incoming electricity as it enters the VFD, so it may be used alone, or it may be used in combination with a ac line reactor when more power-quality improvements are needed. Compared to the ac line reactor, the dc link choke produces less voltage drop for the same amount THDi mitigation, and it tends to be smaller and less expensive. If the VFD cabinet has built-in wiring terminals, it is possible to externally retrofit a dc link choke after the fact, if it is found to be needed when the system is placed in service.

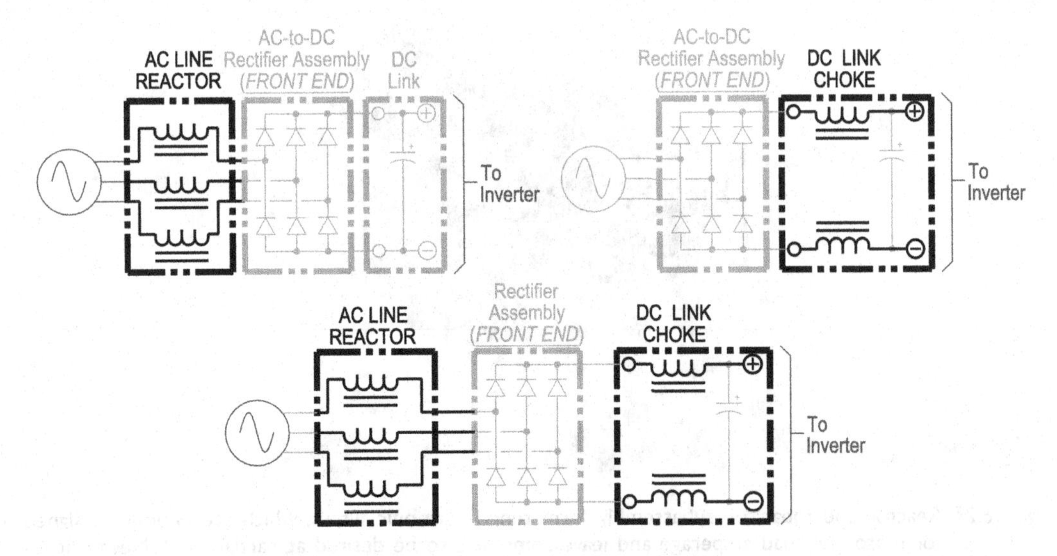

Figure 6.20 AC line reactor (upper left), dc link choke (upper right), and combined link choke and line reactor (bottom center) are used to mitigate harmonic distortion of current and voltage.

AC line reactor

The ac line reactor is a passive inductor that is placed before the VFD. Three identical inductors are preferred, one for each phase, although a single three-phase reactor can be built on a common iron core. A downside of the three-phase reactor is that its common-mode inductance renders it ineffective at canceling high-frequency noise. AC line reactors coup led with the dc link capacitors essentially form a low-pass filter to the VFD, removing high-frequency noise and protecting against voltage spikes. The intended purpose of the ac line reactor is similar to the dc link choke, in that it serves to mitigate source voltage and current spikes and sags, and it reduces the THD possibly to levels that meet the IEEE 519 power-quality requirements.

The ac line reactor may be used alone, or in combination with the dc link reactor when better power-quality is needed. A line reactor also provides some protection to the rectifier assembly from lightning surges or voltage spikes that occur during capacitor switching, while the dc link reactor does not. In addition, a line reactor used in conjunction with a dc link choke will extend the life of the dc link filtering capacitors. Line reactors also help in plants having unequal phase voltages.

The ac line reactors are typically larger and more expensive than the dc link reactors for the same THDi harmonic mitigation, and, its ac reactance drops voltage to the rectifier, and consequently, the dc link and inverter voltage. To solve existing problems, line reactors can be added before the front end at any time after the original installation, assuming there is floor space available, and taking care to avoid an objectionable voltage drop.

Tuned harmonic filters

A tuned harmonic filter is a resonant LC circuit installed in parallel with the supply. It functions to shunt one VFD harmonic from returning to the electrical source, averting the associated system voltage distortion. Tuned harmonic filters use inductors and capacitors that are designed to localize the harmonics in the VFD vicinity (Figure 6.21). The filter seesaws the harmonic current between the filter and the VFD (Figure 6.22 step 5), in a similar way that

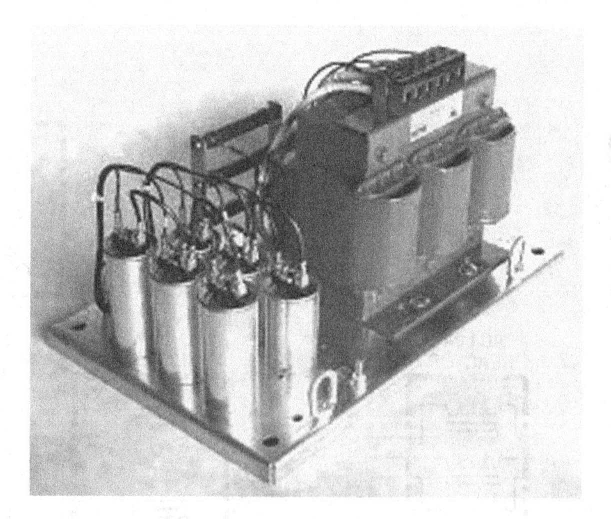

Figure 6.21 Reactor and capacitor subassembly form tuned harmonic filters, which are typically designed for motor full-load amperage and leave something to be desired at partial loads. Newer adaptive technoolgies are capable of mitigating harmonics down to about 20Hz. (Courtesy of MTE Corporation.)

power-factor-correction capacitors circulate reactive-power amperage between a motor's inductance and the power-factor capacitors.

Because each filter is matched to a unique impedance at one load, significant load variations on large VFDs may require that filter elements be cut in or out over the operating range, which can use newer adaptive technologies (Figure 6.23). This compensation is obtained by switching shunt capacitors and inductors in or out of the filter circuit. The use of tuned filters requires an analysis to determine the frequency and magnitude of the worst harmonics, and then the filters are designed to address the worst harmonics. As an example, a six-pulse

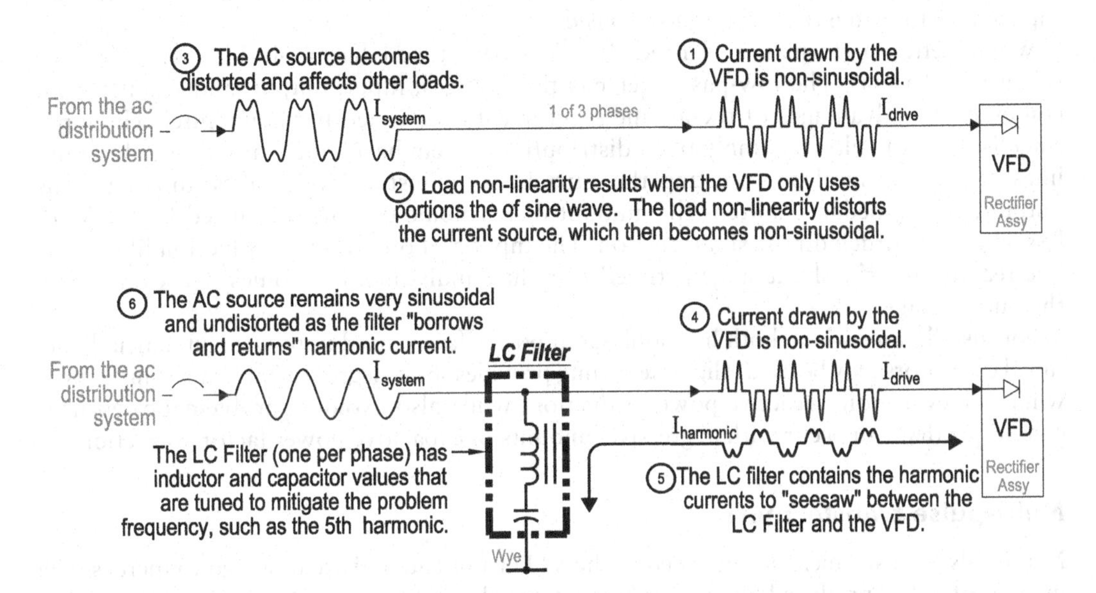

Figure 6.22 Tuned filters are installed to localize the harmonic currents near the non-linear load and to prevent them from reaching the electrical-distribution system.

Figure 6.23 The Matrix® AP harmonic filter (right) is a passive filter installed on the line side of a variable-frequency drive. This filter allows 6-pulse rectifier assemblies to meet the IEEE-519 requirements for harmonic distortion, at loads varying from as low as 30% and up to 100% of full motor rating. The filter uses adaptive passive technology (AP) to adjust the filter impedance in response to changing electrical loads. (Courtesy of the MTE Corporation.)

rectifier commonly has high 5th, 7th, and 11th harmonics, so three sets of LC filters can be ganged into a single module, where each filter is tuned to address its individual harmonic. Each filter may provide additional mitigation to higher-order multiples of that frequency, if they exist.

The best practice is to install the tuned harmonic filters as close as practical to the offending load. Harmonic filters can be dedicated to a single VFD, or larger tuned-filter banks can serve a motor-control center supplying a group of VFDs. For large motors experiencing big load swings, automatic compensation should be employed based on the degree of harmonics occurring at different loads, so that the compensation incrementally inserts or removes capacitance to match the instantaneous need.

When better performance is needed, a low-pass tuned harmonic filter can be used (Figure 6.24 right). The low-pass filter has three stages; input, output, and shunt, and it combines the advantages of the ac line reactor with the tuned harmonic filter. The input stage isolates the filter assembly from distribution-system harmonics caused by other non-linear loads, while also protecting the capacitors against voltage spikes and sags. The output-stage impedance softens the amount of harmonic currents produced by the VFD, lessening the burden on the shunt section. The input and output stages function like the ac line reactor and the shunt stage is tuned to reduce individual harmonics that escape past the output stage.

For installations where both harmonics and power factor need improvement, partially de-tuned capacitors may be used. Slight de-tuning provides some degree of harmonics mitigation, while achieving some reactive-power reduction, while also avoiding problematic resonant conditions that can occur with aggressive amounts of capacitive power-factor correction.

Multi-pulse transformers

Multi-pulse transformers are installed in the VFD front end and are designed to increase the number of pulses produced by the rectifier section. The additional pulses increase the number of humps, which reduces the magnitude of the humps and greatly smoothing the dc-link current ripple before any filtering is applied as shown in Figure 6.25. The additional pulses result in more of the ac source sine wave being used, as opposed to just six sharp spikes of current coinciding with each three-phase sinusoidal peak, as would be the case in a conventional three-phase, six-pulse arrangement. As more of the ac-input sinusoidal current is used, the

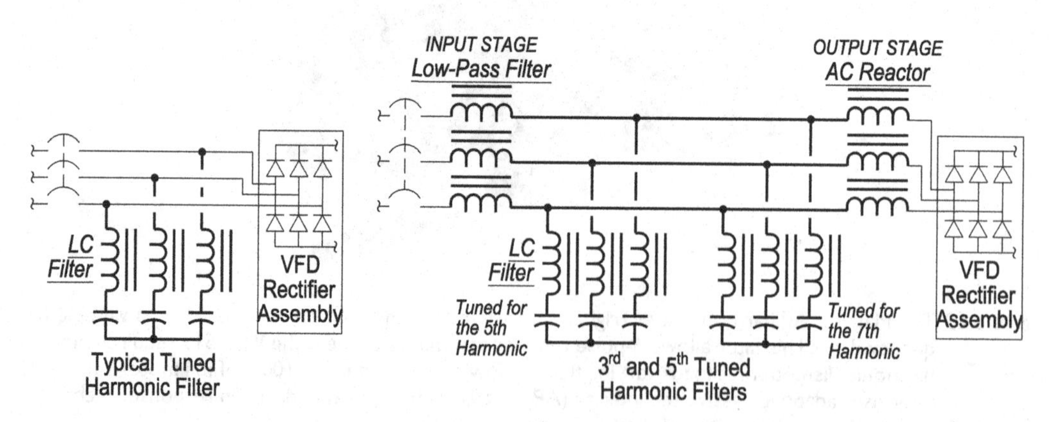

Figure 6.24 Tuned harmonic filters use inductors and capacitors connected in parallel to shunt the harmonic currents. In some cases, a low-pass filter may be added for more harmonic reduction.

associated voltage distortion it causes in the electrical distribution system is reduced. Multi-pulse multi-phase transformers have been designed to produce 12, 18, 24, and 36 pulses. A three-phase, 18-pulse transformer is shown in Figure 6.26 (left) and another transformer having an integral ac line reactor is shown on the right. Multi-pulse transformers can be constructed using delta-delta-wye transformers, or multi-tap three-phase transformers.

Delta-to-delta-wye multi-phase transformer

Multi-phase/multi-pulse transformers use output current-and-voltage phase shifting to produce more sinusoidal peaks from where the diodes can draw current. In the 12-pulse rectifier, the transformer has a set of three primary windings that are magnetically coupled into two

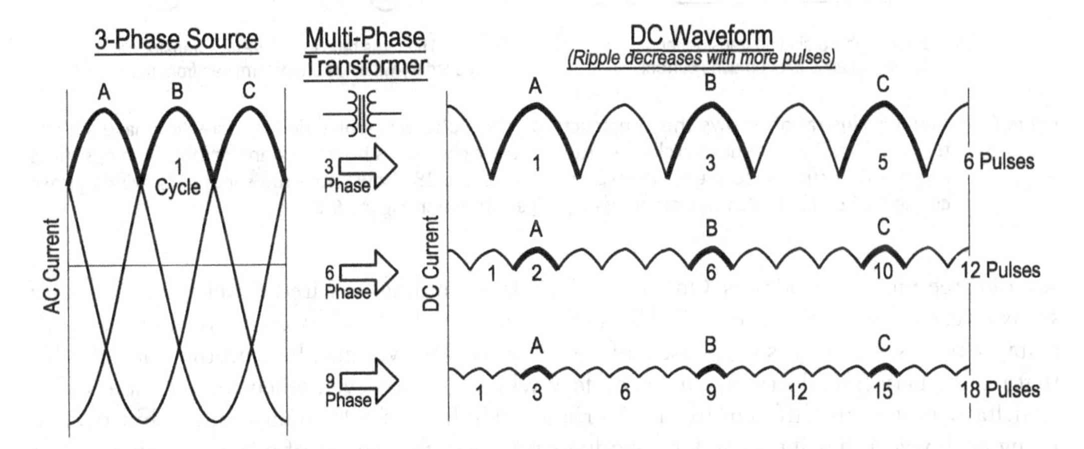

Figure 6.25 Relative comparison of reduced dc ripple by using multi-pulse transformers. Multi-phase/multi-pulse transformers create smoother dc current by using more of the source waveform. This technique results in significantly less distortion of the ac source and less harmonic distortion to the electrical distribution system.

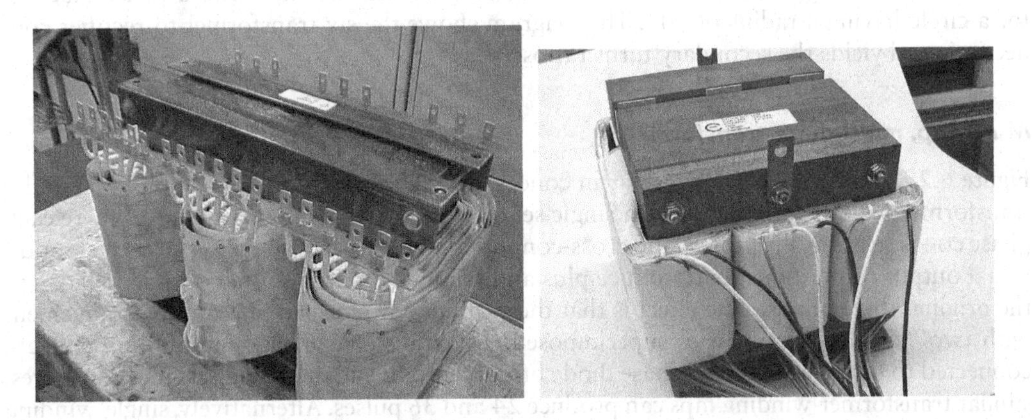

Figure 6.26 Multipulse transformers for full-wave rectification of three-phase electricity. An 18-pulse multi-tap transformer is shown to the left, and a 12-pulse delta-delta-wye transformer with an integral electrostatic shield is shown on the right. (Courtesy of E Craftsmen, Ontario, Canada.)

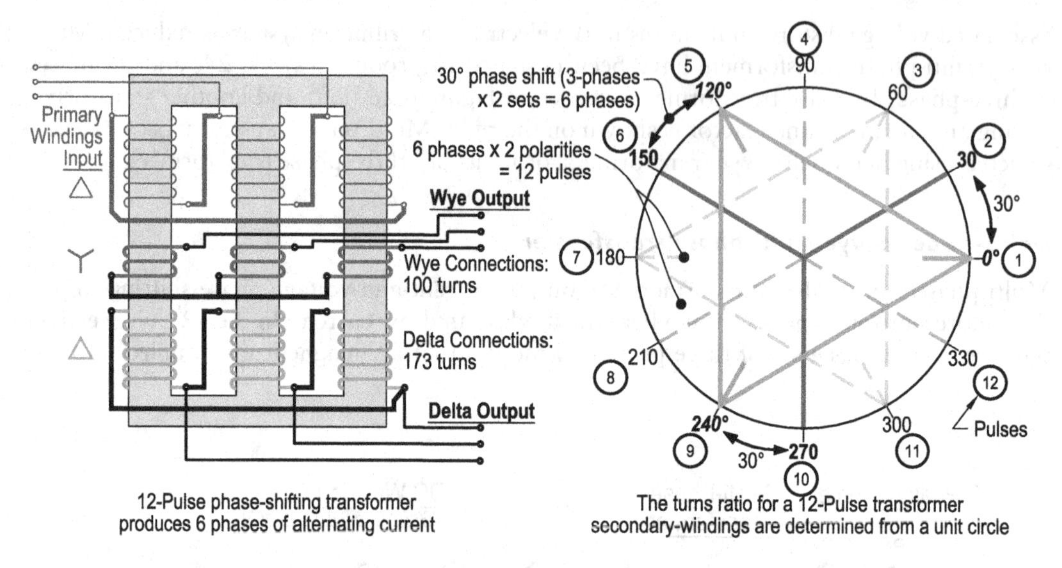

Figure 6.27 The left illustration shows the construction of a 6-phase/12-pulse delta-delta-wye phase shifting transformer. The "100-unit-circle" diagram on the right shows how the transformer-windings turns ratios are calculated to create the 12 pulses. Figure 6.28 shows how the secondary windings are connected to 12 diodes to obtain less ripple as shown in Figure 6.25.

sets of three-phase secondaries. One set of secondary windings is wired as delta and the other set as wye, as shown in Figure 6.27. When the primary and secondary configurations are both delta or both wye, there is no phase shift in the secondary voltage, beyond the normal 180° that occurs between all primary-to-secondary windings. However, when the secondary wye or delta windings are different from the primary windings, a 30° voltage phase shift occurs. Using both wye and delta secondary windings produces three pairs of outputs having *0°* and 30° relative displacements, i.e., *0°* and 30°, *120°* and 150°, and *240°* and 270°. Thus, the two sets of secondaries, one delta and the other wye, have a natural 30° phase shift between each other. The diodes in the full-wave bridge reroute the negative current to the positive terminal, so that the six connections produce a total of 12-pulses. Figure 6.28 shows a schematic representation of a delta-delta-wye phase-shifting 12-pulse transformer, with a phase diagram for a circle having a radius of 100. The diagram shows the six transformer-to-rectifier connections and yields the secondary turns ratios.

Multi-tap, multiphase transformer

Figure 6.29 and Figure 6.30 is similar in concept to Figure 6.28, but for a 9-phase/18-pulse transformer. This arrangement uses a single set of secondary windings per phase, where each phase coil is strategically tapped and cross-connected to the other two phases to provide additional output connections that produce plus and minus 20° phase-shifted outputs relative to the original three phases. The effect is that the windings are formed into a wye arrangement with two "zig-zag" secondaries superimposed. The transformer has nine outputs that are connected to three sets of three-phase diode bridges, producing the 18 pulses. As alternatives, similar transformer-winding taps can produce 24 and 36 pulses. Alternatively, single-winding autotransformers can accomplish the same 18-pulse output, when voltage transformation is not required. Figure 6.31 shows the general arrangements and the typical current waveforms for 6-, 12-, 18-, and 24-pulse rectifier assemblies.

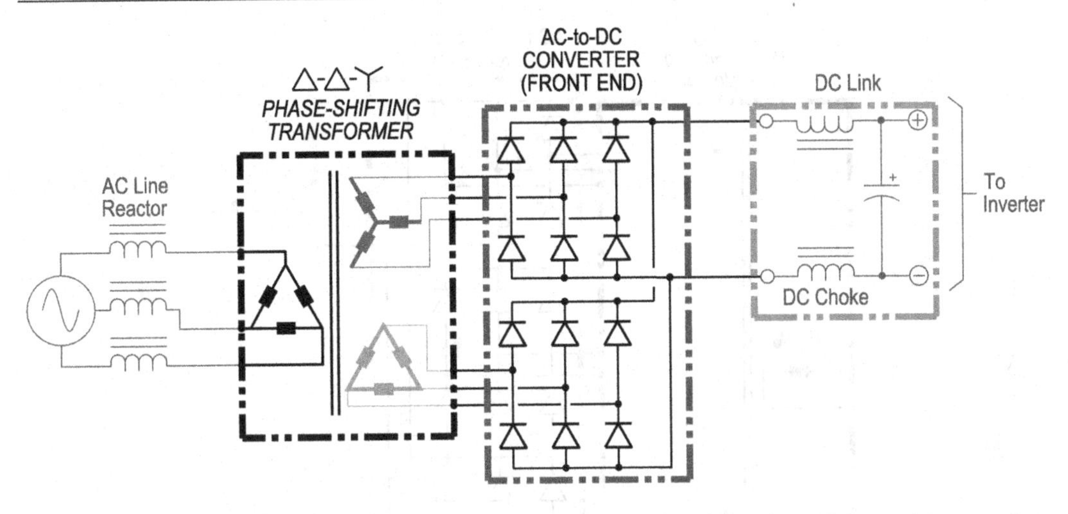

Figure 6.28 The diode wiring for a 12-pulse inverter design using a 6-phase 12-pulse delta-to-delta-wye phase shifting transformer.

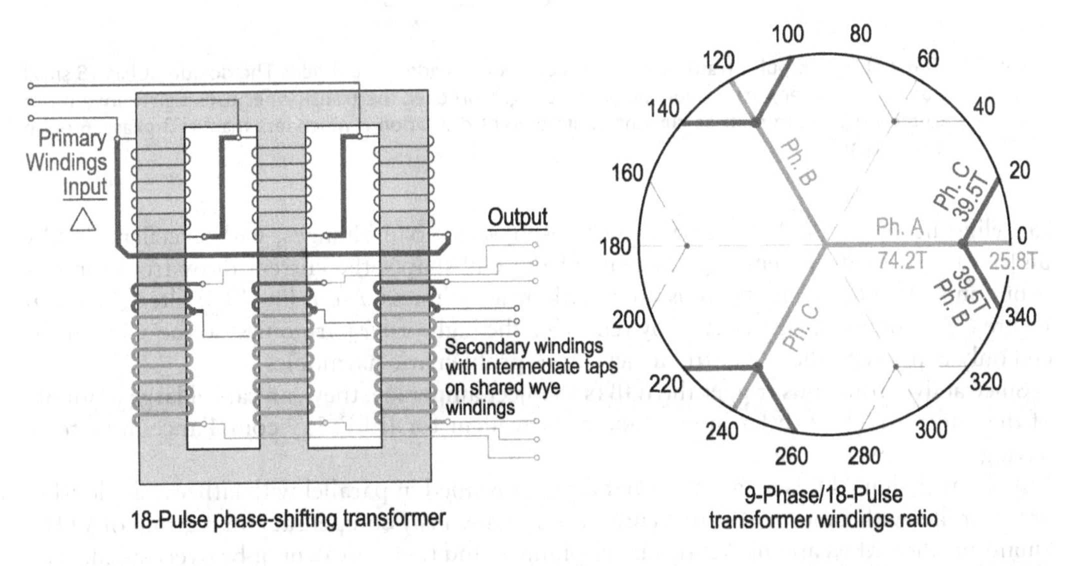

Figure 6.29 A multi-pulse transformer is obtained by tapping the secondary windings as shown in the left illustration, and then interconnecting the tap connections to the other two phases. A circle having a radius of 100 is used to calculate the turns ratios and tap locations from the phase-winding geometry.

Active front ends

Active front ends (AFE) having MOSFET or IGBT dc switching power supplies are used as an alternative to diodes or SCR rectifiers. Active front ends rectify ac to dc using sophisticated gate control to strategically trigger MOSFETs or IGBTs to conduct at just the right time. The wiring arrangement of the active front end is similar to IGBT-based PWM inverters, however, the AFE controller measures the harmonic currents and then uses a microprocessor to trigger the IGBTs in a manner that injects corrective currents in an equal-but-opposite manner as the harmonic amperages. The concept is similar to the active filtering used in noise

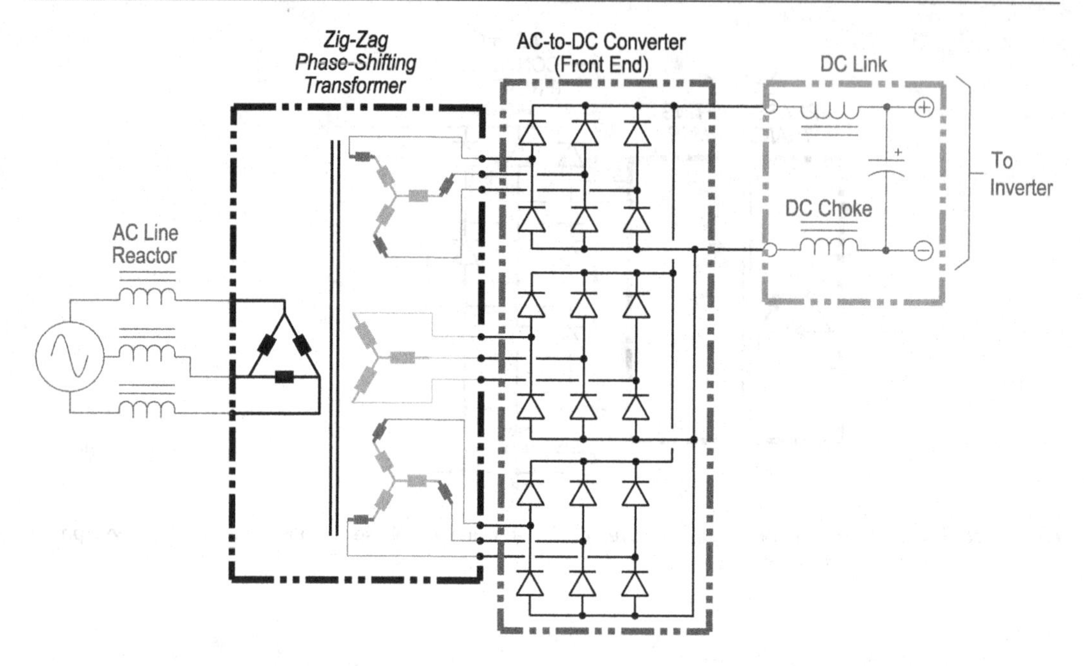

Figure 6.30 The 9-phase, 18-pulse transformer connections are made to 18 diodes. The dc output has 18 small humps having very little ripple before filtering. Moreover, the transformer uses significantly more of the original three-phase current, so its current distortion is much less than its 3-phase, 6-pulse counterpart.

canceling headphones. AFE provides fast response to load changes, with excellent results under varying loads. Essentially, the active front end shapes the current draw from the distribution system to be nearly sinusoidal, with little spikiness. An in-line LCL filter, shown in Figure 6.32, complements the AFE by removing the high-order harmonics, so the active front end only copes with the lower-frequency, more-problematic harmonics.

Since active front ends trigger the IGBTs to inject amperage, they can cancel large amounts of the 2nd through 50th harmonics and are excellent for IEEE 519 compliance in systems having poor power quality.

A shunt-design AFE is a variation that can be installed in parallel with either a single VFD or in parallel with a motor-control center to simultaneously compensate for a group of VFDs. Shunt-installed AFEs are inherently current-limited and therefore cannot be overloaded, even under fault conditions, unlike series-installed AFEs where all amperage passes through the drive. Although shunt AFE filters are more complex and expensive, they may be economical when used with multiple VFDs.

High-power VFDs are apt to use 12- or 18-pulse phase-shift transformers, which are excellent at inherently avoiding harmonics that would otherwise be injected back into the source. The newer PWM switching technologies using IGBTs switch so fast and conduct with little losses so that they are less prone to producing harmonics, especially when compared to earlier generation VFDs. AFE filtering cancels nearly all harmonics without large filtering assemblies.

The active front end has the additional benefit of providing some power-factor correction and can even produce leading power factor to offset a generator's reactive power. Total harmonic distortion is inversely related to the power factor, so as the objectionable THD

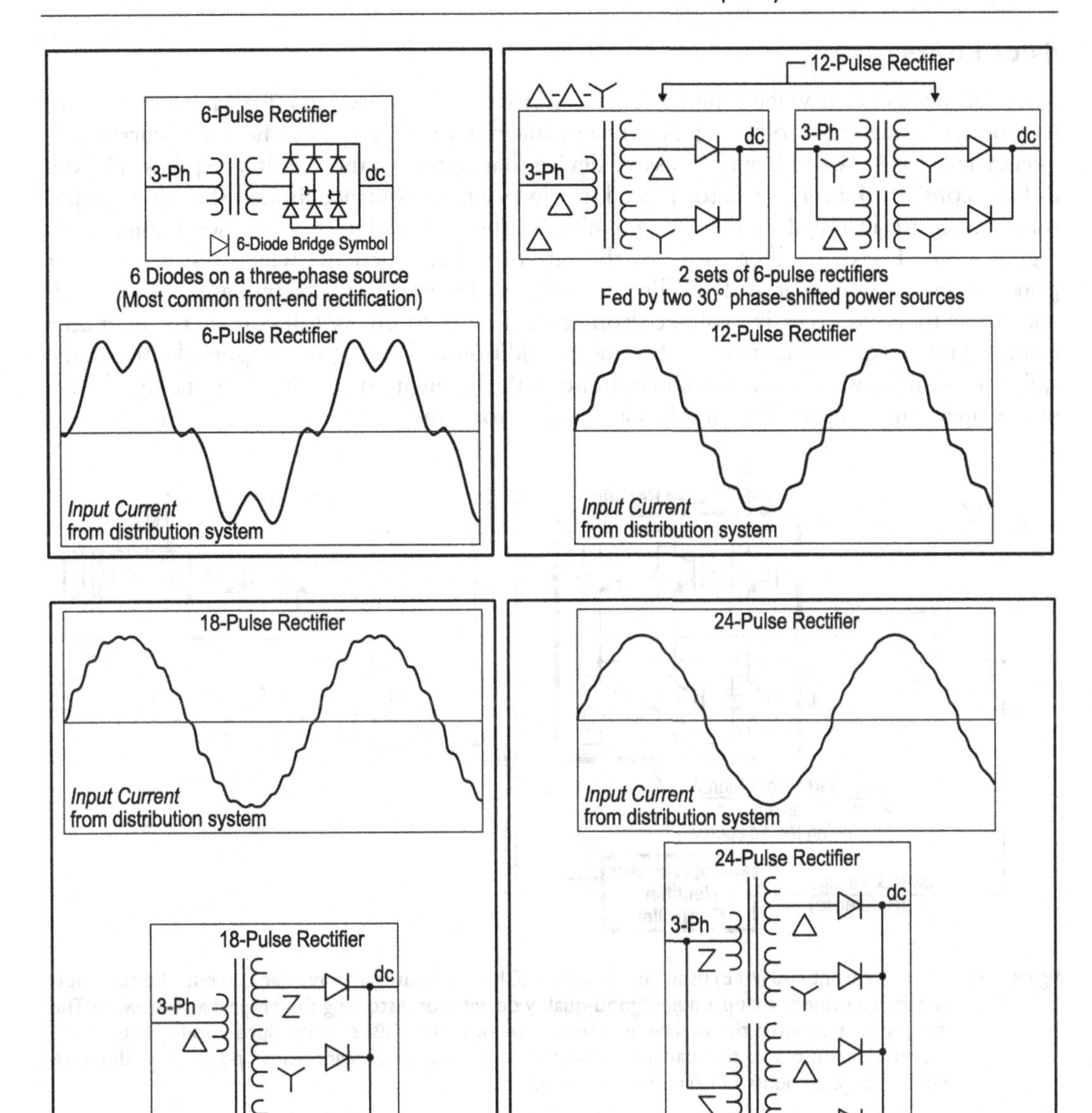

Figure 6.31 Illustrative distortion of motor-supply currents caused by different front-end rectifier assemblies. More pulses make better use of the ac amperage, using more current from the sine wave and resulting in less associated voltage distortion.

increases, the power factor worsens according to the relationship: $pf = \dfrac{\cos\theta}{\sqrt{1+THD^2}}$. The AFE also has "four-quadrant capability," meaning the motor can produce positive or negative torque, giving it regenerative ability during overhauling loads or load reversals, improving energy recovery and providing effective, efficient braking.

dv/dt filters

Reflected waves occur when long cable lengths are installed between a PWM-type VFD and its motor, causing large voltage spikes. The problem is worsened when the PWM carrier frequency is high. A dv/dt filter is a passive device that mitigates those voltage spikes. The dv/dt filter consists of a series reactor placed just in front of resistance-limited shunting capacitors, which are installed as close as possible to the motor (Figure 6.33; see Figure 6.34, Figure 6.35). The dv/dt filter dampens the rate of voltage increase, which results in lower peak voltage at the motor. A dv/dt filter typically is designed for approximately 3% impedance so as to avoid excessive voltage drop at the motor. Applying full voltage to the motor is important for obtaining its rated torque. In addition to limiting the magnitude of voltage spikes to remain within the insulation limits, other benefits of dv/dt filters include lower motor operating temperatures and less audible motor hum.

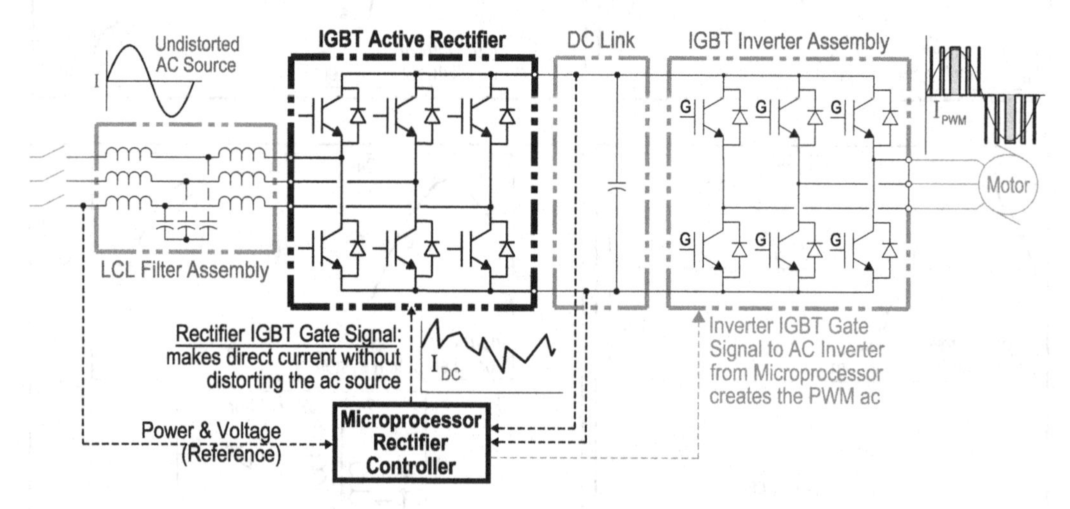

Figure 6.32 The active front end (AFE) uses gate-switched IGBTs to strategically pass ac current. The current is passed in a manner that produces good-quality dc without distorting the original ac sine wave. The inverter section after the dc link also uses gate-switched IGBTs to produce good-quality PWM ac sinusoidal output to the load. This arrangement is capable of regenerative power, in addition to avoiding large amounts of harmonic distortion.

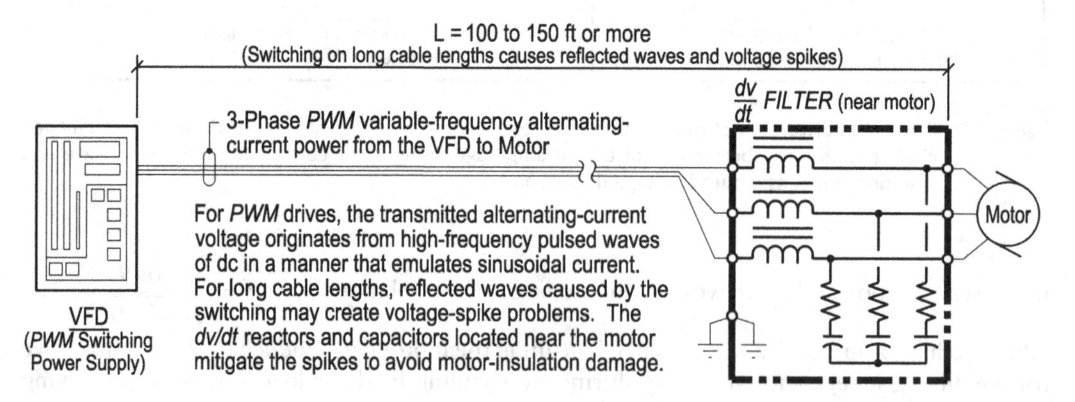

Figure 6.33 A dv/dt filter mitigates voltage spikes caused by PWM-generated ac current transmitted over a long distance.

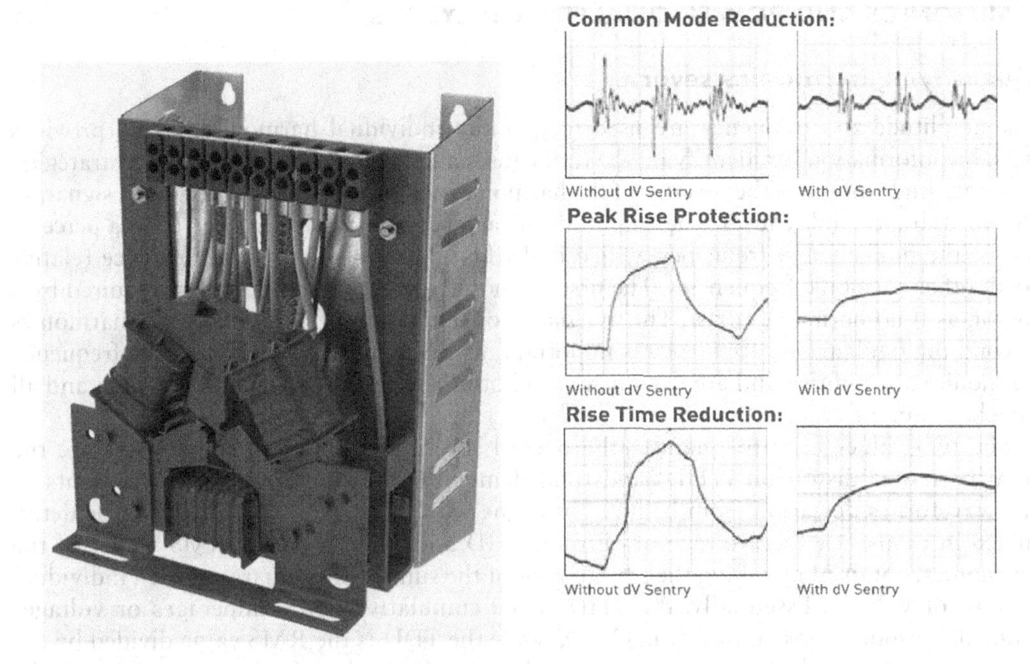

Figure 6.34 The dV Sentry™ is a dv/dt filter that reduces common-mode voltage created by PWM variable-frequency drives, reduces peak voltage spikes, and reduces rise-times, all in one unit. This filter is capable of mitigating over 50% of the common-mode voltage and its associated current imbalances, which occur with PWM inverters supplying motors having long cable runs. The filter eliminates premature bearing failures and the need for shaft grounding. (Courtesy of the MTE Corporation.)

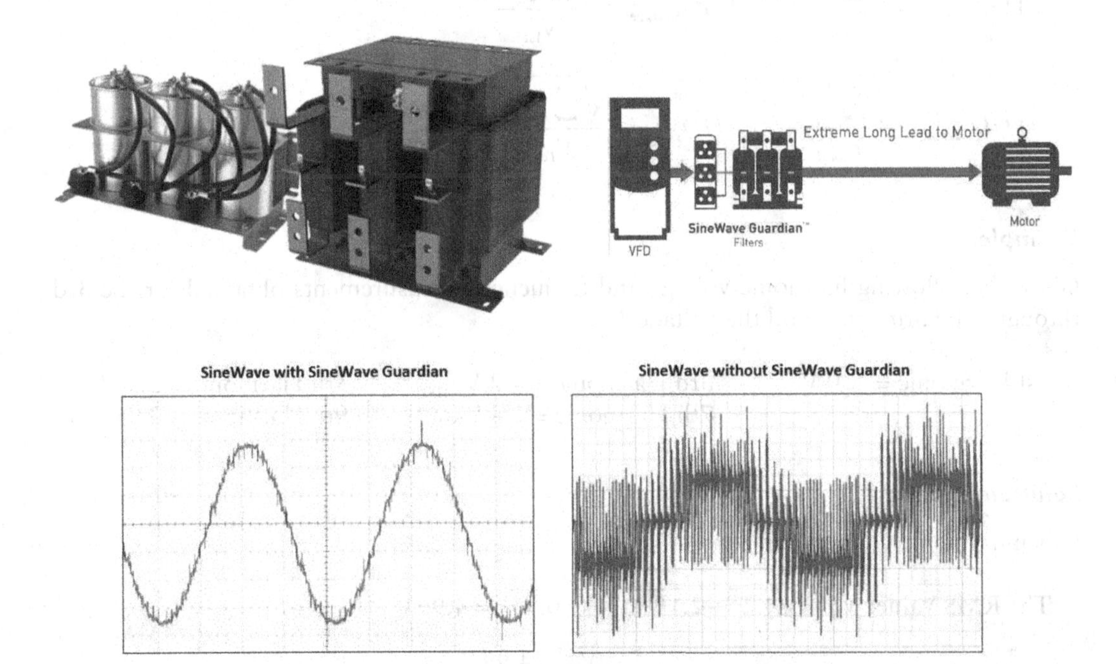

Figure 6.35 Sine-wave filters are used on the motor side of variable-frequency drives using PWM technology and long cable leads. The SineWave Guardian™ filter virtually eliminates high-frequency content and voltage peaks, cleaning the current to be near sinusoidal. Motor life is extended as the filters reduce heating and insulation stress. (Courtesy of MTE Corporation.)

HARMONICS AND POWER QUALITY ANALYZERS

Quantifying harmonics severity

The amplitude and frequency are used to quantify individual harmonics, which provides valuable information for identifying the root causes and for developing mitigation strategies. The amplitude indicates the severity of the harmonic, while its frequency provides a signature that leads to probable causes. The amplitude of each harmonic is typically given as a percentage of the fundamental frequency, which is the first harmonic and is the reference relative to all other harmonic frequencies. The first harmonic represents the electricity required by a motor, as if no harmonics exist. The avoidance or mitigation of all higher-order harmonics beyond the fundamental frequency is important, as the cumulative effects of like-frequency harmonics are additive and are detrimental to both the electrical distribution system and all attached equipment.

The two values used to quantify the overall severity of all harmonic sources are the total harmonic distortion (THD) and total demand distortion (TDD), measurements of the waveform's deviation from being a pure sine wave. The THD and TDD can be determined in terms of either voltage or current. THD and TDD both use RMS values of the harmonics, obtained by taking the square-root of the sum-of-the-squares of each individual current or voltage. Essentially, the THD is the cumulative RMS amperages or voltages from all harmonics for an individual VFD, while the TDD is the RMS value divided by the maximum systemwide demand current or voltage at the "point of common coupling." The equations are:

$$\mathrm{THD}_{\mathrm{Current}} = \frac{\sqrt{\sum_{n=2}^{50} \mathrm{I}_{n}^{2}}}{\mathrm{I}_{1st}} \qquad \mathrm{THD}_{\mathrm{Voltage}} = \frac{\sqrt{\sum_{n=2}^{50} \mathrm{V}_{n}^{2}}}{\mathrm{V}_{1st}}$$

$$TDD_{Current} = \frac{\sqrt{\sum_{n=2}^{50} I_{n}^{2}}}{I_{Total}} \qquad THD_{Voltage} = \frac{\sqrt{\sum_{n=2}^{50} V_{n}^{2}}}{V_{Total}}$$

Example

Given the following harmonic voltage and frequencies measurements obtained for the 3rd through 9th harmonics, find the voltage THD:

1st Harmonic = 120 V

3rd Harmonic = 4.2 V_{RMS}
7th Harmonic = 1.4 V_{RMS}

5th Harmonic = 2.1 V_{RMS}
9th Harmonic = 0.95 V_{RMS}

Solution method I

(when harmonic voltages are given)

The RMS Value: $V_H = \sqrt{4.2^2 + 2.1^2 + 1.4^2 + 0.95^2} = 4.99\,V$

The total harmonic distortion: $THD_V = \frac{V_H}{V_1} = \frac{4.99\,V}{120\,V} \times 100 = 4.16\%_V$

Solution method 2

(when THD is given in percent of the first harmonic)

Individual harmonic distortions, in percent:

$$V_{3rd} = \frac{4.2\ V}{120\ V} \times 100 = 3.5\% \qquad V_{5th} = \frac{2.1\ V}{120\ V} \times 100 = 1.75\%$$

$$V_{7th} = \frac{1.4\ V}{120\ V} \times 100 = 1.17\% \qquad V_{9th} = \frac{2.1\ V}{120\ V} \times 100 = 0.90\%$$

$$THD_V = \sqrt{3.5^2 + 1.75^2 + 1.17^2 + 0.90^2} = 4.16\%_V$$

Note that the individual harmonic distortions in percent can be used directly, since the first harmonic is 100%.

Power-quality measurement

Power quality and energy analyzers are used to quantify levels of harmonics. The analyzers are used to prove compliance of equipment and systems in meeting contractual power-quality requirements and for troubleshooting problematic systems where poor power quality appears to be an issue. In addition to harmonic power analysis, some analyzers have built-in energy-loss calculators that can indicate real and reactive power, power factor, phase imbalances, and even the cost associated with harmonic-energy losses. The Fluke Model 435 is

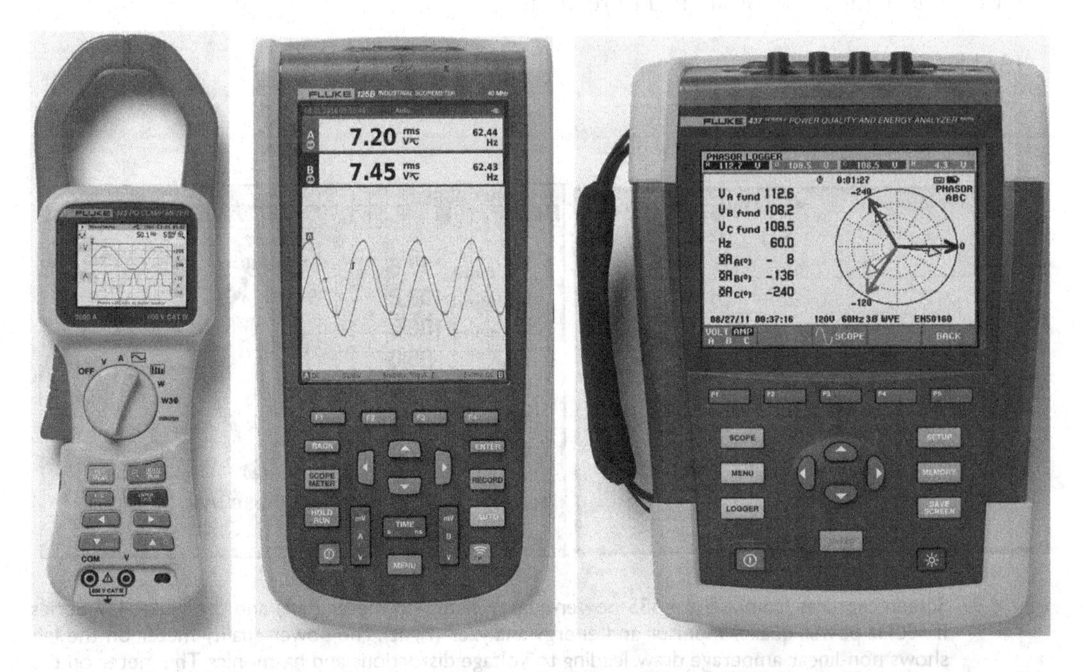

Figure 6.36 Model 345 power-quality clamp-on meter (left) ScopeMeter® 125B handheld oscilloscope, and the Model 437 Series II 400Hz power quality monitor (center) and energy analyzer (right). (Courtesy of Fluke Corporation.)

shown in Figure 6.36. This instrument can analyze power quality in real time and can acquire and store data for later analysis and trending.

Figure 6.37 shows two screens from a power-quality analyzer. The screen on the left shows the sinusoidal applied voltage in the upper half of the screen in the time domain. The lower half shows the amperage drawn by the load, which in this case is from a single-phase full-wave bridge rectifier having capacitive filtering like the circuit shown in Figure 6.15.

This screen on the left is indicating the following:

$V = 121.7\ VAC$

$f = 60.0Hz$

$I_{RMS} = 8.78\ A$ (Non-sinusoidal, spiked current)

$CF = 1.4$ (Voltage Crest Factor)

$V_{peak} = 1.4 \times 121.7 = 170V$

$CF = 3.6$ (Current Crest Factor)

$I_{peak} = 3.6 \times 8.78A = 31.6A_{peak}$

The screenshot on the right shows the harmonics information but displayed in the frequency domain. The frequency-domain bar graph shows the percent contribution of each individual harmonic relative to the fundamental frequency of 60Hz. This screen indicates that the 3rd, 5th, and 7th harmonics are the most significant, tapering in amplitude as the frequency increases. It also has small, but measurable harmonics at the 9th, 11th, 13th, 15th, and 21st. Each harmonic frequency and its associated magnitude is displayed in percentage relative to the fundamental 60Hz frequency as the user scrolls through the frequencies.

In this screen, the cursor is aligned to read the

3rd harmonic

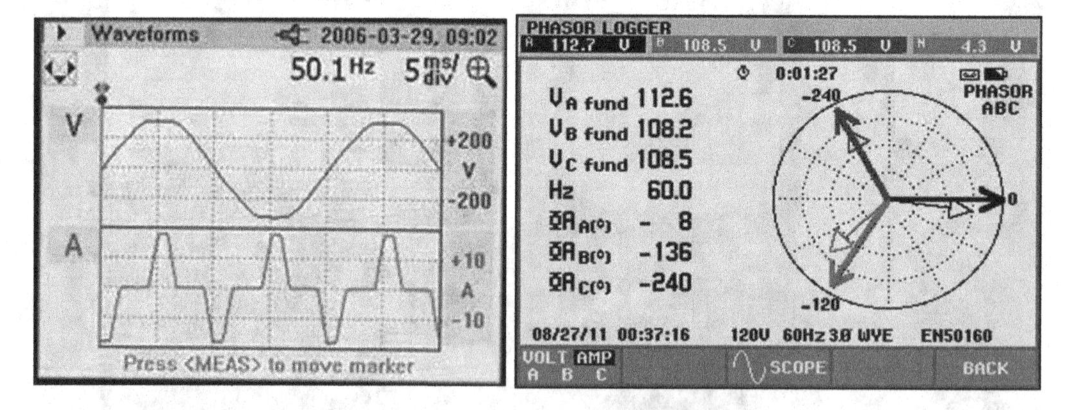

Figure 6.37 Screen captures from a Fluke 435 power-quality clamp-on meter (left) and the Fluke 437 Series II 400Hz power quality monitor and energy analyzer (right). The power-quality meter on the left shows non-linear amperage draw, leading to voltage distortions and harmonics. The meter on the right shows the voltages and phasor-angles from an imbalance of load impedance on each of the three phases. (Courtesy of Fluke Corporation.)

$f = 179.9Hz$

$I_{all\ harmonics} = 5.13A_{RMS}$

$I_{3rd} = 1.29A$

3rd Harmonic Distortion = 25.2%r (reference to the total harmonic current: 1.29A/ 0.252 = 5.12A)

% harmonic THD = 30.9%RMS for all harmonics combined

CONCLUSION

Variable-speed drives are useful for reducing the speeds of motors to obtain better system performance, higher system efficiency, and economical operation. There are several technologies available today, which vary in cost, complexity, and performance, where the selection of the most appropriate technology is often based on the motor size and the stiffness of the electrical distribution system. A downside of VFDs is their production of harmonic currents that distort the electrical distribution system. The harmonics occur at multiples of the fundamental system frequency, and they produce problems with both the motors they serve and the distribution source that feeds them, which affects all other attached loads. If the motors are large and the harmonics significant, more expensive VFD designs can be used to either avoid or mitigate objectionable levels of harmonics. In other cases, passive or active filters can be employed to mitigate the offending harmonics from propagating their deleterious effects into the source and other loads. Frequency-spectrum analyses can be obtained from power-quality meters to determine severity, troubleshoot problematic systems, and formulate corrective actions.

QUESTIONS

1. Describe the principle of operation of a variable-frequency drive (VFD).
2. A 60Hz four-pole 100-HP motor operates at 460V and has 2.8% slip, a power factor of 0.87 and an efficiency of 94%. What is the motor's full-load speed, full-load operating amperage, and full-load torque? Calculate the reduced motor speed, the motor power, the new speed, and the if the motor is driven by a variable-frequency drive operating at 45Hz, operates by holding volt/Hertz ratio constant at 7.67 V/Hz. Motor equations are:

$$HP = \frac{Tn}{63{,}025} \qquad f = \frac{pn}{120} \qquad P_e = \sqrt{3}VIpf \qquad \eta_{mtr} = \frac{BHP \times 0.746\, kW/HP}{P_e} \qquad T \propto V^2$$

$$\%\ Slip = \frac{n_{sync} - n_{full\ load}}{n_{sync}}$$

3. What are the three main sections of a VFD? Sketch a simplified block diagram showing how a VFD works.
4. Sketch the bridge circuit for an older SCR-based inverter and describe how it provides approximated ac electricity.

5. List six types of electronic variable-speed technologies. Briefly list in bulleted format the salient features of each technology and any disadvantages it may have.
6. Describe pulse-width modulation (PWM). What electronic devices/technologies are used for PWM-based drives? Provide sketches to support your description.
7. What are non-linear loads? Provide some examples describing how non-linear amperages are drawn from the system. How do non-linear loads affect the voltage source and the electrical distribution system?
8. What are harmonics? What are some problems that result from harmonics?
9. List the classifications/sources of harmonics. Briefly list the causes.
10. What are reflected waves? Where are systems prone to problems from reflected waves?
11. What is voltage notching? What causes voltage notching? Which technology is prone to voltage notching?
12. List six techniques used for mitigation of harmonics. List the type of harmonics problem the technology addresses.
13. What are multi-pulse transformers? How are they used in VFD front ends and how are they effective at mitigating the harmonics problems from reaching objectionable levels? Sketch the ac current draw by a 6-pulse and 24-pulse rectifier assemblies, as an example.
14. What is an active front end? How does the active front end work?
15. What are dv/dt filters? Where are dv/dt filters used? How can a design be modified to avoid needing dv/dt filters?
16. How are harmonic distortions quantified? What is the difference between total harmonic distortion (THD) and total demand distortion (TDD)? How can power quality be measured, and how are power-quality measurements useful?

NOTES

1. The inductive reactance in the motor windings is directly related to the frequency ($X_L = 2\pi fL$). At lower frequency, the inductive reactance decreases, contributing to lower the circuit impedance and more current. Following Ohm's Law ($V = IZ$), more current leads to more I^2R heat.
2. "Notch" is defined by the IEEE 519 standard as a disturbance in the normal voltage waveform lasting less than half a cycle, which is initially of opposite polarity than the waveform, and is thus subtracted from the normal waveform. Figure 6.18 shows notching.
3. An ac voltmeter produces an output voltage equal to the average voltage when it is passed through a rectifier. For a single-diode half-wave rectifier the dc voltage is $V_{dc} = V_{pk} / \pi = \sqrt{2} \cdot \frac{V_{rms}}{\pi} = 0.45 \cdot V_{rms}$. For a 120Vac source, $V_{dc} = 0.45 \cdot 120\, V_{rms} = 54V$, not the 60V that might be expected when using half the wave.

Chapter 7

Boiler controls

Boiler controls vary widely depending on size, heat source, steam pressure, and primary function. Generalizing a boiler control scheme is difficult, as there are a large number of techniques and devices to accomplish the control. Steam boilers are classified by the ASME Boiler and Pressure Vessel Code as being either heating or power types. Heating boilers generate lower-pressure saturated steam for transporting thermal energy in the form of latent heat, while power boilers generate high-energy superheated steam for running turbines. The ASME classifies boilers as low pressure when the safety-relief valves are set at or below 15 psig, and as high pressure when the safety valves are above 15 psig. The ASME also classifies boilers as unfired when they use steam, high-temperature hot water, or waste-heat gases as the heat source, instead of fuel, such as steam-to-steam contaminated evaporators as an example. Contaminated evaporators are used where oil-heater tube leaks could introduce oil into the condensate returns and boiler feed water, causing poor heat transfer and eventual tube damage to high-pressure boilers. In other plant types, such as diesel or gas-turbine plants, steam may be generated by heat-recovery steam generators, where waste energy within the combustion gases from a topping cycle is used to generate saturated steam for HVAC systems or superheated steam for a Rankine bottoming cycle. The point is that the boiler-control approach, the control systems, and the components vary substantially depending on the service and specific needs of the plant.

Boiler controls use a mix of combinational logic and sequential logic. The combinational logic includes permissives and interlocks, where two-state devices determine whether the operating parameters are within safe and acceptable limits, at which point the boiler is allowed to fire. The steam drum low-low level and loss-of-flame cut-out switches are two examples of combinational logic interlocks. The sequential logic includes the procedural steps followed as the boiler cycles through its normal operations, such as pre-purge, followed by burner light-off, followed by proof-of-flame, etc.

As an example, the Code of Federal Regulations for Shipping, 46 CFR Subpart 62, discusses "Vital System Automation." Subpart 62.35-20 includes the requirements for "Oil-fired main boilers." Some regulation-requirement examples for boiler control are listed below:

> (d) Programming control. The programming control must provide a programmed sequence of interlocks for the safe ignition and normal shutdown of the boiler burners. The programming control must prevent ignition if unsafe conditions exist and must include the following minimum sequence of events and interlocks:
>
> (1) Prepurge. Boilers must undergo a continuous purge of the combustion chamber and convection spaces to make sure of a minimum of 5 changes of air. The purge must not be less than 15 seconds in duration and must occur immediately prior to the trial for ignition of the initial burner of a boiler. All registers and dampers must be open and an air flow of at least 25 percent of the full load volumetric air flow must be proven before

the purge period commences. The prepurge must be complete before trial for ignition of the initial burner.

NOTE: A pre-purge may not be not required immediately after a complete post-purge.

(2) Trial for ignition and ignition.
  (i) Only one burner per boiler is to be in trial for ignition at any time.
  (ii) Total boiler air flow during light off must be sufficient to prevent pocketing and explosive accumulations of combustible gases.
  (iii) The burner igniter must be in position and proven energized before admission of fuel to the boiler. The igniter must remain energized until the burner flame is established and stable, or until the trial for ignition period ends.
  (iv) The trial for ignition period must be as short as practical for the specific installation, but must not exceed 15 seconds.
  (v) Failure of the burner to ignite during a trial for ignition must automatically actuate the burner safety trip controls.

(3) Post-purge.
  (i) Immediately after normal shutdown of the boiler, an automatic purge of the boiler equal to the volume and duration of the pre-purge must occur.
  (ii) Following boiler safety trip control operation, the air flow to the boiler must not automatically increase. Post purge in such cases must be under manual control.

Fuel-fired boiler control systems include the following subsystems:

- Burner management system and flame-safeguard system
- Combustion control
- Drum-level and boiler feedwater regulation
- Draft control and air-pollution mitigation
- Superheat temperature control
- Fuel selection and pollution mitigation

BURNER-MANAGEMENT SYSTEM AND FLAME SAFEGUARD SYSTEM (FSS)

The burner-management system (BMS) controls the sequencing and actuation of the boiler fuel burners. The burner management system works hand-in-hand with the flame safeguard system (FSS) and in conjunction with, but separately from, the combustion-control system. The burner-management operations are summarized in the block diagram in Figure 7.1. The major BMS functions include boiler/burner light-off, shutdown, and on-off burner sequencing during load changes for boilers having multiple-burners, which are summarized in Figure 7.2.

For light-off, the BMS controls the sequencing for furnace pre-purge, the burner light-off attempt, proof-of-flame, and release to modulate. For shutdown, the BMS shuts the fuel, performs a post-purge, and secures the combustion air. When a boiler has multiple burners, the burner management system automatically cuts-in and cuts-out burners as required to adjust the heat input. In multiple-boiler installations, the burner management system has provision to adjust biasing and load-sharing between individual boilers. The BMS is distinguished from the combustion-control system, in that the burner-management system steps through the mechanics of cycling the burners on and off as needed, and it enforces safe operating procedures; whereas the combustion-control system modulates the fuel to match the steam load, while simultaneously adjusting the relative amount of air-to-fuel for good combustion.

The flame-safeguard system is a set of permissive controls that automatically shuts burners upon unsafe conditions. The main purpose of the FSS is to prevent furnace explosions by

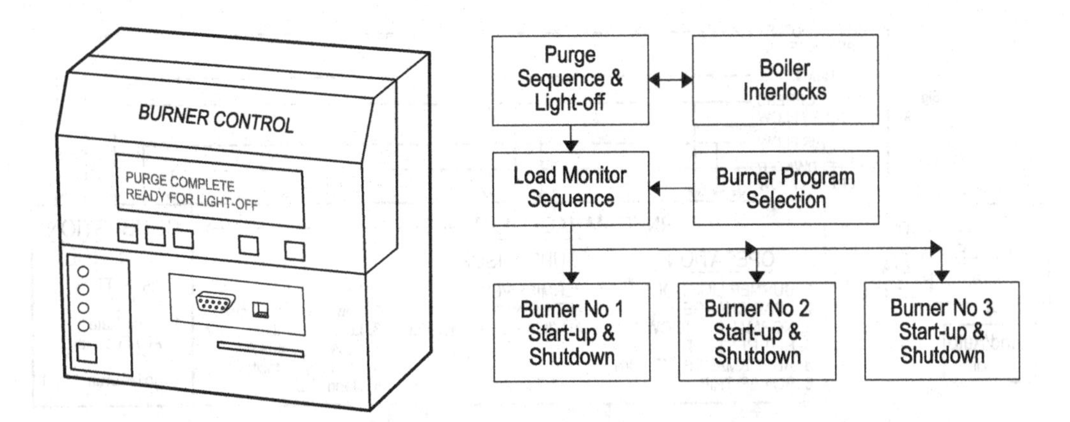

Figure 7.1 An electronic flame-safeguard system checks the status of the boiler interlocks and permits firing when conditions are safe. The burner-management system purges the furnace of combustible gases, lights off burners, performs both controlled and safety shutdowns, and sequences multiple burners in matching the heat input to the boiler load.

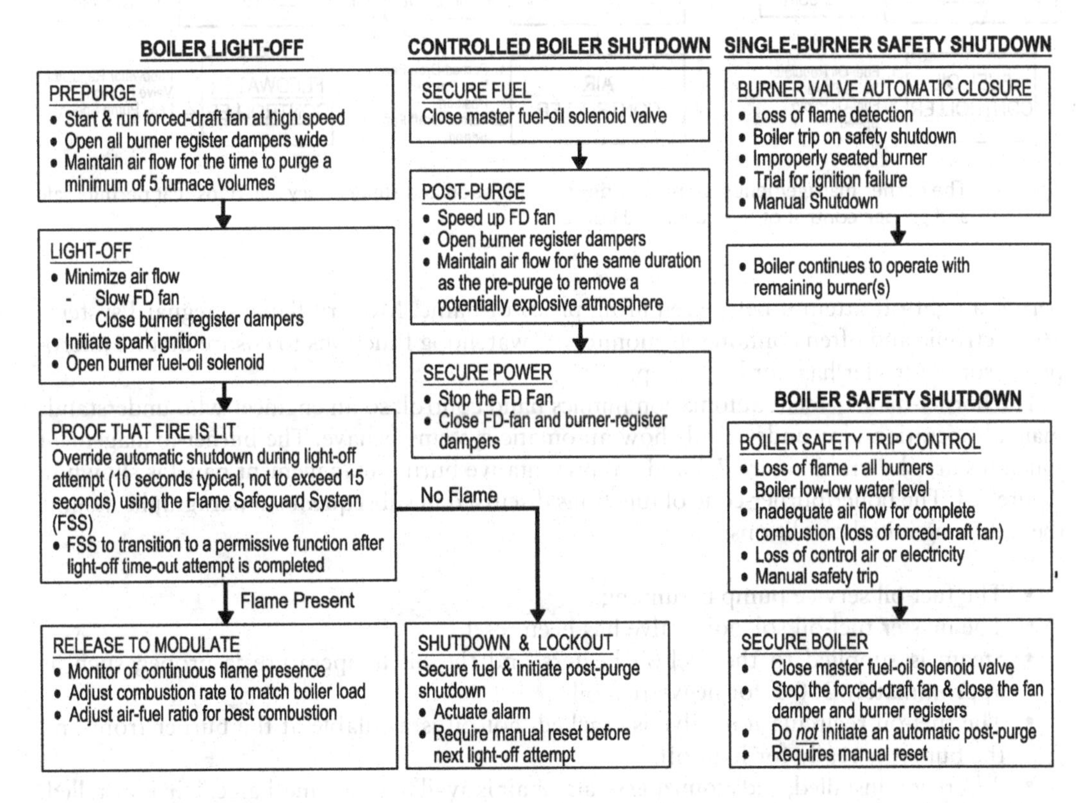

Figure 7.2 These simplified flow diagrams show various boiler and burner management light-off and shutdown actions.

securing the fuel. The FSS ensures that a successful pre-purge operation has been completed before light-off can be attempted, otherwise it trips the main fuel solenoid valve and initiates an alarm. After light-off, the FSS uses a flame scanner to ensure the presence of a fire to avoid pumping unburned fuel into the furnace. The flame safeguard has a built-in time delay to

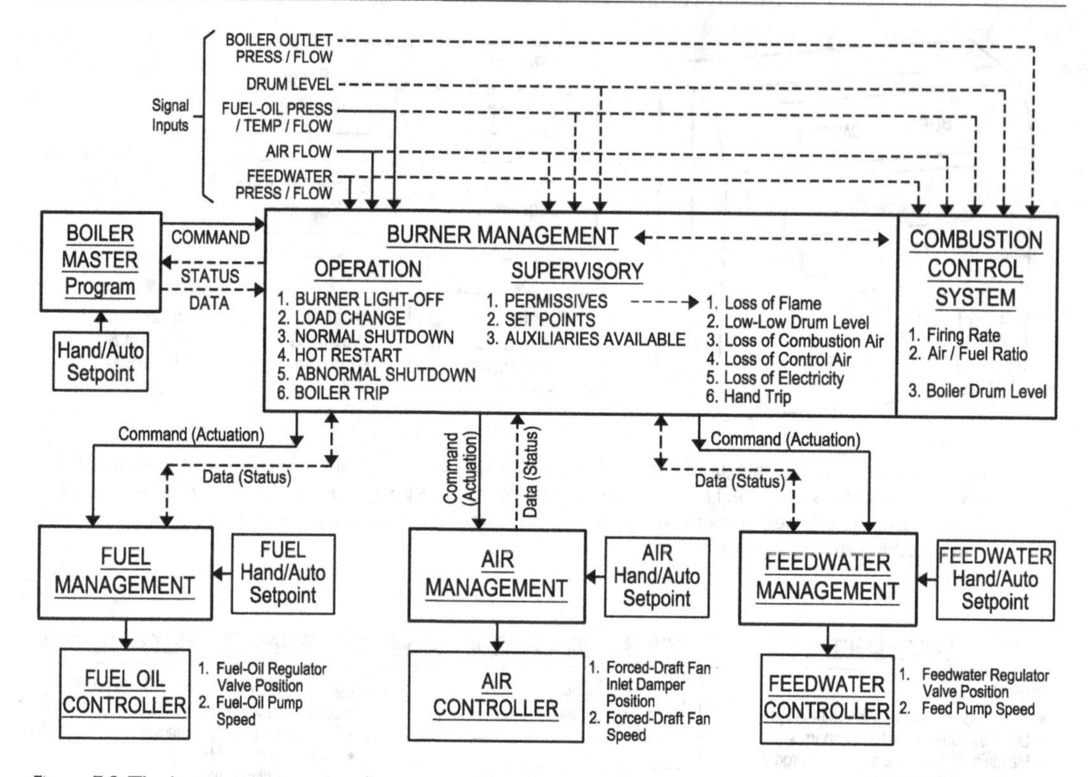

Figure 7.3 The burner management system provides the operation and supervisory functions that permits safe and proper control of the boiler and burners.

permit a light-off attempt before requiring proof-of-flame. Modern flame safeguard systems are electronic and often contain self-monitoring watchdog functions to ensure that its microprocessor controller has not locked up.

The burner-management automation mimics hand control, so an engineer who understands manual operation also understands how automatic systems behave. The burner-management functions are shown in Figure 7.3 and a representative burner-management panel is shown in Figure 7.4. The boiler/burner-control functions discussed in subsequent subparagraphs assume the following initial conditions:

- The fuel-oil service pump is running.
- The master fuel-oil solenoid valve has been reset.
- Steam is supplied to the fuel-oil heaters and the oil temperature is proper, such as approximately 200°F for heavy fuel oil.
- The fuel-oil recirculation valve is cracked, hot oil is available at the burner front and the burner is ready for light off.
- A burner is installed, and atomizing steam or air is available, or a mechanical tip is installed.
- The forced-draft fan is running.

Initial light-off procedures:

1. Check Permissives and Interlocks
2. Boiler Pre-Purge
3. Ignition Attempt

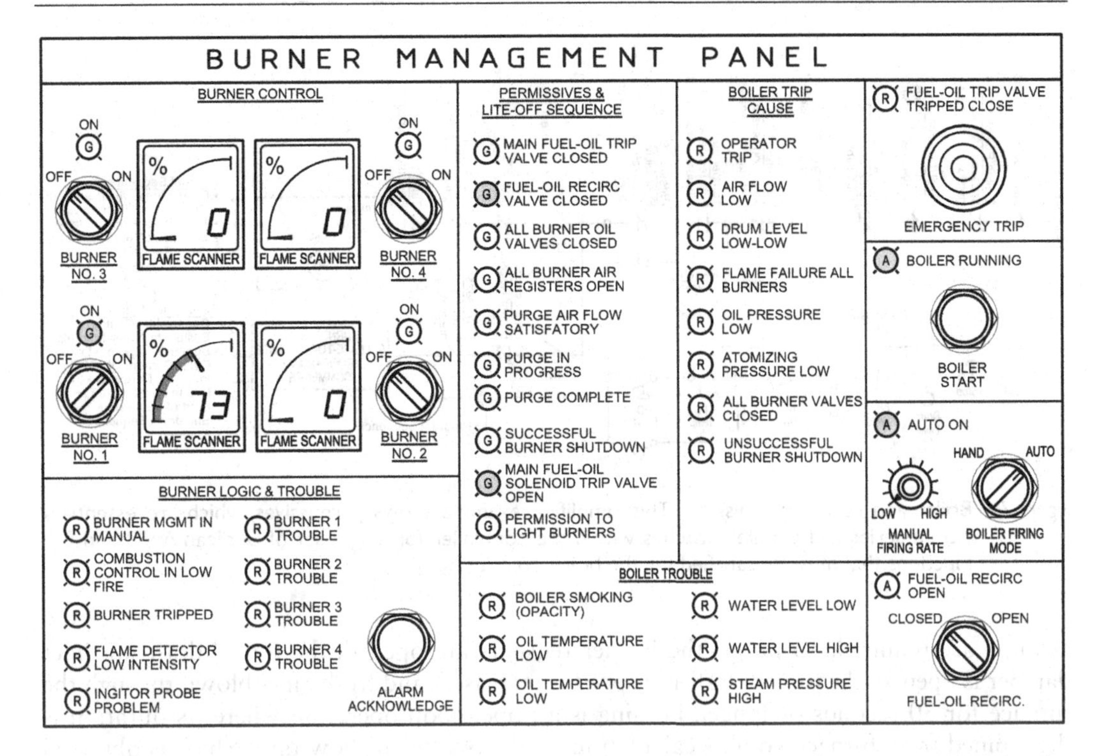

Figure 7.4 A burner management supervisory control panel for an automated boiler provides an interface for the operator.

4. Flame Detection
5. Release to Modulate

Operation and shut-off procedures:

6. Burner Normal Cycling
7. Burner Operation During Load Changes
8. Burner Management Permissives
9. Burner Safety Shutdown
10. Boiler Trips

Check permissives and interlocks

The burner-management system checks that boiler "permissives" are within limits. Some permissives include verification that the steam-drum water level is proper, the forced-draft fan is operating, the fan inlet-damper position is correct, fuel-oil pressure is adequate, atomizing steam pressure is available, a pre-purge cycle has been successfully completed, and that other necessary safety parameters are within acceptable limits. Figure 7.5 shows some permissives in a PLC ladder diagram rung and in a wiring diagram, as an example.

Boiler pre-purge

The purpose of pre-purging is to ensure that there are no explosive mixtures of fuel leftover in the furnace. Before attempting light-off, roughly five to seven furnace-volumes of fresh-air

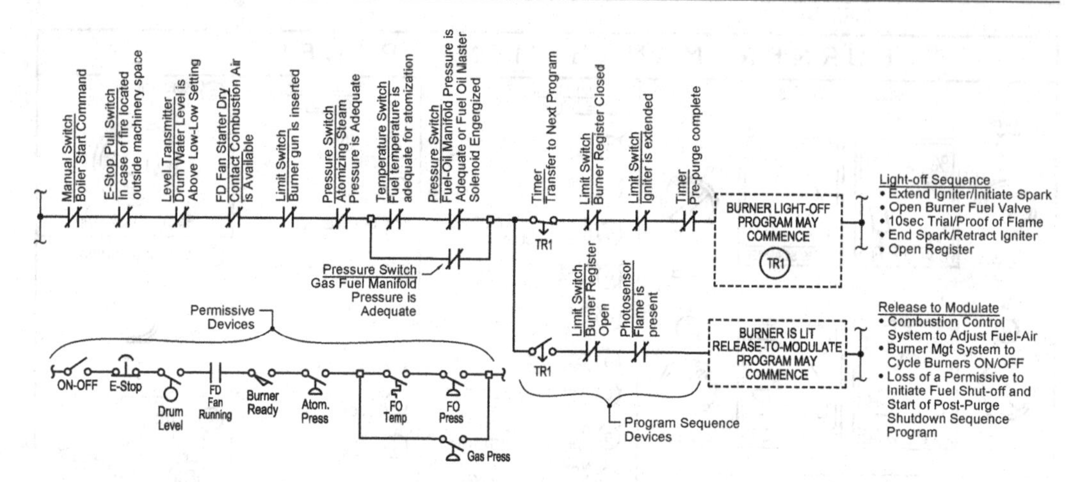

Figure 7.5 Boiler and burner permissives. This simplified example shows permissives, which are essentially closed series and parallel switches within the controller, forming a set of Boolean AND and OR functions that must be satisfied for the boiler to operate.

changes are required. To purge, the burner registers are opened, the forced-draft fan inlet damper is opened, the forced-draft fan speed is increased, and fresh air is blown through the furnace for 30 seconds or longer. Purging is an open-loop operation where its duration is determined from furnace volume calculations and from the air flow rate, which is obtained from the forced-draft-fan curves. At the completion of the purge cycle, the fans are slowed, the burner air registers are closed, and the forced-draft fan damper is positioned to its minimum open position, in preparation for light-off.

Ignition attempt

Prior to ignition, the burner is placed at the minimum firing rate of minimum fuel and least amount of combustion air. During the ignition process, an arc igniter is inserted into the burner by a linear actuator, stopping adjacent to the burner tip. An induction coil produces a high-voltage electric spark at the igniter tip. At this time, the local burner fuel-oil solenoid valve opens, fuel is sprayed, and ignition occurs. With the air flow cut back, it is possible for a small puff of black smoke to occur before the combustion-control system takes over. The flame safeguard system (FSS) begins actively sensing for the presence of a flame, which must be established during a trial period of approximately 10 seconds. At FSS timeout, the fuel will either be secured if no flame is sensed, or the burner will be released to the combustion-control system if all parameters are in good order.

Large boilers are often lit-off with the combustion-control system in "HAND," and the boiler is manually kept at low fire. Intermittent low-fire operation is desirable for bringing the boiler up to pressure slowly, especially when the boiler is cold. When the boiler is up to pressure, the burner is shifted to "AUTO" and the combustion-control system modulates the firing rate in response to the steam pressure. When multiple boilers are on line, a boiler master simultaneously coordinates the firing rates of both boilers, based on the setpoint of the boiler master. If a boiler is prematurely switched to automatic while its steam pressure is low, the combustion-control system would drive the burner to full fire in its attempt to attain the setpoint pressure. Fast firing a cold boiler is stressful to its tubes and refractory and needs to be avoided.

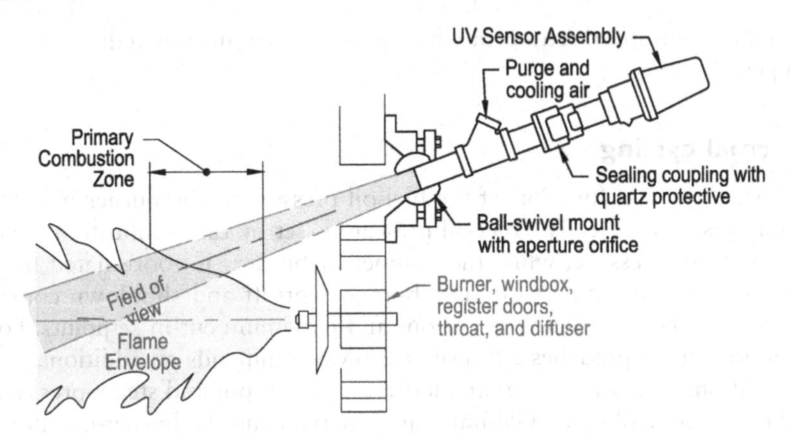

Figure 7.6 The UV-scanner is sighted to measure the flame intensity as part of the proof-of-flame interlock system.

Flame detection

For larger boilers, proof-of-flame is directly sensed using either ultraviolet-light or infrared sensors aligned to observe for the presence of a flame, as shown in Figure 7.6. For industrial applications, ultraviolet flame detectors are more reliable, as they sense the visible light of the fire, whereas infrared flame detectors sense heat, and can mistakenly interpret hot objects in the furnace as a flame. For small single-burner boilers, a pyrostat or other type of thermal detector relay may be indirectly used as the primary flame safety control. The pyrostat detector relay is sometimes referred to as a stack relay or "protecto" relay and is located to sense the boiler flue temperature. The protecto relay contains a bimetallic helix that responds to the hot exhaust-gas temperature that would be present with a flame. If the pyrostat relay does not sense heat, it deenergizes the burner within a period of roughly 60 seconds. A pyrostat relay requires manual resetting before the next light-off attempt can be made.

Release to modulate

When proof-of-flame is first sensed, the fuel will be at the low-fire setting and the BMS will open the burner air register while the fan damper is at its minimum position. For a small boiler, the burner-management system releases both the air and fuel to freely modulate via the combustion-control system. For large boilers, the combustion-control system modulates the air flow via the forced-draft fan inlet damper in coordination with the fuel to obtain good combustion. Typically, the air-fuel ratio is biased or "trimmed" by the operator to avoid smoking, in a fully metered system. When the boiler finally comes up to pressure, it is placed on line by switching the combustion-control system to "AUTO," which then takes full control of the firing rate.

As the steam load and firing rate increases, the combustion air dampers open to meet the combustion requirements. Large boilers often have multiple-speed motors, so that the damper position becomes the input signal to bump the fan to the next speed. When the fan speed changes, the combustion-control system readjusts the damper position to set the correct air-flow rate.

Some boiler burners use variable-speed drives for the forced-draft fans. The combustion-control system on those boilers sets the air flow by modulating the fan speed. Eventually, the fan reaches a minimum speed setting, at which point further air-flow reductions are made via the fan dampers.

Burner normal cycling

The burner firing rate is a function of the fuel-oil pressure at the burner manifold and the number of burners in service. The fuel-oil pressure is set by the combustion-control system in response to steam pressure, while the number of burners is coordinated by the burner management system. The signal to initiate burner light-off and shutdown comes from the fuel-oil pressure, which has minimum cut-out and maximum cut-in setpoints. For instance, when fuel-oil pressure approaches 220 psig, the BMS commands an additional burner to be lit-off. The additional burner leads to an increase in heat input and steam pressure, followed by the combustion control system commensurately reducing the burner-manifold fuel pressure to remain within the normal setpoint range. Likewise, when the burner fuel-oil manifold pressure drops to around 80 psig, the burner management system commands a burner to be secured. Figure 7.7 shows the burner light-off sequence, where the burner lowest and furthest from the stack is the first burner to fire, while the uppermost burner closest to the stack is last.

The burner light-off sequence includes insertion of the igniter adjacent to the burner nozzle, commencement of a high-voltage spark, energization of the local burner fuel solenoid valve, and the burner register door to open when the fire is lit. The flame scanner is enabled as a permissive and the combustion control system modulates the fuel and air in response to

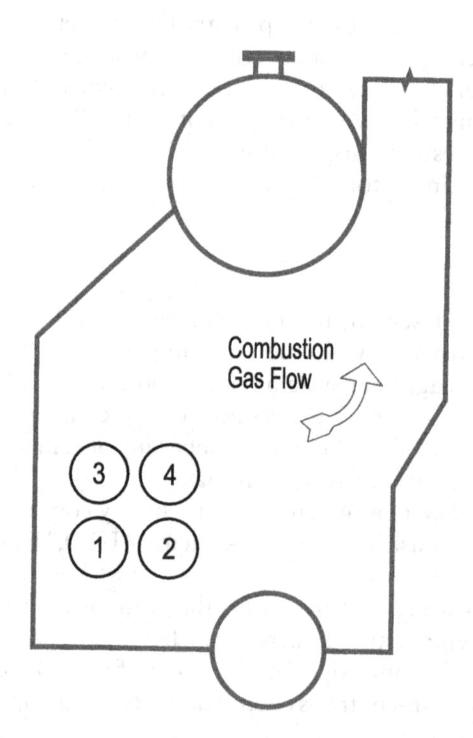

Figure 7.7 The burner light-off sequence starts from the lowest burner furthest from the stack, and sequentially follows a pattern of lowest and furthest until all burners are firing. The burner shutdown sequence is opposite that of light-off, securing the burner that is closest to the stack, and then working toward the lowest furthest burner.

boiler pressure while the additional burner is in service. A burner is secured when the BMS deenergizes the fuel-oil solenoid valve and a linear actuator closes the burner air-register door.

Boiler burner turndown is a function of the fuel-pressure and number of burners in service. When operating four burners between 80-220 psi, the boiler turndown is about 11:1, calculated as.

4 ***burners*** **× 220** ***psi at highest fire***) ÷ (1 ***burner*** **×** ***lowest fire***) **= 11:1** ***turndown***

To obtain even lower firing rates at very low loads, smaller burner tips may be installed.

On small boilers, the burner management system will cycle the last burner off on excessive steam pressure and cycle the burner back on after the pressure drops. However, for industrial boilers, the last burner is usually not secured automatically. Consequently, if the fuel input from only one burner exceeds the steam requirements, the boiler pressure will creep up until the safety valves lift. Before the safeties lift, the operator needs to act, which might include manually securing the fires. Misadjusted air-fuel ratios on a cross-limited combustion-control system might prevent the fuel from reaching its lowest limit, so the corrective action could consist of simply enriching the air-fuel ratio to allow the fuel pressure to go lower.

Boiler operation during load changes

When a boiler is released to modulate, the combustion-control system is given the authority to adjust the fuel and air in response to steam demand by measuring the steam pressure, as the primary input signal. The combustion-control system positions the fuel-oil regulating valve to modulate the heat input, and it correspondingly positions the forced-draft fan damper to provide efficient combustion without smoking. The fan motor speed may be coordinated with the fan-damper position for improved air regulation.

In multiple-burner boilers, when the fuel-oil manifold pressure rises too high during increasing steam demand, which occurs at approximately 220 psig, the burner management system will light the next burner in the sequence. Conversely, when the burner-front fuel-oil manifold pressure drops too low during low steam demand, at approximately 80 psig, the burner-management system will secure a burner. The combustion-control system the readjusts the fuel pressure to match the load.

Burner management permissives

The amount of boiler controls and firing permissives beyond the code minimum requirements will vary with the level of boiler sophistication. Larger boilers may incorporate switches to indicate that fuel-oil pressure is available, forced-draft fan air is available, atomizing steam/air pressure is available, the fuel oil temperature is correct for atomization, the burner is inserted, the igniter has been extended and retracted, the burner air register was closed before light-off and opened afterwards, etc.

Burner safety shutdown

When the flame safeguard system senses a loss of flame, it secures the burner by deenergizing the burner fuel solenoid valve, stopping fuel to the burner, and it sounds an alarm. Assuming no other burner is firing, the forced-draft fan continues to run as part of a furnace post-purge sequence. After the post-purge is complete, the burner air register closes, the forced-draft fan is slowed, and the fan damper goes to its minimum-open position. The boiler is then ready for a hot-boiler light-off attempt. On some small single-burner boilers, the BMS may stop the forced draft fan after post-purge, and the burner may need to be manually reset. On other small boilers, the BMS may make several light-off attempts before locking out the burner.

During any burner light-off attempt for boilers having multiple burners, the burner fuel is secured, and an alarm is actuated if ignition is not sensed before the FSS times out. If the last burner trips on loss of flame, the main fuel-oil solenoid closes, and post purging of the furnace is required. Tripping of the boiler main fuel solenoid requires manual resetting before relight-off can be attempted. Whenever all burners trip, the plant operators should assume that a combustible fuel/air mixture exists in the furnace along with a large enough heat to ignite the mixture. Care should be taken during post- or pre-purging before a subsequent light-off attempt is made, with the operator's understanding that adding combustion air to a hot, fuel-rich furnace could result in an explosion. Pre-purging is always required before any light-off attempt.

Boiler trips

When a single burner trips on loss of flame-scanner signal, the local burner solenoid is deenergized, the burner fuel valve closes, and the boiler continues to fire on the other burners. However, when serious boiler safety permissives are not met, such as during low-low level in the steam drum, the entire boiler is tripped. When a boiler trips, the master fuel-oil solenoid valve closes, securing fuel to all burners and an alarm is actuated. The combustion air is also secured, but the boiler is inhibited from automatically going through a post-purge cycle, as unburned fuel should be assumed to be in the hot furnace. In addition to tripping on low-low drum level, boilers also trip on loss of forced-draft air flow, loss of flame detection, loss of control air, loss of control electricity, and by hand trip. The boiler safety shutdown sequence is shown in Figure 7.2.

Two actions need to be taken by the operator when a boiler trips, first to open the superheater vent valves, if present, when multiple boilers are in service, and second to crack the fuel-oil recirculation valve, if heavy oil is used. The superheater vent is important for maintaining steam flow to avoid overheating damage to the superheater, and the fuel-oil recirculation valve is important for keeping hot oil at the burner front for a subsequent light-off attempt. Although these functions are important, they may not be incorporated into the boiler automation system.

COMBUSTION-CONTROL SYSTEM

The combustion control system takes over after the burner management system has completed its operations and released the boiler to modulate. The safety interlock switches that are incorporated as part of the burner-management and flame-safeguard systems continue to protect the boiler against unsafe conditions, however at this point the combustion control system handles the burner modulation. The combustion-control system serves two main functions:

1. Adjust the heat input released by the fuel to match the heat output in the form of steam production. Heat input is accomplished by setting the fuel-oil pressure at the burner front, cutting burners in and out, and by changing the fuel-pump speed.
2. Modulate the amounts of air and fuel relative to each other for the most complete, smokeless, and efficient combustion at all firing rates. Air flow adjustments are accomplished by setting the fan damper position and by changing the fan speed.

Control symbols

Familiarity with the control diagrams used for plant systems is helpful in understanding the combustion-control system. Functional control-diagram symbols used in the power industry are based on conventions developed by the Scientific Apparatus Makers Association (SAMA) and American National Standards Institute ANSI/ISA-5.1, and they are used to document control strategies. The SAMA methods tend to be used for boiler functional diagrams due to the higher level of control-element detail and the visualization they provide. The newer ANSI symbols have similarities and differences and are used in many industries. The International Society of Automation (ISA) standards are another option for presenting control diagrams. Figure 7.8 shows symbols used in SAMA functional diagrams, Figure 7.9 shows control-device designations, and Figure 7.10 shows a simplified example using both ISA and its equivalent SAMA functional diagram. In general, the ISA symbology tends to be better suited for piping and instrumentation diagrams (P&IDs), while the SAMA diagrams are better suited to indicate logic functions and control.

Control Symbols

Symbol	Meaning	Symbol	Meaning
% or K	PROPORTIONAL ACTION	±	BIAS
$\int$	INTEGRAL ACTION	f(x)	POWER DEVICE (VALVE, DRIVE, etc.)
Σ	SUMMING ACTION	Σ/n	AVERAGING
Δ	DIFFERENCE (SUBTRACTING) ACTION	Σ+%	SUM PLUS PROPORTIONAL
H/A	HAND/AUTO SELECTOR STATION	$\Sigma + \int$	SUM PLUS INTEGRAL
⊁	HIGH LIMITING		
⊀	LOW LIMITING		
>	HIGH SELECT	(PT)	PRESSURE TRANSMITTER
<	LOW SELECT	(DPT)	DIFFERENTIAL-PRESSURE TRANSMITTER
d/dt	DERIVATIVE (RATE)	(FT)	FLOW TRANSMITTER
T	TRANSFER		

Figure 7.8 SAMA signal-processing symbols used for boiler-control diagrams.

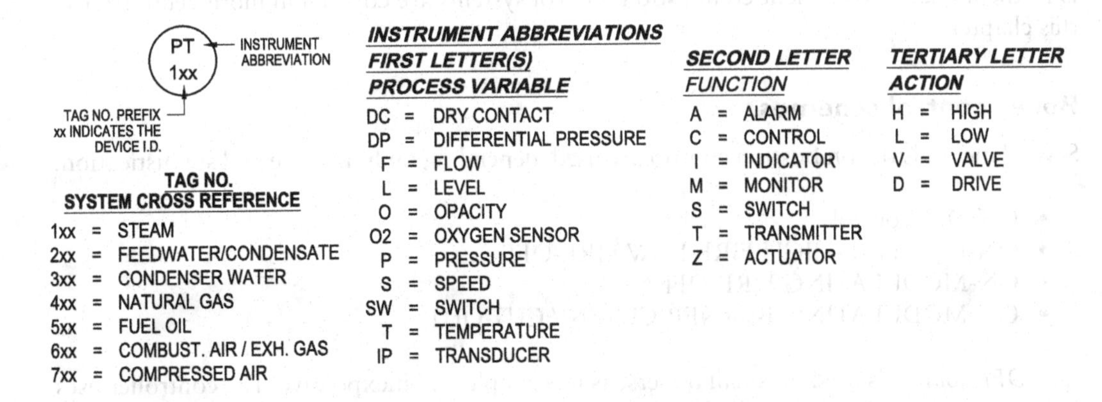

Figure 7.9 Control-diagram symbol-labelling examples used in process-and-instrumentation diagrams.

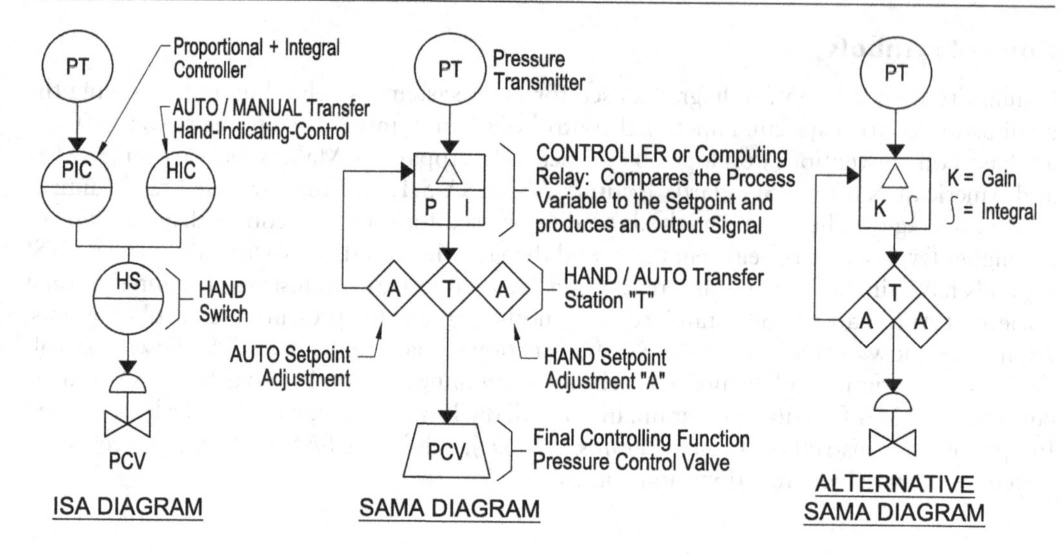

Figure 7.10 A simplified example of an ISA functional diagram (left) compared to equivalent SAMA diagrams.

Single-element and two-element combustion control systems

To match the fuel's heat input to the output in terms of steam flow, the boiler-outlet steam pressure forms the primary indication of the boiler demand. Under steady-state conditions, the firing rate is adjusted so that the steam pressure is at the setpoint and is constant, however, without changing the firing rate, the steam pressure would drop if the load were to increase. In a *single-element* combustion-control system, the steam pressure is the measured parameter to control the firing rate, but in operation there is a delay between a load change, which happens immediately, and the eventual variation in boiler pressure. The delay in response causes excursions from the setpoint during transient conditions.

A *two-element* combustion-control system is used on more sophisticated boilers for better precision. In the two-element system, the steam pressure is still the primary control variable, but the steam flow is added as a second element. The steam flow provides immediate indication that the load has changed. The second element provides *anticipatory control* by sensing the load directly, which in this case is flow. Anticipatory control provides prompt firing-rate response to minimize pressure fluctuations, before the primary element realizes that a change is forthcoming. Two element combustion-control systems are covered in more detail later in this chapter.

Boiler control schemes

Several methods of combustion control are used, depending on boiler size and sophistication:

- ON-OFF control
- ON-LOW FIRE-HIGH FIRE-LOW FIRE-OFF
- ON-MODULATING FIRE-OFF
- ON-MODULATING-BURNER CUT-IN / CUT-OUT

ON-OFF control is used on small boilers, as it is simple and inexpensive. The controller uses a pressure-actuated switch, called a "pressuretrol," so that when the boiler pressure drops, a command signal causes the boiler to light off and stay lit until the pressuretrol high-limit setpoint is reached, at which point the burner is secured. The shutdown and light-off cycles require post-purge and pre-purge operations, causing an inefficient chilling effect as purge air flows through the boiler. This sequence causes a time delay to be introduced between the call-for-steam and burner light-off. ON-OFF control results in large pressure swings, which are usually acceptable for heating services, but may be objectionable for other applications.

The *ON/Low-fire/High-fire/Low-fire/OFF control* is used on moderately small boilers. The control is similar to ON-OFF control; but there are two discreet firing rates after the boiler lights-off at low fire. The burner will switch between low and high fire based on pressure, remaining lit for extended periods of time. Toggling between low and high fire avoids the inefficiencies of the post-purge/pre-purge cycles, along with the associated steam-pressure variations. If demand drops to very low levels where the steam pressure continues to rise at low fire, a high-pressure limit switch will command the burner into a shut-off and post-purge cycle to avoid over pressurization.

The *ON/Modulating fire/OFF control* is used on moderate-size boilers. This burner finely modulates the firing rate between the full-turndown and highest settings. Instead of using pressure switches with built-in differential, this control scheme uses a steam-pressure transmitter to smoothly adjust the firing rate so as to precisely match the steaming rate. The boiler only cycles off during the extended periods of full turndown where the firing rate exceeds the steam load and steam pressure becomes too high.

In very large industrial boilers, multiple burners are used, where all firing burners modulate simultaneously based on steam pressure. Larger variations of loads are accommodated by cutting burners in or out, via the burner management system or by hand. Burners are cut in or out based on the fuel pressure to the atomizers. When the load drops, the oil pressure is reduced, and when it reaches around 80 psig, a burner is cut out and the combustion-control system readjusts the fuel pressure to maintain the required heat input. Conversely, when the oil-manifold pressure rises to approximately 220 psig, an additional burner is cut in and the fuel rate is readjusted. Generally, the first burner stays lit continuously when there is an adequate base load. The operator may install a smaller burner tip if the steam pressure tends to creep up too high, or he may intervene by securing the fires to avoid lifting the safety-relief valves. Low and high steam-pressure alarms may be included to assist in boiler supervision.

Two methods are used to control the air-fuel ratio settings of modulating burners to obtain complete efficient combustion:

1. Single-point positioning for unmetered simultaneous adjustment of fuel and air
2. Full metering control using either series or parallel positioning

Series versus parallel combustion-control systems

Original combustion control systems used a series arrangement when adjusting the fuel and air to maintain good combustion. In the series arrangement, the steam pressure is used as the input signal, which is passed through a boiler master to produce a firing-rate command. In the air-follows-fuel series arrangement, the firing rate command is sent first to the fuel controller, which produces a matching signal to the air controller, as the fuel responds. Conversely, in the fuel-follows-air series arrangement, the firing rate command is sent first to the air controller, which produces a corresponding signal to the fuel controller, as the air responds. The problem with the series combustion control system is that air-follows-fuel produces rich combustion and smoking during load increases, and the fuel-follows-air provides rich combustion and smoking during load decreases, as shown in the block diagrams in Figure 7.11.

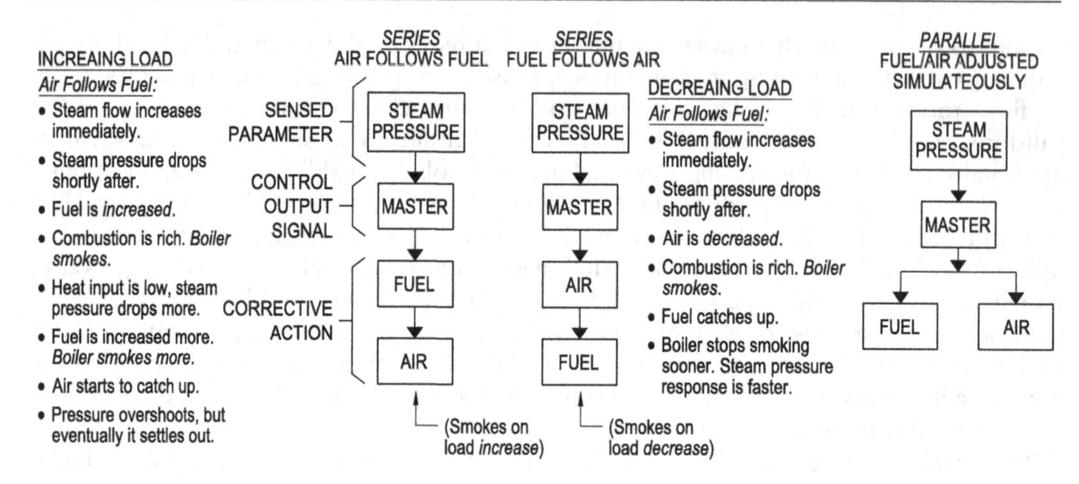

Figure 7.11 Series- and parallel-positioning combustion-control systems. The series combustion-control system by itself is an outdated control type for large boilers. For small boilers, a single actuator simultaneously positions both fuel and air in a control mode called single-point positioning. For large boilers, the fuel and air settings are adjusted simultaneously, but include independent biasing of the air, enabling fine-tune adjustments to avoid smoking. Fuel- and air-flow-measurement signals can be cross connected making the control "full metering," so that combustion remains lean during transients.

In an attempt to eliminate smoking, parallel combustion-control systems were developed, where the boiler-master signal is sent simultaneously to both the fuel and air controllers. This scheme is good in theory and an improvement, but due to the compressibility of air, the control results in a delay in air-flow changes, causing some smoking during load increases. In practice, a boiler fireman could manually override the air damper position during load increases, such as when a ship goes from a stop bell to full ahead.

Parallel positioning using single-point positioning

Single-point positioning is used on smaller auxiliary or heating boilers. In this arrangement, the firing-rate command is sent to a single actuator, such as a modutrol positioning motor as shown in Figure 7.12 and Figure 7.13. The modutrol motor uses a resistor bridge along with two potentiometers, one in the controller and the other in the motor. When the input setpoint is reached, the bridge circuit is balanced, and the motor remains in a fixed position. When the controller-input signal deviates from the setpoint, its new potentiometer voltage imbalances the modutrol bridge circuit, and its amplified signal drives the motor in the direction to correct the change. When the motor's position-feedback potentiometer rebalances the bridge circuit, the motor stops at the new setpoint.

In general, single-point positioned combustion-control systems require annual tuning of the boiler at discrete points over the entire firing range. Tuning is obtained by adjusting linkage-arm lengths to adjust the gain, and then fine-tune adjusting each fuel-valve position relative to each air-damper position, while measuring the stack products. One method of adjustment uses a series of set screws or a series of adjustable dogs to form a cam and follower arrangement. The cam positions the fuel-oil regulating valve relative to the fan-damper position. Tune-up is often done at ten discrete firing rates between full turndown and maximum fire, while optimizing combustion by measuring the stack-gas products. Experience is helpful, as optimum tuning of a boiler on a cold day can result in insufficient combustion air and smoking on a hot humid day.

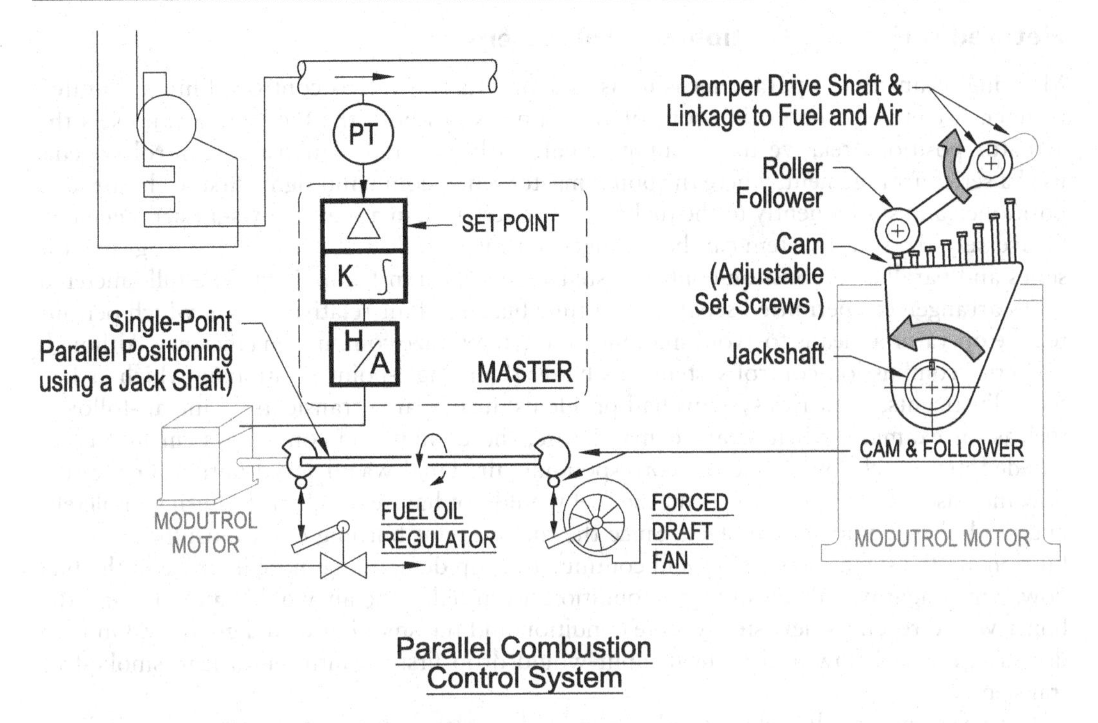

Figure 7.12 Single-point parallel-positioning combustion-control system using a jack shaft and modulating motor.

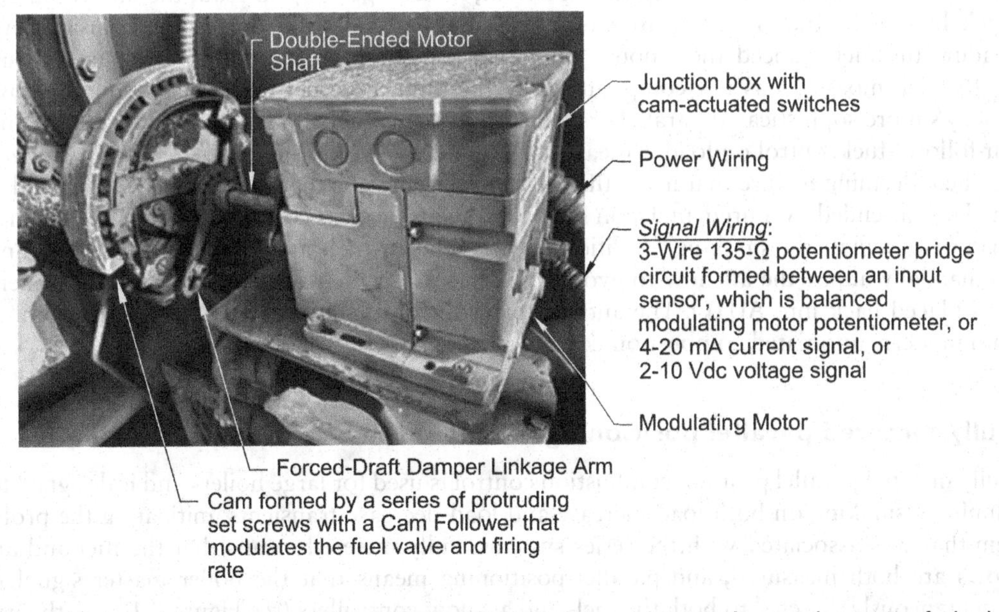

Figure 7.13 A modulating motor with a built-in spring return is used to simultaneously position fuel-valve and air-damper linkages. The motor's rotary shaft movement is translated into linear fuel valve positioning using a cam and follower arrangement. (Courtesy of the U.S. Merchant Marine Academy.)

Metered series combustion-control systems

Metering means that a parameter is measured, and its flowrate is controlled in a calibrated manner. A metered combustion control system measures both the fuel and air and sets the actuator positions relative to the input signal. Early metered combustion-control systems used a series arrangement, where the boiler master sent a command signal first to the air-flow positioner, and subsequently to the fuel-regulator actuator, in a fuel-follows-air arrangement. Conversely, the series system can be arranged for air to follow the fuel. Block diagrams for series and parallel combustion-control systems are shown in Figure 7.11. The fully metered series arrangement permitted separate real-time biasing of air relative to fuel, which permitted the operator a means to avoid smoking and to maximize combustion efficiency. Although the series combustion-control system was better than single-point positioning which lacked fine adjustments, the series systems had problems during large transients. Using air-follows-fuel as an example, when steam demand rose, the combustion-control system first commanded more fuel flow, while the corresponding air signal was delayed until after the fuel flow increased. During the transient, the fuel would not have enough air to burn completely, and the boiler would smoke black. Since the fuel was not burning due to insufficient combustion air, the steam pressure would continue to drop demanding more increase in the fuel flow, which aggravated the smoking condition. Eventually, the air would catch up, and the boiler would reach its new steady state condition and the smoking would go away. On load decrease, the fuel flow and its heat input would drop first, resulting in a lean smoke-free transient.

To prevent potentially severe smoking on load increase, the series systems were modified for the air to lead the fuel, instead of following. On load increase, the boiler master's signal first commanded the air flow to increase, resulting in lean firing during the transient. Shortly thereafter, the fuel flow would catch up, increasing the heat input to match the new steady state load, avoiding the black smoke. However, on a large drop in load, the air would cut back before the fuel, resulting in a fuel-rich smoking condition during the transient. Air leading the fuel reduced the amount of smoke compared to the fuel-follows-air systems, as the rich mixture was not compounded by the unburned fuel from the lagging air flow. Today's more sophisticated parallel systems use air-leads-fuel control on load increases, and air-follows-fuel control on load decreases to completely avoid smoking.

The redeeming feature that made these systems tolerable is that the systems were installed in plants attended by a fireman. Upon large load swings, it was common practice for the fireman to *anticipate* the smoking condition, to place the air-fuel ratio control into HAND, and to manually adjust the air flow to avoid smoking. When the transient was over, the system was placed back into AUTO. The anticipatory control of the fireman is now incorporated into modern automated combustion-control systems.

Fully metered parallel positioning

Fully metered parallel position combustion control is used for large boilers and is designed to eliminate smoking on both load-increase and load-decrease transients, mitigating the problem that was associated with the series systems. Fully metered means that the fuel and air flows are both measured, and parallel positioning means that the boiler master signal is simultaneously directed to both the fuel- and air-flow controllers (see Figure 7.14). In theory, smoking should be avoided by the parallel adjustment of both air and fuel; however, in practice air is compressible, so there is always a delay in its change. Consequently, the system tends to exhibit some of the characteristics of air-follows-fuel, along with some nuisance smoking that occurs during load increases.

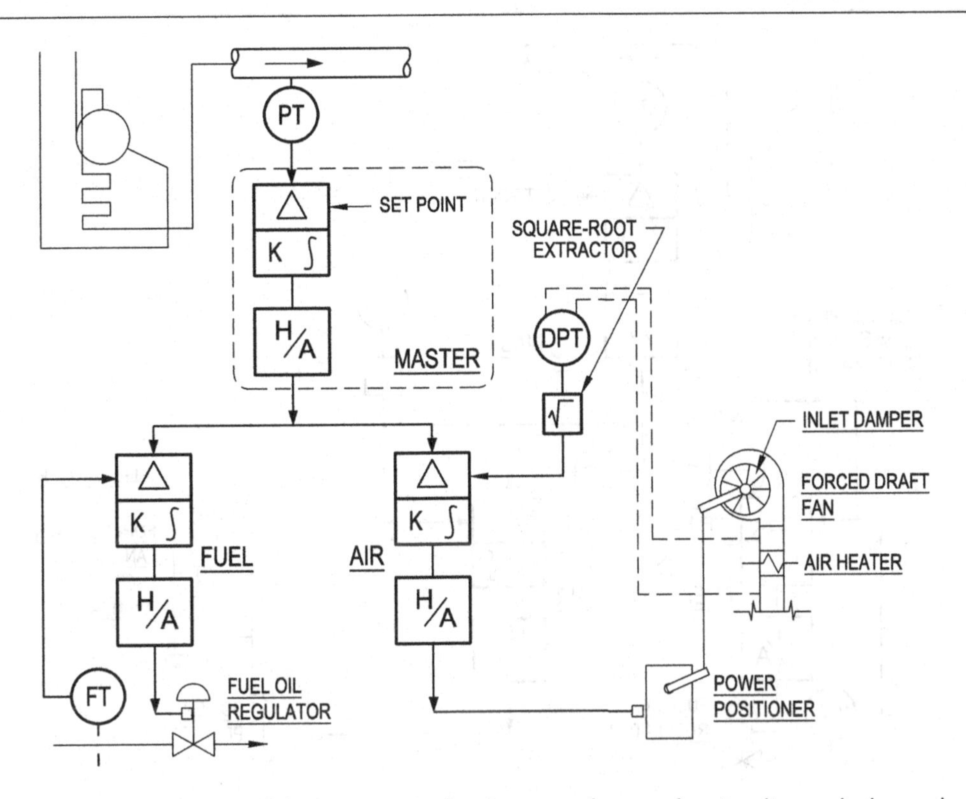

Figure 7.14 Fully metered, parallel-positioning combustion-control system function diagram. In theory, the fuel and air are measured as feedback signals and used simultaneously to adjust the firing rate. The air-fuel ratio can be fine-tuned in fully metered systems to avoid smoking.

Fully metered parallel positioning with cross-limiting

To avoid the smoking that occurs during load increases, *cross-limiting* is added to the combustion-control system, as shown in Figure 7.15. Cross-limiting uses high- or low-select relays, sometimes referred to as auctioneers, to choose the preferred output signal from one of the two input signals. Figure 7.16 (upper) shows a pneumatic high-select relay, where the higher of two input signals is permitted to pass to the output, and Figure 7.16 (lower) shows a low-select relay, where the lower of two input signals passes to the output.

The benefit of cross-limiting is to keep the combustion process slightly lean during all transients. Cross-limiting causes the combustion-control system to behave as the air-leads-fuel series controller when the boiler load increases, and to behave as air-follows-fuel when load decreases. Each select has two inputs; one command signal from the boiler master and the other from the cross-limiting feedback signal from the opposite air- or fuel- flow transmitter. For example, the low select relay passes the lesser input signal from the boiler master or the air-flow transmitter to the fuel regulator. In a similar but opposite manner, the high-select relay passes the greater of the boiler master signal or the cross-limiting fuel-flow transmitter to the air-damper controller. To summarize, when the boiler master sends a signal to increase the firing-rate, the low select passes the smaller air-flow signal to the fuel controller, temporarily ignoring the boiler-master command to increase the fuel. Simultaneously, the increasing boiler-master signal becomes larger than the lower fuel-flow signal, so the high select passes the command for the forced-draft damper to open, increasing the air flow. As the air flow

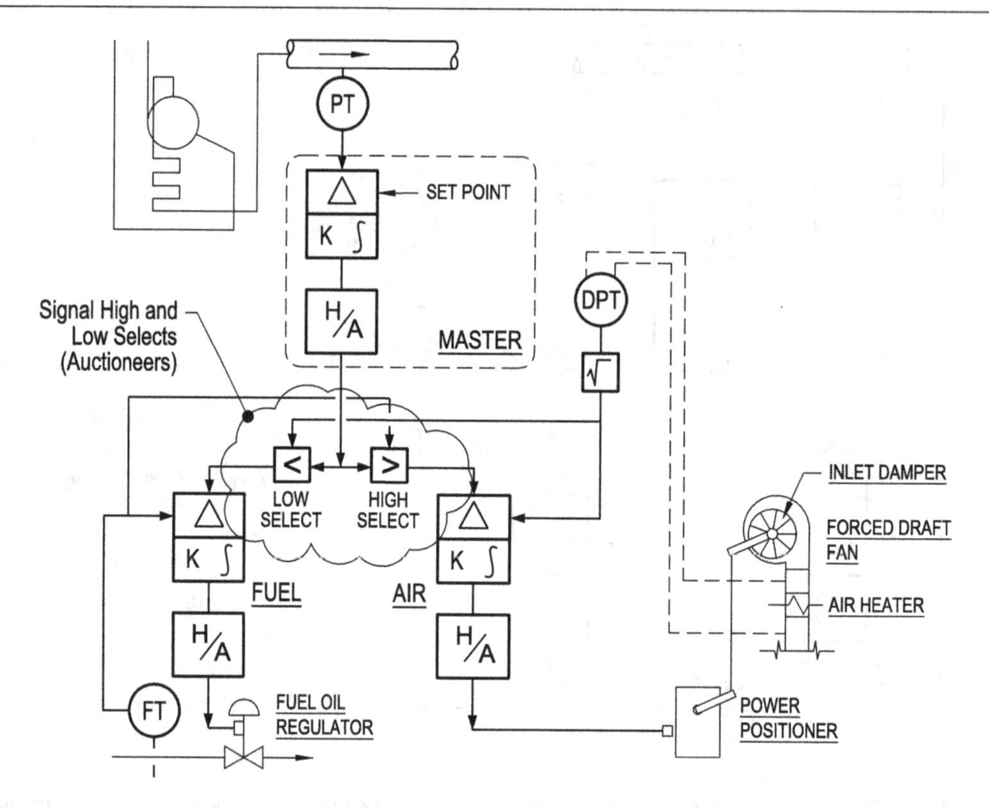

Figure 7.15 The fully metered parallel combustion-control system uses *cross limiting* obtained from high and low selects, which maintains lean combustion during both increasing and decreasing loads.

increases, its increasing transmitter signal passes through the low select, causing the fuel flow to follow the air. This action causes the boiler to remain lean during increasing loads. Similar but opposite actions occur for decreasing loads where the fuel low-select decreases first, and the air follows, also maintaining lean transient conditions. In combustion-control systems where the air system dynamics naturally result in a delayed response, the high-select device is sometimes omitted.

For cross-limiting to work, both the fuel and air flowrates must be metered. Often the fuel flowrate is measured directly, and the air flow is measured indirectly, such as by using the pressure drop across the air heater or boiler windbox. Since pressure drop due to flow is proportional to the flow squared, the air flowrate can be estimated by taking the square root of its pressure drop using a square-root extractor. Figure 7.17 shows the Hagan pneumatic systems square-root extractor, which uses a parabolically cut cam and follower, as one method of deriving the air flow rate.

Two-element combustion-control system (parallel positioning with cross-limiting)

Two-element control is used in applications where large steam-flow variations may be experienced and fast combustion control response with better pressure regulation is desired, such

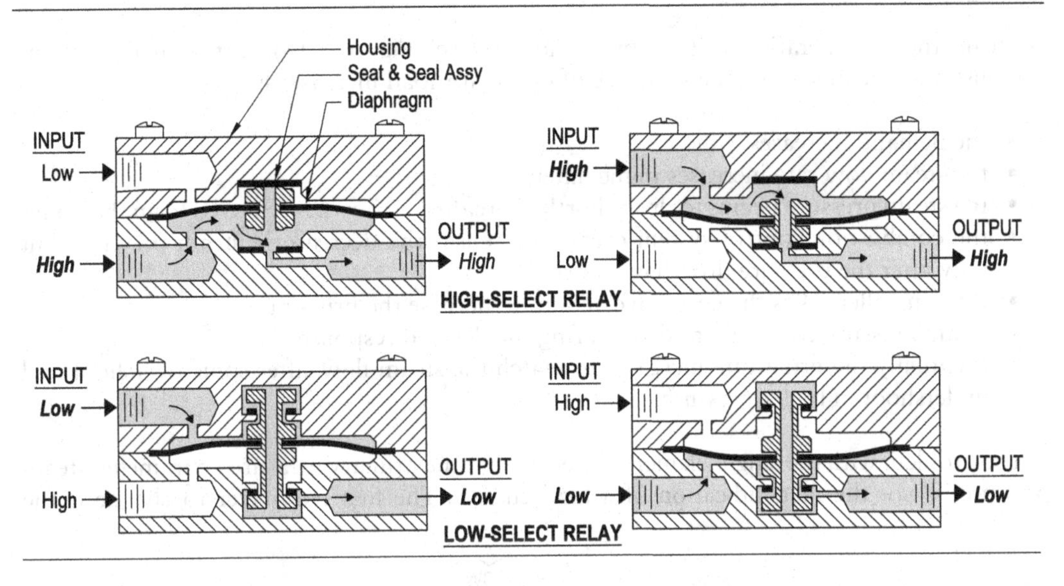

Figure 7.16 A pneumatic high select auctioneer is shown in the upper two illustrations. The higher-pressure input signal deflects the diaphragm, aligning the higher air pressure to the output. A diaphragm supporting a hollow double-seating valve is key to its operation. The low-select auctioneer shown in the bottom two illustrations works using similar principles, but it passes the lower of two signals.

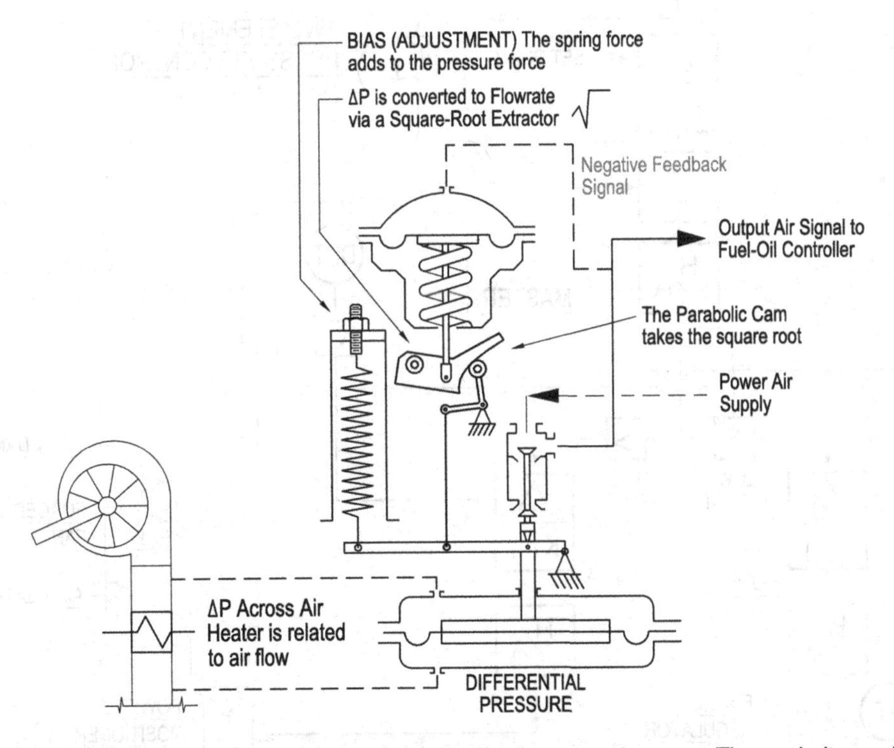

Figure 7.17 An air-flow transmitter used in the Hagan combustion-control systems. The parabolic cam behaves as a square-root extractor to convert the air-heater pressure drop into an air-flowrate measurement. A spring is used as a bias adjustment to add spring force to the pneumatic forces.

as steam turbine operation in the power industry (see Figure 7.18). In the single-element combustion-control system, the sequence of events for load increase is:

- the demand increases
- the steam flow rate increases immediately
- the steam pressure begins to drop shortly thereafter, as steam is removed from the boiler
- the combustion-control system eventually senses the decrease in steam pressure, but only after the pressure drop has occurred
- the controller takes the corrective action to increase the firing rate
- steam pressure continues to drop during the delayed response
- eventually the firing rate increases to match the steam flow, after some overshoot and undershoot in boiler pressure.

The problem with the single-element system is that the primary measured parameter, steam pressure, is *not* the first indication of a load change. The first parameter to change is the

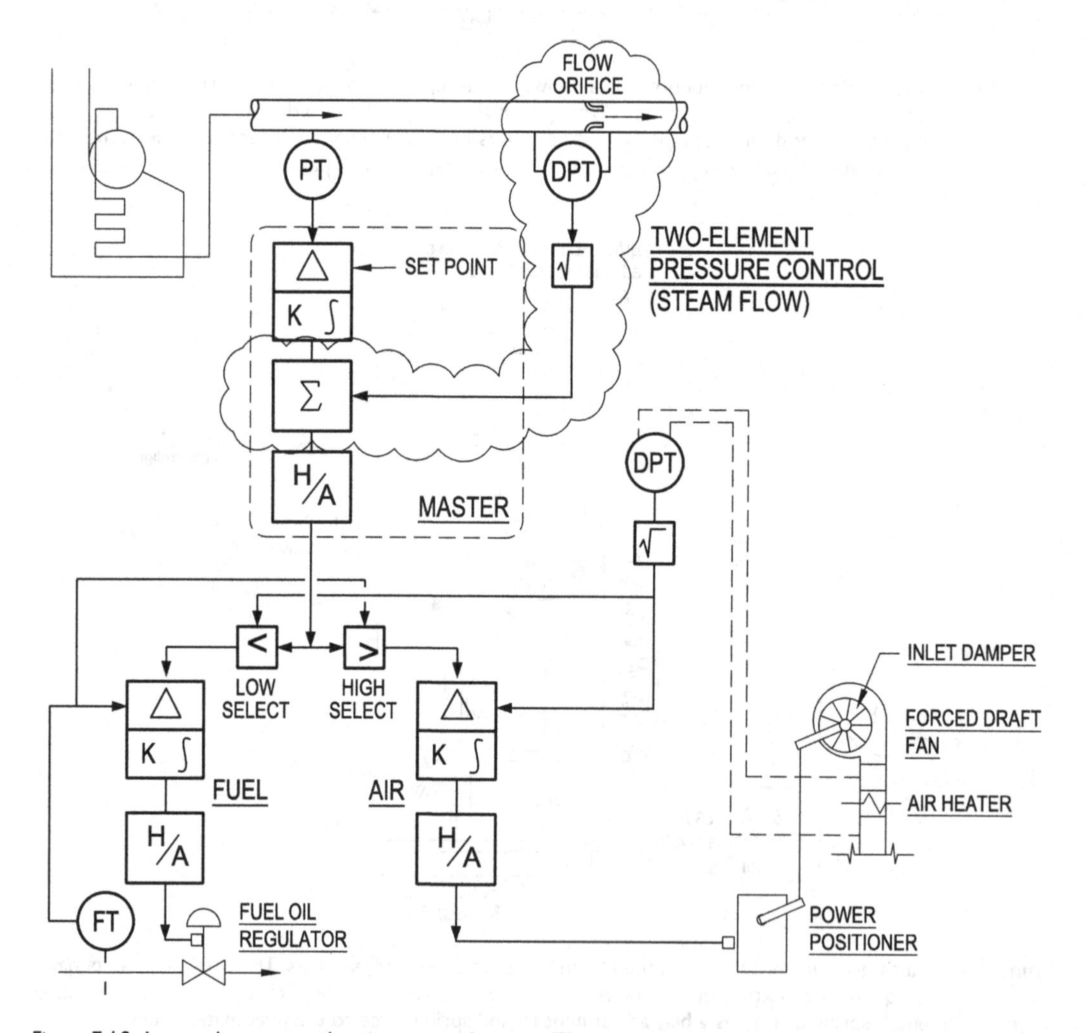

Figure 7.18 A two-element combustion-control system. The primary element is steam pressure, and anticipatory control is obtained from the second element, which is steam flow. The second element results in immediate controller response during transients, before the pressure has had enough time to change and be sensed.

steam flow, itself, before the pressure responds. A new flow rate indicates immediately that the combustion-control system will shortly be calling for a firing-rate change. Steam flow, as the second element, is added as an anticipatory function to command the firing rate to respond before the primary element observes any deviation has occurred. The second element uses a steam-flow transmitter, whose signal is added into the first-element pressure-transmitter signal, biased to provide an immediate and overriding corrective action. After steady state conditions are reached, the second element is no longer required, and its biasing signal is removed from the control signal. Two-element combustion-control systems require tuning to obtain a fast and accurate response, while minimizing overshoot and hunting.

Two-element combustion-control system parallel positioning with cross-limiting and O2 trim

The last function added to a combustion-control system is O2 trim, which is used to maximize the boiler efficiency (see Figure 7.19). The purpose of the O2 trim is to minimize the amount of combustion air while achieving complete combustion. O2 trim requires a metered

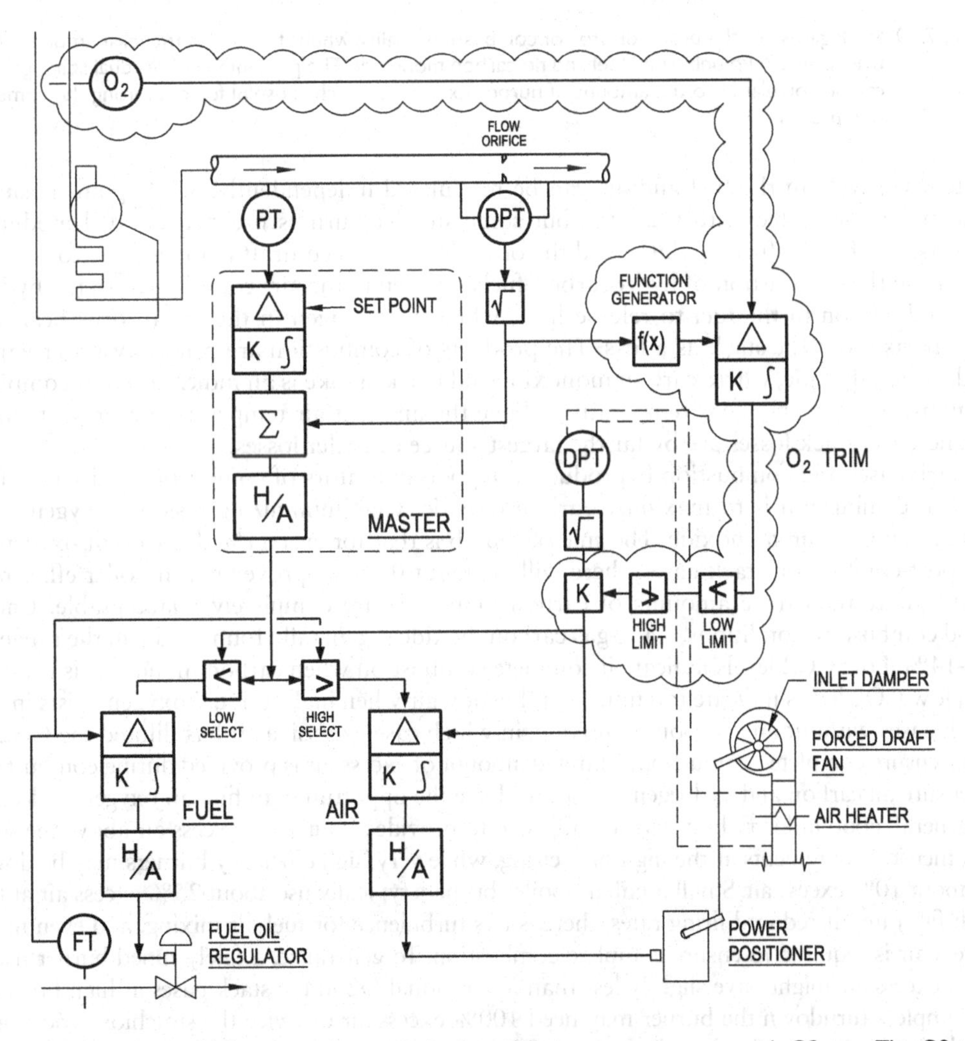

Figure 7.19 Two-element fully metered cross limited combustion-control system with O2 trim. The O2 trim control strategy permits automatic fine-tune adjustments of the fuel-air ratio to minimize excess air.

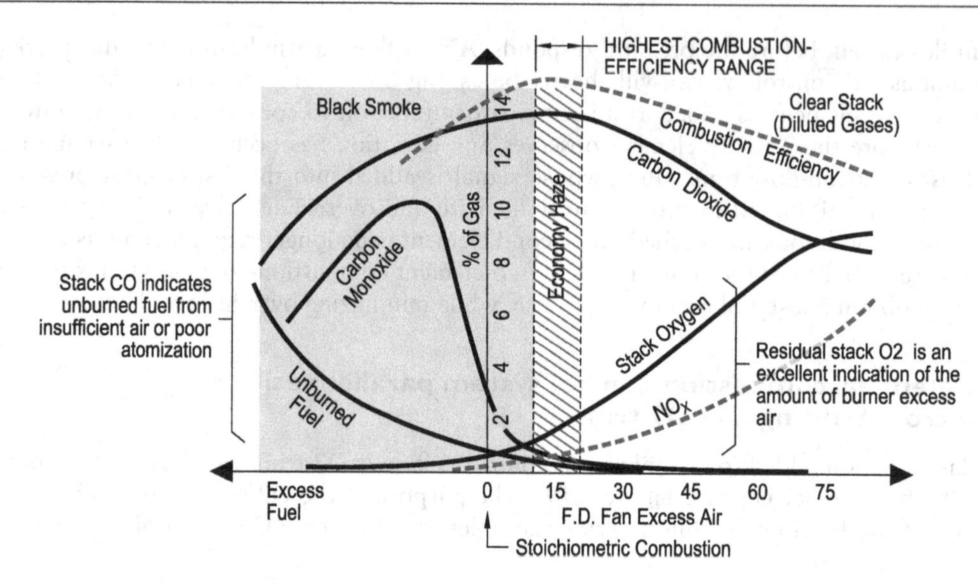

Figure 7.20 Stack gases are the best indicator of combustion quality, where the goal is the least amount of air that results in no unburned fuel and no carbon monoxide. The percentage of residual stack-gas O2 can be correlated to the amount of burner excess air, which is useful for optimizing the combustion process.

system where both the fuel and air can be modulated independently. Single-point position systems are not fitted with O2 trim, but some smaller burners use two parallel modutrol motors, one for the fuel regulator and the other for the forced-draft damper actuator.

During the combustion of hydrocarbon fuels, oxygen in the air[1] combines with the hydrogen and carbon in the fuel to release heat, while the nitrogen in the air absorbs heat and transports it out the stack as a loss. The products of combustion are primarily water vapor and carbon dioxide, where carbon monoxide and black smoke is an indication of incomplete combustion. Any stack-gas temperature above the ambient air temperature represents inefficiency, and stack losses are by far the largest source of boiler losses.

Stack gases and combustion byproducts are good indicators of combustion efficiency. The goal of combustion is to maximize carbon dioxide while minimizing residual oxygen with virtually no carbon monoxide. The rule of thumb is that for every 1% decrease of oxygen in the post-combustion stack gases, there will be about 0.5% improvement in boiler efficiency, where more than trace amounts of carbon monoxide are completely unacceptable. Under good combustion conditions, stack-gas carbon dioxide is generally found to be in the range of 12–14%. Low CO2 levels indicate incomplete combustion when carbon monoxide is present, and low CO2 levels indicate too much combustion air when high residual oxygen exists in the stack gases without any carbon monoxide, in which case excessive air has diluted the CO2.

To ensure complete combustion, a limited amount of excess air is provided during combustion to ensure all carbon and hydrogen fuel atoms have the opportunity to find an oxygen molecule. In general, good industrial burners are adjusted to provide about 15% excess air above the stoichiometric requirements at the high-fire setting, while very high efficiency burners may be down to about 10% excess air. Small auxiliary boiler burners typically use about 25% excess air at the high-fire rate. At reduced firing rates, there is less turbulence for fuel-air mixing, and even more excess air is required to ensure complete combustion. To generalize, a finely tuned burner using 15% excess air might have slightly less than 3% residual O2 in the stack gases at high fire, and at complete turndown the burner may need 100% excess air or twice the stoichiometric value, producing stack gases having about 10% O2 content. Consequently, the fuel-air ratio setpoint and the resulting stack-gas O2 content is a function of the burner firing rate (see Figure 7.20).

Figure 7.21 In-situ zirconium-oxygen O2 sensor and analyzer for fine tuning boiler fuel/air ratios and maximizing efficiency. (Courtesy of Emerson.)

To fine-tune adjust the air-fuel ratio, O2 trim is added. The O2 sensor uses a zirconium-oxide cell with a ceramic heater and thermocouple (see Figure 7.21). When the cell is heated to 1470°F (800°C), it generates a small voltage directly related to the oxygen concentration in the gases, on a wet basis. The flue gases pass through a filter to avoid cell contamination. Calibration gases can be injected into a space behind the filter to permit in-situ calibration. As an alternative technology, tunable diode-laser-absorption spectroscopy (TDLAS) can be used for O2 measurement. The O2 analog-input signal is received by a controller, whose purpose is to make fine-tune adjustments of the combustion air. The O2-trim setpoint is characterized by a function generator, which accounts for the burner firing rate and type of fuel used. The O2-trim controller indirectly uses the steam flow as its firing-rate signal, and it calculates the air-to-fuel ratio setpoint to minimize excess-air while obtaining complete combustion.

To avoid over-correction and hijacking of the fuel-air ratio controller, high and low limits are included to restrict the amount of deviation permitted by the trim control. The high and low limits are similar to the high and low selects used in cross-limiting; however, they permit only small deviations from a setpoint baseline. When O2 trim is used, the controller must be conservatively tuned relative to the boiler master to minimize controller interactions. O2 trim automatically optimizes excess-air adjustments, but used in conjunction with stack temperature, it can provide troubleshooting and alarming functions. Typically, O2 trim is reserved for large boilers.

Figure 7.22 shows the relationship between excess air, combustion efficiency, and stack gases. Figure 7.23 shows boiler efficiency as a function of stack gases and temperature. This information can be used to indicate efficiency in real time and in a data-acquisition system for trending and troubleshooting.

7.3 BOILER FEEDWATER SYSTEMS AND CONTROL

Ultimately feedwater-control systems are designed to balance the mass of steam leaving the boiler with the replacement mass of incoming feed water, and it does so by using water level as its primary control parameter. Small-boiler feedwater systems tend to use constant-speed motor-driven feed pumps, where feedwater is either intermittently controlled via on-off pump

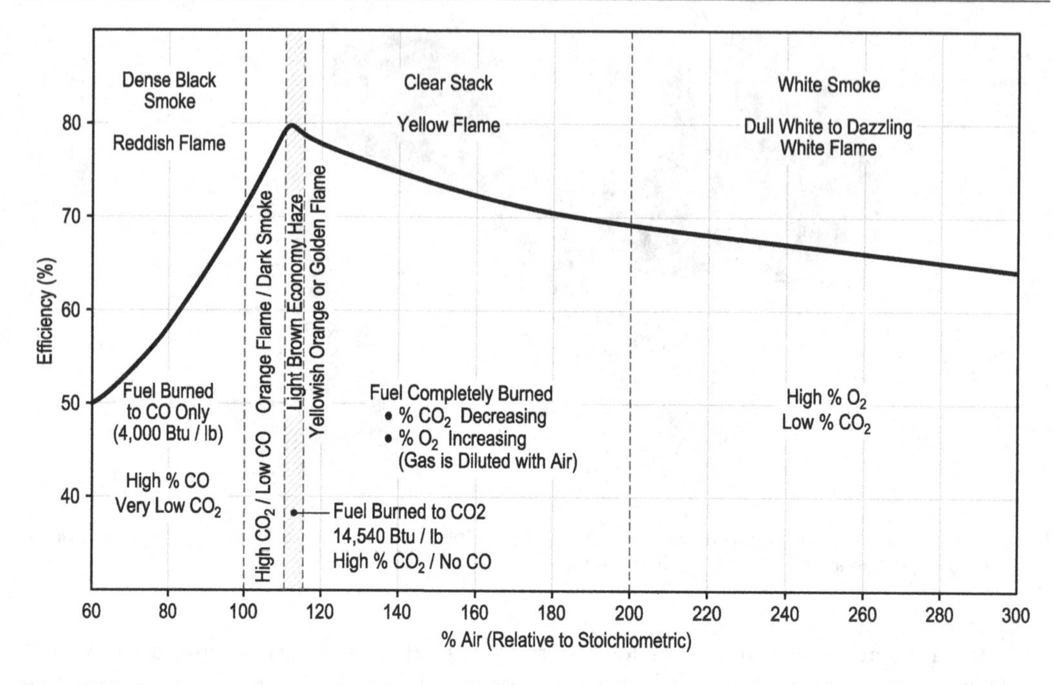

Figure 7.22 Combustion flame and stack-gas conditions with various amounts of air.

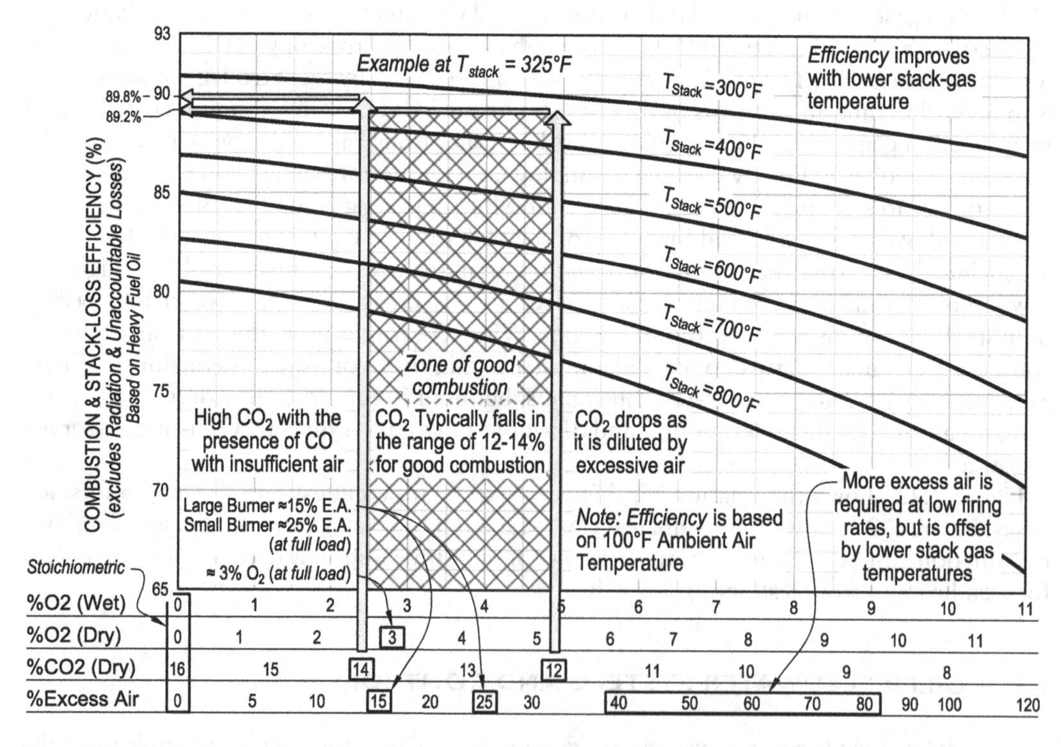

Figure 7.23 Combustion efficiency can be estimated by measuring the stack-gas temperature and the amount of excess air or stack-gas combustion products. This graph is provided as an example and pertains *only* to *heavy fuel oil*. The graph does not include tramp air, radiation and unaccountable losses, as described in ASME PTC 4.1 for determining boiler efficiency.

operation, or it is continuously controlled via a modulating feedwater-regulating valve by a constantly running pump. Feed pumps for large boilers operate at variable speed from either a steam turbine or VFD-driven motor, in which case feedwater flow is a function of both the pump speed and the system pressure, as well as the position of its feedwater-regulating valve. Some interaction will exist between pump-speed control and feedwater-regulator position. The objective of the feedwater-control system is to maintain a fixed and stable steam-drum level, having minimum interaction between the drum-level and combustion-control systems during any fluctuations in feedwater and steam pressure.

Single-element feedwater regulators

Single-element feedwater systems use drum water level as the only control parameter, where level measurements can be done with electrical probes, float controls, thermohydraulic regulators, and thermostatic regulators, as examples. Float and probe-type level controllers are shown in Figure 7.24. Single-element devices directly measure the drum water level and can be configured to intermittently start and stop feed pumps, open and close feed valves, or to modulate feedwater-valve positions, when the pumps run continuously. Better quality feedwater regulators tend to use differential pressure between a static reference leg and a variable leg that rises and falls with water level. The differential-pressure level regulators are used in single-element systems, but unlike other types of controllers, these are conducive to multi-element feedwater systems, which provide faster, more accurate response.

Figure 7.25 shows a single-element thermohydraulic feedwater system, which is flawed in drum-level-control philosophy, but is useful for understanding feedwater-level measurement, control methods, and strategies. The system uses a tube-within-a-tube arrange inclined at an angle, called the generator. The upper sensing-line is a static connection containing steam and is insulated to slow down its condensation. The steam temperature corresponds to its saturation

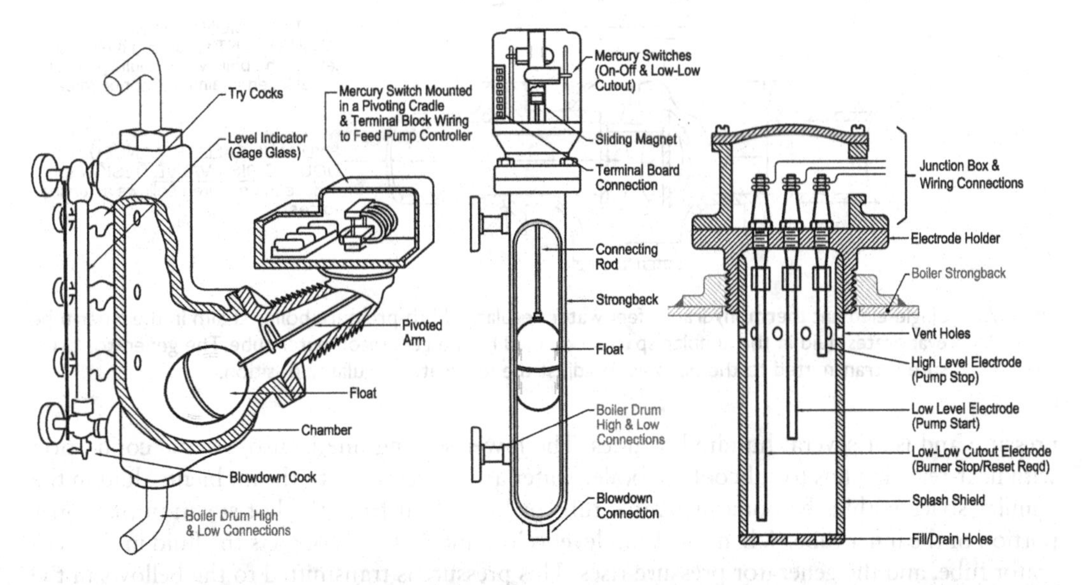

Figure 7.24 Float and electrode feedwater level-control devices. The float is buoyant, so it rises or drops with water level. The float is linked to a switch to provide pump on-off or valve open-close operation, or it is linked to a 0-200-ohm potentiometer for modulating control. The electrode level controller uses probes that become submerged as the boiler level rises. Once submerged, electricity is conducted through the boiler water and the signal is used to start and stop the feedwater pump. A third probe provides a low-low water signal for boiler safe shutdown.

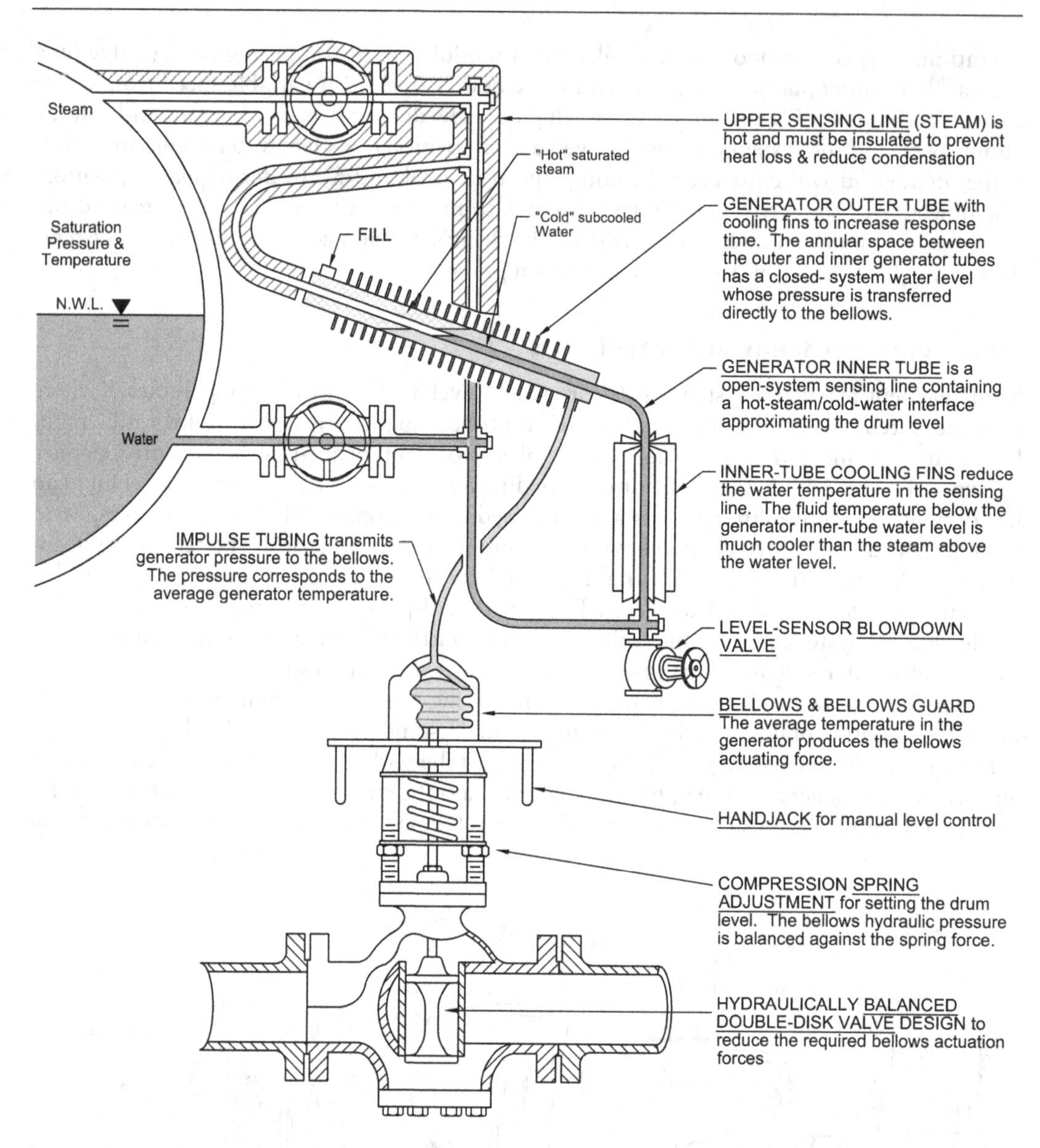

Figure 7.25 Single-element thermohydraulic feedwater regulator. High-pressure boiler steam in the inner tube evaporates fluid in the annular space contained by the generator outer tube. The generator pressure is transmitted to the bellows to adjust the feedwater regulator position.

pressure and is at several hundred degrees. The lower sensing line is also a static connection with heat-rejecting fins to subcool the boiler water as it rejects heat to the ambient. Fluid in the annular space within the outer generator tube picks up heat from the hot steam in the upper portion of the inner tube. When the drum level is low, more steam contacts the fluid in the generator tube, and the generator pressure rises. This pressure is transmitted to the bellows in the feedwater regulating valve causing it to open and to pass more water into the drum.

Conversely, when the drum level is too high, less steam contacts the generator fluid and the pressure on the bellows drops, closing the regulating valve. To increase sensitivity, the generator is inclined at a steep angle so that small changes in drum level result in large movements of liquid within the generator. Cooling fins surround the generator outer tube to increase heat transfer, which improves response time when the level goes high. When the inner tube contains a high-drum level of cooler water, the fins on the generator tube reject the heat faster, lowering the generator pressure and actuating the bellows more quickly. This regulator is suitable only for relatively small heating boilers operating at fairly steady loads, but a fundamental problem exists with the regulator's fail-safe condition. With boiler feedwater, the preferred valve fail-safe position should be to open and overfill the boiler, rather than to run it dry. The problem with this regulator is that if the generator fluid is lost, there will be no pressure on the bellows and the valve closes, risking dry firing the boiler.

Figure 7.26 shows a differential-pressure transmitter used for level control. The transmitter uses a very sensitive diaphragm that measures pressure in inches of water between an unchanging static reference leg and a variable measurement leg that fluctuates with drum level. Upper and lower connections are made so that any drum-pressure variations are transmitted equally to both sides of the sensing diaphragm, cancelling internal pressures that would otherwise rupture it. To avoid over pressurizing one side of the differential-pressure diaphragm when placing the sensor in service, a three-valve arrangement having two isolating valves and one bypass is used. The bypass valve is opened before either sensing valve. In this arrangement, full pressure is applied equally to both sides of the diaphragm when either sensing valve is opened, avoiding damage. The bypass is closed after both sensing valves are open, placing the transmitter in operation.

Differential-pressure-based level sensors can experience errors if the liquid density varies, which can occur with high-pressure boilers that undergo sudden pressure swings, so that hot condensate in the reference leg flashes causing a transient false reading. Constant head chambers or temperature-equalizing columns help reduce these errors by permitting some degree of subcooling in the reference leg to reduce flashing, and to mitigate drum turbulence from reaching the sensor.

It is interesting to note that since the constant head chamber is located outside the drum, the water at the sensor will be subcooled and denser than in the drum, introducing variations at

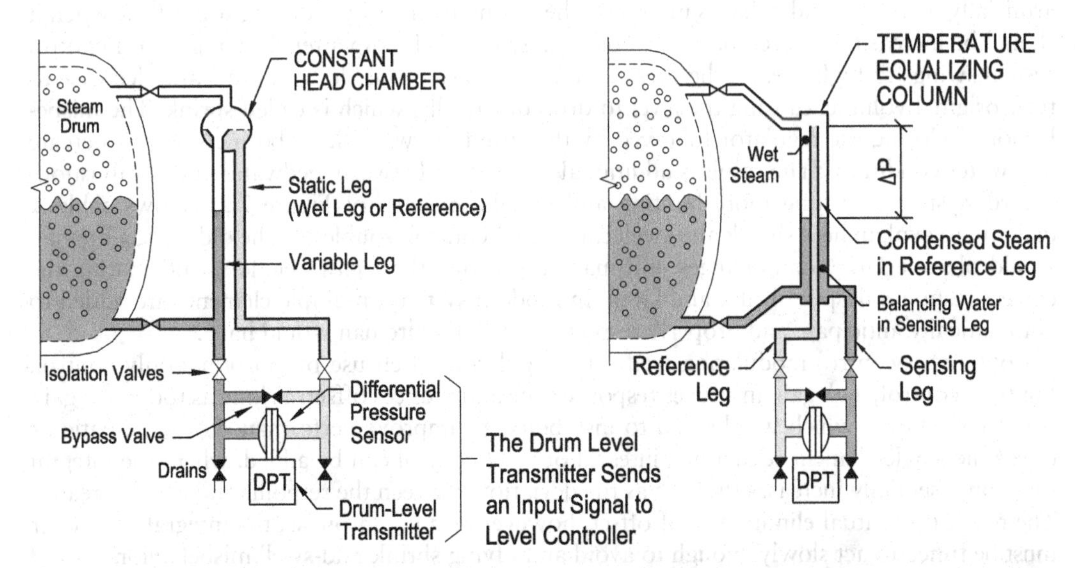

Figure 7.26 The differential-pressure transmitter is probably the best method of obtaining continuous level control. The level transmitter uses a very sensitive diaphragm to measure the difference in water-column height between a static reference leg and a variable water-level leg. The differential pressure correlates to level. This arrangement is used extensively in older pneumatic level controllers, but today is especially handy for creating 4-20mA analog-input signals to electronic controllers. A constant head chamber is shown on the left, and a variation which uses a temperature-equalizing column is on the right.

off-design pressures. To compensate for density differences between hotter water in the drum and cooler water in the level sensor, boiler level-indicating gage glasses are installed slightly below the drum centerline, so that its mid-gage-glass level aligns with the true mid-drum level for intuitive calibration. To compensate for this effect, more complex multi-variable sensors can be used, which would measure differential pressure for determining the water level, and drum pressure and temperature for calculating density for correcting the level measurements. Multivariable sensors may be justified in very large utility boilers, where carryover is more prone and can result in damage to the superheater over time.

Problems with intemittent single-element control

Level switches used with on-off control introduce variations that can be objectionable for some applications. With on-off operation, cold feedwater surges when the low-level limit is reached, and then abruptly stops when it reaches the high-level setpoint. Since the feedwater regulation system and combustion-control system are not linked, there can be significant interaction between the two. For instance, when the boiler is firing at steady state and the feedwater turns on, the sudden in-rush of cold water causes the drum pressure to collapse and the combustion-control system swings to high fire to catch up. After the feedwater is stopped, there is often pressure overshoot as thermal inertia causes the combustion control to lag. The result is hunting and cycling of the combustion-control system.

The second problem associated with single-element feedwater systems is a phenomenon called shrink and swell, which occurs during steam-demand transients. For example, if the steam flow suddenly increases, more steam, i.e., water, is removed from the drum. Intuitively, more feedwater is needed. However, before the water level drops enough to be sensed, the steam pressure decreases. The drum-pressure drop causes the steam bubbles entrained in the boiler water to enlarge. The result is that the drum level rises even though water in the form of steam was removed. This condition is called swell. Consequently, the feedwater controller senses an artificially high level and behaves in exactly the wrong manner by reducing water flow when it should be increasing it. Eventually the boiler pressure catches up after the combustion-control system swings to high fire. As the pressure rises, the steam bubbles compress and decrease to their original volume, causing the level to drop drastically, which is called shrink. The misbehavior of closing the regulator followed by the shrinking water level both conspire to create low-water conditions. The result is additional interaction between feedwater and combustion-control systems, creating more hunting and possible instability. Figure 7.27 shows a block diagram to explain how shrink and swell cause level-control troubles. In the old days, a fireman would observe large load changes and manually adjust the feedwater flow, anticipating the correct response despite shrink and swell. In modern systems, multiple elements are added to automatically anticipate and properly respond, just as the fireman would have.

Continuous-control modulating feedwater regulators often use proportional-only control. For tight control, high gain improves response time and reduces offset, as long as too much gain is not introduced, which would lead to instability. To improve performance of pneumatic or electronic single-element regulators, integral or reset control can be added, where the integral function essentially increases the gain as the deviation between the setpoint and level increases. The result is eventual elimination of offset, however as a compromise, the integral controller must be tuned to act slowly enough to avoid amplifying shrink-and-swell misbehavior.

Multi-element feedwater regulation systems

To address the shrink-and-swell problems associated with single-element feedwater systems, two-element or three-element controllers may be used. All feedwater control systems use

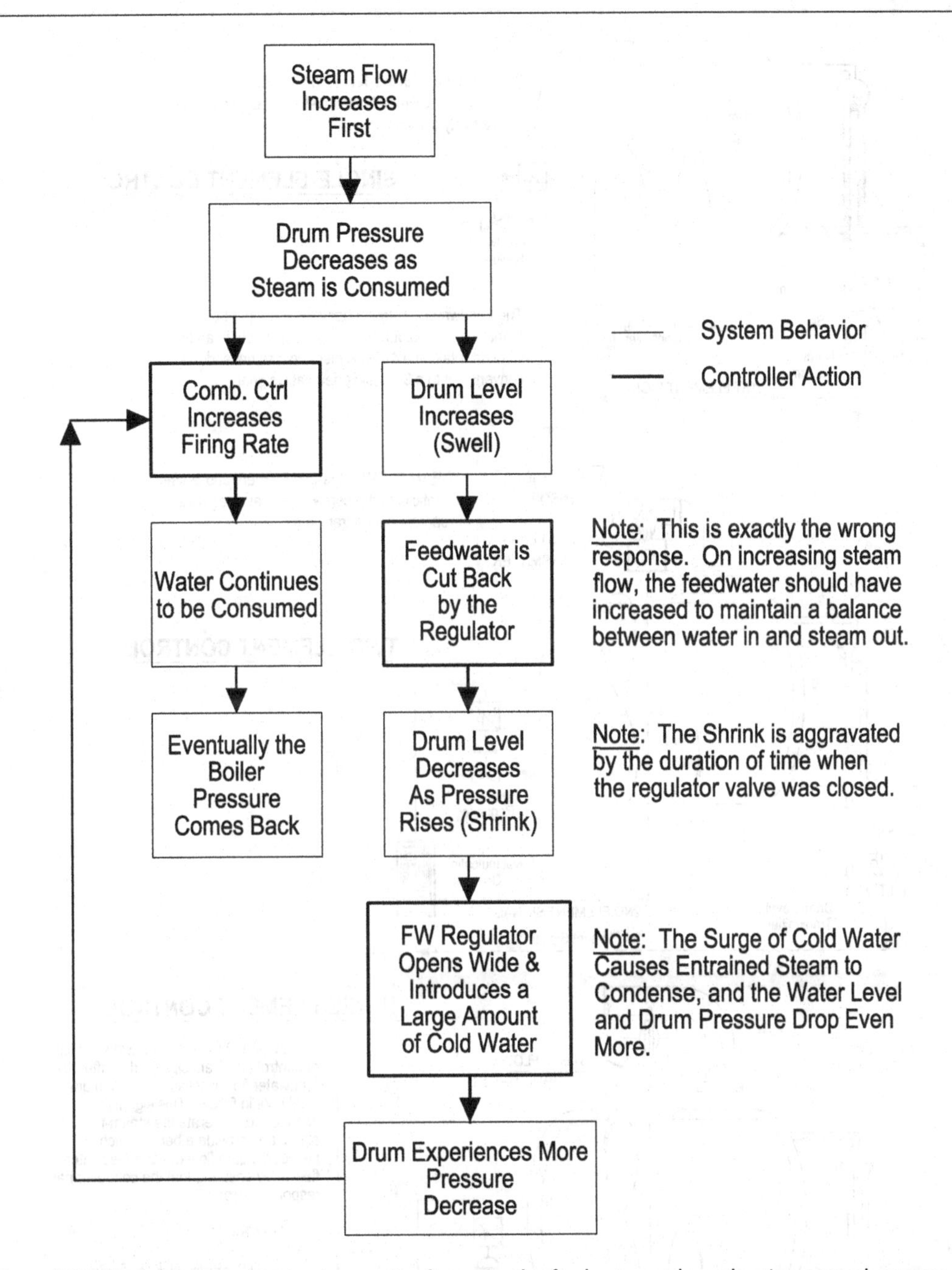

Figure 7.27 Block diagram showing interaction between the feedwater and combustion-control systems on load increase. The interaction results in low-water tendencies, delayed response, and hunting, as a single-element feedwater system experiences shrink and swell during the transient. Conversely, a large load decrease would cause the opposite problems, resulting in high-level misbehavior.

drum level as the primary control element, but the two-element feedwater system adds a steam-flow function to augment the level transmitter during transients. The third element acknowledges that the feedwater flow has responded correctly. During steady state conditions, the controller ignores the steam-flow and feedwater-flow signals. Figure 7.28 and Figure 7.29 show single-, two-, and three-element feedwater control systems.

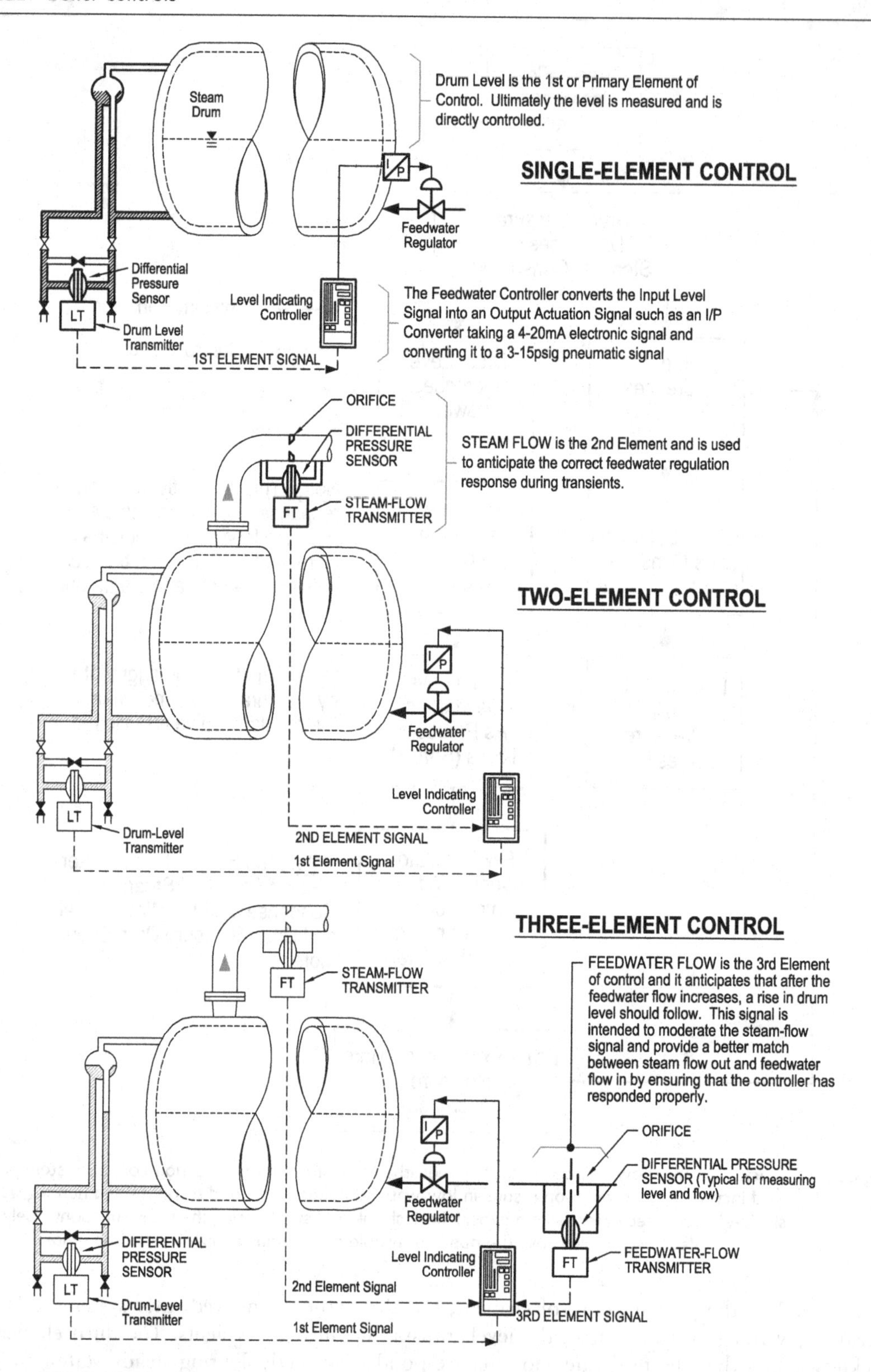

Figure 7.28 Single-, two-element, and three-element feedwater regulation systems. The second and third elements provide anticipatory control during transients to mitigate shrink-and-swell problems and to provide faster more accurate response.

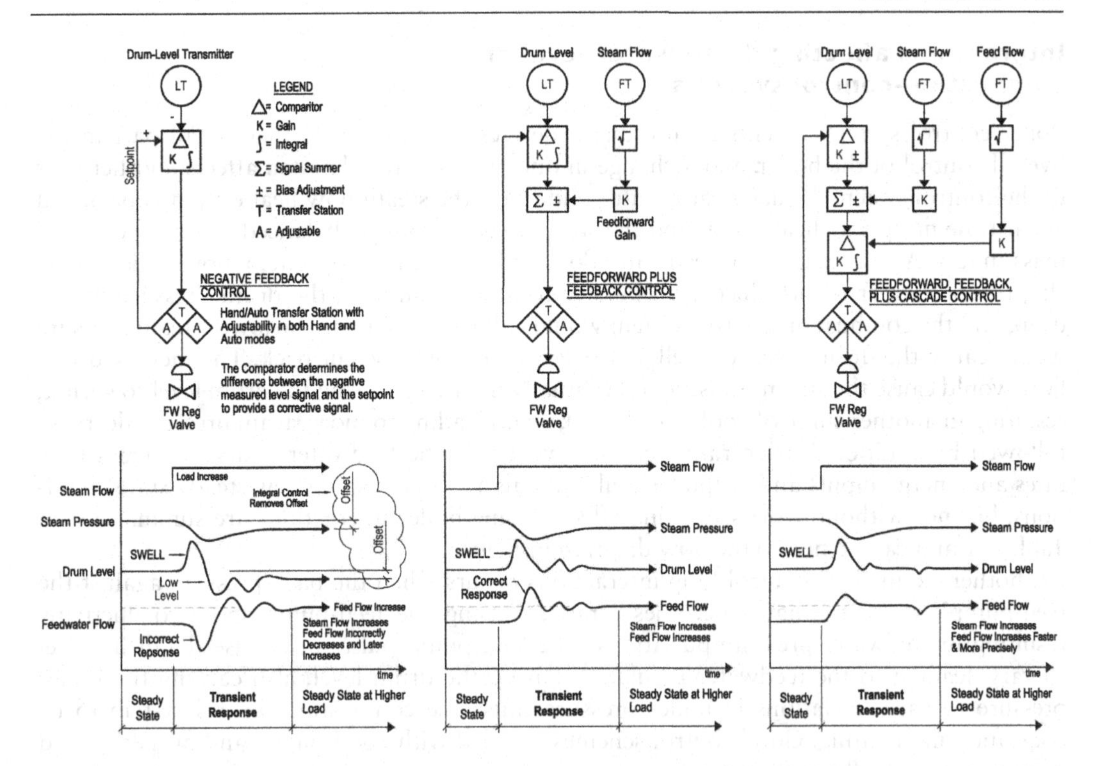

Figure 7.29 Single-, two-, and three-element feedwater control schemes. The relative system behaviors are shown in the response versus time curves for an increasing-load transient.

The two-element feedwater controller uses steam flow as the second element. The steam flow changes without delay during a transient, before the steam pressure or drum level has reacted, consequently, steam flow is the best indication to anticipate that the feedwater flow needs to change. Furthermore, unlike the single-element level controller which reacts incorrectly at first due to shrink and swell, the steam-flow determines immediately and correctly whether more or less feedwater is needed, as shown in the response curves in Figure 7.29. After the second element is added, the controller is tuned so that the steam flow adds a strong enough signal to the level controller to command the feedwater regulator to respond correctly. To work well, the control signal should be linearly related to the control-valve flow rate, and it is desirable that the feedwater-regulator pressure drop is fairly constant and kept within the range where the valve has good controllability. Boiler feed pumps operating with constant-pressure and constant-differential pressure governors are discussed later in this chapter, as pump pressure affects the feedwater flow through the regulator. Nicely, the two-element feedwater controller uses the same steam-flow transmitter signal used by the two-element combustion-control system.

The three-element feedwater system uses drum level and steam flow as the first and second elements, but also adds feedwater flow into the control scheme as a third element. Essentially, the third element provides feedback to account for unpredictability in the control-valve throughput, which may have to accommodate variations in the valve-disk and actuator design, friction in the moving parts, and pressure deviations in the feed-pump discharge and boiler steam drum. Feedwater flow as a feedback signal closes the open-loop nature of the regulating valve, ensuring that the corrective action has indeed taken place. There are several ways to configure three-element control systems, and the feedforward and feedback with cascade control is shown in Figure 7.29.

Interactions affecting drum water level and combustion-control systems

Boiler control systems essentially use dumb devices working to do their small part in the overall control of the boiler, and a change in one process variable often affects another part of the control system. Under steady-state conditions, the steam-flow heat output is balanced against the firing-rate heat input, and the steam mass outflow is balanced against feedwater mass inflow. As an example of feedwater-flow and firing-rate loop interactions, if a sudden slug of cold water is introduced during steady-state conditions, the drum pressure would drop, and the combustion-control system would call for more fuel. The lower steam pressure would cause the drum level to swell, and feedwater would be cut back. The increased fuel flow would cause the steam pressure to eventually increase, causing the drum level to shrink, resulting in another surge of cooler feedwater, again leading to more steam pressure decrease followed by additional firing rate increase. Eventually the feedwater and steam mass flow rates and energy inputs and outputs should all balance, establishing new steady state conditions, but not without some seesawing effects. Some boiler interactions are summarized in Table 7.1 and can be tied to the flow diagram in Figure 7.30.

Another example of control-loop interactions occurs when the back pressure against the feed-pump's turbine exhaust becomes erratic, causing the feed-pump speed to fluctuate, resulting in feedwater pressure pulsations. The feed pump pulsations cause the drum level to vary, leading to the feedwater regulator chasing the drum level, also causing the boiler pressure to vary. Variations in boiler pressure cause the combustion-control system to be constantly modulating. Good control schemes coupled with good tuning and properly sized devices reduce the fluctuations and the disruptive interactions.

Feed pump pressure control

Pump control systems and control valves are discussed in Chapters 8 and 9 respectively, and their behaviors affect the drum-level control systems. Pumps may be operated at constant speed using motors, or at variable speeds using steam turbines or motors with variable-frequency

Table 7.1 Some boiler control parameters and their interacting disturbances

Control parameter	*Disturbances that interact with the control parameter*
Boiler Water Level	• the position of the feedwater regulator valve • the feed pump discharge pressure • the boiler pressure, and • steam flow/firing rate particularly during transients producing shrink and swell
Feedwater Regulator Position	• boiler pressure • feed pump discharge pressure, for example, a low boiler pressure and high feed pump pressure would result in a regulating valve being mostly closed
Feed Pump Discharge Pressure	• feed pump speed, which is affected by steam flow to the feed pump turbine • steam drum pressure, which is influenced by the boiler steam flow • position of the feedwater regulating valve • amount of back pressure on the exhaust of the steam turbine
Auxiliary-Exhaust/Back-Pressure System: Turbine back pressure affects turbine/feed pump speed	• when the back pressure increases, the turbine slows down, and the feed pump discharge pressure drops. A high back-pressure trip can shut the pump down. • when the back pressure decreases, the turbine speeds up and the feed pump discharge pressure rises. Some pumps have an overspeed trip mechanism that can shut the pump down.

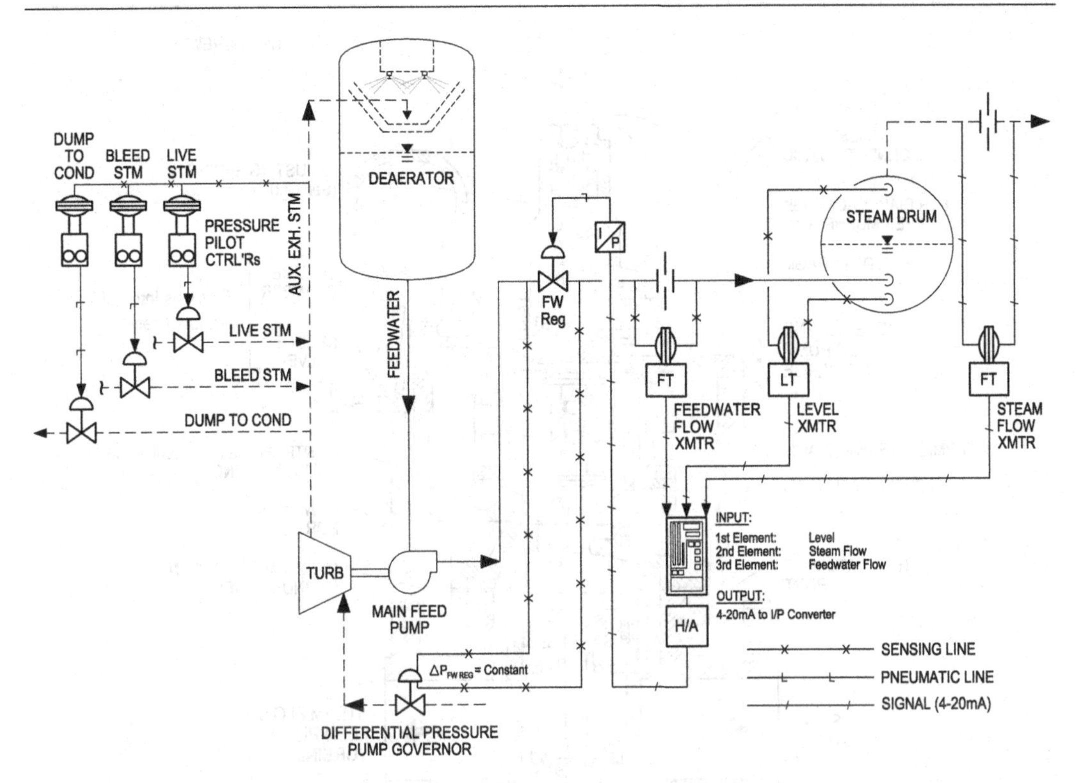

Figure 7.30 Interactions that affect the boiler water level occur from the feedwater regulator, feed-pump governor, and turbine auxiliary-exhaust system. A three-element feedwater regulator is shown in a system using an excess-pressure (XP) governor to set a constant pressure differential across the regulator valve for reliable control under all loads.

drives. The speed of the pump affects the feedwater-pump-discharge pressure, and the pump pressure affects the flow through the feedwater regulator at any valve-stem position, so it stands to reason that the boiler water level is affected by the interaction between the turbine steam flow, the pump speed, and the feedwater control valve.

Small boilers typically use constant-speed motor-driven feed pumps, through pump on-off operation or through a simple modulating feedwater regulator. Often the modulating feedwater regulator is motor actuated using a modutrol motor such as shown in Figure 7.13, or for larger systems, by using a 4-20mA signal to an I/P converter and a pneumatically controlled regulator.

Two feed pump control schemes are used for variable speed applications, a constant-pressure governor to maintain a fixed feed pump discharge pressure (Figure 7.31), or a constant differential-pressure governor to maintain a fixed pressure drop across the feedwater regulator (Figure 7.32). The constant-pressure governor senses the pump discharge pressure and then modulates the turbine steam flow to set the feedwater pump discharge pressure at one value. The operator adjusts the discharge pressure to be greater than the steam-drum pressure so that flow occurs. Problems with this arrangement occur in applications where the boiler-drum pressure varies. As an example, if the boiler-drum pressure increases and reaches the feed-pump pressure, no feedwater flow will occur, and a low-water-level condition will ensue. Conversely, if the drum pressure drops substantially, without readjustment by the operator, the feedwater valve will be nearly closed and will tend to hunt and may wire draw

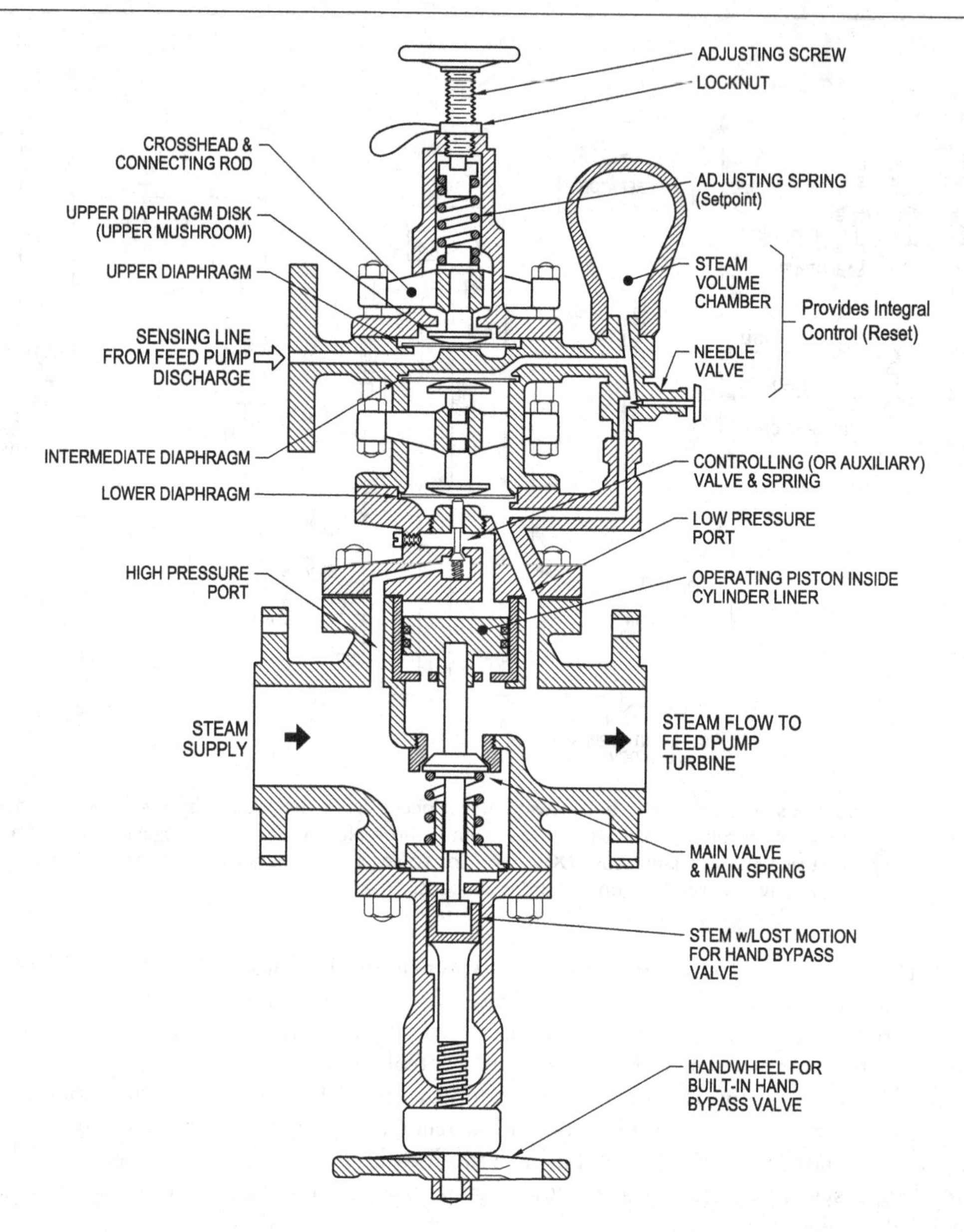

Figure 7.31 Constant-pressure feed-pump governor, similar to the Leslie CP governor, controls steam to the turbine. The upper handwheel and adjusting spring provides the setpoint, the needle valve and volume chamber adds integral control, and the operating piston provides strong hydraulic-actuation forces. The lower handwheel is part of the built-in hand bypass valve, useful for warmup or when the governor becomes erratic. (Courtesy of CIRCOR | Leslie.)

the seats. Boilers having a superheater and large load variations are prone to these problems, as the superheater outlet pressure varies substantially with its load-dependent pressure drop. In practice, these constant-pressure feedwater systems require manual readjustments as boiler load changes occur.

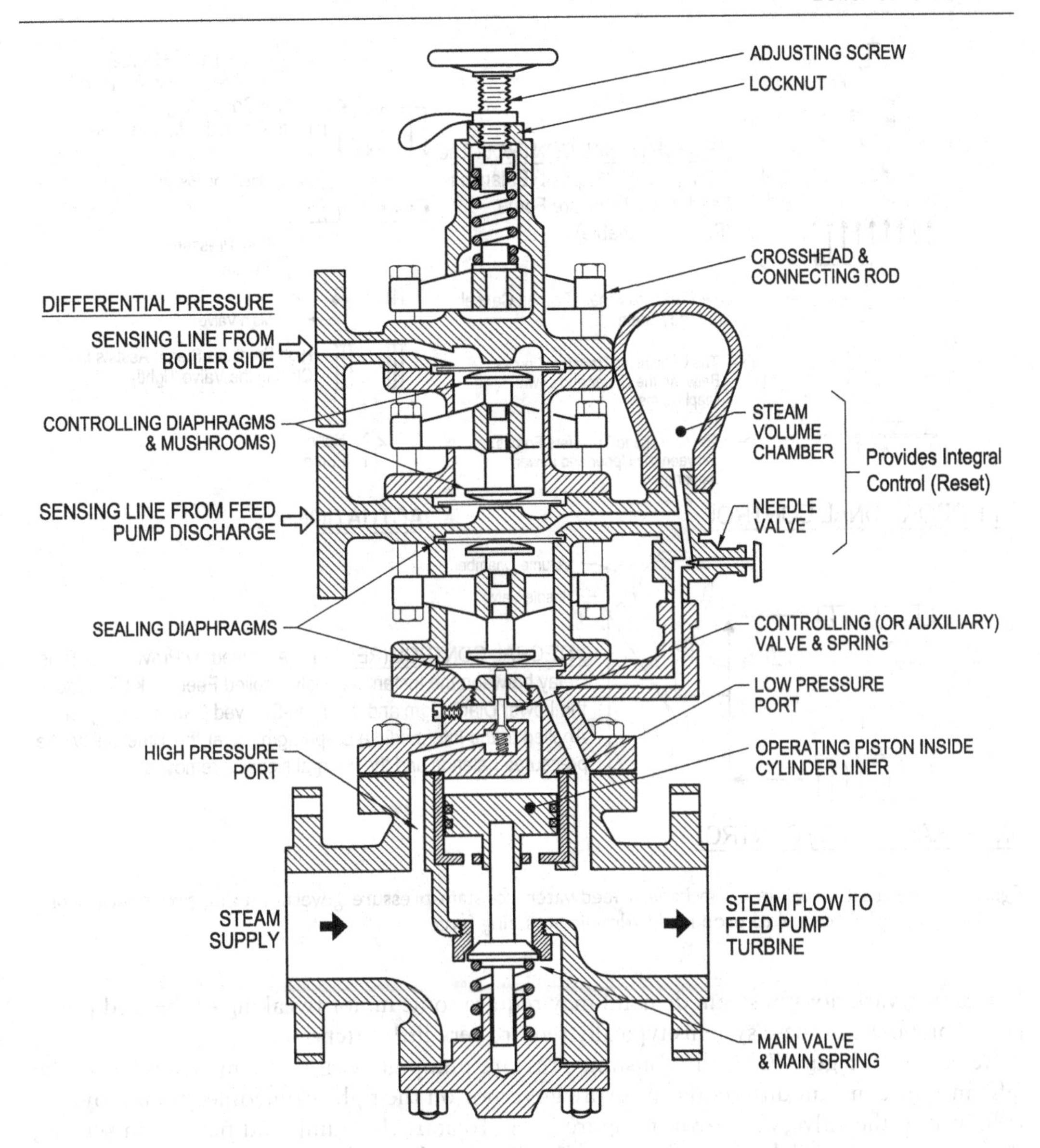

Figure 7.32 The Leslie XP (excess pressure) governor adjusts the feed-pump speed to maintain constant-differential pressure across the feedwater regulator, at all boiler-pressure variations. The built-in hand bypass valve is not included in this illustration, but can be added. (Courtesy of CIRCOR | Leslie.)

The constant differential-pressure governor is similar in construction, except it has two connections, one on the feed-pump side of the feedwater regulating valve and the other on the boiler side. The governor uses internal diaphragms to create the valve positioning forces that are shown in Figure 7.33, and spring force via an adjusting screw to change the setpoint. The intention is to modulate the turbine steam flow so as to always maintain the feedwater pressure moderately greater than the boiler drum pressure. Constant differential across the regulator ensures the potential to fill the boiler is always available regardless of boiler pressure. The advantage to this system is that as the boiler pressure varies, the feedwater-pump discharge pressure tracks the variation while maintaining the setpoint pressure differential.

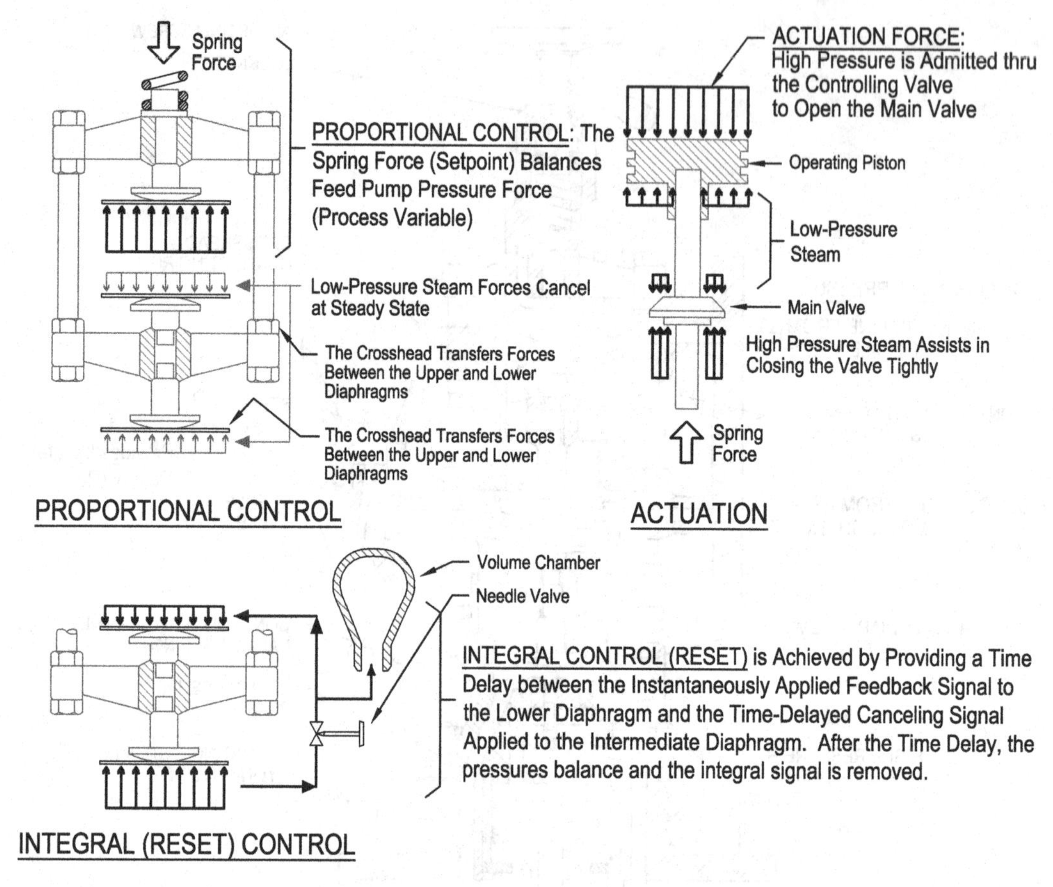

Figure 7.33 Internal forces in a mechanical feedwater constant-pressure governor using proportional plus integral control to produced hydraulic actuating forces.

In practice, variations in steam demand may require some minor tweaking of the feed pump governor, however, these systems typically require very little attention.

Referring to Figure 7.34, the constant-pressure pump governor system is shown on the left, and the constant differential-pressure governor on the right. Sometimes a hand bypass is built into the valve, as shown in Figure 7.31. To start the pump and place it in service, the pump suction and discharge valves are lined up, with the check valve preventing reverse flow. The feed pump recirculation valve is opened to avoid deadheading the pump. The turbine exhaust is lined up, while the governor is backed off to the no-flow position. The pump is slowly rolled by cracking the hand bypass valve until steam flow is heard and the pump shaft is observed to begin rotation. The operator then allows time for the turbine to warm up while listening for water hammer and ensuring lubrication is flowing. After the turbine is warmed, the governor handwheel is screwed down until the governor takes control, at which point the hand bypass is closed and the turbine brought up to speed using the governor.

In practice the governor is adjusted so that the feed pump discharge pressure results in the feedwater regulating valve being 50% open under steady-state conditions. At the 50% setting, the feedwater regulator has plenty of open-or-close excursion travel, permitting good control during large load changes. The 50% setting can be observed using the valve-stem

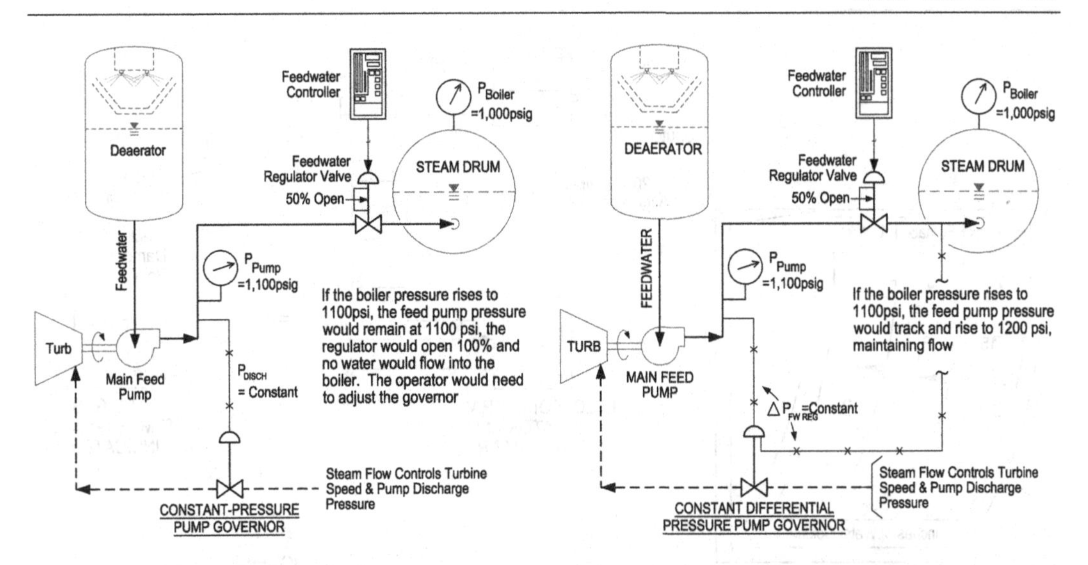

Figure 7.34 The constant pressure governor has a single pressure-sensing connection from the pump discharge (left), while the constant-differential-pressure governor senses pressure from both sides of the feedwater regulator (right).

Vernier scale, by observing the pneumatic signal to the actuator, which would be at about 9psig for a 3-15psi actuator, or by observing the controller position feedback signal.

FORCED DRAFT, INDUCED DRAFT, AND FURNACE DRAFT CONTROL

Apart from combustion-control systems having O2 trim to automatically tweak air-fuel ratios, all combustion-control systems are adjusted to set the air-fuel ratio by establishing delicate balances between mechanical components. As operating parameters change, air-fuel-ratio deviations may occur. For instance, the air-fuel ratio controller balances the fuel-oil regulating valve position with the forced-draft fan damper position for any given fuel-oil service-pump and forced-draft-fan speed. For small boilers, the burner is tuned once by adjusting the fuel relative to the air at multiple discrete firing rates over the entire range, while large boilers have the capability of biasing the fuel or air signal so as to correct for deviations and optimize performance. The fuel-air mixture is a function of several factors that may affect tuning, such as ambient temperature and humidity, consistent fuel oil supply pressure to the regulator valve, and furnace pressure, which is directly affected by the stack draft. For short stacks, stack draft may not affect the burner much, and no special provision may be required; however, for tall-stack systems, stack draft may be controlled using barometric dampers, boiler-outlet dampers, variable-speed forced-draft fans, or induced-draft fans. Some stack monitoring and control devices are shown in Figure 7.35.

Stack draft or stack effect is the natural buoyant effect of hot gases rising because of their lower density compared to the surrounding outside air. Draft provides a natural circulation that assists and reduces the load on the forced-draft fan. Draft is opposed by stack-gas fluid-flow friction. In some cases, when stack draft is inadequate, such as very large industrial boilers having exhaust systems that turn downward, induced-draft fans are installed on the outlet to assist draft in pulling out combustion gases.

Pressure-fired boilers do not have an induced draft fan, and the furnace is operated at positive pressure, at least at high loads. Tight boiler-casing construction is mandatory to avoid

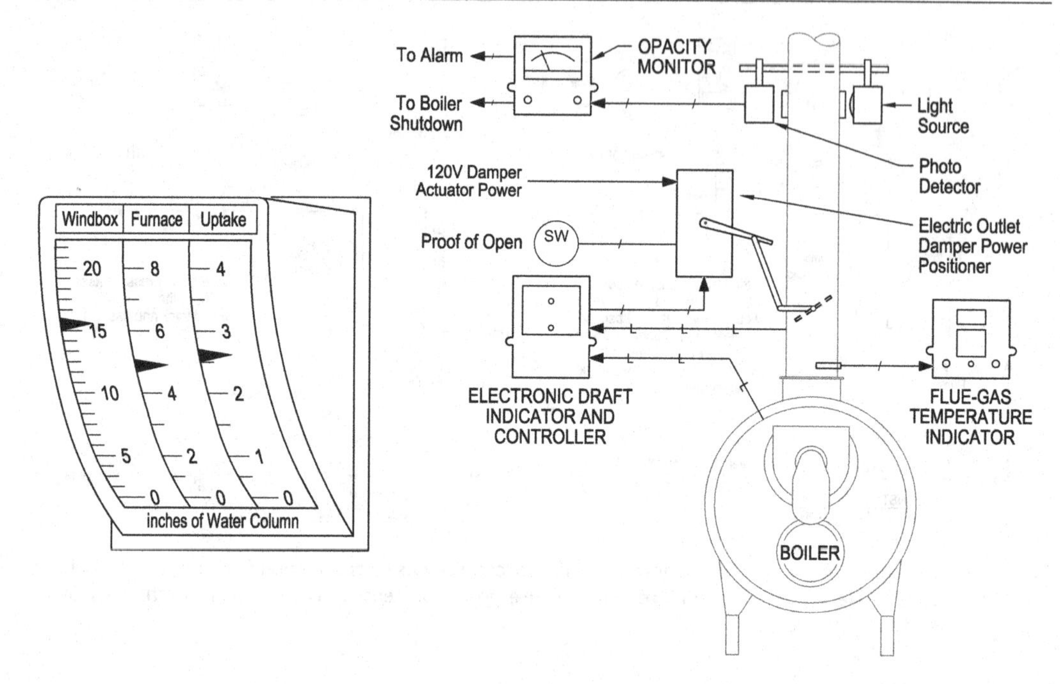

Figure 7.35 A three-module vertical scale draft indicator (left) shows windbox, furnace, and uptake draft in inches of water column. Some commonly installed stack devices (right) include smoke opacity monitors, draft controllers with damper actuators, and flue-gas temperature indicators.

leaking combustion gases into working spaces and to avoid the chilling effect of tramp air when the furnace pressure goes negative. Balanced-draft boilers are used where positive pressure would cause gas leakage, such as stoker-fired boiler which are open to the boiler room, and typically have a forced draft fan for under-and over-fire combustion air. The induced-draft fan pulls enough gas so as to maintain the furnace at a slight negative pressure.

Barometric dampers are passive devices that blend cool ambient air into the hot stack gases to reduce the stack effect. The barometric damper uses a counterweight to close the damper balanced against negative internal pressure that tends to open it. At low firing rates, the damper may be fully closed, but at higher firing rates with large stack effect, the barometric damper opens, mixing cool air into the hot gases. This device is used for small boilers when stack-draft control is required.

Outlet dampers are used when there is significant stack effect, and the damper is used to provide a consistent outlet pressure under any ambient conditions, so that the burner is firing into predictable furnace pressures. Doing so results in more consistent air flow through the burner over the entire range of firing rates. Outlet dampers may be of single-bladed design for small boilers, or multi-bladed design for larger boilers. The multi-blade design may use parallel blades which are simpler in construction but non-linear, or more expensive opposed blades that have better flow characteristics. Linear behavior is desirable for matching simpler control signals, but good linearity can be engineered into the actuator-to-damper linkage geometry as an option.

In some applications where boilers tend to operate at low loads and cycle on/off, a non-modulating stack damper may be installed. It is fully open when firing and completely closed when the boiler cycles off, so as to reduce tramp-air flow through the boiler from natural-convection and to retain heat. In installations using stack-outlet dampers, it is mandatory

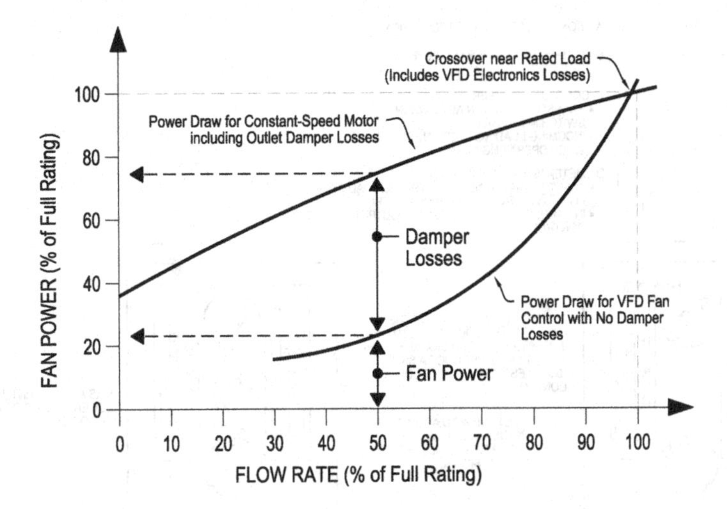

Figure 7.36 These two curves show the power requirements for constant-speed damper-controlled fan systems compared to speed-controlled systems. The upper curve shows the fan power consumption when using an outlet damper to control air flow. The lower curve shows power consumption when using variable speed to modulate air flow, where the fan output matches the system friction. The upper power curve includes the system's required fan power plus the damper losses, while the lower curve consumes only the fan power to overcome the system's friction.

that a proof-of-open switch be connected into the boiler permissive circuit to maintain safe operating conditions.

Variable-speed forced-draft fans provide a good option for energy savings and are used to control the combustion-air flow for larger boilers. Variable-speed operation eliminates the fan-power losses associated with air flow overcoming pressure losses across the damper, as shown in Figure 7.36. The speed setpoint is obtained from the boiler master, possibly cross-limited with the fuel, and in conjunction with an air-flow feedback signal. In some cases, a forced-draft damper is not used, but where high turndown is required, both the variable-speed drive and the forced-draft damper control are needed to work in a coordinated manner. At high boiler demand, the damper is commanded to the fully open position, and air flow is controlled solely by modulating the fan speed, where air flow follows the affinity laws.[2] As the steam demand and air-flow requirements decrease, the fan response becomes lazy and less predictable, and eventually the fan may not be able to develop the right amount of flow. As low air flowrates are approached, the controller begins to throttle the fan damper.3[3] At some point, the controller fixes the fan at a minimum speed and the air flow is then controlled solely by the damper position. When variable-speed forced-draft fan control is used, modulating stack dampers to control furnace pressure may not be required, as good stack draft aids flow and reduces the fan power consumption.

Where induced-draft fans are required, the forced draft fan pushes the air through the windbox and into the furnace, while the induced-draft fan pulls air through the boiler. Induced-draft fans may be required where downdraft occurs or where the stack draft is insufficient to maintain a negative pressure. Figure 7.37 a hand-auto operating station used for fuel, air, and feedwater flow control. Figure 7.38 shows an electronic controller used as a boiler master to control firing rate and to balance steam production from multiple boilers, and coupled with other controllers in Figure 7.39 to form a complete boiler control system. Figure 7.40 shows a conventional multispeed forced-draft fan system, where the speed is manually adjusted to mitigate damper losses at all firing rates, while providing good modulation of air flow.

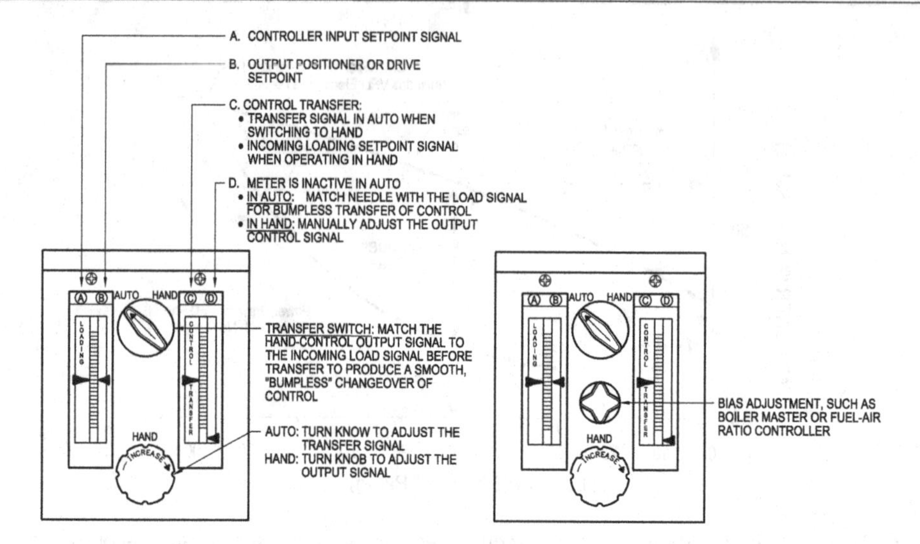

Figure 7.37 Hand/auto station with adjustment knob for manual control and for "bumpless" transfer between AUTO and HAND control (left). In pneumatic control systems, the HAND adjustment is simply a pressure-control valve where its output is the control signal. Some controllers contained a bias or trim adjustment (right) for fine-tuning the signal

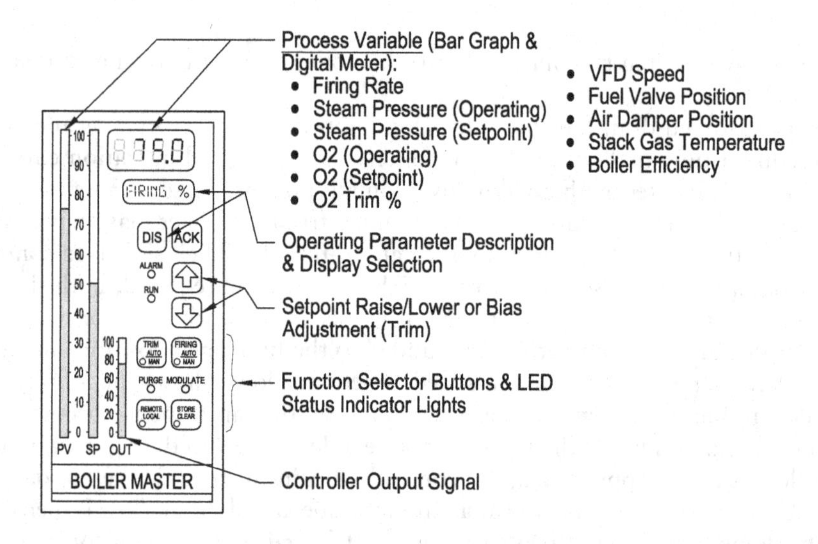

Figure 7.38 PLC-based multi-loop controller used as a boiler master. The operator can scroll through displays to monitor parameters and to change setpoints using the up/down arrows. LED lights may be used to indicate status and alarm, and alarm acknowledgement (ACK) is included. The alarm history is retained in memory for troubleshooting, using "first-out" programming.

SUPERHEATER TEMPERATURE CONTROL AND DESUPERHEATED STEAM

Superheaters are used in power generating plants to increase the steam's available energy so as to do more turbine work per pound of steam and increase the plant efficiency. Early boilers generated steam at much lower pressures than today, and the superheat temperature was uncontrolled, except by boiler design. Superheaters built with a large amount of

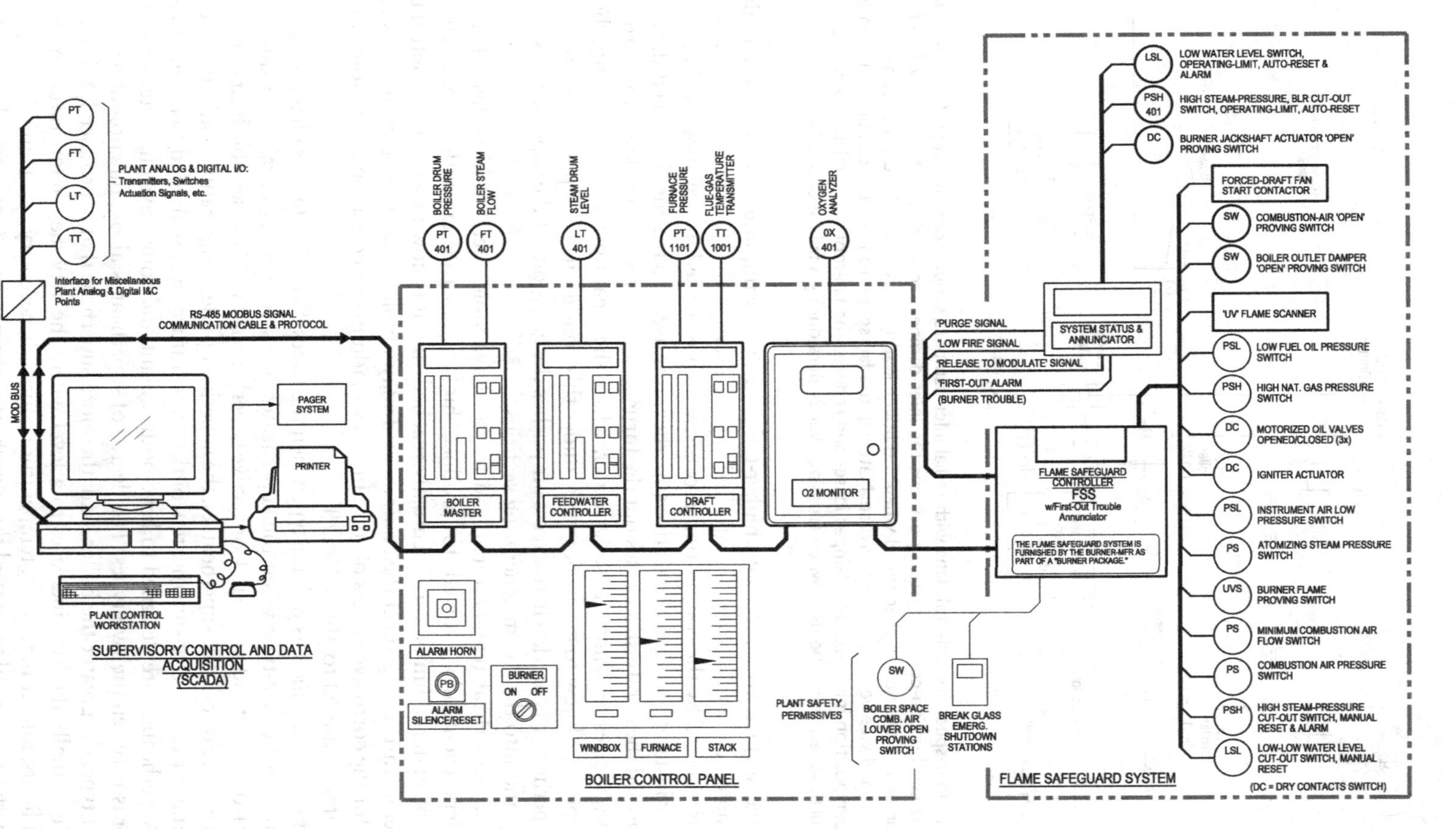

Figure 7.39 Boiler-burner control with Flame Safeguard System (FSS) showing signal and data-link connections to the loop controllers and a Supervisory Control and Data Acquisition System (SCADA) used for plant-wide monitoring and control.

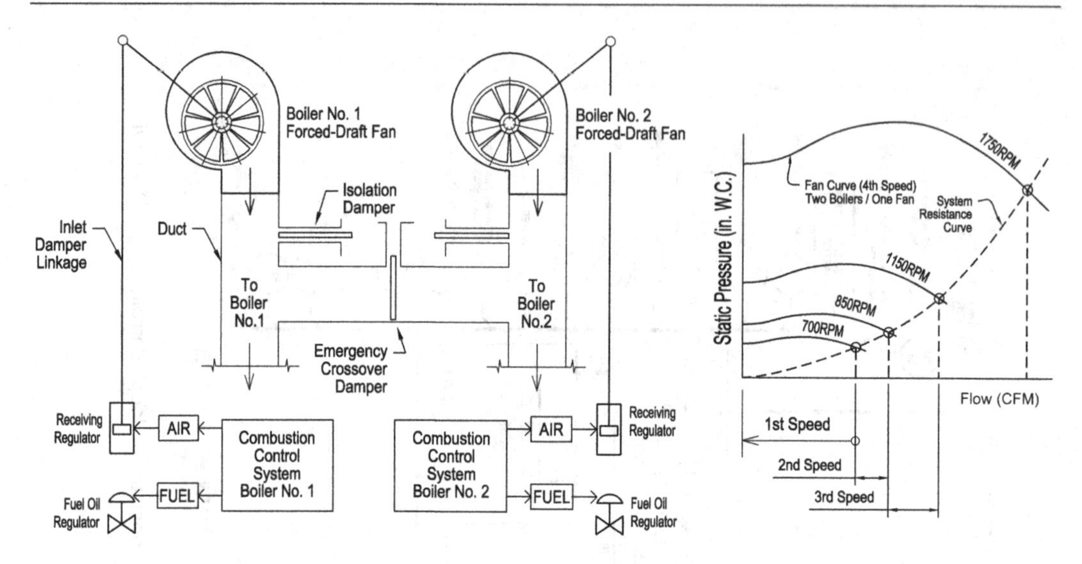

Figure 7.40 Two four-speed forced-draft fans with radial inlet dampers supply combustion air for two boilers. The combustion-control system positions the inlet dampers to modulate the air flow. The fan is bumped to the next higher speed when the boiler load increases and the damper is nearly wide open, and vice versa on load decrease. Matching the fan speed to keep the damper in the mostly open position minimizes the damper losses and reduces the fan-motor electrical power draw. The fourth speed would be used when crossing over boilers during a failure of one fan.

heat-exchange surface area produced higher-temperature steam, and vice versa for boilers with small superheaters. For the uncontrolled superheaters of the day, the steam-outlet temperature was affected by the combustion process, which was used in adjusting the boiler air-fuel ratio. For instance, when a convective superheater is used, increasing the forced-draft fan excess air pushed the combustion gases out the furnace sooner, resulting in less steam production and hotter gases reaching the superheater, where the steam picked up more superheat. The boiler operator knew to cut back the air under this condition. The opposite is true when a radiant superheater is used, where excess air pushes the gases out of the superheater section prematurely and into the steam generating sections.

Since higher superheat temperatures lead to more available energy and higher-efficiency plants, boilers eventually evolved to produce higher superheater outlet temperatures. However, tube material limits were eventually reached, and it is now necessary to control the superheat temperature to protect the equipment. Today's boilers use chrome-moly steels and accounting for the corrosive effects of sodium and vanadium salt contaminants in some fuels, the superheater is limited to about 1050°F.

Superheaters are classified as radiant-type when the tubes "see" the direct radiant heat from the flame, and convective-type when the superheater has screen tubes installed to shield the superheater from direct radiant heat. Convective superheaters "feel" the sensible heat of the hot combustion gases. For radiant superheaters, the superheat outlet temperature tends to decrease with load, and for convective superheaters outlet temperature rises with load (see Figure 7.41 left). Some superheaters are designed to have both radiant and convective characteristics, and the behavior is between the two types. The degree of superheat is also a function of load, and regardless of type, the greater the boiler load, the more superheat that is produced. With modern boilers, eventually the load increases to a level where the superheat temperature matches the material limits, and at that point, temperature control is necessary (see Figure 7.41 right).[4]

Several methods are used to control the superheater temperature. One method is to extract a portion of the steam part way through the superheater tube passes and divert it through a submerged-tube heat exchanger located beneath the boiler water level. This heat exchanger

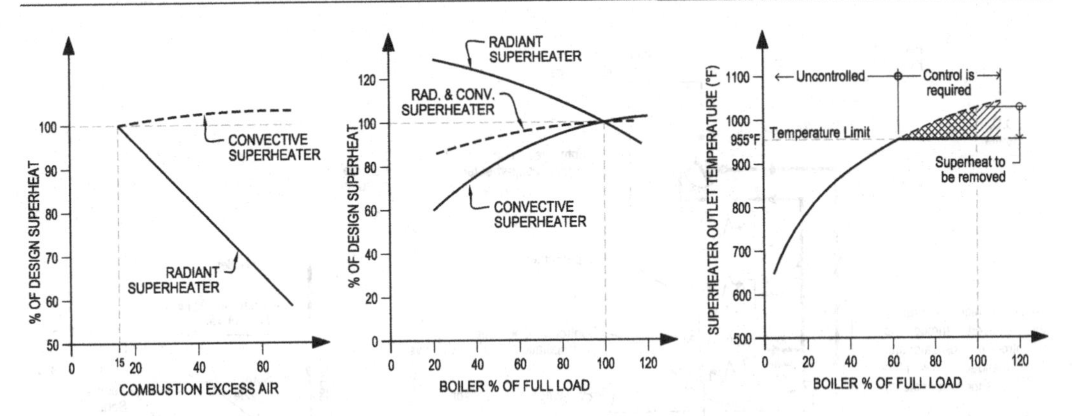

Figure 7.41 (Left): Superheater-outlet temperature behavior as a function of boiler load for radiant and convective superheaters. (Right): Convective-superheater outlet temperature as a function of load. The superheater is uncontrolled at low loads, but excessive superheat temperatures at high loads mandates control be added.

is called a control desuperheater. A temperature-actuated flow-control valve is used to divert the correct amount of steam through the control desuperheater, where an orifice in the superheater header is installed to create enough pressure difference for steam to flow. The controller senses the superheater-outlet temperature and modulates the correct amount of diverted flow through the control valve before blending the cooler steam back.

Another method of superheat temperature control is an attemporating-type desuperheater. The attemporator sprays feedwater into the steam flow, where the water absorbs sensible and latent heat of vaporization, while at the same time cooling excess temperature from the steam. The controller measures the attemporator-outlet temperature and throttles the water flow to obtain the correct amount of superheat.

The superheat control signal comes from measurements of the superheater-outlet temperature as the primary element. In some large boilers that undergo large sudden load changes, the steam flow rate is added as a second element for feedforward anticipatory control. In the case where steam flow is rapidly decreased, the longer residence time in the superheater coupled with the thermal inertia of very hot combustion gases, can cause excessive superheater temperature. The signal from a steam-flow transmitter can be used to proactively initiate more control desuperheating before the high steam temperature occurs.

Superheated steam is used for high-energy steam-turbine applications to improve energy efficiency; however, saturated steam is much preferred for transporting large amounts of thermal energy for heating applications. Heating primarily uses the steam's latent heat, where a large amount of thermal energy is transferred very effectively via superior convective-film heat transfer that occurs during the condensation phase change. Recuperative heat exchangers without extended-surface fins are relatively poor at removing superheat from vapor before it reaches saturation, resulting in oversized heat exchangers, sometimes requiring a dedicated desuperheating section. Auxiliary steam used for heating services is often desuperheated to improve its heat transfer characteristics, so as to reduce the heat-exchanger size.

Either the submerged-tube or the attemporating type of desuperheater is used to create auxiliary steam, as shown in Figure 7.42. Both types maintain some amount of superheat, where the goal is to approach saturation. The submerged-tube desuperheater is an uncontrolled heat exchanger located below the boiler water level, where superheated steam rejects heat back into the boiler for energy recovery. Following second law principles and because the desuperheater is at a lower pressure than the boiler drums, the desuperheated steam even if it approaches the drum temperature, is always slightly superheated. As an uncontrolled heat exchanger, the steam will closely approach the drum temperature at low loads, but at

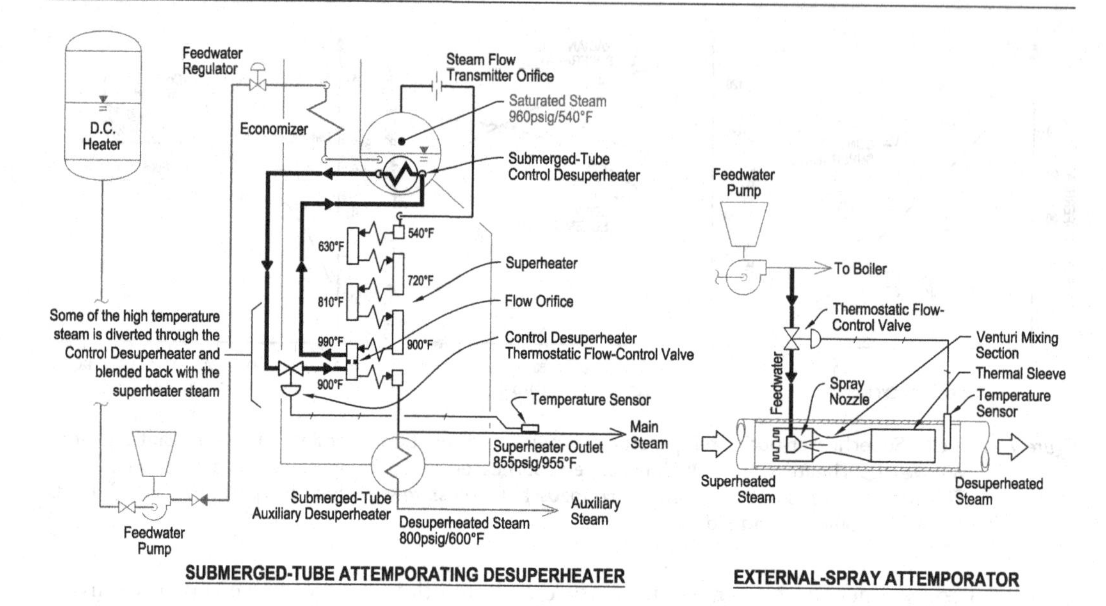

Figure 7.42 Two methods of superheater temperature control. A submerged-tube control desuperheater is shown on the left and an external-spray attemporator is shown on the right.

high loads, the steam may be 50-80°F higher than the drum temperature. The system naturally self-adjusts if the boiler pressure varies.

The auxiliary-steam attemporating desuperheater controls the temperature to within 30-50°F above saturation by spraying water into the steam. A slight amount of superheat is necessary, because pressure and temperature are not independent within the saturated region, so accurately controlling the water flow would otherwise be difficult. In boilers where the steam pressure varies substantially, temperature control alone is insufficient for ensuring that the steam is dry, so temperature and pressure must be characterized into the control scheme to maintain a slight amount of superheat. In high-pressure steam applications, throttling processes to produce low-pressure heating steam result in significant amounts of adiabatic superheating, and attemporating desuperheaters are sometimes installed to make low-pressure heaters more effective, so that the heat-exchanger physical size can be reduced. Figure 7.43 shows an instrumentation and controls diagram for a large industrial boiler, combining all required systems.

Figure 7.44 shows a method of superheat temperature control used in large utility boilers. In this method, steam generation and superheater-outlet temperature are both modulated by the combustion-control system, by adjusting the flame size and by tilting the burners in response to superheater outlet temperature. When the burners are tilted downward, the combustion gases have more residence time in the furnace, evaporating more steam in the water walls, and the superheater experiences lower-temperature gases creating less superheat. When the burners are tilted upward, combustion gases are pushed out of the furnace earlier with less heat absorbed by the waterwall generating tubes, and the superheater experiences hotter combustion gases. The combustion-control system simultaneously modulates the firing rate to maintain steam flow and the burner tilt to maintain the superheat.

FUEL SELECTION AND POLLUTION MITIGATION

There are several strategies for mitigating pollution, where devices may be tied into the boiler controls, alarms, or shutdown-permissives. Much of the pollution control is accomplished by equipment design or post combustion stack-gas treatment. The major types of pollution are oxides

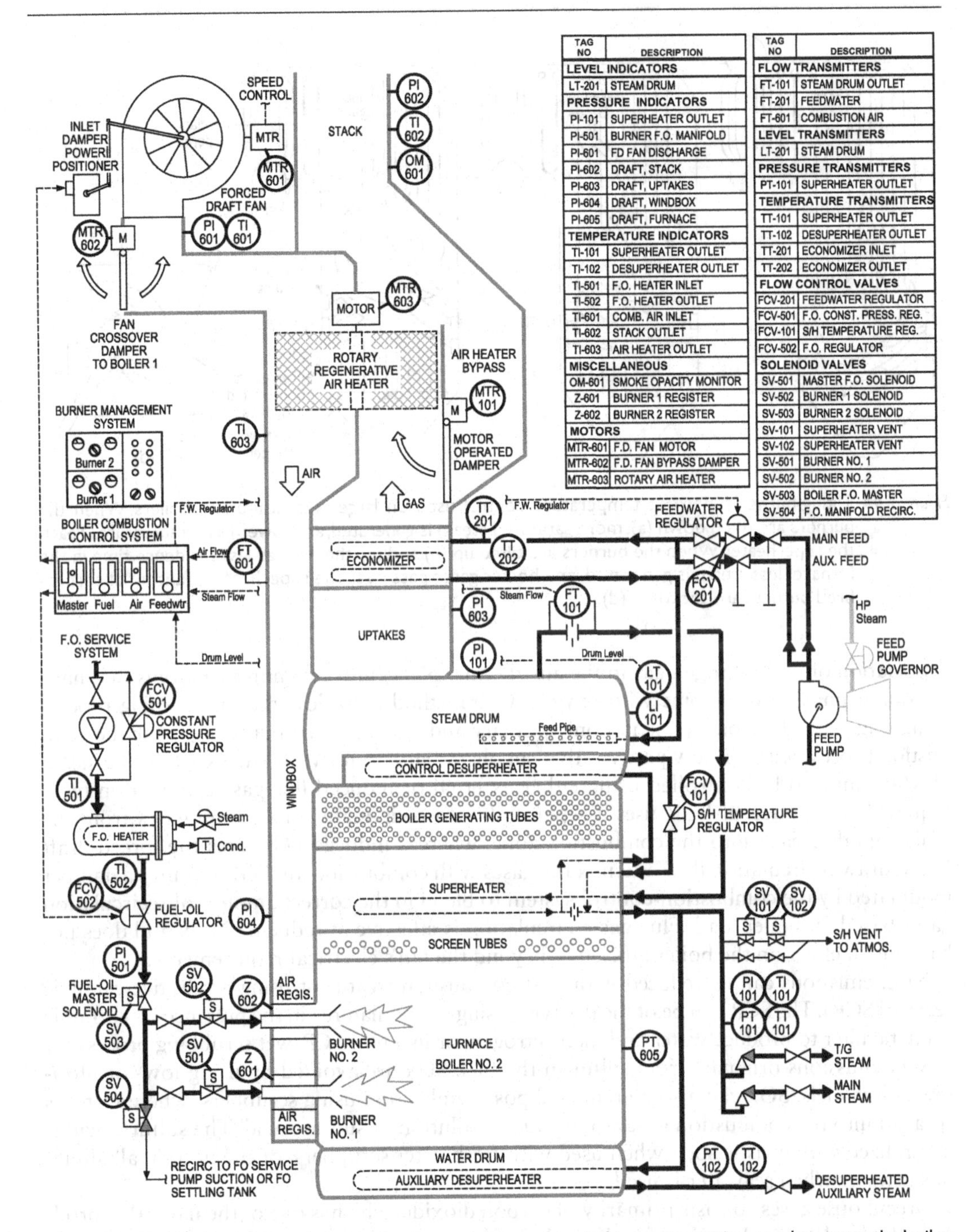

Figure 7.43 This block diagram shows boiler components and typical instrumentation, used to control a boiler that has a Ljungstrom rotary-regenerative air heater and burns heavy fuel oil.

of nitrogen (NOx), oxides of sulfur (SOx), greenhouse gases (primarily CO2), volatile organic compounds (VOCs or unburned hydrocarbons), carbon monoxide (CO—which is toxic), and particulate matter (PM). SOx and NOx lead to sulfuric and nitric acids, constituents of acid rain.

NOx is controlled by modern burner design, which functions to slow down the chemical reaction of combustion. The idea is to reduce the peak flame temperature to avoid the chemical

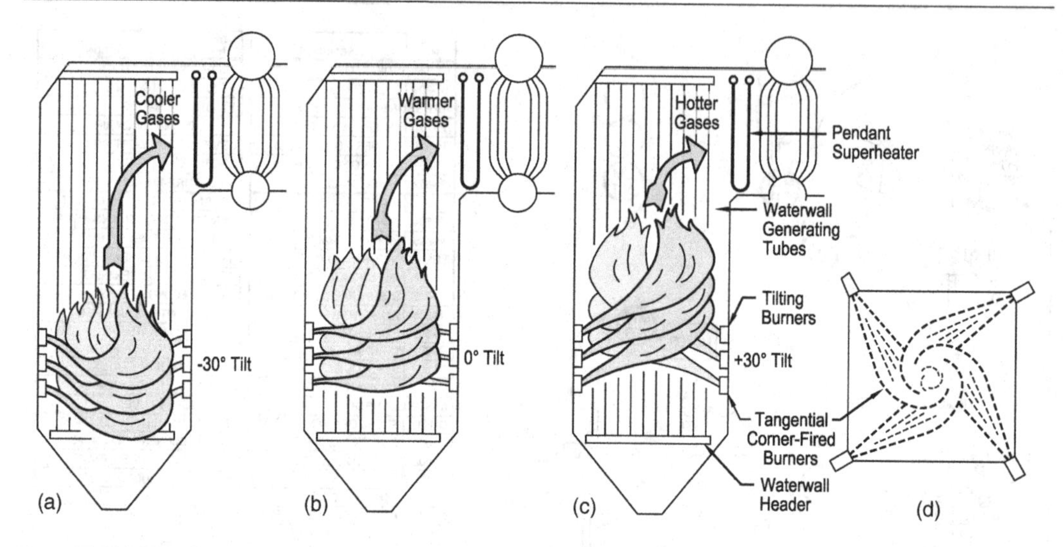

Figure 7.44 Tilting-burner superheat temperature control used on large electrical-utility boilers. When the burners are tilted down (a), more saturated steam is generated, and lower gas temperatures reach the superheater. When the burners are tilted up (b) and (c), the fire has less residence time in the furnace, less steam is generated, and hotter gases pass over the superheater. Tangential corner-fired burners are shown in (d).

dissociation of inert nitrogen gas in the air (thermal NOx) into two nitrogen atoms that have the opportunity to recombine with oxygen. One method is to slow the fuel-burning process using staged combustion having not only primary and secondary air, but to complete the combustion process with air-flow patterns that supply tertiary air for very-low-excess-air burners.

NOx can also be controlled using either external or induced flue-gas recirculation techniques (FGR). External FGR uses a separate fan to recirculate combustion gases from the boiler uptakes back into the combustion zone, whereas induced FGR uses the forced-draft fan to draw recirculation flow and mix the gases with combustion air. FGR requires a damper modulated by the combustion-control system to blend in the correct amount of recirculation gases based on boiler load. Flue-gas recirculation is effective at reducing NOx and does not have much effect on the boiler efficiency, beyond the fan's electrical requirements.

NOx emissions can be reduced with post combustion treatment using a selective catalytic reactor (SCR). The SCR can be of the dry type, using ammonia/urea in the presence of a ceramic catalytic filter to produce water and inert nitrogen, or in a non-SCR wet scrubbing process.

SOx emissions originate from sulfur in the fuel. SOx are avoided by using low- or ultra-low-sulfur fuels. SOx can also be removed post-combustion using scrubbers, where water is sprayed into the combustion gases to produce a solution of sulfuric acid. The sulfuric acid is neutralized with caustic soda when used with freshwater scrubbers, or the natural alkalinity of saltwater when seawater is used.

Greenhouse gases consists primarily of carbon dioxide, which is one of the natural byproducts of complete combustion of hydrocarbon fuels. CO2 can be reduced using clean-burning lower-carbon fuels, such as natural gas, and the inclusion of high-efficiency equipment that results in lower fuel consumption.

Volatile-organic compounds (VOCs) are airborne hydrocarbons from unburned fuels. VOCs react with NOx in the presence of sunlight to produce ground-level ozone (O_3) as a secondary pollutant. Ground-level ozone is the main component of smog. VOCs can be formed by poor combustion, but they more commonly come from vapors emitted by fuels. Most steam boilers use good burners to avoid producing VOCs, but some chemical plants require post-combustion thermal oxidizers in an after-burner to decompose the VOCs at high temperatures. Flue-gas recirculation, while reducing NOx, can actually increase VOC emissions, if done incorrectly.

Carbon monoxide (CO) is a product of incomplete combustion. CO is avoided by good burner design, enough excess air, and effective fuel-air mixing techniques. In addition to significantly affecting the combustion efficiency, carbon monoxide is highly toxic and can cause chemical asphyxiation even at low levels given a long-enough exposure.

Particulate matter (PM) is the solid and liquid particles suspended in the exhaust gases, many of which are hazardous. For boilers, the largest sources of PM come from ash, which are fuel contaminants that do not burn, and soot from unburned carbon, as well as other compounds. PM can be abrasive, corrosive, toxic to plants and animals, and harmful to human respiratory systems. PM levels are worse when burning heavy fuel oils and coal. Electrostatic precipitators, scrubbers, and baghouses are used for utility boilers, to filter the gases, and smaller boilers can benefit from burning clean fuels. In some installations, opacity monitors are installed for detecting and alarming smoking conditions (see Figure 7.45). Some units are self-calibrating during the boiler off cycle, when the stack is known to be clear.

For large boiler installations, continuous emission monitoring (CEM) and reporting may be required by federal or local regulations. Like the O2 trim combustion-control tuning technologies discussed earlier in this chapter, stacks may be fitted with gas analyzers that measure unburned hydrocarbons, CO, CO_2, NO, NO_2, NH_3, SO_2, H_2O, etc. Many CEM analyzers directly measure the gases using non-contact opto-electronic technology, based on the wavelength-specific light absorption of different gases in the measuring path across the stack outlet. The light beam is split and directed to detectors configured to measure the desired gas.

Fuel selection is another method of mitigating emissions. Fossil fuel options consist of natural gas, light distillate oils, heavy fuel oils, and coal. Natural gas has a very high percentage of methane, depending on where it was extracted. Methane is the cleanest-burning fossil fuel, having the largest heating value per pound and producing the least amount of CO2 per pound. Generally, the more carbon in the fuel, the heavier it is, and the more CO2 it will produce. When accounting for the percentage of carbon and hydrogen and the differences in heating value, heavy fuel oil produces nearly 40% more CO2 than natural gas per btu of released energy. Heavy fuels are more prone to having sulfur, ash, and other contaminants.

Gas fuels, light distillates, and heavy fuel oils can be burned in boilers or internal-combustion engines. Heavy fuel oil requires heating to be atomized in a burner nozzle or engine injector, and it reaches its autoignition temperature in a diesel cycle. Natural gas does not reach its autoignition temperature and is commonly mixed with intake air in a similar manner as a carburetor in

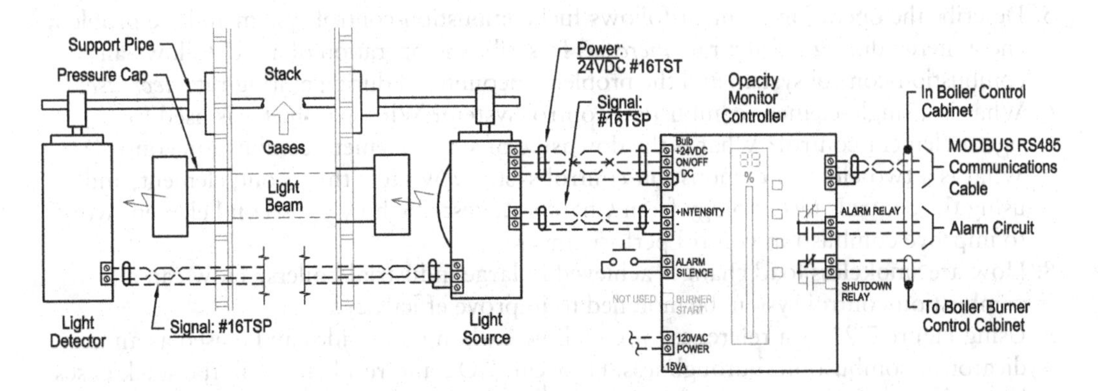

Figure 7.45 The general arrangement and wiring are shown for an opacity monitor that detects smoke, acquires and transmits data, and alarm alarms when smoking is excessive.

an automotive engine, using either spark plugs or fuel injectors for pilot ignition. Natural gas burns cleaner, but its density is less than half that of heavy fuel oil. When carried in the liquid state for use as a ship's fuel, it needs about twice the storage volume, even when considering its larger heating value. When stored as a cryogenic liquid, the fuel system must handle the boil-off as the tanks absorb ambient heat, where the natural boil-off typically contributes as a fuel in the combustion process. When starting from its liquid state, natural gas requires a heat source to force its boil-off, often using waste heat from the combustion process. The burner management system may have a fuel selector switch, which may align the fuel tanks and blend liquid fuels while maintaining correct atomizing temperature, or to switch over to gas.

DEFINE:

1. Series combustion control
2. Single-point, parallel positioning
3. Parallel positioning with *full metering*
4. Cross-limiting
5. High & low selects (auctioneers)
6. Oxygen trim
7. High and low limits
8. The relationship between the pressure drop across a flow restriction and flow rate
9. Signal biasing
10. The difference between excess air and excess O2.
11. Two reasons that forced-draft-fan inlet dampers are located on the fan inlet instead of the outlet

QUESTIONS

1. List the operations that are supervised and controlled by the burner-management system, describing the functions for both a single- and multi-burner boiler.
2. List some of the checks, permissives, and interlocks provided by the burner-management system.
3. What are the two *main functions* a boiler combustion-control system?
4. List the four types of boiler burner control schemes. Provide a brief description of their operation and the applications where they are appropriate.
5. Describe the operation of an air-follows-fuel combustion-control system and the problem encountered during a firing-rate *increase*. Describe the operation of a fuel-follows-air combustion-control system and the problem encountered during a firing-rate *decrease*.
6. What is a single-element combustion-control system? What parameter is used for signal-element control? What is the downside of single-element combustion control?
7. What is a two-element combustion-control system? What is the second element, and using the controls terminology from Chapter 2, describe how the second element works to improve combustion-control performance?
8. How are smokeless load changes achieved in large industrial boilers? How can a combustion-control system be modified to improve efficiency?
9. Using Figure 7.21 as a reference, discuss how carbon monoxide can be used as an indicator of combustion thoroughness. How can CO2 and residual O2 in the stack gases be used as an indicator of combustion efficiency for tuning a boiler burner? What levels of stack-gas O2, CO2, and CO are indicative of good combustion?

10. Estimate the combustion efficiency using the graph in Fgure 7.22 for heavy fuel oil for an ambient air temperature of 100°F, stack-gas temperature measurement of 400°F, and a residual O2 measurement of 4% on a dry basis. What recommendations might be made to improve the boiler efficiency?
11. List the parameters used in single-element, two-element, and three-element feedwater regulators, in the order of precedence. Using control terminology, what is the purpose of the second element? What is the purpose of the third element?
12. How is the proper feed-pump discharge pressure or differential pressure determined?
13. How is boiler steam-drum water level typically measured in a modern boiler. What are two options for producing a control signal from this measurement technique?
14. How does a steam-flow transmitter work? What two control applications make use of the steam flow transmitter?
15. Describe the auxiliary exhaust system make-up and dump steam system for a plant having cheaper turbine-bleed steam, more expensive live steam, and excess-pressure dump. Describe deadband and its importance in this system.
16. Why is feed-pump discharge-pressure control important to the operation of an industrial boiler? Describe the difference between a *constant-pressure* feedwater system and a *constant-differential-pressure* feedwater system. What is the advantage of the constant-differential-pressure feedwater system?
17. Why is constant furnace draft important in a boiler combustion-control system? Describe two methods of controlling boiler furnace draft. Why is it advantageous to control forced-draft fan air flow using a radial inlet damper?
18. Describe three methods of controlling superheater-outlet temperature in an industrial boiler. What is the measured parameter? What is the "controlled quantity" for each method?

NOTES

1. Neglecting trace elements, air consists primarily of 21% oxygen and 79% nitrogen by weight (or mass), which correlates to 4.76 volumes of air containing 1.00 volume of oxygen and 3.76 volumes of nitrogen. Chemical equations are balanced on a volume basis, but the theoretical amount of oxygen/air is determined on a mass basis using the molecular weights. A stoichiometric analysis yields the theoretical amount of combustion air required to obtain perfect 100% combustion without any residual O2 in the stack gases. To ensure real-world complete combustion, excess air beyond stoichiometric requirements is added, since it would otherwise be statistically unlikely for all oxygen molecules to find a hydrogen or carbon atom.
2. Fan affinity laws for speed variations: the flow varies directly with fan speed, the developed pressure varies as the square of the fan speed, and the power varies as the cube of the fan speed. As an example, if the fan is reduced to 70% of its rated speed, 70% of the air flow would be produced, 49% of the discharge pressure would be developed, but the only 34% of the power would be drawn. Fans are discussed in Chapter 8.
3. Several things conspire against VFD fan systems operating at low speeds. First, the motor's built-in fan may not provide enough cooling air flow and the motor may overheat, and second, the fan developed pressure may drop to the point where overcoming the system losses may be difficult, slow to respond, and the fan may hunt.
4. Note that radially oriented inlet dampers are used on large industrial fans. The radial orientation induces a pre-whirl of flow into the impeller eye, which reduces inertial losses and making the fan more efficient. The inlet location tends to starve the fan by reducing air density, resulting in less power consumption for the same mass flow of air, while at the same time virtually eliminating the possibility of fan surging, which would negatively affect combustion.

Chapter 8

Pump and fan controls

Pumps are used to create pressure and move fluids. They are classified as dynamic or positive displacement. Dynamic pumps include centrifugal pumps, and positive displacement include reciprocating, gear, and screw pumps. Centrifugal pumps impart kinetic energy, in the form of velocity, onto the fluid by an impeller, where some of the velocity is converted into pressure in the progressively enlarging portion of the casing called the volute, or in diffuser vanes, and some of the kinetic energy is used to move the fluid. Centrifugal pumps are classified by the impeller construction, such as radial, mixed-flow, or axial, where the impeller design affects the pump's operating characteristics. In general, radial pumps are better suited for lower-flow, higher-head applications, axial pumps are suited for high-flow, low-head applications, and mixed-flow pumps are used for applications between the two. Discharge valves can be throttled to control the flowrate, but it results in some suction and discharge recirculation losses at off-design conditions. With radial centrifugal pumps, it is usually possible to operate at shut off head with the discharge valve completely closed for brief periods without causing damage; however, prolonged operation at shutoff leads to overheating of the fluid, cavitation, and potential seal damage.

Positive-displacement pumps work by trapping fluid within a cavity and progressing the volume toward the discharge, where it is expressed. Except for some internal recirculation of fluid from the discharge back to the suction through close clearances of moving parts, called slippage, positive-displacement pumps tend to move a constant volume during each pumping cycle. Positive displacement pumps have the potential to produce very high pressures and are therefore not well-suited for direct throttling flow control, but typically use constant-pressure regulating valves to recirculate excess flow from the discharge back to the suction and relief valves to avoid damage.

PUMP CURVES

Pump curves are important, as they describe the operating behavior of pumps, allowing an engineer to select an appropriate model and size for any application. The manufacturer produces the pump curves by operating the pump in a minimized closed-loop system while accurately measuring the flow rate, developed pressure, power, and other parameters. The test engineer throttles the discharge valve to obtain different flow rates from maximum flow to shut-off, repeating the measurement-and-calculation process until a series of points defines the entire curve. The manufacturer calculates and plots the total developed head,[1] power, efficiency, and net positive suction head required (see Figure 8.1)[2]. It cannot be stressed enough that, in service, the pump always operates at some point on the pump curve.

Figure 8.2 shows normalized capacity versus head and power pump curves. Normalized curves are obtained by comparing the actual pump parameter to the value at the best

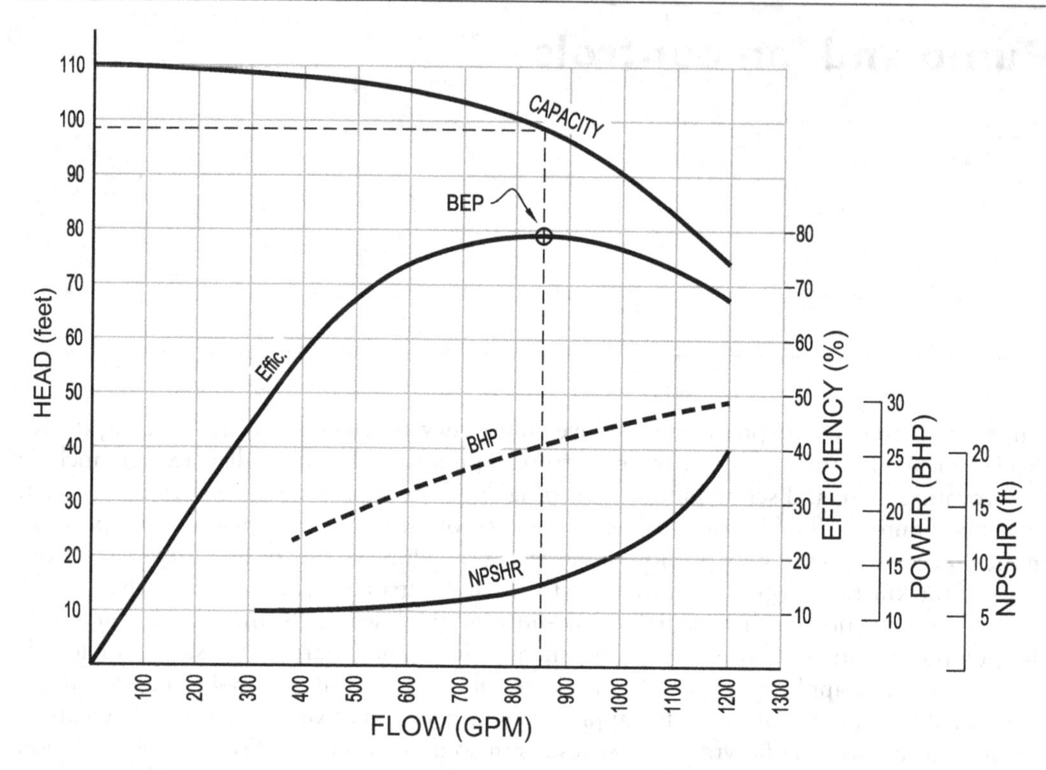

Figure 8.1 A typical pump curve indicates total-developed head, power, efficiency, and NPSHR plotted over the entire flow range from shut-off to maximum flow. Often, multiple scales are provided so all parameters conveniently fit on a single graph. In service, the pump always operates on its curve.

efficiency point (BEP), allowing pumps that operate at completely different values to be plotted together for comparison. The curves indicate that radial pumps tend to have flatter capacity curves where pumping power decreases as the flow is throttled, while axial flow pumps have steep capacity curves and pumping power that increases sharply as flow is throttled. The trends show that throttling flow control can work well with radial centrifugal pumps, while recirculation control via a constant-pressure regulator coupled with a relief valve may be preferred when the pump curve is steep.

The other factor that is important for understanding pump operation and control is the system curve. The system curve is produced by the plant design engineer, which consists of a parabolic frictional curve that intersects with the pump curve, defining the operating point. In selecting a pump, the Bernoulli head terms are calculated around two reference points, typically the source and destination, based on worst-case values, such as the lowest pressure and tank level at the source, and the relief-valve pressure setting and highest tank level at the destination. Since the Bernoulli terms are based on fixed values, the sum of the pressure, elevation, and velocity heads is called the static head. In addition to using the worst-case operating parameters, and additional flow margin of 10-25% is typically included during design to allow for pump wear over time or to account for uncertainty. The inflated values are used as the pump design operating point, and actual pump selections usually exceed the system requirements. The frictional head loss is estimated as being proportional to the flow squared,[3] or a parabolic relationship, as shown in Figure 8.3. To compound the oversizing, the next-larger standard size pump impeller is often selected to avoid additional costs to trim the impeller to better match the design operating point.

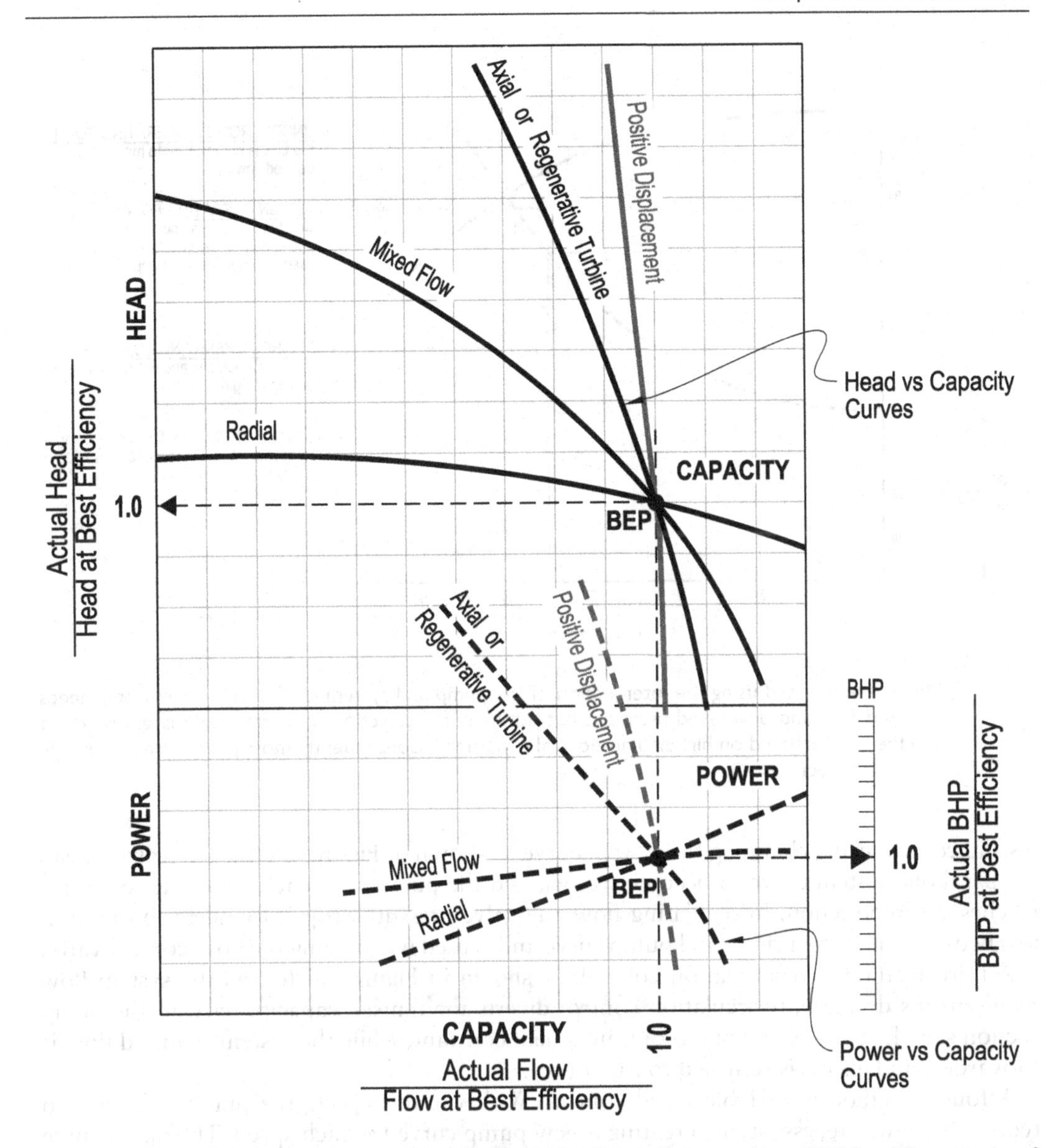

Figure 8.2 These pump curves are normalized relative to their best efficiency points (BEP) for different types of centrifugal pumps and a positive-displacement pump. The normalized curves permit comparison of the generic behaviors between different types of pumps. Understanding the pump behavior is useful for determining an appropriate control scheme.

Several options exist for controlling pump operation, including uncontrolled, throttle control, recirculation control, and variable-speed, whose pump curves are shown in Figure 8.4. In the uncontrolled system, the pump operating point is limited by the natural system resistance, which varies in a parabolic manner. As the flow increases, the piping losses increase, until the system resistance is matched by the pump developed head. In uncontrolled systems, changes in the static pressure component of TDH translate to variations in flow rate, as shown with the system curve moving up or down.

A second method of pump control is throttle control, where a pump discharge valve is closed-in until the desired flow is obtained. Essentially, the valve introduces more flow

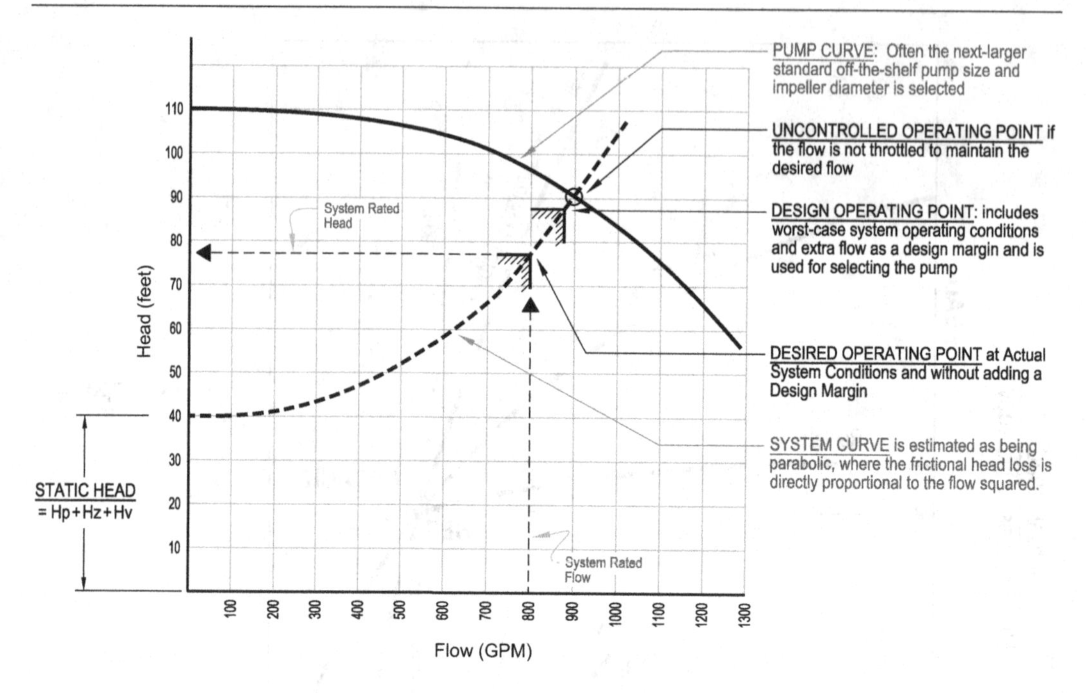

Figure 8.3 Pumps are selected using the intersection of the pump and system curves. The system only needs its rated flow and developed pressure, but during pump-selection, an extra flow margin is added, and the head is based on higher-than-normal conditions. Consequently, most pumps are larger than the service requires.

resistance, artificially changing the system curve as shown in Figure 8.4 (b). The valve causes the parabolic system curve to become steeper, and the pump runs back on its curve until it reaches the intersection. In regulating flow, the valve pressure drop is found from the head loss between the system curve and pump curve and is used when sizing the flow-control valve.

A third method is recirculation control, as shown in Figure 8.4 (c). As the system flow requirements decrease, recirculation control diverts the unused capacity back to the pump suction side. Essentially, the pump output is held constant, while the system-required flow is forwarded, and excess is returned to the source.

A fourth method is variable-speed control. With variable-speed, the pump is slowed to reduce the flow rate, essentially creating a new pump curve for each speed. The pump curve shifts downward until its intersection with the system curve matches the desired flow rate, as shown in Figure 8.4 (d). The reduced-speed pump curves follow the pump affinity laws for speed changes for the equations that follow:

Equation 1. Pump Affinity Laws based on speed

- Flow: $$\frac{Q_2}{Q_1} = \frac{n_2}{n_1}$$
- Developed Pressure: $$\frac{H_2}{H_1} = \left(\frac{Q_2}{Q_1}\right)^2 = \left(\frac{n_2}{n_1}\right)^2$$
- Power Consumption: $$\frac{P_2}{P_1} = \left(\frac{Q_2}{Q_1}\right)^3 = \left(\frac{n_2}{n_1}\right)^3$$

where n is speed (rpm), Q is flow (gpm), H is head (ft), or pressure (psi), and P is power (BHP).

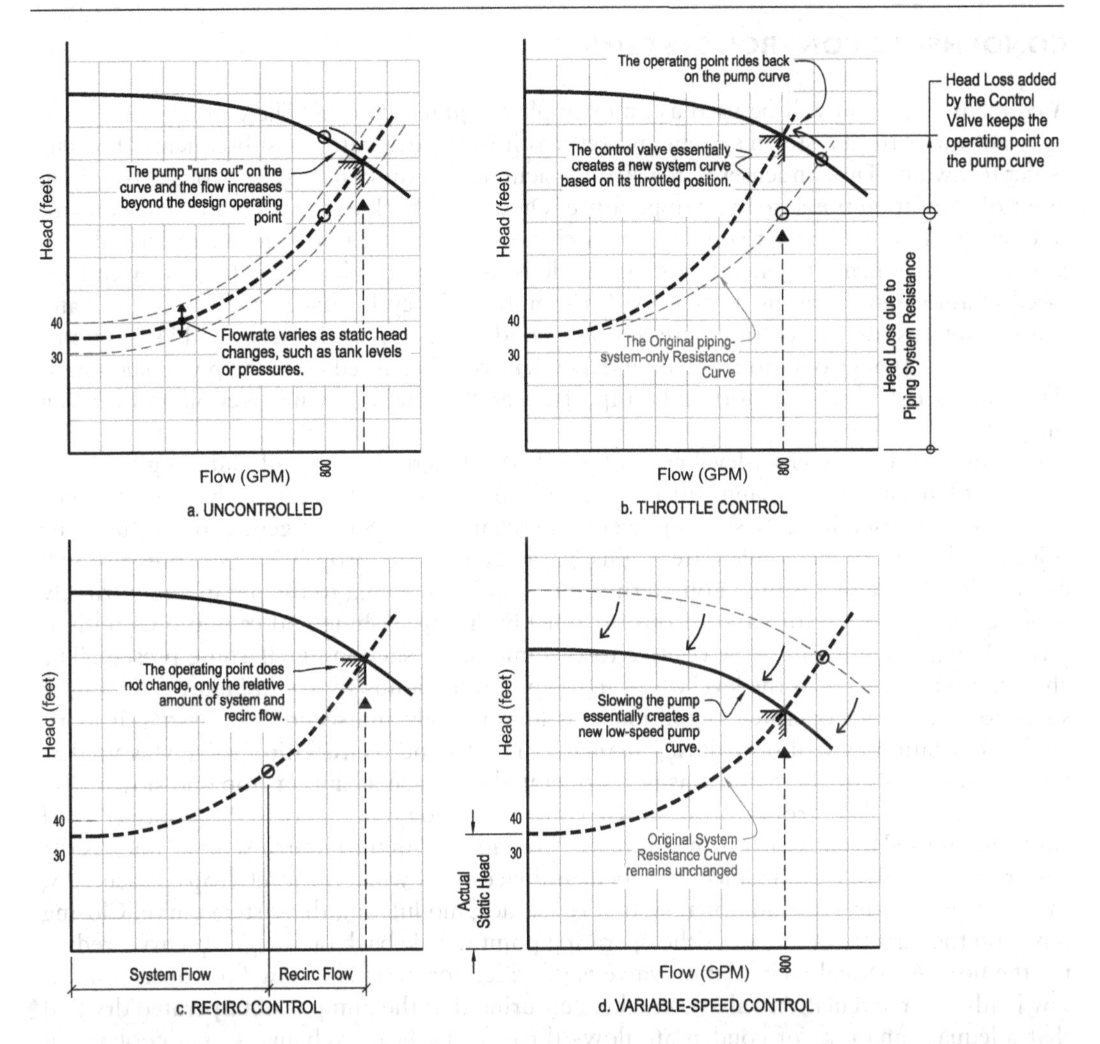

Figure 8.4 Pump and system curves for different control schemes. (a) Uncontrolled system: the operating point runs out on the pump curve until it intersects with the system resistance curve. (b) Throttle control: a discharge valve is choked to induce a pressure drop and additional system resistance. (c) Recirculation control: The pump flowrate and operating point is held constant by recirculating excess flow. (d) Variable-speed control: the pump is slowed, producing less flow and less developed head to exactly match the system requirements.

The affinity laws show the potential energy savings that can be obtained if a pump can be operated at reduced speed for extended periods of time. As an example, if a pump can be operated at 80% of full-rated speed, the developed head would be 0.80^2 or 64% of its original pressure, and the power usage would be 0.80^3 or 51% of its original consumption. A nearly 15% energy savings can be realized simply by operating a pump at 95% of its rated speed, well within the design margins used during pump selection.

Variable speed can be obtained several ways, some which have become outdated. Some methods for varying speed include steam-turbine drives, DC motors, AC multiple speed motors, AC wound-rotor induction motors, variable-frequency drives (VFDs), mechanical / hydraulic / magnetic clutch drives. Today, for nearly all plant applications, VFDs are the preferred option. Variable-frequency drives benefits and problems are discussed in Chapter 6.

CONDENSATE CONTROL SYSTEMS

A condensate system will be used as an example for pump-control. The condensate system recovers water from a steam system and returns it to the deaerator to subsequently become boiler feedwater. The condenser is at a strong vacuum, and the condensate starts at its boiling point of saturation pressure and temperature. On its way back to the deaerator, the condensate serves as the coolant for condensing air-ejector and gland leak-off steam, and to subcool heater drains. When a regenerative steam cycle is used, the condensate is heated by turbine bleed steam when operating at high loads to improve the cycle efficiency. Four condensate-control options are shown in Figure 8.5. For condensate applications, low NPHSR pumps are needed to avoid cavitation, and an independent vent line is led from the pump suction to the condenser shell so as to form a U-trap of water, flooding the pump suction and making the pump self-priming.

One method of condenser-level control is called submergence control. Submergence control is outdated and not recommended, but it aids in understanding pump behavior. This type of control was used in early steam-powered merchant ships. Submergence control uses the principle that when the condensate begins at saturation, any drop in suction pressure will cause some degree of flashing. Long-term cavitation is damaging to the pump and normally avoided, but pump performance drops off quickly during flashing and cavitation, continuously changing the pump curve relative to the amount of suction-line flashing present. The changing pump curves as the level drops in the pump suction is shown in Figure 8.5a. Older ships got by, as they operated at full load for long periods and secured the system in port, so that cavitation existed only during maneuvering. A hand recirculation valve was used at lower loads to ensure the pump was never operated dry, such as during stop and slow bells.

The next type of control is throttle-valve control, as shown in Figure 8.5b. A hotwell level controller is used to choke in on an automatic valve located in the pump discharge line. As the level rises, the valve is commanded to open, and vice versa when the level drops. Essentially, the throttle valve inserts additional system resistance, modulating the system curve. Closing down on the throttle valve causes the operating point to ride back on the pump curve, reducing the flow. A manual recirculation valve is provided for periods of low-flow operation. At low loads, the recirculation valve is cracked, ensuring that the pump is not operated dry and that adequate amounts of condensate flows through the heat exchangers as a coolant. In some plants, the recirculation valve is automated to open when low-flow conditions occur. Like any control scheme, the best option is to base the control on the closest parameter to the process, in this case flow. Because of the predictability of the pump curve, some plants use the pump discharge pressure to avoid the expense of a differential-pressure or flow transmitter. The thought is that as a throttle valve closes, the operating point rides back on the pump curve, and the developed pressure increases. This temptation should be avoided, as deviations to the system static head or pump wear over time will reduce pump discharge pressure, eventually causing the control system to become ineffective. This example shows the importance of understanding the system behavior when design control solutions.

An advantage of throttle control with radial centrifugal pumps is a slight energy savings which occurs because the pump consumes less power[4] at reduced flows. Another advantage is that the NPSHR and any cavitation decreases quickly as the flow decreases. Reduced pump efficiency and an imbalance of impeller radial forces as fluid emerges unequally from the vanes are disadvantages that occur during low flow, as the operating point shifts away from the best efficiency point. The radial loads cause higher bearing loads, higher shaft fatigue bending loads, and more vibration.

Recirculation control is another method commonly used for condensate systems (Figure 8.5c). In this control method, the hotwell level signal is sent to a recirculation valve

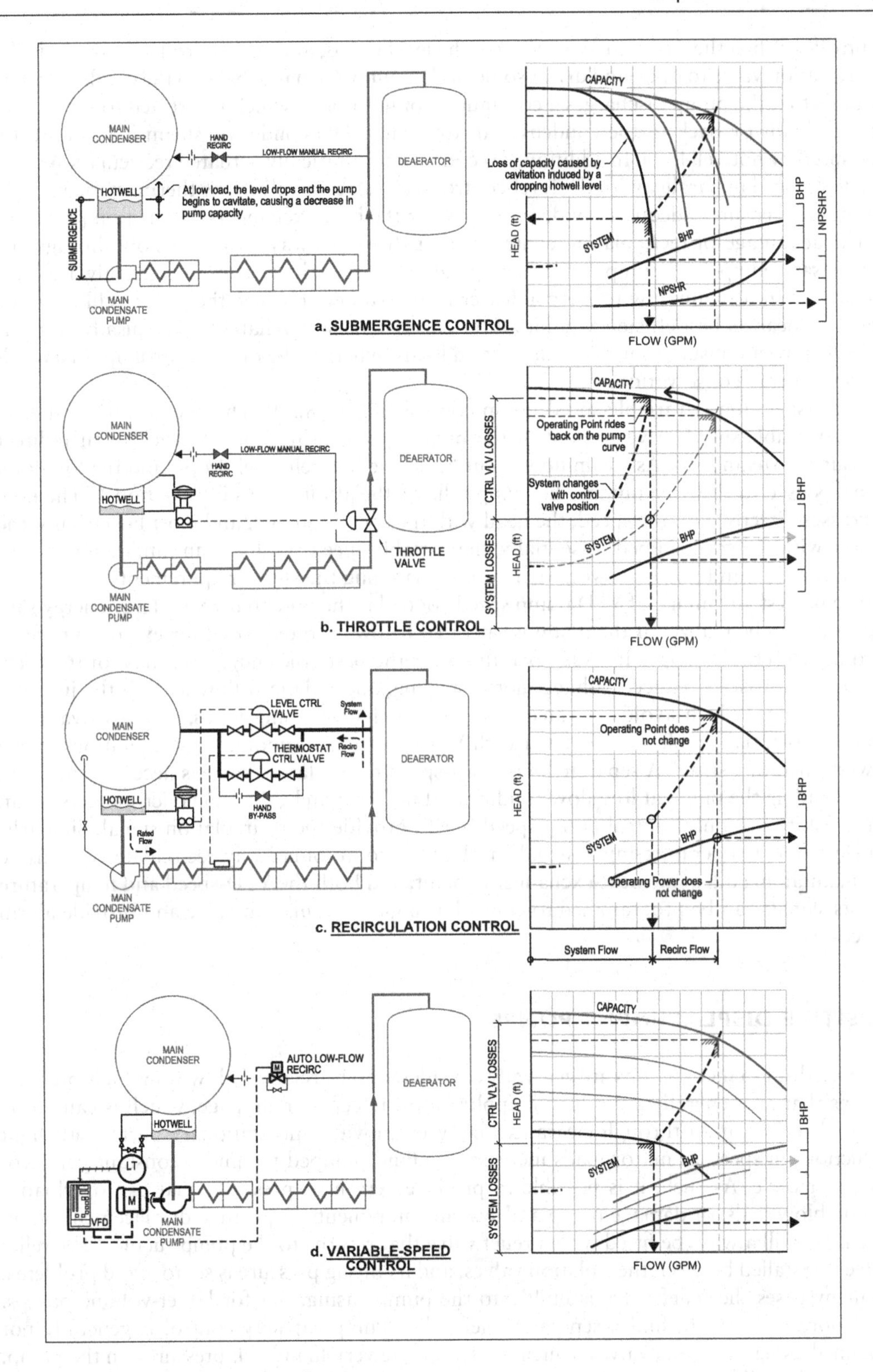

Figure 8.5 Condensate system control options.

controller. When the steam load is reduced, the level drops, and the controller commands the recirculation valve to open and divert some of the pump discharge back to the condenser. The result is that the pump discharges a constant amount of flow, which is directed to the system at high loads or back to the condenser at low loads. The condenser steam load indirectly measured by hotwell level. In addition, a second thermostatically actuated recirculation valve is often placed in parallel with the level-control recirculation valve. The thermostatic recirculation valve ensures adequate condensate flow for the heat exchangers to function properly.

The advantage of recirculation control is that there is always constant flow through the condensate pump. The pump cannot be operated dry and the heat exchangers always have coolant, even if steam flow to the condenser is interrupted. Further, the pump is likely to be operated near its best efficiency point all the time, increasing reliability. The disadvantage is that the power consumption is constant at all loads, whereas the other control options result in some energy conservation.

The last option is variable-speed pump control (Figure 8.5d). This method uses an electronic variable-speed drive to control the pump speed, using hotwell level as an indirect measure of steam load. As steam flow is cut back, the hotwell level drops, and the low-level signal is used to slow the pump. The pump follows the affinity laws in Equation 1. The flow decreases directly with the speed, the head with the speed squared, and most beneficially, the power with the speed cubed. Essentially, the variable speed results in an infinite number of pump capacity and power curves, each exactly matching the steam requirements.

The biggest advantage of VFD pump speed control is the potentially very large energy savings that can be realized if the pump is operated below its rated speed for extended periods of time. Other advantages include operating near the best efficiency point most of the time, electronic soft-start reduced-voltage motor starting, and balanced flow leaving the impeller circumference, which avoids elevated bearing and shaft-bending loads. The disadvantages are any potential problems associated with VFDs, such as harmonics and limitations to the lowest operating speed. A separate low-motor-speed recirculation valve is necessary to avoid dead heading the pump at low-flow conditions, but fortunately, the VFD microprocessor can multitask to not only control pump speed but to provide the recirculation signal. Similarly, an electronic temperature-input signal can be used to automatically recirculate condensate for maintaining adequate heat-exchanger coolant, and both the VFD-speed and temperature inputs signals can be programmed to control a single recirculation valve and provide alarm functions.

POSITIVE-DISPLACEMENT PUMPS

Positive-displacement pumps are generally considered to be relatively low-flow, high-pressure pumps that are generally good for oil applications. Except for slippage, which is caused by internal recirculation through clearances between moving and stationary parts, and slight reductions in speed as motor loads increase, the fluid pumped volume is constant per revolution or cycle. As such, it is possible to produce very high pressures that can overload a motor, blow seals, or damage pump and system components. A positive-displacement-pump discharge will always be fitted with a relief valve that returns to the pump suction. The relief valve is installed between the isolation valves, and its lifting pressure is set to avoid problems. In many cases the relief valve is built into the pump casing, but for larger-volume pumps, it is more common to find a separate relief valve. Pump capacity control is generally not obtained using throttling valves, which tend to create very high back pressures on the pump. In general, the best way to achieve capacity control is to use a constant-pressure regulating valve that recirculates excess flow back to the pump suction or source. The constant pressure

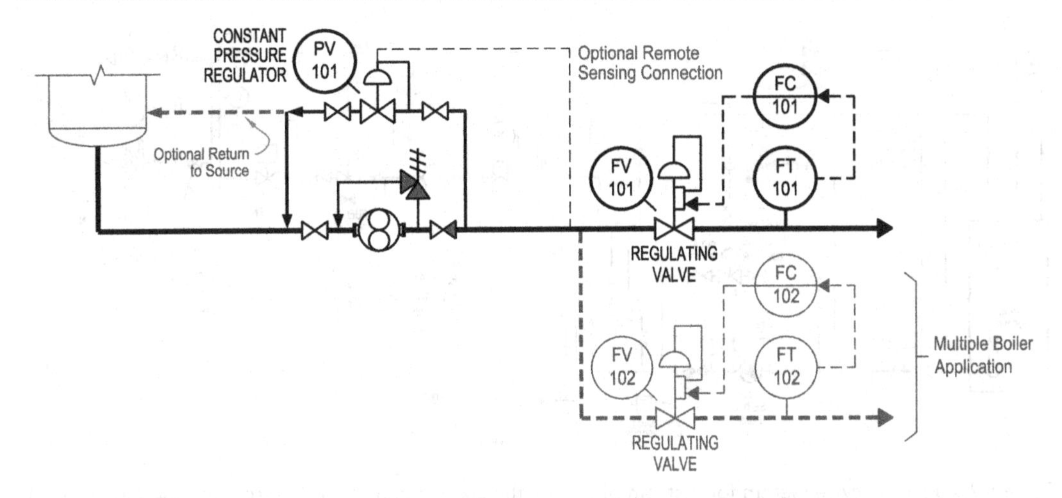

Figure 8.6 A constant-pressure regulator (backpressure regulator) maintains the pump discharge pressure fixed, and diverts excess flow back to the pump suction or source. As the regulating control valve opens and system flow increases, the recirculated flow is automatically reduced. The constant-pressure regulator functions in a similar manner as a relief valve, but is used as a control device, rather than as a safety device.

regulator is a backpressure regulating valve. It senses pressure on its inlet side and opens to reduced excess pressure. In some cases, it is a self-contained device, and in other cases it has a sensing line to remotely control pressure at the equipment needing a fixed value. Figure 8.6 shows a constant-pressure regulator used to accurately set the inlet pressure to a flow-regulating valve. Maintaining constant pressure at the regulator inlet provides flow-rate predictability and allows a combustion-control system to maintain a consistent balance between air and fuel flow.

Figure 8.7 shows a schematic diagram of a heavy-fuel-oil-service system used on steam ships, where the constant pressure regulating valve (CPR) maintains correct fuel-oil pressure at the inlet to the fuel-oil-regulating valve. The combustion-control system blindly sets the fuel-oil regulating valve position based on demand, so for good control, the inlet pressure to the valve must be constant.

The system takes suction from the settling tanks through a low suction valve. If water contaminates the tank, it is possible to switch to the high suction valve. The tank valves have reach rods that extend outside the machinery space to permit remote closure in the event of fire. The tanks are steam-heated to around 100°F to reduce the oil viscosity enough to be pumped. Oil passes to the fuel-oil service pump through a coarse-mesh duplex low-pressure suction strainer, which can be changed and cleaned in service. As a positive-displacement pump, a relief valve is fitted from the discharge back to the suction to protect against excessive pressure, even if the pump is incorrectly lined up. The constant-pressure regulator is often located near the boiler operating console and is adjusted to recirculate excess pressure back to the pump suction. Its oil-pressure-sensing line is near the fuel-oil regulator inlet, which needs constant pressure to reliably set the fuel-to-air ratio adjustment. The fuel passes through fuel-oil heaters, which sets temperature to around 200°F. High temperature reduces the oil viscosity to meet proper atomization levels, and then flows to the fuel-oil regulator by way of the fine-mesh duplex high-pressure discharge strainer. Because the heater can be isolated and steam heat applied, the fuel side is fitted with a relief valve to burp excessive pressures. The fuel oil regulator is positioned by the combustion control system, and a

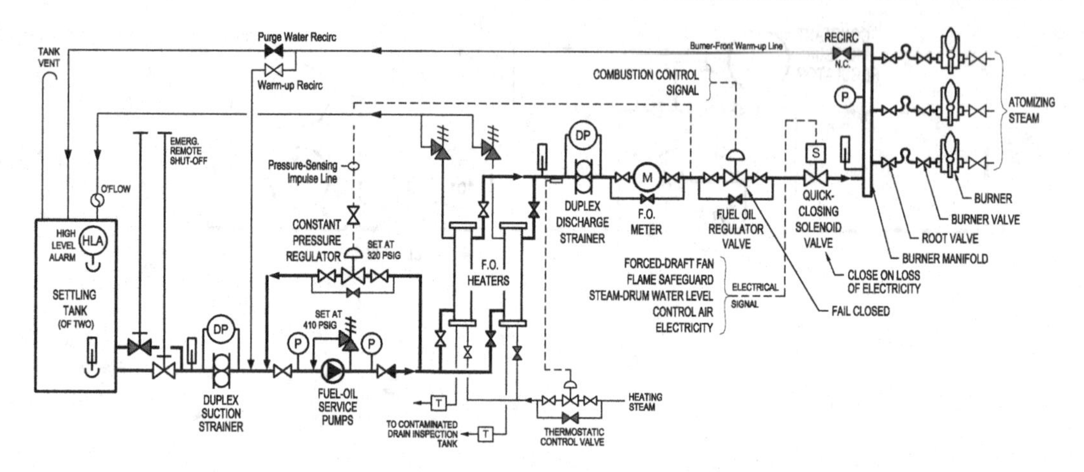

Figure 8.7 Fuel-oil service system for a steam plant. For the combustion control system to function properly, the fuel-oil pressure must be held constant. A constant-pressure regulating valve (CPR) senses pressure at the fuel-oil regulator inlet and fixes the pressure by recirculating excess flow back to the pump suction. Note that for simplicity, redundant standby fuel-oil service pumps and multiple boilers are not shown.

quick-closing solenoid valve is part of the boiler permissives, controlled by the burner management system. The solenoid valve is of the normally closed type, and for safety reasons will fail-closed on loss of electricity for any reason. At the manifold, fuel is distributed to the individual burners at atomizing temperature. A burner-front recirculation valve at the manifold can be cracked to maintain hot oil availability to the burners, during pre-purge or post-purge cycles. Normally burner-front fuel is recirculated back to the pump suction, however, if the lines become contaminated with water, recirculation can be redirected back to a settling tank to clear the water.

Often the fuel-oil service pump motor is designed for multiple speeds. When boiler steam loads are low, the pump is operated at the lowest speed to avoid large amounts of recirculated flow and the associated power consumption, and when the fuel consumption increases enough, the pump is sped up to meet the burner-flow requirements. Low pressure at the fuel-oil regulating valve is an indication that the pump should be sped up.

Figure 8.8 shows a fuel-oil service system for a dual-fuel large slow-speed diesel engine used for shipboard propulsion. This system begins with a fuel-oil centrifugal purifying pre-processing system that draws fuel from a settling tank and transfers it to a day tank. The purifiers are designed to remove water, dense solids, and the catalytic fines used in the refinery cracking process, which would be abrasive to the close-tolerances inside engine fuel pumps and injectors. The fuel-oil booster system draws from the day tank and circulates fuel in a loop via the mixing tank. Fuel is pumped through the fuel-oil heaters, a viscometer, also called a viscosimeter, and the engine injection pumps. The viscometer measures the viscosity and sends a signal to a control valve to adjust the amount of steam to the fuel-oil heaters (see Chapter 1 regarding instrumentation). As heated fuel is consumed by the engine, cooler fuel is drawn from the day tank and blended into the loop within the mixing tank. The purpose of the mixing tank is to permit the smooth in-service changeover between heavy fuel oil and light diesel oil, used when arriving into nonattainment port areas or entering maritime emission control areas (ECAs). As light, thinner oil is slowly blended with the heavy oil, the need for steam heat is reduced. The viscometer senses the fuel viscosity and commands the heater steam valve to open or close in response.

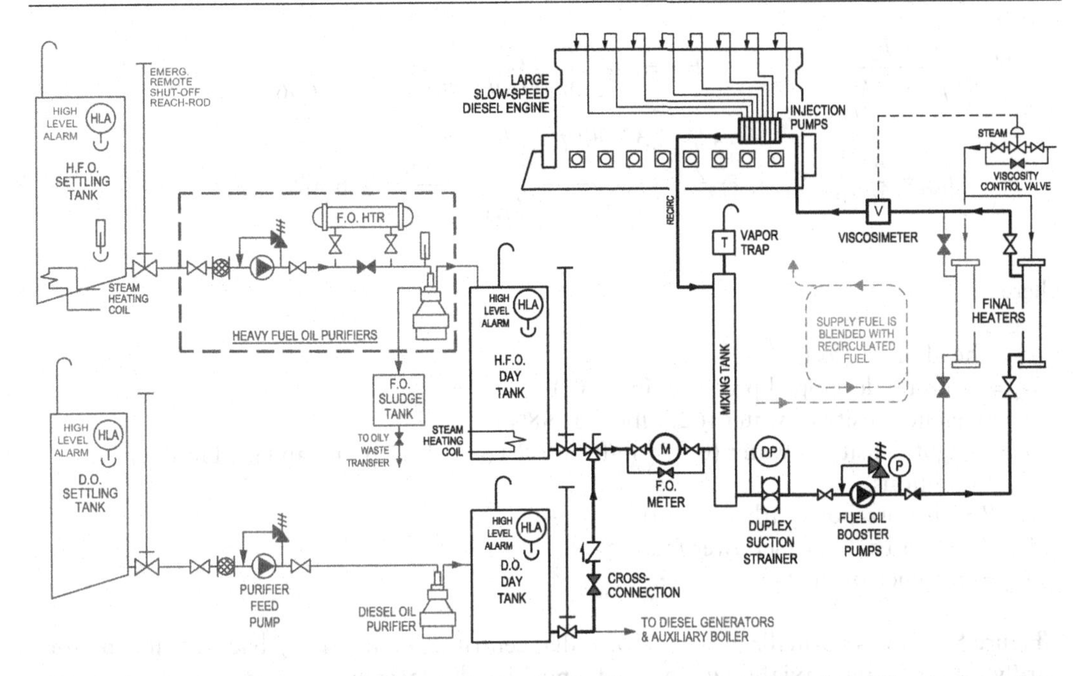

Figure 8.8 A typical fuel-oil service system used on a dual-fuel large slow-speed diesel plant for ship propulsion.

Table 8.1 ASME differentiation between fans, blowers, and compressors

Equipment	*Developed pressure increase ratio*	*Pressure rise*
Fan	0 to 1.11	Up to 1136 mm W.C.=44.72 in.W.C.= 1.61 psid
Blower	1.11 to 1.20	Up to 2066 mm W.C.=81.34 in.W.C.= 2.93 psid
Compressor	More than 1.20	

FANS, FAN CURVES, AND FAN CONTROL

Air moving devices are classified as fans, blowers, or compressors. Typically, fans and blowers are used in ventilation and industrial processes to overcome resistance in ducts, dampers, heat exchangers, etc. Compressors are devices that produce higher pressures that serve as a potential-energy source to power pneumatic tools or to position actuators. The ASME classifies these machines as shown in Table 8.1. Essentially, when the developed pressure is low, the air density does not vary much, and the fan calculations are more like pump calculations for incompressible fluids. However, when the discharge pressure is high, the air density increases quite a bit, heat of compression is produced, and the compressor calculations are more complicated and thermodynamic in nature.

Fan curves are very similar to pump curves, but given the compressible nature and lower viscosity of air, there are definite differences, including prominent droop for many fans (Figure 8.10 left). In general, the fan volumetric capacity is plotted against fan static pressure. Static pressure is directly measured in inches of water column, rather than measured as head in feet of air. The flow and developed pressure are useful in calculating fan power. The fan equations are:

$$H_a = \frac{\rho_w}{\rho_a}\frac{h_w}{12\,in/ft} \qquad AHP = \frac{\rho_a H_a \times CFM}{33{,}000\,ft \cdot lbf/min/HP} = \frac{h_w \times CFM}{6362}$$

$$\eta_{fan} = \frac{AHP}{BHP} \qquad \frac{12\,in/ft \cdot 33{,}000\,ft \cdot lbf/min/HP}{\rho_w = 62.3\,lb/ft^3} = 6362$$

where

H_a = head of air (feet)
h_w = fan total developed pressure (in. W.C.)
ρ_w = weight density of water (62.3 lb/ft^3 at 68°F)
ρ_a = weight density of water (0.075 lb/ft^3 at 68°F and 14.7 psia standard fan inlet conditions)
AHP = air horsepower (theoretical)
BHP = shaft brake horsepower (actual)
η_{fan} = efficiency of the fan

Figure 8.9 shows normalized curves for radial centrifugal fans having backwardly and forwardly inclined vanes, axial flow fans, and a positive-displacement air compressor as a reference. The fan curves exhibit many similarities to pump curves, but often have droop, where the peak pressure is not at dead head. Instead, the peak pressure defines the point of instability. The fan operating point should be to the right of the point of instability (Figure 8.10). Intuitively, as the fan discharge resistance increases, its associated back-pressure is also expected to increase, and the flow is expected to decrease. However, at low flow rates, some fans experience a counterintuitive pressure drop against the increasing system resistance in a phenomenon called droop (Figure 8.11).

Figure 8.12 shows the progression of system curves as a discharge damper closes while modulating flow. As the damper closes, the system resistance increases, the fan rides back on the curve decreasing flow and increasing developed pressure, as would be expected (Figure 8.12b). Eventually, the damper closes enough to put the fan into the unstable operating range, where the fan pressure drops as the damper closes, instead of increasing (Figure 8.12c). At low discharge pressure, the fan curve has two possible operating flowrates, as shown as Q1 and Q2 in Figure 8.11, at which point the fan will surge. Surging occurs as the flow separates from the impeller vane in a condition known as stall. The flow separation and stall occurs when the angle of attack of the relative air flow over the vane exceeds the critical angle of attack. Stall can lead to surging, noise, vibration, fatigue, inefficiency, difficult pressure/flow measurement, and poor system behavior. Instability and surge are a low flow phenomenon and even small amounts would be intolerable in systems requiring fine control, such as combustion-air fans.

FAN CONTROL

Several methods are used for reducing flow where variable fan output is required, such as combustion-control systems. Flow control methods include bypass or recirculation control, discharge-damper throttling, inlet-damper throttling, inlet-damper throttling using radial vanes, multi-speed motors, and variable-frequency-drive fan-speed control. Figure 8.13 shows generic power-consumption relationships for radial fans using different control schemes.

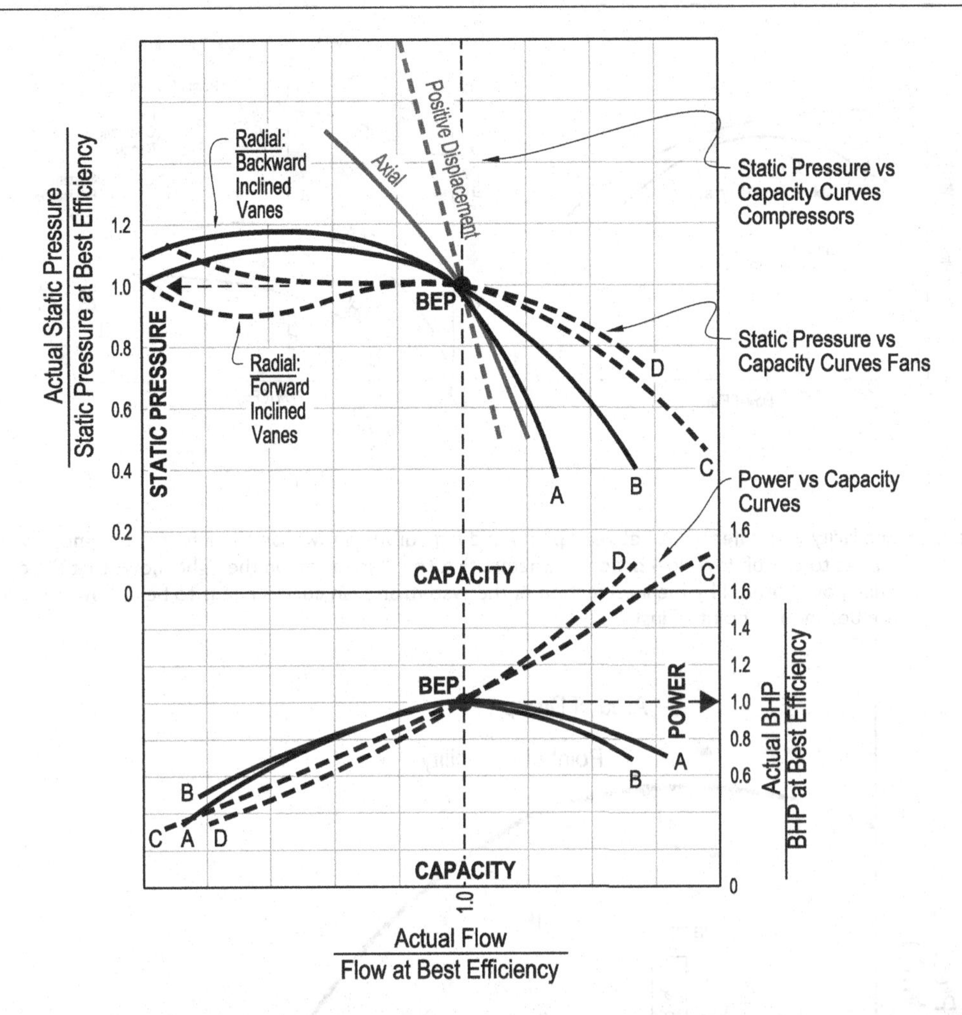

Figure 8.9 Normalized curves for fans of different types. All values used to form the curves are scaled relative to the value at the best-efficiency point (BEP). The static pressure in inches of water column and power in BHP are plotted against flow in CFM. Normalizing the curves allows the relative behaviors and trends for fans of different types to be compared.

Bypass control uses a recirculation duct and damper from the fan discharge back to its inlet, as shown in Figure 8.14. The recirculation damper controls the system flow by diverting excess flow back to the inlet. This arrangement fixes the operating point on the fan curve, where needed flow goes to the system, and the surplus is recirculated. This arrangement avoids riding back on the curve, so instability and surging never occurs. Because the fan-operating point is fixed, the power consumption at reduced flow never changes. Figure 8.13 shows that the bypass/recirculation power consumption is constant, and there are no energy savings to be had.

Discharge-damper throttling is also used to modulate flow, where a damper on the outlet is throttled to increase the system resistance. Radial fans with relatively flat fan curves, like radial centrifugal pumps, have the characteristic where the power decreases as the flow is reduced, as shown in Figure 8.9 power curves (c) and (d) and Figure 8.13. However, when the fan curve exhibits droop and the flow decreases enough, it is possible for the fan to become unstable and surge as describe in Figure 8.10, Figure 8.11, and Figure 8.12. Fan curves

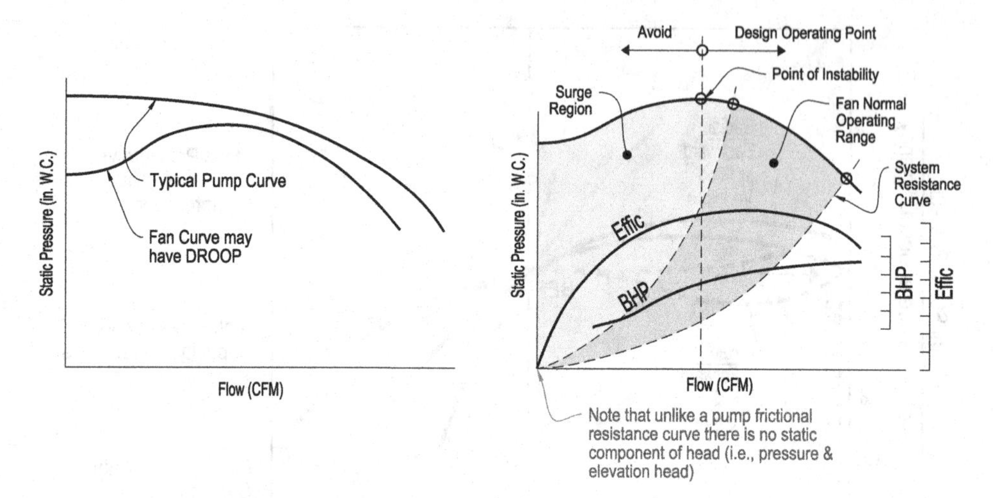

Figure 8.10 Similarity and differences between pump and fan curves, shown on the left. It is common for fan curves to exhibit low-flow droop characteristics. The illustration on the right shows that the operating point formed by the intersection of the system and fan curves needs to be at flow rates that are beyond the point of instability.

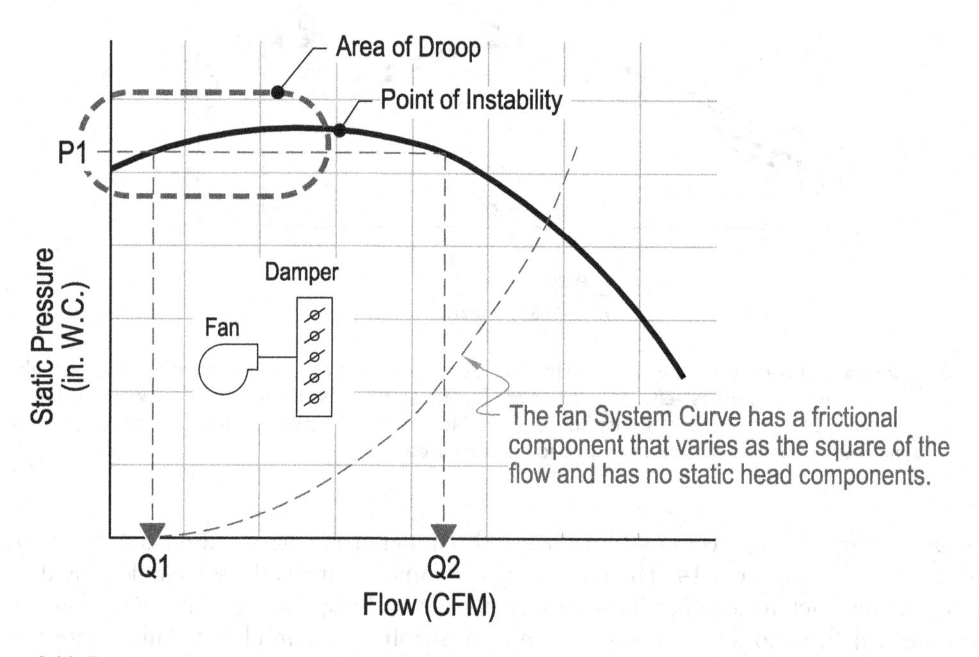

Figure 8.11 Fan droop characteristics. When droop exists, there will be two flow rates that experience the same fan developed pressure, one above and one below the point of instability. Essentially, the fan becomes "confused" and "does not know" whether to operate at the high or low flow. The result is fan surging between two flowrates, Q1 and Q2.

need to be examined when controlling applications operating at low-flow conditions, and in applications such as combustion control, where any surging is completely unacceptable. Shutter-type dampers are typically used for flow control and may be of the parallel-blade or opposed-blade design (Figure 8.15). The actuating mechanism for the parallel shutter dampers is simpler and less expensive, but it produces higher losses and the damper tends to open

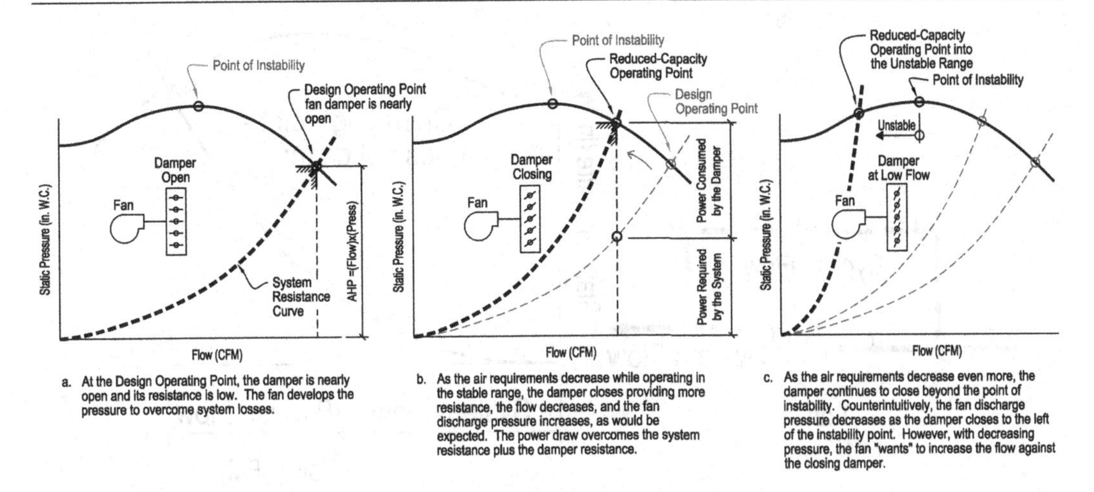

Figure 8.12 Fan operation can eventually go into the unstable range, as a discharge damper closes to reduce flow.

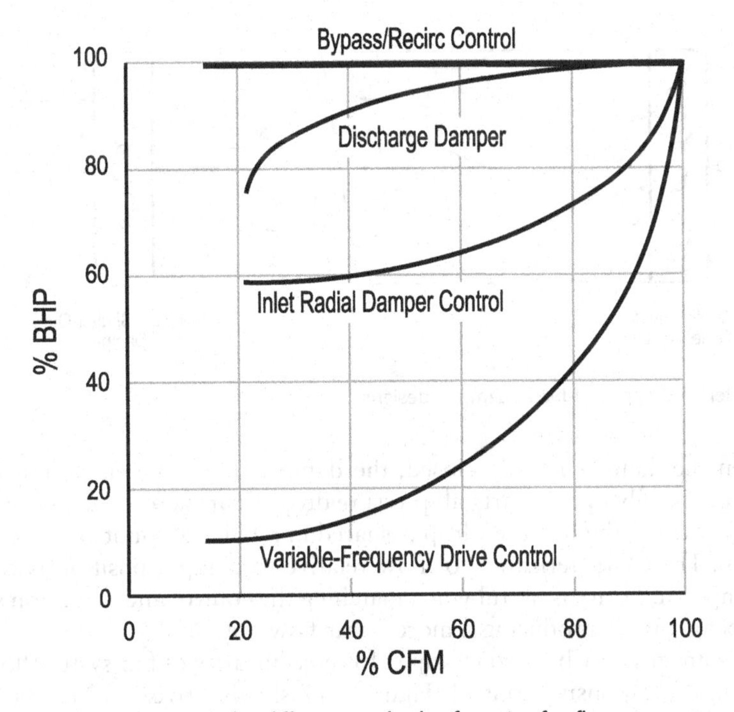

Figure 8.13 Power consumption trends for different methods of varying fan flowrates.

fast, giving less control authority at the more-open positions. The opposed shutter damper requires a more complex actuating mechanism, is more expensive, but it provides better control authority over a larger flow range. In operation, the opposed damper provides slightly less resistance and power losses,

When dampers are installed in a system, the dampers contribute to the total system resistance, which also include pressure losses across ductwork, fittings, filters, and heat exchanger coils. The system pressure drop distorts the damper pressure drop (Figure 8.16). As an

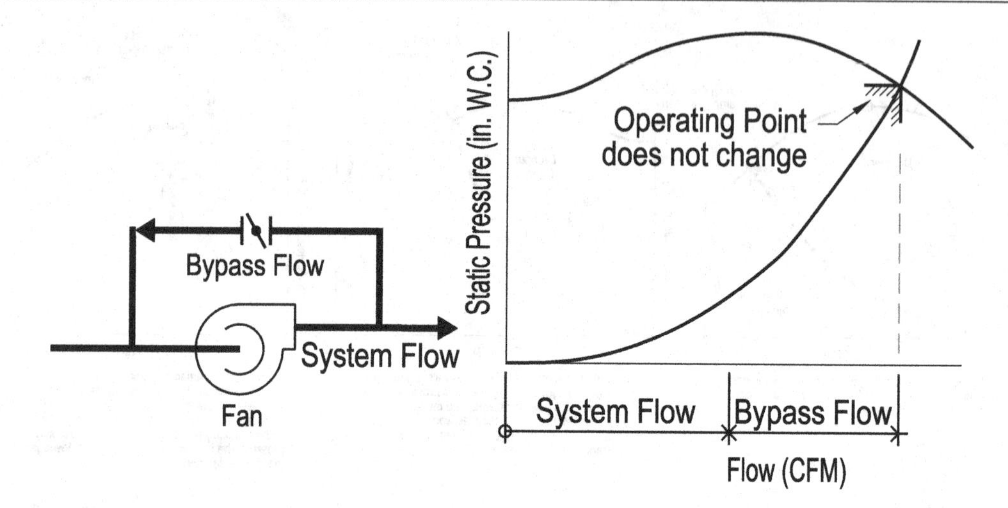

Figure 8.14 Bypass control to modulate system flow.

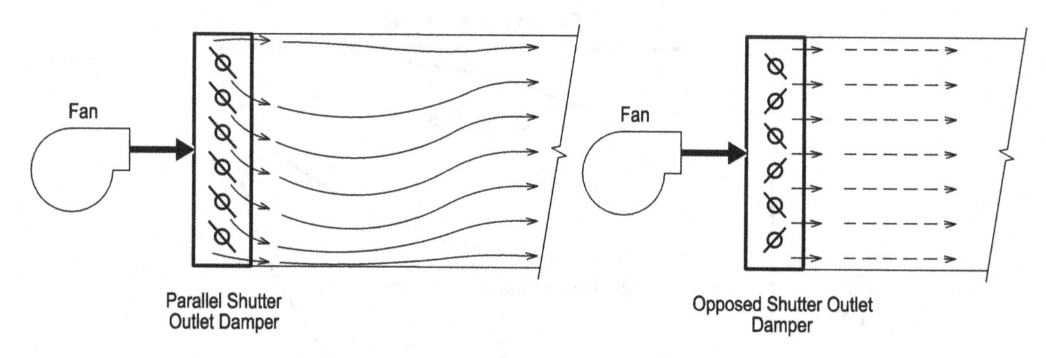

Figure 8.15 Parallel- and opposed-blade damper designs.

example, when the damper is fully closed, the damper drops the entire fan pressure, but when the damper is fully open, nearly all pressure drop occurs across the system. At intermediate damper positions, the pressure drop is shared in relative amounts between the damper and the system. The characterization of flow relative to damper position is called damper authority. Damper authority is useful when matching the control and actuation signals to the damper flow behavior for producing a more linear flow output.

Linkage mechanisms may be used to adjust for non-linearity of fan-system flow to provide for smoother more responsive control. Figure 8.17 shows driver and driven arm lengths and shaft angles that can be adjusted to change the flow characteristics in an advantageous manner. For electronic control systems that operate devices like the rotary actuator shown in Figure 8.18, a mathematical function may be programmed into the controller to produce linear flow characteristics.

Inlet dampers may also be used to control flow as shown in Figure 8.19. Inlet dampers provide the advantages of significantly reducing power losses at partial flowrates, and they result in much less likelihood for surging. The advantages occur because throttling the inlet starves the fan of air, dropping the inlet pressure, which in turn lowers the air density entering the fan. Lower air density translates to reduced power consumption with little likelihood of surging. Surging is reduced because the fan discharges into a lower

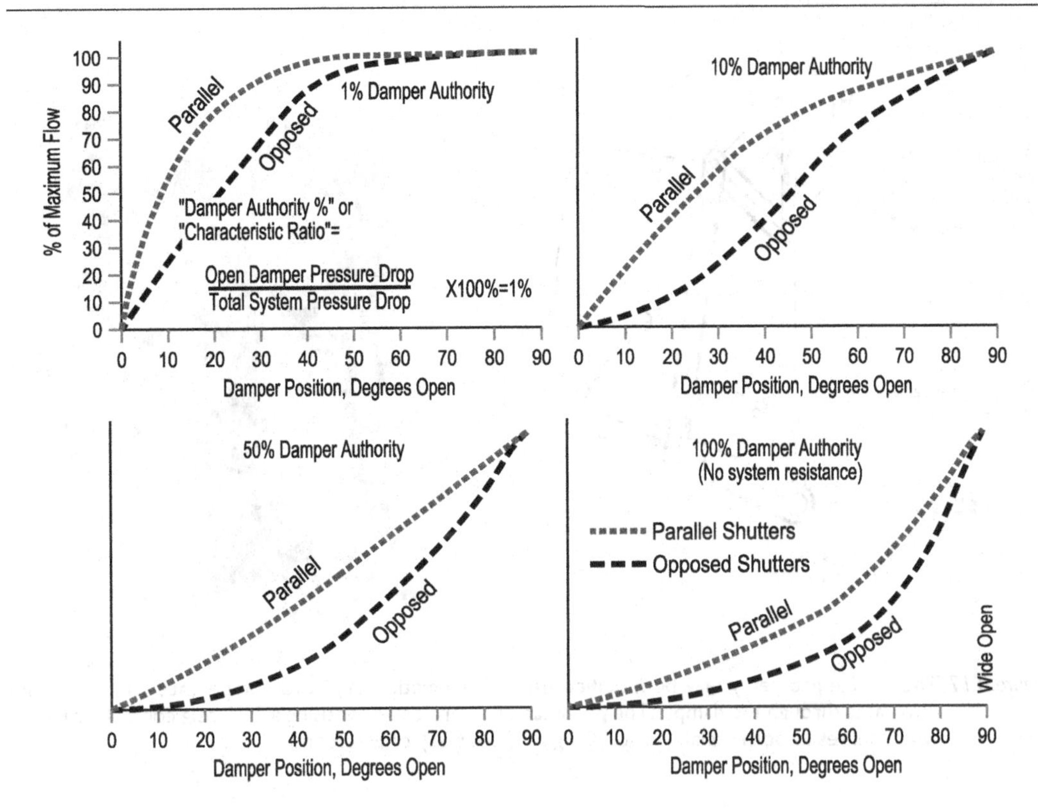

Figure 8.16 The damper flow characteristics change relative to the percent of damper-open position. The amount of damper resistance compared to the amount of system resistance affects the air-flow rate, which is referred to as *damper authority*.

pressure system, lacking a high discharge pressure drop. This arrangement has little potential to produce high fan discharge pressure against a flow-reversing component near the fan outlet.

Centrifugal fans having backwardly inclined vanes coupled with inlet shutter dampers that are severely throttled can be susceptible to a special case of instability called inlet rotating stall. Ideally, during rotation, each vane passageway formed between adjacent impeller vanes, also called cells, should evenly create flow and pressure around the scroll periphery. However, at very low flow rates, the fan flow is developed by just a few cells, so the scroll pressure can exceed the pressure within the cells having little-or-no flow. At that condition, reverse flow back to the impeller eye occurs through the low-flow cells, where the reverse flow emerges at the vane inlet as a "bubble" that jumps to the preceding cell. The effect is rotating stall, which can lead to surging. Surging in fans can be very problematic, but for dynamic compressors, surging can be damaging.

Inlet radial vanes are often used for larger variable-volume fans, such as combustion-air forced-draft fans. Advantages of radial inlet dampers include improved fan efficiency at partial loads, finer flow-control characteristics, and the virtual elimination of stall or surging. The positioning mechanisms for inlet radial dampers are more complex and expensive, but the expense is justified in large fans where energy costs are significant, typically above 75HP, assuming the fan will be operated for long periods at partial flow.

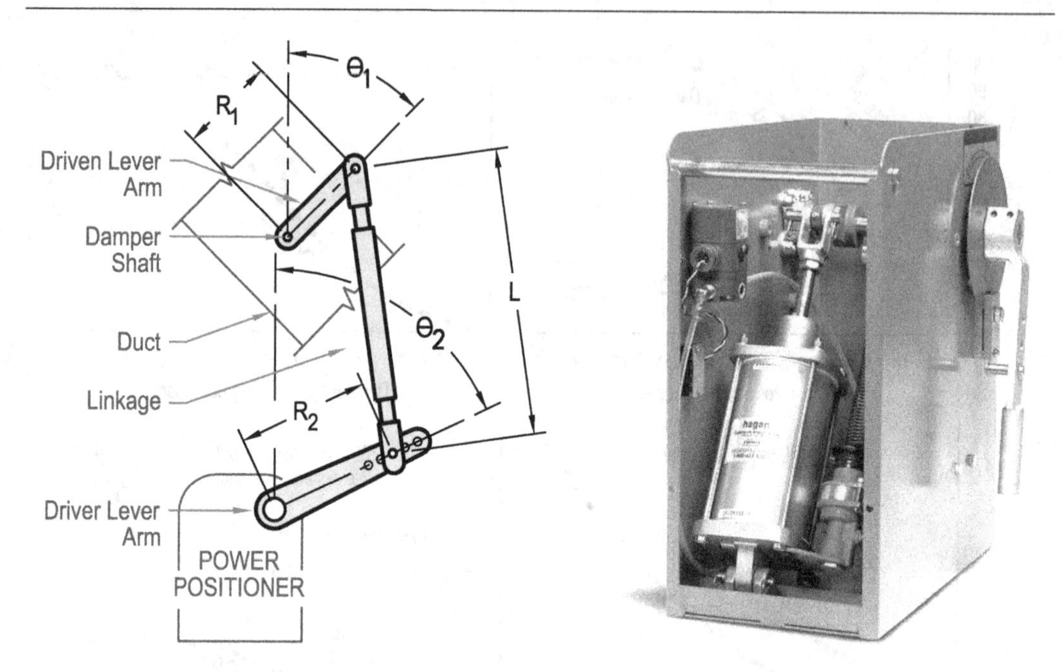

Figure 8.17 The linkage geometry may be modified using a kinematic analysis to compensate for non-linear flow rates through the damper. The pneumatic power positioner using a 3-15 psig control-air input, which serves as both signal and actuating air. (Courtesy of Emerson.)

Inlet vanes are oriented to produce a prewhirl of air into the impeller eye, spinning the air in the same direction as impeller rotation. The prewhirl reduces inertial entrance losses, and the inlet location of the damper "starves" the fan, which reduces air density. Both effects combine for significantly less power consumption. Further, the radial inlet vanes essentially behave as if to create an infinite number of fan characteristic curves corresponding to the instantaneous damper position (Figure 8.20). Droop tends to flatten or be eliminated as the inlet vanes close. The low-low fan-curves avoid stall and surge by removing the point of instability, in most applications.

Figure 8.21 shows a pneumatic power positioner actuation system used on a fan having an inlet radial-vane damper. The illustration is shown with a command signal to increase fan air flow. In this case, the I/P converter receives an increasing 4-20mA electronic command signal from a controller and it outputs an increasing 3-15 psig pneumatic signal to the power positioner's receiver diaphragm. The receiver diaphragm pushes down against a restoring spring to shift a spool valve. The spool valve directs large control-air flow to the power positioner cylinder, pushing the power piston upward. At the same time, the spool valve allows trapped air on the top of the power piston to bleed off, avoiding hydraulic lock. The power positioner linkage opens the fan inlet damper vanes, increasing air flow. As the linkage moves upward, an inclined bar attached to the piston rod follows upward. The bar causes a cam mechanism to pivot and move the restoring linkage upward to compress the spring that opposes the diaphragm pressure force. The upward movement of the diaphragm is the negative feedback that shifts the spool valve into a neutral position, where control air neither enters nor leaves the power piston cylinder. The air pneumatic pressure on both sides of the power piston locks the actuator at the new setpoint.

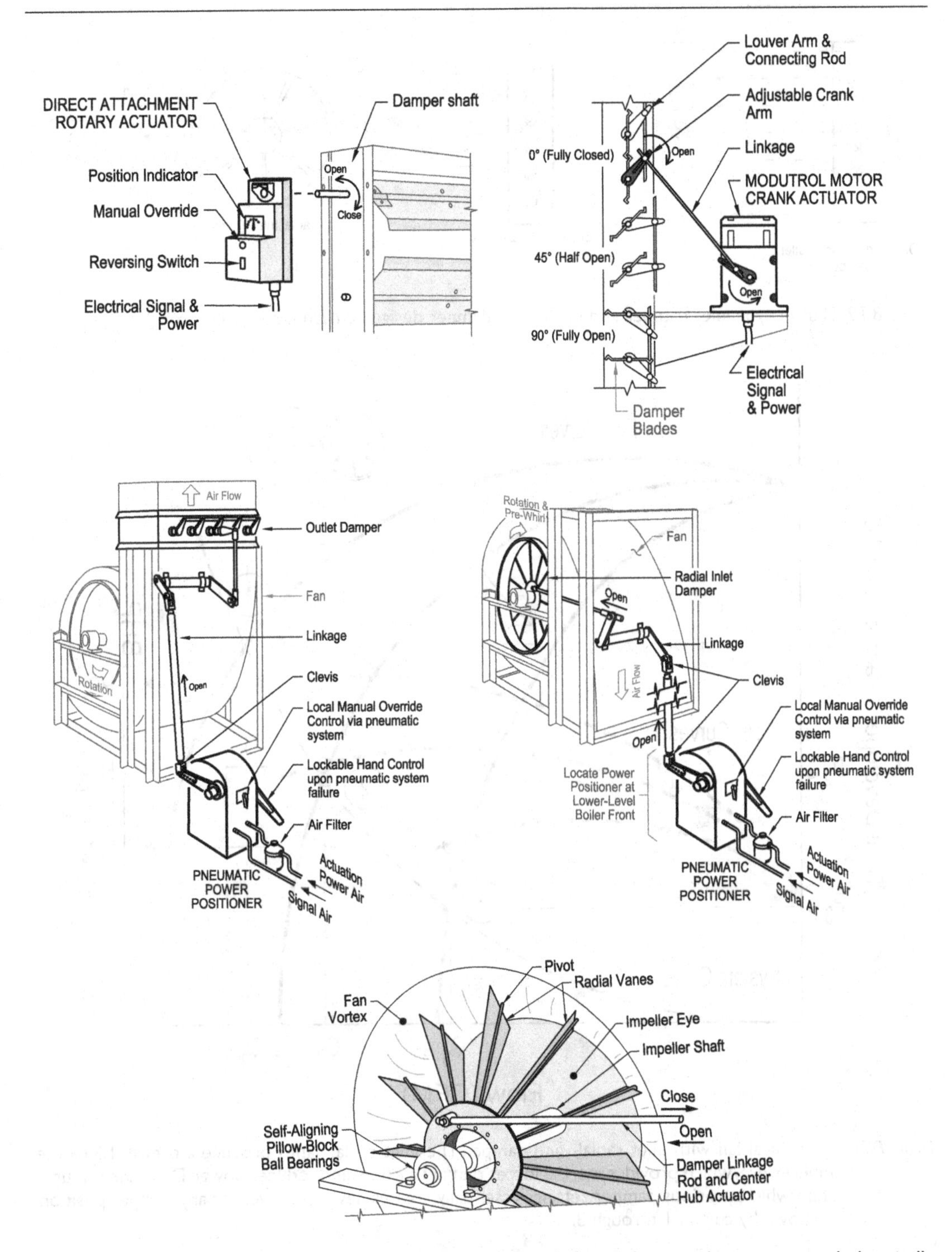

Figure 8.18 Damper and actuator arrangements. The upper left figure shows a direct-connected electrically powered rotary actuator. The upper right figure shows an electrically powered crank/linkage actuator using a modutrol motor.

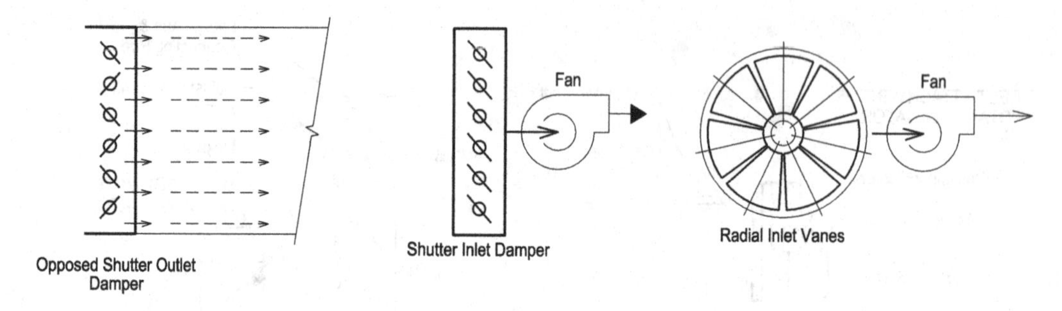

Figure 8.19 Shutter-type inlet damper and radial inlet damper designs for fan capacity control.

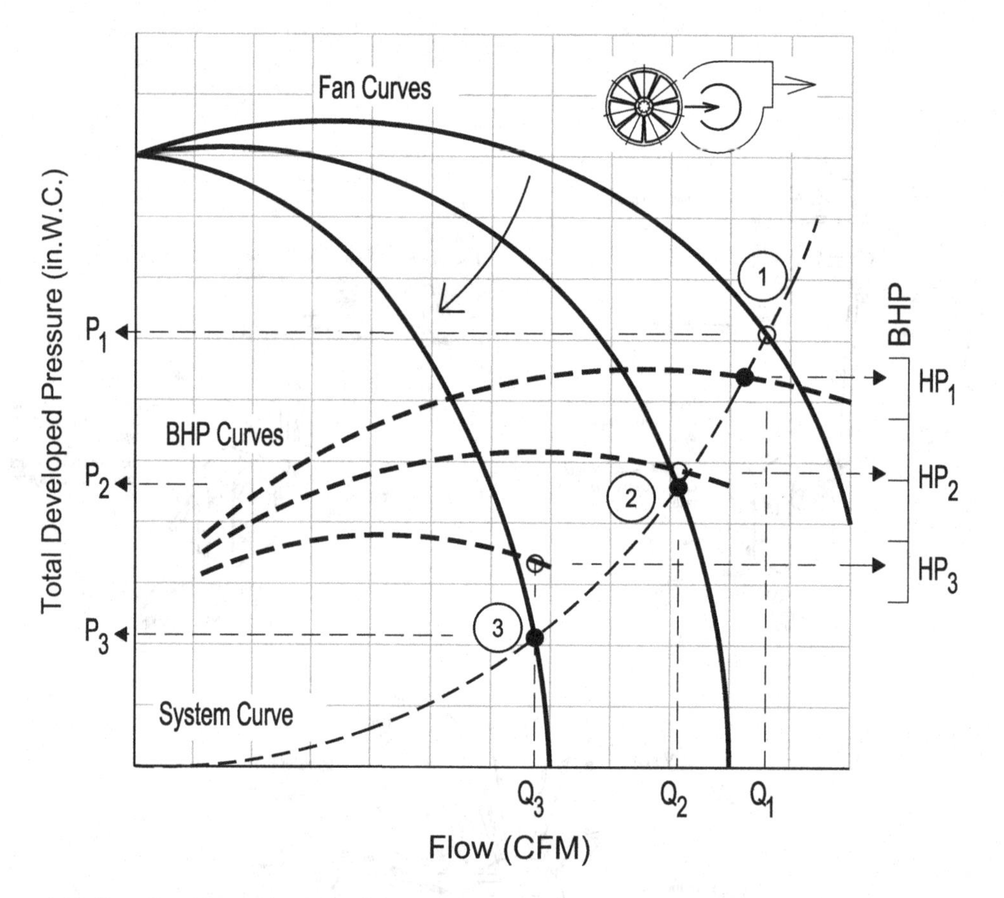

Figure 8.20 Centrifugal fan with inlet radial-vane damper. The inlet radial vanes produce a prewhirl into the impeller eye, which provides inertial forces that aid air flow and reduces power. Different amounts of prewhirl at various damper settings essentially creates new fan curves for any damper position, as shown by curves 1 through 3.

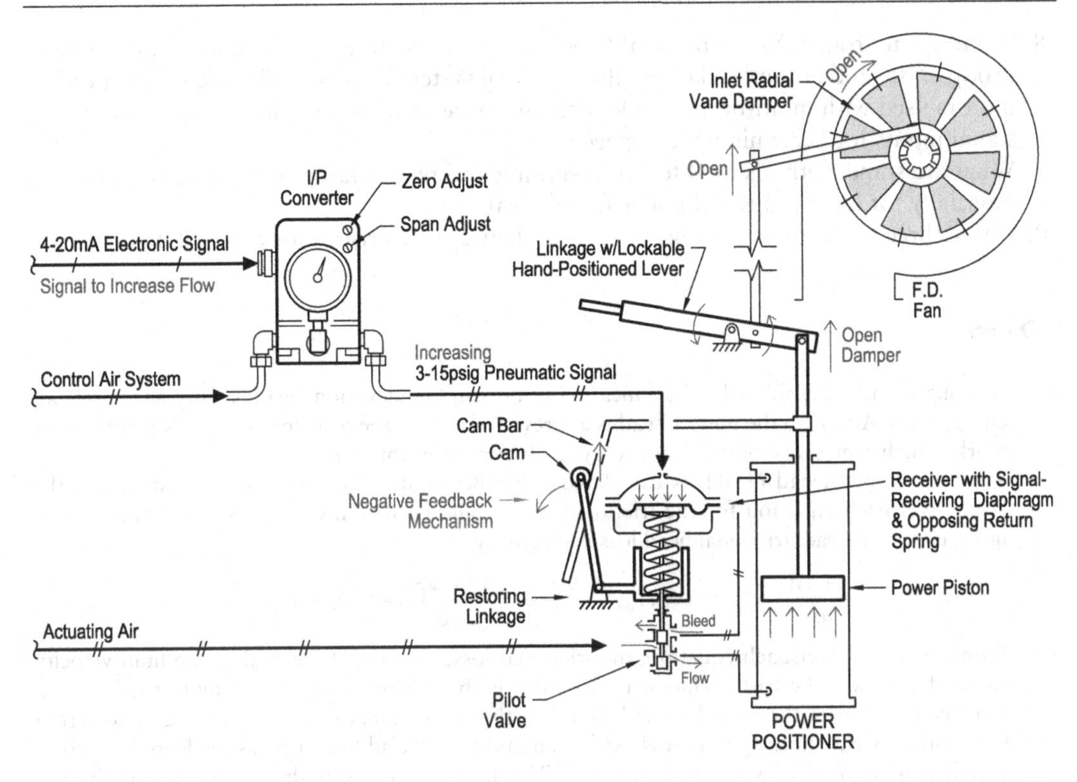

Figure 8.21 A radial inlet-vane damper and its pneumatic power positioner is shown with an I/P converter. The system is responding to an increasing air-flow signal from its controller. Negative feedback is provided by a cam and restoring linkage to stop the positioner at the desired setting. This arrangement is typical for the receiving regulator used with combustion-air forced-draft fans.

QUESTIONS

1. What are pump curves and what are they used for? What four parameters are often plotted against flowrate?
2. What is a system curve? What Bernoulli equation parameters comprise the system curve? Compare the static head and frictional head. What is the approximate shape of the frictional head curve? Why is the system curve so important to pump curves?
3. Write the pump affinity laws for speed variations. In an uncontrolled system, how does flow vary relative to speed? In an uncontrolled system, how does pressure/head and power vary with capacity?
4. Including uncontrolled, list four types of pump control schemes. How is the operating point on the pump curve affected by each control scheme?
5. Using condensate control as an example, describe the operation for varying flow using submergence control, throttle control, recirculation control, and variable-speed control.
6. For a rotary screw, positive-displacement pump used in a boiler fuel-oil system, describe the purpose of the relief valve, the constant-pressure regulator (CPR valve), and the fuel-oil regulating valve. What is the control signal input for each? Why is the CPR valve so important to the system operation?
7. In a dual-fuel plant, using the option of light and heavy fuel oil, how is the transition of fuels seamlessly accomplished?

8. What is fan droop? What undesirable behaviors can occur at low flow conditions when droop exists and an outlet-damper flow-control system is fitted? Why are multi-speed motors used with multi-burner boilers having large turndown? What are good control-signal inputs for determining fan speed?
9. What are some methods used to avoid surging and reduce fan power consumption, especially for large industrial boiler forced-draft fans?
10. Name three fan-damper arrangements and their characteristics for controlling air flow.

NOTES

1. Although head is considered to be a measure of pressure or elevation, head is more accurately an energy term. Although the unit of head is abbreviated as feet, the true unit of head is ft-lbf/lbm or work, which represents mechanical energy (ft-lb_f) on a per pound basis.
2. Total Developed Head (TDH) is calculated using Bernoulli's Equation for the system and the Darcy-Weisbach equation for the frictional losses. The head terms are pressure head, velocity head, elevation head, frictional head loss respectively:

$$H_{pressure} = \frac{144 \cdot \Delta P}{\rho},\ H_Z = \Delta Z,\ H_{Vel} = \frac{\Delta(V^2)}{2 \cdot g_c} = \frac{V_2^2 - V_1^2}{2 \cdot g_c},\ \text{and}\ h_L = f\frac{L}{D}\frac{v^2}{2g_c}$$

3. From the Darcy-Weisbach equation, the frictional losses are largely related to the fluid velocity squared, however, the friction factor is also affected by velocity to a lesser extent. It is accepted practice to estimate the head loss as being directly proportional to the velocity squared. From the continuity equation ($Q=vA$), it is extrapolated that the head loss or pressure drop is approximately related to the flow squared ($\Delta p \propto Q^2$). This flow- and pressure-drop relationship is important in understanding and predicting flow behavior.
4. Theoretical pumping power is fundamentally the flow rate multiplied by the pressure (or head) and then converted into horsepower or kW. The brake power would be determined from the pump efficiency.

Chapter 9

Control valves

Control valves are devices that modulate flow rate, pressure, temperature, or level. These are final control devices and understanding their operation, nuances, and options is important for integrating them into control systems. Control valves range from the self-contained, self-actuated devices to the externally powered, pilot-actuated remote-sensing devices. The separately powered systems can be entirely pneumatic-control systems, or they may use electronic controls interfaced with pneumatic or electrically powered actuators, which are much more conducive to large devices using remote-sensing and control. Proper sizing of the control valve and selection of its flow characteristics is important for long reliable operation. Typical plant applications include:

- Pressure-reducing valves
- Back-pressure regulators
- Pressure-switching devices
- Pressure-relief or vacuum breakers
- Low-level makeup or high-level dump
- Flow regulation or flow bypass for temperature control

SELF-CONTAINED PRESSURE REDUCING VALVES

Self-contained reducing valves provide control without any external power source, and tend to be reliable, low-maintenance devices. In their simplest form, these control valves function by establishing a balance between an applied adjusting-spring force and feedback from hydraulic forces developed by fluid pressure applied against a diaphragm. Figure 9.1 shows a simple pressure-reducing valve. Fluid enters from the left and flows to the outlet at the right. Internal ports are drilled within the body to direct the downstream low-pressure fluid against the underside of a diaphragm, where the outlet pressure pushes upward against the diaphragm. Without any adjusting-spring force, the controlling spring pushes the control-valve disk upward, closing the valve with assistance from the inlet high pressure fluid. Screwing down on the handwheel increases the adjusting spring compression, eventually opening the control valve and starting flow. The downstream pressure builds until a force balance is reached between the spring force on top and diaphragm pressure force underneath. The force balance is shown in Figure 9.1 (right). The balancing pressure is the valve setpoint. The setpoint can be changed by turning the adjusting screw until the desired output pressure is obtained.

A small vent in the bonnet prevents accidental trapping of fluid and any influence it might have on the top of the diaphragm. This valve operates using simple proportional-only control, without any ability for in-service gain adjustment. As the operator screws down on the

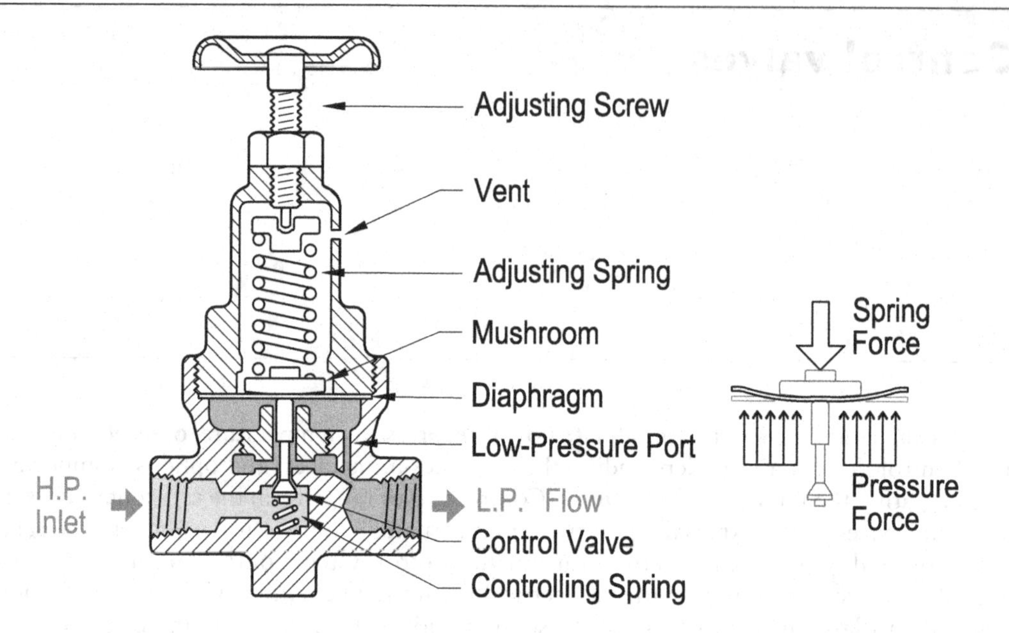

Figure 9.1 Leslie LC series simple pressure-reducing valve. The adjusting spring force pushes down while pressure is directed through internal ports to push up on the underside of the diaphragm, to obtain a force balance. Screwing down on the adjusting screw increases the spring compression, raising the output-pressure setpoint. (Adapted from CIRCOR | Leslie.)

handwheel, spring compression increases, and the output pressure needs to increase proportionally until it balances the diaphragm pressure force. The range and sensitivity are a function of the spring rate (k_{sp}). The gain or sensitivity can be increased by using a softer, more compliant spring. *Offset* under load variations is certain, and occasional readjustment may be required for some applications.

Figure 9.2 shows a self-contained, self-actuated pressure-reducing valve. A small auxiliary valve is similar to the controlling valve used in the simple reducing valve shown in Figure 9.1, however, the high-pressure fluid is ported through the auxiliary valve to be used as a static pressure source against the top of a piston. The high-pressure fluid pushes against the piston to create the main-valve driving force. The main spring serves to move the main valve upward to the closed position, where inlet-pressure also assists in sealing against leak-by in the upward-seating valve. The low outlet pressure is ported to the underside of a thin steel diaphragm, whose force is balanced by the adjusting spring compression. The force balance is shown in Figure 9.2 (right). The handwheel adjustment is similar to the simple valve, where screwing down on the handwheel increases spring compression and the setpoint. Like the simple reducing valve, this valve is proportional only, but capable of larger flowrates. Once it is set at a steady-state load, this valve tends to maintain very tight control, however, it is prone to *offset* during transient conditions.

SELF-CONTAINED TEMPERATURE REGULATING VALVES

Figure 9.3 and Figure 9.4 show two styles of self-contained thermostatic control valves used with steam heated systems. Figure 9.3 is a control valve common in hot-water heaters or fuel-oil heaters, where the oil grade is constant, and a viscometer is not needed. The valve is fitted

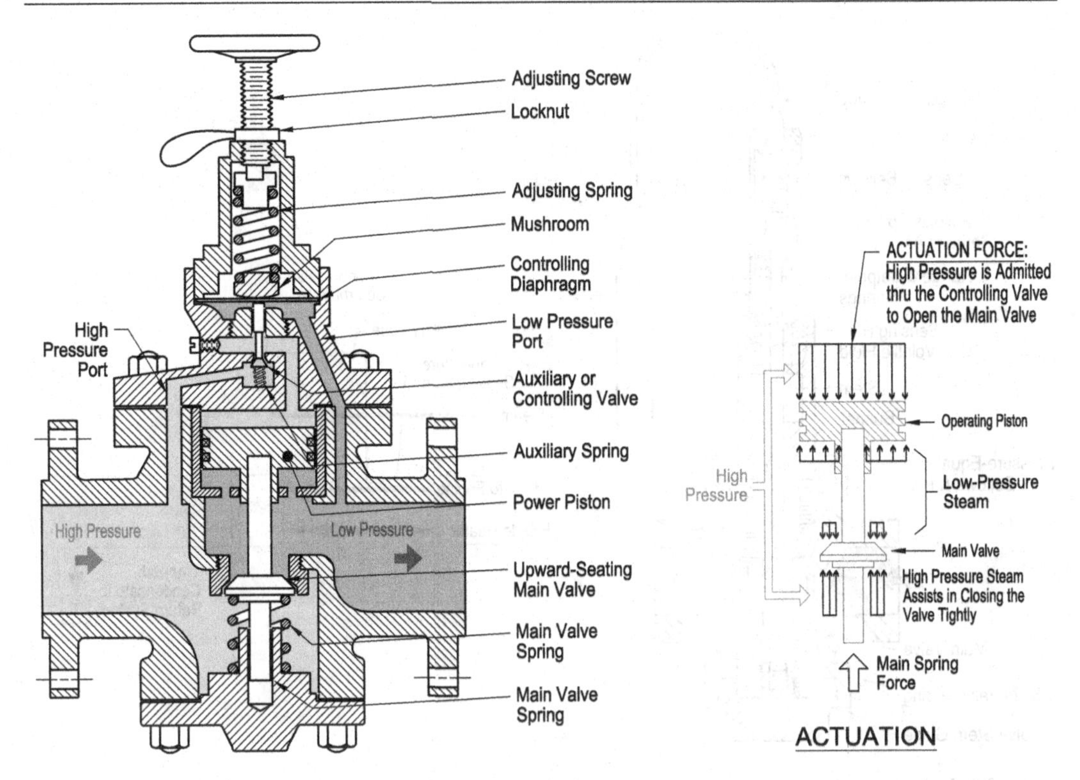

Figure 9.2 Self-contained, self-actuated pressure reducing valve. (Adapted from CIRCOR | Leslie.)

with a sensing bulb containing a volatile liquid that is placed in the process fluid stream, whose temperature is being controlled. As the process temperature rises, the bulb temperature and its corresponding saturation pressure also rises. The increasing bulb pressure is transmitted through a capillary tube to an actuating bellows, which applies a downward force to close the downward-seating steam valve, cutting back the heat input. Temperature adjustments are made by turning the bellows cover. When the cover is screwed down or clockwise, the adjusting-spring compression is increased, and it takes more bulb temperature/ pressure to expand the bellows, raising the setpoint. The main valve stem has an internal port that connects an equalizing bellows located on top of the valve disk with the high-pressure located on the underside of the disk. The internal port equalizes the high-pressure between both the disk bottom and top, creating a hydraulically balanced disk design. Balancing of the disk forces smooths stem movement, improving control response, and the upper process-sensing bellows does not need to create a very large pressure force to tightly close the main valve.

If the bulb charge is lost, the valve fails open, and the process temperature is limited by the steam saturation temperature and process flow rate. This valve is typically used for heating systems, but it can be converted to a cooling service by exchanging the main valve with an upward-seating disk and seat.

Figure 9.4 shows a self-contained steam thermostatic control valve used in larger-flow applications. This valve is essentially a dual-function pressure and temperature control valve, having a self-contained pressure reducing valve with a linkage attached to a thermostatic mechanism, which biases the pressure-limit-adjusting spring. The bulb senses temperature, which is transmitted as a pressure signal to a larger rubber diaphragm whose force is

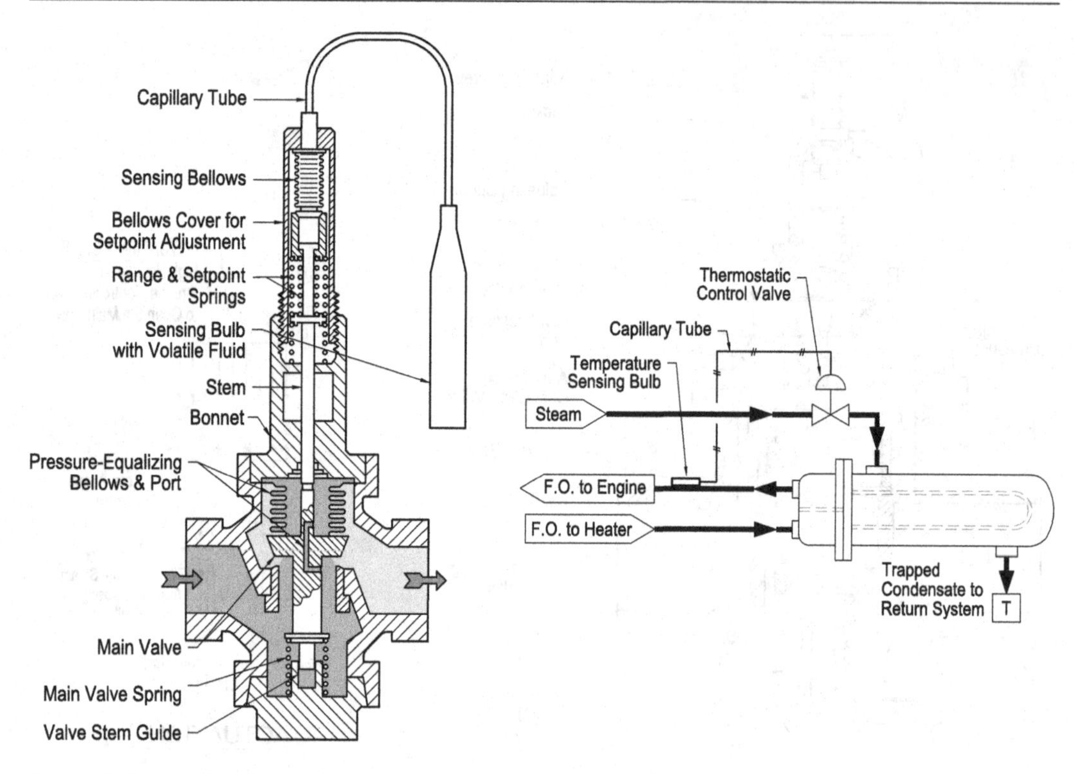

Figure 9.3 A thermostatic temperature control valve senses the fuel oil temperature leaving the heater, to modulate steam flow. (Adapted from CIRCOR | Leslie.)

balanced against an external spring. The valve has three adjustments; pressure, which uses the pressure-limiting compression spring; temperature, which is set using the temperature-adjusting spring and lever mechanism; and proportional band or gain, which is adjusted using the blade spring and slider mechanism, similar to the one shown on the right side of Figure 9.9.

Figure 9.5 shows a pneumatically actuated rotary three-way valve and its application as a temperature-controlled valve in an engine jacket-water cooling system.

PNEUMATIC PILOT-ACTUATED CONTROL VALVES

Pneumatic pilot-actuated valves are preferred for higher-flow applications, where the internal diaphragm and piston for self-actuation would otherwise become too large and cumbersome, or where more precise control is required. The pilot-actuated valve also has the ability of installing the sensing pilot controller close to the process-measurement for ease of adjusting the setpoint, and locating the control valve far away, where process is manipulated. For instance, a steam plant uses the deaerator as a surge tank and its level is used for determining if the system contains the correct volume of water. The makeup and dump pilot controllers are typically located near the deaerator sight glass for ease of adjustment. However, the makeup and dump control valves are located near the condenser, where the making up and dumping actions occur.

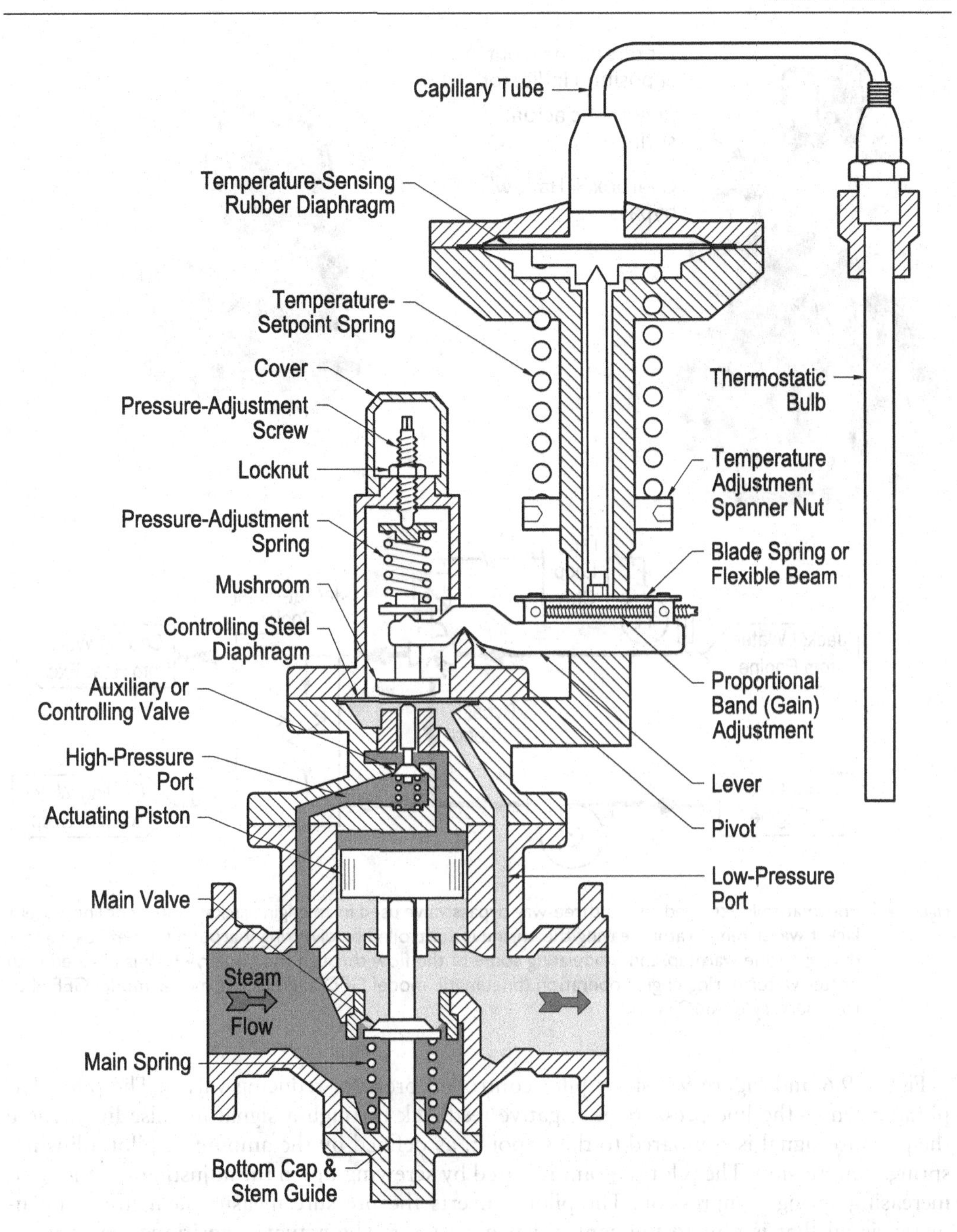

Figure 9.4 The Leslie *LTC Duo-Matic* is a combination pressure reducing and temperature-control valve. This valve has a flexible beam between the sensing bulb and actuating mushroom, used for adjusting the temperature-to-pressure gain, which is detailed in Figure 9.9. (Adapted from CIRCOR | Leslie.)

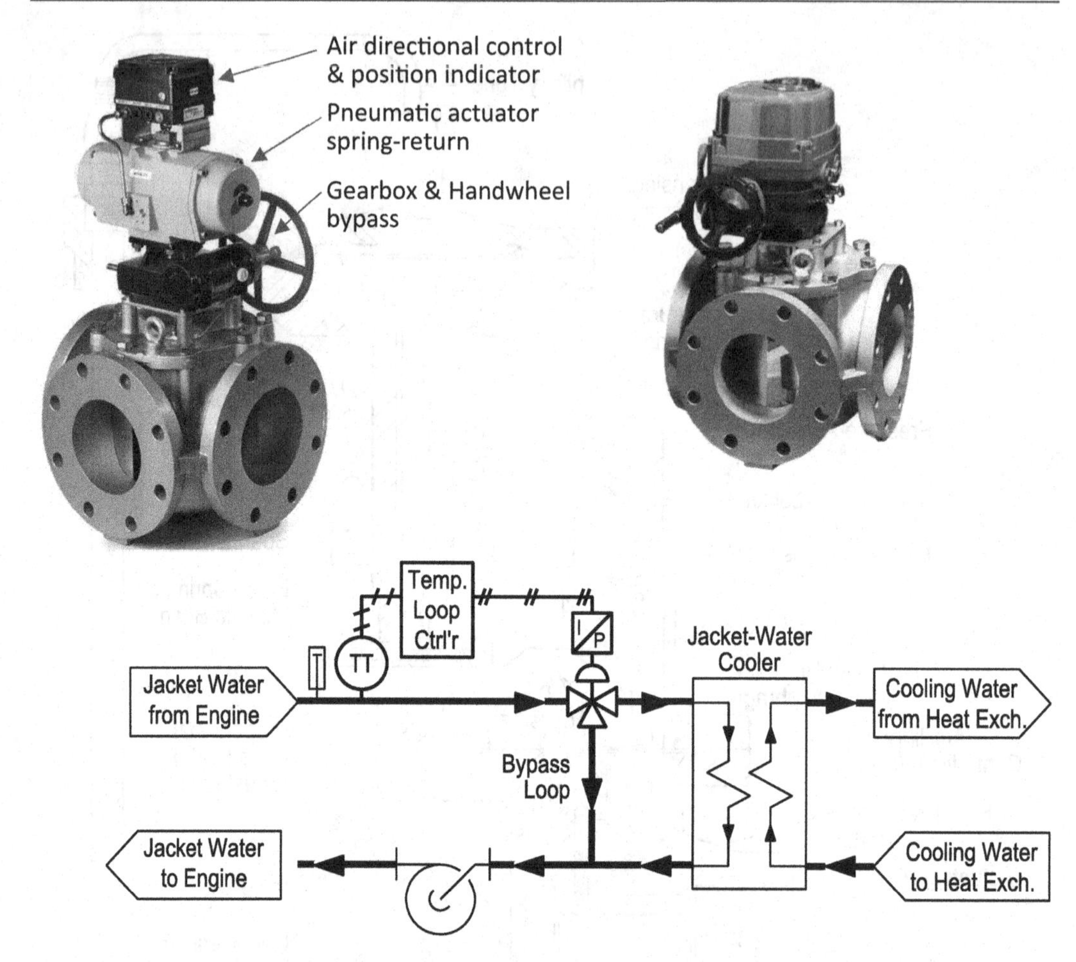

Figure 9.5 Pneumatically actuated rotary three-way bypass valve used in an engine jacket-water cooling system. Jacket water temperature leaving the engine is controlled by diverting it around the heat exchanger during engine warmup, and modulating some of the flow through the cooler, which is blended with hotter water during engine operation (pneumatic model GPD and electric motor model GEF photos courtesy of AMOT).

Figure 9.6 and Figure 9.7 show pilot-controlled pressure-reducing valves. The pilot diaphragm senses the line pressure as negative feedback through a signal impulse line, where the pressure signal is compared to the setpoint, as defined by the amount of pilot-adjusting spring compression. The pilot setpoint is raised by screwing up on the adjusting spanner nut, increasing spring compression. The pilot converts the pressure measurement into a pneumatic signal that is sent to the control-valve actuator. The actuator positions the valve to match the process pressure to the pilot setpoint. The pilot output usually operates between 3-15 psig, but some systems use 0-30 psig. In Figure 9.6, high outlet pressure is transmitted as a signal to the direct acting pilot actuator, which moves down, opening the signal-air pilot valve, causing an increasing signal to the control-valve actuator, moving its stem down to close the valve. In this manner, the control valve uses *air to close*, and *the valve fails open upon loss of control air.*

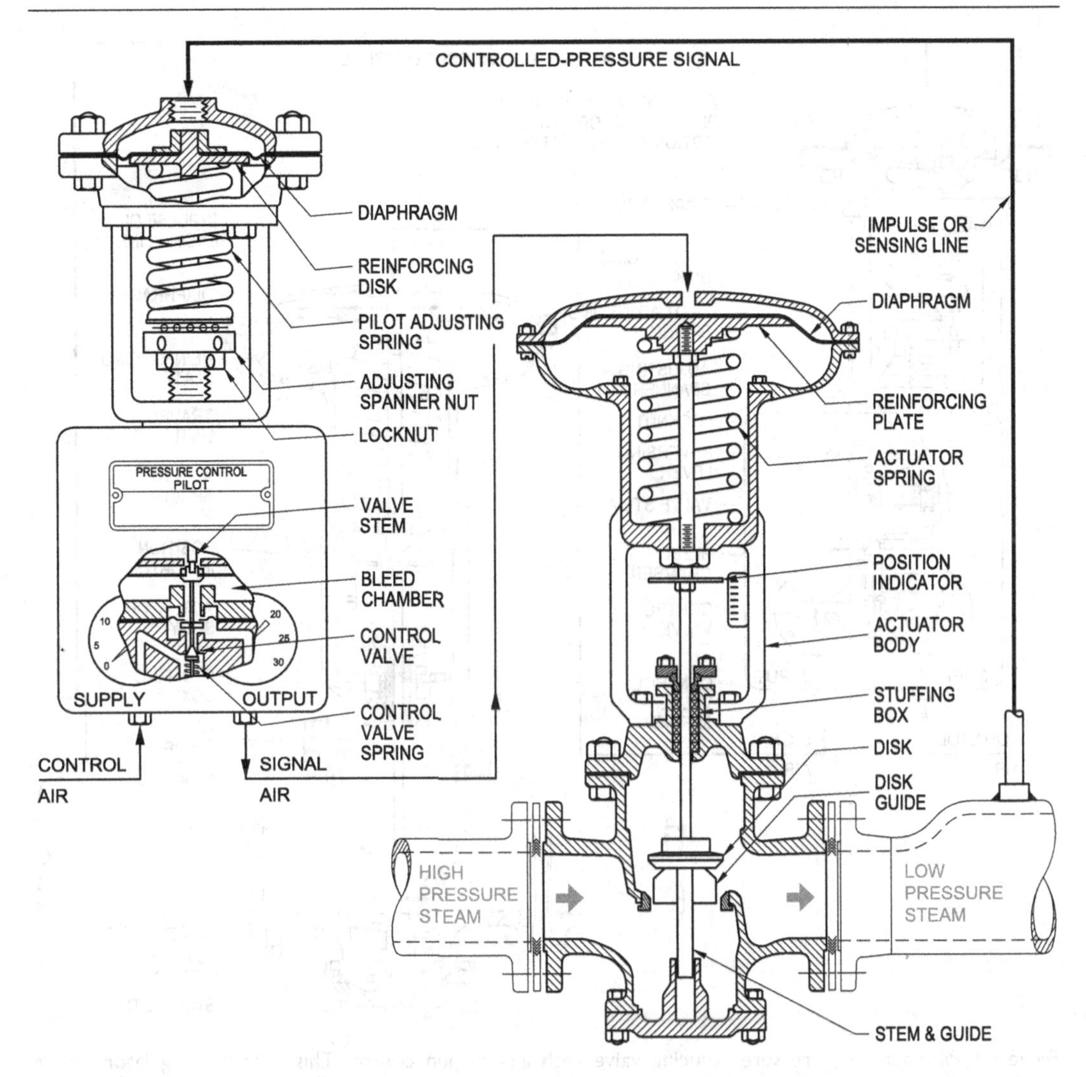

Figure 9.6 Pilot-actuated pressure-reducing valve with air-to-close control. This pressure regulator uses a direct-acting pilot, a direct-acting valve actuator, and a downward seating valve disk. On loss of control air, this valve fails open.

Figure 9.7 is like Figure 9.6, but instead it uses a reverse-acting pilot controller and an upward seating valve disk. In this arrangement, signal-*air pressure is used to open the valve*, and *the valve fails closed upon loss of control air*.

PNEUMATIC PILOT CONTROLLER

The pneumatic pilot controller is a force-balance device that senses the process, compares its measurement to a setpoint, and produces a corrective load-air signal that is sent to the control valve actuator. The pilot controller consists of a spring-adjusted diaphragm-actuator assembly mounted on top of a pilot controller (Figure 9.8). The actuator balances the sensing-line-pressure forces against the adjusting spring to create the setpoint. The two forces balance each other to position a pneumatic bleed nozzle disk up or down, essentially setting its

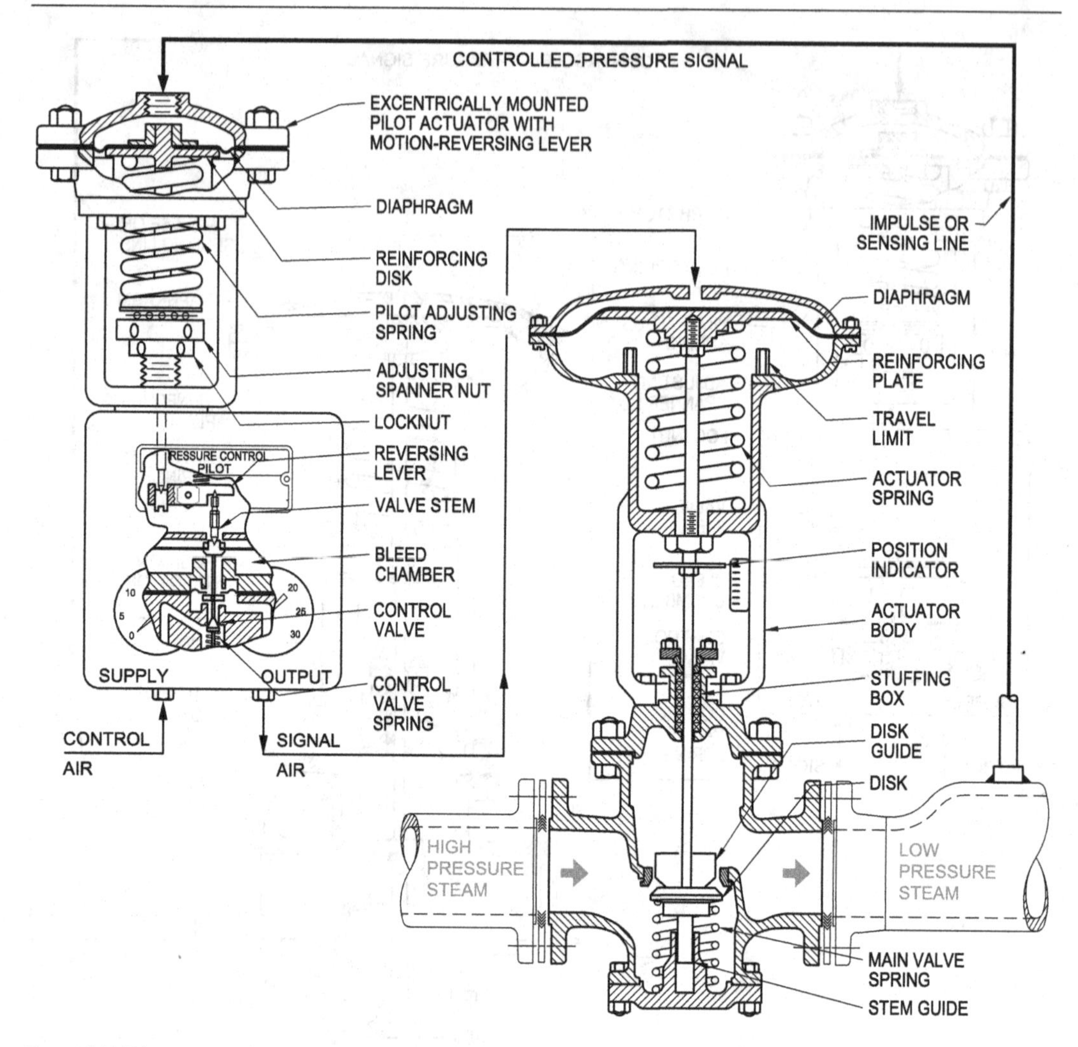

Figure 9.7 Pilot-actuated pressure-reducing valve with air-to-open control. This pressure regulator uses a reverse-acting pilot, a reverse-acting valve actuator, and an upward seating valve disk. On loss of control air, this valve fails closed.

location where controller output signal air is trapped when the signal is correct, bled when the signal is too high, or increased when the signal is too low. The pilot valve itself is a nozzle mounted in the flexible diaphragm. The output signal pressure acts on the underside of the nozzle diaphragm to position the nozzle for the correct output signal.

At the correct output signal, the nozzle seals against the nozzle disk, and the pilot valve closes, resulting in a static signal-air pressure. For the controller shown in Figure 9.8, if the sensed pressure drops, the actuator stem moves up, the nozzle disk lifts uncovering the nozzle, the signal air pressure bleeds past the nozzle, and the output signal pressure drops. This arrangement describes a direct-acting pilot controller.

In some instances, reverse action is required from the controller. In those cases, the actuator on top of the pilot controller is shifted to the side, and a motion-reversing lever is installed. A reverse-acting pilot controller mechanism is shown in Figure 9.15.

In many cases the pilot-controller output signal has enough flow to power the control-valve pneumatic actuator. In other cases, the pilot air flow is too low, and the control valve

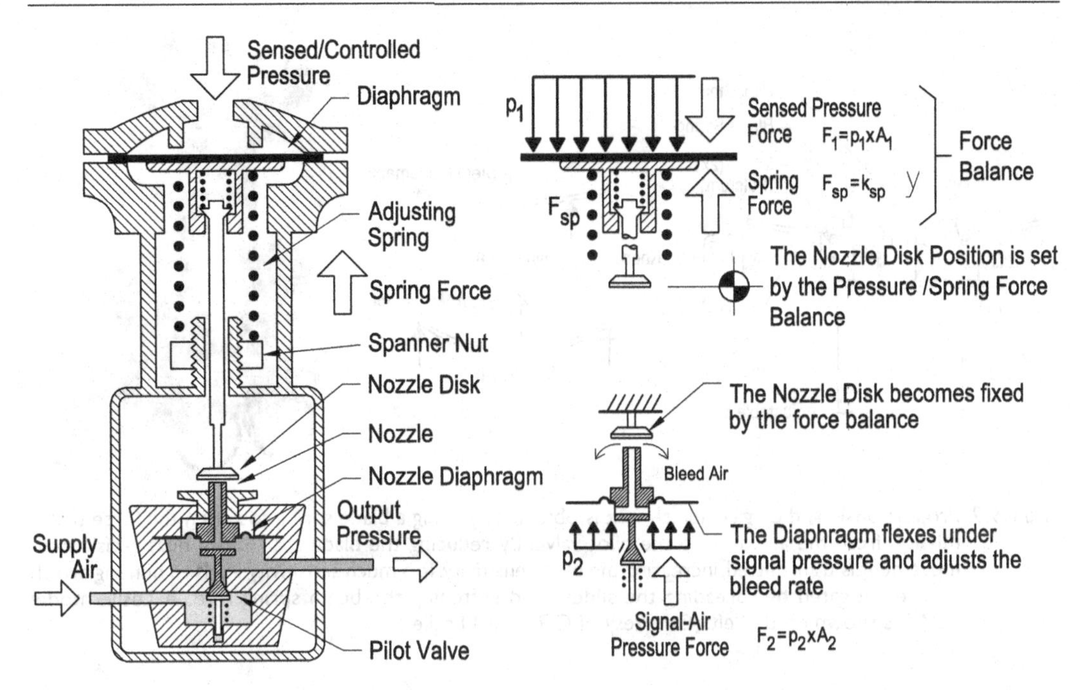

Figure 9.8 Pilot controller operation. Sensed pressure on the upper diaphragm opens a pilot valve sending an increasing 3–15 psi air signal to the controlled device. On sensed pressure drop, the stem moves up by spring force and opens a nozzle disk, bleeding and dropping signal-air pressure.

response is too slow for satisfactory operation. In those cases, a volume-booster relay may be installed. The volume-booster relay uses the small pilot signal to position a larger spool valve, designed to permit large actuation-air flows. The amplifier relay shown in Figure 9.14 is an example of a volume booster.

Some controllers are fitted with a proportional band or gain adjustment as shown in Figure 9.9. Proportional band adjustment is beneficial in control stations having significant time lags between load-changes and the corrective actions that tend to make the system overreact and hunt. Time lags may occur from system inertia, long distances between control elements, or other causes. Proportional band is calibrated during large transients by moving the sliders closer to stiffen the blade spring until hunting occurs. At that point, the proportional band is widened by moving the sliders outward beyond the point where the hunting stops. The system is observed for correct operation during various subsequent load changes, and further adjustments can be made, if needed.

PNEUMATIC ACTUATORS AND POSITIONERS

Pneumatic actuators are used on larger valves to create greater actuation forces or where more precise control is needed. The actuators function by accepting loading air signals from either a pilot controller or an electronic current-to-pneumatic (I/P) converter. The signal load-air is applied against a large rubber diaphragm or piston in the actuator. The load-air pressure pushes against the diaphragm to slide the valve stem. The rubber diaphragm is reinforced by a stiffening plate, which connects to the actuator stem. The rubber between the plate edge and diaphragm casing behaves like a hinge, permitting stem movement. A spring seat is provided for adjusting the spring compressive pre-load, and adjustable stops are provided for

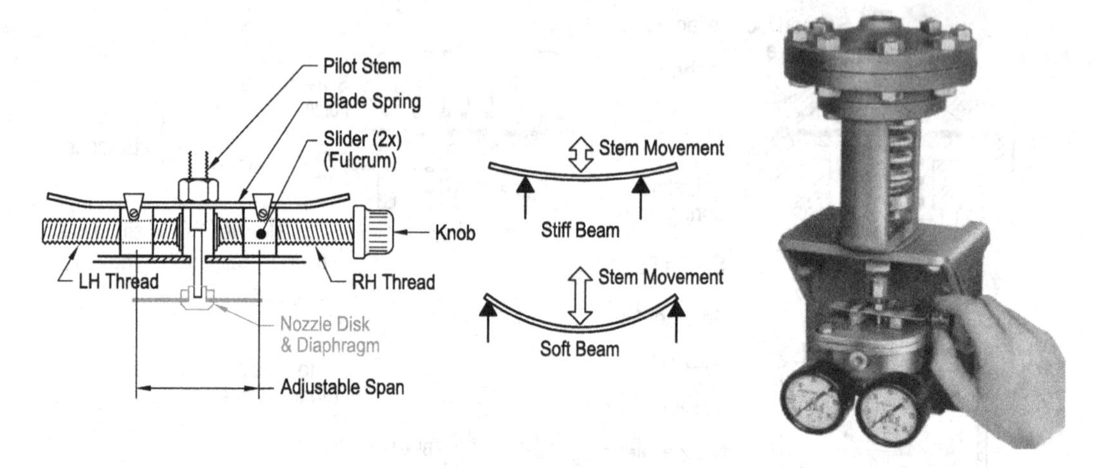

Figure 9.9 Proportional band or gain adjustment is obtained by using a blade spring to soften the force transmission from the actuator to the pilot valve. By reducing the blade span, more nozzle-disk stem movement is transmitted, increasing the pilot sensitivity. Too much sensitivity leads to hunting, which can be mitigated by spreading the sliders and softening the beam spring rate. A Leslie model PDAP is shown on the left. (Courtesy of CIRCOR | Leslie.)

setting the maximum permissible stem travel within the manufacturer limits. The stem travel needs to be limited, since extra movement will not allow any more than wide open flow, but the extra motion will delay the actuator response when operating near full open.

Pneumatic actuators can be direct acting or reverse acting (Figure 9.10 and Figure 9.11). In the direct-acting actuator, load air is applied on top of the diaphragm to slide the stem downward. For the reverse-acting actuator, load air is applied underneath the diaphragm to lift the stem. The reverse-acting actuator construction has a slightly more complex spring-seat design, and an O-ring seal is required between the stem and the diaphragm housing to contain the load-air pressure.

Typical control-valve practice includes isolation valves and a hand bypass; however, some pneumatic actuators can be procured with a built-in hand bypass as shown in Figure 9.12. In manual operation, the handwheel directly positions the diaphragm and stem, and the diaphragm travel becomes restricted. When transitioning back to automatic operation, it is important that the bypass handwheel is positioned to permit the full stem travel.

Essentially, a pneumatic actuator by itself is an open-loop device, where it is hoped that the command signal moved the actuator to the correct operating position. However, the open loop does not account for any sloppiness, stickiness, or hysteresis in the mechanics, and some variations are natural. For installations where very precise control is required, negative feedback is added to provide closed-loop control to ensure that the proper corrective action indeed took place. Actuator negative feedback is obtained using linkages that produce a stem movement signal, which is added back into the original control command, when the proper response has occurred.

Figure 9.13 and Figure 9.14 show a positioner and feedback linkage. The bellows receives the load-air signal from a controller. The bellows causes a quadrant lever to pivot around a vertical axis. The quadrant positions a nozzle-and-flapper arrangement on an amplifier-relay, which acts as a volume booster. The amplifier relay forwards the boosted air signal to the actuator chamber to position the valve stem. Movement of the stem's mechanical linkage pivots the quadrant beam around its horizontal axis to counteract the nozzle-flapper

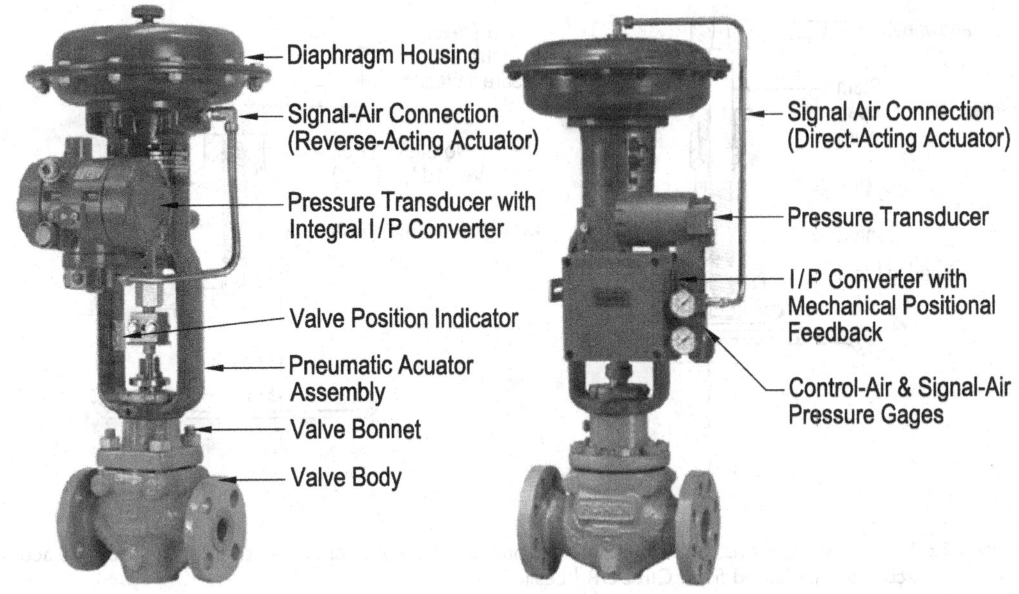

Figure 9.10 The two electro-pneumatic control valves are using integral 4-20mA transmitters to create their own 3–15 psi actuator signals. The left valve applies the pneumatic signal to a reverse-acting actuator, and the right valve applies the pneumatic signal to a direct-acting actuator. The box on the righthand valve contains a mechanical positioner that provides proportional negative feedback for precise positioning. (Courtesy of Emerson.)

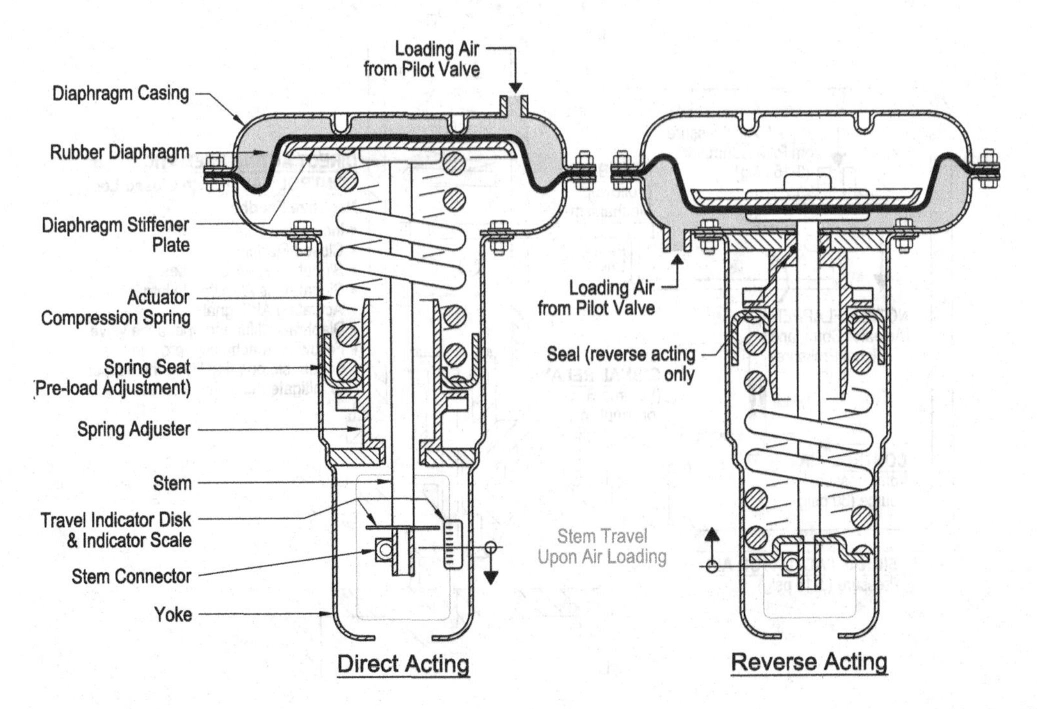

Figure 9.11 Pneumatic actuators, direct acting (left) and reverse acting (right). (Adapted from CIRCOR | Leslie.)

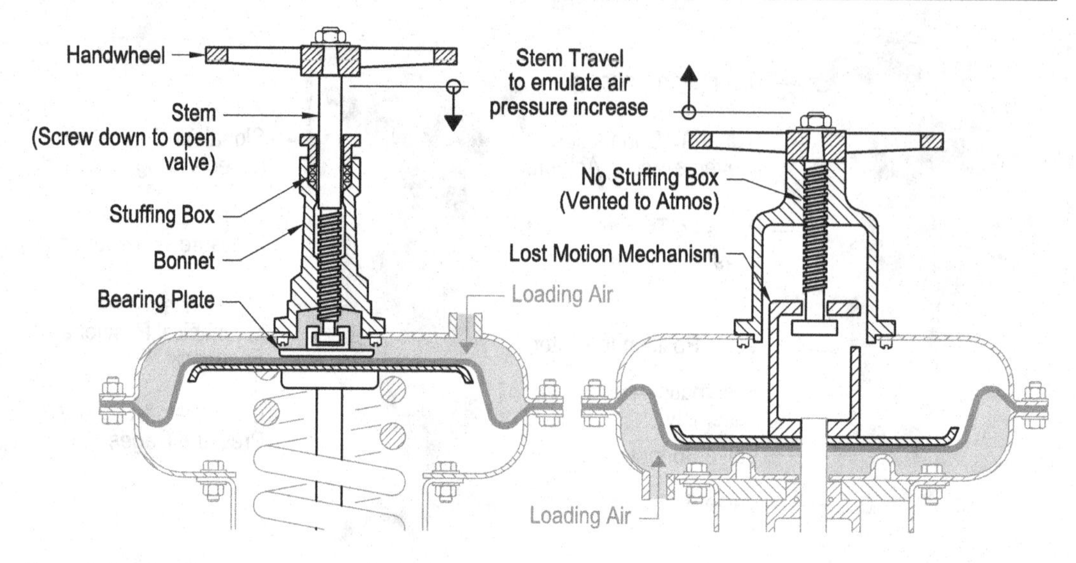

Figure 9.12 Hand bypass options can be built into the pneumatic positioners for both direct and reverse acting actuators. (Adapted from CIRCOR | Leslie.)

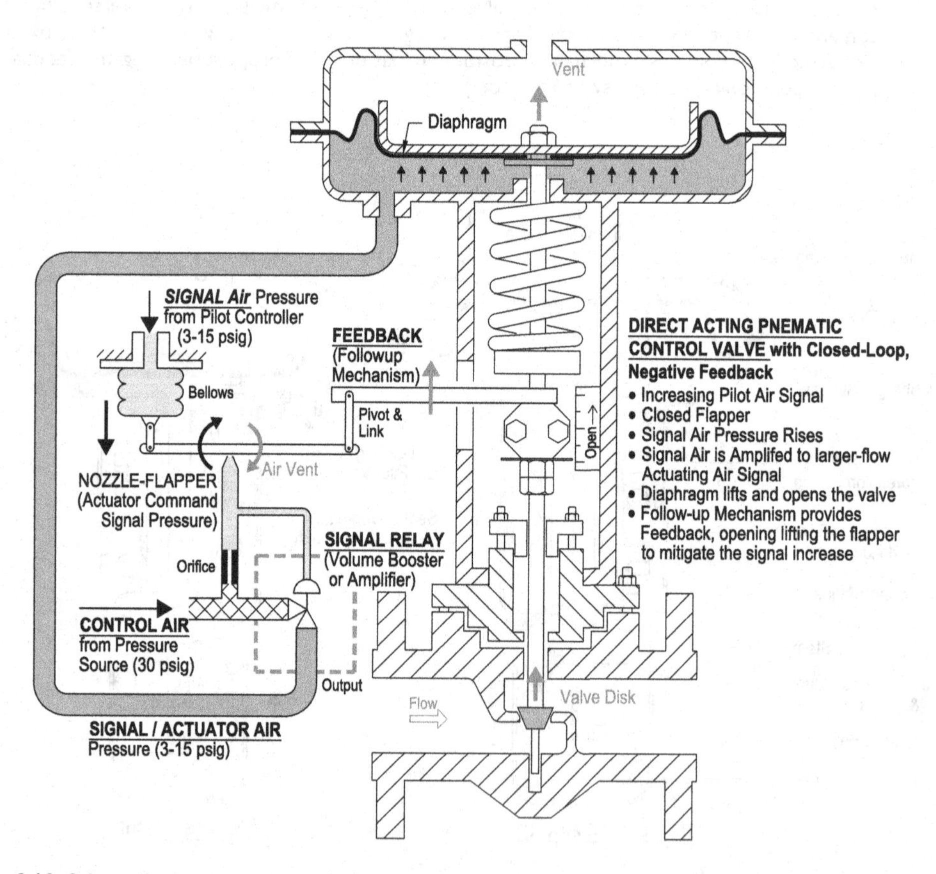

Figure 9.13 Schematic representation of a direct-acting closed-loop pneumatic valve actuator, which provides negative feedback on the valve position. (Adapted from Emerson.)

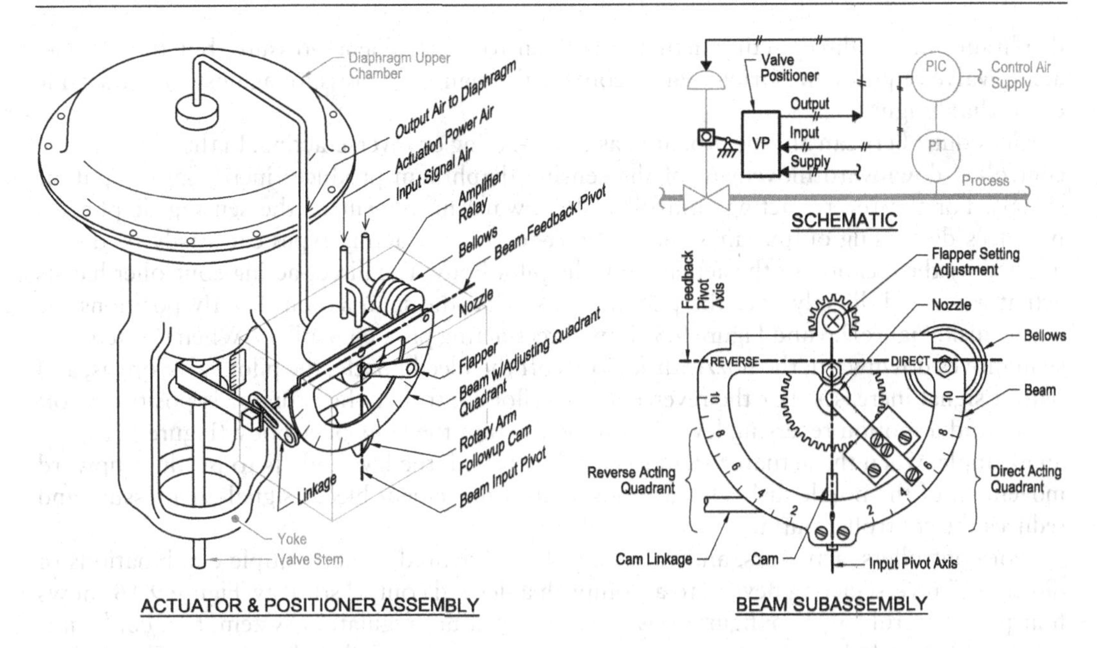

Figure 9.14 A Fisher single-acting pneumatic positioner for sliding-stem control valves. The positioner receives a pneumatic input signal from a controller and adjusts the control-air pressure to the actuator. The beam assembly provides a mechanical negative-feedback signal to ensure the corrective action has taken place. (Courtesy of Emerson.)

movement after the corrective action has indeed taken place, providing stem positional feedback. The mechanism resets the nozzle flapper position, ensuring that the valve indeed moved in response to the signal. Gain is adjusted by rotating the flapper around the quadrant, changing the relative amount of vertical pivoting from the signal input relative to the amount of horizontal pivoting during stem-movement feedback. This positioner can be used for reverse-acting actuators by swinging the flapper to the left quadrant. Boiler feedwater regulators and fuel-oil regulators are good examples where precise control is important and where feedback positioners are recommended, as fluctuations can result in unnecessary high or low water alarms, or unnecessary smoking or fuel inefficiencies.

DIRECT ACTING/REVERSE ACTING AND FAIL-SAFE POSITION

In selecting control valves, the valve itself may be downward seating, like a conventional globe valve, or it may be upward seating, where inlet pressure helps seal the disk to the seat. The upward seating valve construction is a little more complicated, in that it needs a spring under the valve plug to push it closed and the actuator may need to overcome the high-pressure-side hydraulic forces to open the valve. Although the terminology is not used, the downward-seating valve can be thought of as direct-acting, and upward-seating valve can be thought of as reverse acting.

The actuators themselves are classified as direct acting or reverse acting. An actuator that slides the stem downward, as in a conventional globe valve, is classified as a direct-acting actuator (see Figure 9.11 left). When actuating air pressure enters underneath the diaphragm and slides the stem upward, it is reverse-acting (Figure 9.11 right). The underside of the diaphragm in the direct-acting actuator is open to the ambient pressure, but the reverse-acting valve requires an O-ring seal to contain the signal air pressure against the underside of the diaphragm.

Pilot controllers can also be procured as direct-acting or reverse-acting. In the direct-acting controller, downward movement of the sensing diaphragm produces increasing output-air signals. For the reverse-acting controller, downward movement of the sensing diaphragm produces decreasing output-air signals. Direct- or reverse-acting types can easily be distinguished by the location of the actuator on the pilot body. The direct-acting controller has its actuator located directly over the pilot valve, so that the stem action directly positions the nozzle disk. Figure 9.6 and Figure 9.8 show direct-acting pilot controllers. When the actuator stem moves downward, the nozzle disk closes off the bleed nozzle, the pilot valve opens, and output signal increases. For the reverse-acting pilot controller, the actuator is positioned off center and a motion-reversing lever is installed within the controller body (Figure 9.15). As an example, when the actuator stem moves downward, the lever pivots to produce upward movement of the nozzle disk. Upward nozzle-disk movement bleeds signal-air pressure and reduces the controller output.

Pilot controllers, actuators, and valves can be configured using multiple combinations of direct- and reverse-acting devices to accomplish a desired control strategy. Figure 9.16 shows four pilot control valve configurations used in a pressure-regulation system, two configurations produce a fail-open response and two produce a failed-closed response. The desired actuation direction of the final control element is typically governed by failsafe considerations. Failsafe means that upon loss of actuation power, the control valve should move to the safest position for personnel and the process. As an example, if a pneumatic feedwater regulator were to lose compressed air, it should be designed to fail open. While high- or low-water casualties are both bad, a low-water casualty can produce more immediate and more catastrophic failures than a high-water casualty. As another example, a boiler fuel-oil

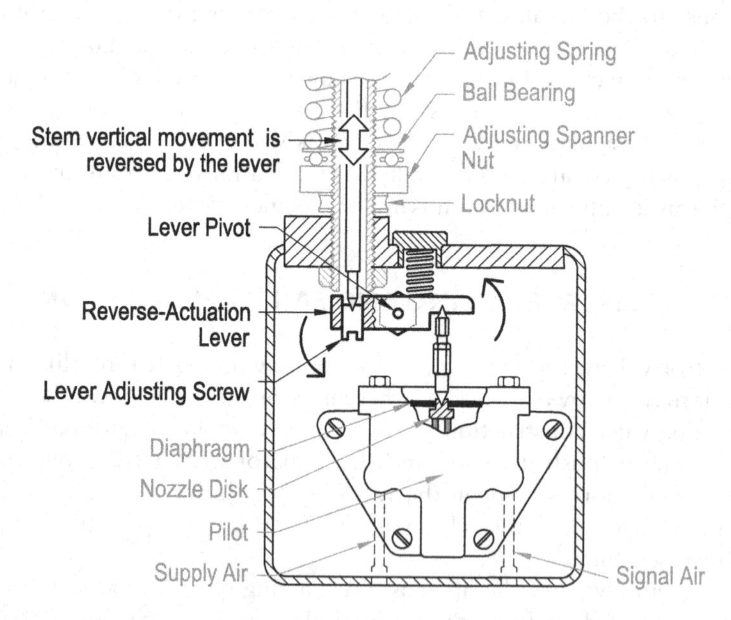

Figure 9.15 Reversing pilot controller. The actuator is shifted off-center, and a lever is installed to pivot and reverse the actuator stem motion. (Adapted from CIRCOR | Leslie.)

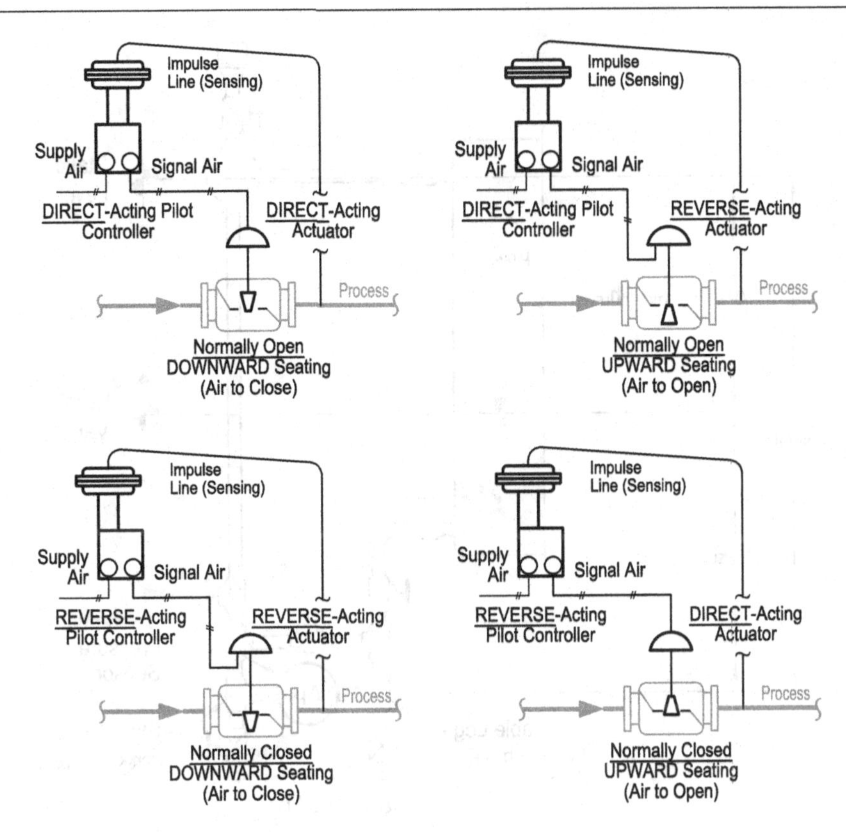

Figure 9.16 Fail-open and fail-close configurations for control valves using combinations of direct- and reverse-acting pilot controllers and actuators, and upward and downward seating valves. (Adapted from CIRCOR | Leslie.)

regulator must be configured to fail closed. Spraying unburned fuel into the furnace upon loss of the control system could create a possibly explosive condition and is very unsafe.

LEVEL CONTROL

Continuous level control is often accomplished accurately and reliably by differential pressure measurement, although other techniques are available. Tank differential pressure measurement compares the pressure between a static leg and a variable leg. The static leg is a reference leg that allows tank pressure variations to occur without affecting the level measurement. The static leg is a high connection on the tank that is filled with liquid to its overflow level, thus creating a fixed-height or static column of fluid. The variable leg is formed by making a tank connection below the lowest anticipated water level. The variable leg, which moves with level changes, applies the fluid height to the opposite side of the Δp sensor. The differential pressure sensing device must be physically located below the lowest measurement level. Figure 9.17 shows the hydrostatics that produce the pressure-differential measurement used for determining tank liquid level. The figure shows that the tank internal pressure is equal, opposite, and cancels across the Δp sensor leaving only the height difference (Δh) between the static and variable legs, so that the pressure differential is $\Delta p = \rho g \Delta h = \lambda \Delta h$.

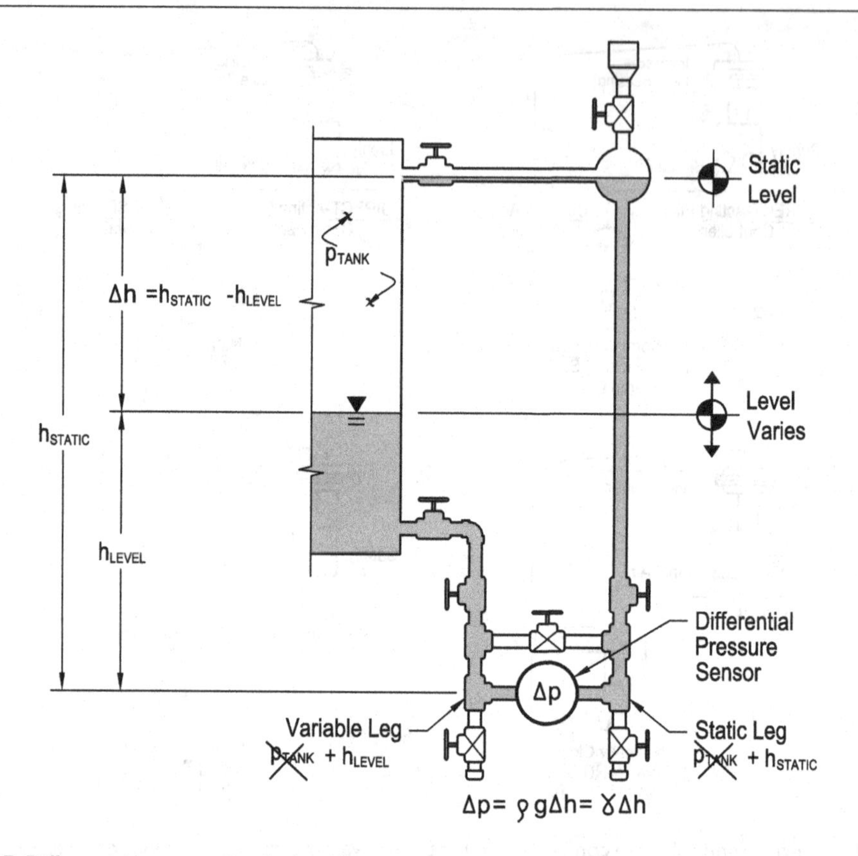

Figure 9.17 Differential pressure is obtained between static and variable legs. The tank pressure is applied equally to both sides of the Δp sensor, and varying tank-pressures are effectively cancelled, leaving only the difference in liquid heights.

In cases where the tank is vented to the atmosphere, the static-leg connection is not necessary, and the static-leg connection can be left open to the ambient air. Both the tank fluid-level and sensor static connections are exposed to the same pressure. In those installations, the pressure resulting from the tank-level is adequate for level control.

Figure 9.18 shows a differential-pressure pilot level controller. The actuator is fitted with two connections, one to the upper side of the sensing diaphragm, and the other to the underside of the diaphragm. The differential pressure arrangement shown is direct acting when the variable leg is connected to the upper side of the diaphragm. The lower side senses the static-leg pressure. As the level rises, the diaphragm deflects downward, and the output load-air signal increases.

Figure 9.19 and Figure 9.20 show some of many level-control options for pressure or vacuum tanks. Figure 9.19 shows a low-level makeup arrangement, where a low tank level commands a control valve to open and fill the tank. Combinations of direct and reversing devices produce the failsafe direction of the makeup valve, which will either fail open and flood the tank (Figure 9.19 left) or fail closed and empty the tank (Figure 9.19 right) upon loss of signal air. Figure 9.20 shows a high-level dump arrangement, where a high-tank level causes a control valve to open and dump fluid from the system. Combinations of direct- and reverse-acting devices can result in open or closed failsafe conditions upon loss of signal air.

Figure 9.21 shows a condenser hotwell level-control system for a steam plant. Refer to Chapter 8 for more discussion regarding condensate pumping systems. This system is

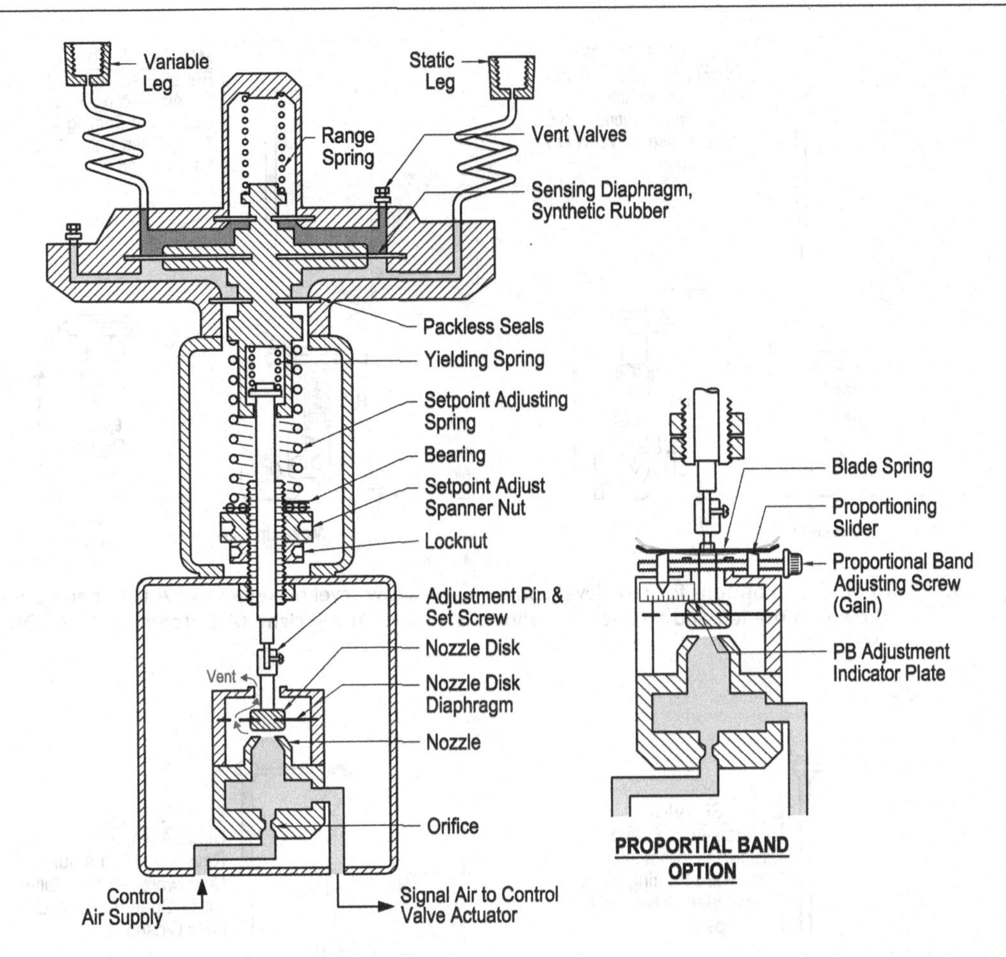

Figure 9.18 Differential-pressure pilot controller used for level-control applications. (Adapted from CIRCOR | Leslie.)

important for low-flow conditions to avoid operating the condensate pump dry. At low steaming loads, the control valve diverts condensate to recirculate back to the condenser in lieu of being forwarded to the deaerator, thus maintaining continuous flow, even before the steam turbine is placed in operation. When the pump draws condensate faster than steam enters, the level begins to drop. The control system recognizes the decreasing level and begins to open the recirculation valve. To avoid recirculating too much condensate, an orifice is usually installed in the recirculation line. The orifice is sized to permit enough condensate flow through the pump impeller to avoid cavitation as the motor power heats the liquid churning within. It is also common practice to install a thermostatically controlled recirculation valve, whose purpose is to ensure that there is enough coolant through the condensate system steam heaters. High condensate temperature indicates insufficient cooling, and more condensate is recirculated. Good practice is to crack the hand recirculation valve at low flow conditions to mitigate delays in the control-valve response, and to close the hand valve when steady flow conditions are reached.

Where tank or system levels are to be held constant, makeup and dump pilot controllers are commonly ganged together to address both high and low disturbances. Figure 9.22 shows a distilled water makeup and dump system used in steam plants. The makeup valve adds

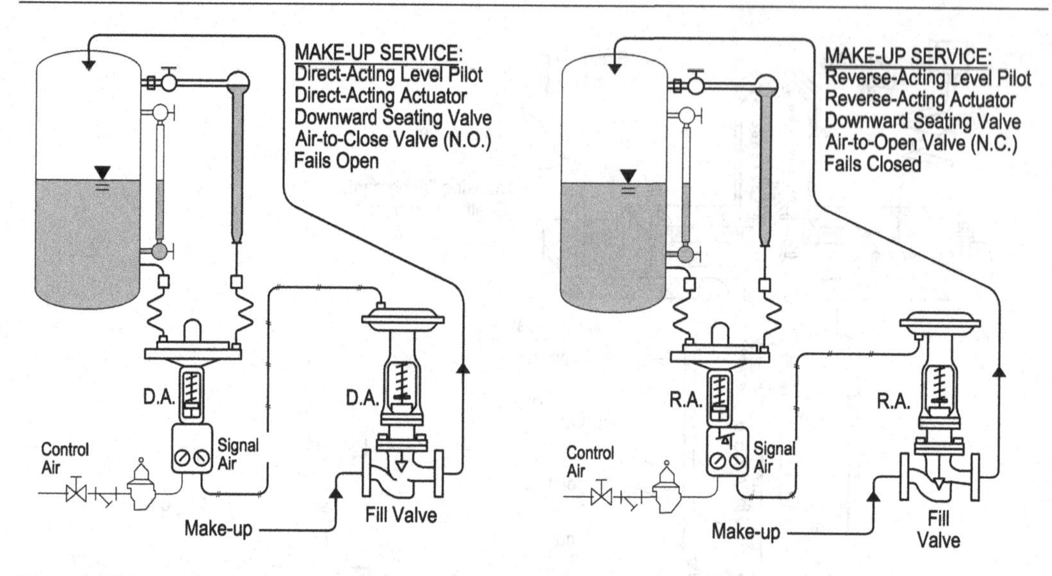

Figure 9.19 Two of several options for tank level control using a low-level makeup valve. A fail-open option is shown on the left and a fail-closed option is shown on the right. (Adapted from CIRCOR | Leslie.)

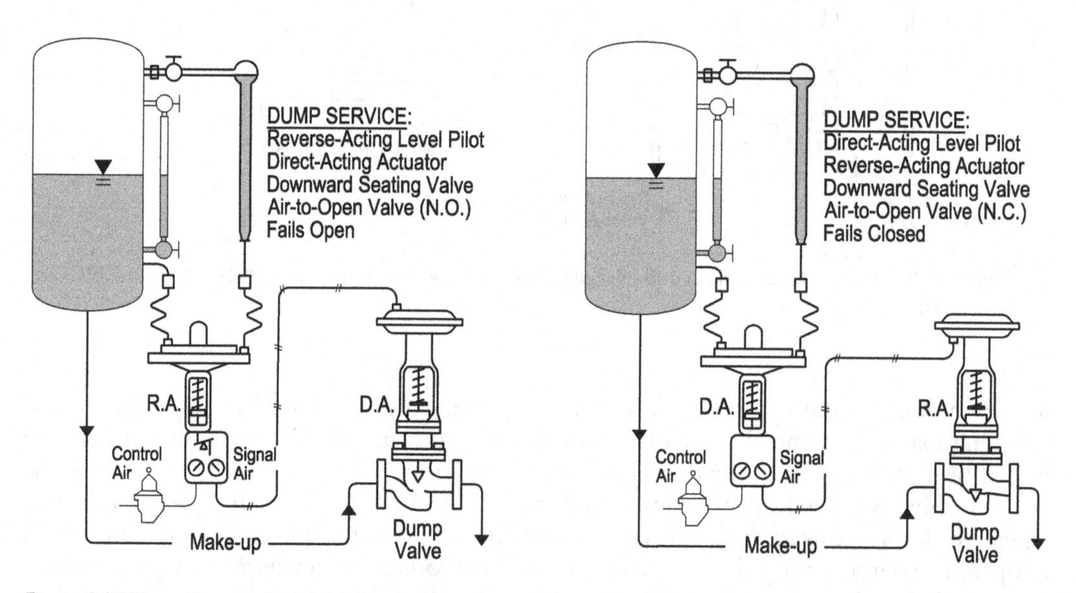

Figure 9.20 Two of several options for tank level control using a high-level dump valve. The fail-open option is shown on the left and the fail-closed option is shown on the right. (Adapted from CIRCOR | Leslie.)

water to compensate for losses, and the dump valve removes excess water from the system. In this arrangement, the pilot controllers share the same sensing connections. The two pilots are located adjacent to each other, and in sight of the level gage glass for ease of adjustment. The pneumatically actuated valves would be located remotely, convenient to the system layout, and to minimize piping runs. High- and low-level alarms would be installed on the tank to alert the operators to conditions beyond the normal range, and typically the deaerator would

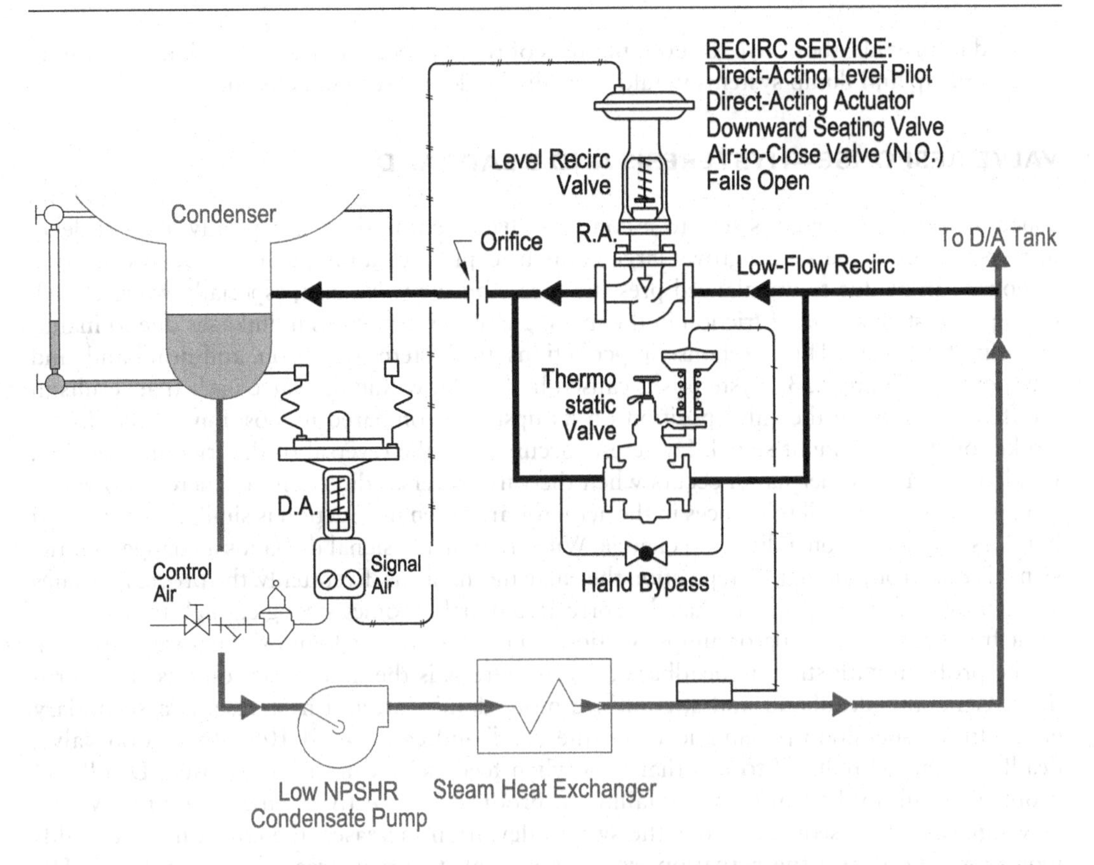

Figure 9.21 Level control using a low-flow condensate recirculation system that senses hotwell-level. Note the thermostatic recirculation valve is added to ensure enough coolant flows through the heat exchanger.

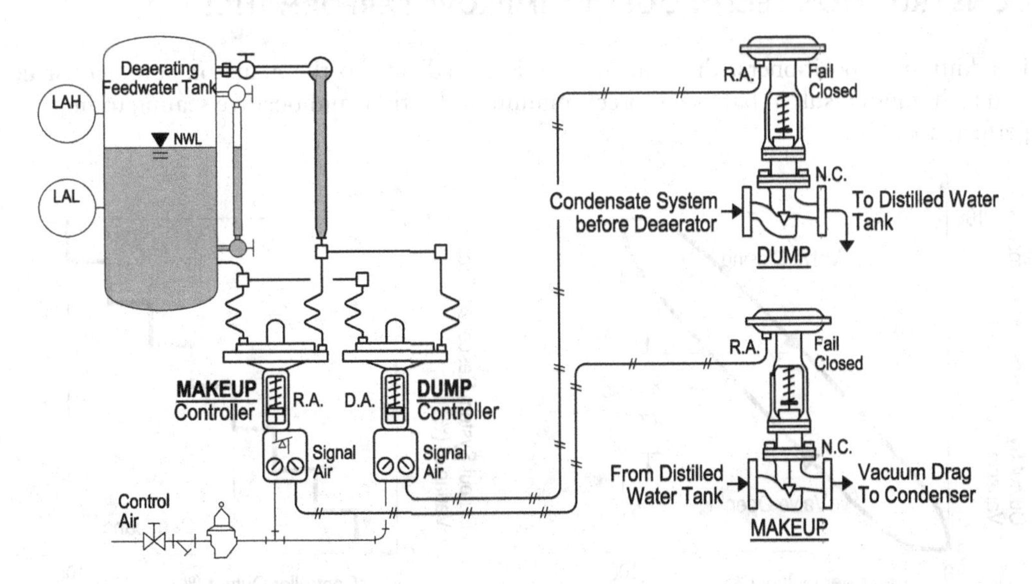

Figure 9.22 Distilled water makeup and dump system used on a steam plant. The same sensing connections are input into both controllers, one controller for replacing water losses and the other for addressing excessive water conditions.

be sized to provide about five to seven minutes of reserve feedwater at the designed steaming rate. Makeup and dump systems would typically be designed to fail closed.

VALVE RESPONSE—HYSTERESIS AND DEADBAND

When a pneumatic signal is sent to a control-valve actuator, there is typically a small delay in response, until the signal grows large enough to produce an adjustment. Delays in valve response can be due to unbalanced pressure forces on the valve disk, especially when closed, the slip-and-stick action of friction in the moving parts, or looseness in linkages due to manufacturing tolerances. Three response imperfections are hysteresis, stiction, and deadband, and are shown in Figure 9.23. Hysteresis is caused by friction within the valve itself that results in a difference between the valve position on its upstroke compared its position on the downstroke for the same input signal. Hysteresis occurs from the reversal of the friction direction. Deadband, on the other hand, occurs when the valve reverses direction but there is no valve-stem movement until all tolerances in the actuator are taken up. Stiction is similar to deadband but does not depend on a direction change. With stiction, the signal demands a change, but the signal is not strong enough to reposition the valve mechanisms. Eventually, the process deviates sufficiently from its setpoint so that the corrective signal becomes strong enough to overcome static friction, and the actuator jumps. Stiction is a result of the stick-and-slip nature of friction.

The problem with stiction, deadband, and hysteresis is the non-linear response they produce. Generally, deadband and stiction are more dominant, and hysteresis is a secondary effect. In an open-loop pneumatic valve, the deadband can exceed 10%. In a good valve, deadband can be reduced to less than 1% when feedback positioners are used. Deadband problems result in the control valve failing to produce a corrective change upon receiving a new signal, and subsequently after the system deviation increases, the controller then adds more correction until the actuation signal is enough to break free of the deadband. The delayed overreaction force often causes overshoot.

CONSTRUCTION TECHNIQUES TO IMPROVE PERFORMANCE

In addition to positioners, that add mechanical feedback to the actuator, some construction techniques result in balanced forces, minimized friction, and positive sealing to improve performance.

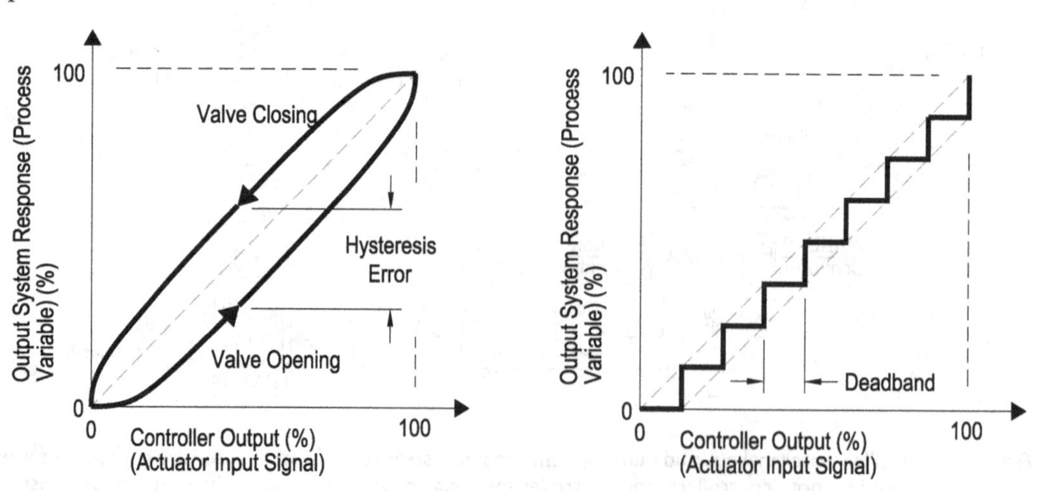

Figure 9.23 Hysteresis (left) and deadband (right) adversely affect system response.

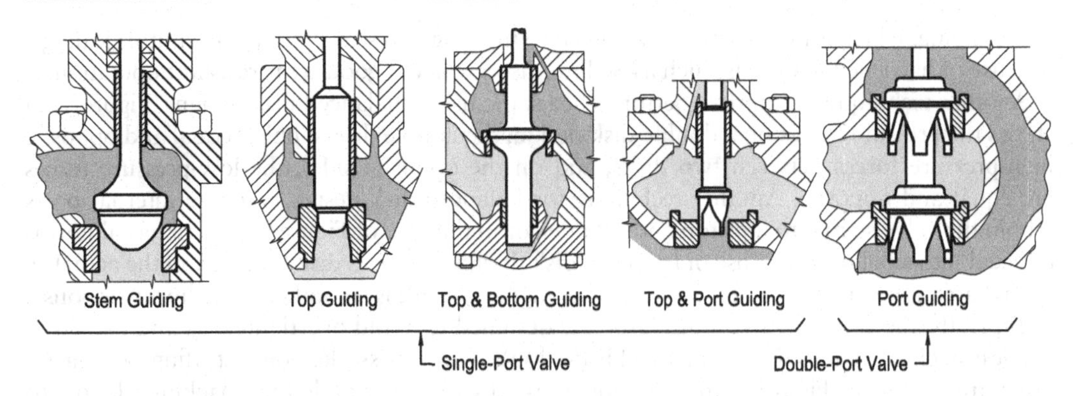

Figure 9.24 Plug guiding methods. Good disk-to-seat alignment results in better shut-off sealing, longer valve-trim life, and higher reliability. (Adapted from CIRCOR | Leslie.)

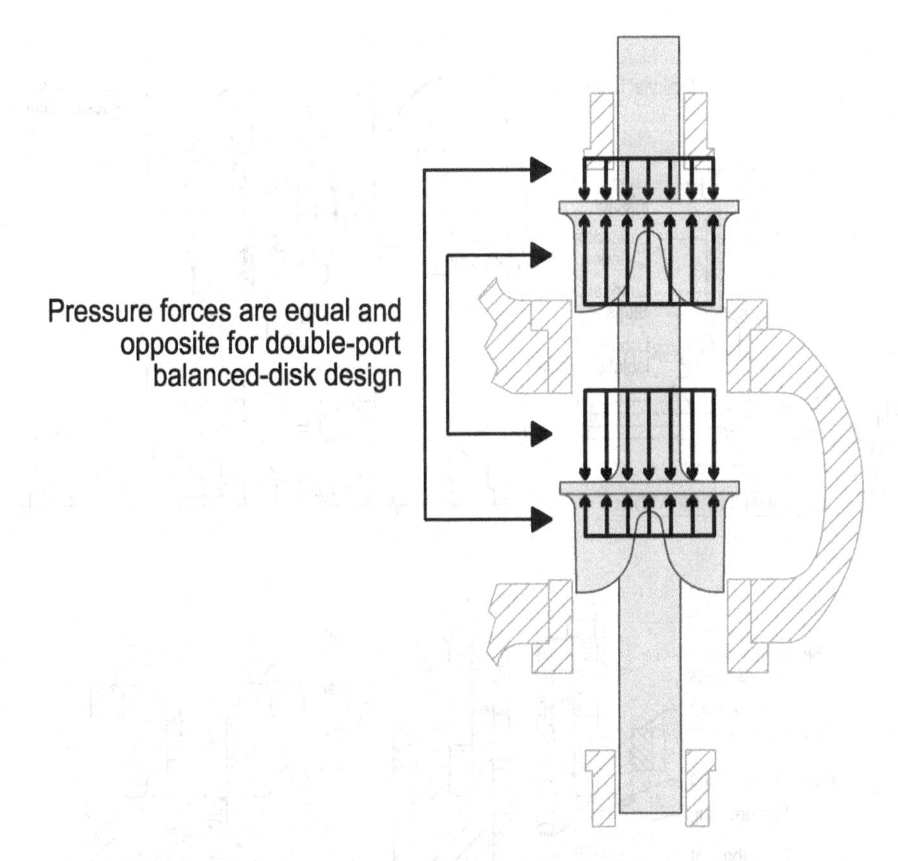

Figure 9.25 Balanced-disk design. Actuator forces only need to overcome friction when internal hydraulic forces cancel each other, resulting in smoother valve operation.

One technique to improve performance is to guide the valve plug so that its movement is completely perpendicular to the valve seat. Some techniques for guiding the plug are shown in Figure 9.24, including stem guiding, top guiding, top-and-bottom guiding, top-and-port guiding, port guiding, and double-port guiding. Precise seating of the plug against the seat avoids wiredrawing, which is fluid friction that cuts into the plug and seat, resulting in shutoff leakage. The use of hardened materials, such as Stellite, also produces good wear-resistant operation.

Another technique to improve performance is the use of balanced disk designs. Figure 9.25 shows the pressure forces on a double-disk design, the double disk supplies equal and opposite high-pressure forces between two disks, and on the opposite sides, the low-pressure forces are equal and opposite. Another technique for balanced-disk design is to use internal ports within the valve stem into a piston and cylinder, or in the case of Figure 9.3, the internal port is directed into a bellows. The use of balanced disk designs removes disk forces from the actuator, so that actuator can focus its on valve positioning. The result is smoother, more linear response.

Typically, the control-valve stuffing-box is designed to avoid overtightening and uses low-friction packing materials, such as PTFE (Teflon), where possible. Some stuffing-box variations are shown in Figure 9.26. It is not unusual to use spring-loaded packing glands, to control the packing compression forces. The figure shows Belleville washers used as compression springs outside the stuffing box and a helical spring, installed within the gland. In some control valves, an external lubricator can be installed. The lubricator uses an injector that

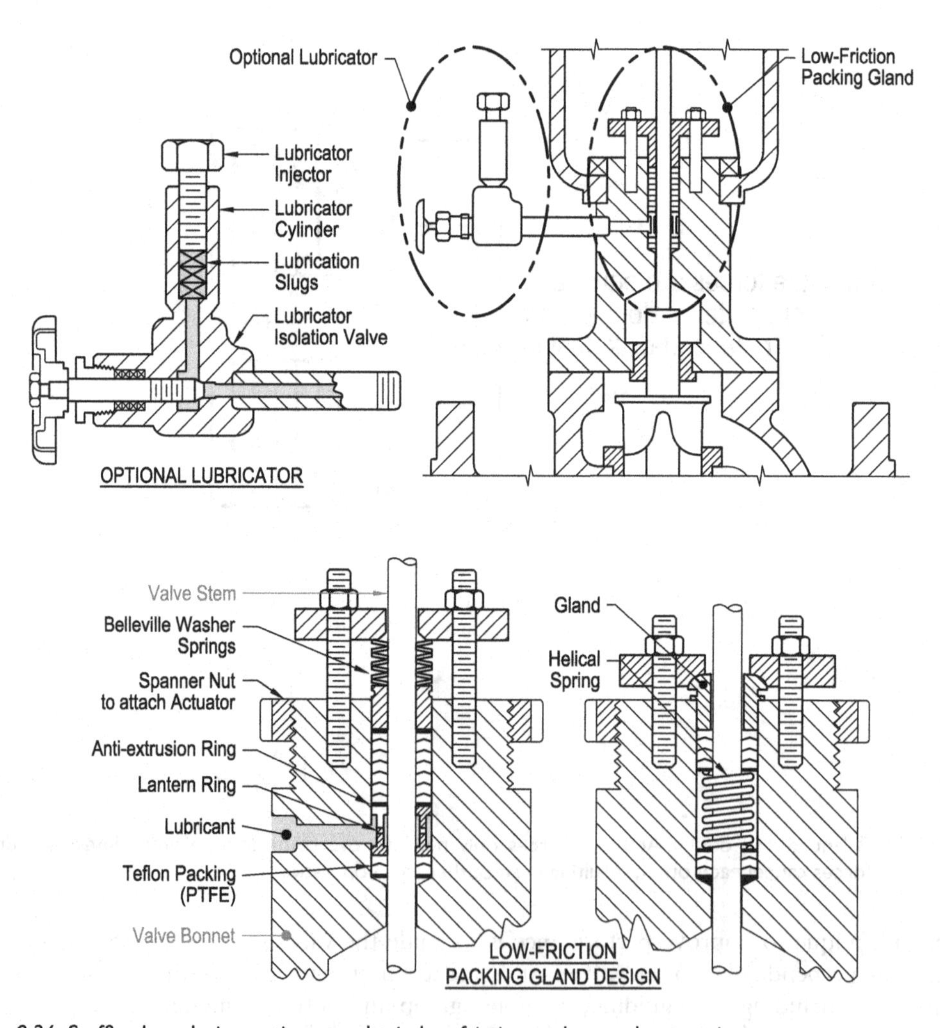

Figure 9.26 Stuffing box design options to obtain low friction and smooth operation.

compresses slugs of lubricant inserted into a cylinder. A hand valve is opened, and lubricant is extruded through a lantern ring located in the stuffing box by screwing down on the injector bolt-head. After the lubricant is injected, the hand valve is closed.

SIZING CONTROL VALVES FOR LIQUID APPLICATIONS

Sizing control valves typically begins with a thermodynamic analysis to determine the mass flow requirements. However, real-world devices, such as control valves, are selected on a volume basis. As discussed in Chapter 8, the relationship between flow and system head loss is close to a parabolic relationship. This parabolic relationship where $\Delta p \propto Flow^2$ applies to control valves as well. Manufacturers use a valve coefficient or Cv to describe valve size, where Cv is the constant of proportionality in the parabolic relationship. The Cv is defined as the constant of proportionality that produces a 1 psi drop when the flow rate is 1 gpm. Cv accounts for fluid density and is determined at multiple positions between mostly closed to fully open. The Cv relationships are:

$$Q = Cv \cdot \sqrt{\frac{\Delta p}{s.g.}} \quad or \quad Cv = Q \cdot \sqrt{\frac{s.g.}{\Delta p}} \quad or \quad Cv = \frac{\dot{m}_{wtr}}{500 \cdot s.g.} \cdot \sqrt{\frac{s.g.}{\Delta p}}$$

where Q is the flow rate in gpm
Cv is the valve flow coefficient, corresponding to the flow rate in gpm that produces a 1 psi pressure drop
Δp is the pressure drop in psi
$\dot{m}_{wtr}$ is the mass flow of the liquid in lbm/hr
$s.g.$ is the liquid specific gravity

To further understand control valve behavior, the Cv value is different for every valve-stem position between fully closed to fully open, so the Cv needs to be characterized over the entire range of valve positions. Moreover, the control valve flow characteristics can be modified simply by changing the valve trim, i.e., by using different shaped disks and seats. A single valve body can be fitted with multiple trim options that produce very different behaviors. The manufacturer performs factory tests to measure the Cv values at valve-position increments and the measured flow rates represent the control valve only, without any influence from an attached piping system. The testing yields the inherent characteristics of the valve, which does not include any other influence, such as an attached piping-system. Table 9.1 shows design values at various disk positions as an example.

Figure 9.27 shows the generalized inherent characteristics for the three most commonly used control valves; quick-opening, linear, and equal-percentage, and Figure 9.28 shows the flow characteristics for various cages forming the valve seats. The inherent characteristics of ball and butterfly valves are included, as these types of valves are also used in control applications. The figure also illustrates three examples of valve-plug shapes that produce those characteristics; however, there are plug-shape variations that produce these flow characteristics along with other features such as balanced forces, noise-reducing flutes, or staged-pressure drops to avoid cavitation damage. Note that these curves show only the valve performance, without any external influence from the system for which it is intended, and performance variations will occur depending on the exact make and model valve.

The true behavior of the control valve is a function of the interaction between its inherent flow and pressure-drop characteristics and the piping system in which it is installed.

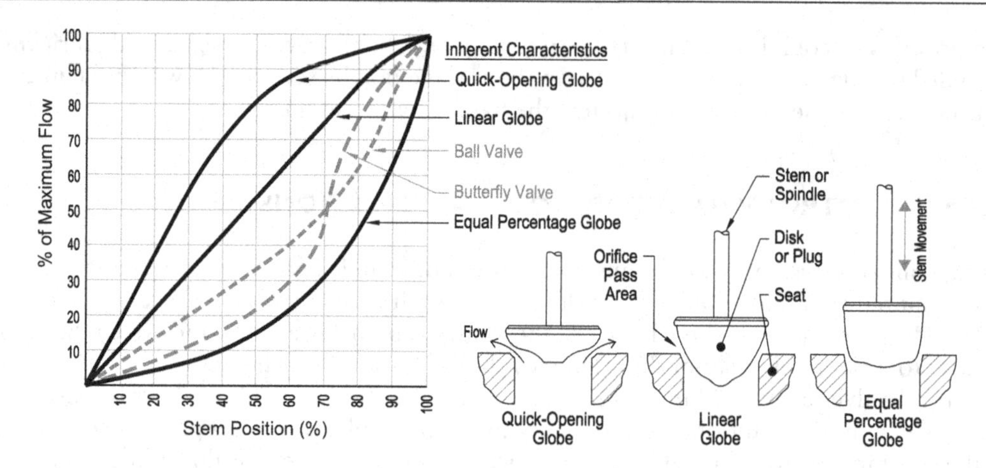

Figure 9.27 Inherent flow characteristics for different valve plugs. When placed in service, the control valve inherent characteristics interact with the system dynamics to produce an installation curve.

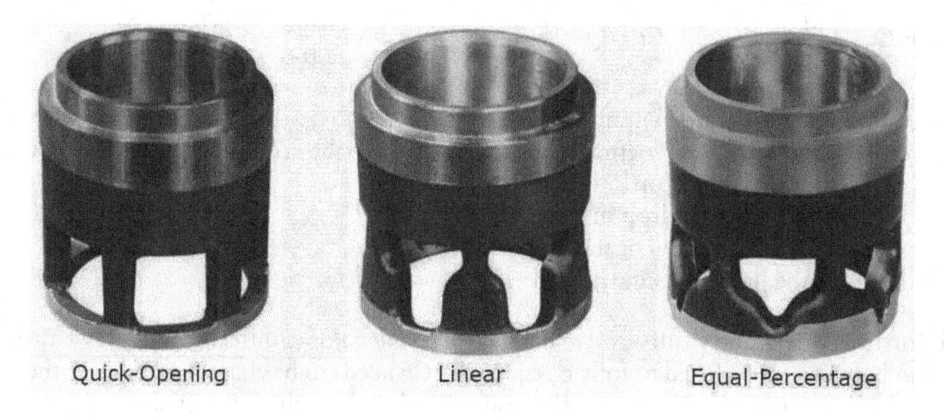

Figure 9.28 Inherent flow characteristics obtained using different cage openings with globe-type plugs. (Courtesy of Emerson from Fisher Controls International LLC, *Control Valve Handbook*, 5th edition, Emerson, 2019.)

Combining the control valve characteristics together with the system behavior yields the *installation curve*. The installation curve represents the true behavior of the valve when it is installed in the system, versus its behavior in a test fixture. The goal of selecting the trim is to produce accurate control, control-loop stability, and a control process whose pressure, temperature, level or flow control varies as linearly as possible. Ultimately, the engineer needs to understand the system and its behavior at different flow rates to select the disk shape. Also, the valve size is based on the *Cv* required at a strategic stem position, such as, with the valve designed to be 50% open for a variable-speed pump or 85% open for a constant-speed pump, at maximum flow conditions.

Linear valves have the inherent flow characteristics that are represented by a straight line when flowrate is plotted against stem travel, assuming that the pressure drop across the valve is constant. In this case, the valve gain K_v has a constant slope, and the control sensitivity is similar when the valve is nearly closed or nearly full open. When the valve is half open, the flow will be at half flow, and likewise for all other stem positions.

Table 9.1 Control valve sizing table for Fisher Design ED, ANSI Class 125–600 using a Linear Cage

Linear																*Linear characteristic*
Valve Size, NPS	*Port Diameter*		*Maximum Travel* [2]		*Flow Coeffi-cient*	*Valve Opening – percent of total Travel*										
	Mm	Inches	mm	Inches		10	20	30	40	50	60	70	80	90	100	$F_{L(1)}$
					C_v	3.21	5.50	8.18	10.9	13.2	15.0	16.6	18.6	19.9	20.6	0.84
1 & 1-1/4	33.3	1.3125	19	0.75	K_v	2.78	4.76	7.08	9.43	11.4	13.0	14.6	16.1	17.2	17.8	...
					X_T	0.340	0.644	0.494	0.509	0.532	0.580	0.610	0.620	0.628	0.636	...
	47.6	1.875	19	0.75	C_V	4.23	7.84	11.8	15.8	20.4	25.3	30.3	34.7	37.2	39.2	0.82
					K_V	3.66	6.78	10.2	13.7	17.6	21.9	26.2	30.0	32.2	33.9	...
					X_T	0.656	0.709	0.758	0.799	0.738	0.729	0.708	0.686	0.683	0.656	...
1-1/2					F_d	0.30	0.37	0.41	0.44	0.44	0.41	0.38	0.35	0.34	0.34	...
	33.3	1.3125	19	0.75	C_V	2.92	5.70	9.05	12.5	15.6	18.5	21.1	23.9	26.8	29.2	0.91
					K_V	2.53	4.93	7.83	10.8	13.5	16.0	18.3	20.7	23.2	25.3	...
					X_T	0.690	0.651	0.633	0.634	0.650	0.666	0.708	0.718	0.737	0.733	...
	58.7	2.3125	29	1.125	C_V	7.87	16.0	24.9	33.4	42.1	51.8	62.0	68.1	70.6	72.9	0.77
					K_V	6.81	13.8	21.5	28.9	36.4	44.8	53.6	58.9	61.1	63.1	...
					X_T	0.641	0.720	0.728	0.767	0.793	0.754	0.683	0.658	0.652	0.638	...
2					F_d	0.30	0.35	0.36	0.37	0.37	0.36	0.35	0.35	0.34	0.33	...
	33.3	1.3125	19	0.75	C_V	3.53	6.36	9.92	13.3	16.5	19.7	22.7	25.6	29.3	33.3	0.87
					K_V	3.05	5.50	8.58	11.5	14.3	17.0	19.6	22.1	25.3	28.8	...
					X_T	0.456	0.529	0.549	0.582	0.611	0.633	0.671	0.723	0.727	0.694	...
	73.0	2.875	38	1.5	C_V	9.34	21.6	35.5	49.5	62.7	74.1	83.6	93.5	102	108	0.81
					K_V	8.08	18.7	30.7	42.8	54.2	64.1	72.3	80.9	88.2	93.4	...
					X_T	0.680	0.660	0.644	0.669	0.674	0.706	0.716	0.687	0.658	0.641	...
2-1/2					F_d	0.27	0.33	0.35	0.36	0.35	0.34	0.32	0.29	0.27	0.27	...
	47.6	1.875	19	0.75	C_V	4.10	8.09	12.3	16.7	21.1	26.8	33.7	41.3	49.2	57.0	0.84
					K_V	3.55	7.00	10.6	14.4	18.3	23.2	29.2	35.7	42.6	49.3	...
					X_T	0.668	0.646	0.684	0.688	0.698	0.694	0.678	0.668	0.669	0.666	...

(continued)

Table 9.1 (Cont.)

Linear															*Linear characteristic*
Valve Size, NPS	*Port Diameter*		*Maximum Travel* (2)		*Flow Coeffi-cient*	*Valve Opening – percent of total Travel*									
	Mm	Inches	mm	Inches		10	20	30	40	50	60	70	80	90	100	$F_{L(1)}$
	87.3	3.4375	38	1.5	C_V	14.5	32.9	52.1	70.4	88.5	105	118	133	142	148	0.82
					K_V	12.5	28.5	45.1	60.9	76.6	90.8	102	115	123	128	...
					X_T	0.671	0.699	0.697	0.720	0.733	0.718	0.707	0.650	0.630	0.620	...
3					F_d	0.26	0.32	0.35	0.36	0.36	0.36	0.36	0.28	0.29	0.30	...
	58.7	2.3125	29	1.125	C_V	8.06	16.9	26.7	37.5	49.0	61.4	73.8	85.3	94.7	102	0.85
					K_V	6.97	14.6	23.1	32.4	42.4	53.1	63.8	73.8	81.9	88.2	...
					X_T	0.592	0.614	0.662	0.672	0.674	0.676	0.694	0.722	0.736	0.732	...
	111.1	4.375	51	2	C_V	23.3	50.3	78.1	105	127	152	181	203	223	236	0.82
					K_V	20.2	43.5	67.6	90.8	110	131	157	176	193	204	...
					X_T	0.692	0.714	0.720	0.731	0.764	0.757	0.748	0.762	0.732	0.688	...
4					F_d	0.31	0.36	0.38	0.38	0.37	0.35	0.32	0.30	0.27	0.28	...
	73.0	2.875	38	1.5	C_V	9.77	22.6	37.2	51.8	65.7	77.5	87.5	97.9	107	113	0.84
					K_V	8.45	19.5	32.2	44.8	56.8	67.0	75.7	84.7	92.6	97.7	...
					X_T	0.926	0.926	0.899	0.873	0.904	0.919	0.962	0.937	0.891	0.872	...

Notes: refer to ISA-75.01.01 for a description of using XT, FL, and Fd coefficients for valve performance calculations. Refer to manufacturer catalogs as valve trim and valve coefficients are subject to change.

Source: Adapted from Emerson.

Quick-opening valves have a rather flattened valve disk that is shaped to provide large flowrate changes when the valve is nearly closed. Typically, a quick-opening valve may pass 80-90% of the total flow when it reaches 50% of the stem lift. This valve behavior approaches on-off operation, and the valve gain (K_v) near shutoff is typically too high for fine modulating control. Quick-opening valves may be appropriate for tank-overflow or deluge applications, where its operation is intermittent, and the valve needs to react quickly and decisively.

Equal-percentage valves have contoured plugs that produce logarithmic characteristics so that the percent *change* of valve-stem position produces a corresponding equal percentage of *flow change*. The mathematical relationship between valve lift and flowrate/orifice size is:

$$Q(l) = Cv_{\max} R^{(l-1)} \sqrt{\frac{\Delta p}{s.g.}} \quad or \quad Q(l) = \frac{R^h}{R} \cdot Q_{\max}$$

where $Q(l)$ – flow at the valve lift position
l – valve lift in decimal percent, where $0 \leq l \leq 1.0$
R – Rangeability—the ratio of the maximum to minimum controllable flowrate

Equal-percentage valves have gain that is low and relatively insensitive when the valve is barely open, and the gain/responsiveness increases rapidly at the high flows near the upper excursion of stem travel. The exact performance of a valve is a function of its rangeability, and the graph in Figure 9.29 shows curves based on several rangeability values relative to a linear characteristic. The table values on the right are provided for an equal-percentage valve having a rangeability 50:1.

The manufacturer measures the control-valve inherent characteristics using a fixed differential pressure across the valve under laboratory test conditions. In real plant systems, the actual pressure drop across the control valve often varies, in which case the Cv becomes dynamic as the system interacts with the valve. Figure 9.30 shows a centrifugal pump curve

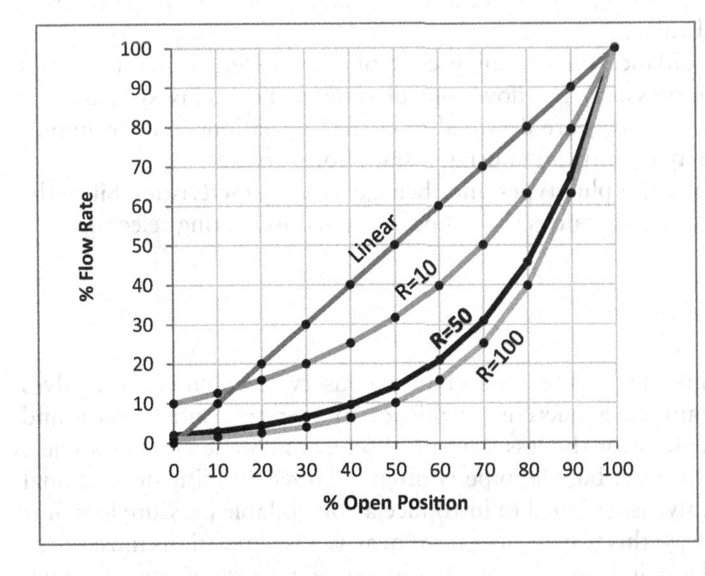

Valve Stem Position (% Open)	Flow Rate (% of Full Open Flow)	% Increase in Flow from Previous
0%	2.00	
10%	2.96	48%
20%	4.37	48%
30%	6.47	48%
40%	9.56	48%
50%	14.14	48%
60%	20.91	48%
70%	30.92	48%
80%	45.73	48%
90%	67.62	48%
100%	100.00	48%

based on Rangeability = 50

Figure 9.29 The mathematics behind an equal-percentage valve is shown in the table. The calculations are shown for a rangeability of 50, meaning a turndown ratio of 50:1, representing the minimum controllable flowrate just before shutoff.

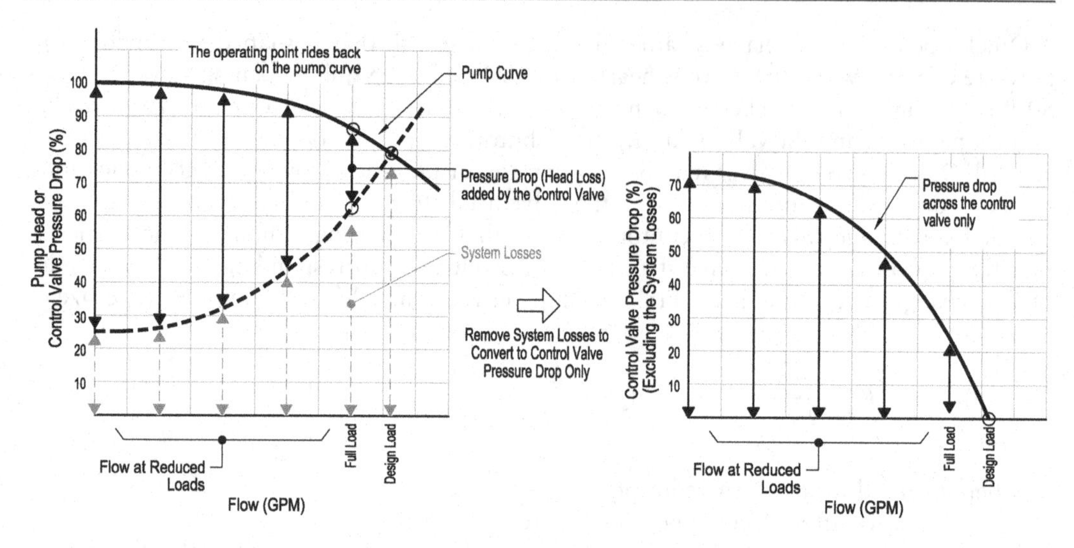

Figure 9.30 Pressure-drop or head-loss requirements are shown for a throttle-valve capacity-controlled pumping system. The pump curve on the left shows the parabolic system-resistance curve, which intersects with the pump curve. The control valve resistance at reduced flowrates adds to the system curve, which keeps the operating point on the pump curve as the valve chokes down. The curve on the right shows the head loss across only the control valve. This non-linearity makes an equal-percentage valve a good option for straightening out the installation curve and making the system response more linear.

with a discharge valve being throttled to modulate flow. The operating point on the pump curve at any flow rate is the intersection of the pump curve with the system-resistance curve *plus* the control-valve pressure drop (Figure 9.30, left). The illustration on the right side of Figure 9.30 shows only the control-valve pressure drop, which is very non-linear. To be reliable, repeatable, and robust, the control-valve flow characteristics should compensate for the system non-linearity as much as possible, to produce a near-linear response for the composite effect of the system and control valve.

Table 9.2 provides general guidance for matching control-valve inherent characteristics with various temperature, level, pressure, and flow-control systems. For many systems, this guidance is satisfactory, as control valves are designed to correct deviations, but for important systems, better selections may be warranted using installation curves.

Table 9.3 shows some control-valve plug types and their generic characteristics. Since the data is product-specific, manufacturer catalogs should be consulted in making selections.

INSTALLATION CURVE

There is often a tendency to think bigger is better and more conservative, but control valves need to be sized properly. In many cases, there is a tendency to oversize control valves and rely on the controller to compensate for the less-than-ideal selection. Some engineers select the valve body to match the pipe size, but the pipe is often enlarged to mitigate frictional losses. Conversely, the control valve is designed to introduce a controllable pressure loss. It is not unusual for valves that are sized this way to operate at nearly closed positions most of the time, and the oversized valves lack fineness of control. In practice, oversizing control valves is probably the biggest source of operational problems. The increased sensitivity caused by operating over a very small span of stem travel due to oversizing often leads to instability and

Table 9.2 General guidance for selecting control-valve flow characteristics

Temperature control systems

CONTROL VALVE APPLICATION	RECOMMENDED INHERENT CHARACTERISTIC
Temperature Control—all applications	Equal-Percentage

LEVEL CONTROL SYSTEMS

CONTROL VALVE PRESSURE DROP	RECOMMENDED INHERENT CHARACTERISTIC
Constant ΔP	Linear
ΔP decreases with increasing load ΔP Max Load > 0.20 x ΔP Min Load	Linear
ΔP decreases with increasing load ΔP Max Load < 0.20 x ΔP Min Load	Equal-Percentage
ΔP increases with increasing load ΔP Max Load < 2.0 x ΔP Min Load	Linear
ΔP increases with increasing load ΔP Max Load > 2.0 x ΔP Min Load	Quick Opening

PRESSURE CONTROL SYSTEMS

CONTROL VALVE FLUID TYPE & PRESSURE DROP	RECOMMENDED INHERENT CHARACTERISTIC
Liquids	Equal-Percentage
Gas, small volume past control valve, typically less than 10-ft of pipe length	Equal-Percentage
Gas, large volume past control valve, typically receiver or greater than 100-ft of pipe length ΔP decreases with load ΔP Max load > 0.20 x ΔP Min load	Linear
Gas, large volume past control valve, typically receiver or greater than 100-ft of pipe length ΔP increases with load ΔP Max load < 0.20 x ΔP Min load	Equal-Percentage

FLOW CONTROL SYSTEMS

FLOW MEASUREMENT SIGNAL	LOCATION OF THE CONTROL VALVE	WIDE RANGE OF FLOW SETPOINTS	NARROW RANGE OF FLOW SETPOINTS, *AND* LARGE ΔP CHANGES WITH INCREASING LOAD
Signal is Proportional to Flow	Throttling	Linear	Equal-Percentage
	Bypass	Linear	Equal-Percentage
Signal is Proportional to Flow Squared	Throttling	Linear	Equal-Percentage
	Bypass	Equal-Percentage	Equal-Percentage

Source: Courtesy of Emerson: Fisher Conrtols International LLC, *Control Valve Sourcebook—Oil and Gas*, Emerson, 2013.

Table 9.3 Partial representation of control valve properties for illustrative use

Valve plug type	*Size range*	*ANSI valve class (Max)*	*Valve size (NPT)*	*Approx wide-open Cv*	*Typical pressure differential*	*Approximate rangeability*
Globe	1"-24"	2500	1"	12	100 psid	20:1 to 50:1
			1.5"	30	100 psid	
			2"	47	100 psid	
			2.5"	63	60 psid	
			3"	100	45 psid	
			4"	160	25 psid	
			6"	400	12 psid	
90° Rotary Segmented Ball/ Plug Valve	1"-24"	600	1"	24	200 psid	Up to 300:1
			1.5"	55		
			2"	77		
			3"	207		
			4"	350		
			6"	507		
90° Rotary Ball	1"-12"	900	1"	110	50 psid	Up to 300:1
			1.5"	350		
			2"	600		
			3"	1330		
			4"	2420		
Butterfly & High-Performance Butterfly	2"-96"	600	2"	159	75 psid	10:1 to 35:1
			2.5"	266		
			3"	457		
			4"	860		
			5"	1320		
			6"	2020		
			8"	3540		
Double-Seated or Balanced Globe	1–14"	600	1"	19	150 psid	50:1 to 100:1
			2"	34		
			2.5"	60		
			3"	84		
			4"	211		

Note: Refer to manufacturer catalogs for actual performance.

Source: Adapted from Emerson.

hunting, and possible wire-drawing of the disk and seat. In practice, control-valve bodies are usually, but not always, one or two NPT pipe sizes smaller than the line size, to provide fineness of control over a large travel range. Although less common, undersized control valves should also be avoided, or the valve may not permit the required flow to pass at the system maximum load. Additionally, the valve inherent characteristics need to be appropriately matched to the system dynamics to provide stable in-service operation. It is important that the engineer selecting the control valve understands the plant system and its typical operating scenarios, and it may be prudent to produce installation curves for important systems to make the best valve selection.

Figure 9.27 shows manufacturer-developed inherent characteristics for different control-valve types before the valve is installed into a system, and the right-side curve in Figure 9.30 shows the system losses required from the control valve. In practice, the behaviors of the valve and system can be married together to provide an installation curve, where the goal is to provide near-linear system response. Table 9.2 provides general guidance that is usually adequate for good operation. However, in some cases, it may be

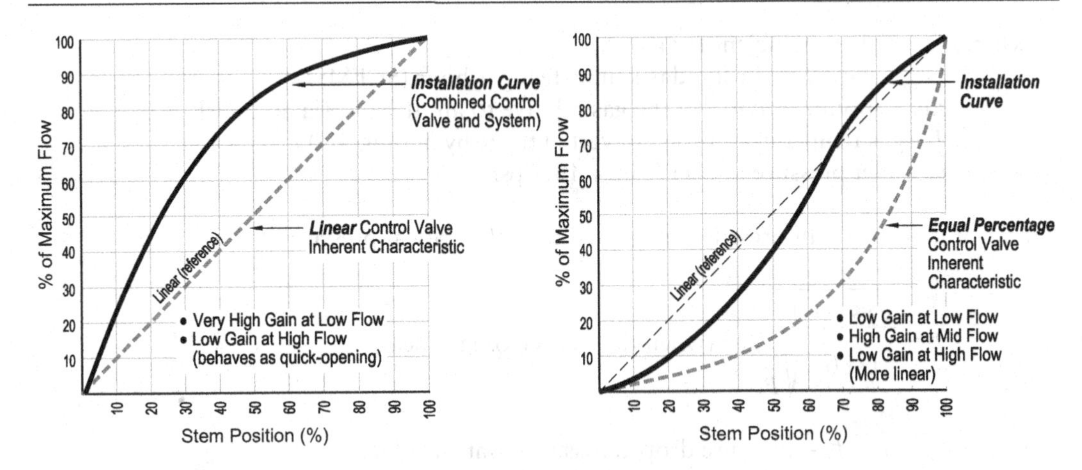

Figure 9.31 Two installation curves are provided for the same piping system having moderate frictional losses. The installation curve is formed by combining the control-valve characteristics with the piping-system curve. The left installation curve combines a linear valve with the system to approach quick-acting behavior, possibly for emergency cooling. The right installation curve is formed using an equal percentage valve, which results in more linear operation. This more-linear installation curve exhibits precise characteristics at low and high flow rates, with faster response at mid-flow.

desirable to more precisely determine the valve characteristic when it is installed in the system. Installation curves can be produced by using the manufacturer table values of Cv at different stem positions combined with the pump/system curves at corresponding flows. The installation curve is of interest between the normal minimum and maximum system flow requirements. In general, the control valve selection should be for a valve whose Cv corresponds to 75-85% open at peak flowrate is and 10-20% open at minimum flow. Figure 9.31 shows the resulting installation curve for the same system using either a linear or equal-percentage valve.

SIZING CONTROL VALVES FOR COMPRESSIBLE-FLOW APPLICATIONS

The Cv values in tables are based on incompressible fluids and would result in some error when used with compressible fluids, such as steam, air, or gases. To obtain more accurate selections when controlling compressible fluids, significant changes in density as the pressure drops need to be included. Different manufacturers handle the calculations differently, and they provide good guidelines for selecting valves for compressible fluid services. One method is to calculate an equivalent liquid Cv value, based on whether the fluid is air, gas or steam, and based on whether the pressure drop produces choked flow or not. The following equations provide good estimates; however, the manufacturer should be consulted for important systems when high accuracy is required:

For air or gases operating beyond the critical pressure drop (choked flow[1]):

$$C_v = Q_{scfh} \cdot \frac{\sqrt{s.g. \cdot (T+460)}}{660 \cdot p_i} \text{ (when the gas is close to ideal)}$$

where: C_g – valve coefficient for gas
Q_{scfh} – flow rate in standard cubic feet per hour (SCFMx60)
$s.g.$ – specific gravity[2] of the gas relative to air at 14.7psia and 60°F
T – gas temperature in °F, converted to °R by adding 460°
p_i – inlet pressure in psia (psig + 14.7 psi)

For air or gases at small pressure drops, less than the critical pressure:

$$C_v = Q_{scfh} \cdot \frac{\sqrt{s.g. \cdot (T+460)}}{1360 \cdot \sqrt{\frac{\Delta p}{Po}}} \quad \text{(when the gas is close to ideal)}$$

where: $\Delta p = p_i - p_o$ – pressure drop across the control valve.
p_o – outlet pressure in psia (psig + 14.7 psi)

For near-100% quality saturated steam operating beyond the critical pressure:

$$Cv = \frac{\dot{m}_{stm}}{1.61 \cdot p_i} \quad \text{where } Cs \text{ – valve coefficient for steam}$$

For 100%-quality saturated steam less than the critical pressure drop:

$$Cv = \frac{\dot{m}_{stm}}{2.1 \cdot \sqrt{(p_i + p_o) \cdot \Delta p}} \quad \text{where } C_s \text{ – valve coefficient for steam}$$

Cv Correction factor for superheated steam:

$$Cv = Cv_{sat} \cdot (1 + 0.00065°S / H)$$

Cv Correction factor for wet steam:

$$Cv = Cv_{sat} \cdot \sqrt{x_{stm}} \quad \text{where } x_{stm} \text{ – steam quality in decimal percent}$$

Cv is the valve coefficient using US Customary units which is the flow in GPM to produce 1 psi drop.

Kv is the valve coefficient using SI units which is the flow in m³/h to produce 1 bar drop.

X_T. is the valve pressure-ratio factor without any fittings, given by the manufacturer and used with other system-related factors for more accurate CV calculations involving air, gases, or steam.

F_L is the pressure recovery factor, which is an alternative method of presenting the valve recovery factor, K_m., where $K_m = F_L^2$. Valve recovery factors, K_m. or F_L, are appropriate for liquid service only in predicting whether flashing and cavitation will occur.

F_d valve style modifier for full-size trim, used for calculating Reynolds number through the valve. The valve style modifier is the ratio of hydraulic diameter of a single flow passage to the diameter of a circular orifice. F_d is given by the manufacturer as a function of travel, since the flow area and hydraulic diameter changes with valve position.

FLASHING AND CAVITATION

Flashing and cavitation are "high-temperature" phenomena associated with *liquids* only and are shown in Figure 9.32. High temperature is defined as being close to the boiling temperature and pressure. Flashing and cavitation can cause very high local pressures, noise and vibration, mechanical damage to parts, and wear to the valve trim and body.

The pressure drop across an orifice is useful for understanding control valve behavior, where a control valve is essentially an adjustable orifice. As the liquid approaches the restriction, the flow streamlines converge inward as shown in Figure 9.33. The flow continues to converge past the obstruction, so that the minimum flow area occurs downstream at the "vena contracta." Fluid follows the continuity equation ($Q=vA$), so that as the flow area reduces, the velocity increases. In accordance with Bernoulli's principle, the flow reaches maximum velocity at the vena contracta, and the pressure drops to its minimum value. Beyond the vena contracta, the streamlines expand, the flow area increases, and the velocity decelerates until the streamlines again reach the pipe wall. As the fluid decelerates, some of the pressure energy dropped at the vena contracta is recovered, where the velocity energy is converted back into pressure. However, a permanent unrecoverable pressure loss also takes place across the restriction. In accordance with the continuity equation, the velocity downstream of the orifice will be the same as upstream.

The fluid is considered to be "hot" when its temperature is near saturation. In the orifice, it is possible that the pressure drop at the vena contracta could dip below the vapor pressure or boiling pressure of the fluid. When this happens, a portion of the fluid flashes into vapor. Cavitation ensues if the pressure recovery implodes the vapor bubbles back to its liquid state.

There are three cases to consider when flashing or cavitation may be a problem; no cavitation, flashing only, and flashing followed by cavitation.

Case 1: If the liquid temperature is low and the pressure at the vena contract does not dip below the boiling point, no flashing occurs, and cavitation will not be a problem (Figure 9.34).

Case 2: If the fluid pressure at the vena contracta drops well below the vapor pressure but does not rise above the vapor pressure during recovery, then flashing occurs without cavitation (Figure 9.35). Flashing can damage the valve trim.

Case 3: If the fluid at the vena contracta drops below the vapor pressure, and later the fluid pressure rises above the vapor pressure during recovery, then flashing occurs followed by the imploding of the vapor bubbles, which is cavitation (Figure 9.36). This condition is damaging to the valve if it is severe enough, prolonged, and the materials are not hardened and erosion resistant.

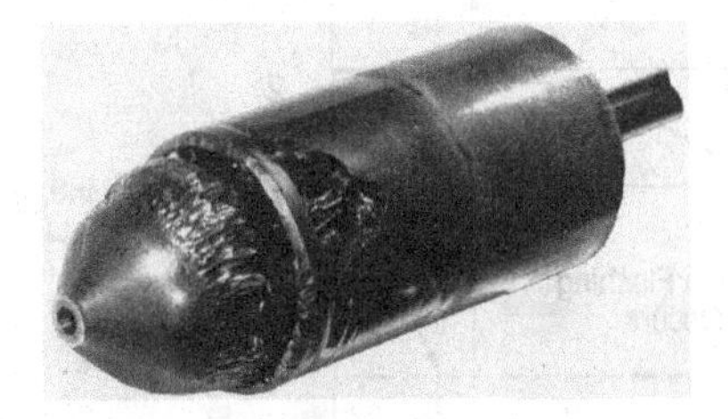

Figure 9.32 A valve plug having severe flashing damage is shown on the left, and a valve plug with its seat exhibiting severe cavitation damage is on the right. (Courtesy of Emerson from Fisher Controls International LLC, *Control Valve Handbook*, 5th edition, Emerson, 2019.)

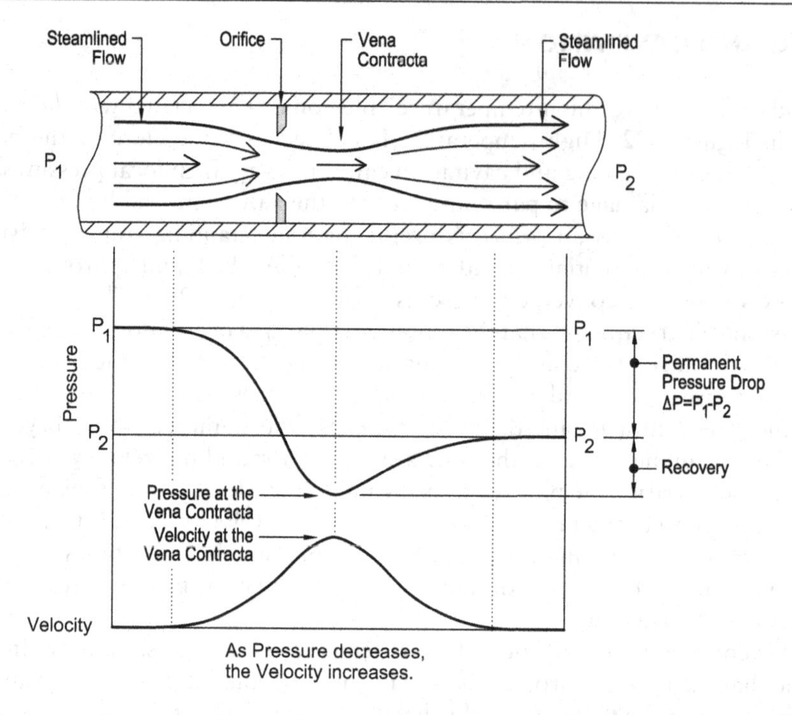

Figure 9.33 Pressure-velocity relationships of an incompressible liquid moving past a flow restriction.

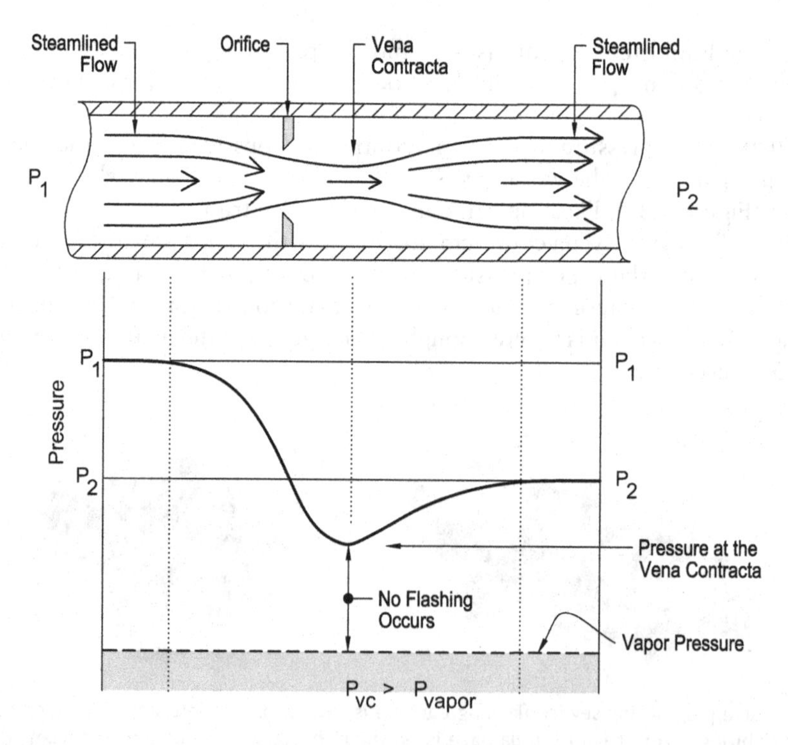

Figure 9.34 Case 1: The pressure at the vena contracta is higher than the vapor pressure and no flashing or cavitation occurs.

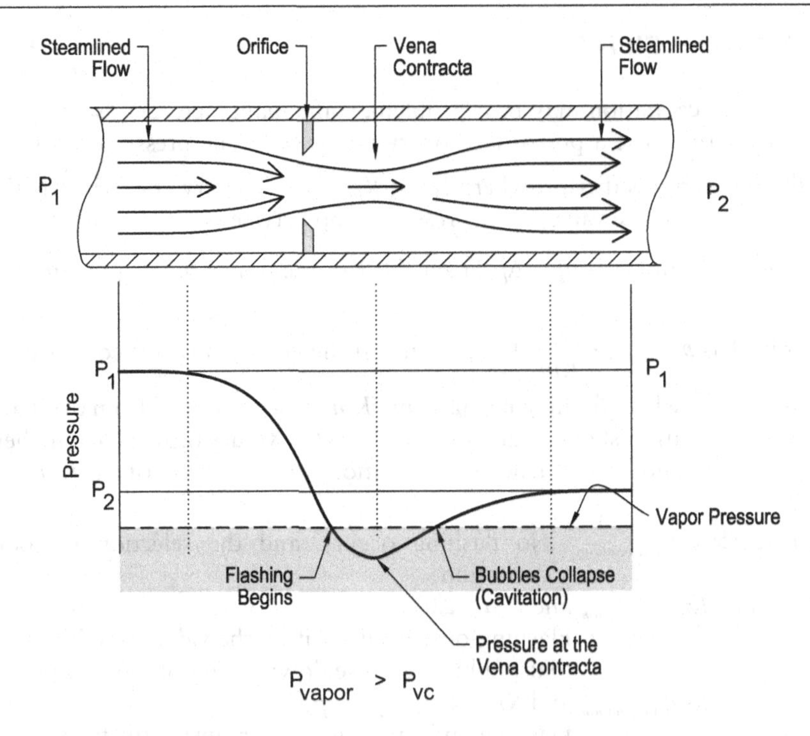

Figure 9.35 Case 2: Flashing occurs when the pressure at the vena contracta drops below the vapor pressure without enough recovery to exceed the vapor pressure. Flashing occurs without cavitation, and damage can be expected.

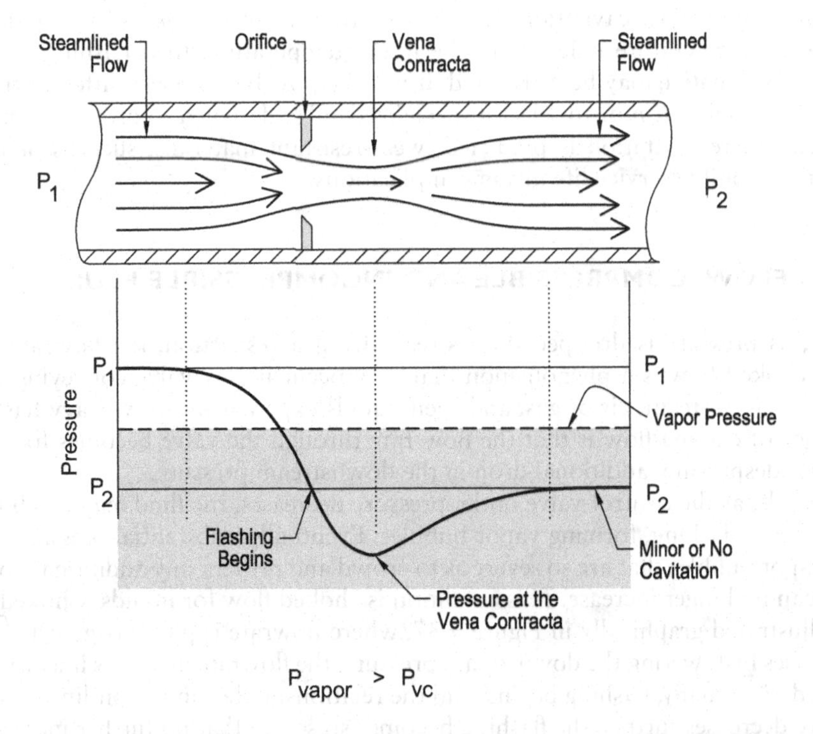

Figure 9.36 Case 3: The pressure drops below the vapor pressure causing flashing. Later, during recovery, the pressure rises above the vapor pressure, collapsing the vapor bubbles. The implosion of bubbles is cavitation and is abrasive to the oxide coatings that protect metal surfaces. Pitting and erosion damage can be expected.

RECOVERY COEFFICIENT

To predict whether cavitation may be a problem, some manufacturers use a recovery coefficient (Km) and others use a pressure-recovery factor (F_L). The pressure recovery factor is related to the recovery coefficient, where $F_L = \sqrt{K_m}$. The recovery coefficient is the ratio of the pressure drop across the valve to the pressure drop between the inlet and vena contracta, where $K_m = \frac{p_1 - p_2}{p_1 - p_{v.c.}}$ and $p_1 - p_2 = \Delta p$ From the Km relationship, the pressure at the vena contract is solved as $p_{v.c.} = p_1 - \frac{\Delta p}{Km}$. The pressure at the vena contract is compared to the saturation pressure (p_{sat}) where flashing begins using Km values obtained from the manufacturer catalog data. The equation shows that valves with higher values of Km have higher pressure at the vena contracta and less tendency for cavitation. Three cases exist for Km.

- *Case 1*: $Km > Km_{\min caviation}$ No flashing occurs, and the selection is good against cavitation
- *Case 2*: $Km < Km_{\min caviation}$ and $Km > 1.0$
  Flashing-only occurs within the valve trim, but it is possible that bubbles collapse downstream of the valve
- *Case 3*: $Km < Km_{\min caviation}$ and $Km < 1.0 \rightarrow p_2 > p_{vapor}$
  Flashing and cavitation occurs inside the trim

The equation $\Delta p_{\max\ cavitation} = Km \cdot (p_1 - p_{vapor})$ is convenient for selecting control valves. If the Δp_{max_cav} calculation predicts that the liquid is subcooled at the vena contracta, then the selection is good against cavitation. However, if the calculated $\Delta p_{max\ cavitation}$ indicates that the saturation temperature is less than the liquid temperature, then flashing is inevitable, and another alternative may be warranted. It may be possible to select different trim having a satisfactorily higher Km or to select a specialized valve designed to drop pressure in small increments. Where cavitation is predicted, wear-resistant materials, such as Stellite[3], may provide an acceptable service life in some applications.

CHOKED FLOW: COMPRESSIBLE AND INCOMPRESSIBLE FLUIDS

Intuitively, as pressure is dropped downstream from a restriction, the flowrate increases. However, choked flow is a phenomenon that may occur in any valve conveying compressible fluids, such as steam, air or gas, and even liquids experiencing very heavy flashing. One consequence of choked flow is that the flow rate through the valve becomes fixed and will not increase despite any additional drop in the downstream pressure.

With liquids, as the control-valve outlet pressure decreases, the fluid may reach its boiling point and begin flashing, forming vapor bubbles. Eventually, substantial pressure drops will produce vapor bubbles that are so severe as to crowd and prevent any additional flow, so that flow rate can no longer increase. This condition is choked flow for liquids. Choked flow of a liquid is illustrated graphically in Figure 9.37, where flowrate is plotted against $\sqrt{\Delta p}$ As the drop increases by lowering the downstream pressure, the flowrate increases linearly, as would be expected. Eventually, flashing begins, and the relationship becomes non-linear. As the outlet pressure decreases further, the flashing becomes so severe that no further increase in flow can occur. At that point choked flow is reached.

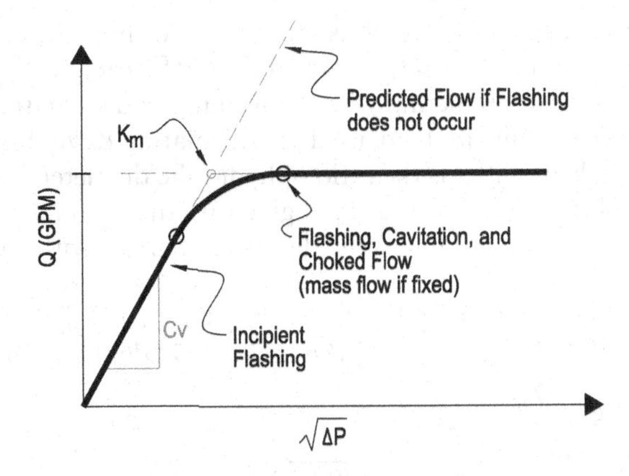

Figure 9.37 Flow rate plotted against $\sqrt{\Delta p}$. When the outlet pressure drops enough, liquid flashes into vapor, and flowrate tapers. When the outlet pressure decreases substantially, full cavitation is reached, and the flowrate at that valve position becomes fixed. The flow rate can be increased by opening the valve more, but it is no longer influenced by pressure difference. Eventual damage to the valve trim can occur from the cavitation.

Choked flow also occurs and is common with compressible fluids, such as steam, air and gases, although the mechanism causing choked-flow is different, and there is no associated cavitation. As shown in pressure-velocity profile curves in Figure 9.33, as the outlet pressure drops, the velocity increases, following Bernoulli's equation. Eventually, the critical pressure ratio of the fluid is reached. The critical pressure is the point where the fluid reaches the speed of sound within that fluid, forming a standing normal shock wave. No further velocity increase can occur beyond sonic velocity, despite any additional outlet-pressure decrease. In accordance with the continuity equation, when the velocity and area is fixed, there can be no further increase in the flow rate, and choked flow occurs. At the point of choked flow, the flowrate can only be increased by opening the valve more.

Whether choked flow occurs from severe flashing of a liquid or whether choked flow occurs from a compressible gas reaching its critical pressure ratio, valve manufacturers have documentation that can be consulted to select valves to meet these extraordinary conditions.

SELECTING THE VALVE CV USING MANUFACTURER PRODUCT DATA

Example problem

Select a feedwater regulator valve having *linear* trim. The regulator shall be installed in a 2.5" NPT line flowing 150 gpm into a boiler whose operating pressure is 1000 psig. The feedwater comes from a deaerator, which operates at 35 psig and saturation temperature. The feed pump is variable-speed turbine-driven and adjusted to maintain a constant differential pressure of 50 psid at all flow rates.

Kv is the valve coefficient using SI units which is the flow in m³/h to produce 1 bar drop.

X_T. is the valve pressure-ratio factor without any fittings. XT is given by the manufacturer and used with system-related parameters to obtain more accurate C_V calculations for air, gases, or steam services.

F_L is the pressure recovery factor, which is an alternative method of presenting the valve recovery factor, *Km.*, where $K_m = F_L^2$. Valve recovery factors, *Km.* or F_L, are appropriate only for liquid service in predicting whether flashing and cavitation will occur.

F_d valve style modifier for full-size trim, used for calculating Reynolds number through the valve. The valve style modifier is the ratio of hydraulic diameter of a single flow passage to the diameter of a circular orifice. F_d is given by the manufacturer as a function of travel, since the flow area and hydraulic diameter changes with valve position.

Valve Inlet Pressure: $p_1 = 1000\ psig + psig = 50\ psid = 1050\ \ psig$

Feedwater Properties: $T_{sat\ \ at\ 50\ psia} = 281°F$ and $\gamma = 57.9 lbm\ /\ ft^3$ (from steam tables)

Specific gravity: $s.g. = \dfrac{\gamma_{liq}}{62.4} = 0.923$

Valve Coefficient: $C_v = Q\sqrt{\dfrac{s.g.}{\Delta p}} = 150\sqrt{\dfrac{0.923}{50}}$ $\qquad Cv\ =\ 20.38$

If the pump operates at constant speed, the valve should be selected to be between 70-85% open. Since this system uses a variable-speed pump, the Cv should be selected for a valve that is approximately 50% open.

Using Table 9.1, the 1-1/2" full-port control valve has a Cv of 20.4 at 50% open and is a good choice. This valve has a port diameter of 1.875" and a maximum travel of 0.75". Note that a 2-1/2" *reduced port* valve with a Cv of 21.1 at 50% open would also work while about 48% open, however, the larger body valve would be more expensive and not necessary.

Now check the valve selection for cavitation: $F_L = 0.82 \quad K_m = F_L^2 = 0.672$

Pressure at the vena contracta: $p_{v.c.} = p_1 - \dfrac{\Delta p}{km} = 1065\,psia - \dfrac{50\,psid}{0.672}$ $\qquad p_{v.c.} = 990psia$

Verify vena-contracta temperature is subcooled: $\lfloor T_{sat}\ @990\,psia = 543°F \rfloor > \lfloor T_{fw} = 281°F \rfloor$

T_{sat} at vena-contracta exceeds T_{actual} --> No Cavitation

Note that although the feedwater started at saturation, the very high feed-pump discharge pressure produces a very subcooled condition, so it is not surprising that this valve is far from cavitating.

ANSI VALVE CLASS

Valve-body strength requirements must be considered when selecting control valves. One system used for valve body strength is the ANSI class, which also standardizes flange dimensions. A larger class number indicates a stronger thicker-wall valve. General guidance for selecting cast-iron and steel valve[4] classes is plotted in Figure 9.38 using table values published in the ANSI standards. The figure indicates the valve-class limits are based on both temperature and pressure. Intuitively, the higher the operating temperature, the lower the material strength, and the lower the allowable pressure will be.

As an example, if the maximum steam conditions are 1000psig/800°F, a 900# class valve is the appropriate selection. However, a 600# class valve is adequate for a feedwater regulator valve operating at 1300 psig and 300°F. Cast-iron valves are less expensive than forged-steel, but they have lower pressure and temperature limits and are more susceptible to brittle failures. For economy, cast-iron valves are appropriate for lower energy systems

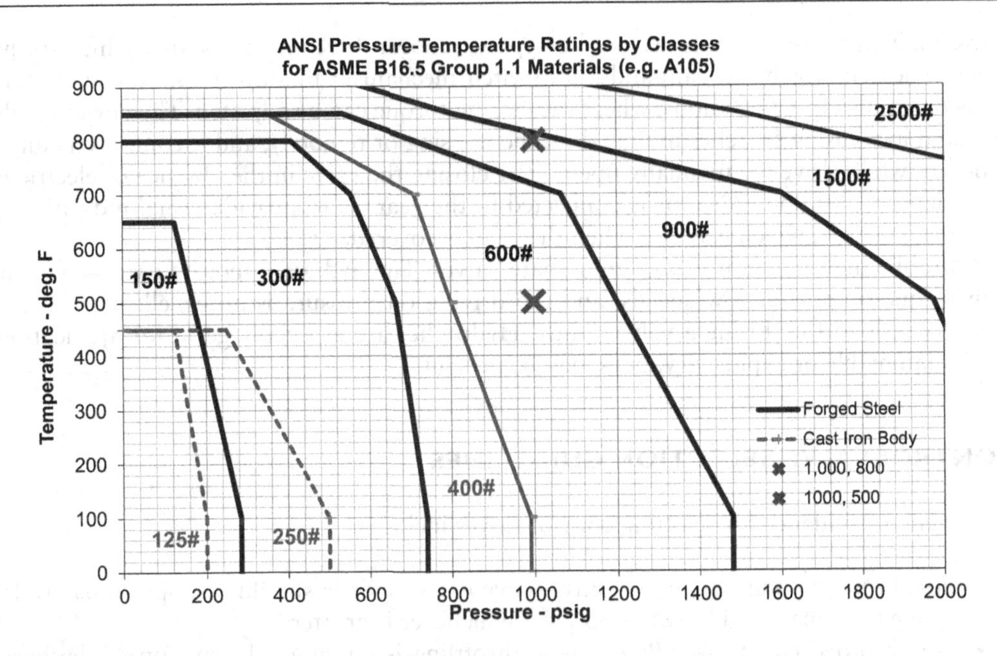

Figure 9.38 ANSI classes for cast-iron and steel-body valves, showing the operating pressure-temperature limits. The maximum permissible temperature for small cast-iron valves is 450°F, and larger cast-iron valves are limited to even lower temperatures. Material specifications should be referenced for pressure-temperature ratings for other materials and their maximum permissible temperature limits. Note that class 125# cast iron and 150# and class 250# cast iron and 300# have the same bolt patterns and can be accidentally or intentionally interchanged. Refer to the ANSI standards.

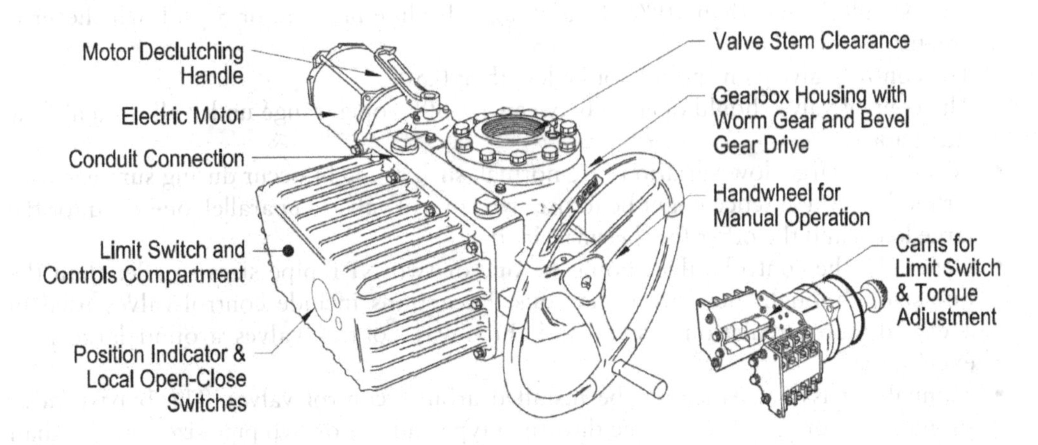

Figure 9.39 Limitorque motor-driven actuator for automatic or remote open-close operation. (Images provided courtesy of Flowserve U.S., Inc., all rights reserved.)

MOTOR ACTUATORS

Motor actuators are also used in plant automation. The motors are often used for open-close operation for remote control or automatic sequencing of a process, but in some instances, they are used in modulating applications. Figure 9.39 shows a large Limitorque motor-driven valve actuator. This valve actuator uses high-gear-ratio worm gears into a bevel-gear set to convert a high-speed, low-torque, low-horsepower motor drive into a low speed, high torque driver. The actuator has a cam-and-limit-switch mechanism to stop the motor at its fully opened or fully closed positions, where the closing torque can be adjusted. The device can be operated local using switches mounted on the housing or remotely, and it contains a clutch-and-handwheel drive for manually opening or closing the valve in the absence of electricity. A mechanical position indicator is connected to the gear set to provide visual indication of the valve position, which is useful when the stem is covered.

Hydraulic motors are also used to actuate valves. The hydraulic motors naturally create high torque outputs at low speeds, and the driving oil pressure is an excellent indication of actuator load for closing a valve tightly. Hydraulics are especially good for applications where flammable or explosive vapors may be present.

CONTROL VALVE SELECTION GUIDELINES

Some general guidance for selecting control valves includes:

- For throttled pump flow the control-valve pressure drop should be approximately 1/3 of the frictional head losses or 15 psid, whichever is greater
- For throttled compressor flow, where throttling is on either the suction or discharge side, the control-valve pressure drop should be about 5% of the suction absolute pressure or 1/2 of the frictional pressure losses, whichever is greater
- For throttling transfer flow between pressure vessels, the control-valve pressure drop should be about 10% of the low-side pressure, or 1/2 of the frictional pressure drop, whichever is greater
- For throttled flow to steam turbines or associated equipment, the control-valve pressure drop should be less than 10% of the design absolute pressure or 5 psid, whichever is greater
- The control-valve gain should not be less than 0.50.
- The control valve should operate between 10–80%-open range under all normal flow excursions.
- Where very large flow variations are normal, such as might occur during summer winter mode, good practice might be to use two control valves in parallel, one sized for the large loads and the other for the small loads.
- Generally, the control valve should be one or two NPT pipe sizes smaller than the line size, although there are exceptions. Exceptions include control-valves used in a closed-loop circulation system and three-way control valves around large heat exchangers.
- Manual bypass valves should be installed around control valves. The bypass valve should be a non-gate valve of the throttling type and one or two pipe sizes smaller than the main line to provide good control.
- The valve-trim inherent characteristics should be selected to shift the installation curve to be closest to linear for most modulating control applications. Table guidance for particular control applications is typically adequate.
- In cases where flashing or cavitation may be of concern, valves designed for multiple internal pressure drops can be selected.
- In cases when pressure drops are very large, if may be necessary to install two control valves in series and drop the pressure in stages.
- When bolting cast-iron valves with 150# or 300# flanges, the flange raised faces should be machined off to avoid breaking the more-brittle cast-iron flange during tightening.

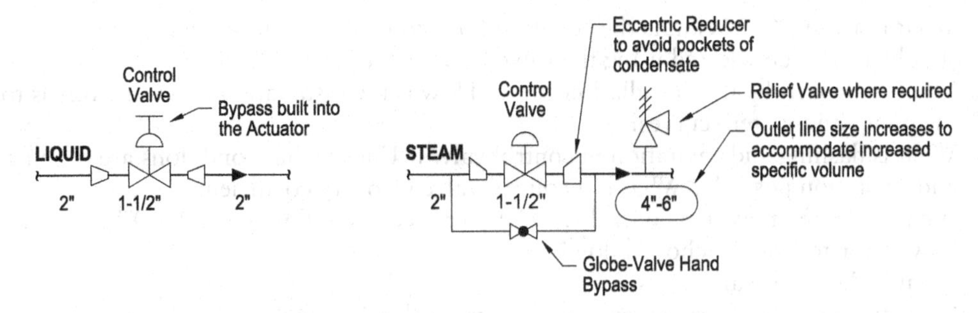

Figure 9.40 Although there are some exceptions, the control valve body is typically smaller than the main line size. Eccentric reducers are usually used with steam valves, when water hammer can be a problem.

- When the fluid is compressible, expansion of the gas/vapor results in a larger specific volume and a corresponding increase in fluid velocity. To keep pipe flows within reasonable-velocity ranges, the outlet piping is often increased to be one or more pipe sizes larger than the supply pipe.
- In steam lines operating near saturation, it sometimes good practice to use an eccentric reducer with the flat-side down and pitched in the direction of flow to avoid a pocket that may trap condensate that could lead to water hammer (Figure 9.40).
- Dropping steam pressure results in adiabatic superheating, which could affect heat exchanger design.
- In addition to control valves, traps are required at the outlets of steam heat exchangers to ensure transfer of all latent heat

QUESTIONS

1. List six plant applications that use control valves.
2. In general terms, describe how self-contained pressure-reducing valves work.
3. Using the self-contained regulating valve in Figure 9.2, describe the functions of the low- and high-pressure ports, the controlling diaphragm, the auxiliary or controlling valve, the power piston and main valve, and the adjusting spring. How is the output setpoint pressure raised with this valve? What disadvantage occurs with this valve under varying loads?
4. How does a self-contained temperature-regulating valve work?
5. How do pneumatic pilot-actuated control valves work to maintain a constant outlet pressure? Discuss the operation of the pilot.
6. What is the meaning of direct-acting and reverse-acting for pneumatic pilots and actuators? What is meant by a valve's fail-safe position? How can pilot-actuated control valves be configured to fail open or fail closed on the loss of control-air pressure?
7. How is level control accomplished with pilot controllers? What is the measured parameter in an open tank? What is the measured parameter in a pressurized tank?
8. Describe hysteresis and deadband with respect to a single control valve.
9. Discuss how the shape of the valve disk and seat affect control-valve flow. What are three inherent flow characteristics for control valves?
10. Using Table 9.2, what are the probable recommended inherent characteristics for a temperature-control valve, for a constant-differential-pressure control valve, for liquids, and for sudden very large requirements of large flows.

11. What is a valve Cv? What is the equation for a control-valve flow coefficient? How should a valve be selected using the valve flow coefficient?
12. What is a control-valve installation curve? How is the installation curve analogous to the pump-and-system curves?
13. What is flashing and cavitation in control valves? Under what conditions are flashing and cavitation possible? What is a control-valve recovery coefficient?
14. What is choked flow in a control valve? Where is choked flow possible? What are the flow characteristics of choked flow?
15. What is the ANSI valve classification?
16. What types of actuators are used with control valves?
17. What is control-valve mechanical-positional feedback and what is its benefits?

NOTES

1. The critical pressure ratio for determining choked flow is calculated as $\frac{p_{cr}}{p_o} = \left(\frac{2}{k+1}\right)^{\frac{k}{k+1}}$, where k is the ratio of specific heats ($k = \frac{c_p}{c_v}$). For air, $k=1.4$. and $p_{cr} = 0.528p_o$. For the wet steam, the critical pressure ratio is 0.577 ($k_{sat} = 1.135$) and for superheated steam, $p_{cr}=0.546$ ($k_{s/h}=1.3$).
2. The density of air at STP is $0.074887 lb_m/ft^3$. Note that pressure and temperature need to be in absolute units when using the ideal gas equations.
3. Stellite is a trademarked cobalt-chromium alloy containing tungsten or molybdenum and a small amount of carbon. Stellite is designed specifically to be very wear resistant and used in valve-trim for applications prone to wire-drawing damage.
4. 125# and 250# are the valve classes for cast-iron-body valves, and the others are for ductile iron or steel. Cast-iron is a brittle material and is limited to lower temperature/pressure installations.

Chapter 10

Speed, load, and alternator control

Speed control of rotating equipment is used to maintain constant speed, to limit machinery speed from becoming excessive, and to provide automatic shutdown when safe limits are exceeded. Speed droop is a good indicator when matching engine input power to the connected load, and it provides an indirect signal input to a control system, as speeds tend to drop as loads increase. Speed-control devices are classified as:

- Speed-limiting governors,
- Constant-speed governors,
- Overspeed trip devices.

In addition to providing speed limiting, speed setpoint, and overspeed protection, these devices may incorporate other automatic shutdown features, such as protection against low lubricating oil pressure or high exhaust backpressure from a steam turbine. This chapter covers speed-control devices for engines, turbines,[1] and electrical alternators.

SPEED-LIMITING GOVERNORS

A speed-limiting governor is used to throttle back a prime-mover's energy input without interrupting service when its excess-speed setpoint is reached. Speed limitations avoid the hazardous conditions of high centrifugal forces that could exceed the structural strength of the rotating machinery. As an example, speed-limiting governors are used in shipboard main propulsion steam turbines.[2] During periods of rough weather, a ship's propeller may go in and out of the water, cyclically loading and unloading the main engine. Sudden unloading causes the engine to overspeed. Tripping a propulsion engine is not an option, as it is critical that a ship continuously makes way in severe weather. Typically, speed limiting is actuated between 10–15% higher than the engine's normal full speed.[3] Later, when the engine speed drops as the propeller digs into the sea, the throttle resumes its normal control of speed. Figure 10.1 shows a schematic representation of a steam-turbine speed-limiting governor. Two limiting functions are provided by this device; the first sets the speed at which overspeed limiting begins, and the second sets the low lube-oil-shutdown pressure.

Main steam is piped in series through the speed-limiting governor valve and the main-engine throttle valve. The throttle normally controls speed, while the speed-limiting governor is a wide-open permissive, protecting against abnormal conditions. The speed-limiting governor uses the bearing lube-oil-pressure as a hydraulic source for opening the governor valve. In some plants the bearing lube oil comes from a gravity tank through an orifice designed to set the oil pressure to the bearings, and in other plants the bearing oil is supplied directly from a pressure-regulating valve. Both methods produce a continuous source of constant-pressure

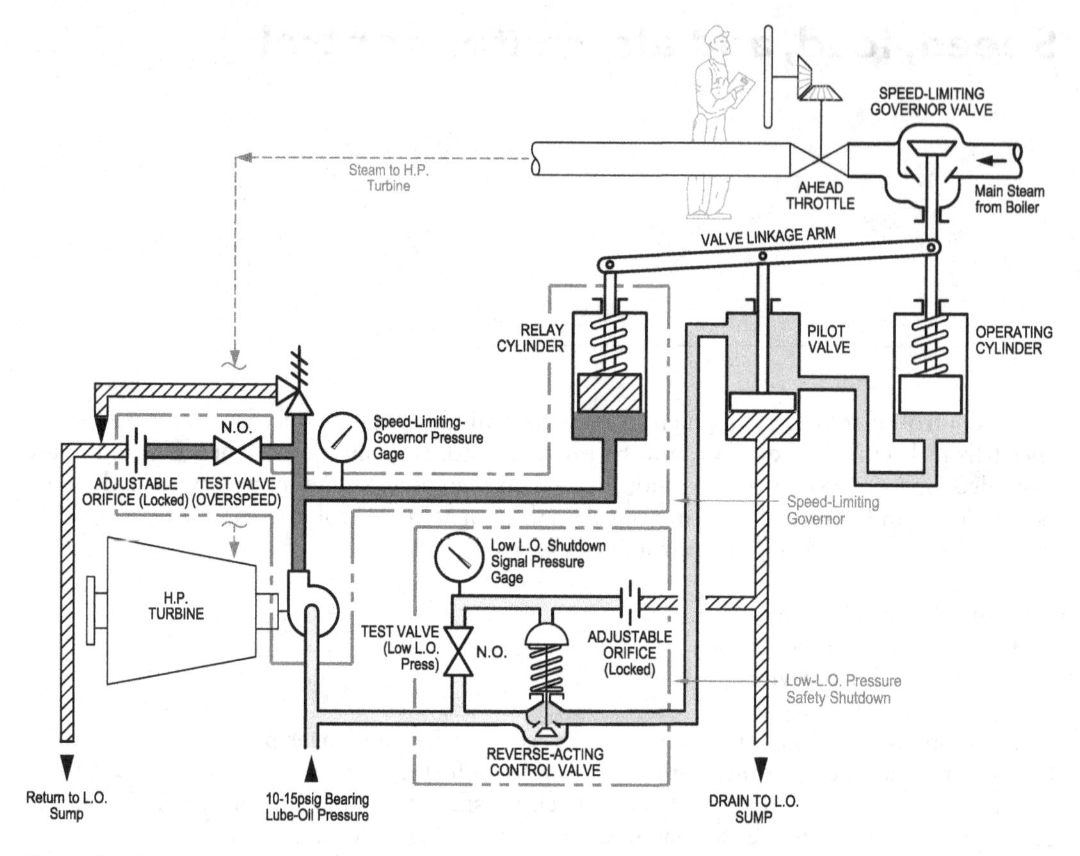

Figure 10.1 A shipboard main-propulsion steam turbine speed-limiting governor having a built-in low lube-oil trip.

oil to the turbine bearings, in the range of 10–15psig, for lubrication. The primary function of the system is machine lubrication, but the second function of the bearing oil is to provide the speed-limiting governor's hydraulic control signal and actuation force.

Speed-limiting protection

Referring to Figure 10.1, a crude centrifugal pump is built into the end of the steam turbine rotor near the thrust bearing. The pump produces the control input signal by stepping up the bearing oil pressure in accordance with the pump affinity laws. The pump pressure is the control input signal and is proportional to the square of the rotor-speed.[4] This signal pressure is directed to the relay cylinder, which is connected to the governor valve via a floating linkage arm. When the rotor speed and the corresponding pump pressure is low, the relay cylinder piston moves its linkage arm down, lowering the pilot-valve piston. The pilot valve aligns to direct oil to the operating cylinder, which lifts the speed-limiting governor valve to open wide. The governor valve remains fully open when below the overspeed setpoint, thus serving as a permissive. The higher the rotor speed, the faster the pump spins, the higher the oil pressure into the relay cylinder, eventually forcing the relay piston to move up. Initially, the valve-linkage arm pivots around a pin on the operating-cylinder valve stem as shown in the left side of Figure 10.2. The linkage arm lifts the pilot valve until the pilot is aligned to spill operating-cylinder oil to the drain, at which point the operating-cylinder spring closes

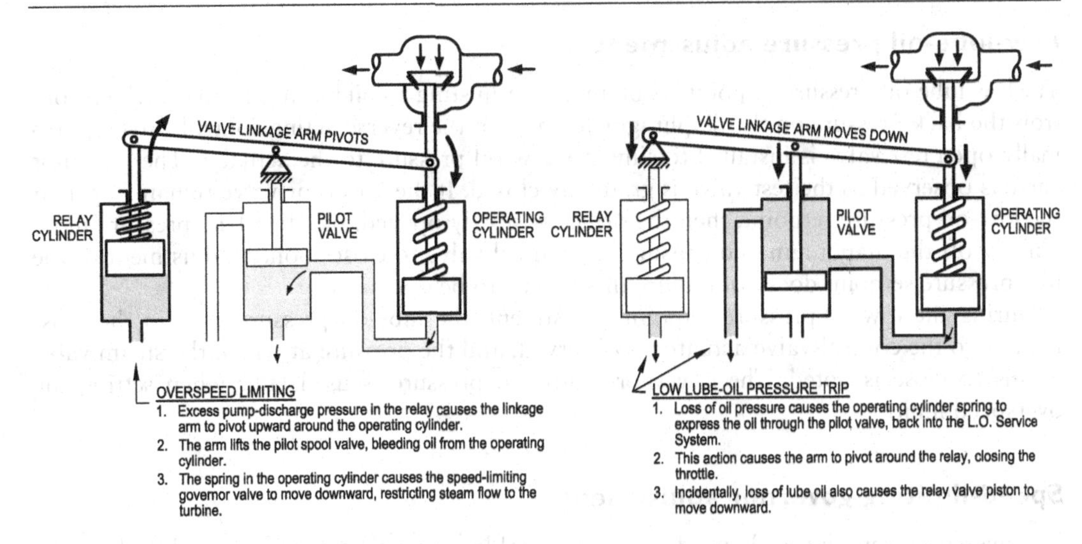

Figure 10.2 Actuation for speed-limitation (left) and actuation for low lube-oil pressure shutdown (right).

down on the speed-limiting governor valve, slowing the turbine and protecting against overspeed. As the operating-cylinder piston moves down, the pilot-valve piston also moves down, providing negative feedback until the pilot-valve port is again covered. Covering the pilot-valve port inhibits further closure of the governor valve and any further speed reduction. As the turbine slows to safe levels as the propeller submerges back into the sea, the process is reversed until the governor valve is fully open. The relay piston moves down, aligning bearing lube-oil pressure to the operating cylinder, causing the governor valve to reopen. To summarize, turbine overspeed causes the oil-pump speed to increase, raising the oil-pressure signal to the relay, which lifts a pilot valve piston causing oil to drain from the operating cylinder, thus closing down on the speed-limiting valve and slowing the turbine.

Because the system uses lube-oil pressure for its hydraulic actuation, this governor also serves as a low-lube-oil-pressure safety shutdown device, designed to avoid severe damage to the main-engine bearings and reduction-gear meshes, as shown on the right side of Figure 10.2.

Low-lube-oil-pressure protection

For loss-of-lube-oil protection, lube oil to the bearings serves as both the pressure-input signal directed to the diaphragm of a reverse-acting control valve and the actuating force to lift the operating-cylinder piston, when the reverse-acting pressure-control valve is aligned through the pilot valve as shown in Figure 10.1. Lube-oil pressure that is directed to the underside of the operating-cylinder piston provides the hydraulic force to lift the piston and fully open the governor valve. The hydraulic force is opposed by spring and gravity forces to ensure positive and failsafe closure of the valve if lube-oil pressure is lost. The bearing lube-oil pressure serves as both the control input signal and the actuation force. Upon loss of lube-oil in the operating cylinder, the governor valve moves downward, cutting back steam from reaching the ahead throttle. The oil pressure opens the control valve when adequate lubrication is present. The reverse-acting control-valve closes by spring force upon loss of oil pressure to its actuator, dropping the oil pressure to the operating cylinder. The operating-cylinder spring completely closes the speed-limiting governor valve; ultimately stopping the turbine.

Low-lube-oil pressure adjustment

The low lube-oil pressure setpoint is obtained by adjusting an oil bleed-off orifice, which controls the back pressure on the diaphragm actuator of the reverse-acting control valve. A normally open test valve is installed to simulate low oil pressure to the actuator. The governor valve is observed as the test valve is gradually closed. If the governor valve remains open at the low-oil-pressure setpoint, then the orifice is slowly opened, bleeding backpressure more quickly off the diaphragm and causing the control valve to close sooner. Adjustment of the low-pressure setpoint does not require the turbine to be operating.

During the low-oil-pressure setpoint adjustment, the lube-oil pressure gage in the sensing line to the control-valve actuator is observed, and the pressure at which the steam valve begins to close is noted. The governor-shutdown pressure is used later when setting the overspeed limit.

Speed-limiting governor adjustment

The overspeed setpoint is changed via an adjustable orifice that continuously bleeds pump flow from the line leading to the relay cylinder. Changing the orifice area changes the system resistance, essentially changing the piping "system curve." The setpoint is obtained by observing and plotting various pump-discharge pressures versus the corresponding rotor speeds so as to produce a "system curve" based on that orifice position. According to the affinity laws, pump flowrate is directly related to rotational speed, and the developed pressure is directly related to the square of the speed/flow, as shown in the upper graph in Figure 10.3. Eventually the pressure reaches its maximum value at full throttle, which should be 10–15% lower than the desired overspeed limit. Because the overspeed setpoint cannot be reached under normal conditions, the curve can only be plotted up to the turbine maximum speed. To predict the lube-oil pressure that would close the governor at 15% overspeed, the speed-versus-pressure curve is extrapolated, as shown in the lower curves in Figure 10.3. The governor shutdown pressure is the value that was noted during the low-lube-oil pressure calibration process. Linear extrapolation is used, which is accurate over the short extrapolation distance but slightly conservative. Ideally, the straight-line extrapolation should cause the governor valve to close as the lube-oil shutdown pressure is attained when the RPM reaches or is slightly below the 115% of the turbine rated speed. The overspeed setting is obtained by trial and error. When the predicted overspeed RPM and pressure do not correspond, the orifice is adjusted, and the trial repeated.

Trial 1 in Figure 10.3 shows the extrapolated pressure exceeds the governor-close pressure at 115% of the full-power speed, and consequently, the turbine will prematurely slow before reaching its required overspeed setpoint. In that case, the orifice would be opened slightly, requiring the turbine to rotate faster, and to develop more oil pressure. After making an adjustment, the test is run again, and the speed-versus-pressure graph is reproduced. Conversely, Trial 2 shows that if the extrapolated pressure is too low to close the valve at the 15% overspeed value, the rotor overspeed will be excessive before speed limiting begins. The corrective adjustment is to open the orifice slightly. The testing and adjustments are repeated until the 115% extrapolated speed predicts the lube-oil pressure that closes the governor.

CONSTANT-SPEED GOVERNORS

Constant-speed governors modulate the energy input to an electrical alternator's prime mover for producing the synchronous speed, setting the required frequency[5]. Early governor

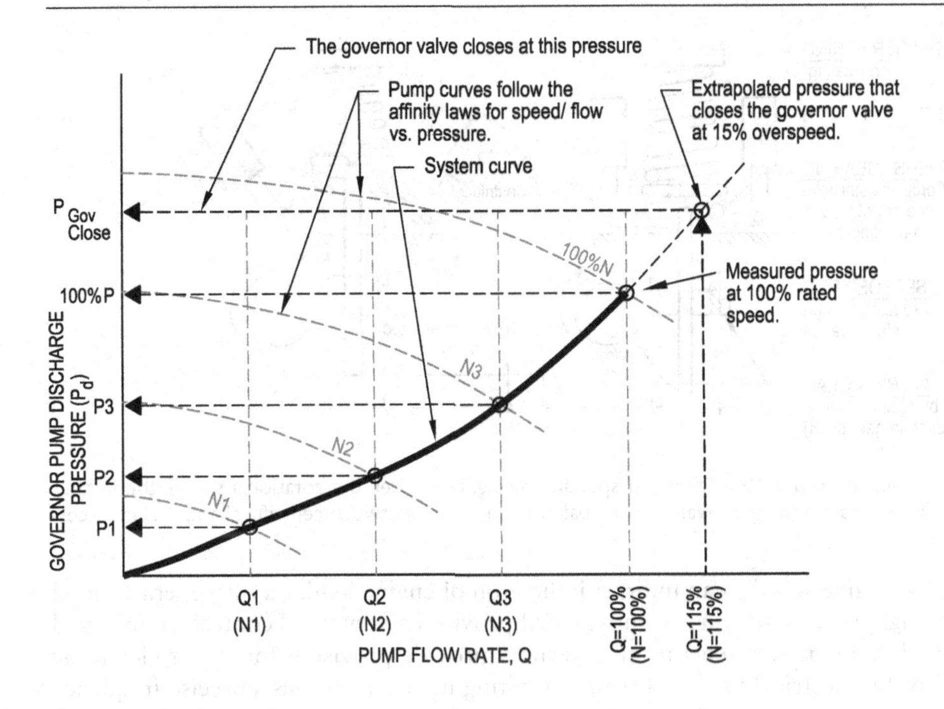

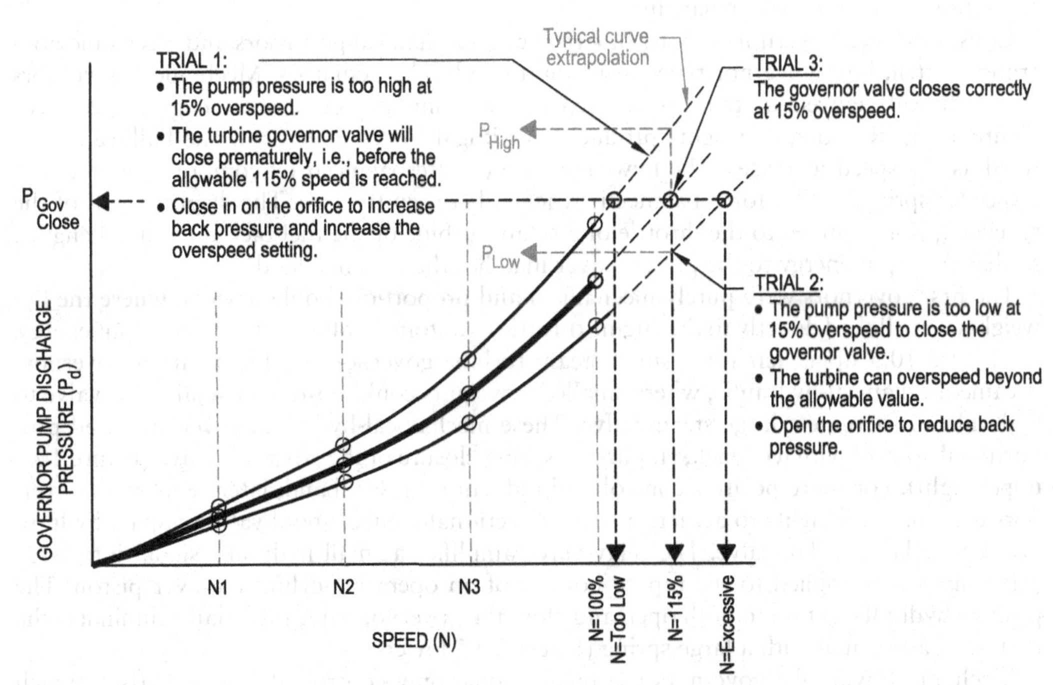

Figure 10.3 The upper flow-versus-pressure pump curves show the application of the pump affinity laws for producing higher pressures at higher flowrates, and how the speed is extrapolated using pump curves. The lower curves show the procedure for adjusting the governor overspeed setpoint. Speed-versus-pressure points are plotted and then extrapolated to predict the speed where the governor valve closure begins.

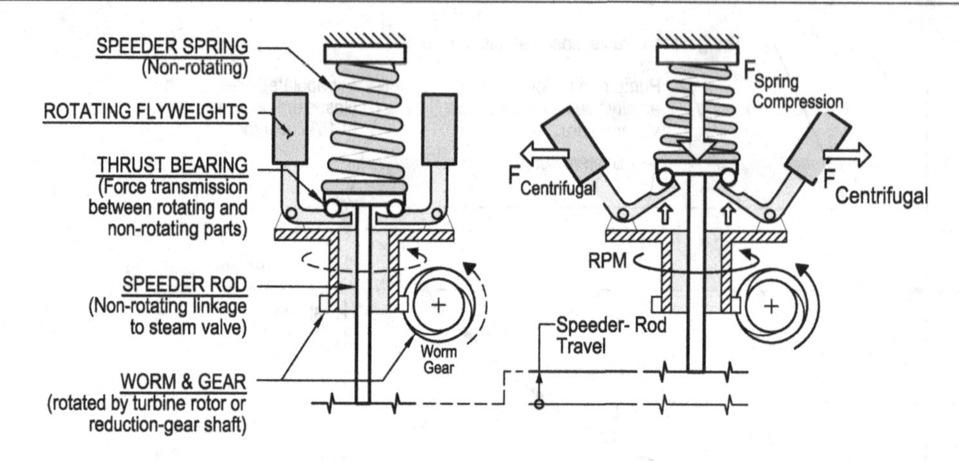

Figure 10.4 Mechanical governor flyweights and speeder spring. The faster the rotational speed, the further the flyweights are flung outward, until a balance is reached between centrifugal and spring forces.

designs used the engine speed as an indirect indication of engine load. As the generator load increases, the engine slows, so consequently, speed provides a very good control-input signal to indicate load. Some newer-generation governors provide provision for direct electronic measurement of the electrical load and speed, resulting in faster response, precise frequency regulation, and better load balancing.

Constant-speed governors consist of two types, mechanical governors and electronic governors. Often, both governor types incorporate hydraulic actuators. Mechanical governors use "ballhead" flyweights that are rotated by the prime mover to sense engine speed (see Figure 10.4). Rotating flyweights produce centrifugal forces that throw the ballheads outward as the speed increases. The flyweights are counterbalanced by compression forces in a speeder spring, until a force balance is reached between the two. The displacement of the speeder spring is linked to the throttle of a steam turbine or the fuel rack on a diesel engine, so that the input energy to the prime mover matches the attached load.

The first governors were purely mechanical and proportional-only devices, where the flyweights were linked directly to the engine throttle to automatically modulate the input energy, see Figure 10.5 upper left for a small steam turbine governor. Next-generation governors became mechanical-hydraulic, where smaller flyweights could position a small pilot valve to hydraulically actuate a large steam valve. These mechanical-hydraulic governors used proportional control with feedback, employing spring closure of the governor valve (Figure 10.5 upper right). For more positive control and reduction of deadband, later-generation governors use small flyweights to actuate a small directional-control spool valve coupled hydraulically to a larger relay valve. The relay valve amplifies a small hydraulic signal into large actuating forces applied to the top or bottom of an operating-cylinder power piston. The positive hydraulic action to both open and close the governor valve essentially eliminates the hysteresis associated with a large spring (Figure 10.5 lower).

Mechanical-hydraulic governors use proportional-only control (P), having offset, which is called droop, or they can add integral gain (PI) for zero-droop or isochronous operation. Many governors use centrifugal forces from flyweights to sense engine speed, where engine speed is directly related to generator frequency; but nowadays, electronic speed sensors provide faster and more accurate speed measurement and governor response, and they can be used with or instead of the flyweights. In general, control systems behave best when the controlled variable is measured most directly. Today, it is possible to electronically measure the generator's electrical output power (kW), and to incorporate it into the governor's

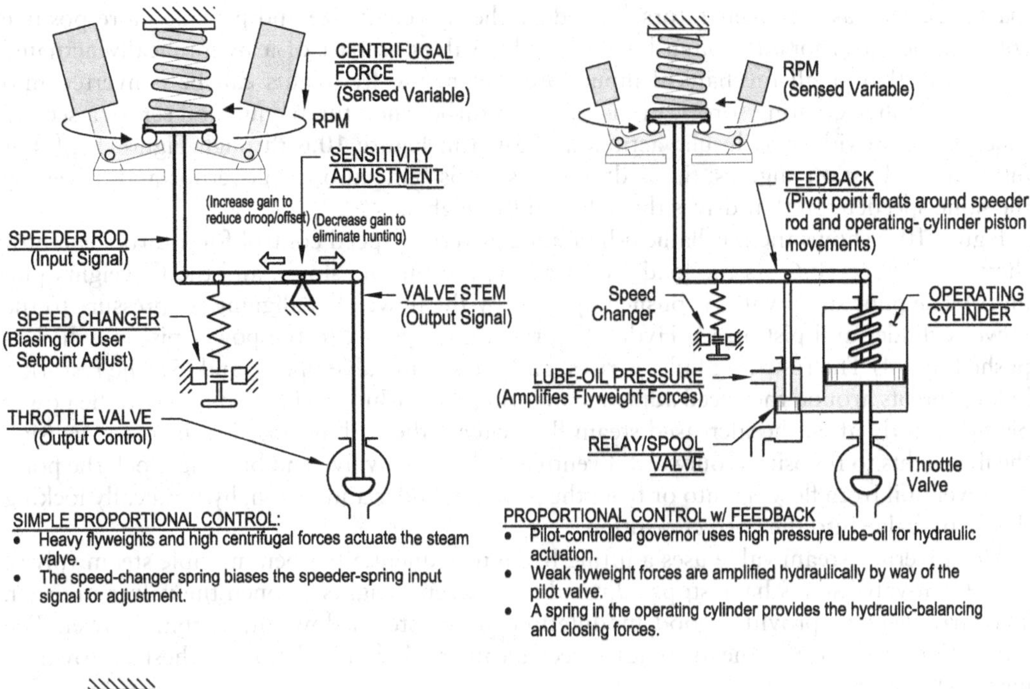

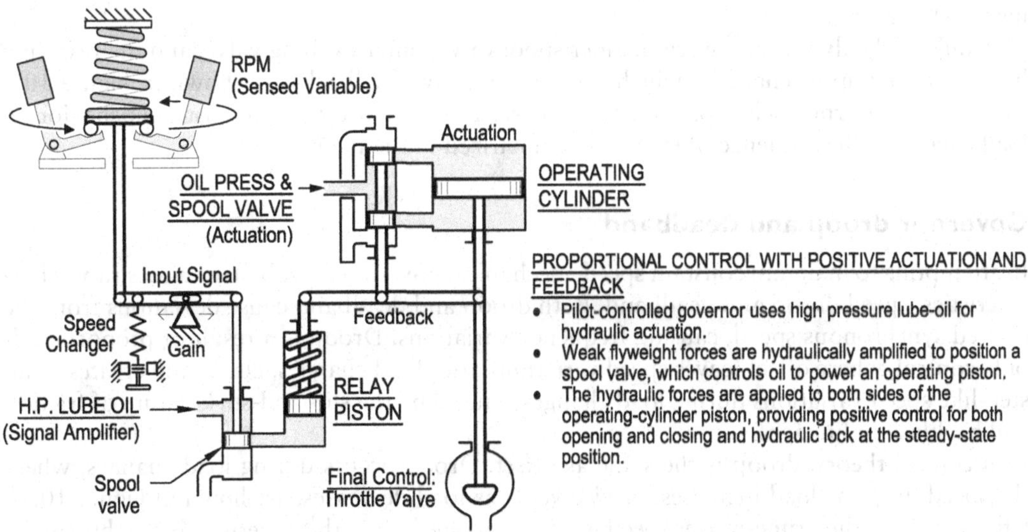

Figure 10.5 Three variations of mechanical-governors for steam-turbine throttle valves.

energy-input control signal, improving response and load sharing. The speed input signal is used to control frequency, while the power measurement input signal is used to speed up the governor's response during transients and to balance the loads when generators are paralleled.

Mechanical-hydraulic governors

Although mechanical-governor flyweights are directly linked to the energy input for small non-condensing auxiliary turbines, the flyweights become objectionably large for big machines, such as turbogenerators. To reduce the flyweight size and provide more positive control, the governors are often linked to pilot valves as part of a hydraulically actuated system, so that small mechanical input forces from the flyweights can be converted into large hydraulic actuating forces. In the case of a turbogenerator, the lube-oil serves a second function of providing hydraulic action, as shown in Figure 10.6 through Figure 10.9. For internal-combustion engines, the hydraulic system is built into the governor and driven by the same speeder rod that drives the ballhead flyweights.

Figure 10.7 shows the mechanical-hydraulic governor speed control for a turbogenerator shown in Figure 10.6. As the load increases, (1) the turbine slows, and the flyweights pull in; (2) The pilot spool valve is pushed upward by the flyweights aligning oil pressure to the power cylinder and piston; (3) Hydraulic pressure is applied to the power piston, which is pushed up; (4) The restoring linkage rises, and the steam valve opens; and (5) The restoring linkage pivots around the speed adjustment, lifting the bushing to block oil flow to the power piston. Shortly after, the increased steam flow causes the turbine speed to increase, causing the flyweights to reposition outward. Eventually, the spool valve and bushing block the ports to prevent oil from flowing into or from the power cylinder and piston, hydraulically locking the steam valve into the correct position.

The governor steam valve uses a lifting beam to sequentially open multiple steam valves. The steam-valve stems have stops adjusted to different heights to open the valves in order. This arrangement provides good modulation of the steam flow and turbine power. The sequential valves for some turbines direct steam to individual nozzle chests, providing higher efficiency.

Details of the flyweight governor, inner spool valve, outer bushing valve, and the restoring linkage follow-up mechanism, which provides negative feedback, are shown in Figure 10.8 to illustrate governor behavior on load increase, and in Figure 10.9 to illustrate behavior on load decrease. The sequence of events is summarized in the figures.

Governor droop and deadband

In attempting to maintain constant speed, mechanical governors exhibit two undesirable characteristics, speed droop and deadband. Both droop and deadband cause deviations from the desired synchronous speed, causing frequency variations. Droop is a result of proportional-only control, where offset leaves speed variations after load changes. Deadband creates small step-like speed variations during load changes, caused by the slip-and-stick nature of friction and backlash in linkages.

In control theory, droop is the same as offset. Droop occurs during load changes, where the speed drops as load increases, or vice versa as load decreases, as shown in Figure 10.10 left. Intuitively, the prime-mover speed needs to change before the governor flyweights can be repositioned, which is followed by a change of the engine fuel-rack or steam-turbine-throttle-valve position. Outward flyweight motion causes the governor linkages to move, eventually modulating the energy input in the direction that corrects the speed deviation. A reduced-speed signal indicates that the load has increased, and this signal is needed for the controller to "know" that more input energy is required; and the residual speed deviation is droop. Governor performance can be described in terms of sensitivity, or fineness of control, which is often measured in percent droop. The droop calculation is similar to the percent-slip calculation used for induction motors. Percent droop is often an adjustable governor characteristic and is calculated between the no-load and full-load speeds (see Equation 1a and b: percent speed droop is a governor characteristic (a) and frequency is a function of engine speed (b)). In a well-designed governor system, droop will be nearly linear when the governor linkage and fuel rack mechanism are properly matched.

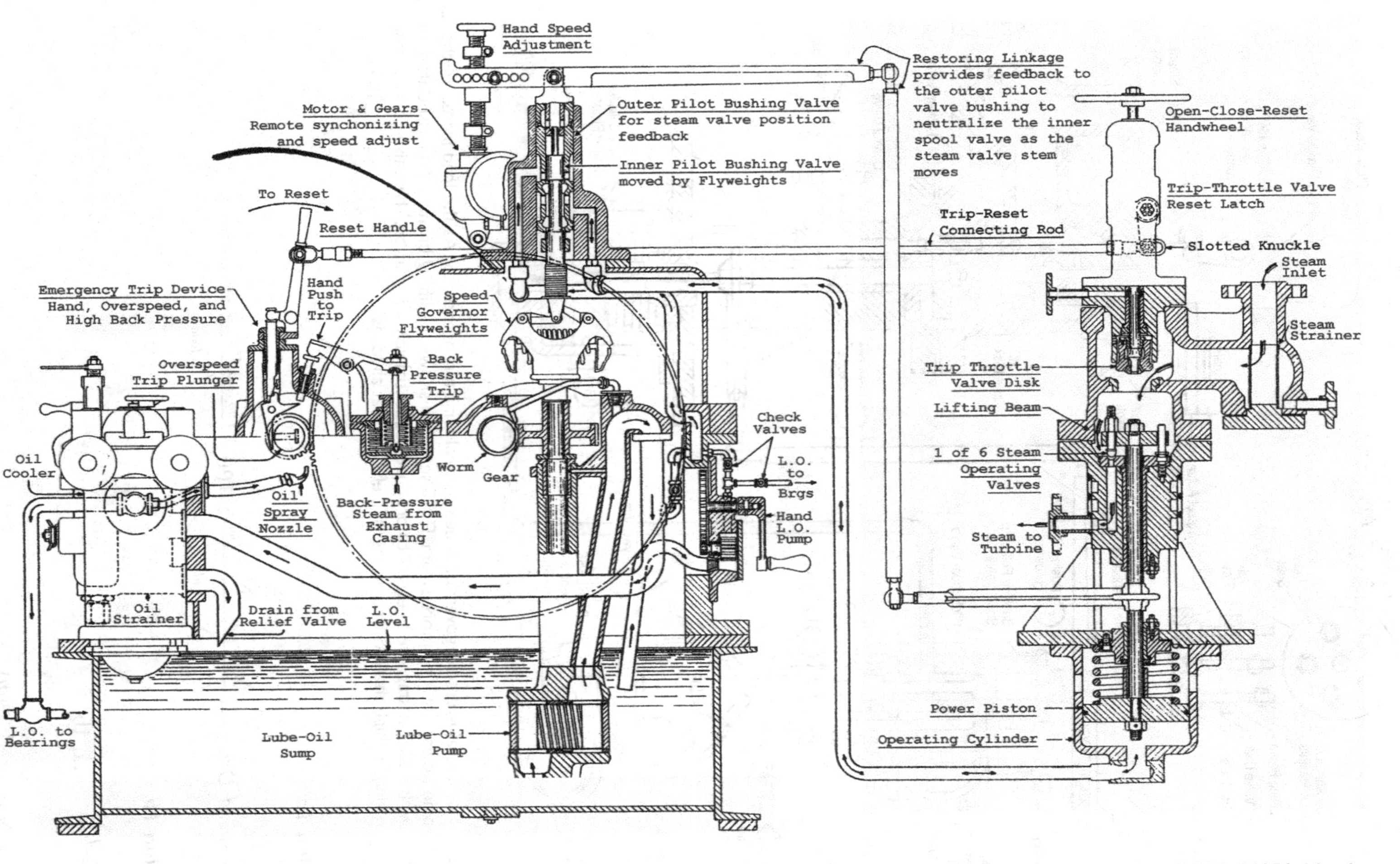

Figure 10.6 Turbogenerator mechanical-hydraulic constant-speed governor. (Adapted from U.S. Navy, *Machinist's Mate 3 and 2, NAVTRA 14151*, Naval Education and Training, 2004.)

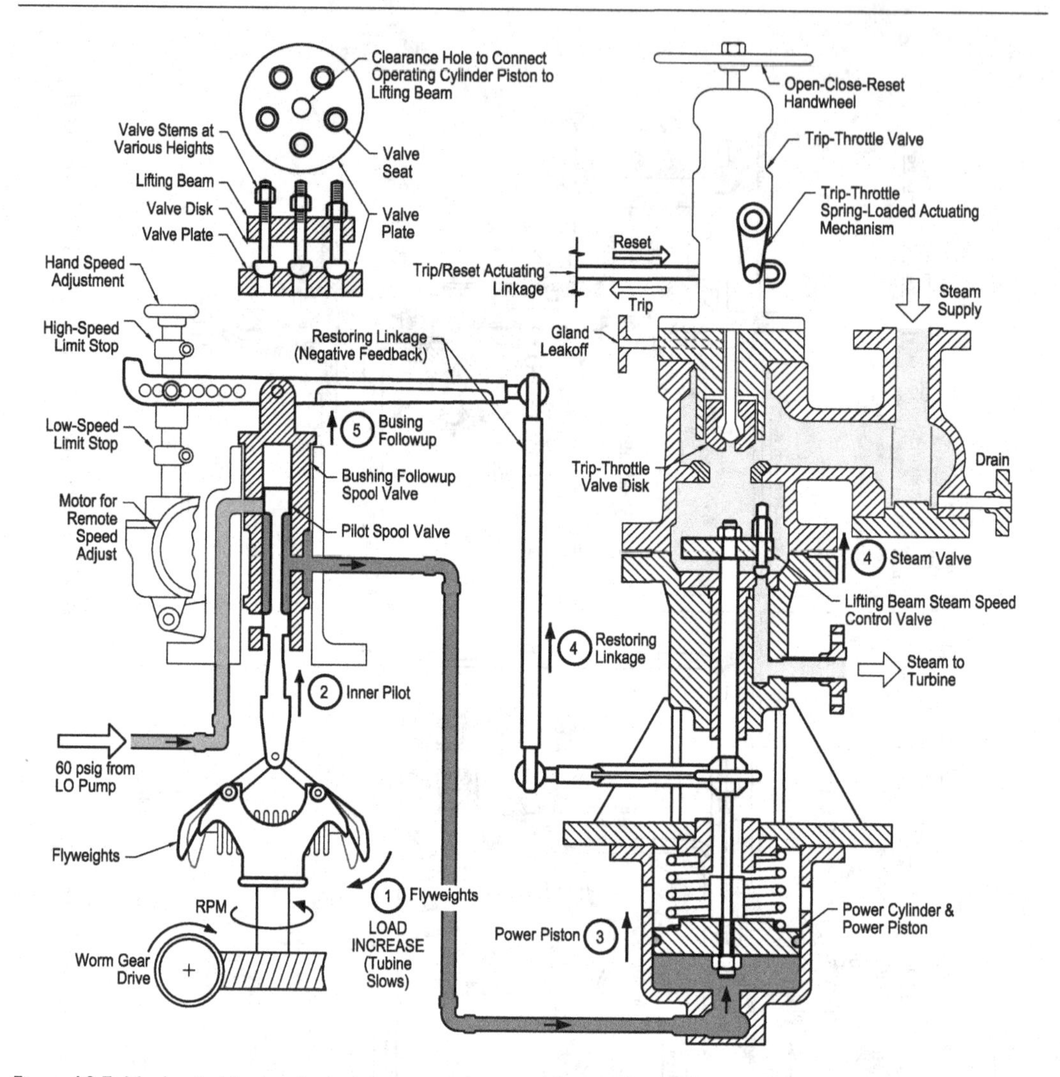

Figure 10.7 Mechanical-hydraulic governor speed control for a turbogenerator. As the load increases, the turbine slows, requiring more steam. The inset detail shows the lifting beam and its associated valves with stems at staggered heights to produce sequential valve operation. This valve control scheme maintains high thermodynamic turbine efficiency at all loads.

Equation 1a

$$\%\ Speed\ Droop = \frac{N_{no\ load} - N_{full\ load}}{N_{no\ load}} \times 100 = \frac{f_{no\ load} - f_{full\ load}}{f_{no\ load}} \times 100$$

Equation 1b

$$f = \frac{pN}{120} \quad \text{(Eq.1b)}$$

where N = speed (rpm)
f = frequency (Hz)
p = number of poles

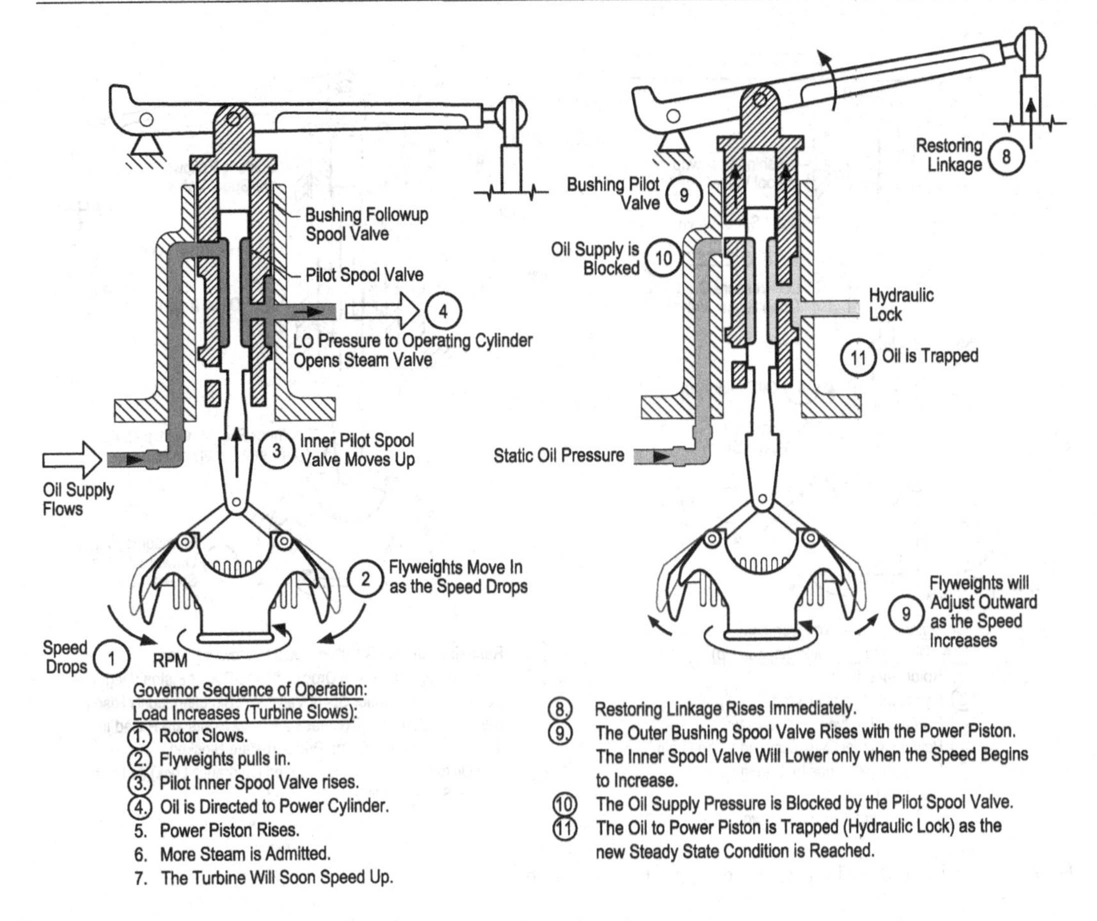

Figure 10.8 Pilot and bushing action during a load decrease.

In mechanical governors, droop always exists, however, the amount of droop is typically adjusted by changing the gain/proportional band[6]. In general, tight speed control requires high gain or a small proportional band, which is obtained by decreasing the droop setting. Governor sensitivity and response improves when the gain is high, but excessive gain results in instability or hunting. Conversely, greater stability occurs when the gain is reduced. Typically, alternators fitted with drooping governors are adjusted for 3–5% droop, to provide good, stable response without too much frequency variation after load changes.

Deadband is the amount of engine speed change that occurs before the energy input is "triggered" to reposition after a load change (see Figure 10.10 right). The problem with mechanical governors is that internal friction results in some "stickiness" in the governor moving parts, and the stickiness results in stick-and-slip frictional variations. Any backlash clearances in the linkages also contributes to deadband. As an example, when a small load is added, the turbine will slow a small amount. When the turbine slows, the governor should open to put in more energy. However, if the turbine slows only a small amount, internal friction could inhibit the governor valve from repositioning. After the speed change is large enough, the extra input-signal would cause the governor to jump to a new position. The amount of speed jump is deadband.

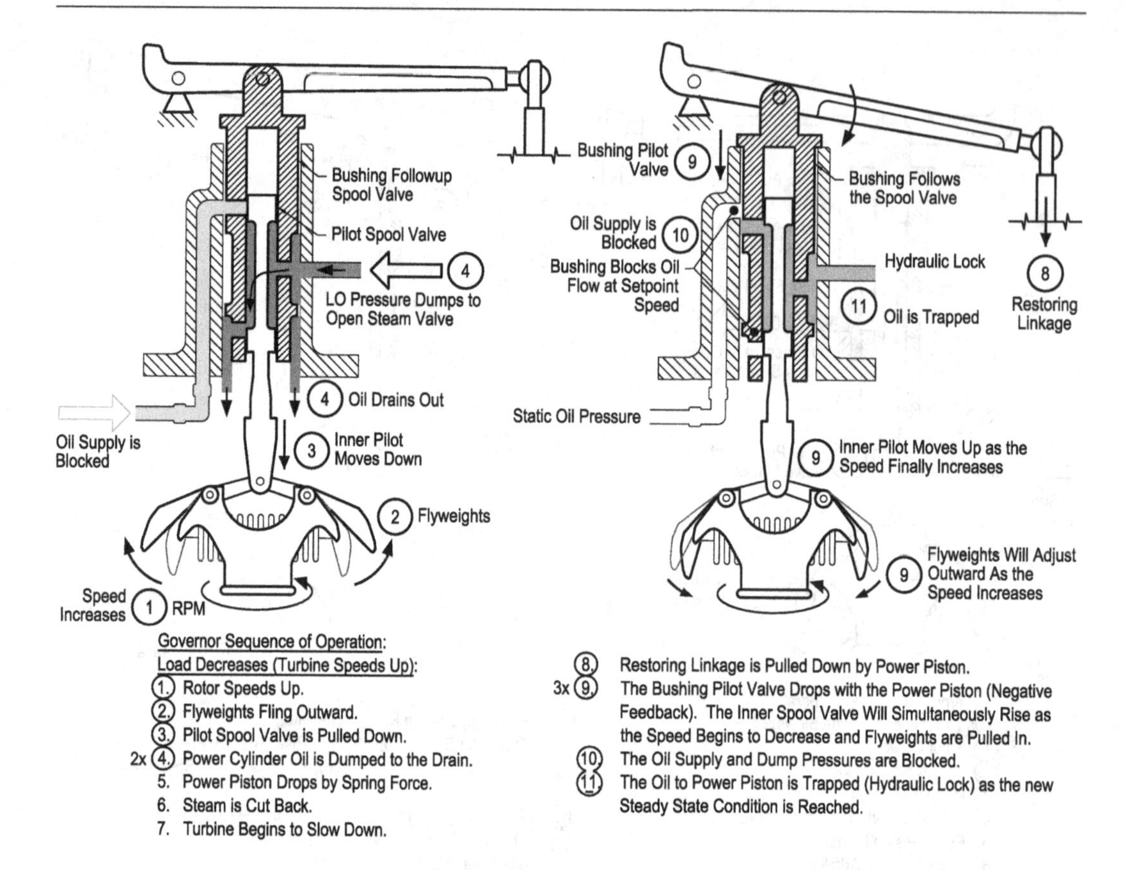

Figure 10.9 Pilot and bushing action during a load increase.

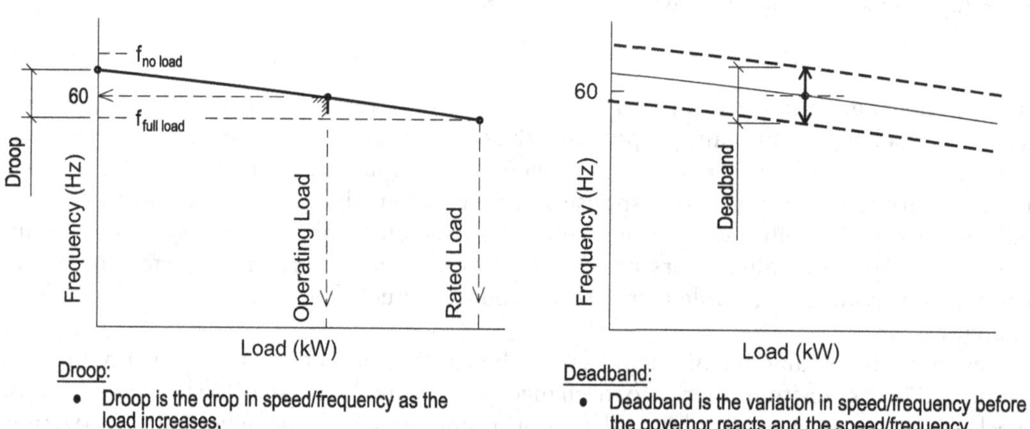

Droop:

- Droop is the drop in speed/frequency as the load increases,
- droop is the same as offset,
- percent droop is calculated between no load and full load.
- The greater the droop, the steeper the curve.

Deadband:

- Deadband is the variation in speed/frequency before the governor reacts and the speed/frequency modulated,
- Deadband results from internal friction within the governor and backlash due to linkage clearances.

Figure 10.10 Droop (left) is the decrease in prime-mover speed between no load and full load. In a well-designed system, droop is nearly linear. Deadband (right) is the variation in speed as the load changes before the governor reacts.

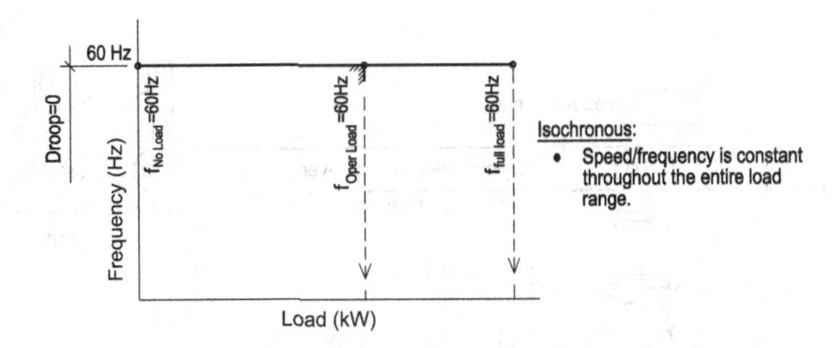

Figure 10.11 Isochronous governor behavior. Prime-mover speed and generator frequency is constant for all loads.

Governor compensation and isochronous control

In addition to having drooping characteristics, mechanical-hydraulic governors can also be designed for isochronous operation, so that engine speed and corresponding alternator frequency is held completely constant regardless of load (Figure 10.11). Isochronous governors have two actions; first, droop that occurs as the speed changes before the fuel is added, followed by droop removal, which is also called temporary droop. Temporary droop occurs in response to the observed engine-fuel usage change, and when done correctly, it returns the flyweights to their original positions and consequently the engine to its original speed. To obtain isochronous operation, a more complex hydraulic circuit adds temporary droop during follow up of the fuel setpoint as it responds to a load change. In control terminology, the temporary droop is called integral control or reset, so that proportional-plus-integral control (PI) is obtained. In internal-combustion engine governors, the temporary speed droop is called compensation. The purpose of compensation is to remove offset and return the engine to its original speed after a load change, i.e., to produce isochronous speed control. Compensation effectively repositions the flyweights to its exact setpoint position, and the engine driving the flyweights therefore returns to its exact isochronous speed setpoint.

As an alternative, electronic governors can measure both alternator speed and attached load and use circuits designed to return the frequency to exactly 60Hz. The governor controller effectively adds a form of temporary droop by initially over-adjusting the actuator in response to how far off the speed is from the setpoint. Load control provides fast response and is mandatory when isochronous generators are paralleled. Figure 10.12 shows a governor that introduces temporary droop for isochronous speed control.

In the hydraulic governor, it is common that the governor has its own self-contained hydraulic-oil system, including a sump, a small gear-type positive-displacement pump, and a built-in pressure control valve. The oil pump is gear driven from the engine and provides high actuation forces with low friction.

Referring to Figure 10.12, the functions of the mechanical-hydraulic isochronous-governor components are:

- *Hydraulic Pump*: driven by the engine. It is a gear-type positive-displacement pump that provides hydraulic-oil pressure to the circuit. When fitted with a four-check-valve arrangement, the pump and governor can be used with reversing engines. The pump is fitted with a spring-loaded control valve providing a means to set the hydraulic pressure by recirculating excess pressure.
- *Accumulator*: device that stores hydraulic energy. In this case, the accumulator absorbs and mitigates any pressure fluctuations caused by the hydraulic pump as its gears discharge into the hydraulic circuit.

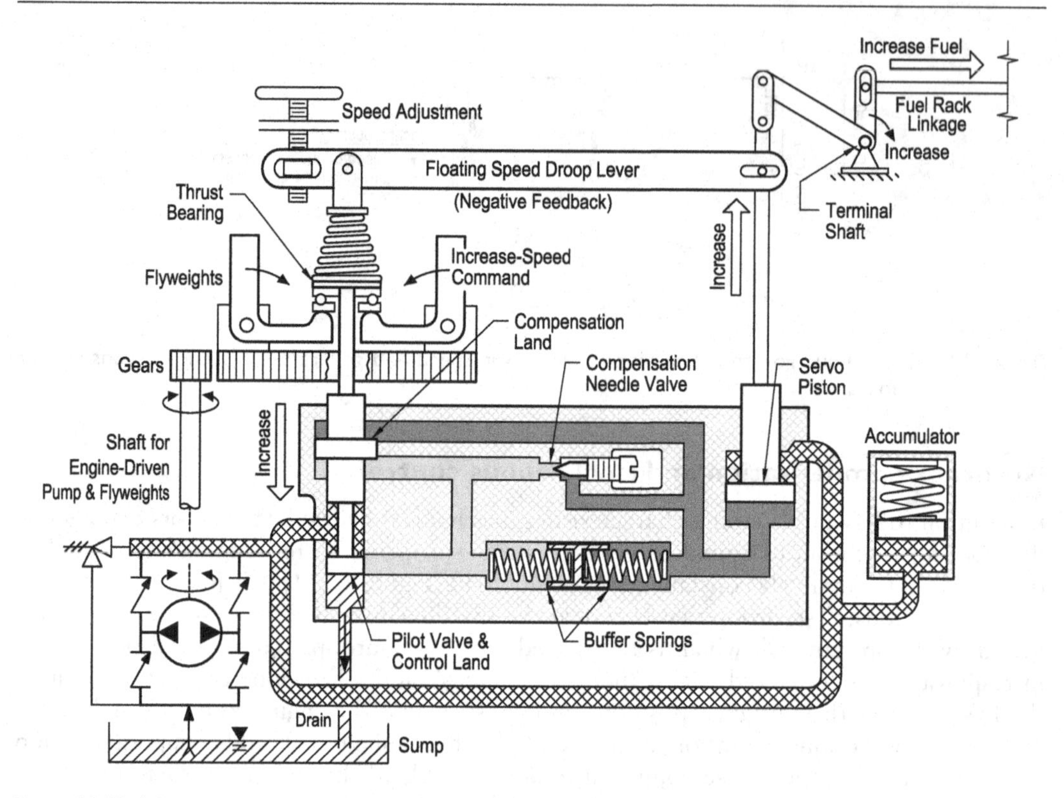

Figure 10.12 Schematic representation of a mechanical-hydraulic governor, configurable for drooping or isochronous operation (simplified from Woodward Inc., EGB governor manual).

- *Drive Gear*: consists of the engine-driven pinon and gear assembly that rotates the flyweight assembly and pump. The gears provide the engine-speed input signal into the governor assembly.
- *Flyweights*: sense the engine speed as part of the input-control signal.
- *Speeder Spring*: provides the compression force that counteracts flyweight centrifugal force. The speeder-spring displacement becomes the input signal into the governor pilot valve through the speeder rod.
- *Thrust Bearing*: The controller error-signal comparator between the desired speeder-spring setpoint and the negative-feedback flyweight speed measurement. The thrust bearing sums the actual speed, sensed as flyweight centrifugal force, with the desired speed, which is set by the speeder-spring compression force, to produce the error signal.
- *Pilot Valve and Control Land*: Part of the actuator assembly. The pilot valve receives the speed-error signal through the speeder-rod movement, and its control land is moved to direct hydraulic oil to/from the servo piston. Hydraulic oil to/from the servo piston only flows when the control land is off center. Hydraulic locking of the fuel rack is obtained when the control land is centered, and oil is trapped beneath the servo piston.
- *Servo Piston (also called the Power Piston)*: Acts through mechanical linkages to control the fuel-rack position. Hydraulic oil pressure actuates the piston both up and down. The piston is of the differential type, with more exposed area on the piston bottom than on the top. Pump discharge pressure is continuously directed to the top. When the pilot-valve control land moves down to direct oil to the buffer springs, the pump

hydraulic pressure is transmitted to the large-area underside of the piston, forcing it up and increasing fuel. The upward force occurs because the same pressure acting on the small area of the servo piston top land simultaneously acts on the larger piston area of the underside. Conversely, when the pilot-valve control land moves up, the servo-piston pressure is bled past the compensation land to the sump, forcing the servo piston down to cut back on fuel.

- *Buffer Springs and Piston*: Used to adjust the hydraulic-circuit gain by varying the buffer-spring compression. Buffer spring compression reduces hunting to provide stability.
- *Compensation Land*: Transmits the force differential between the hydraulic action that occurs immediately to move the servo, and the time-delayed feedback, which acknowledges that the signal has been received by the piston. Differential-pressure forces across the compensation land act opposite to the thrust-bearing force, which moderates the signal caused by the error. Essentially, this action is integral gain, which eliminates offset and achieves isochronous speed control.
- *Compensation Needle Valve*: determines whether the governor will operate in droop-ing or isochronous mode. When the needle valve is wide open, hydraulic pressure is applied simultaneously to the top and bottom of the compensation land, canceling the integral gain. The effect is proportional-only control with droop. When the needle valve is cracked open, a time delay is created between the immediate application of hydraulic pressure on the bottom of the compensation land and delayed application of pressure to the top of the compensation land. The speed-control action is initiated quickly, but the integral feedback from the hydraulic pressure to the top land is delayed from moving the speeder rod opposite to the speeder-spring force, assisting the pilot valve to return to its center when the fuel correction has been made. When the flyweights reach the same position before and after a speed change, the speed will be constant.
- *Terminal Shaft*: The rotational-output shaft from the governor that is connected to the fuel-rack linkage.
- *Speed Adjustment Handwheel*: A biasing signal to add or subtract speeder-spring compression. The speed adjustment changes the amount of flyweight centrifugal force required to reposition the pilot valve, and ultimately the engine-speed setpoint. For alternators, a motor-actuated adjustment is installed for remote paralleling and load sharing.
- *Motor Speed Adjustment (Figure 10.13)*: Installed in lieu of the handwheel for remote operation of the speed control.
- *Speed-Droop Feedback Lever*: Provides negative feedback from the servo piston movement for proportional control. Upward movement of the servo piston lessens the speeder-spring compression, allowing flyweights to move outward.

Two-element governor

An improved method of obtaining isochronous operation is to use a two-element governor. The primary element is the mechanical-hydraulic flyweight controller, which addresses both frequency and load. The second element is an electronic-hydraulic control that responds directly to the alternator's instantaneous load. The second element superimposes its corrective signal onto the primary-element signal. The second element is useful for setting the engine's power output, and for load balancing two isochronous generators in parallel. The load can be measured using kW for an alternator or torque and speed for mechanical loads. In the case of the alternator, load measurement alone is insufficient, since it does not address frequency. However, the second element is effective because it senses the electrical-load changes

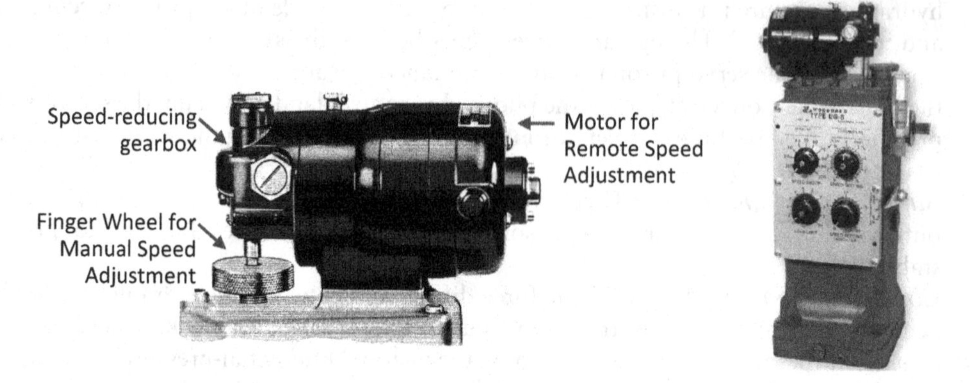

Figure 10.13 A motor used with Woodward constant-speed-governors provides remote engine-speed and generator-frequency adjustment, synchronizing, and kW balancing between alternators. The motor provides remote control, while the knurled finger wheel is used for local manual adjustment. (Courtesy of Woodward, Inc. from technical manual 03505 and shown with the Type UG-8 governor.)

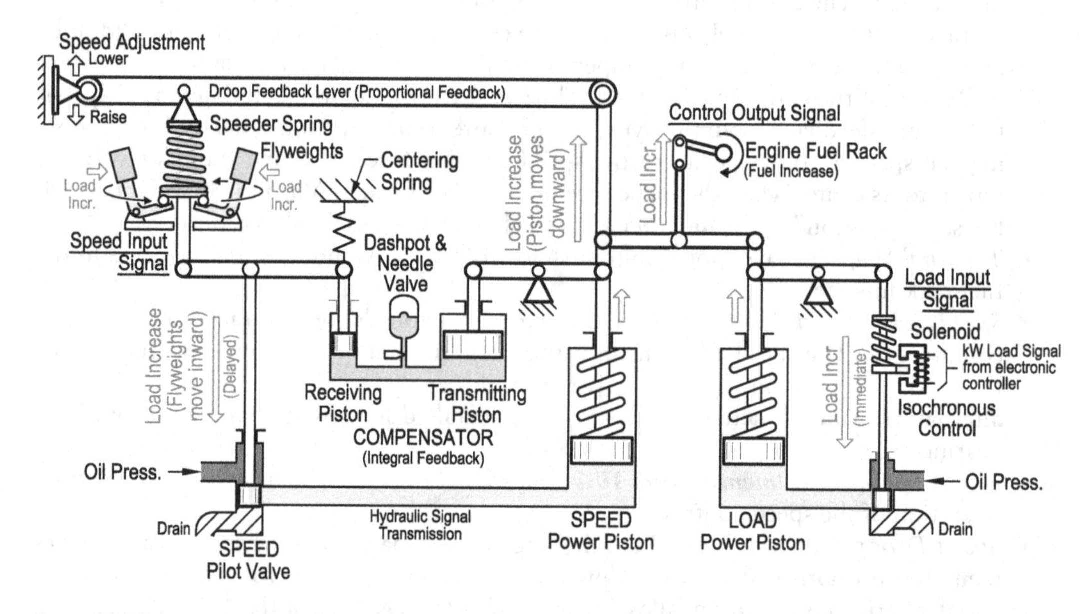

Figure 10.14 A constant-speed governor uses a flyweight for speed control as shown to the left of the fuel rack. The flyweight governor experiences proportional-control droop and delays in producing the speed adjustment. A needle-valve-and-dashpot compensator adds integral feedback, permitting isochronous operation. The load pilot valve and power piston shown to the right of the fuel rack provides Immediate, anticipatory load adjustment for faster governor response.

immediately, before the machine's inertia, hydraulic viscosity, and friction in the linkages are overcome, all which delay the speed change. Thus, the electronic load-sensing control provides an immediate, anticipatory corrective action to the fuel rack, before the speed-change signal is sensed and addressed.

Figure 10.14 shows a two-element constant-speed governor, having mechanical-hydraulic speed control, shown to the left of the fuel rack, and electro-hydraulic load sensing, shown to the right of the fuel rack. The two sensing elements are independent of each other. In this governor, the flyweight speed-sensing control is the primary element since it measures both load and frequency, and the load-sensing control is the second element, providing anticipatory response. The mechanical-hydraulic governor can be configured for either drooping or isochronous operation. Load control provides faster governor response. If electronic speed measurement is added into the controller, the governor can provide precise load sharing, along with the ability to set frequency very precisely, whether operated singly or in parallel. When the electronic controller directly measures load and speed, its alternator can be paralleled with another isochronous generator or a utility's "infinite bus" while in isochronous mode. In that case, speed-sensing control through the mechanical flyweights provides the redundant backup, and it, too, can be operated in drooping or isochronous mode, with the restriction that it cannot be paralleled in isochronous mode.

Load-sensing element

The load-sensing element is shown in the right-hand side of the two-element governor shown in Figure 10.14. Electrical load variations are sensed and addressed immediately by the load-sensing governor before the engine slows. The electronic controller measures changes in the generator kW, which are applied as an input signal to the governor. The controller converts the kW signal into a corresponding electrical signal, which is applied to a solenoid. The load-control solenoid repositions the load pilot valve in proportion to the required load change. As the kW increases, the pilot valve moves down. High-pressure oil applied to the underside of the load power piston pivots the fuel-rack linkage up and provides more fuel to speed up the engine. As the load power piston responds by moving up to increase the fuel supply, the solenoid linkage provides negative feedback by closing off the pilot valve, hydraulically locking the load power piston and fuel rack. In isochronous mode, electronic speed measurement is added to produce constant frequency, in conjunction with the fuel adjustments needed for load sharing.

For non-electrical loads, mechanical load measurements are required when paralleling engines, such as when using two ship-propulsion diesel engines clutched to drive a single propeller shaft through a double-input gearset. Mechanical loads can be measured using two methods to calculate torque; a speed sensor, coupled with either a strain-gage torque meter or with an angle-of-twist meter.[7] In the propulsion-engine application, the electronic engine-load signal could also be used to feather the blades of a controllable-pitch propeller, to maintain a constant engine-power output at any engine speed. When multiple engines are used to drive a common load, it is common practice to set one engine as the master, and the other engines as slaves in a lead-lag arrangement. The master provides the setpoint signals to the slaves as required to produce its predetermined share of load contribution.

Flyweight speed-sensing element

Speed control is produced via a mechanical-hydraulic linkage, originating from the "ball-head" flyweights. The flyweights are gear-driven by the engine. Actuation is delayed by inertia, friction in the linkages, and fluid viscosity. Governor response happens only after the load change has produced a speed change. Since load changes occur immediately, the flyweights serve provide a reliable back-up to the electronic load-sensing element. Like the electronic load-sensing element, the mechanical-hydraulic speed controller can be configured for either isochronous or drooping operation. Referring to Figure 10.14, to obtain droop mode, the needle valve to the compensator dashpot is completely closed, so that any changes in pressure in the transmitting piston are instantly directed to the receiving piston.

When the load increases in droop mode, the engine slows, the ballheads pull in, and the linkage drops around the pin located at the receiving piston/centering spring, and the speed pilot valve moves down. High-pressure oil is directed to the speed power piston, which moves up, shifting the fuel rack upward, increasing the fuel flow. The droop feedback lever follows up control actuation by easing some compression off the speeder spring, thereby establishing a new balance between the ballhead centrifugal force and speeder-spring compression force. The result is droop. Droop is required when an engine having a mechanical governor is paralleled with an isochronous bus. The speed can be changed using the speed adjustment, which essentially repositions the droop-feedback lever fulcrum located to the left of the speeder spring. Moving the droop-feedback linkage down increases the speeder-spring compression, which commands the engine to run faster, throwing the ballheads outward.

In isochronous mode, the needle valve in the compensator is cracked a small amount. The needle valve slows the flow of fluid into or from a dashpot, providing integral feedback (reset) for PI control. Integral control removes offset, or the speed droop that would otherwise occur. The governor operation experiences similar linkage movements as when in droop mode. However, as the speed power piston moves up, a linkage on the power piston causes the compensator transmitting piston to move down. Immediately the oil is flows from the transmitting piston into the compensator receiving piston, lifting the linkage against a centering spring. Upward movement of the receiving piston closes off the speed pilot valve port, which moderates the control response. As steady-state speed approaches, the centering spring pushes the oil into the dashpot until eventually the centering spring/receiving piston are restored to the original setpoint position that existed before a load change occurred. For the linkage between the centering spring and speeder spring to be in the same position as it was prior to the load change, the ballheads must be in the same position, which only occurs when the engine speed matches the original isochronous setpoint speed.

Electronic constant speed governor

The electronic governor is a speed-control device that works without the use of any direct connection between stationary and rotating parts, and thus eliminating the need for more complicated flyweights and linkages (Figure 10.15). The electronic sensors provide extremely accurate speed measurements and more robust control with less deadband (see Figure 10.24 for an electronic panel used for electrical generators). Electronic constant speed governors use either a passive magnetic proximity sensor as the source of speed measurement (Figure 10.16, Figure 10.17, and Figure 10.28) or an active inductive proximity sensor (Figure 10.18). The sensor face is near the gear-teeth tips that are rotated by the prime mover. The speed of the engine is determined by the number of gear teeth detected per minute divided by the number of teeth around the circumference, and modified, if required, by any reduction-gear ratio between the engine shaft and the governor speeder rod.

As its principle of operation, the speed pickup senses the magnet's changing reluctance. The sensor has a permanent magnet surrounded by an electromagnetic coil in the detector face. The magnetic sensor is a passive device, so it only needs signal wiring. As the ferrous gear tooth approaches the probe, the magnetic reluctance drops, and more magnetic lines of flux are produced by the magnet. Later, when the probe sees high-reluctance air as the tooth moves past, the reluctance increases and the magnetic lines of flux decrease. Essentially, the changing magnetic-field strength through the coil creates a generator effect, producing one AC cycle of voltage in the coil for each tooth that approaches and leaves the probe. The result is a signal of varying voltage amplitude and frequency, so the faster the speed, the greater the RMS voltage. To work properly, it is necessary that the sensor proximity and gear speed are capable of generating 1.5V at the lowest service speed.

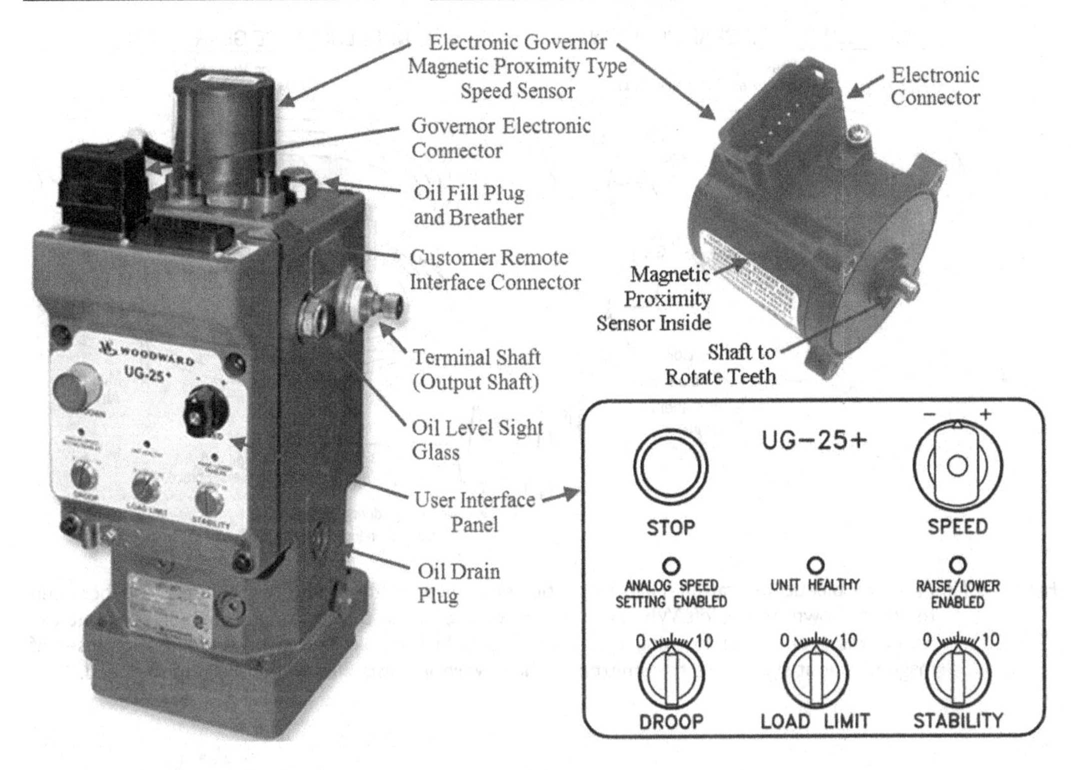

Figure 10.15 The Woodward electronic governor uses a magnetic pickup to accurately sense engines speed. (Adapted from Woodward, Inc. product manual 26579.)

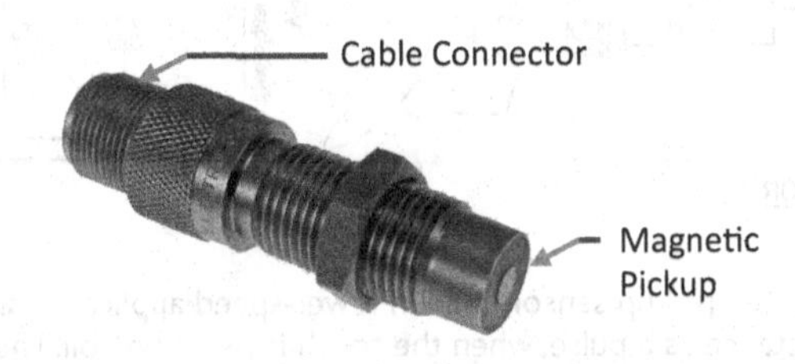

Figure 10.16 Woodward magnetic speed-sensor pickup. (Adapted from Woodward, Inc. product specification 03225.)

The inductive proximity speed sensor (Figure 10.18) is like the magnetic proximity sensor, but it works on the principle that a coil's inductive reactance through is low when an air core is used, but high when an iron core is used. Alternating current is applied to a coil to create an oscillating magnetic field, which in turn induces eddy currents into a close-by conductor, such as a gear tooth. The inductive proximity sensor is an active device that powers an oscillator to induce an AC magnetic field in a coil located at the probe's detector face. When a tooth moves by the probe, eddy currents are induced in the conductive material of the tooth, and the eddy currents in turn create their own magnetic fields. Interaction from the induced magnetic field in the tooth distorts the magnetic field in the sensor coil, which triggers a spike of signal voltage. Essentially, as teeth approach and leave the sensor, the inductive reactance

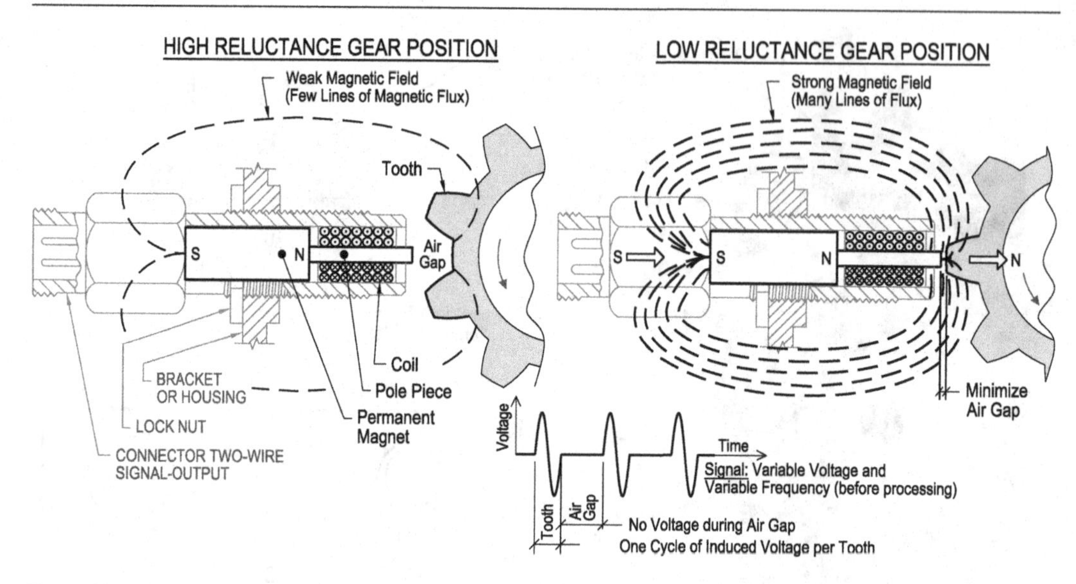

Figure 10.17 Magnetic pickup operation. High magnetic reluctance occurs when the pole piece is between teeth, as shown on the left. When the pole piece is close to the tooth, as shown on the right, the iron in the gear causes low magnetic reluctance, which results in a stronger field. The pulses of magnetic-input signals are transmitted to the governor controller and correlated to speed.

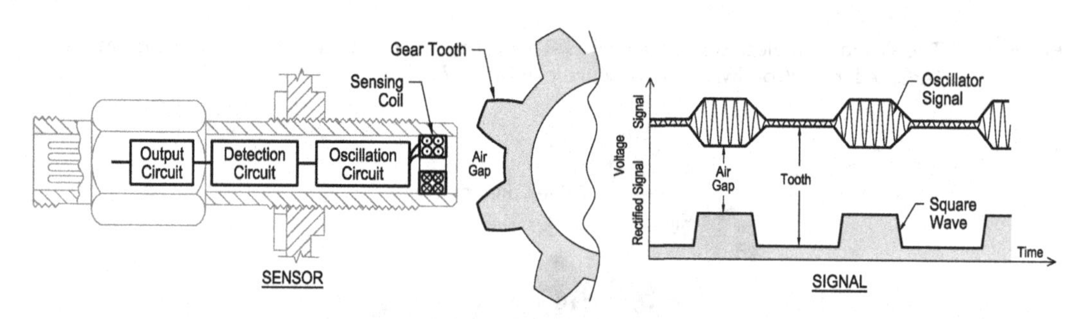

Figure 10.18 Inductive-type speed pickup sensor, used in lower-speed applications. A sensing coil measures an increase in inductance as a pulse, when the tooth is near the coil. The pulse signals are transmitted to the governor controller, and correlated to speed.

through the sensing coil changes, intermittently providing more oscillator-circuit impedance when the tooth is near, followed by less impedance when the air gap is near. The changing impedance produces a square-wave signal.

The inductive-type sensor is an active device, so it requires separate power wiring for the oscillator and coil circuits, and separate wiring for the signal. Depending on the sensor model, the wiring may either be sink or source connected to produce the square wave. Because of hysteresis effects, the inductive-type sensor is better suited for low-speed applications compared to the magnetic proximity sensor. In some devices, the electronic oscillator circuit is located within the sensor body, as shown, while in other devices, the electronics is external to the sensor, decreasing the probe size. The signal is converted to speed based on the number of teeth passing the sensor per minute. If the speed pick-up is

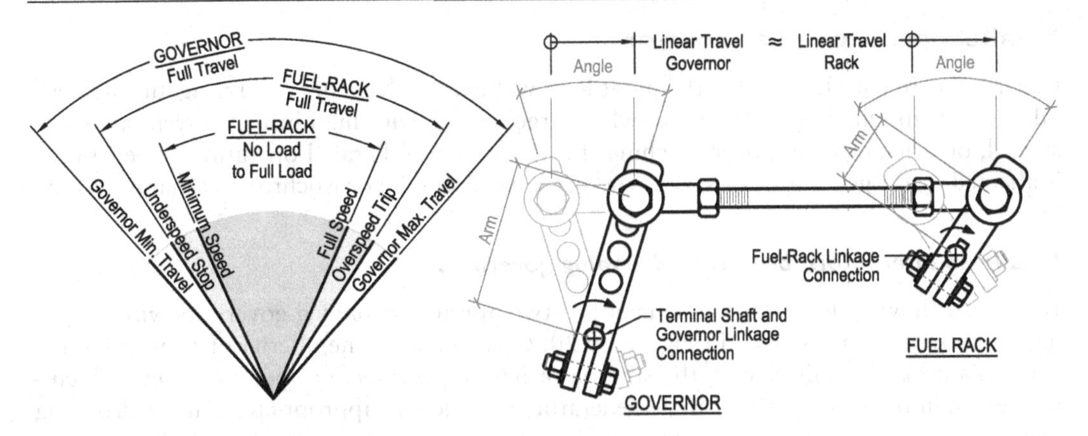

Figure 10.19 The governor linkage must be adjusted to obtain the full range of fuel-rack motion between the engine's physical stops. (Adapted from Woodward, Inc. product manual 26579.)

sensed inside the governor, the engine-to-speeder-rod reduction-gear ratio must be included in the conversion.

$$f(Hz) = \frac{No.of\ Teeth \times Gear\ RPM}{60\,{}^{sec}\!/_{min}} \times Rotor\text{-}to\text{-}Governor\ Gear\ Ratio$$

Governor linkage

Figure 10.19 shows governor-linkage settings on the left coordinated with the governor and fuel-rack linkage movements on the right. The goal is to best-fit the full fuel-rack motion between the engine-stop position at the low end and the overspeed-trip position at the high end. Figure 10.19 (right) shows the governor linkage connected at its maximum arm length, so that small angular movements of the governor terminal shaft are amplified to produce large movements of the fuel-rack. Conversely, if the governor linkage connection at the terminal shaft were moved inward toward the center of the terminal shaft, then large governor movements would produce small fuel-rack movements.

To set the governor control system to work satisfactorily, the linkage adjustments need to properly match the governor terminal shaft range of movement to the fuel-rack range of movement, from the no-fuel stop position to the rated load position. The adjustment needs to include some overtravel margin for the mechanical stops. To ensure that the physical stops on the engine rack mechanism can be reached, the correct angular span between the underspeed fuel-stop setting and the governor minimum travel needs to be properly set, and the correct angular span between the overspeed trip and the governor maximum travel also needs to be properly set. Some fuel-rack motion above the rated engine speed is needed to allow adequate engine acceleration during variations at high loads; and some fuel-rack span below the no-load setting is required to permit deceleration during changes at low loads without reaching the no-fuel setting. Enough range of motion is required to eventually reach the no-fuel rack position to shut down the engine. If the no-load-to-full-load governor movement is too large relative to fuel-rack movement, then the fuel-rack will over-respond and the engine is likely to hunt and surge. If the governor movement is too small, then the engine may not shut down or may not reach its rated output.

Parallel operation of alternators

Generators are paralleled when the plant load will exceed the generator rating, during periods when high reliability is required, when large motors with high inrush current are to be started, or when changing over machines for maintenance. Parallel operation of generators is affected by the governor types, i.e., whether it has drooping or isochronous characteristics.

Paralleled alternators, both with drooping governors

Four cases showing load-sharing behavior of two alternators having governors with drooping characteristics are shown in Figure 10.20. Case 1 shows one alternator supporting the entire plant load of 1,000kW. If this alternator has ample reserve to permit the in-rush current of additional loads, then single-generator operation is appropriate. With a drooping governor, some variation in frequency will be experienced as the plant load changes, and manual readjustment is required if 60 Hz is to be maintained.

Case 2 shows two paralleled alternators having identical ratings, however, the governor for Alternator A has less droop, i.e., tighter speed control, than Alternator B. When two

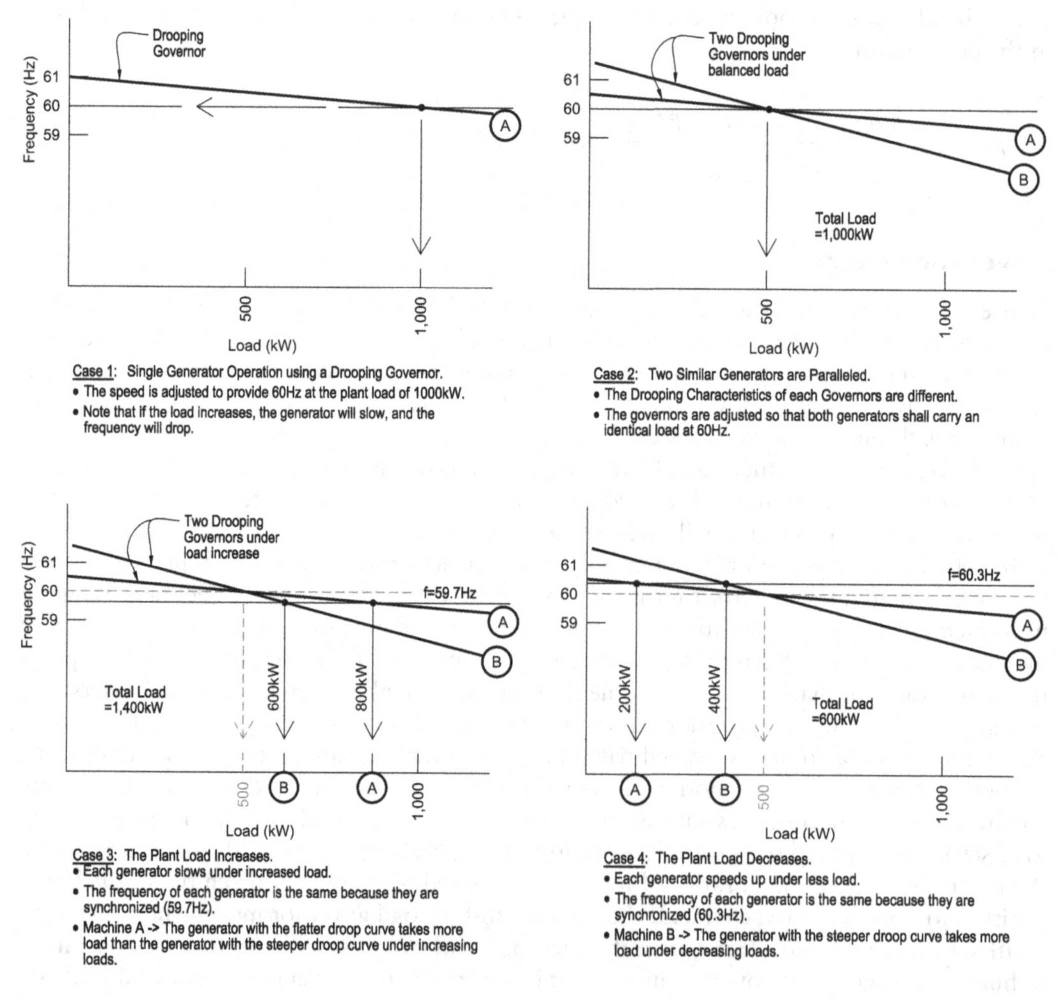

Figure 10.20 Parallel operation of two generators having drooping governors.

alternators having the same rating are paralleled, the loads should be shared equally[8]. In this case, each alternator would be balanced to take 500kW of the 1000kW total load. Ideally, both governors should have identical droop characteristics; in which case, each alternator would provide the same output at all loads. However, in practice, it is likely that the governor droop characteristics will be somewhat different.

Case 3 shows the load-sharing behavior of these two machines when the attached electrical load increases. When the plant electrical load increases, the prime-mover speed and generator frequency drops because both alternators have drooping governors. Regardless of load, the frequency and operating speed must always be the same for both machines, as the generators are "magnetically locked" together in synchronization. In this figure, both machines A and B drop in speed until they both settle at 59.7Hz. The figure shows that Alternator A, with the flatter droop curve, picks up more of the load increase than Alternator B, which has less droop. To rebalance the loads after a change, Machine B would need to be sped up while Machine A is slowed a little until the loads are balanced, and then both machines might need to be tweaked to reestablish 60Hz.

Case 4 is similar to Case 3, but it shows the load-sharing behavior when plant load decreases. Following a similar thought process to Case 3, the result is opposite. On electrical load decrease, the bus frequency increases to 60.3Hz, while both alternators are magnetically locked in synchronization at the same speed. Upon load decrease, Alternator B has the steeper droop curve and it picks up more load than Alternator A, with the flatter curve. Again, if load balancing and 60Hz frequency is desired, the governors of both machines need to be adjusted.

In summary, when governors for paralleled alternators have different amounts of droop, the machine having a flatter droop curve will pick up more load upon load increase, and the machine having a steeper droop curve will take more load upon load decrease.

Paralleled alternators, one isochronous and the other with a drooping governor

Original isochronous-governor designs were mechanical-hydraulic, which achieved constant speed by introducing "temporary droop," like the governor shown in Figure 10.12. Case 5 in Figure 10.21 shows that using two alternators having speed-based isochronous governors, with both generators in parallel isochronous operation, will cause electrical-load sharing problems. The machine adjusted for a slightly higher speed looks at the lower-speed bus and adds more energy to go faster. The machine adjusted for a slightly lower speed looks at the higher-speed bus and cuts back its input energy to go slower. The result is that the slower-speed generator sheds all load, while the higher-speed generator picks up all load. Eventually the lower-speed generator becomes motorized and trips on reverse power. In cases where the setpoint speeds of the two machines are nearly identical, the governors wind up swinging the loads back and forth in an unstable manner.

Case 6 in Figure 10.22 shows an alternator with a drooping governor paralleled with an isochronous machine. In this case, isochronous Alternator A maintains the bus frequency at a constant 60Hz, regardless of load. Because the bus frequency never varies, Alternator B with the drooping governor always operates at constant speed. Consequently, at constant speed, the load on Alternator B never changes, and Alternator A picks up all load variations.

If it is desired to balance the loads after a load change, the governor for Alternator B is raised or lowered, changing its prime-mover energy input. Because the bus frequency is held constant by the isochronous machine, the energy-input changes to Alternator B will only vary its load-sharing contribution. No adjustments should be required of Alternator A.

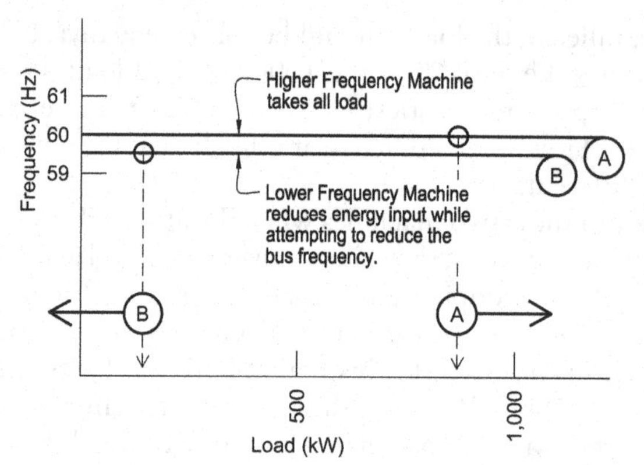

Case 5: Parallel-Generator Operation with two Isochronous-Governor Generators.

- The governor which is adjusted for the higher speed will look at the slower bus frequency and determine that the generator needs to go faster. That generator would eventually take all of the load (Generator A).
- The generator governor having the lower-speed setting (Generator B) would look at the faster bus frequency and determine that it needs to cut the energy input in an attempt to reduce the bus frequency. That generator will eventually motorize and trip off line.
- If it was possible to exactly match the speeds of each machine, there would be no way for each machine to "know" how much load to share. The loads would swing back and forth between the generators in an unstable manner.

Figure 10.21 Attempts to parallel two alternators, both having speed-based isochronous governors, will result in the lower-speed generator transferring its entire load to the higher-speed generator. The lower speed generator will eventually become motorized and will be tripped via the reverse-power relay.

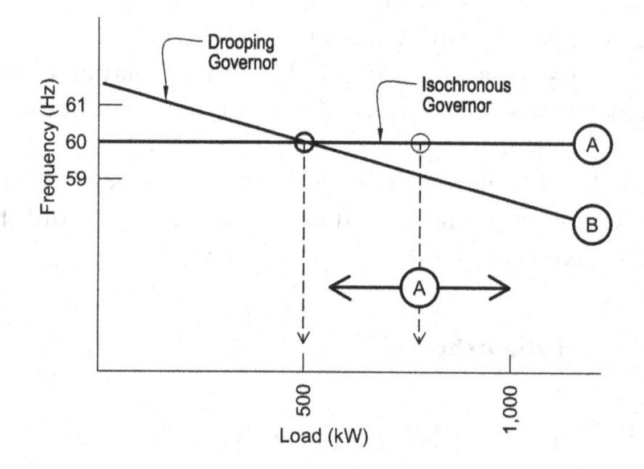

Case 6: Parallel-Generator Operation with an Isochronous Governor and Drooping Governor.

- The Isochronous Governor forces the bus to remain at a constant 60Hz frequency at all loads.
- Machine B (Drooping Governor) always operates at one frequency. Consequently, the speed is always the same, and therefore, the load on Machine B never changes.
- Machine A (Isochronous) picks up all load variations, while always maintaining 60Hz.

Figure 10.22 Parallel operation of a drooping-governor alternator with an isochronous machine. The isochronous machine fixes the bus frequency at exactly 60Hz. Since the speed of the drooping alternator remains fixed, its load-sharing contribution to the bus never changes. Consequently, the isochronous machine picks up all load swings.

Case 7 in Figure 10.23 shows two alternators, each fitted with load-based isochronous generators. These machines are capable of equally sharing all loads while maintaining a constant 60 Hz, because they directly sense load, rather than sensing speed as a function of load. The speed sensors maintain 60Hz, while the load sensors balance the real power. The automatic voltage-regulators balance the reactive power.

Other alternator controls

When an electrical alternator is operated singly, it picks up the entire real and reactive components of the attached load. The governor establishes the bus frequency by controlling its prime-mover speed, and its automatic voltage regulator sets the voltage. However, when more than one alternator is required for powering the attached loads, the alternators need to be synchronized, paralleled, and then both the real and reactive loads need to be shared between machines.

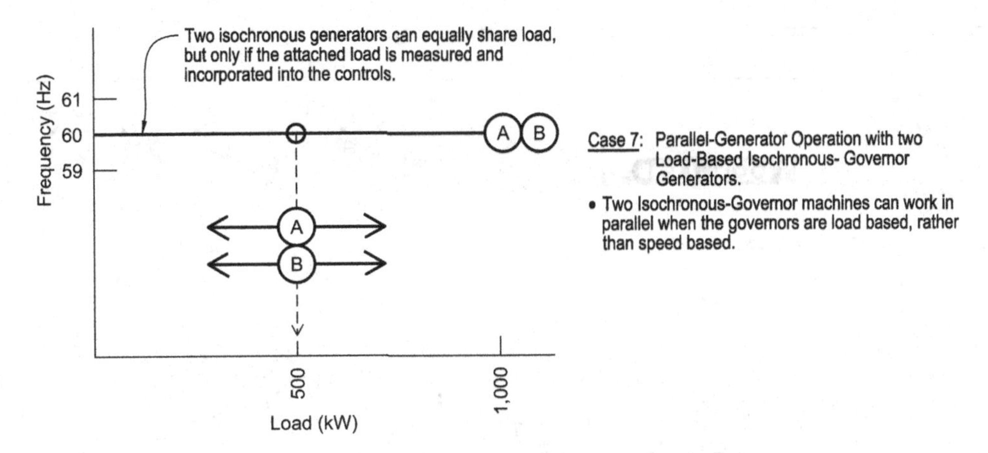

Figure 10.23 Parallel operation of two isochronous generators, each having speed-based frequency control and load-based power sharing. These machines are capable of equally balancing all loads while maintaining a constant 60Hz.

For synchronizing and balancing loads, the generators must have

- the same number of phases
- the same phase sequence
- the same voltage
- the same frequency
- a means of closing the circuit breaker at the proper time

And after synchronizing is complete, the following adjustments are needed:

- Real power—via the governor
- Reactive power—via the voltage regulator
- Frequency—via both governors

Real power is the power that is consumed by the electrical system, in kW, and is directly related to the fuel consumption. Real power is calculated from: P=3VIpf Where the power factor represents the phase shift between the voltage and current from the inductive components of the attached loads. Real power and frequency are both adjusted by the governor setpoint.

Reactive power is the power to supply the magnetic components of inductive loads, such as motors and transformers. The reactive component is the amperage that makes the magnetic fields. That current is returned back to the system when the magnetic fields collapse. The reactive power is adjusted between alternators using the automatic voltage regulator (AVR). The AVR modulates the alternator field current. Raising the voltage results in more field current, leading to more alternator output voltage and current, and consequently, more-lagging power factor. Adjusting the AVRs of both machines balances the reactive loads and sets the desired bus voltage.

When the governors and automatic voltage regulators are properly adjusted, the real power, reactive power, power factor and amperages for both machines will be identical to each other, while maintaining the bus at 60Hz.

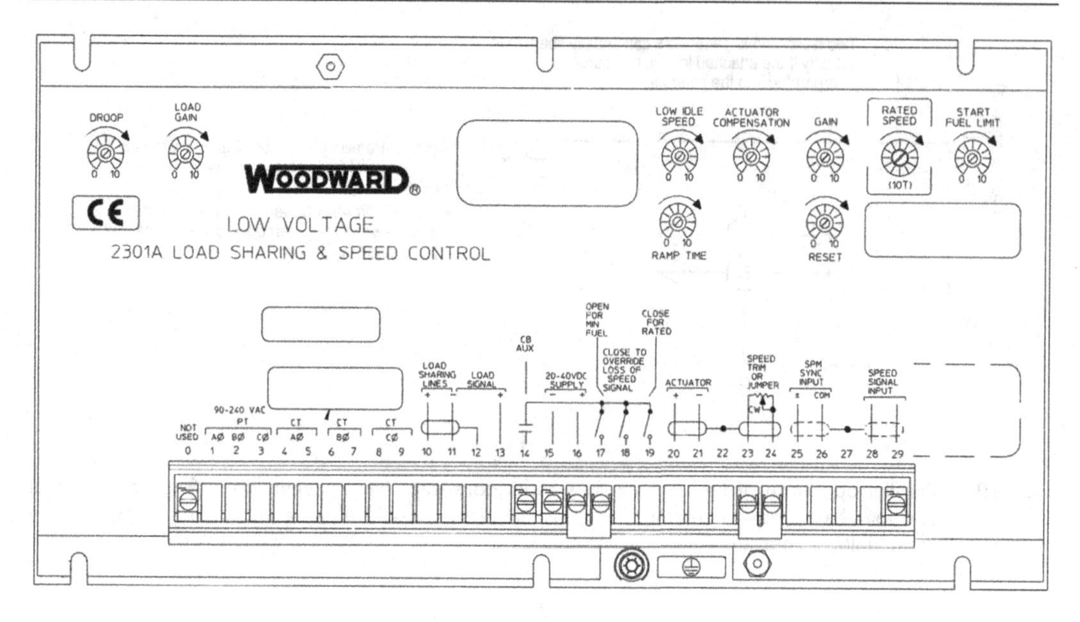

Figure 10.24 Woodward electronic load-sharing and speed controller. (Adapted from Woodward, Inc. technical manual 82389.)

OVERSPEED TRIP DEVICES

An automatic safety shutdown installed on many rotating prime movers is the overspeed trip mechanism. The overspeed trip disables the source of input energy and allows the prime mover to coast to a stop. Overspeed trips are used in conjunction with governors, where overspeed will create a hazard and interrupted service can be tolerated. Overspeed trips consist of mechanical and electronic types.

Mechanical overspeed trip

The mechanical overspeed trip uses a weighted plunger mounted within a radial hole in the prime-mover's rotor (Figure 10.25). The center of gravity of the plunger is located eccentrically outboard from the center of shaft rotation; so that as the rotor spins faster, the centrifugal force builds, flinging the plunger outward. The centrifugal force is balanced by compression-spring force. When the centrifugal force increases enough, the plunger begins to protrude from the rotor surface until it eventually strikes a tripping arm and latching mechanism. A latched spring-loaded trip-throttle valve for a steam turbine is shown in Figure 10.27, where the tripping arm initiates the trip-throttle valve closure. A trip-throttle valve is shown in Figure 10.27. Once unlatched, the power spring in the trip-throttle valve slams its disk closed, allowing the machine to safely stop.

As an alternative to the latched trip-throttle valve, the centrifugal overspeed tripping mechanism can be linked to open a pressure-unloading valve in the hydraulic governor circuit. When actuated, the linkage opens a valve, draining the hydraulic pressure that keeps the governor valve open.

Referring to Figure 10.27, after the valve trips, it requires manual resetting by the plant operator before the machine can be placed back into service. To manually reset the trip-throttle valve, the handwheel is turned in the "close" direction. However, since the valve is

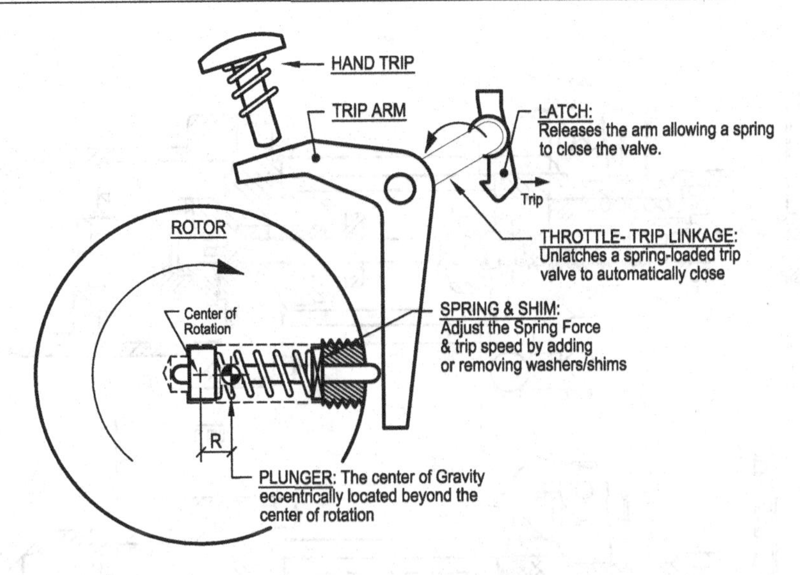

Figure 10.25 Overspeed trip mechanism. Refer to Figure 10.27 for the operation of its associated trip-throttle valve.

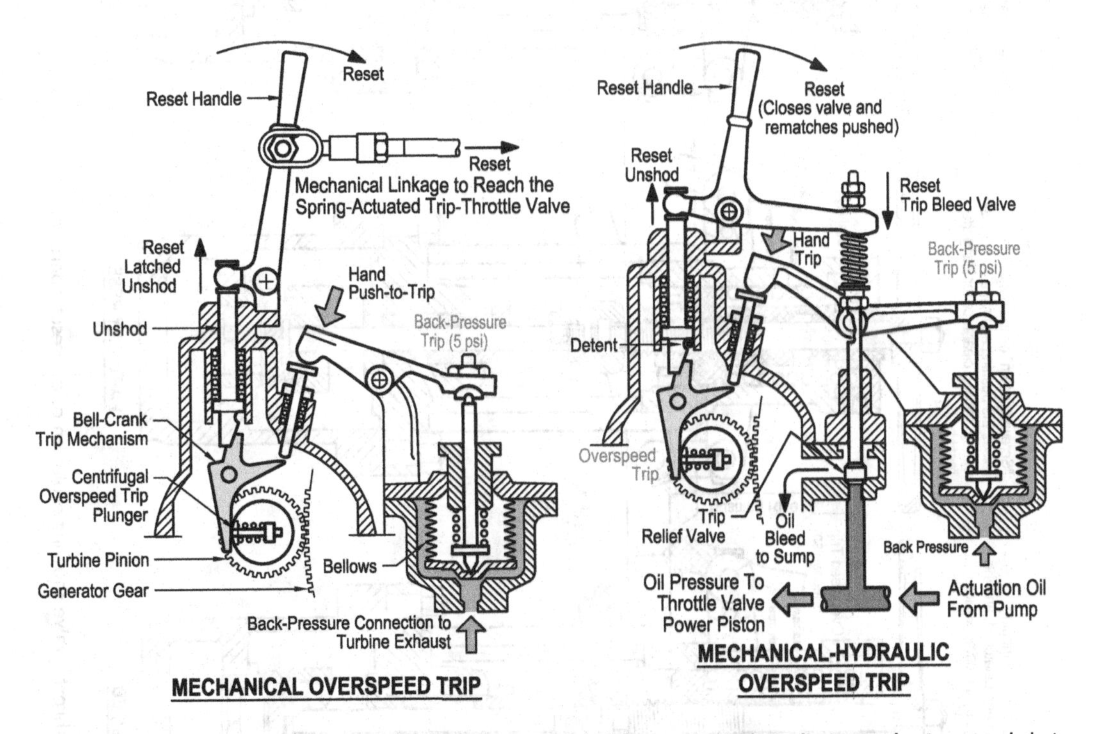

Figure 10.26 Two variations of high-back-pressure-trip, hand-trip, and overspeed-trip mechanisms, and their associated reset mechanisms. The trip on the left mechanically unlatches a spring-actuated steam valve, while the mechanism on the right bleeds the lube-oil pressure that operates a hydraulic governor.

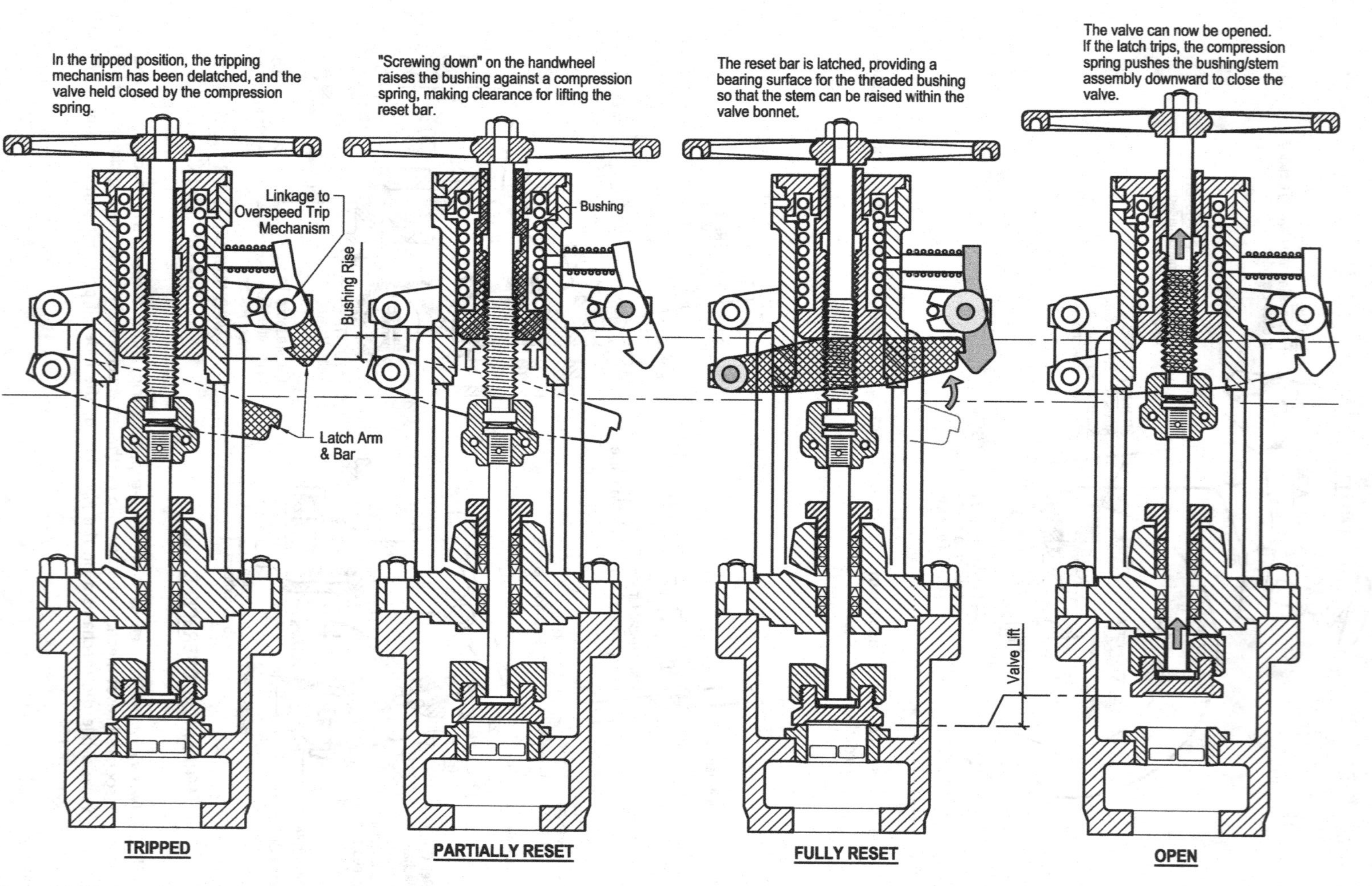

Figure 10.27 Trip throttle valve showing evolution of the reset operation.

already closed, this action instead causes an unrestrained threaded bushing to ride up on the valve-stem. The bushing, in turn, compresses and re-energizes the power spring that is used to quick-close the valve. Raising the bushing makes clearance so that the latch bar can be lifted for resetting. The latch bar is lifted manually by a lever handle attached to the tripping arm linkage, thus engaging the latch, as shown by the reset handle in Figure 10.26.

To open the trip-throttle valve after it has been reset, the handwheel is turned in the "open" direction. Initially, the bushing is driven downward by the valve-stem threads, where it soon bears against the latched bar. The bar prevents the bushing from moving further down the valve stem. At that point, the turning threads of the valve stem pull the stem through the bushing and the stem rises, opening the throttle valve. During normal startup, the valve is cracked to the point where the turbine comes up to speed slowly. Slow startup allows the operator to observe for abnormal conditions, such as excessive vibrations, water hammer, etc., and the operator has the option to choke in on the valve to mitigate the problem, or to hand-trip the valve. Once the governor takes control, the trip-throttle valve is opened fully, at which time it serves as a safety permissive.

Electronic overspeed trip

As an alternative to the centrifugal trip, it is possible to use electronic speed indicators to provide the trip-actuating signal. For reliability, the overspeed trip sensors must be independent from the governor speed sensor, and best practice would be to use three magnetic pickups (Figure 10.28) in a two-out-of-three "voting scheme" (2oo3) to ensure that the unit will not shut down because of transients or failure of a single pickup. The two-out-of-three voting scheme is shown in Figure 10.29. Electronic governors and trips are particularly conducive to high-speed machines, where mechanical trip devices tend to be less accurate and where dangerous overspeed can occur during a sudden loss-of-load; an application requiring extremely fast-acting overspeed tripping action. The electronic overspeed trip is easier to calibrate, and does not require the same trial-and-error methods as is necessary for adjusting the spring compression in a mechanical trip. Electronic trips can be obtained with accuracies up to 0.1% of the speed. Testing of the overspeed trip can be done using a frequency generator to simulator overspeed, and often the test/calibration function is built into the electronic controller.

Figure 10.30 shows a schematic arrangement for a highly reliable, fail-safe electronic overspeed trip system. No single fault in the power supply, controller, or dump-valve manifold will result in an interruption of service. High reliability is obtained using multiple power supplies, redundant two-out-of-three-voting speed-sensing modules, and a 2oo3 dump-valve

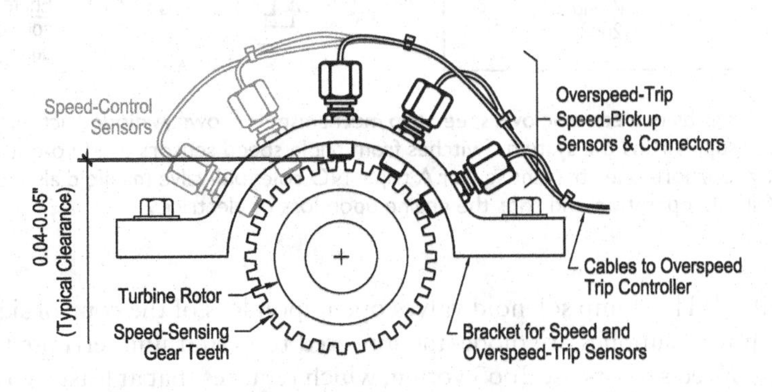

Figure 10.28 Electronic speed-pickup sensors and mounting bracket.

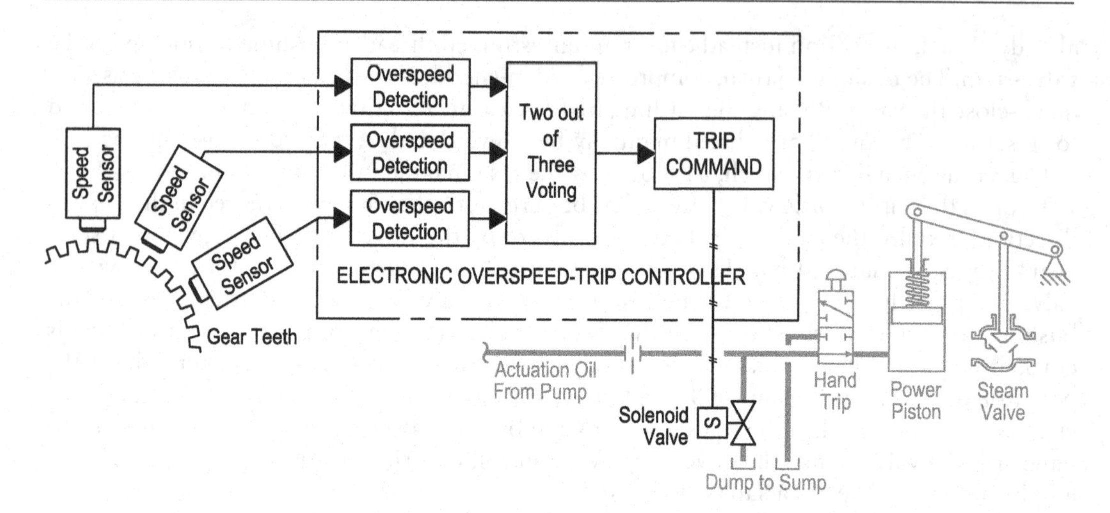

Figure 10.29 Electronic-hydraulic overspeed trip system. Signals from three speed sensors are input into a two-out-of-three voting system, for high reliability. Lube-oil pressure used for actuation is bled from the system to trip the engine.

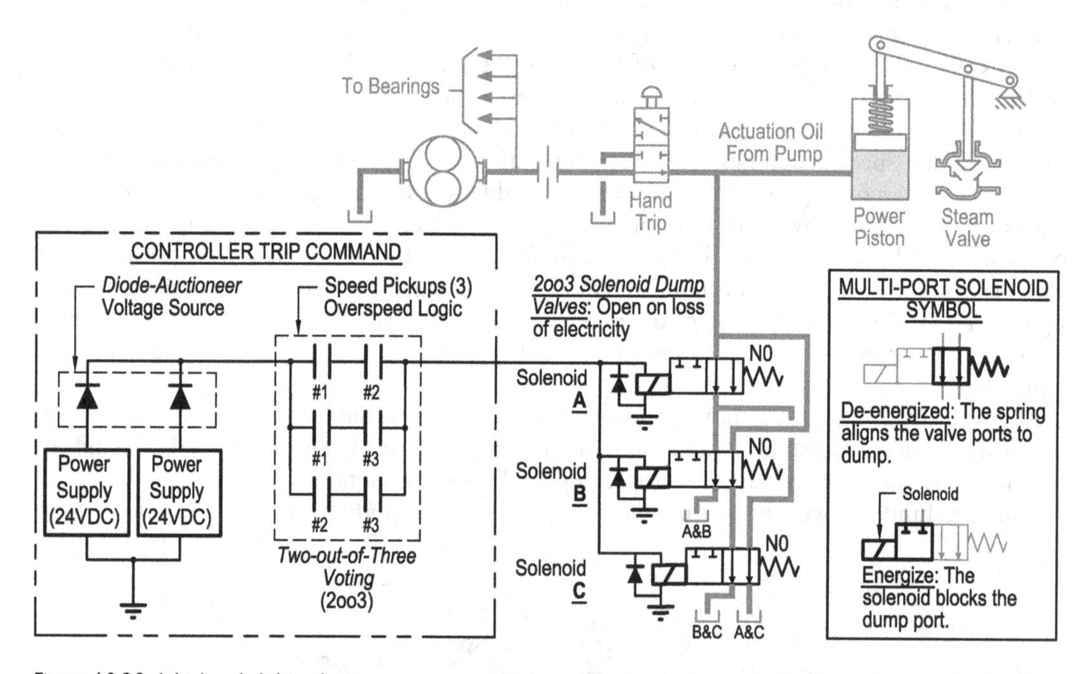

Figure 10.30 A high-reliability electronic overspeed trip mechanism is shown. A diode-auctioneered redundant power supply runs the system. Switches from triple speed sensors use two-out-of-three (2oo3) voting to ignore one aberrant sensor. A triple NO solenoid-valve manifold also uses 2oo3 voting for fail-safe operation, and trips the engine upon loss of electricity.

hydraulic manifold. The dump solenoid valves open upon loss of the control signal or loss of electricity. The power supplies are diode-auctioneered to permit uninterrupted power if one fails. The triplex speed sensors use 2oo3 voting, which requires that at least two pickups must simultaneously sense an overspeed condition before the trip mechanism is actuated. This arrangement eliminates nuisance tripping from an occasional spurious transient. Likewise, three solenoid valves are connected in a similar 2oo3 voting arrangement, so that the system continues to function even if one solenoid valve or coil is defective.

Other safety shutdown devices

In addition to overspeed and hand trips, low-lube-oil-pressure and steam-turbine-exhaust excess-backpressure trips are commonly incorporated into the trip-throttle valve linkage. Because the lube-oil pump is commonly driven by the turbine rotor, it cannot develop pressure until the rotor is turning. However, to start the turbine, lube-oil pressure is needed as a permissive to allow the trip-throttle valve to open. Consequently, one of several methods needs to be employed before the turbine can be started:

- the plant operator pries on the low-oil-trip mechanism, simulating that oil pressure is present until the machine comes up to speed, or
- the low-pressure oil trip is blocked by a bar, simulating oil pressure. As the turbine comes up to speed and pressure is developed, the bar drops by gravity, placing the low lube-oil-pressure trip in service, or
- a hand lube-oil pump is used to create pressure, until the turbine develops its own pressure, or
- a motor-driven lube-oil pump provides the initial lube-oil pressure. As the turbine comes up to speed, the rotor-driven pump actuates a pressure switch, cutting-out the electrical pump.

To secure the turbine, the best practice is to remove load from the turbine and then to hand trip the trip-throttle valve at a low steam-flow rate. Quick closure avoids wiredrawing[9] of the trip-throttle valve, which can damage the valve disk and seat. After the valve is tripped, the valve handwheel should be "closed," which is really "resetting" the bushing position in preparation for the next startup. Since the power spring holds the valve tightly closed, the handwheel should be backed off from being tight, to avoid binding from thermal contraction of parts as the valve cools and to allow the plant operators to easily determine whether the valve was left in the tripped or reset position.

QUESTIONS

1. What is a speed-limiting governor? Where might a speed-limiting governor be used?
2. What is a constant-speed governor? Where might a constant-speed governor be used?
3. As an example, how does a shipboard speed-limiting governor work for a propulsion steam turbine? What secondary control feature is incorporated into this governor?
4. Describe how an alternator mechanical constant-speed governor works. What inherent undesirable trait occurs with the mechanical constant-speed governor? For large machines, how are high actuation forces obtained?
5. What is governor droop, and where does it come from? What is governor deadband, and what is its cause?
6. What is an isochronous governor? Why is it that two generators both having mechanical isochronous governors cannot be paralleled?
7. How are electrical load changes shared between two generators when one generator has a drooping governor and the other has an isochronous governor?
8. How does an electronic speed-control governor work?

9. Why is it possible that generators having modern electronically controlled isochronous governors can be paralleled and share loads equally, even after load changes have occurred?
10. How do mechanical overspeed-trip mechanisms work? How would a mechanical overspeed trip mechanism be recalibrated, if its trip setpoint was too low?
11. What is a trip-throttle valve used in a generator mechanism? How is the trip throttle valve reset? How is the trip-throttle valve actuated? What is wiredrawing in a steam valve?
12. How does an electronic overspeed device work?
13. What is two-out-of-three (2oo3) voting? How does 2oo3 voting increase reliability?

NOTES

1. Electric motor speed control is not covered in this chapter. Motor speed variations can be obtained using dc electricity, multiple-pole windings, wound-rotor construction, variable-frequency drives, and other methods as discussed in Chapter 6.
2. A ship's main engine may be a steam turbine, gas turbine, or an internal-combustion engine, such as a diesel engine. Each has provision for avoiding dangerous overspeeding.
3. "ABS Rules for Building and Classing Steel Vessels," 4-2-4/7. 1 (Part 4, Chapter 2, Section 4, Paragraph 7.1) limits a propulsion turbine to be no more than 15% overspeed for shipboard applications.
4. From the affinity laws, flow is directly related to pump speed, and the developed pressure varies with the speed squared.
5. Standard utility frequencies are either 50Hz or 60Hz. 60Hz is used in North America, while 50Hz is common in Europe, with worldwide use divided between one or the other. Other frequencies are used for non-utility purposes, such as 400Hz for military or aviation electronics or 25Hz for some electric subways.
6. Note when adjusting governor controllers, manufacturers may refer to gain as droop, offset, sensitivity, or proportional band, where proportional band is the inverse of gain.
7. The angle of twist is calculated from: $\theta = \frac{TL}{JG}$ and the horsepower is calculated from: $HP = \frac{Tn}{63025}$
8. Electrical loads should be shared in proportion to the generator ratings when not identical.
9. Wiredrawing is high-velocity fluid-flow friction in steam valves that can gouge the sealing surfaces and prevent tight valve closure. Wire drawing is exacerbated by operating the valve in the throttled position and when the steam contains moisture. Minor wiredrawing can be repaired by lapping with valve compound, but severe wiredrawing requires replacement of the disk and seat.

Chapter 11

Programmable-logic controllers and operation

PLC HISTORY

Today PLCs are found in many industrial, commercial, and even residential applications. However, before the late 1960s, PLCs did not exist, and industrial control primarily used mechanical, pneumatic, hydraulic, and relay-based electrical controllers, where the electrical relays performed the on-off logic functions to produce output signals to start and stop equipment. In automated plant processes, complex control and sequencing functions were performed by small low-voltage control relays interacting with sensors; and a control cabinet could be large and contain dozens of relays with a complicated array of wires. The programming language for those relay-based controllers was a wiring diagram, and changing the logic consisted of adding or removing control relays and rewiring the controller.

In the late 1960s, the General Motors Corporation solicited industry for a standardized programmable machine controller. GM's original requirements included a solid-state electronic system having flexible programming, easy maintainability, competitive costs when compared to conventional control-relay systems, easily understood programming language using familiar relay-logic symbology, functionality in a harsh industrial environment, and modular design for exchange of components and expandability. Dick Morley is credited as being the father of programmable-logic controllers through his company Bedford Associates and later the Modicon Company (Modular Digital Control). By the mid-1970s, the Modicon PLCs became the first microprocessor-based, distributed-control systems using algorithms for continuous digital-and-analog control. Later the company developed the Modbus industrial communications network, which allowed the direct interface of PLCs to computers. Modbus became an industry standard due to its high reliability, and its topology is so robust that it is still one of several communication topologies and protocols used today. The first PLC programming devices were large non-portable dedicated terminals. Later, handheld programmers were developed for field troubleshooting and in-situ reprogramming, and now personal computer software is used for factory programming and in-situ troubleshooting. The Modicon brand is still around and is owned by the Schneider Electric company.

Today there are several PLC programming languages in use, including ladder-logic programming, which is similar to the original programming language. Functional block diagram programming methods, such as the Siemens SIMATIC language, are also used as well as some high-level programming languages. Some feel block-diagram programming is more intuitive when reprogramming may be required by plant operators who may not completely understand ladder diagrams. Chapters 13 and 15 focus around the Allen-Bradley/Rockwell Automation PLCs and ladder-diagram programming language.

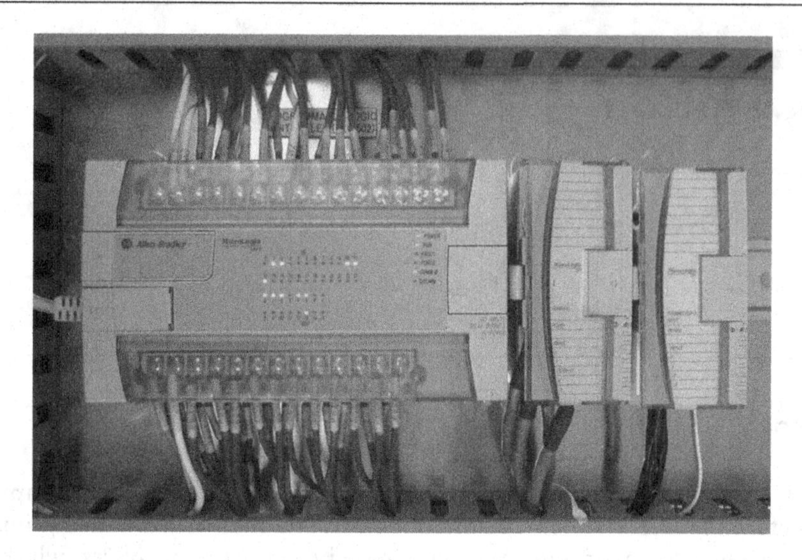

Figure 11.1 PLC with two additional I/O expansion modules attached.

MODERN PLCS

Programmable-Logic Controllers are microprocessor-based systems that receive input signals from electromechanical sensors, process the input signals, and then send output signals to perform control or monitoring functions. PLCs perform real-time control of plant systems.

Figure 11.1 shows a PLC on the left with two I/O expansion modules on the right. The wiring entering on the top side are digital inputs and the wiring on the bottom include an external dc power source and digital outputs. Power and communications span from the PLC to the I/O modules through a ribbon cable. An Ethernet connection for programming, networking, or connecting to an HMI device can be seen on the left.

Essentially, the PLC has many similarities to personal computers, with some big differences. Like a computer, the PLC uses a microprocessor to do the control; however, the PLC microprocessor often uses "reduced instruction set computing" technology (RISC). RISC technology provides only the minimum amount of circuit instructions to reliably carry out only the intended control functions, so that the microprocessor runs faster than if it ran unnecessary routines or if it was operating high-level programs. In addition, the PLC is typically programmed once to perform its control functions, and then it runs the same program repeatedly. Compared to microcomputers, PLCs are typically small, they often do not have a monitor and keyboard, and they tend to have large numbers of input and output connections. Essentially the PLC repetitively polls for input states, executes a program based on the input values, and sends updated output signals to perform the control. User input is usually limited, such as setpoint adjustments.

PLCs come in many shapes, sizes, and configurations, but all work in a similar manner. Some are physically very small and limited in functionality, while others are rack-mounted, modular, and expandable to attach additional input, output, control, memory, and communication modules. Smaller units are sometimes called "bricks" and may contain the dc power supply within the unit and have a limited number of input and output points. Larger PLCs may be of the "shoebox" type, where the PLC is plugged into a backplane printed-circuit board, and expansion modules providing I/O functions are also plugged into the backplane, guided within a carrier frame. These units may or may not include their own dedicated dc power supply, and the size of the carrier frame and backplane limits the number of expansion modules. Still other PLCs may be mounted on a DIN rail, where additional modules may be snapped onto the rail and daisy-chained from one module to the next using a ribbon cable, and where expandability is a function of the PLC model limitations.

PLCS IN POWER PLANT CONTROL

In the simplest form, PLCs may be electronic single-loop controllers, where the PLC is dedicated to one function and the quantity of input and output signals may be very limited. In other instances, the PLC may be a multi-loop controller, where the PLC is dedicated for multiple control functions oftentimes provided as part of a complete system from a single manufacturer. The multi-loop controller generally has more input and output connections. In more extensive arrangements, one PLC may control entire systems, equipment, and devices in large plants.

In many plants, multiple PLCs are used, where the PLCs may be ganged together in a central location or distributed to remote locations close to the equipment. These systems are referred to as Distributed Control Systems (DCS) or Supervisory Control and Data Acquisition systems (SCADA). These multi-PLC systems are linked through a data bus, which in turn may link to a central computer. The SCADA computers often have graphic displays that indicate plant parameters, often superimposed onto a flow diagram. The SCADA system provides real-time operational status, such as running/standby equipment, pressures, temperatures, flows, levels, alarm status, etc. It also can archive data for historical trending, used for equipment maintenance, troubleshooting, and determination of efficiency.

Inputs and outputs consist of digital or analog signals. Examples of digital signals are open-close signals from temperature, pressure, or level switches. Analog signals are continuous signals ranging from a low-range level to a high-range level. Analog inputs are obtained from transmitters, such as the 4–20mA signals commonly obtained from temperature, pressure, or level transmitters; or the signals may be voltage-based, or even directly converted into PLC words, such as from the millivolt signals received from thermocouples. Figure 11.2 shows a block diagram for a PLC used for industrial control.

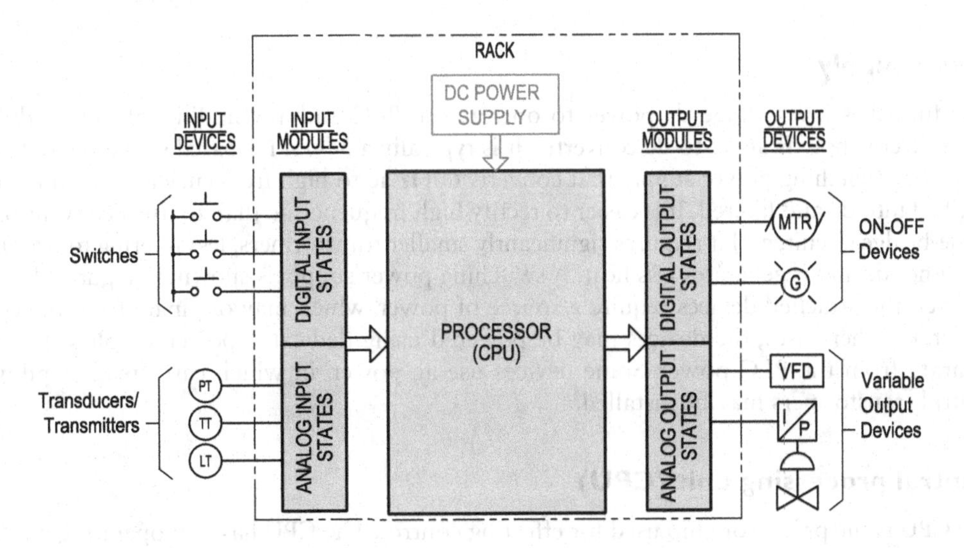

Figure 11.2 Block diagram of PLC components.

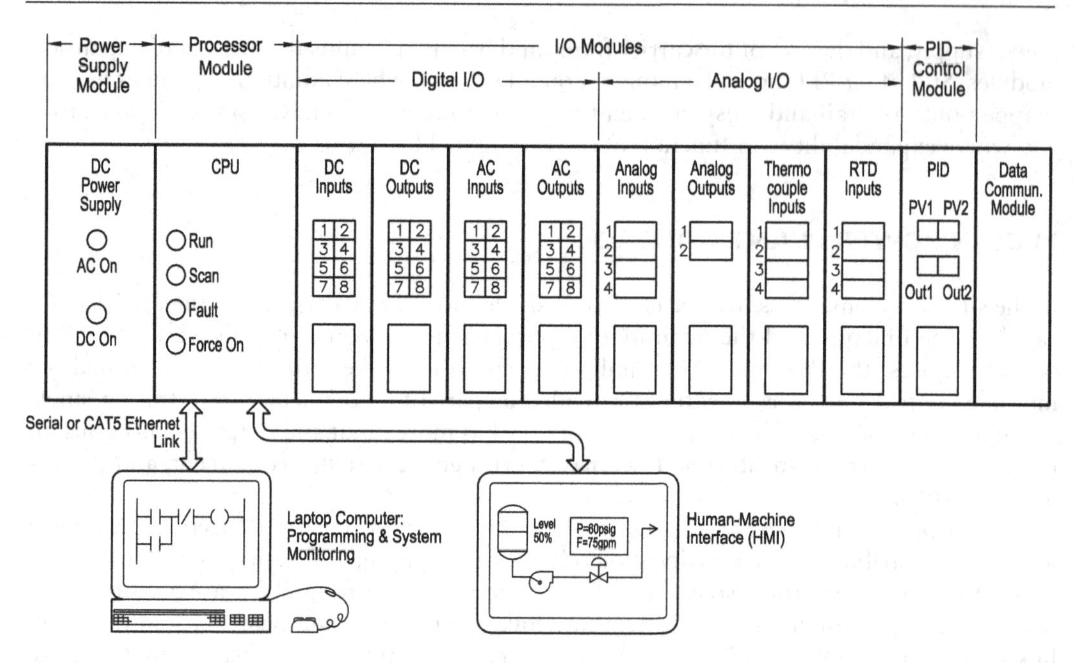

Figure 11.3 The main parts of a PLC rack include the dc power supply, the CPU, digital-and-analog inputs and outputs, and other specialty modules, such as thermocouples, RTDs, PIDs, memory, etc. For plant-wide applications, the PLC will have data communication capability, either as a dedicated module or as part of its CPU. PLCs are often attached to HMI devices for user input and monitoring.

PARTS OF THE PLC

The major parts of the PLC consist of the processor (CPU), a power source, input devices, output devices, cabling, human-machine interfaces (HMI), and an external programming device such as a laptop PC or hand-held portable programmer. The CPU and all connected I/O devices are referred to as the rack, which is a useful representation when addressing multiple PLCs in a plantwide control system, which are shown in Figure 11.3 and Figure 11.5.

Power supply

This furnishes low-voltage dc power to operate the PLC and any additional I/O modules. The power supply is an ac-to-dc converter. It is typically a MOSFET-based single-phase high-frequency switching power supply that converts 60Hz ac to high frequency ac, which is then rectified into dc and filtered. It is easier to rectify high frequency ac than to use a conventional diode-bridge circuit, and it requires significantly smaller transformers, less filtering to produce smoother dc and it generates less heat. A switching power supply is shown, in Figure 11.4.

Often the attached devices require a source of power, which may originate from the controller. In other cases, the devices may be powered using dedicated power supplies that are separate from the PLC power. Some devices use ac power, in which case small stepdown control transformers may be installed.

Central processing unit (CPU)

The CPU is the processor chip used for effecting control. The CPU has two operating modes, programming and run. In the programming mode, the CPU receives the program from a

Figure 11.4 Front and back sides of a PLC external switching power supply.

computer terminal. In the run mode, the CPU executes the program and controls the processes. The CPU receives input-state signals from sensing devices, processes the signals during the program execution, and then sends output signals to the control devices. The central processor unit organizes, controls, and supervises all operations within the PLC. The CPU also performs housekeeping functions, such as internal self-diagnosis of components, data-integrity and parity checking, and communication functions between modules and other PLCs in the distributed control system.

Memory

Memory, such as read-only memory (ROM) and random-access memory (RAM), is used by microcomputers and PLCs to retain the states of functions and devices. ROM is retentive memory that is maintained during a power-off cycle, and RAM is volatile that requires power to be maintained. The memory can be classified as *program-executive memory* (ROM), *system memory* (RAM), *I/O status memory* (RAM), *data memory* (RAM), and *user memory* (RAM or EPROM). These types of memory are allocated to addresses within the PLC memory chips. The *executive memory* is programmed once by the manufacturer, and the memory executes the program and does the scans. The *system memory* is where the operating system is loaded and where temporary information is stored during program execution, including error codes. *I/O memory* is where the input-and-output-device states are stored and modified, including virtual devices. *Data memory* is where results of mathematics, timers, counters, accumulators, and process parameters are stored. The *user memory* is where the

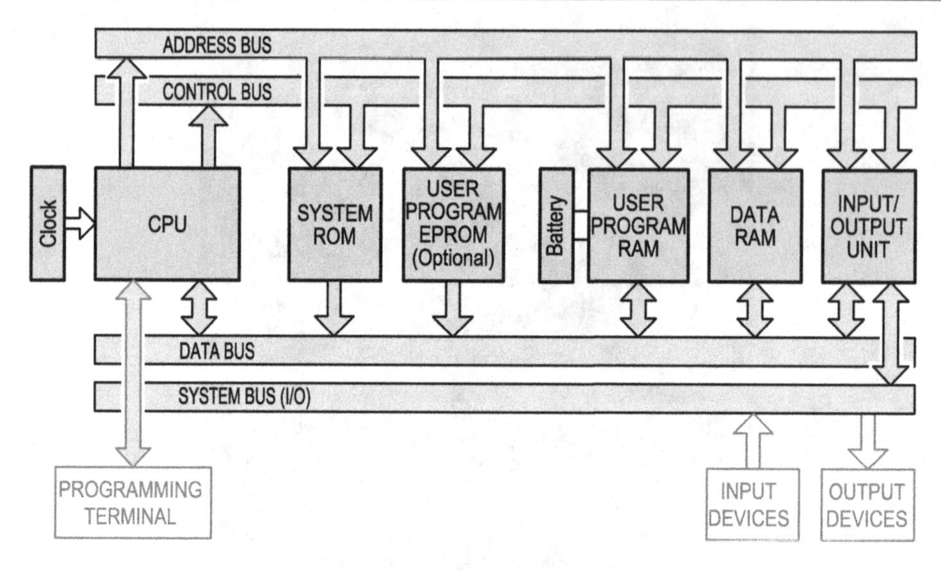

Figure 11.5 Block diagram of the PLC architecture.

Figure 11.6 EPROM chip.

program is stored and where the program deposits and recovers data values occurring during the program scans. The user memory also includes storage of ASCII messages, subroutines, and program functions that may be called as part of the main program.

ROM

Read-Only Memory (ROM) is permanent memory and is not easily erased or changed, i.e., it is non-volatile memory and not lost during a power-off cycle. The ROM chip contains the PLC startup instructions and the basic operating system, and in some cases, it contains the entire controller program. To change the ROM program, the ROM chip may be "burned-in" or a replacement chip containing with the modified program may need to be substituted. The ROM chip essentially breathes life into the PLC in starting up properly

When the PLC starts, it first accesses read-only memory. The ROM performs a computer self-diagnosis and verifies the status of connected hardware devices, such as I/O, communications modules, etc. The ROM then proceeds to boot-up the PLC by directing the processor to access additional steps, often contained in an EPROM chip or a nonvolatile RAM chip. Eventually the computer loads functions and programs into the random-access memory to perform sophisticated tasks.

In some PLCs, all the boot-up instructions and the entire control program may be hard coded into a PROM chip, or *Programmable* Read Only Memory integrated-circuit chip.

RAM is volatile memory, meaning that when a machine is de-energized, the contents of memory are lost. PLCs that use RAM may use a back-up battery, a supercapacitor, or nowadays, flash memory to retain the contents of RAM during power interruption.

RAM

Random-Access Memory (RAM) is sometimes called volatile memory. RAM is fast, but it loses its contents upon loss-of-power, hence, it is volatile. Many PLCs have a battery backup to retain the contents of volatile memory during power interruptions. The CPU uses RAM for its operation, such as storing input status, performing calculations, formulating output commands, etc.

There are two types of RAM, static (SRAM) and dynamic (DRAM). After receiving data, static RAM holds it as long as power is maintained, whereas dynamic RAM must be periodically refreshed. SDRAM is faster, more stable, consumes less power, but is more expensive. In a microcomputer, SRAM is used for cache memory and DRAM for main memory.

EPROM

The difference between an *Erasable-Programmable* ROM (EPROM) chip and the ROM chip is that it can be reprogrammed more easily than the ROM chip. Early versions of EPROM chips were burned in with the code using UV light through a crystal window on the chip, which was later covered with a light-proof tape (see Figure 11.6). The EPROM is hard coded, and its programming cannot be changed unless the chip is removed from the circuit board and mounted in a "chip burner." Newer EPROM chips are electrically erasable and do not require removal and burn-in.

EEPROM

Electrically Erasable PROM (EEPROM) chips are used on newer PLCs. These chips can be directly programmed through a laptop connection to the PLC, where the code is seamlessly uploaded onto the chip without needing its removal and replacement. EEPROM chips are extensively used in modern PLCs.

Nonvolative RAM

This is a category of RAM that stores and retains data when power is lost. NVRAM in its simplest form may consist of battery-backup conventional static RAM, and its retention time is limited only by battery life. Other NVRAM options use flash memory to provide nearly indefinite retention times. Some PLCs store their operating programs in the PROM chips and other PLCs use NVRAM chips.

Input devices

Input devices are the digital and analog devices installed on the systems being controlled to receive status information from the field. Digital devices are simply the two-state switches, whereas analog devices are the transmitters carrying the continuously varying control signals. These devices provide input information to the PLC, so it can execute its program and derive appropriate output responses. Figure 11.7 shows examples of both digital and analog input and output devices.

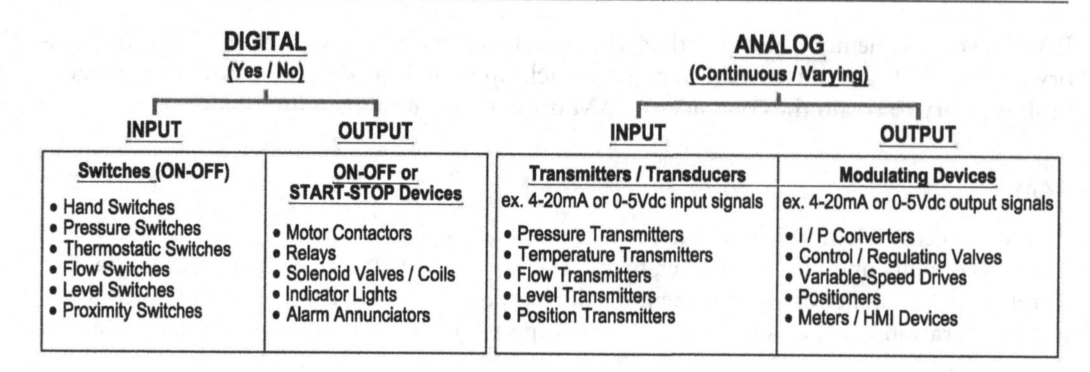

Figure 11.7 Input and output devices and their signals.

Digital input devices

These are essentially are switches having two states; either open or closed, so as to perform yes/no or true/false functions. The switches may be hand-operated, or they may be automatically actuated from a sensed parameter, for automatic control. An example of a digital-input device is a thermostat, which has a bellows that is linked to make or break switch contacts upon reaching a setpoint. Switches can be actuated from pressure, temperature, flow, level, proximity, force, or position.

Analog input devices

These consist of transmitters or transducers that make continuous measurements, or they can even be hand-actuated devices such as Vernier switches or potentiometers that create variable input values. Transmitters/sensors measure magnitudes, not just one value. Transmitters may measure temperature, pressure, differential pressure, flow, liquid level, and proximity. Proximity devices may use contact, inductive, capacitive, or optical methods. Transmitter measurements are converted to electrical signals such as 4–20mA signals, 0-5VDV, or other signals so that the measurement can be transmitted and ultimately imported into the PLC. The transmitter is an analog-input device and the corresponding 4–20mA transmission is its analog signal. The PLC analog input module receives the 4–20mA signal and the converts it into a binary number that represents the magnitude of the input signal. The PLC program analyzes the signal and computes an output action, after the signal is converted into machine language. A block diagram showing analog input and output signal transmission is shown in Figure 11.8.

Output devices

Like the inputs, PLCs have outputs that are either digital or analog.

Digital Output Devices are essentially two-state devices that are either turned on or off. Examples of digital outputs are motor-starter contactors, relay coils, solenoids, status indicator lights, alarm horns, etc., where the device is either operating or not.

Analog Output Devices are any devices that have a variable output or modulate. Examples of analog-output devices include regulating valves using I/P converters and operating at throttled positions, variable-frequency drives producing changing motor speeds, digital meters, positioners, etc. Human-machine interface devices and computer monitors that mimic meter indications are other examples of output devices.

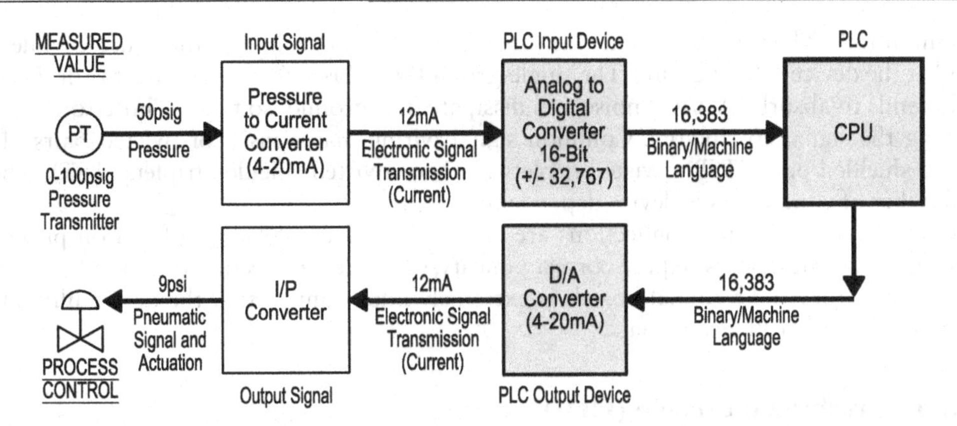

Figure 11.8 Block diagram showing the analog-input measurements and signal conversions leading to the PLC for pressure control, and the PLC analog-output signals that are transmitted to the process-control valve. The values represent control being maintained at 50 psig using a 0–100-psig transmitter.

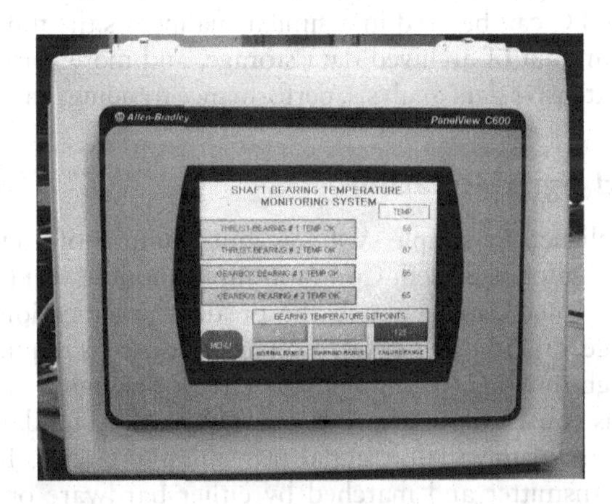

Figure 11.9 Human-machine interface (HMI) with touchscreen menus and switches for entering setpoints and initiating control.

Cabling and interfaces

Cabling is the means of connecting field devices to the PLC controller. Digital input and output signals are transmitted on a pair of wires, as they represent the two states of open-or-closed inputs or on-or-off outputs. These devices use a fixed voltage and are relatively immune to noise and interferences. They consist of a pair of wires, or a twisted-pair cable. Generally, shielding to mitigate noise is not required.

Analog signals are generally low-voltage and low-amperage variable electrical transmissions, such as 4–20mA. Because the signals are small, the wire gage is small, often in the range of 16–20AWG. The cabling may be exposed to stray electromagnetic signals, or noise, through inductive and capacitive coupling with other conductors or EMI/RFI emitting devices that are near the cable. To minimize noise coupling, analog signal wires are twisted together. When twisted, the current induced by electromagnetic-noise in one direction in one conductor is balanced by simultaneously induced current in the opposite direction of the other conductor, essentially canceling the noise. In many instances, the analog-signal cable is fabricated with a "shield" that surrounds the conductors. The shield can be formed from fine strands of wires intertwined into a braid around the conductors, or from a conductive aluminum foil. When used, the shield is often connected to ground at the source, while the shield at the device is left floating. The single-ground shield is sometimes called the drain. The shield tends to absorb EMI/RFI noise and dissipate it to ground, so there is little-to-no noise reaching the signal conductors. Common signal wiring may consist of twisted pairs (TP), twisted-shielded pairs (TSP), twisted-triplets (TT), or twisted-shielded triplets (TST), where the number of conductors is device dependent.

In PLCs, carefully made connections are required for some devices to function properly. For example, some devices require correct polarity, or a solid connection to ground. Labeling of the conductors and terminal boards is extremely important so that the device ultimately lands on the correct PLC terminal.

Human-machine interface (HMI)

The HMI is a screen interface used to display system information. Often, the HMI is an LCD display having a touchscreen, although older models use touchpad switches (Figure 11.9). The touchscreen allows the user to scroll through pages of information, but also permits the user to modify set points or to remotely start and stop equipment. Like the HMI, a microcomputer can be linked to a PLC. The PC can be used in a similar manner as the touchscreen, but the hard drive allows extensive amount of archived data storage, and more sophisticated programs on the PC can be used for extensive data analysis, performance trending, and efficiency calculations.

Control signals and signal isolation

Digital input is accomplished by simple switches, but a source of voltage is needed for the PLC to sense when the switch is closed. Common input-module voltages are 12/24Vdc and 120/240Vac, where the voltage must match the product specifications. In some cases, the PLC provides the voltage, and in other cases, a separate source is required. In either case, the voltage is passed through the PLC to complete the circuit through a digital-input terminal.

Analog-signal options consist of 4–20mA or one of a several standard voltages, where the signal is produced as part of the analog signal transmission circuit. The analog signals are often selected at the transmitter and matched by either hardware or software switches at the input module. The analog signal is converted into a binary word through and analog-to-digital converter for use by the program.

Because of power quality issues, and even to protect the PLC electronics from a local I/O fault, incoming signals are typically isolated from the processor. Opto-isolators are used for inputs and opto-isolators or reed switches, which are small relays, are used for digital outputs. Opto-isolators use light-emitting diodes (LEDs) to change the electricity into light, where a second circuit in the input module converts the light back into an electronic signal that is used by the processor. The LED optical coupling isolates stray voltages and noise on the input signal from reaching the PLC electronics, and the robust nature of the LED is very forgiving for wiring errors or faults that might otherwise damage the PLC's sensitive electronics, if the wiring was directly connected to the PLC. Usually, a single damaged input can be isolated from all other PLC inputs, so that a device can be reconnected and readdressed to a spare terminal after input fault is corrected.

GENERAL PLC TROUBLESHOOTING

Probably the biggest source of PLC control-related problems involves peripheral devices. The CPU is responsible for handling the assembly of input-signal data, the analysis and control program, and the assembly of the output data. In general, the PLC will provide LED or LCD visual indication for the status of all digital incoming and outgoing signals. There will also be locations where a multimeter may be attached for testing continuity or the presence of voltage on the I/O.

Some troubleshooting procedures include:

- Look for obvious problems, such as burned components or connections, physical damage, broken wires, etc.
- Look for blown fuses or tripped circuit breakers. Check that the fault is cleared before resetting the protection.
- Measure for the presence a correct voltage level. Check amperage and determine if within limits.
- Observe for high ambient temperatures and obstruction of any ventilation systems or filters.
- Transient voltage surges can cause problems. Use a digital multimeter with a peak-hold function to capture transients that may result in intermittent problems.

NETWORKING THE PLC

Many PLCs can be networked together to communicate with each other and share data and device states, or to just pass through information to other devices or to a SCADA system. Some networking systems are Bus, Ring, Star, and Mesh networks, and there is similarity to computer networking systems. Networking can be accomplished with twisted-pair conductors, CAT5 or 6 cabling, coaxial cable, or fiber-optic cable radiating from a single device or daisy-chaining from one unit to the next. In the daisy-chain arrangement, it is possible to return the end-or-run device back to the source for redundancy. Networking software and protocols are required to solicit for data, package the data, and decode the data as part of the communication process. In many cases, dedicated integrated circuit cards or communication modules, such as routers, may be needed to accomplish the communication. While some systems use the Modbus RS485 cabling and data-transmission protocol originating from the early days of PLCs, newer CAT5 or CAT6 Ethernet cabling and data transmission protocols provide options. Guidance for wiring and configurating communications is provided by manufacturer instructions.

Bus topology is a network configuration where all nodes are connected in parallel to a common communication bus and all nodes receive the same information from all other nodes. Each node then determines whether to use the information if it is intended for them, or to ignore the information, if it is not. Each node is directly connected to a common bus cable through a network interface module using a drop cable, and each drop must have its own terminating resistor. This topology tends to use less cable than other topologies. Nodes are free to talk directly with each other without any control limitations from a master server, and consequently, these networks are useful in distributed control systems, where individual nodes perform local control, but information is exchanged with other controllers as needed. This topology is conducive to expansion or reconfiguration, however, a single break in the bus can be very disruptive to the entire control system.

Star topology uses a central switch, hub, or computer which acts as the common-point conduit for routing communications for all connected nodes. The switch becomes the server and the nodes the clients. The switch may be passive, meaning that the node initiating the communication must be able to ignore its echo as the message is retransmitted back to all nodes, or active, where the retransmission prevents echoing the message back to its source node. The start topology limits line failure to any one node experiencing a communication problem, while all other nodes will continue to function, however without communication with the damaged node. This network requires the expense of more cables and if the server/hub crashes, then the entire intercommunication system is disabled.

Ring topology is a network where each node is connected in series with the next until the last node connects back to the source in a peer-to-peer arrangement. This topology is easily expandable by breaking a new node into the ring, as along as the total circumference of cable does not exceed the latency requirements of the network protocol, i.e., the time it takes for information packets to traverse the entire cable length. Cable redundancy is built into the ring to protect remainder of the control system against a single node failure or cable break. Technology exists to identify the node location for a break in closed-loop continuity.

Hierarchal topology is a variant of the ring topology, however, each node connected to the central hub has its own devices that it communicates with. This topology is recommended for larger sized local area networks, when communication traffic might affect the system response.

Fully connected *mesh technology* uses cable connections from each node to every other node in the network, and data can be routed directly to any other node with less delay. Mesh networks are complex and expensive when more than a small number of nodes exist that are all interconnected.

PROGRAMMING THE PLC

Traditional motor controllers use discrete electrical devices mounted in a cabinet that not only contains the motor-starter and overload-protection devices, but also the control relays and timers that produce the control. External sensing devices, such as pressure switches, are connected to energize or de-energize control-relay coils, or they are wired as permissives as part of a circuit. The coil then toggles normally-open (NO) or normally-closed (NC) contacts. When control relays are configured properly, decision-making logic is performed, and very sophisticated control can be obtained. Because the circuits take the appearance of rungs on a ladder, these relay-logic schematic drawings are also known as ladder diagrams.

There are several different languages that may be used to program a PLC, from BASIC, C, PASCAL, C++, Java, etc. However, since industrial PLC control schemes produce similar functions as the traditional logic used in the electrical ladder diagrams, the most common industrial PLC programming language is ladder-logic. Some PLCs use combinations of ladder-diagram programming with a more sophisticated language. An example might be that when an action occurs, one of the PLCs responses may be to supply a message on a computer screen through a user-defined function generated by the higher-level language. In addition, more complex functions, such as proportional-plus-integral-plus-derivative control (PID) may be available within the PLC software. During operation, the user may be given the ability to tune the system for good response by modifying variables in the program.

Other PLC programming languages include Functional Block Diagrams, Structured Text, and Instructional List, and Sequential Function Chart languages. This text will focus mostly on ladder diagram, which is probably the most common language in industrial use.

PROGRAM EXECUTION

Understanding relay ladder logic is half the challenge in PLC programming, as the language is very similar in appearance and function. But there is a big difference, as electricity conducts through all circuits having continuity so that real-world relays and contactors are energized nearly instantly, except for short time delays as switches swing positions. On the other hand, PLC ladder diagrams behave like sequential events in a computer program. As with all computer programs, the PLC program flows sequentially from top-to-bottom, and it also flows from left-to-right. The importance is that rungs on an electromechanical control-relay wiring diagram can be placed in any order, but the rung order in the PLC program can be as important as the ladder logic. In some cases, a misplaced rung can erroneously cause devices to lock out or misbehave. To further exacerbate the programming challenges, intermittent problems sometimes occur based on the exact instant in program execution when an input is toggled. For instance, an input change occurring near the end of the sequence may change an output that subsequently disables a preceding input on the ladder. If the misplaced rung is close to the rung producing the trouble, it is possible that the problem may occur so seldomly, so as to be baffling. If the offending rungs are separated by a larger amount of code, the problem may surface frequently enough to force more aggressive troubleshooting.

The program execution follows three main discrete steps, with a fourth set of housekeeping tasks for reliability and communications (see Figure 11.10).

One scan is the time it takes to execute the three steps plus the housekeeping, where the input scan must be fully completed. Considering that it is possible for an input to change just after the input scan begins, the worst-case scan time is twice the time to cycle through all steps. The importance of the scan time is that for a guarantee that an input change will be seen by the PLC, the input must remain active for the complete duration of at least one complete *input* scan. Scan times can fall between less than a millisecond and several hundred milliseconds, depending on the program complexity and the number of attached devices. The scan time is very important for very fast applications, such as some manufacturing systems;

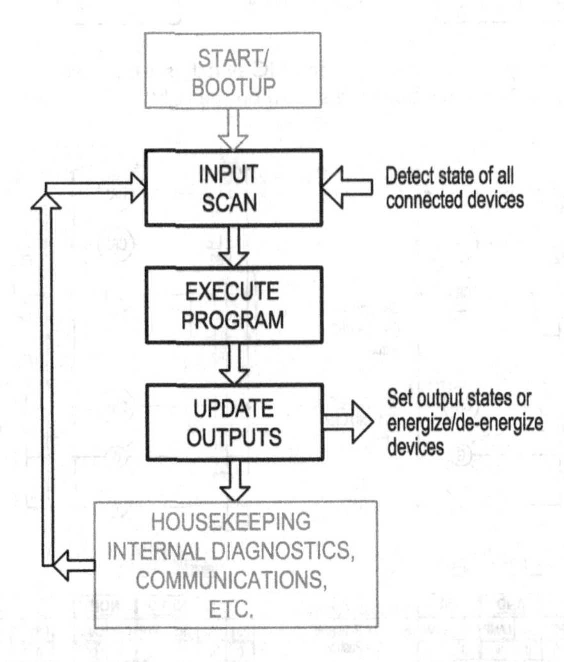

Figure 11.10 The PLC program scan. Fundamentally the program polls all input devices, placing input measurements into memory. Then the PLC executes the program and stores its results into output memory. Finally, the PLC program updates all of the outputs, using the values obtained from the output memory to operate the controlled devices. The scan also includes "housekeeping" functions, such as internal diagnostics and communication to other PLCs or SCADA systems.

however, scan times tend to be much less of a problem in plant automation systems, which tend to be slow-acting in comparison.

Another trait that makes the PLC program different from a wiring diagram is the fact that the PLC ladder diagram refers indirectly to the attached real-world devices. The PLC never "sees" the device, but instead its program refers to a computer "bit" that represents the device state. More specifically, the "components" on the PLC ladder diagram are in fact computer instructions, and the instructions look for true/false values stored in a computer address.

Relay logic

Understanding digital theory, motor-controller functionality, and ladder logic provides a good background for programming a PLC, so reviewing those chapters might be helpful. The relay wiring diagrams that produce the various Boolean functions are summarized in Figure 11.11 through Figure 11.13.

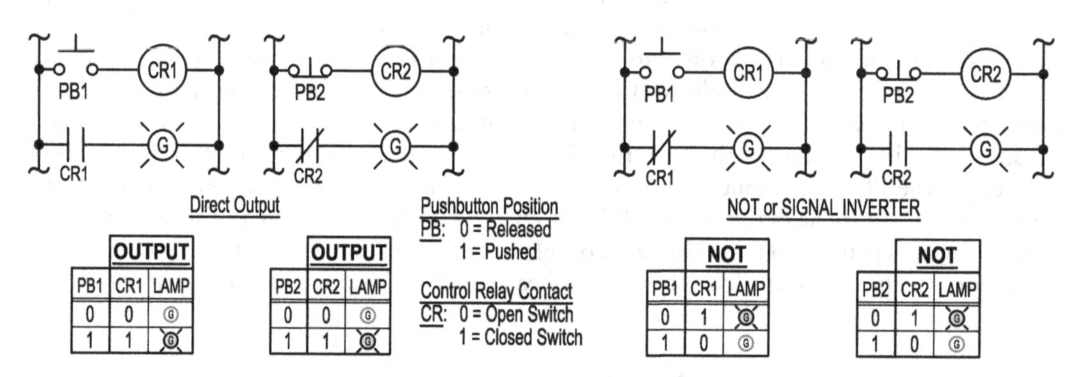

Figure 11.11 Direct-acting relay output using NO and NC switches are shown on the left, and inverting contacts to produce NOT functions are shown on the right.

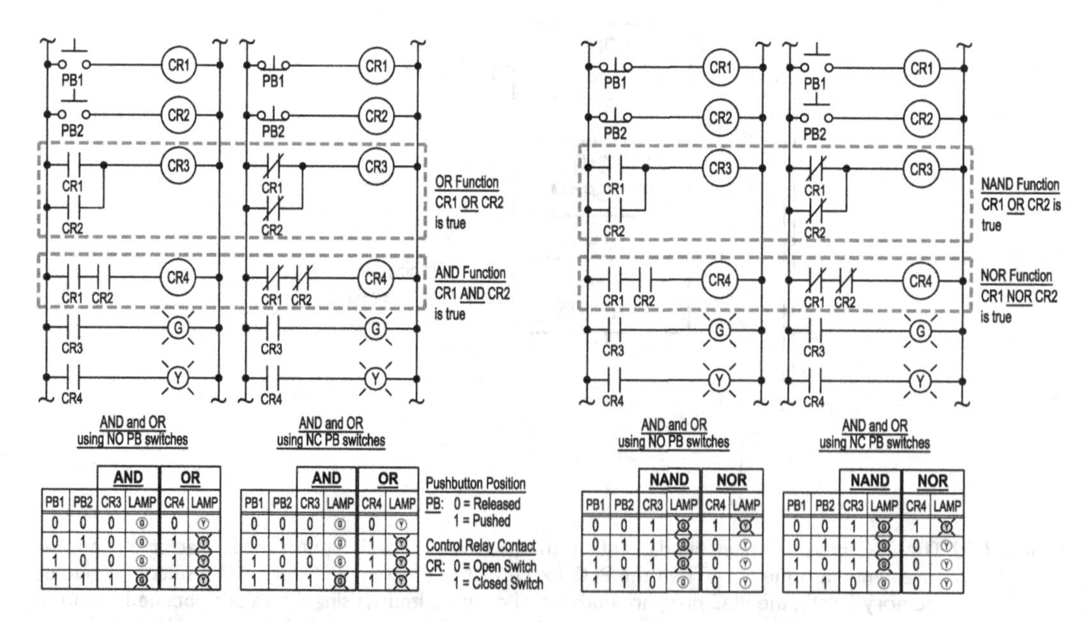

Figure 11.12 Relay logic can be used to produce AND, OR, NAND, NOR control functions, including DeMorgan equivalents using NO and NC pushbutton alternatives.

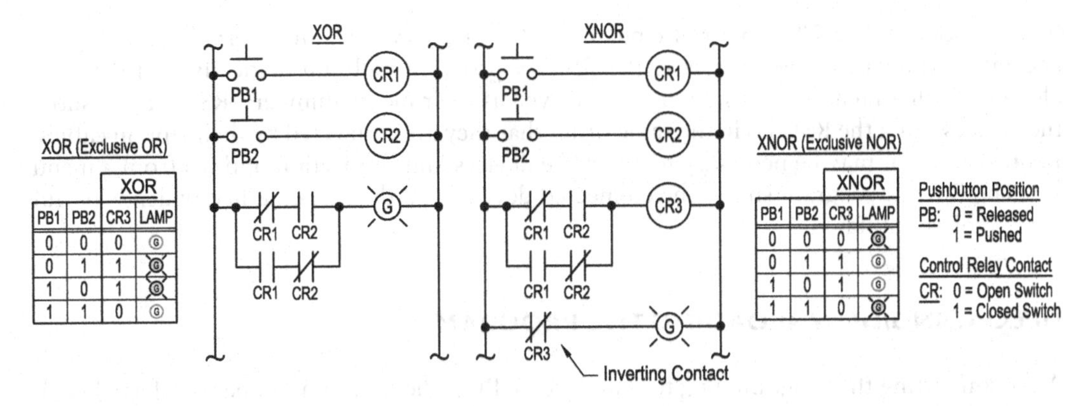

Figure 11.13 Relay logic produces *Exclusive OR* (XOR) and *Exclusive NOR* (XNOR) functions.

FUNCTIONS WITHIN PLCS

PLCs produce the automatic control of systems by measuring variables, comparing those values to a desired setpoint, and then outputting a control signal. For small applications, the output may apply combinational logic simply to open and close a valve, or to start and stop a motor. For more extensive sequential-logic control, PLCs have instructions that emulate their discrete relay counterparts for latching relays, on-delay and off-delay timers, and counters. The PLC includes mathematical functions that accomplish the same tasks that complicated mechanical mechanisms did in the past, and for calculating required outputs.

In applications requiring fine continuous control, closed-loop or negative feedback is provided where the sensed variable is compared to a user setpoint to determine an error. The PLC corrective output signal may be 4–20mA, 0–5Vdc, or another signal whose value is relative to the amount of error, and the correction is applied in a manner to eliminate the error. To provide sufficient actuation forces, the 4–20mA or 0–5Vdc output signal is often converted to a 3–15 psi or a 0–30 psi pneumatic signal to position a valve actuator or a power cylinder. The current-to-pneumatic converter is referred to as an I/P converter, which is essentially a signal amplifier. Proportional-only control leaves offset, or a deviation from the setpoint after a process disturbance has taken place, because the corrective signal is generated only after observing that a deviation exists. If the system can tolerate some level variations and the response speed is satisfactory, then proportional-only control is adequate. To reduce offset, the gain or sensitivity of the proportional control can be increased. Gain is the amplification of the output-signal relative to the input-signal. Gain can only be increased within limits constrained by the system dynamics, and controls that are too sensitive result in instability or hunting. PID mathematical instructions can be programmed into the PLC to improve the control. The integral and derivative functions add behavior that removes offset and speeds the response. This topic is discussed in Chapter 2 on "Control Terminology and Theory."

USING ALLEN-BRADLEY/ROCKWELL AUTOMATION RSLOGIX 500 SOFTWARE

The first step in programming a PLC is to establish communications with the PLC and all attached modules. The Allen Bradley (A-B) communication software is called RSLinx. The programming terminal is cable-connected to the PLC using an RS232 (COM port) cable and

9-pin D-connector, a USB-to-serial emulator cable and driver, or an Ethernet cable in later generation models. In the best case, the RSLinx program will automatically find the PLC along with all attached expansion devices, as well as their model numbers. RSLinx will show the devices using the RSWho in the same order that they are connected by their ribbon cables. In other cases, it may be necessary to select the devices and connection options from a menu list. The RSLinx program shows the connected devices and their availability, which is useful for troubleshooting.

UPLOADING/DOWNLOADING THE PROGRAM

After connecting the programming terminal to the PLC, the program can be transferred back and forth between the two devices. The A-B terminology considers the PLC to be the important device and the programming terminal is the remote program source used for maintenance and archiving. From the PLC perspective, downloading is the transfer of program files into the PLC. Uploading is the retrieval of the program from the PLC back to the programming terminal. However, some PLC manufacturers reverse this terminology.

RSLogix is the software containing the PLC ladder-logic programming language. Common practice is to annotate the rungs of code and to label the rung devices as the program is being authored. Commenting assists in understanding and debugging the program and in making future modifications. Thorough and detailed commenting is good practice.

To download the program, the PLC typically must be toggled out-of-operation by removing it from run mode and setting to the program mode, although some PLC models are capable of online updating while in run mode. The programming software contains the commands and prompts for downloading. After code modifications are complete, the program is downloaded while the PLC is offline. Afterwards, the PLC is placed back into run mode. In run mode, it is possible for the user to go online and observe the program operation and resulting bit states in real time. While online, it is also possible to apply external forces to override program bit settings, that is, switches can be forced on or off to directly control devices. The online features are excellent for troubleshooting purposes. An LED indicator on the PLC will illuminate whenever user forces are in effect, this indication serves as a warning that some control features have been overridden and non-functional.

It is possible to upload or retrieve the program from an operating PLC back into the programming terminal. In most PLCs, the annotations and comments embedded in the original source code is not downloaded to conserve PLC memory and increase speed. Thus, uploading the program results in the loss of all annotation. However, some software packages will compare the uploaded software against the original source code, if it is available on the programming terminal, which is useful in identifying any changes that may have been incorporated while in service. However, if the original program file is unavailable, the uploaded retrieval will lack the original comments.

ADDRESSING THE PLC DEVICES

Addressing the PLC is done using the device location in the rack and the type of data associated with it. Essentially, the address is the wiring terminal connection, and the type of data is described by its data file, where the information is stored in PLC memory. Numbering always begins with 0.

For a rack consisting of a PLC followed by a digital-input module, a digital-output module, and a combination analog input-and-output module, the PLC would be device 0, and the remaining three expansion modules would be numbered 1 through 3, respectively. The data files are labeled by their type and the file number, which are:

- **O0** - Output
- **I1** - Input
- **S2** - Status
- **B3** - Binary (bit)
- **T4** - Timer
- **C5** - Counter
- **R6** - Control
- **N7** - Integer
- **F8** - Float
- **L** - Long
- **MG** - Message
- **PD** - PID
- **PLS** - Programmable Limit Switch
- **RI** - Routing Information
- **RIX** - Extended Routing Information

Not only are the data files used for addressing, but each data file has pre-assigned memory limits, based on the number of bits and words required.

The O0 and I1 data files are the only two that are associated with real PLC hardware. As such, they are both physically constrained by the number of terminal connections on the PLC and expansion modules. Consequently, the zero and one in the O0 output and I1 input files can be omitted when addressing memory locations. The O0 and I1 files represent two-state inputs and outputs, such as input switches that can be either open or closed, or output coils that are either energized or deenergized. These data files contain single bits as part of a "word" of information that can have a value of either 0 or 1. All other data files beyond O0 and I1 are either part of virtual devices that exist only in memory or they are data types that have certain memory requirements.

The "B" files represent two-state binary files where the information consists of only a single bit, having a value of either 0 or 1. The B3 data file is just like the O0 and I1 files, but is limited to virtual devices. Since there are no hardware restrictions, it is possible to exceed the number of virtual devices permitted by the B3 file, in which case additional user-defined binary data files can be created. For example, a B9 file can be added to extend the quantity of available bits of memory. For that matter, if PID control is needed, a PD10 data file can be added, and so on. A total of 256 data files can be created, including the nine defaults.

Referring to the digital-input device in Figure 11.14 as an example for creating a binary I/O data address:

- The first letter represents the data type: *I* for input or *O* for output, followed by a colon. The data type defines the data file where the input open/closed state is stored in memory, which in this case is the I1 data file. The outputs are stored the O0 data file.
- The next number represents the location of the I/O device in the rack: The PLC is at location 0, the first expansion module is at location 1, and so on. While 1 physically corresponds to the device location, it also corresponds to the "word" address in the PLC memory that contains the information.
- The slash indicates that the information is binary, meaning it contains two states of information having values of 0 or 1. Digital inputs and outputs take only one bit of information in memory.

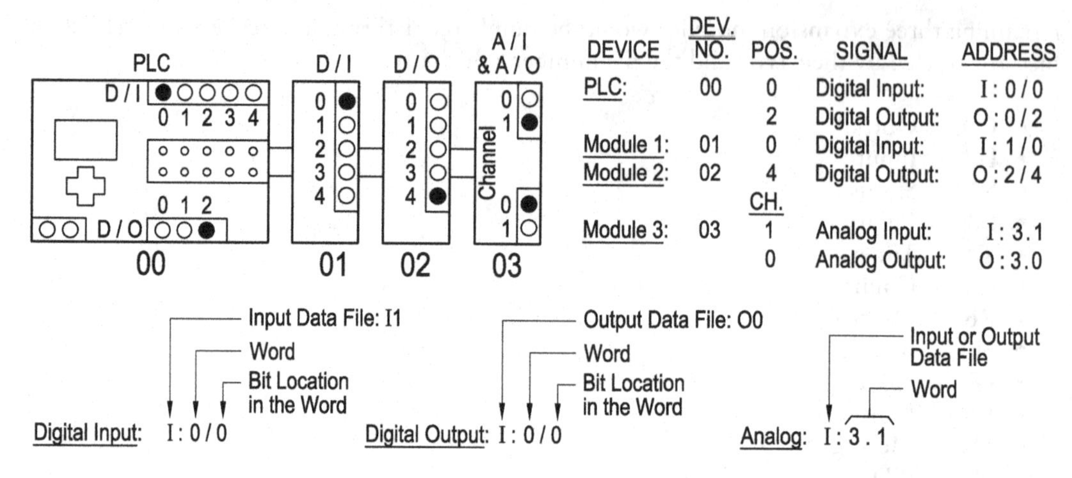

Figure 11.14 Examples of I/O addressing. Addresses relate the physical connections of the PLC and expansion modules to the program's data file, word, and bit locations in memory.

- The last number represents the screw-terminal position on the device: If the input module has 16 terminals, the number corresponds to the screw position, numbered from 0 to 15, and it simultaneously represents the bit location in the 16-bit "word" stored in memory, also numbered from 0 to 15.

The address is I:1/4, where the bit state is stored in the input data file I1 in the fifth bit location from the right in word 1.

Referring to the analog-input device in Figure 11.14 as an example, the analog I/O data addresses are created as follows:

- The first letter represents the data type: *I* for input and O for output, followed by a colon. This letter represents the data file where the value will be stored in PLC memory.
- The next number represents the location in the rack: The PLC is at location 0, and the analog I/O expansion module is the third device beyond the PLC, which is location 3.
- The dot indicates that the information is analog: Analog data takes the entire word, or all 16 bits in a two-byte word.
- The last 0 indicates the channel terminal on the A/O device and it corresponds to the word in PLC memory.

The address is I:3.0, where 3.0 is the word in the I1 data file.

QUESTIONS

1. What is a PLC?
2. Where and how are PLC's used in power-plant operations?
3. What are the major components of the PLC and provide a brief explanation of what the components do.? How does a PLC work?
4. What is the BIOS, ROM, EEPROM, RAM, and non-volatile RAM?
5. Sketch and describe the PLC scan sequence.
6. What are the four types of signals used with PLCs?

7. What are the differences between digital and analog signals?
8. What types of variables are measured using PLCs?
9. What are typical digital-*input* signals? What are typical digital-*output* signals? What do PLCs do with the input signals?
10. What are typical analog output signals? How are the analog output signals converted into high actuation forces?
11. What is the programming "language" most commonly used for PLC controllers for shipboard power plants?

Chapter 12

Wiring PLCs and I/O devices

PLC connections consist of the power source, a dc power supply, and input/output-device wiring. The PLC and I/O modules operate from dc voltage, but the source in most installations originates from ac electricity. In most cases, the source is single-phase 120V 60Hz, which can be supplied directly or stepped using a transformer, and then it is converted to dc electricity, which is used by the PLC microprocessor and the signal expansion modules. In some cases, the I/O field devices use a separate source of power, which can be ac or dc and of various voltage levels.

PLC POWER-SUPPLY SYSTEMS

Power-supply types can be unregulated, linear, or switching. Unregulated power supplies lack the feedback circuitry that controls the output voltage, which decreases as the amperage draw increases. If the unregulated power supply meets the load current while maintaining the minimum voltage requirement of the load, then the unregulated supply is a cheaper alternative. However, it is important to size the power supply adequately, as the power supply naturally increases the output current as the voltage drops to meet the power required by the load, which can cause overheating of the supply or load. The unregulated supply should have built-in capacitors to stiffen the dc voltage during transients. The regulated power supply includes feedback circuitry to fix the output voltage.

A linear power supply typically uses a stepdown transformer to set the output voltage, coupled with a diode-based bridge rectifier assembly. Linear power supplies are used where large in-rush current exists. They tend to have large capacitors for filtering, removing ripple, and stiffening the voltage. They are relatively large, less efficient, and produce more heat, which needs to be rejected. Figure 12.1 shows a PLC power supply using and stepdown transformer and an ac-to-dc power supply.

Switched-mode power supplies are regulated, MOSFET based, high-frequency-switching power supplies, which regulates its output voltage by varying the amount of on-to-off time. These supplies are about one-quarter of the size and weight of a linear power supply, and can be nearly twice as efficient, but they pass some of the ac source noise and create some electromagnetic interference and harmonics. Switching power supplies have a larger range of input voltages than a linear supply, which might function well between 85–265 VAC and either 50 or 60 Hz.

Switching power supplies can be obtained from single-phase or three phase power. They should have built-in current-limiting or shutdown features, in the event of a shorted output. Some products include a digital output that can be used by the PLC program to monitor for overvoltage, undervoltage, or overheat. Overheat can initiate an alarm, start additional cooling, or execute a safe shutdown sequence.

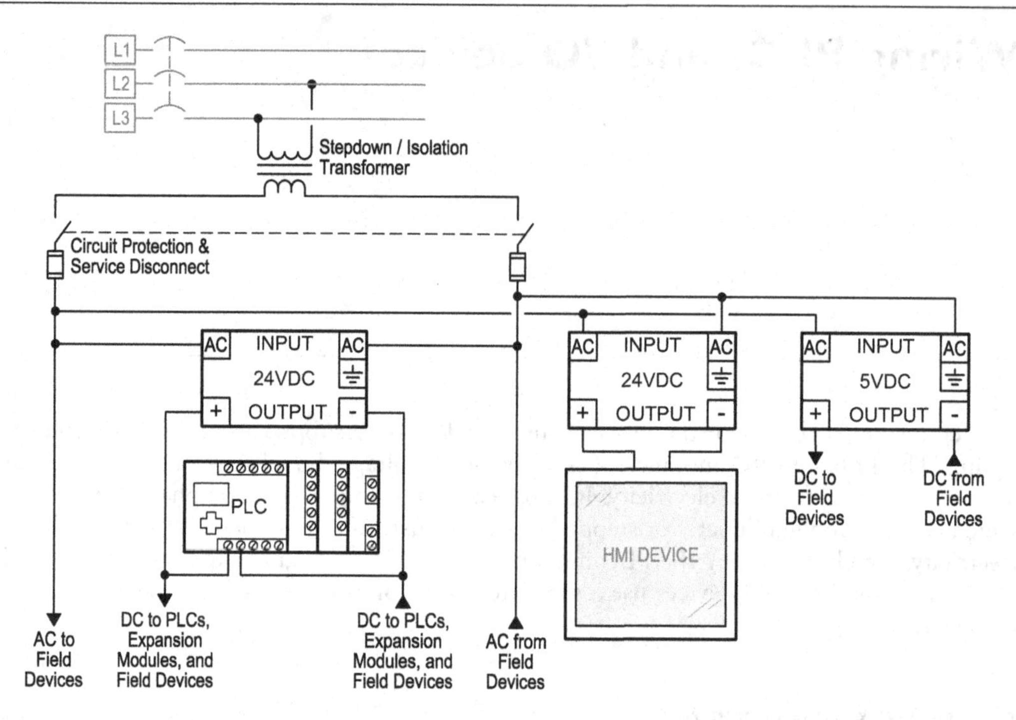

Figure 12.1 PLC System: A single source of ac power is applied to three dc supplies, which serve the PLC and its I/O expansion modules, the HMI device, and I/O devices. An isolation transformer provides the correct voltage and mitigates harmonics problems.

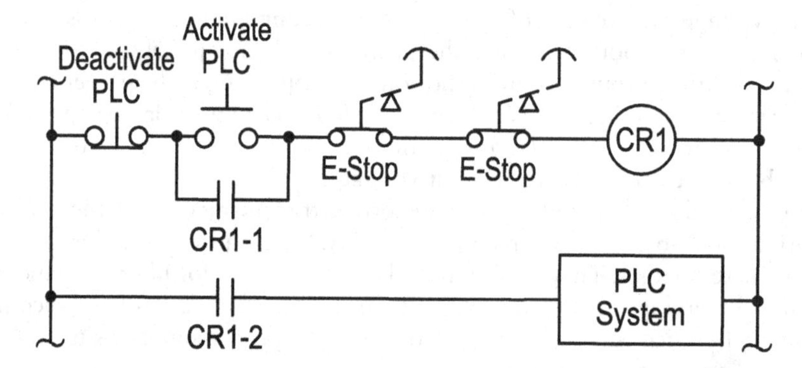

Figure 12.2 A simple PLC activation and deactivation system. Emergency-stop switches can be used to interrupt controller operation when unsafe conditions are observed. Note that failure of the CR1 coil prevents PLC operation without any fault indication. See Figure 12.3 for a high-reliability alternative.

A fused-disconnect switch or circuit breaker should be installed for circuit-protection and starting and stopping the PLC system. In cases where safety is a concern, an emergency stop system may be warranted. Figure 12.2 shows E-Stop switches connected in series with a control relay which acts as a permissive to the PLC system. When safety is concerned, it is important that the E-Stop switch be hardwired to kill power, rather than a PLC instruction that may not execute. This circuit contains activation/deactivation switches to control the system independently from the circuit protection disconnect switch.

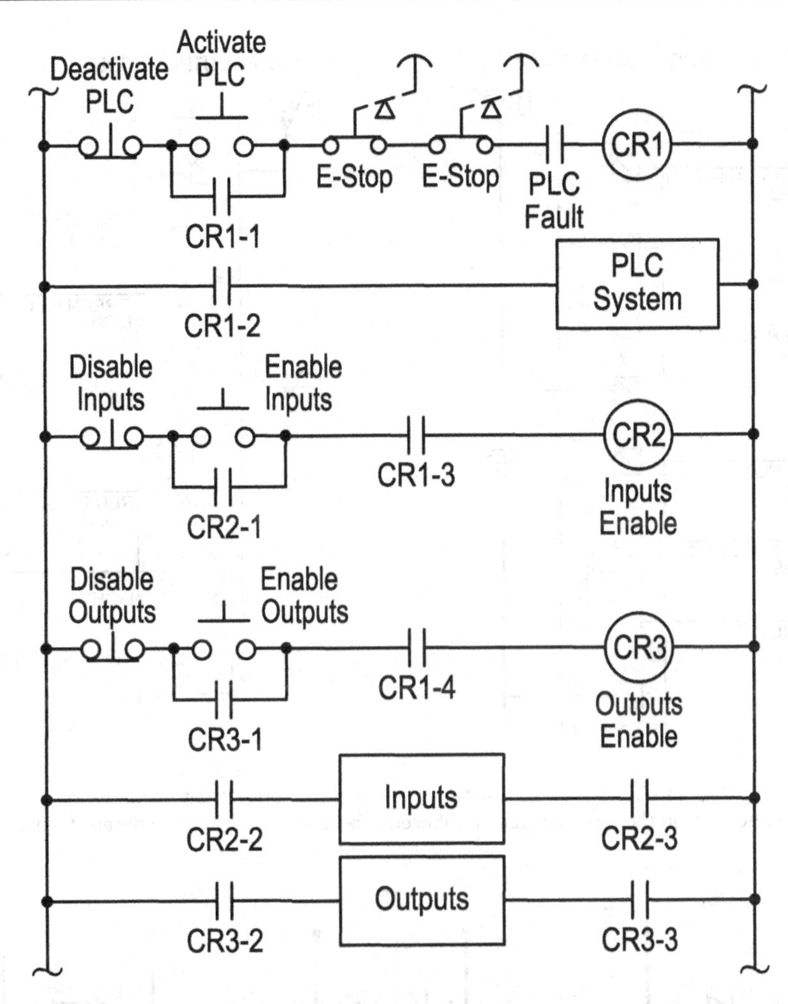

Figure 12.3 Enable and disable circuits can be added to permit PLC operations without running any inputs and outputs, which is a feature useful for troubleshooting without exercising the controlled equipment. This circuit is arranged to incrementally apply power to the PLC first, and then to the inputs and outputs, shown in Figure 12.4. For safety, a fault in the PLC can be used to automatically secure the system.

Figure 12.3 shows a higher level of control where enable/disable switches are added to the input and output circuits, in addition to the activate/deactivate switches that start the PLC system. This arrangement provides a progressive manner of starting the PLC systems, and enables CPU operation while all inputs and outputs are deenergized. Optionally, a PLC-fault contact can be added into the deactivate circuit to kill power in the event of a problem with the CPU, such as memory errors or loss of communication, or processor lockup. Note that if the CR1 coil were to burn out, the PLC system would be completely inoperative, and troubleshooting might be challenging.

Figure 13.34 shows a high-reliability alternative for the E-Stop circuit. The circuit makes use of AND, OR, NOR, and XOR functions to require two control relays to both be deenergized in order to trip the circuit and indicate an emergency shutdown occurred. The failure of one relay initiates an alarm buzzer and relay-failure indicator, where the audible alarm can be silenced while leaving the lamp lit.

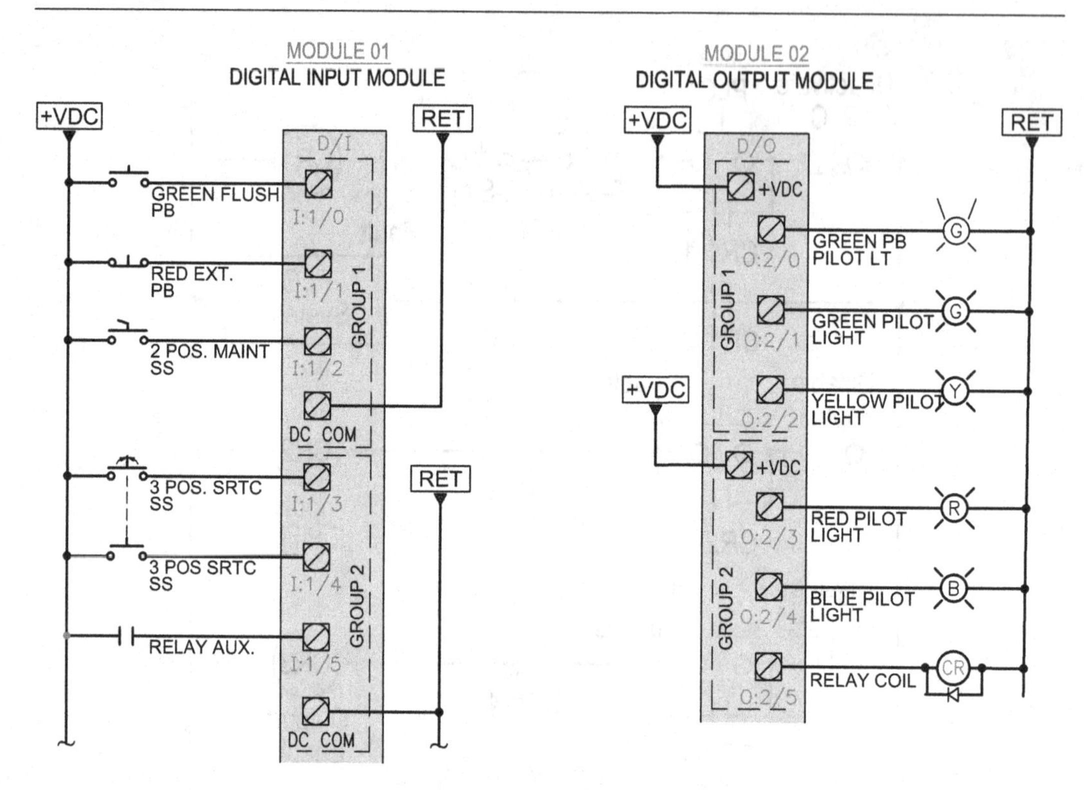

Figure 12.4 Wiring diagrams are shown for digital input and digital output expansion modules. Groups of I/O connections on the modules permit different operating voltages for different field devices.

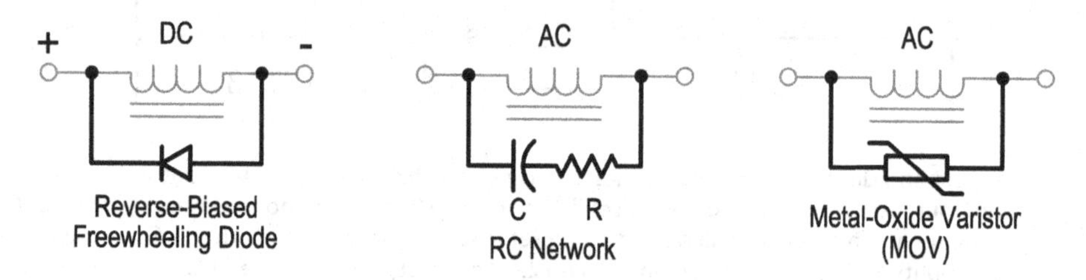

Figure 12.5 Surge suppression spanning across inductive loads mitigates potentially damaging voltage spikes during transients.

Although not shown in the figures, some PLC systems use either an uninterruptible power supply (UPS) or a dynamic sag corrector. The UPS is essentially a battery-backup system maintained in a ready state by a constant trickle charge. The UPS keeps the system functioning until the operator can correct the loss of power or perform a controlled shutdown of equipment. The dynamic sag corrector is essentially a set of capacitors that maintain enough power to ride through a transient sag in voltage or a brief power interruption, typically designed for five seconds.

Most output modules have built in surge suppression to avoid damage from high voltage spikes. However, when loads are known to have significant inductive properties, additional surge suppression is recommended. Surge suppression is recommended for relay coils, motor contactor coils, solenoid coils, motors, etc., especially when using dc electricity, as shown in Figure 12.5. Sometime surge suppression is recommended for small control-input switches in series with the inductive loads, to prolong switch life. The surge suppression device is connected in parallel around the inductor.

PLC INPUT AND OUTPUT DEVICES

Solid-state digital input and digital output devices

Digital input and output devices may use conventional dry-contact switches to make or break a circuit, non-polarity dependent solid-state relays using TRIACs, or they may use direct-current solid-state transistor-based sensors, which have precise wiring requirements and is the topic of this section. Field-installed transistor-based sensors have two distinct components, power for the sensor electronics to operate, and signals for control inputs or actuation outputs. Solid-state digital-input and digital-output devices come in two, three, and four-wire configurations.

IEC 61131-2 defines three types of digital input sensors, all of which use binary 1 or 0 signals in the PLC I/O module:

Type 1: *Two-wire Mechanical Switching Contacts* are devices, such as pushbuttons and relays, developed in an era preceding semiconductor switches. Type 1 devices can be used in place of three-wire Type 3 PNP semiconductor sensors, but cannot be used in place of two-wire Type 2 semiconductor devices.

Type 2: *Two-wire Semiconductor Sensors* transmit both power and signal over the same wires. Power is continuously drawn from the PLC I/O module to operate the sensor circuitry. Type 2 sensors are restricted to modules having few channels.

Type 3: *Three-wire Semiconductor Sensors* derive their operating power from a separate source than from the I/O module itself. Their lack of power draw from the I/O module permits a large number of connected signals, producing little heat in the module. Type 3 sensors can be used with Type 1, 2, or 3 input modules.

In the two-wire device, the same conductors that power the device also carry superimposed amperage that forms the digital opened-or-closed-switch signal. The biggest problem with a two-wire sensor is that the PLC I/O module continuously supplies the sensor with quiescent or "overhead" current, including their associated I^2R losses which occur in the PLC module. A design concern with two-wire devices is that the quiescent current must be kept lower than the switching threshold of the digital module. The relatively large operating current when using multiple two-wire devices limits the number of channels that can be connected to a PLC I/O module, as more devices cause the module to run hotter.

A three-wire sensor uses two separate conductors, one for device power and the other for signal, with the third conductor being used to return both the power and signal amperages. The sensor circuitry is operated from the PLC direct-current power supply through a brown conductor, and the sensor produces the amperage used for the output signal. A black conductor is dedicated for the signal and is wired directly to a terminal on the I/O module, where the wiring terminal defines the PLC memory address. Since the signal current is very small, very little amperage and heat is imposed on the module. The third, blue wire is the shared conductor, completing the circuit for both power and signal. A simplified schematic is shown in Figure 12.6 for two- and three-wire sensors. It is the Type 3, three-wire semiconductor sensors that require careful coordination of polarity, which is referred to as sourcing and sinking.

Four-wire sensors use signal conductors that are independent from the power conductors. Four-wire sensors are used for high-power devices, or where the field device circuit uses a different voltage than the PLC power supply.

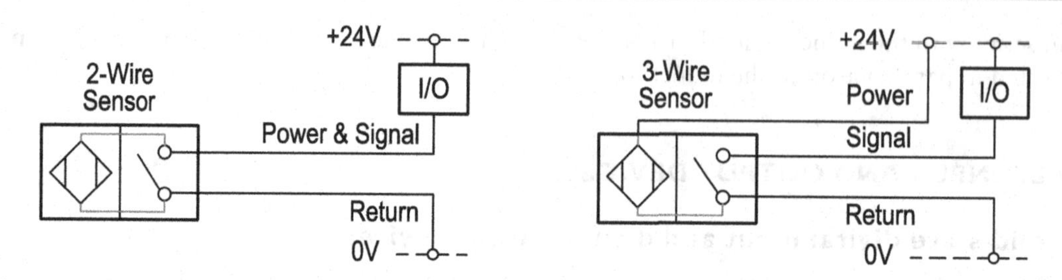

Figure 12.6 (Left) Two-wire devices carry both power and signals on a single pair of wires for very low-power sensors. (Right) For three-wire devices power and signals are carried on two separate wires, but use a shared return for completing both circuits. Three-wire sensors are the most common.

For PLCs, sourcing and sinking concepts are important when using polarity-dependent direct-current solid-state devices in three-wire systems, which have transistors that need to be installed with forward-biased polarity. The standard that establishes *sinking/sourcing* terminology is the IEC 61131-2 *Industrial-Process Measurement and Control—Programmable Controllers—Part 2 Equipment Requirements and Tests.* This IEC standard defines current sinking as the property of receiving current, and current sourcing as the property of receiving current, and goes on to reference sourcing and sinking to the *PLC input/output modules.* It should be noted that some manufacturers use terminology and wire colors that vary from convention, so engineers should always refer to the manufacturer instructions.

Sourcing and sinking

Sourcing and sinking terminology applies to PLC digital I/O modules and their associated field devices. Sourcing and sinking are limited to devices meeting all of the following conditions:

- Digital devices, using solid-state electronic switching
- Direct-current power and signals
- Transistor-based sensor technology, having strict dc polarity requirements
- Three-wire devices, using separate wires for power and signals with a shared return
- PLC digital input and output modules that are polarity dependent.

Many input and output field devices do not have polarity requirements. For example, PLC *dc-input-signal* amperage can flow in either direction through the "dry contacts" of a thermostat, or in either direction from a PLC *dc output signal* directed through non-LED lamps or solenoid coils. However, some three-wire *transistor-based electronic field devices use dc* electricity, where the relative polarity between the devices and the power source matters. The circuit will not work if a polarity-dependent device is not matched to the I/O polarity, just as a diode or transistor wired in the reverse-biased direction will not conduct. The terminology used to correctly match the polarities of *transistor-based* I/O devices to PLC modules is called *sourcing* and *sinking*.

Figure 12.7 and Figure 12.8 show an analogy of sourcing and sinking for a conventional ac circuit wired two ways.[1] Although placing the switch in either the hot or neutral legs does not affect functionality, the circuit on the left shows that voltage does not exist at the load until the switch is closed, while the circuit on the right shows that voltage is continuously applied through the load, regardless as to whether the circuit is energized or not. The figure shows that the terms sourcing and sinking are relative, so that the switch in the left diagram sources current to the load, while the load sinks current to the return. Conversely, the load in the right

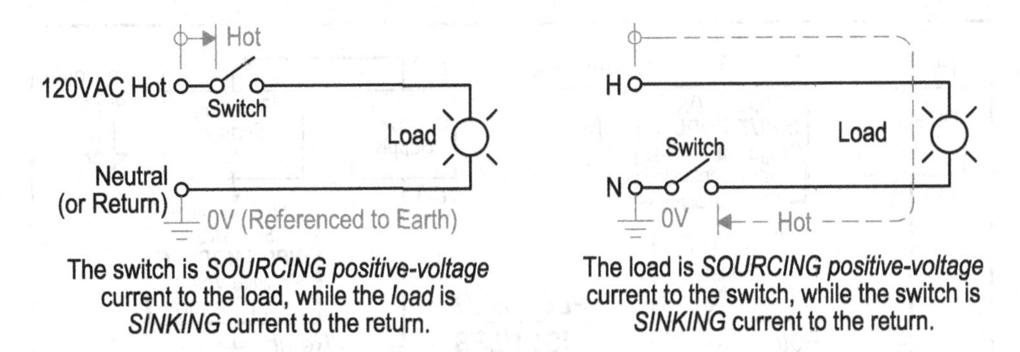

Figure 12.7 *Sourcing* and *sinking* terminology as applied to an *output* device using an ac electricity analogy. Functionally, the circuit works with the switch in either location, but its position in the circuit affects when voltage is available at the load.

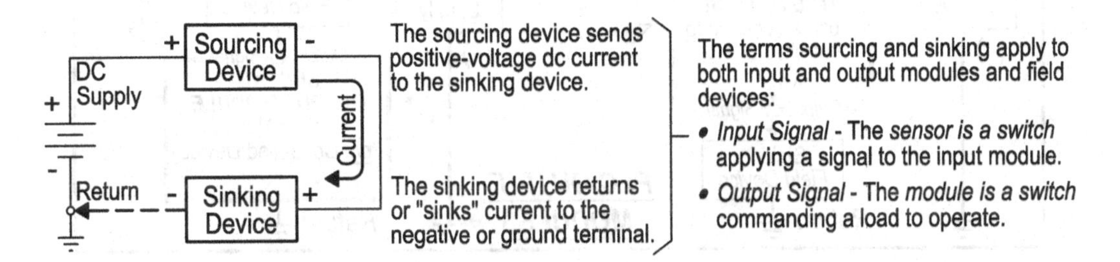

Figure 12.8 The *sourcing* device sends positive-polarity *current flowing away from* and toward the *sinking* device, which completes or sinks the circuit to the negative power supply terminal. Sourcing or sinking terminology applies to both the PLC input and output modules, where the input module behaves as a "load" controlled by a field switch, and conversely, the output module has a built-in transistor-based switch that turns the field-installed output "load" on or off.

diagram sources current to the switch, while the switch sinks current to the return. The perspective of using the switch or the load to define sourcing and sinking affects the terminology, which can cause confusion. For consistency, most PLC manufacturers reference sourcing and sinking to the PLC input/output modules, following IEC 161131-2 conventions.

Correctly sourcing or sinking device connections is necessary to match NPN or PNP transistor-based field devices to the polarity of both the dc power-supply and the module I/O terminals. To complicate the terminology, sinking and sourcing depends not only on the transistor type, but also whether the device is used for digital inputs or outputs (Figure 12.9). Table 12.1 shows the sourcing and sinking relationships for digital input modules, and Table 12.2 shows the relationships for digital output modules. Incorrect selection of field devices relative to the sourced or sinked I/O modules can be avoided by following the manufacturer wiring diagrams, but if not done correctly, hardware replacement will be required, adding schedule delays and expense. Figure 12.10 shows simplified externally powered three-wire *PNP* and *NPN* sensors. For the *PNP* sensor, the *signal* current flows *away* from the sensor, *sourcing* positive voltage into the PLC digital-input module. Conversely, the *NPN* sensor pulls or *sinks* current *from* the PLC module, returning it directly to ground.

Figure 12.11 shows the signal arrangements for sourcing and sinking input modules, but with sensor power omitted for clarity. PNP sensors with sinking input modules are more common in Europe and North America, while NPN sensors coupled with sourcing input modules are more common in Asia. Sinking modules use a separate power source, which supplies the

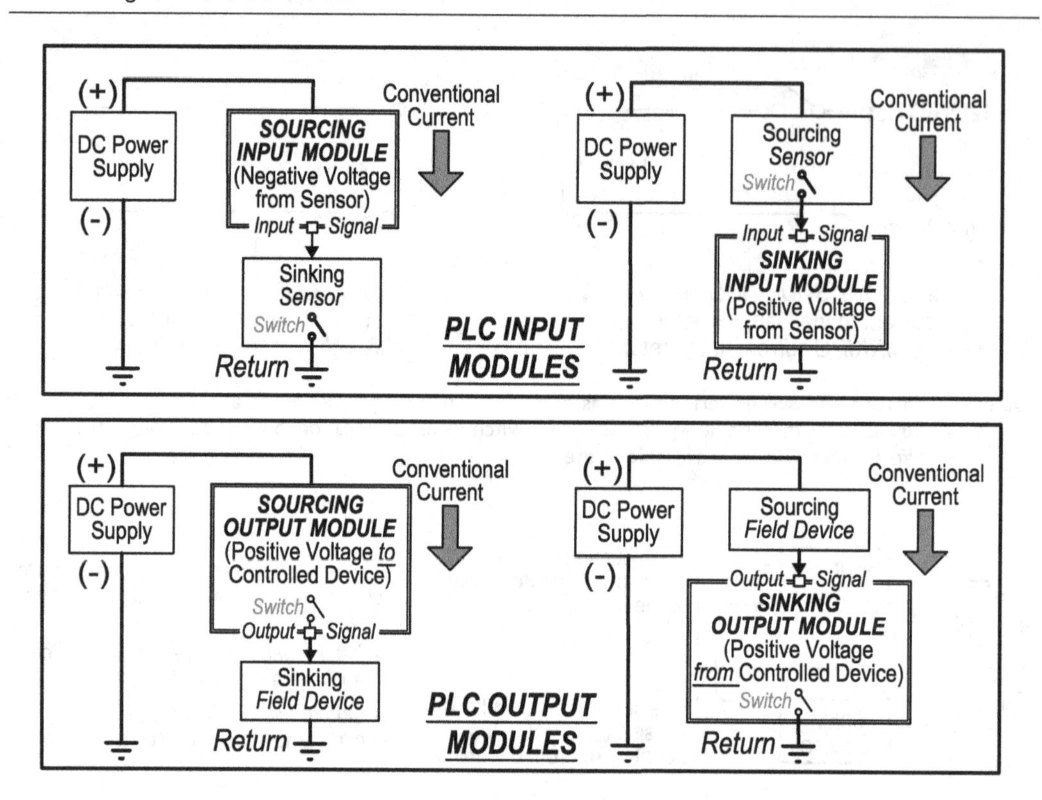

Figure 12.9 Block diagrams showing PLC sourcing and sinking *input modules* (upper) and PLC sourcing and sinking *output modules* (lower). The input sensors are electronically actuated transistor-based switches that send a true or false signal to the PLC, while the output modules use transistor-based switches from the PLC to energize or deenergize the field devices. Sourcing and sinking are commonly referenced to the PLC modules, and NPN or PNP terminology is used to define the input device types.

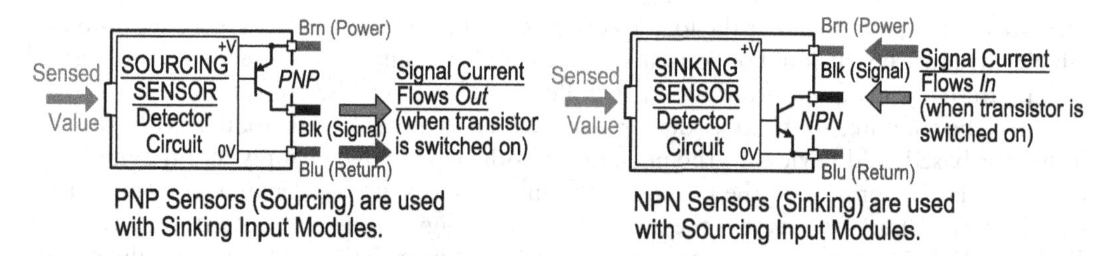

Figure 12.10 Simplified diagrams showing externally powered 3-wire solid-state sensors. The sensor circuit fully biases a transistor to behave as a closed switch, providing a digital-input signal to the PLC input module. For the *PNP* sensor, signal current flows *away*, sourcing current to the input module. Conversely, signal current flows *into* the *NPN* sensor, which sinks module current to the return.

signal into the module in an arrangement called "loop powered." The PNP sensors coupled with sinking-input modules tend to be more vulnerable to short-circuits. However, NPN sensors coupled with sourcing modules tend to be more vulnerable to false signals that inadvertently start equipment, such as if a broken wire shorts to ground. Unplanned starting of equipment is more likely to be considered unsafe.

Table 12.1 Digital-*input* relationships between sourcing and sinking terminology, logic states, module signals, and locations of switches and "loads" in the circuit

Terminology	*Logic states*	*Module signal*	*Switching*
Sourcing Input Module Positive polarity current leaves the module signal terminal *NPN* field device (switch) IEC Negative Logic The *sinking switch* is located between the module and ground	**1** (True) 0 (False)	**0V** +24V	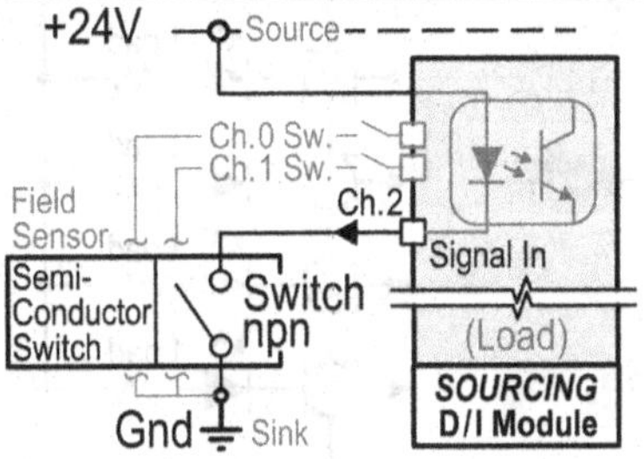
Sinking Input Module Positive polarity current enters the module signal terminal *PNP* field device (switch) IEC Positive Logic The *sourcing switch* is located between the +24V and module.	**1** (True) 0 (False)	**+24V** 0V	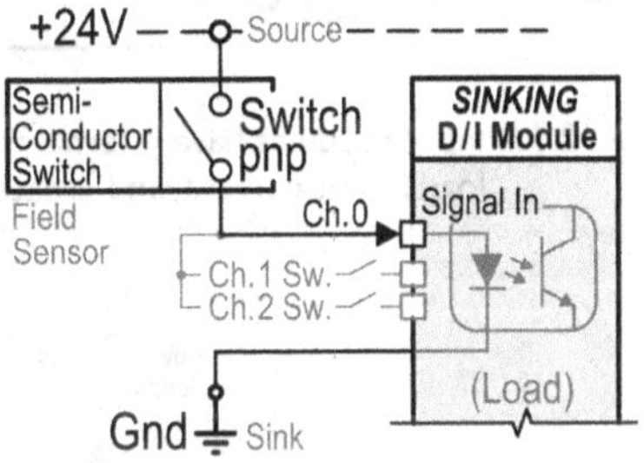
Both Sinking and Sourcing	**0** (False)	**0 amps** (Open)	

Table 12.2 Digital-*output* relationships between sourcing and sinking terminology, logic states, module signals, and locations of switches and "loads" in the circuit

Terminology	*Logic states*	*Module signal*	*Switching*
Sourcing Output Module Positive polarity current leaves the module signal terminal PNP Output Module (switch) IEC Positive Logic The *sinking switch* is located between the module and ground	**1** (True) 0 (False)	**+24V** 0V	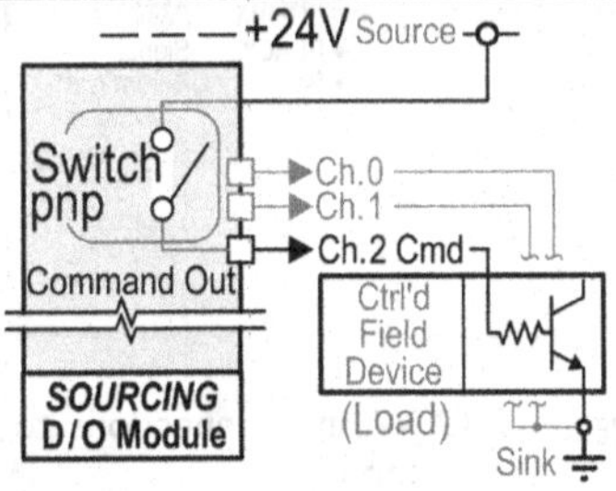
Sinking Output Module Positive polarity current enters the module signal terminal NPN Output Module (switch) IEC Negative Logic The *sourcing switch* is located between the +24V and module.	**1** (True) 0 (False)	**0V** +24V	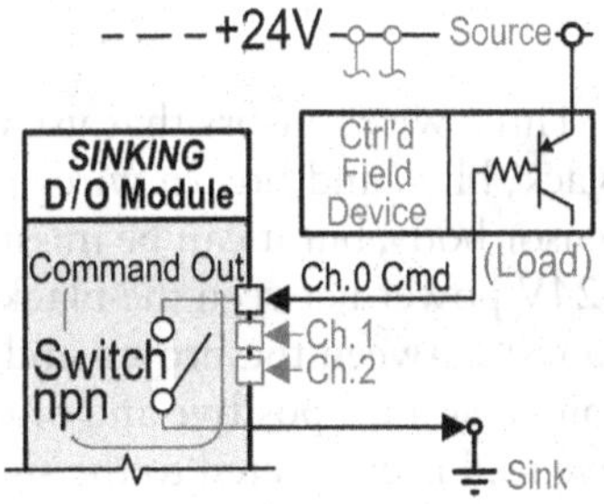
Both Sinking and Sourcing	**0** (False)	**0 amps** (Open)	

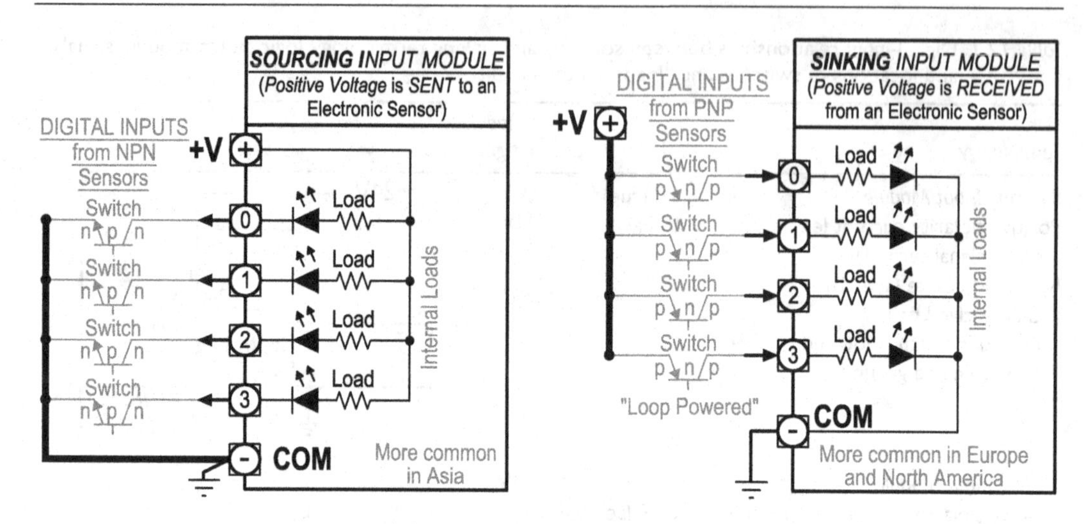

Figure 12.11 Input modules have internal resistances between the source voltage and ground, which behave as loads, limit current, and create the voltage drops that form the input signal.

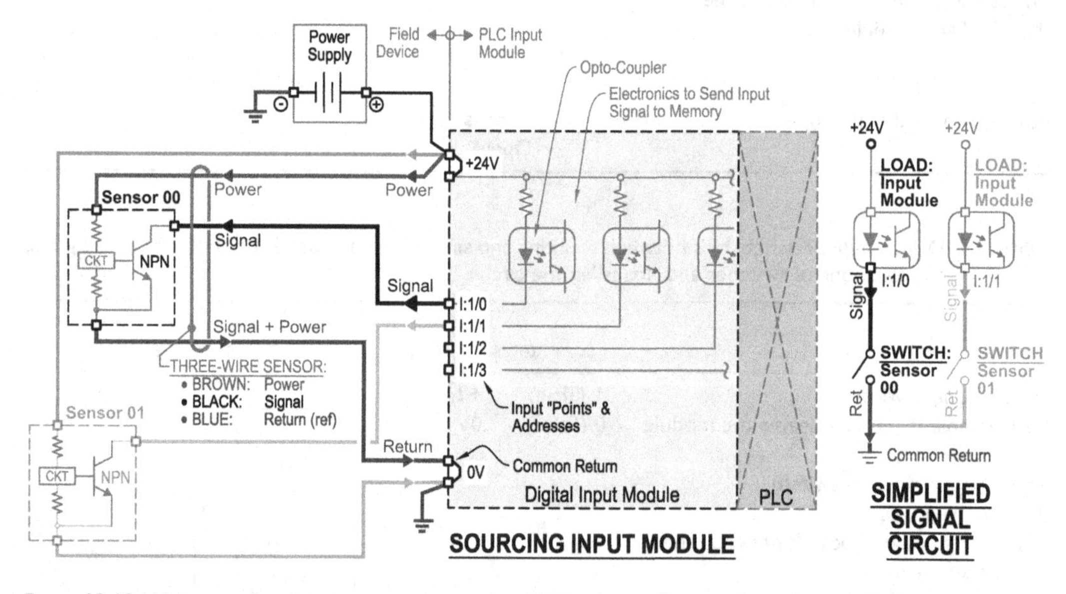

Figure 12.12 Wiring connections to two three-wire *NPN* sensors from a four-channel PLC sourcing input module.

Three-wire sensors that use dc voltage are either of the PNP or NPN type; usually having black, blue, and brown wires (see Figure 12.12). Usually the sensor type is printed onto the sensor body, but it can be missing. The sensor can be tested to determine its type by applying +24V power between the black lead and the blue 0V return lead, and then using a voltmeter to test between the brown and blue leads. In a PNP sensor, brown is typically connected to source-voltage positive and blue to the 0V reference or negative terminal. The voltmeter positive lead is connected to the signal output (brown) and voltmeter negative lead to the sensor 0V reference (blue). The reading should be zero when not sensing. Then the sensor is forced to produce an output. For example, if the sensor is a proximity switch, a piece of steel could be placed close to its pole-face. If the voltmeter reads a signal between 10 to 30V, then it is a PNP type sensor, to be connected to a sinking input module. If the meter stays at 0V, then the sensor is of the NPN type, to be connected to a sourcing input module.

Flexible, polarity-independent transistor-based digital I/O devices

Figure 12.13 shows more robust polarity-independent I/O opto-couplers, constructed with two reverse-paralleled light-emitting diodes, allowing bi-directional current to flow. Bi-directional signal capability eliminates the strict NPN and PNP sensor-polarity requirements, simplifying connections. Figure 12.14 shows all four combinations of polarity-dependent sinking-and-sourcing and input-and-output field-device connections.

Other PLC output-module options

Mechanical relay outputs

Some PLC outputs use miniature mechanical relays. Like the opto-couplers, the relays provide electrical isolation between the field circuit and the sensitive PLC electronics. The relays

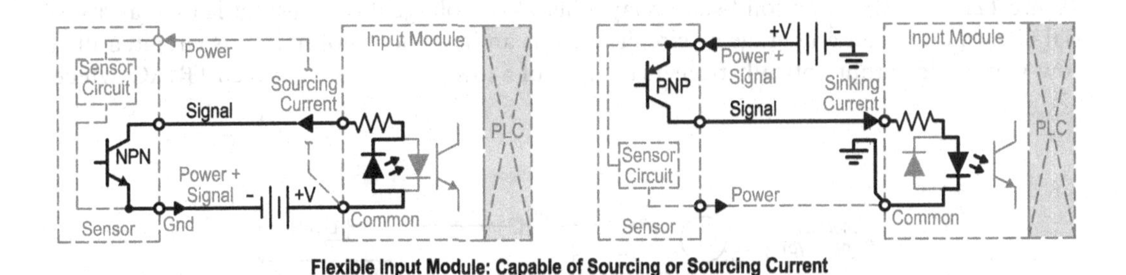

Figure 12.13 Flexible input module design. Strict NPN or PNP sensor-polarity requirements are eliminated when the module has a double-LED optocoupler. This design is becoming more prevalent.

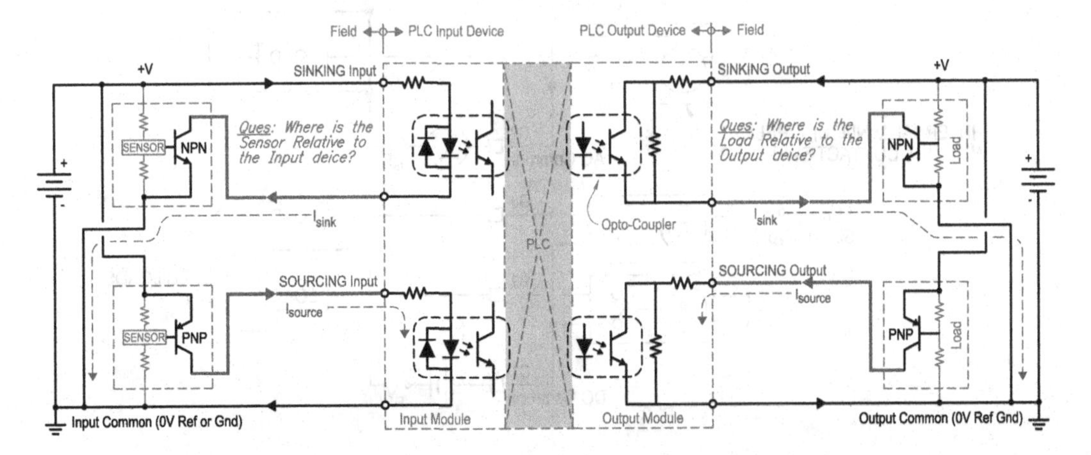

Figure 12.14 NPN or PNP input field devices will work with PLCs having polarity-independent input modules (left). The right circuit shows that either *NPN-* or *PNP-type* output field devices will work with a polarity-dependent output module when it can be wired in differential mode, instead of wiring in the singled-ended mode, which uses a shared source or a shared ground.

tend to be more expensive, they do not switch very fast, and can have switch-bounce problems in small programs controlling PLCs with fast-changing output states. One side of each relay switch is connected to a common terminal, where an external source of ac or dc voltage is applied, and the other side is wired to send output current to the individual loads. Figure 12.15 shows a PLC output where several groups of loads may be powered from different electrical sources for increased design flexibility. In this manner, a single PLC can power several load types simultaneously.

High-power DC outputs

Some PLC dc-output loads may require high amperage, especially if the loads are inductive. Figure 12.16 shows two high-power transistors biased to conduct via an opto-coupled signal, where one is sinked and the other sourced. This circuit also shows a Zener diode which provides the dual function of setting the transistor collector-to-emitter voltage when it is connected in its normal reverse-biased direction, and it also behaves as a freewheeling diode to mitigate collapsing magnetic fields when inductive loads are turned off.

AC solid-state relay output

Figure 12.17 shows an ac solid-state relay where low-voltage direct current is used as a signal to turn on a high-amperage ac circuit, such as an induction motor. In this circuit, a PLC low-voltage dc output optically triggers the gate of a low-power opto-coupled TRIAC, which

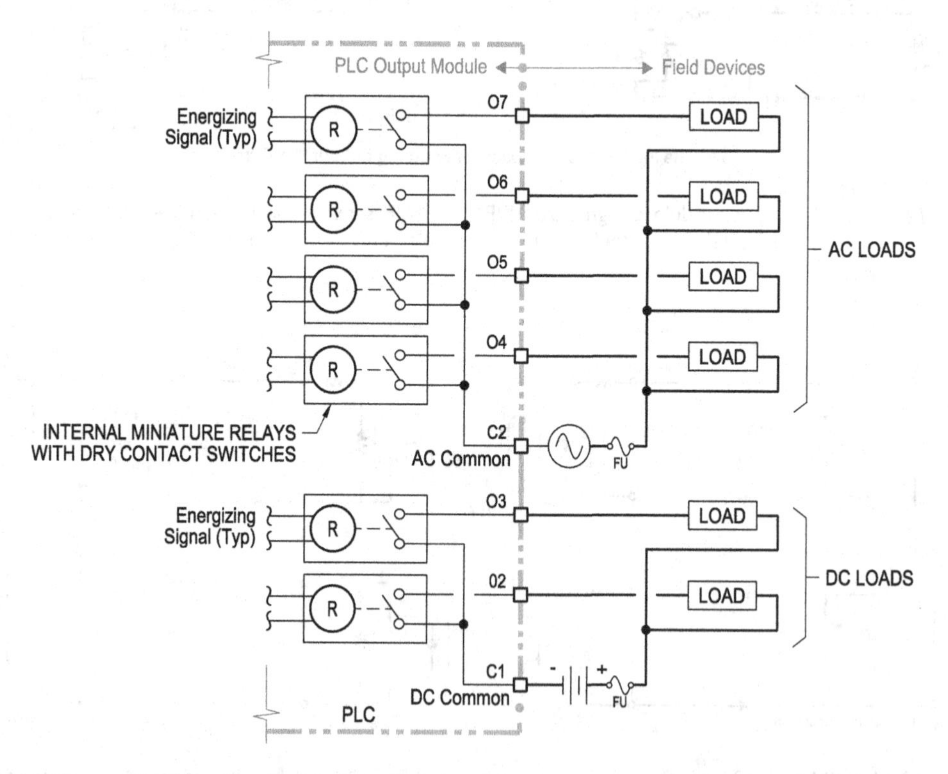

Figure 12.15 Allen-Bradley MicroLogix 1000 relay output provides electrical isolation between higher-voltage ac/dc field devices and PLC electronic circuits.

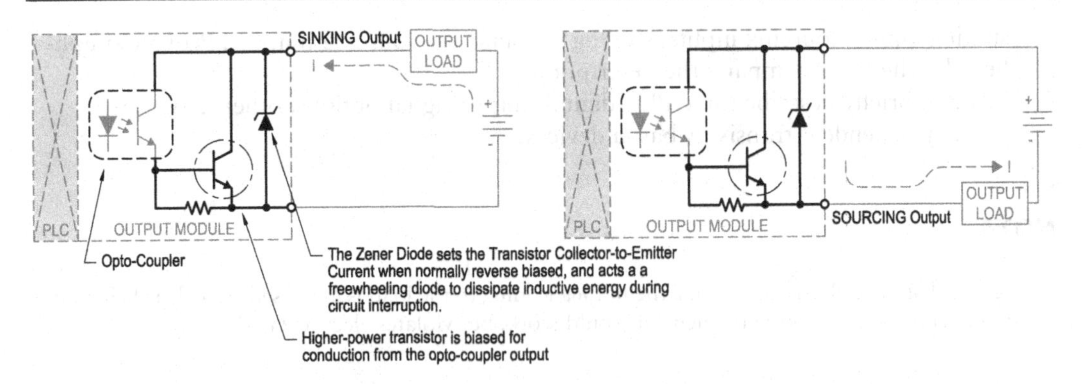

Figure 12.16 A PLC opto-coupled high-amperage output transistor is powering a dc load. A sinking output arrangement is shown on the left and a sourcing-output is shown on the right.

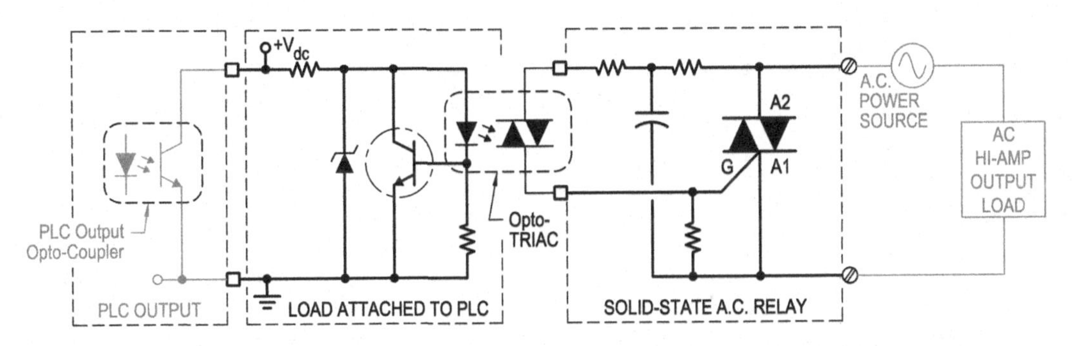

Figure 12.17 The dc output from a PLC is optocoupled to drive a high-power TRIAC, which serves as a solid-state contactor, powering high-amperage ac loads.

is a device that conducts in both polarities. The small opto-coupled TRIAC in turn triggers a larger-power TRIAC, which forms the power portion of the solid-state relay. The correct sink-or-source polarity must be applied into the small opto-TRIAC, so as to trigger the circuit to function.

QUESTIONS

1. What type of voltage is used to power PLC's? How is that voltage obtained?
2. What is an E-Stop switch? Where would E-Stop switches be used?
3. How are digital inputs and digital outputs wired to a PLC? How are the connections defined on an I/O wiring diagram?
4. How are outputs protected when powering inductive loads?
5. Describe two-, three-, and four-wire semiconductor sensors and their wiring requirements.
6. Discuss sourcing and sinking in broad terms, when using transistor-based direct-current input or output devices having strict polarity requirements. Summarize the location

sinking inputs, sourcing inputs, sinking outputs, and sourcing outputs. Provide simplified sketches to accompany the description.

7. List and briefly describe three PLC digital-output-signal options, other than the polarity-dependent transistor-based devices.

NOTE

1. Note that in ac electrical circuits, the hot leg leading to the load is always switched, as being more inherently safe. Breaking the neutral would work, but violates electrical codes.

Chapter 13

Allen-Bradley RSLogix software and ladder-diagram programming

ROCKWELL/ALLEN-BRADLEY PLC SOFTWARE

Allen-Bradley is the brand of PLC equipment manufactured by Rockwell Automation. RSLinx, Boot-DHCP Tool, and RSLogix are three computer applications for establishing communications, writing programs, and transferring program information into the PLC.

PLC communications are required to:

- "download" a computer program from a PC to a PLC to establish control,
- "upload" a program from the PLC into a PC to view the as-installed program and to make program modifications,
- "go online" to observe the program operation in real time for functional verification or troubleshooting,
- insert overriding "forces" into the program instructions to modify its operation, for observing the system reaction and for troubleshooting,
- examine the memory-resident "bit states" that describe the I/O status information.

RSLinx is the software that establishes the basic hardware communication link between a PC and the PLC, where communication is coordinated via the cable type and data transmission protocol. Some common options include serial protocol using PS2/RS232 connections, Modbus protocol using RS485 cabling/connectors, and Ethernet protocol and IP addresses.

The BOOT-DHCP Tool can establish dynamic IP addresses by default for easier connectivity, but for distributed control systems the BOOT-DHCP application is used to define static PLC IP addresses. Unique static addresses are necessary for avoiding data collisions when multiple PLCs are networked into a DCS/SCADA system.

The RSLogix 500 and RSLogix 5000[1] are software packages used for programming the MicroLogix and SLC models of the A-B PLCs and their associated I/O expansion devices; but RSLogix also contains needed addressing features for program communication. RSLogix is fundamentally an IEC-61131[2]-compliant ladder-diagram programming language. The software can be toggled between classic ladder-logic and structured-text programming languages, for increased flexibility. The ladder-diagram language also includes instructions that provide virtual devices[3] that emulate real-world timers, counters, complex mathematical functions, and PID algorithms, and it has the ability to use subroutines for efficient programming and faster execution.

FUNCTIONS WITHIN PLCS

Discrete control/combinational logic

For discrete or digital logic and control, PLC input functions include internal coils, internal switches, timers, counters, etc. In PLC programming, the internal or virtual switches are called "examine-if-closed" contacts (XIC) or "examine-if-open" contacts (XIO). "Examine-if-closed" (XIC) are somewhat analogous to NO relay contacts in electrical systems, and "examine-if-open" (XIO) are somewhat analogous to NC relay contacts. However, it needs to be understood that the symbols represent logical computer instructions, which may or may not correspond exactly with electrical behavior.

For automatic control of systems, PLCs measure system parameters as inputs, compare those measured values to a desired setpoint, and then output an appropriate control signal. In simpler applications, the output can be a discrete open-or-close command to a valve or a start-or-stop command to a motor. Electromechanical motor control devices, such as conventional relays, latching relays, comparators, and mathematical functions, are easily mimicked within the PLC built-in program functions.

Sequential control

Some systems require sequential operations to occur or for equipment to be started and stopped at specific periods of time. Typically, on-delay timers, off-delay timers, and counters have been used in the past to perform recurring or sequential control, and these functions are now available as virtual devices from program instructions. The use of subroutines aids in placing or removing sequencing functions into the program scan. The programming accomplishes sequencing operations in a simpler, more reliable manner than the older electromechanical controllers.

Continuous control

In fine control of complex systems requiring fast, smooth action, the output needs to be modulated based on a calculated output signal. The PLC can incorporate complex mathematical functions, which can emulate older mechanical-control mechanisms used to provide a real-time analog response.

For applications requiring continuous modulating control, closed-loop negative feedback compares measured values to a user-defined setpoint to determine an error. A corrective signal is typically transmitted as 4–20mA or as a low-voltage signal.[4] The magnitude of the corrective signal is related to not only the amount of error, which is gain, but the signal can be modified to account for time away from the setpoint (reset) and the speed at which corrective action occurs (rate). The corrective signal is applied in a manner to eliminate the error, and the response is tuned to avoid over- or under reaction to disturbances. To provide sufficient actuation forces at the final control device, the 4–20mA signals are often converted to a corresponding pneumatic signal, which positions actuators. Pneumatic signals are typically 3–15 psi or a 0–30 psi, either of which can create large actuation forces. The current-to-pneumatic converter or I/P converter is more than a signal converter; it is also a force amplifier. In plants lacking control air, the signals can position stepper motors, such as modutrol motors, to position sizeable valves or dampers.

Proportional-only control leaves offset, which is a setpoint deviation after a new steady-state control position is attained following a process disturbance, and occurring because the corrective signal can only be generated from an observed deviation. If a system can tolerate offset or level variations and if the response speed is satisfactory, then the proportional-only control is adequate, straightforward, inexpensive, and recommended. To reduce offset for better control, the gain or sensitivity of the proportional control can be increased in the PLC program. Gain amplifies the output-signal relative to the input-signal. However, gain can only be increased within system behavioral limits, as controls that are too sensitive become unstable and hunt. Built-in PID instructions can improve both control accuracy and response time. The integral (reset) and derivative (rate) control functions add anticipatory behavior that eliminates offset and speeds the response. This topic is discussed in Chapter 2 on "Control Terminology and Theory."

RSLINX, BOOT-DHCP TOOL, AND ESTABLISHING COMMUNICATIONS

RSLinx

The first step in programming a PLC is to establish communication with the PLC and all attached modules. The Allen-Bradley (AB) communication software is called RSLinx. The programming computer is cable-connected to the PLC using an RS232 COM port cable and nine-pin D-connector, or a USB-to-serial emulator cable and driver, or an Ethernet cable in later generation models. In the best case, the RSWho function will automatically find the PLC and all attached devices, including their model numbers, using plug-and-play technology, in the same order they are ribbon-cable connected to the PLC. With older devices, it may be necessary to manually select the devices and their connection options from a menu list. The RSLinx program shows the connected devices and their availability.

RSLinx establishes PC communications to the PLC for programming, browsing automation networks, configuring network devices, observing ladder diagram actions in real time, and troubleshooting problems. PC-to-PLC device communication protocols are selected from the RSLinx program, which are either serial or Ethernet. Modbus or Ethernet protocols are often used for PLC interconnections within a distributed-control or network system. RSWho verifies which devices are connected as well as the connection protocols (see Figure 13.1).

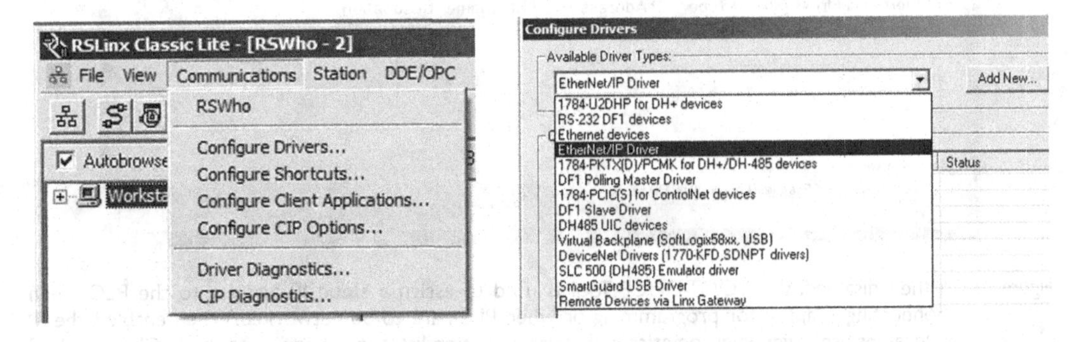

Figure 13.1 RSWho and its driver type submenu are used to establish communications, which need to be consistent with the available connector types and wiring.

BOOTP-DHCP

When configuring the PLC using Ethernet protocol as an example, is possible to set the BOOTP-DHCP so the Ethernet IP addresses are established dynamically in a plug-and-play manner, depending on unused network IP addresses. Conversely, in many plant installations, multiple PLCs and devices need to work synergistically on a single network, and it is advantageous to manually preassign IP addresses. A fixed PLC IP address can also be defined in the PLC using the "BOOT-DHCP Tool" application (see Figure 13.2) and the address coordinated within the ladder-logic program through the RSLogix software. In this configuration, unique static IP addresses are coordinated to avoid communication conflicts in a multi-PLC distributed control system or SCADA systems.

UPLOADING/DOWNLOADING THE PROGRAM

After connecting the computer to the PLC, the program can be transferred back and forth between the two devices. A-B defines the PLC to be the "local device" and the computer to be the "remote server" used for downloading programs, program maintenance, and archiving area. From the PLC's perspective, downloading is the transfer of program files from the PC into the PLC. Uploading is the retrieval of the program from the PLC into the PC. However, this terminology is not universal for all manufacturers.

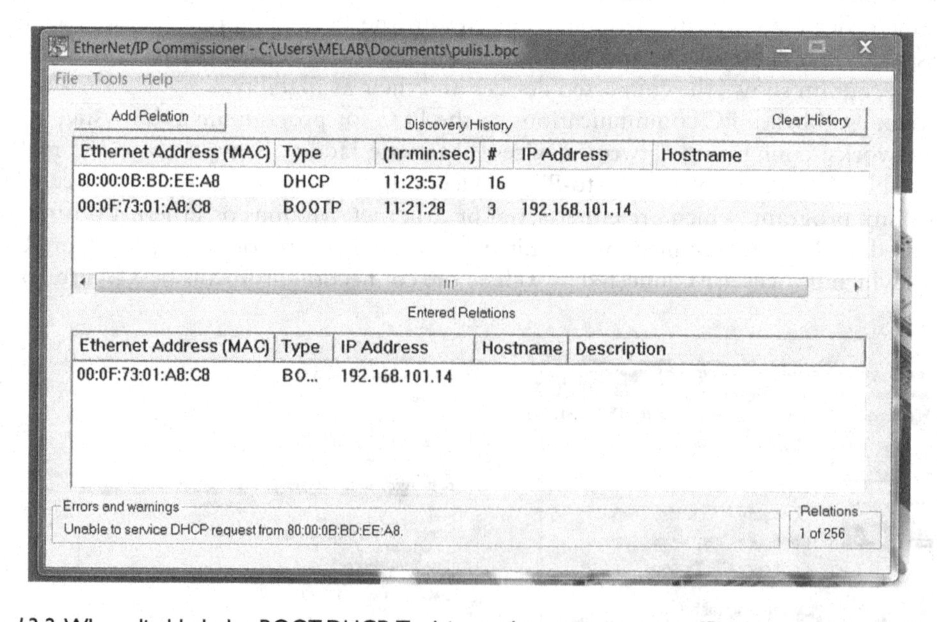

Figure 13.2 When disabled, the BOOT-DHCP Tool is used to assign a static IP address to the PLC when connecting a laptop for programming or when PLCs are to be networked. When enabled, the IP addresses are assigned automatically through a relation list, corresponding to the PLC's local MAC address.

RSLogix is the software containing the PLC ladder-logic programming language. Common practice is to annotate the rungs of code and to label the rung devices as the program is being authored. Commenting assists in understanding and debugging the program and in making future modifications. Clear and thorough commenting during programming is good practice.

While it is possible to retrieve or upload an active program from an operating PLC into the PC, the annotations embedded in the original source code are typically not resident in most PLCs, to conserve memory and increase speed. Thus, all annotations are usually lost during uploading. However, some software packages can compare the uploaded program against the original annotated source code, to identify any changes that may have been incorporated while in service, if the original program is available.

To download the program, the PLC typically must be toggled out-of-operation by removing it from run mode and setting it to the program mode, although some PLC models are capable of "hot" online updates while in run mode. The PC programming software contains the commands and prompts for downloading the PLC program. After code modifications are complete, the program is downloaded while the PLC is offline. Afterwards, the PLC is placed back into run mode, where it is possible for the user to go online and observe the program operation and resulting bit states in real time. While online, it is possible to apply forces to override program bit settings, i.e., bits in a data file can be "forced" on or off for direct control of devices. The online observations of the ladder diagram and application of external forces are excellent tools for troubleshooting. An LED status indicator on the PLC illuminates whenever user forces are in effect; and this indication serves as a warning that some control features have been overridden and are non-functional.

RS LOGIX PROGRAMMING

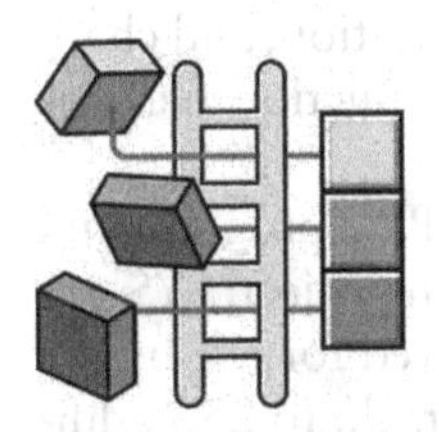

RSLogix is the software used to program the PLC. RSLogix essentially follows ladder-diagram syntax with the addition of complex mathematical functions and the availability to use programming subroutines. The program integrates real-world I/O devices with internal or virtual devices that exist only as part of the PLC memory. Input-device statuses are tracked by the PLC memory bits and are used to determine the output signals. This section discusses the ladder-logic programming associated with the Allen-Bradley RSLogix 500 software.

The RSLogix 500 window is shown in Figure 13.3. The main sections within the RSLogix window include the project window, the ladder-view window, the instructions sections, and online dropdown options. Major programming functions contained within the project folder are:

- *Instruction tabs*: allow instructions to be dragged-and-dropped onto the rungs,
- *Controller*: establishes program communications,
- *Database files*: contains project organizational and annotation information,
- *Program files*: includes programs, subroutines, and functions,
- *Data files*: store the values in memory of inputs, outputs, and variables used by instructions,
- *Forces*: provide manual overrides of I/O data for observing program behavior during troubleshooting.

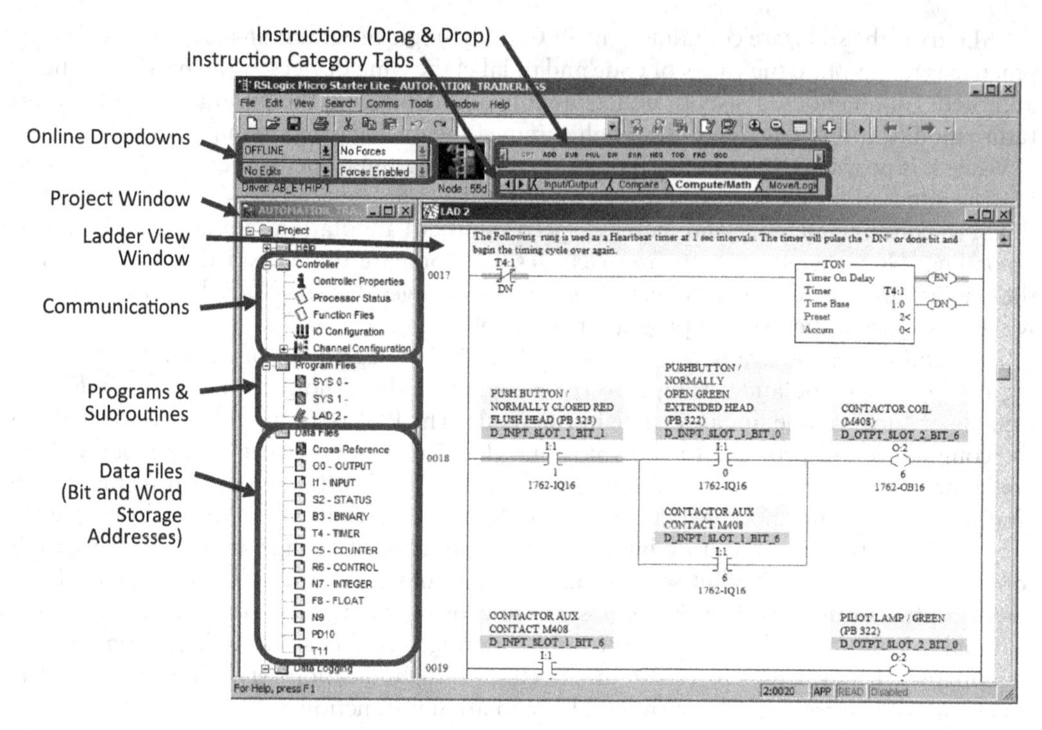

Figure 13.3 The RSLogix 500 programming window.

The *Controller* folder consists of files containing PLC properties, IO configuration, and channel communication for linking the PC to the PLC. Processor status and function files are included in this folder.

The *Program Files* folder can contain up to 256 files associated with the PLC program and subroutines. SYS 0, SYS 1 and LAD 2 are always included by default in each project. SYS 0 is a system file that contains the PLC controller configuration. SYS 1 is reserved for use by the controller and LAD 2 is the main ladder-diagram programmed by the user. Up to 253 additional program-related files are allowed, where the additional files are typically user-defined subroutines.

The *Data Files* folder can contain up to 256 data files. The data files contain operational status information for all main ladder program instructions, internal instructions, and subroutines. A summary of the data-file types is included in Table 13.1.

The *Instruction* tabs and drag and drop instructions are available for connecting real I/O devices, internal devices, functions, actions, etc. An overview summary of instructions and their locations are included in Table 13.2.

TERMS USED WITH PLC LADDER LOGIC PROGRAMMING:

Rack

This is the PLC and the string of devices connected off the PLC. An example of a rack is shown in Figure 13.16. The rack, PLC, and connected expansion modules are used together in creating addresses. The rack includes the power supply and communication modules.

Table 13.1 Default data-file types

Data file		*Stored information*
Cross Reference Report		Cross references of device/instruction type usage by addresses & symbol
O0	Output	Bit status for each PLC output
I1	Input	Bit status for each PLC input
S2	Status	PLC operation information
B3	Binary	Bit status for each internal binary-logic instruction value
T4	Timer	Timer Data (preset time, accumulated time, timer status bits)
C5	Counter	Counter Data (preset count, accumulated count, counter status bits)
R6	Control	Control Data for shift registers and sequencer instructions (length, pointer position, status bits)
N7	Integer	Integer numeric values or bits
F8	Float	Real numeric values
Files 9-255	User Defined	Bits, timers, counters, control, programmable limit switch date (6-word elements), double words, messages, PID word files, etc.

Note: Except for the O0 and I1 files, which are limited by hardware connections, data files can be added for all other types when the available files are completely used.

Table 13.2 RSLogix instructions are organized within the various tab menus

Tab	*Instructions*	*RSLogix drag-and-drop instruction menu*
User	New rung, branch, XIC, XIO, OTE, OTL, OTU, ABL -ASCII test for line, ABSolute	ABL ABS User Bit Timer/Counter Input/Output Compare
Bit	XIC, XIO, OTE, OTL, OTU, One Shots: ONS, OSR, OSF	ONS OSR OSF DDT FBC User Bit Timer/Counter Input/Output Compare
Timer/ Counter	TON, TOF, RTO, CTU, CTD, RES, RTA	TON TOF RTO CTU CTD RES HSC RHC RTA User Bit Timer/Counter Input/Output Compare
Input/ Output	IIM/IOM – immediate input or output w/mask MSG – message REF – I/0 refresh	BTR BTW IIM IOM SVC MSG IIE IID RMP RPI REF DEM CEM User Bit Timer/Counter Input/Output Compare
Compare	LIM – limit test EQU/NEQ -equals (not equal) LES/GRT – less /greater than LEQ/GEQ (<= or >=)	LIM MEQ EQU NEQ LES GRT LEQ GEQ Bit Timer/Counter Input/Output Compare Com
Compute/ Math	ADD, SUB MUL, DIV, SQR, NEG, TOD/FRD – to/from BCD GCD – Gray code	CPT ADD SUB MUL DIV SQR NEG TOD FRD GCD Input/Output Compare Compute/Math Move/Logi
Move/ Logical	MOV – move AND, OR, NOR, NOT – bitwise logical CLR – clear	MOV MVM AND OR XOR NOT CLR Compare Compute/Math Move/Logical File/Misc

Table 13.2 (Cont.)

Tab	*Instructions*	*RSLogix drag-and-drop instruction menu*
File/Misc	COP – file copy SCL – scale PID, PWM CPW – copy word	 COP FLL DDV DLG SCL INT STE STS STD PID PTO PWM UID I Move/Logical File/Misc File Shift/Sequencer Pro
File Shift/ Sequence	BSL/BSR – bit shift left/right SOC/SQL/SOO Sequencer FFL/FFU - FIFO load/unload LFL/LFU – LIFO load/unload	 BSL BSR SQC SQL SQO FFL FFU LFL LFU Move/Logical File/Misc File Shift/Sequencer Pro
Program Control	JMP – jump to label LBL – label JSR – jump to subroutine RET - return	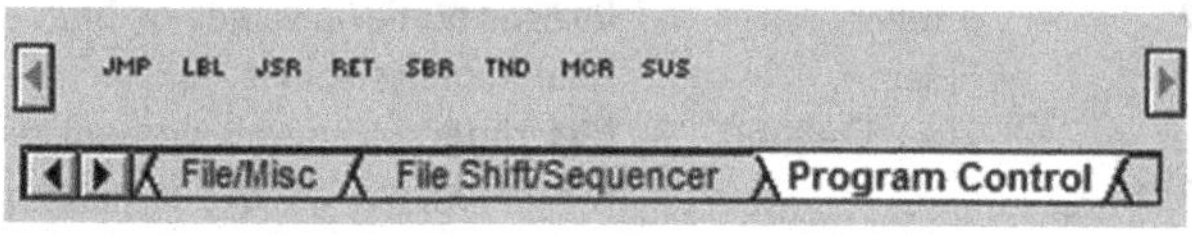 JMP LBL JSR RET SBR TND MCR SUS File/Misc File Shift/Sequencer Program Control
ASCII Control	ABL, ACB, ARD, ARL, AWT, AHL, ACL ASCII read/write/append/ Handshake, clear buffer,	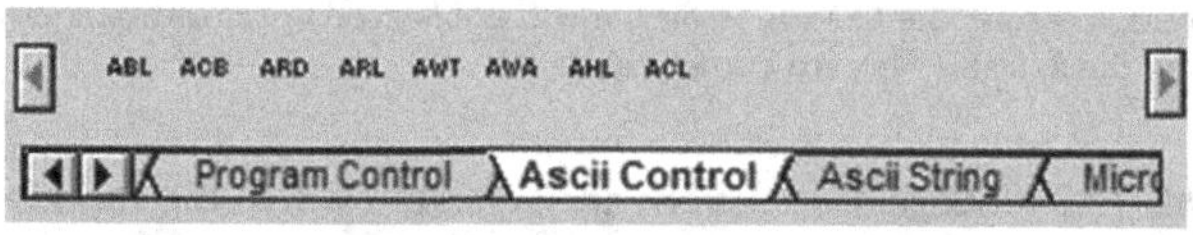 ABL ACB ARD ARL AWT AWA AHL ACL Program Control Ascii Control Ascii String Micr
ASCII String	ACN, ACI, AIC, ASC, ASR concatenate, string to Integer, integer to string, extract, search, compare	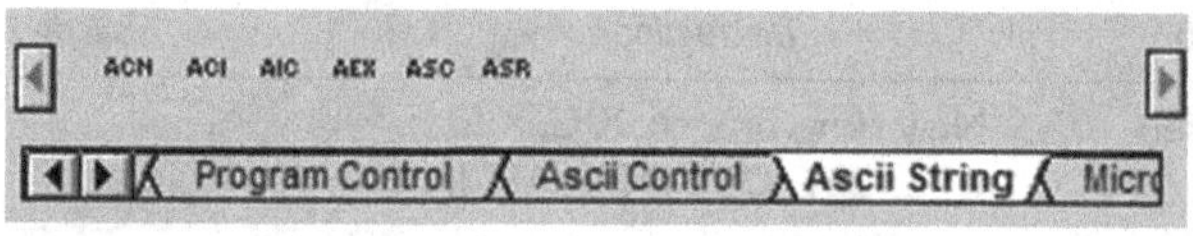 ACN ACI AIC AEX ASC ASR Program Control Ascii Control Ascii String Micr
Micro High-Speed Control	HSC, HSL, RAC, HIS, HSD High-speed counter, load, reset, interrupt enable/ disable	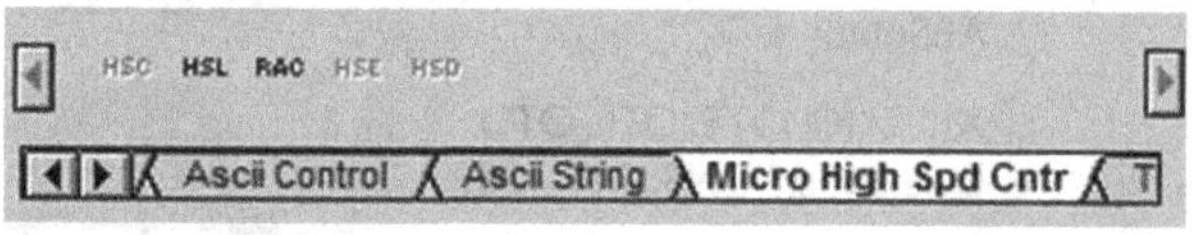 HSC HSL RAC HSE HSD Ascii Control Ascii String Micro High Spd Cntr T
Trig Functions	SIN, COS, TAN, ASN, SCS, ATN Sine, cosine, tangent & inverse trig functions	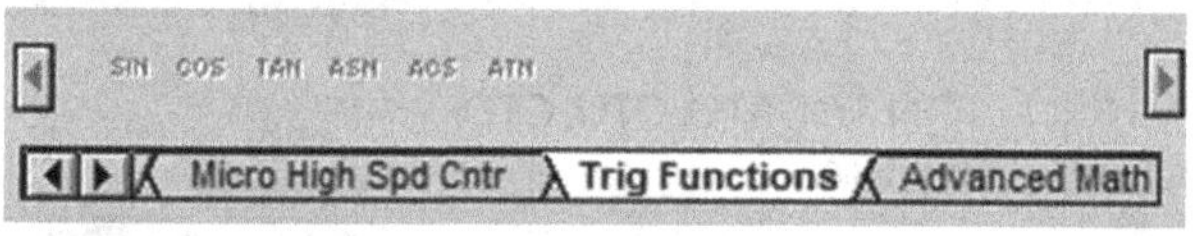 SIN COS TAN ASN ACS ATN Micro High Spd Cntr Trig Functions Advanced Math
Advanced Math	LN, LOG, DEG, RAD, XPY, ABS, SCP, SWF, DCD, ENG DFF Scale, swap, encode, etc.	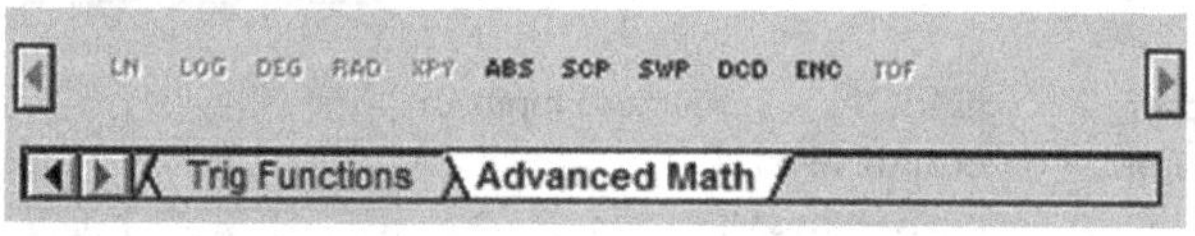 LN LOG DEG RAD XPY ABS SCP SWP DCD ENC TDF Trig Functions Advanced Math

Rung

This is a horizontal section of the ladder diagram where the instructions are placed, and which terminates in an output instruction. The first rung in a ladder program begins at numeral 0000. The rung contains instructions with addresses and annotations. A rung of instructions is shown in Figure 13.4.

Although the programming is very graphical in appearance, the code is really an organized string of textural commands and addresses. The programming code underlying the ladder rung in Figure 13.4 is shown in Figure 13.5 and can be obtained by double-clicking the rung number. "XIC I:1.0/2 XIC I:1.0/1 BST XIC I:1.0/0 NXB XIC B3:0/0 BND XIO B3:1/3 OTE B3:0/0" is the programming code that would have been entered manually in early generations of PLCs to represent the rung. Modern programs automatically translate the graphical code into text during the downloading process.

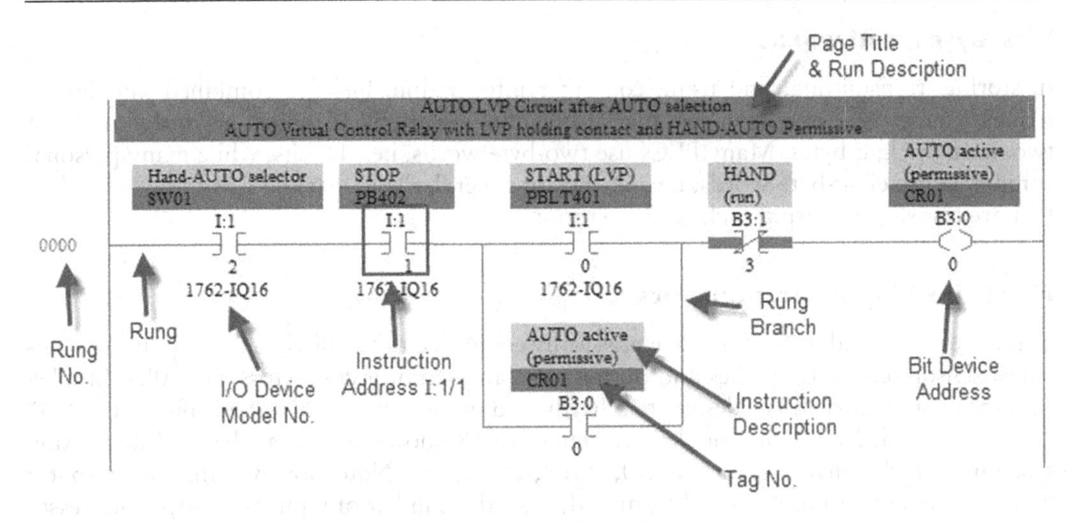

Figure 13.4 A rung is shown, including instructions, a parallel rung branch, and a mix of real and binary I/O devices. Comments, annotations, and device tag numbers assist in programming and debugging.

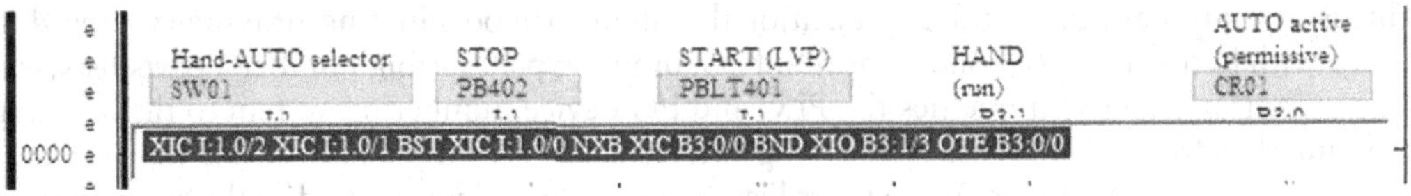

Hand-AUTO selector SW01 | STOP PB402 | START (LVP) PBLT401 | HAND (run) | AUTO active (permissive) CR01

0000

```
XIC I:1.0/2 XIC I:1.0/1 BST XIC I:1.0/0 NXB XIC B3:0/0 BND XIO B3:1/3 OTE B3:0/0
```

Figure 13.5 Double-clicking the rung shows the underlying program code, which consists of instructions and addressing. Legacy PLC programs were coded in this textural manner.

Instruction

The programming language is comprised of individual instructions or functions inserted into a ladder diagram. Bit-type instructions include commands, such as XIO (eXamine If Open), which looks like a NO contact, XIC (eXamine If Closed), which looks like a NC contact, and OTE (OutpuT Energize, which looks like a relay coil. Instructions that use Word data types include timers, counters, sequencers, math functions, etc. Figure 13.6 shows the data files associated with the instruction types.

Bit

This is a two-state storage address within the PLC. A bit has a value of 0 or 1, representing OFF/ON or FALSE/TRUE, and it stores the state of digital inputs from switches or digital outputs, represented as coils. The bit information can originate from external real-world devices or internal virtual devices. Bits assigned to real devices are stored in the I1 data file for inputs or stored in the O0 data file for outputs. Both input and output bits obtained from internal virtual devices are stored in the B3 data files. The B3 input and output bits are easily cross-referenced by the address. For efficiency, individual bits are organized into groups, called words.

Bits, bytes, and words

In storing, transmitting, and using computer information, bits are combined into longer strings, where eight bits form one byte. Depending on the PLC model, a word consists of two, four, or eight bytes. Many PLCs use two-byte words, i.e., 16 bits, while many personal computers use eight-byte words, i.e., 64 bits. In general, the longer the word, the more data that are transferred during each scan operation.

Hardwired inputs and outputs

These are real-world devices that are physically connected to the PLC, where inputs are manual or sensor-actuated switches and outputs are lamps, motor contactors, etc. Allen-Bradley uses the capital letter "I" to designate a hardwired input device, while the capital letter "O" designates a hardwired output device. Figure 13.18 shows the data files and addressing alphanumeric designations, such as O0, I1, B3, T4, C5, etc. Note, however, that the output 0 number and input 1 number can be omitted, since the number of input and output addresses are limited to the quantity of terminal connections on the devices, while all other data-file addresses use both the letter and number.

Line-smart/page smart numbering

This is an optional but useful organizational system for coordinating drawing wiring diagrams with hardware tags, conductor numbers, drawing pages, rung numbers, parts lists, etc. This organizational technique ties the PLC and I/O device numbering system to the drawing page number for ease of cross-referencing information. Essentially, all device tag numbers link the associated hardware to the parts lists, to the wiring diagram, and to the rung number on that diagram.

Address

Addressing is the key to linking real-world or virtual devices to the PLC program. RSLogix addressing is in the form of [DATA TYPE] : [word] / [bit location]. Data types, such as output, input, binary, etc., are indicated by "I" and "O", which include real-world input and output devices, while binary devices are essentially virtual devices existing only in the computer memory, and are indicated by "B3." Other data types include timers (T4), counters (C5), integers (N7), and floating-point numbers (F8).

For I/O data, the word numbering defines the physical module location in the chain of cable PLC devices. The word also defines the memory location where bits of information are stored, and it forms the tag numbers for cross-referencing instructions within the PLC program. The bit represents the screw-terminal number and is corresponded to its memory location within the word.

Using the I:2/5 address as an example; I indicates the data is an input. 2 indicates that the signal originates from the third device mounted in the rack, where the first device is 0 and is the PLC itself. 5 indicates the sixth screw-terminal connection, remembering that terminal numbers begin with 0. Using address O:0/2, as another example, O indicates output data, 0 indicates the first device, which is the PLC itself, and 2 indicates the third wiring connection. It is important to distinguish between the letter O and the numeral 0 when addressing,[5] and to note that the ":" and "/" symbols are separators when manually typing the addresses.

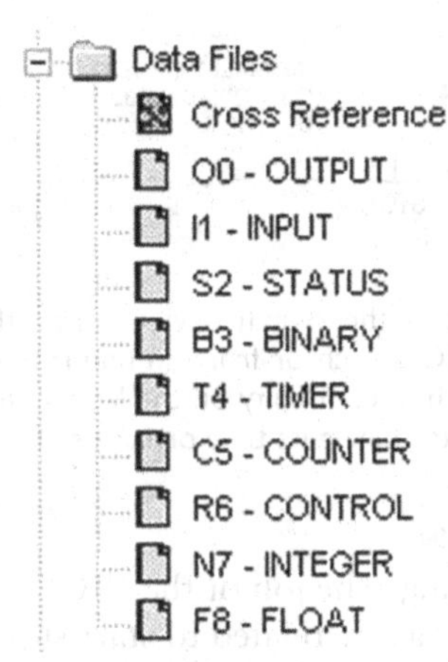

Figure 13.6 RSLogix software contains nine default categories of data files. These categories are used to store status information and variables. Note that except for output (O) and input (I), all other data-file types must include both the letter designation and number when addressing. Some examples of addressing are binary I:1/6, O:2/3, B3:2/5, and timer T4:0. In the case of the T4 timer, the data file contains several bits and words to define its setpoint and status.

Bitbox

This is a convenient way of thinking about binary or two-state memory locations. The bitbox is the virtual location containing a two-state memory bit in the form of 0 or 1, where the value relates to the device as being in a TRUE or FALSE state. Figure 13.18 shows an Input Data File, with all bits set to zero. Each zero represents a false value stored in that bitbox. Each line in that data file has 16 bitboxes that are stored as one word in memory. The bitboxes are subsets of the Word address, numbered 0–15.

Internal contacts and internal coils

These are the "examine-if" switches and "output-energize" instructions that are parts of virtual relays contained within the program and memory, and are analogous to real-world relay coils and contacts. Virtual relay coils and switches are linked in operation, so they share the same memory address. Virtual relays are two-state devices and are therefore referred to as binary or bits. RSLogix uses the "B3" designation to define the internal-relay's nomenclature and to link its 0/1 state to its binary data-file memory location. "B3" indicates that the relay device exists in the virtual PLC world as the result of an instruction, rather than in real world hardware. An example of a binary address is "B3:1/6," where "B3" indicates a virtual device referencing a binary data file, "1" means that the stored value is in the second word in the table, and "6" means the seventh bit in the word contains the addressed value. This address can be accessed by the XIO, XIC, and OTE instructions.

INTERNAL CONTACTS—EXAMINE IF CLOSED (XIC)

B3:6
—] [—
0

The Examine-If-Closed (XIC) instruction is a switch that looks like a NO contact, but the XIC is in fact a software instruction. It is important to remember that instructions do not

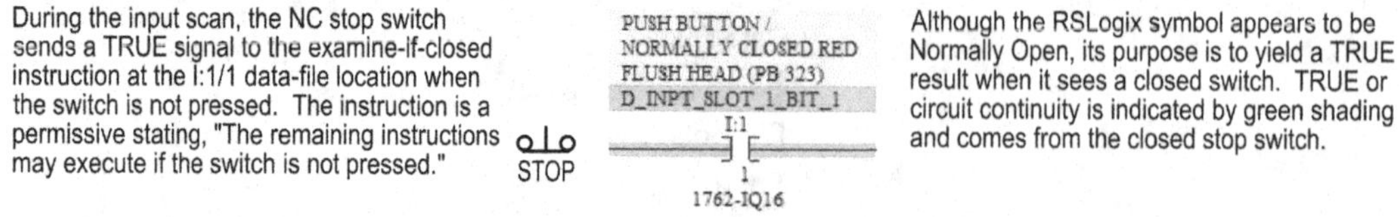

During the input scan, the NC stop switch sends a TRUE signal to the examine-if-closed instruction at the I:1/1 data-file location when the switch is not pressed. The instruction is a permissive stating, "The remaining instructions may execute if the switch is not pressed."

STOP

PUSH BUTTON / NORMALLY CLOSED RED FLUSH HEAD (PB 323) D_INPT_SLOT_1_BIT_1

I:1

1

1762-IQ16

Although the RSLogix symbol appears to be Normally Open, its purpose is to yield a TRUE result when it sees a closed switch. TRUE or circuit continuity is indicated by green shading and comes from the closed stop switch.

Figure 13.7 The XIC instruction looks at the switch device during the input scan. If the switch is closed, whether from a pressed NO switch or from an un-pressed NC switch, the result is 1 or TRUE, which shows as a green path of continuity on the ladder rung. The programmer should ask of the switch, "is it closed," for determining its true or false state.

behave exactly like electrical wiring. The job of the XIO instruction is to look for a value of 1 in the I1 data file address. XIC can be related to start-stop push buttons, selector switches, limit switches, internal bits, etc. Essentially, as the program input scan reaches this instruction, XIC looks at the switch position and sets the corresponding bit state in memory. When the XIC switch is CLOSED or TRUE, the corresponding bit value is set to 1. The XIC instruction addressed as I:1/1in Figure 13.7 says to go to the I1 data table, find the second 16-bit Word at location 1, and then use the second bit state value, which is found at position 1. If the value of the I:1/1 bit is 1 or TRUE, then set the XIC output to be 1 or TRUE, behaving as if the "NO" contact is closed or conducting.

INTERNAL CONTACTS—EXAMINE IF OPEN (XIO)

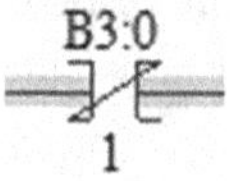

B3:0

1

The Examine-If-Open (XIO) is an instruction that looks like a NC contact, but the XIO is indeed a software instruction. The job of the XIO instruction is to look for a value of 0. XIO can be related to start/stop push buttons, selector switches, limit switches, internal bits, etc. Essentially, as the program reaches this instruction during the input scan, XIC looks at the switch position and sets the corresponding bit state in the memory. When the switch is OPEN or FALSE (0), the corresponding bit value is set to TRUE (1) or ON. During the scan portion of the program, the XIO instruction in Figure 13.8 says to go to the B3 data table, find the first 16-bit word 0, and then use its second bit-box value, which is in the bit 1 location. If the value of the B3:0/1 bit is 0 or FALSE, then set the XIC output to be 1 or TRUE, behaving as if the NC switch is conducting.

To show how this symbol can be counterintuitive, if a normally open STOP switch is wired into the circuit, in place of a normally closed switch that is commonly used in an electrical-diagram, the input scan would see the unpressed NO switch value as 0, but the XIO instruction would instead assign a value of 1 (TRUE), as if the switch was conducting. This XIO instruction combined with an open NO STOP switch is like the saying that "two negatives become a positive." Table 13.3 shows the four relationships between the "examine-if" instructions and the 0/1 logic-state inputs. It shall be stressed again that the input device may set the bit-box value, but it is the value in the bitbox that is seen by the "examine-if" instruction to produce the TRUE or FALSE output. Ultimately it is the bit-box value that is used by the program logic.

During the input scan, the NO switch sends a 0 or FALSE signal to the XIO-instruction data-file location when the switch is not pressed. The instruction states, "It is TRUE that the switch is open and there is no continuity."

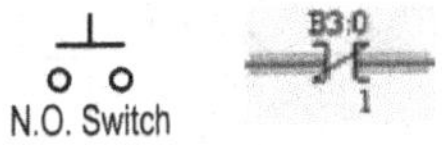

N.O. Switch

B3:0
1

Although the RSLogix symbol appears to be Normally Closed, its purpose is to yield a TRUE result when it sees an open switch. TRUE or circuit continuity is indicated by green shading and comes from observing an open switch.

Figure 13.8 The XIO instruction looks at the switch device during the input scan. If the switch is open, whether from an inactivated NO switch or from an activated NC switch, the result is 1 or TRUE, which is shown as a green path of conduction on the ladder rung. The programmer should ask of the switch, "is it open," for determining its true or false state.

Table 13.3 Truth table summarizing XIC and XIO input instructions

RSLogix I, O, or B3 Data file Bit state	*Instruction status*			
		XIC Examine if closed		*XIO Examine if open*
Logic **0**	FALSE	I:1 6	*TRUE*	B3:0 1
Logic **1**	*TRUE*	I:1 1	FALSE	B3:0 1

INTERNAL COILS—OUTPUT ENERGIZE (OTE)

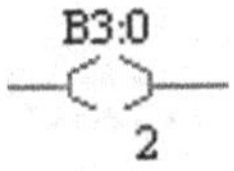

B3:0
2

The OutpuT EnergizE (OTE) is an instruction that simulates a coil, or a device that is turned on or off during the output scan. This device is used for indicator lights, solenoid valves, motor run commands, internal bits, or any two-state output device. Just like when a conductive path leading to a coil energizes a relay, the OTE is TRUE or "energized" when a complete path of "*examine-if*" instructions leading to the OTE device are TRUE, at which point its associated memory O0 output data-file bit is toggled to 1. Figure 13.9 shows that the virtual coil defined by OTE B3:7/6 is true, setting bit 6 in word 7 in the B3 binary data file to 1, emulating an energized coil.

OUTPUT LATCH (OTL)/ OUTPUT UNLATCH (OTU)

Latch and Unlatch are retentive instructions that are usually used together (Figure 13.10). Like real latching relays, once the latching coil is energized, even for an instant, the OTL latch remains permanently toggled until such time that it is unlatched using the OTU instruction. In other words, the OTL is a retentive instruction that once toggled will maintain its latched state, whether or not the rung has a path of true instructions and even if power is interrupted

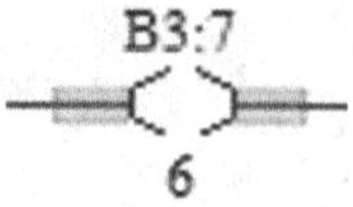

B3:7

6

Figure 13.9 The output energize or OTE instruction emulates a relay coil. The B3 indicates the data-file where its information will be stored, and that the device is binary and virtual, existing only in PLC memory. The single-bit information is stored in the B3 data file, in word-7 and at the number 6-bit position. Memory location B3:7/6 presently has a bit value of 1, as seen by its green highlight.

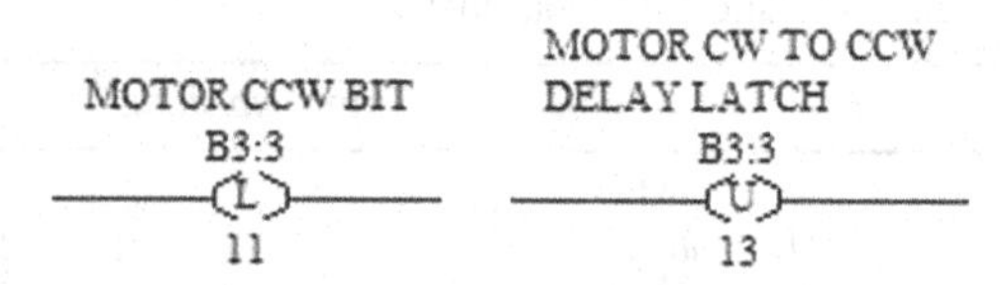

Figure 13.10 Output Latch (OTL) and Unlatch (OTU) instructions emulate latching relays.

and later restored. Once energized, the OTL normally open or normally closed states are toggled indefinitely.

The output unlatch instruction (OTU) is opposite to the OTL output-latch instruction. OTU is used to unlatch the OTL coil. The OTU instruction leaves the coil in the unlatched position, which is analogous the “normal” position of a de-energized relay. Like the OTL instruction, OTU is retentive and remains in the unlatched position indefinitely.

In some cases, one-shot instructions are used with latches, where it only takes a single pulse signal to execute the latching or unlatching action. This technique avoids repetitive signals spanning multiple scans, which could result in miscounts.

ONE-SHOT (ONS)/ONE-SHOT RISING (OSR)/ONE-SHOT FALLING (OSF)

One-shot commands are retentive instructions that trigger an output event to occur only one-time (Figure 13.11). Typically, when the ONR one-shot-run instruction is initiated, the corresponding bit remains ON for only one complete cycle of program scan, resetting to OFF when the preceding ladder rung is executed. To be available for future scans, the one-shot must first be driven false. One-shot instructions are useful for math operations, where a condition may be sensed during many scan cycles, but the one-shot permits only the first instance to trigger the math operation, consequently, the math is done only once. The one-shot may also be used to reset a condition using a single scan, or to read a fixed value from an analog-input value into an output display, where constantly changing meter values might be hard to read. One-shots can also be used to initiate or stop data acquisition.

The ONS it functions as an input instruction and it fires as a single true pulse to the latter part of the rung logic on the rising edge of the logic. The ONS instruction works only on the rung where it is inserted. The ONR and ONF are both outputs and must be the last instruction on a rung. The ONR fires as a single pulse when the preceding logic rises from logic 0 to logic 1, and the ONF fires a single pulse when the preceding logic falls from logic 1 to logic 0. The ONR and ONF instructions toggle a second memory bit that is available for other rungs in the program.

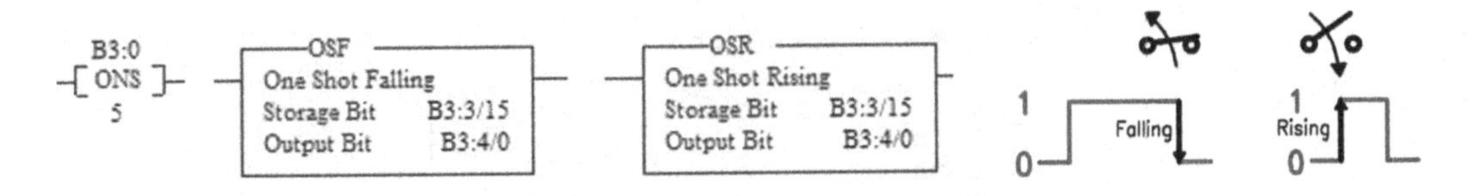

Figure 13.11 One-shot, one-shot falling, and one-shot rising instructions.

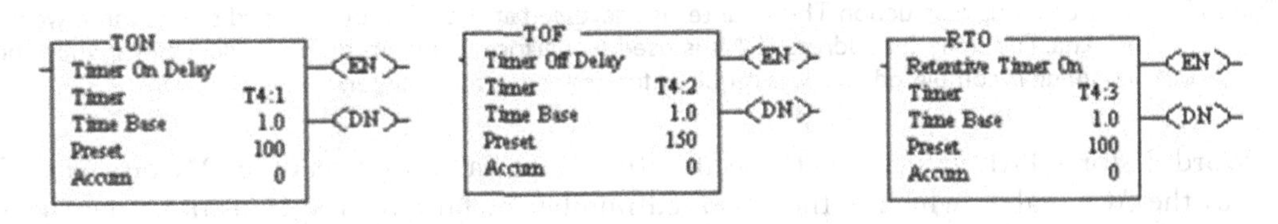

Figure 13.12 Timer instructions, time on delay (TON), timer off delay (TOF), and retentive timer (RTO).

TIMERS—ON-TIME DELAY (TON) AND OFF-TIME DELAY (TOF)

A timer is a T4 instruction that delays the turn-on or turn-off program functions using enable bits (EN) or done bits (DN) after a preset delay period is reached. The timer types include the "timer on delay" (TON) and the "timer off delay" (TOF) (Figure 13.12). RSLogix defines internal timer nomenclature using the T4 letter/number designation, which references an internal timer data file. As an example, T4:7 indicates an internal timer file using the eighth timer in the table.

The timer instruction is a three-word element, where word 1 is the control word, word 2 is the preset value (PRE), and word 3 stores the real-time accumulated value (ACC). The control word contains a count-up enabled bit no. 15 (CU), a count-down enabled bit no. 14 (CD), an updated accumulated value bit no. 10 (UA), an underflow bit no. 11 (UN), and overflow bit no. 12 (OV), a done bit no. 13 (DN).

The programmer specifies the timer address from the T4 data file and enters the delay time as part of the preset and the time base. The time base specifies hundredths, thousandths, or integer seconds. The output signals can be when the timer begins to time, using EN for enable, or when the time delay has expired, using DN for done.

COUNTERS—COUNT UP (CTU) AND (CTD)

A counter is a C5 data-file instruction that toggles output bits on or off after a preset count has been reached. Counters may be count up (CTU) or count down (CTD) (Figure 13.13). The reset instruction (RES) is available to reinitiate the count for starting the next sequence. The count increments at each false-to-true transition.

The counter instruction is a three-word element, where the Word 1 is the control word, Word 2 is the preset value (PRE), and Word 3 stores the accumulated value (ACC). The control word contains binary count-up enabled bit 15 (CU), a count-down enabled bit 14 (CD), an updated accumulated value bit 10 (UA), an underflow bit 11 (UN), and overflow bit 12 (OV), a done bit 13 (DN).

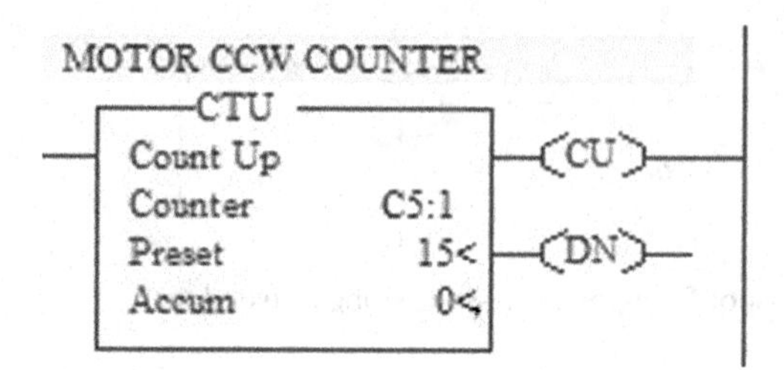

Figure 13.13 The count up instruction. This counter is addressed at the C5:1 location and stores three words of data. The done-bit address (DN) is used to perform program output logic actions when the counting is completed. Reset is required for restarting the counter.

Word 2 stores PRE value, which specifies the count value that sets the DN bit. Word 3 stores the ACC value, which is the observed number of false-to-true transitions that have occurred since the counter was last reset.

SWITCH BOUNCE AND COUNTERS

Some mechanical switches used in fast PLCs can experience inherent switch-bounce problems during closure, where the spring causes make-and-break rebounding of the contacts. In long programs or slow PLCs, the switch bounce may not be a problem. However, in short programs with fast scan rates, it is possible that subsequent input-scans may observe the switch contact in both the closed and open states. In PLC counting operations, the switch bounce is sometimes recognized as several closed-and-open actions occurring during a single actuation of the switch, leading to miscounts. In rare instances, program logic having poor rung sequencing can pick up a badly timed switch bounce that could toggle instructions in an unexpected manner. These problems can result in unpredictable, intermittent, buggy behavior.

Hardware solutions are available to mitigate switch bounce; however, PLC-based solutions may be more expeditious. Some PLC techniques include the insertion of delay timers to accept the switch signal only after the bounce is finished, the use of one-shot instructions, or the use of set-and-reset or latch-and-unlatch instructions.

ARITHMETIC FUNCTIONS AND STATUS FILE (S2)

Arithmetic functions include addition, subtraction, multiplication, and division (Figure 13.14). These functions are outputs located at the end of the rung and are executed when the preceding Boolean instructions are all true. They permit more complex control operations, going beyond the simple Boolean-logic decisions. They retrieve one or more values from a memory register in the data file, perform the mathematical operation, and store the results back in memory. Typically, the values can be integer or real numbers and the results are stored in integer data files (N7) or floating-point data files (F8), respectively.

The ADD instruction takes data stored in Source A and Source B, adds the values, and then stores the result in the destination address. Source A or Source B can be either a constant embedded in the program, however, at least one value must be obtained from a variable stored in a data-file address.

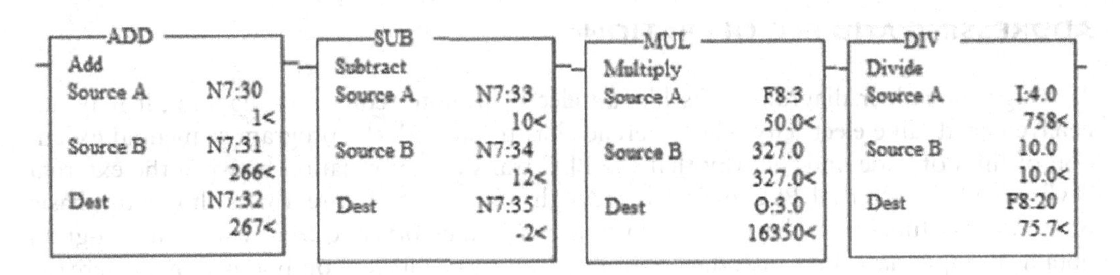

Figure 13.14 Arithmetic functions.

Table 13.4 Syntax for arithmetic instructions and the arithmetic data-file bits

Instruction	*Description when preceding rung instructions leading to math instruction is TRUE*
ADD	Add Source A address to Source B address (A+B).
Addition	Store result at destination address
SUB	Subtract Source B from Source A (AB)
Subtract	Store result at destination address
MUL	Multiply Source A and Source B (AxB)
Multiply	Store result at destination address
DIV	Divide Source A by Source B (A/B)
Division	Store result at destination address
	* Either Source A or Source B may contain a constant, but not both

Instruction	*Carry (C)*	*Overflow (V)*	*Zero (Z)*	*Sign (S)*
ADD	Sets for carry	Sets if underflow	Sets if zero	Sets if minus
Addition	Resets if not	Resets if not	Resets if not	Rests if not
SUB	Sets if borrow	Sets if underflow	Sets if zero	Sets if minus
Subtract	Resets if not	Resets if not	Resets if not	Rests if not
MUL	Always reset	Sets if overflow	Sets if zero	Sets if minus
Multiply		Resets if not	Resets if not	Rests if not
DIV	Sets for carry, if	Sets if divide-by-zero or	Sets if zero	Sets if minus
Division	not reset	result exceeds +/-32,768	Resets if not	Rests if not
		Resets if normal		Undefined for overflow

Status file

The status data file maintains a close relationship between mathematical operations and some of the corresponding status bits (see Table 13.4). The program executes the math instruction then updates the related bits in the S2 status data file. As an example, the status word 0 address would store the following math information: S2:0/0 for Carry (C), S2:0/1 for Overflow (V), S2:0/2 for Zero (Z), S2:0/3 for Sign (S). In addition, the S2:0/5-bit location is reserved if a mathematical overflow or divide-by-zero error occurs. When errors are not handled, they will produce a “fault,” which locks-up the CPU and actuates a trouble light on the PLC. To avoid locking up the CPU, it is possible to examine and catch the fault, unlatching this bit before a program-END statement is reached.

Finally, the S13 and S14 math registers contain the 32-bit word that stores the results of the MULtiplication and DIVision functions. The S:13 register contains the least significant 16-bit word of the result, and the S:14 register contains the most significant 16-bit word.

ADDRESSING AND PLC OPERATION

Although the ladder diagram looks like an electrical motor-controller diagram, it may not behave exactly like electricity. The difference has to do with the program sequential execution of lines of code and the way that the PLC transfers information between the external devices and the internal PLC memory, through addressing. Figure 13.15 shows the basic program scan functions where PLC data is stored in memory, and from where the program obtains its input data. Likewise, the program moves its results into output memory, where the bit-box memory values are used to set the output devices. Addressing is the key to coordinating PLC operation.

Figure 13.16 shows a PLC and expansion module general arrangement. The physical arrangement is used to define I/O addresses used within the PLC, where the first device (the PLC) is number 00, the next connected device is number 01, and so on. The wiring terminals on the devices become the bit number assignment in the memory address for digital devices, or the channel number for analog devices. Conversely, internal or virtual devices have user-defined addresses. These device addresses are typically assigned to the drag-and-drop rung-instruction symbols, and then the PLC memory addresses behave like the hardware-derived addresses.

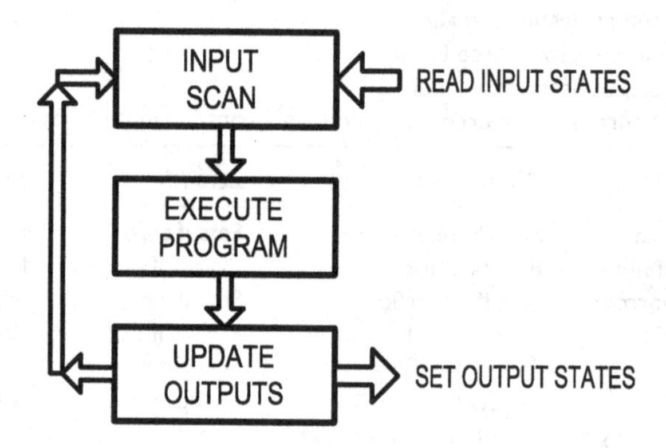

Figure 13.15 The PLC scan fundamentally works by reading all inputs from top down and left to right, and then storing the input-device states in memory. The execute-program scan then takes all input memory values and runs the program logic. The results from the program are stored in output-memory registers. The update-outputs step sends the memory values to the output devices.

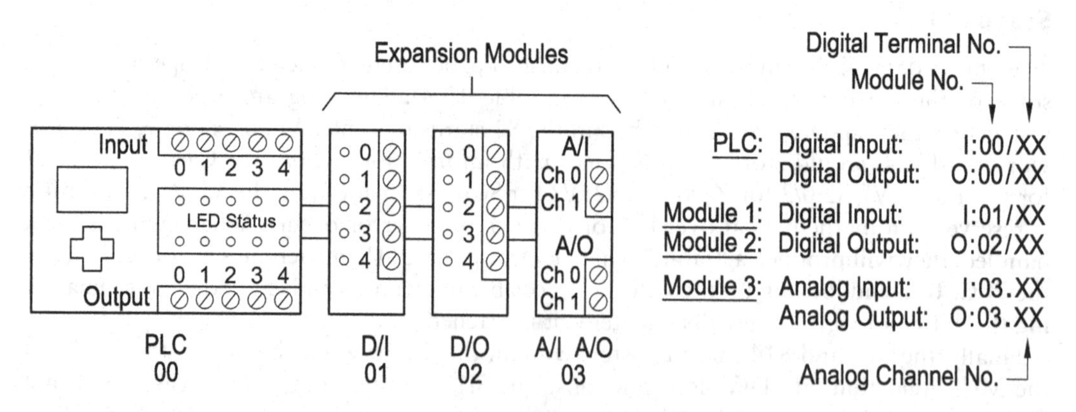

Figure 13.16 PLC rack. Addressing consists of a function "type" followed by the device wiring location. The terminal connections on each module are shown by "XX."

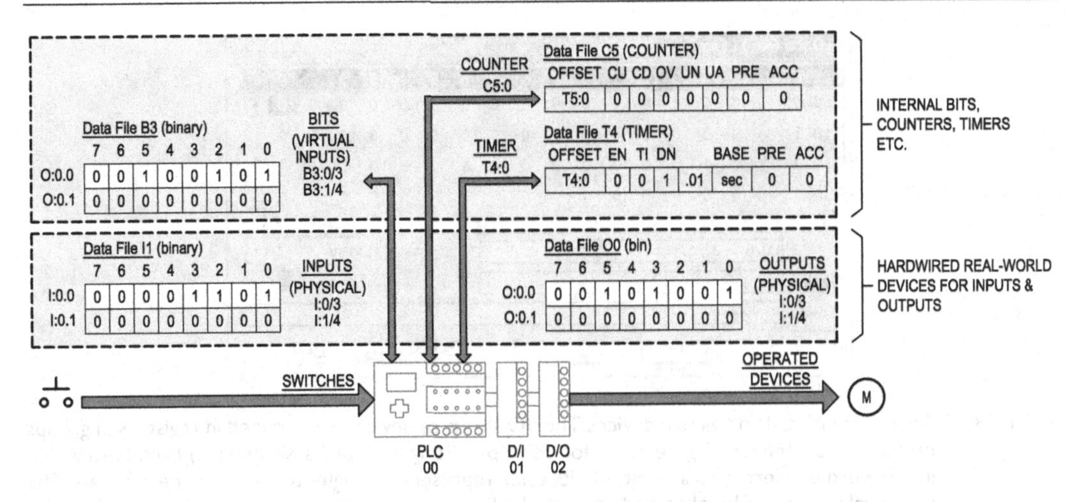

Figure 13.17 The PLC scans all inputs and maps their status values into memory. During program execution, the input-status information is pulled from the input memory and the program creates output values that are routed to output addresses. The program also updates memory addresses used by counters, timers, and other instructions, as needed. After program execution, all output devices are set using the values obtained from the output-memory addresses.

Figure 13.17 shows the PLC taking the input status and mapping it into the appropriate data-file memory location, for later use by the program. This information is called from memory by the program instructions, using the assigned addresses. When an output memory value changes, the output device is toggled to match the bit-box state. Values in the memory locations are used by the program, and outputs are updated during each scan

Figure 13.18 shows the Input Data File and corresponding 16-bit memory locations. As the storage locations for binary devices, each 16-bit word shows the status of 16 individual two-state devices. During operation, a 1 or 0 in each bit location would be used to indicate TRUE or FALSE for the corresponding device. This Data File table can also be manipulated to show which memory addresses are in use, as indicated by an "X," thus showing which addresses are free for program use.

Figure 13.19 shows a functional-block diagram to illustrate the movement of digital I/O data, from the input signal, to the memory address, and ultimately to an output. The inputs are switches whose open-or-close positions are converted into 0-or-1 status bits that are assigned to a bit-box memory address in the I1 data file. The values pulled from the data-file memory are used in the program execution. The program produces results that are stored in the O0 data-file, and technically, it is the bit-box value, which is converted to an output, and which should correspond to the switch position.

PLC INPUT/OUTPUT DATA LINKS

Digital input devices are essentially switches, which may be mechanical or electronic, and digital output devices are loads that are either on or off. Output devices may be lights, or electromechanical, electromagnetic, or electronic devices. I/O devices are typically powered separately and isolated from the PLC's electronics, as shown in Figure 13.20.

Input/output devices are typically linked to PLCs using opto-isolators or opto-couplers in lieu of direct electrical connections. The higher-voltage electrical signals from field devices,

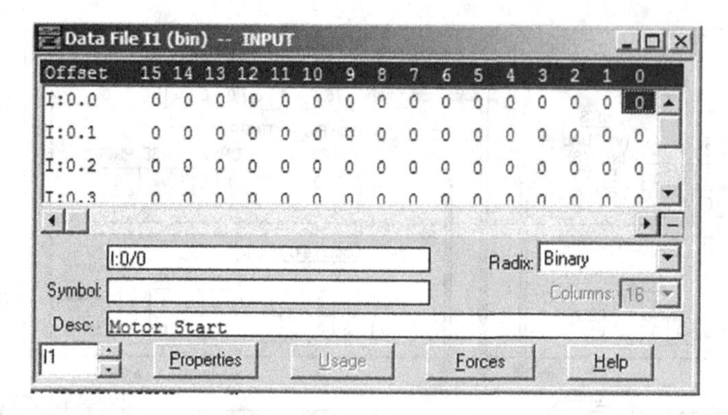

Figure 13.18 Addressing of real and virtual devices. The I/O status of devices is contained in registers in groups called words. (Refer to Figure 13.16 for conceptualizing bit/word assignments.) Each binary digit in the word is referred to as a bit, and its value represents a single- true-or-false device state. The individual values can be observed through the RSLogix software, by double-clicking the data files shown in Figure 13.6.

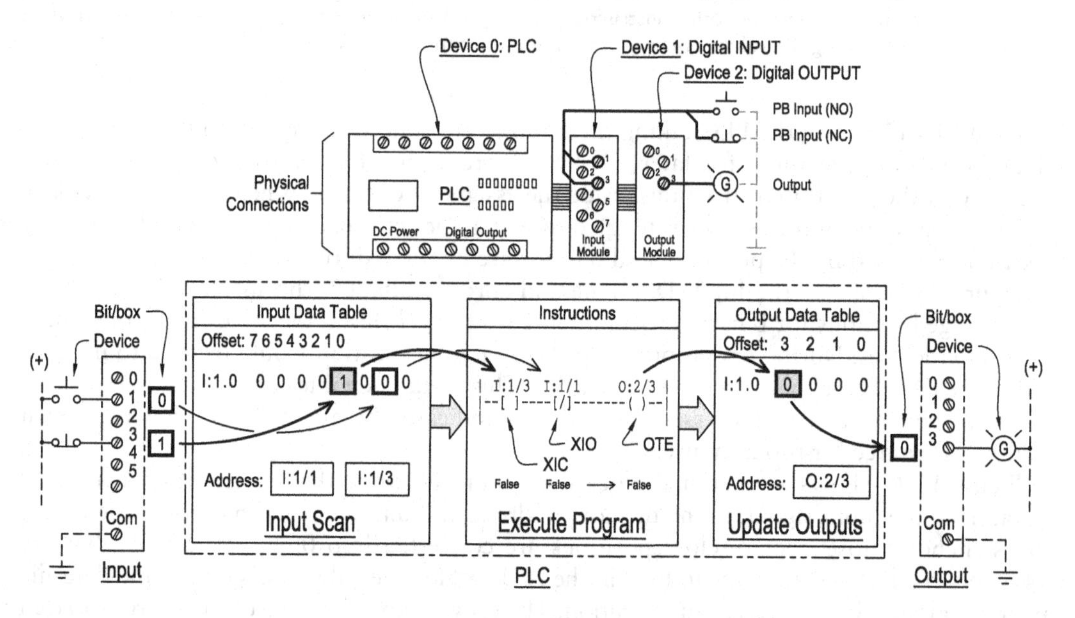

Figure 13.19 This functional-block diagram shows the relationships between the input devices, the mapping and storage of their states in an input data-file table, the use of the stored data during program execution, updating of the output data-file table memory, and the use of the output data file bit-box values to control a device.

which may contain spikes, are transferred into the sensitive lower-voltage PLC internal electronic components using a light, so that the internal PLC electronics are electrically isolated from the external circuit (Figure 13.21). This arrangement prevents abnormal external faults from reaching the PLC electronics.

The opto-isolator uses an infrared-emitting LED interface that is energized from an input-sensor circuit when an input-signal switch is closed. Often, the input-signal power is furnished by the PLC and returned through a PLC "signal ground" connection. The LED illumination is picked up by a photosensitive transistor. The light drives the transistor base into

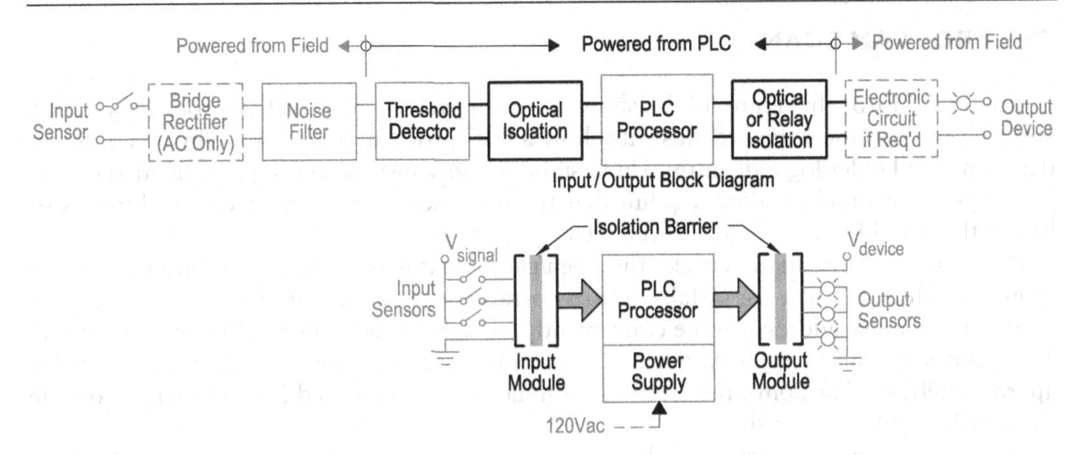

Figure 13.20 Electrical isolation of the PLC electronics from the attached input and output devices using optical couplers or relay separation.

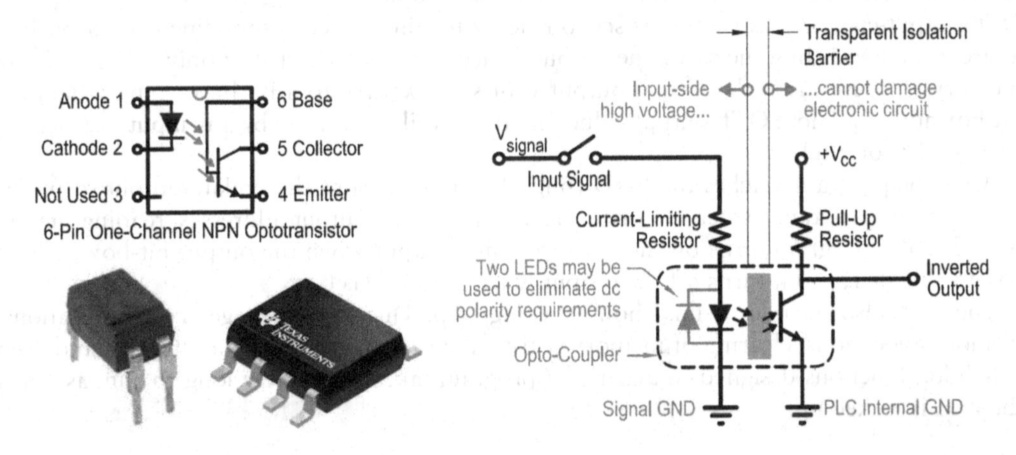

Figure 13.21 Opto-isolators: one-channel and two-channel chips are pictured on the left, accompanied by simplified circuit diagrams. (Courtesy of Texas Instruments.)

a forward-biased or conducting condition. The phototransistor is part of the PLC internal electrical circuit. Transistor conduction causes a corresponding memory bit-box address to take a value of 1 (true). A turned-off transistor is interpreted as 0 (false), which is stored in the memory address.

The advantage of opto-couplers is that robust LEDs, in general, will automatically resume operation when a fault is corrected. Further, if a severe fault causes an LED failure, the problem is usually limited to the affected I/O address, and the remaining PLC I/O circuits tend to function normally. Some solid-state opto-couplers are formed from phototransistors, for either input or output signals, and photodarlingtons using compound push-pull transistors for outputs. Photo-SCRs are used for dc loads and photo-TRIACs for ac loads, where high-gain or high-power outputs are needed. In addition to opto-coupled outputs, miniature low-power mechanical relays having dry-contact switches are common for isolating the internal PLC circuits from the external output circuits. The relays provide isolation but are slower and more susceptible to damage.

THE PROGRAM SCAN

Figure 13.22 through Figure 13.28 show the scan-cycle sequence that produces control. Stepping through the sequence illustrates how a PLC program works and why it is different than a motor ladder-logic diagram. Understanding how instruction sequencing affects program operation should make the point that timing or sequencing errors can lead to misbehavior that would not be seen with relay-based controls.

The input scan essentially checks the open/closed status of switches and inserts a corresponding value of 1 (true) or 0 (false) into a corresponding input data-file memory location. The input-memory address can be conveniently thought of as a bit-box location in a word. The input scan checks each switch in order and toggles the corresponding bit-box value appropriately. At that point, the electrical conductivity is converted into a binary two-state value or bit, which is stored.

The program scan next runs the lines of code starting from the top rung in order from left-to-right and top-to-bottom. If rung sequencing is ever in question, it is possible to double-click the rung number to see the computer code. The program toggles XIC examine if open and XIO examine-if-closed or other input instructions to match the 1/0 states contained in the instruction bit-box address. When any path along the rung is completely true, the OTE output-energize instruction is set to true by toggling its corresponding output bit-box address value to 1. In some cases, the output devices are virtual and exist only in the program memory. In those cases, the virtual output values are written to a B3 bit or binary data-file bit-box address. Those OTE output values are then available for use by any input instruction, such as XIC or XIO.

After the program reaches the END rung, the program runs the update-output scan. For this scan, the program sequentially runs down each of the output addresses, reading the O0 data-file. The program turns on the corresponding output when the output bit-box address has a value of 1, and it turns off the output when the value is 0.

The scan also includes a final housekeeping step. This scan manages communications, memory allocations, clearing of memory registers, parity checks for data integrity, and even watchdog functions designed to alarm if a program takes excessively long to run, as when the PLC is locked up.

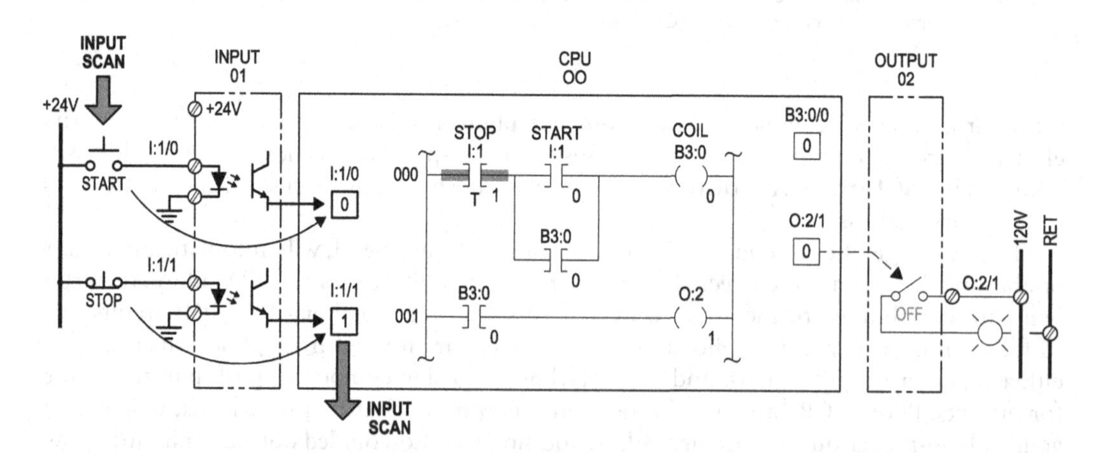

Figure 13.22 Input scan: The status of the input switch position is stored in an input data-file bitbox, where electrical conduction is 1 and lack of conduction is 0. In this case, the normally open start switch address is assigned a value 0 at I:1/0 and the normally closed stop switch is assigned 1 at I:1/1.

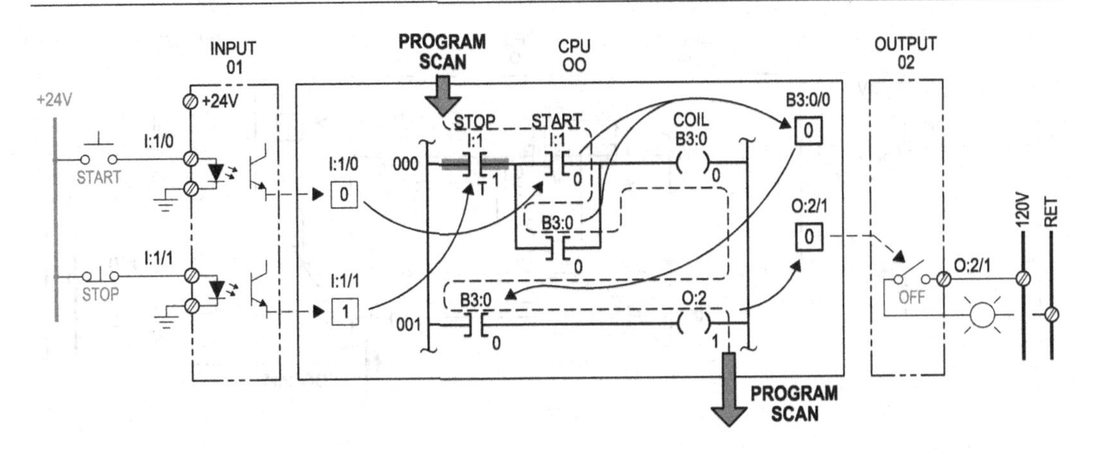

Figure 13.23 Program execution scan: The ladder logic code draws its value from the input-address bitbox. When a rung path has instructions that are all true, the OTE output-energize coil is driven to 1. The instructions are executed from top down and left-to-right. Note that double-clicking the rung number shows the instruction sequential order.

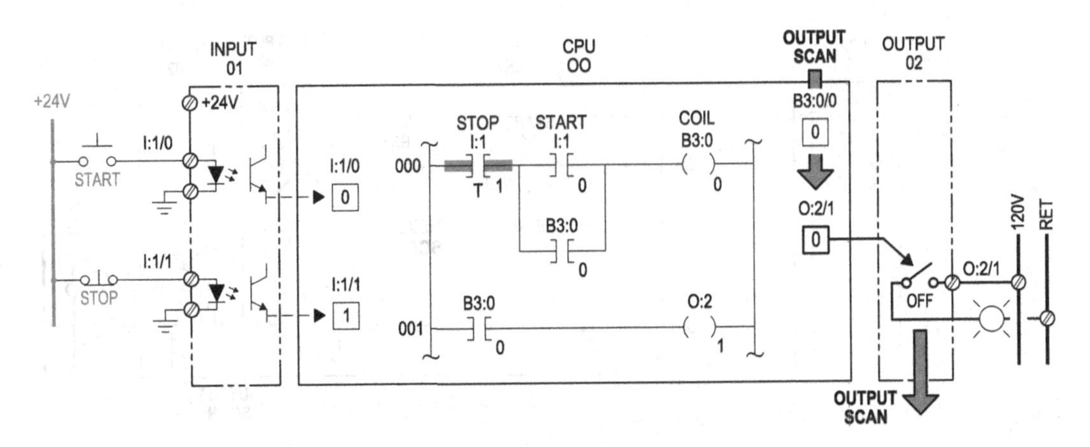

Figure 13.24 Output update scan: The value of 1 in the output data-file bitbox would toggle the output switch to turn on the attached load, while a value of 0 would deenergize the load.

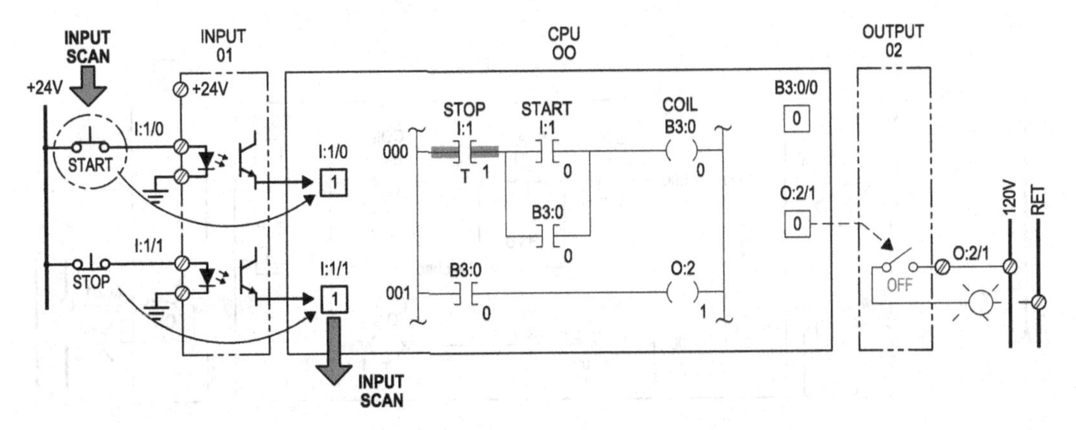

Figure 13.25 *Second input* scan: Pressing the START switch produces electrical continuity into the input. The bitbox at address I:1/0 is then toggled to 1.

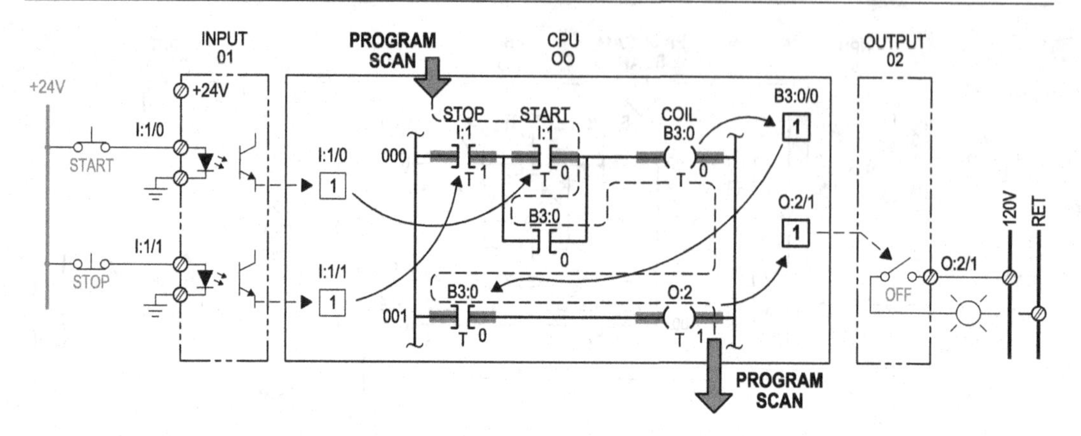

Figure 13.26 *Second program* scan: The I:1/0 bit-box value of 1 causes the stop-start rung of input instructions to have continuity leading to the B3:0/0 OTE, which is then set to 1. The true B3:0/0-XIC instruction on line 001, in turn sets bitbox O:2/1 to a value of 1. Note that the B3:0/0-holding contact remains false for the remainder of this scan, but will be picked up during the next program scan.

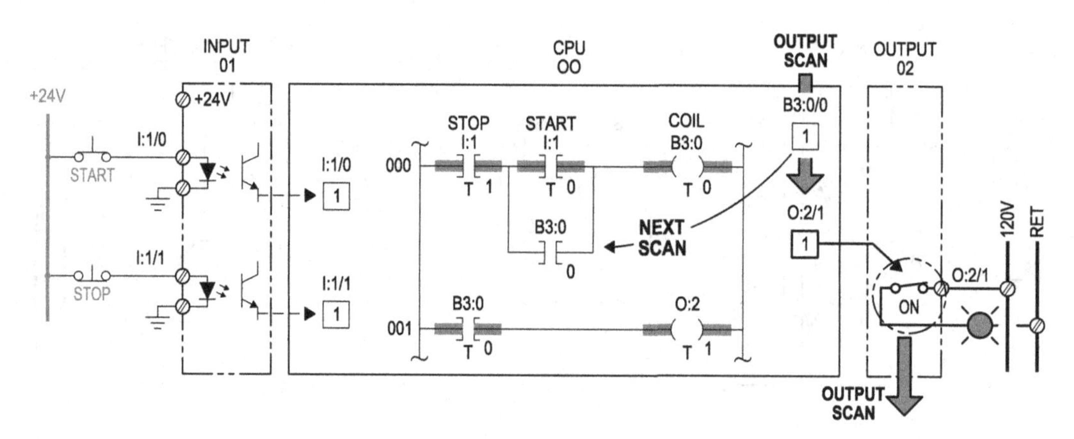

Figure 13.27 *Second output* scan: The output at address O:2/1 is directed to electrically conduct since its address value is 1, and the indicator lamp turns on. The B3:0/0-XIC holding instruction is set to 1 during the next program scan, which occurs in just a few milliseconds.

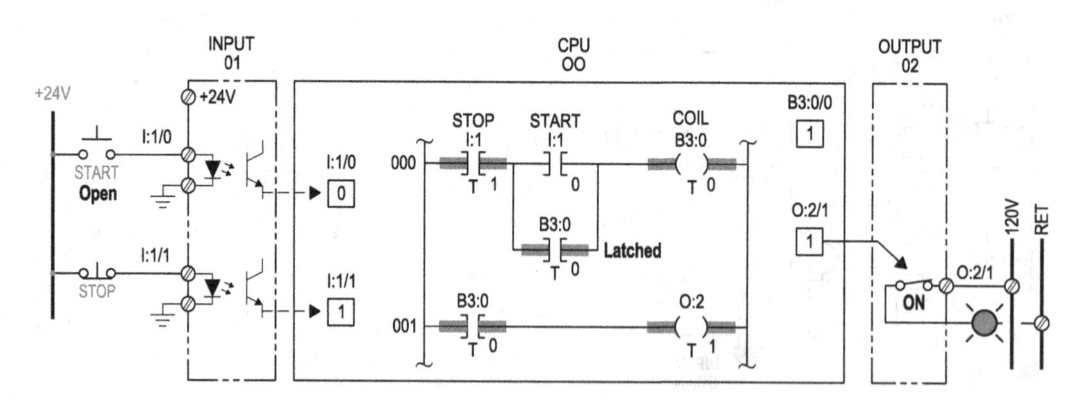

Figure 13.28 The START button is released in the final scan, and its paralleled B3:0/0-XIC holding instruction maintains the B3:0/0-OTE coil at a value of 1. The output will remain energized until the bit at address B3:0/0 returns to 0, such as after pressing the stop button to break continuity.

FLOW CHARTS AND PSEUDOCODE

Flow charts and pseudocode are tools that can be used to describe the basic control logic. These tools are excellent ways for a controls engineer to convey behavioral requirements to the programmer in construction drawings or specifications. Subsequently, the flow chart or pseudocode information is converted into ladder-diagram or any other type of PLC programming language.

Flow charts are graphical representations to help visualize the process. They use start and end terminators, input/outputs, processes, and decision trees. A sample flow chart is shown in Figure 13.29. The flow chart represents the control logic and sequence for the tank-level controller illustrated in Figure 13.30, as an example.

Pseudocode is a loose informal representation of the logic, control sequence, or control code, but it lacks the syntax precision needed by the programming language. Pseudocode is useful for describing the process in plain English and identifying the what-if conditions that affect the proper behavior of the system. Using the sample flow chart and control diagrams in Figure 13.29 and Figure 13.30, the pseudocode might look like the following:

1. Press START: Enable Controls
2. Sense tank level using level switches
   - LSHH indicates excessively high level
   - LSH indicates the normal high-level extents
   - LSL indicates the normal low-level extents
   - LSLL indicates excessively low level
3. Conditional Statements—*IF-THEN-ELSE, WHILE, etc.*
   - *IF*: LSL is TRUE (low limit reached) then start the pump
   - *WHILE*: Pump running is TRUE *AND* LSH is FALSE (level is between limits) *THEN* keep the pump running until upper limit is reached
   - *IF*: LSH is TRUE (upper level is reached) *THEN* stop the pump AND close the valve
   - *WHILE*: Pump running is FALSE *AND* LSH is FALSE (level is between limits) THEN keep the pump off until the lower limit is reached
   - *IF*: LSLL is TRUE (lowest level is reached) *THEN* actuate the alarm *AND* stop the pump *AND* close the valves (to avoid running without adequate differential)
   - *WHEN*: LSLL becomes FALSE, *PERMIT* the control to function properly
   - *IF*: LSHH is TRUE, THEN Actuate the alarm AND stop the pump *AND* close the valves (to void flooding the tank)
4. Press STOP: Stop the pump AND close the valves AND disable the controls

ANTICIPATING PROBLEMS WITH PROGRAM LOGIC

When implementing the logic, the design engineer needs to consider all normal and abnormal conditions or modes of operation that may occur. Referring to the previous example, line 1 has the LSL float switch in series with LSH high, so that both floats need to be low for the pump to start. However, when the level starts to rise, LSL opens and would shut the pump before a significant level-increase has happened, resulting in short cycling of the pump. To keep the pump operating until LSH is reached, a holding contact from CR1 is placed in parallel around LSL on line 2. The holding contact keeps the pump running after LSL opens, until the high-level switch is reached, solving the problem.

Another logic consideration exists with the configuration of the LSLL permissive. Presently, the LSLL permissive simply actuates an alarm. However, if this system performs a critical function, it

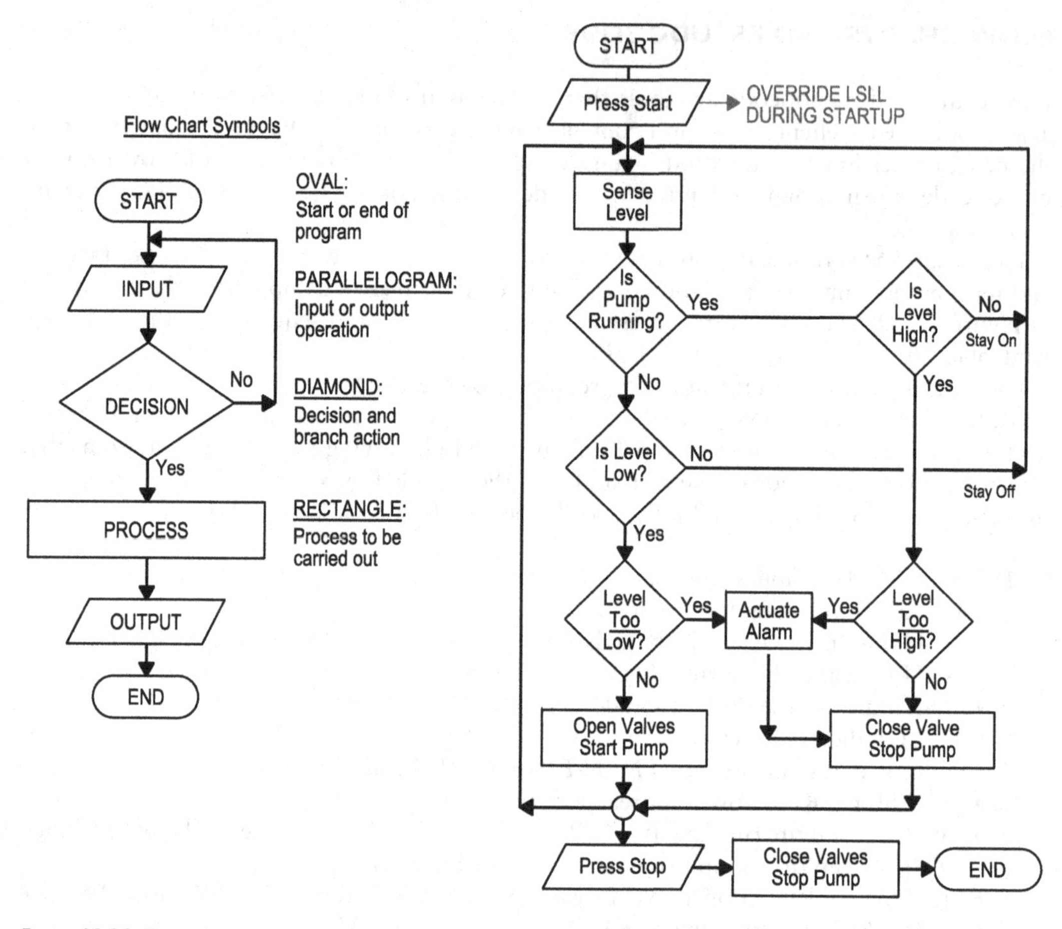

Figure 13.29 Flow chart symbols and a simplified sample flow chart.

might be desirable that the LSLL control should also send a signal to a back-up filling circuit to increase reliability. For that matter, an automatic overflow might be desirable as well.

Using the level control system in Figure 13.30 as another example, the system may function perfectly in a building, but the sloshing of water that occurs during rolling in a shipboard application may cause the float to act prematurely, possibly causing short cycling. One solution may be to reorient the tank, if the float favors one side. Another solution may be to program a short time delay into the level-sensing circuit, so that the start and stop commands are initiated only after the high or low levels have proven themselves to be real. Either of these solutions might avoid redesigning the tank to hold more volume.

Another programming feature to improve reliability might be to include a "stuck-on" or "stuck off" fault circuit. In the case of the tank design, pump down might occur within a minute under peak loads. The designer may opt to install an alarm circuit that initiates after two minutes to alert the operator to potential dry-running operation.

DOCUMENTING THE RSLOGIX PROGRAM

The RSLogix program can be annotated using rung comments, instruction descriptions/tag numbers, and addresses. The annotations can describe how the program works, the purpose

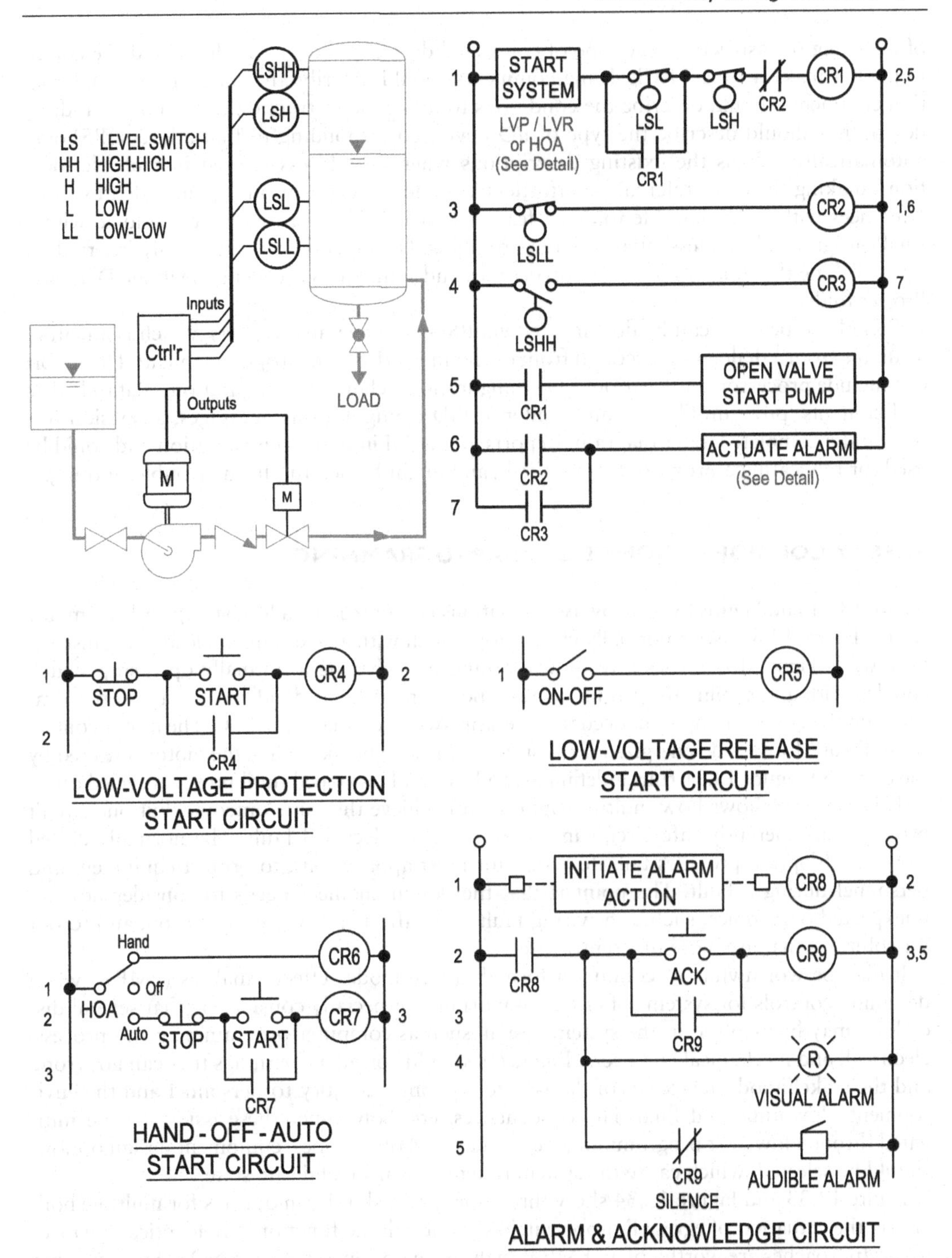

Figure 13.30 Flow diagram (upper left) and ladder logic (upper right) developed from the flow-chart example in Figure 13.29. Note that repetitive circuits using preprogrammed starting options or various alarm-with-acknowledge routines can be easily replicated as standard functions.

of any rung or instruction, the type of connected device, the device address, and the conditions that make a rung TRUE. Rung comments should describe the rung function, instruction comments should describe the conditions to make the instruction true, and the address description should describe the type of I/O device corresponding to that address. RSLogix automatically assigns the existing descriptions when an address is used in multiple locations, making the cross-referencing effortless. All address descriptions are included by cross references within the data-file folder, a feature that is helpful in following the logic within small programs, but is essential in large programs. Documentation is generally inserted by right-clicking the rung numbers or instructions, and summaries are available in the Database Project folder.

Completed projects can be documented via RSLogix-generated reports, which is a feature available from a dialog box accessed from the main File/Reports dropdown menu. The report can include processor information, I/O configurations, channel configurations, data-file lists and contents, program-file lists and contents, PID configurations, messages, a revision history, and a complete ladder diagram. Reports are useful in user documentation and could be used for replicating a program from scratch, and might be helpful during troubleshooting.

SAFETY CONSIDERATIONS DURING PROGRAMMING

Figure 13.31 and Figure 13.32 show two variations for controlling a low-voltage-release motor starter. Figure 13.31 uses a normally closed stop switch with an examine-if-closed PLC instruction, while Figure 13.32 uses a reciprocal arrangement having a normally open stop switch coupled with an examine-if-open instruction. Both circuits provide identical start-stop operations, until an open-circuit fault occurs in the stop-switch wiring to the PLC. The motor contactor in Figure 13.31 is de-energized if the stop-switch wire breaks, while the motor powered by the circuit in Figure 13.31 runs indefinitely, and cannot be stopped until power is cycled off.

This example shows how multiple options can achieve the same outcome, but one circuit provides an inherently safe circuit in the event of an electrical fault. The normally closed stop-switch option provides failsafe action for most applications, to protect equipment and personnel during a fault. The point is that the design engineer needs to consider how an unexpected occurrence, such as a wiring fault, may affect the system operation, and to opt for solutions that are more inherently safe.

It may be worthwhile to conduct a formal failure modes effects analysis (FMEA) when designing controls for systems of critical importance, especially considering that several disciplines may be involved in the system design, such as computer programmers and process, electrical, and mechanical engineers. Engineers should consider the things that can go wrong and their likelihood, the extent of damage to equipment, injury to personnel and the environment, downtime and financial consequences, etc. Sometimes these issues can be mitigated through low-cost programming techniques, and other times, equipment design options should be selected, which cause the system to tend toward its failsafe state.

Figure 13.33 and Figure 13.34 show three emergency-shutdown options for multiple boilers via the actuation of egress E-Stop switches. Although the functions are identical, nuances in the approaches are worthy of an FMEA analysis. In option 1, multiple NO E-Stop switches are wired in parallel, so that actuation of any switch energizes control relay CR1 to break a NC permissive switches in the control circuits to each boiler. For option 1, it is noted that CR1 is deenergized at all times, until any E-Stop switch is actuated.

For option 2 in Figure 13.33 and option 3 in Figure 13.34, it is *stressed* that the shutdown control relays are energized continuously for placing the circuit in operation, and therefore, the contacts are toggled to the alternate state than is shown. Multiple NC E-Stop switches are

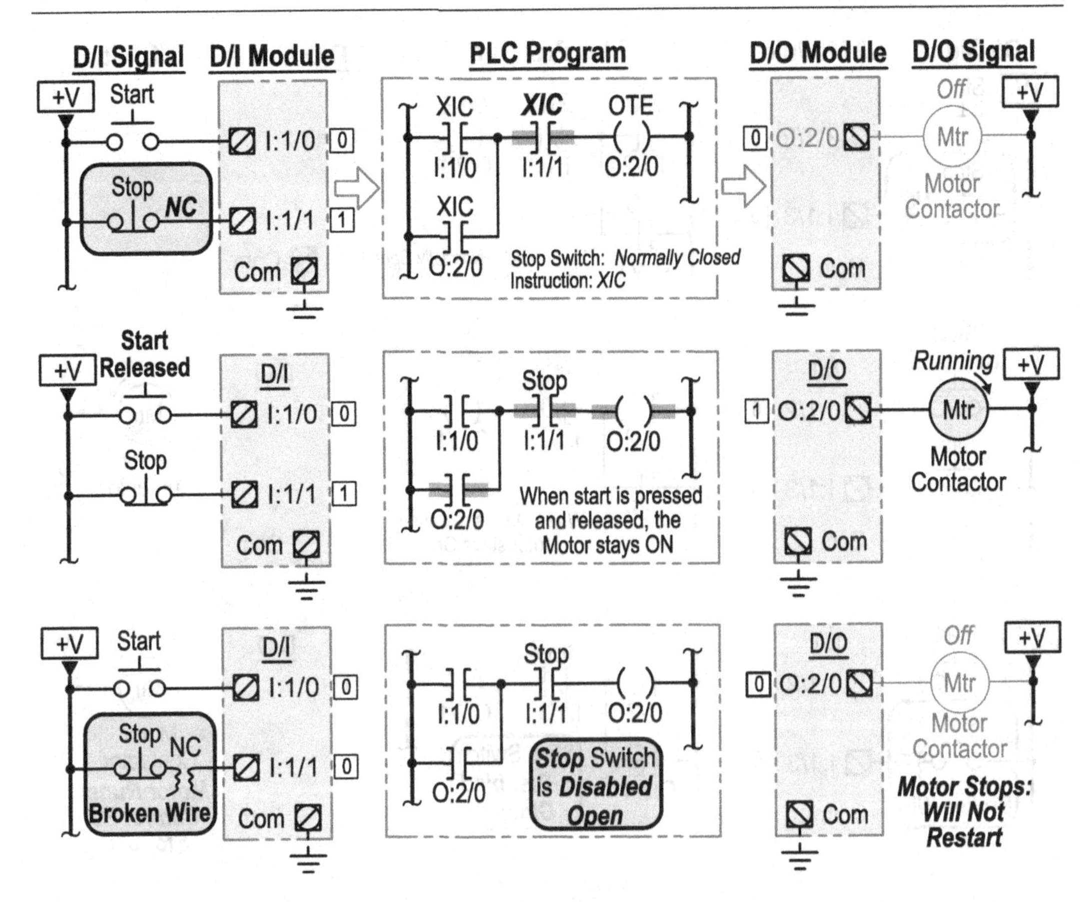

Figure 13.31 This diagram exemplifies preferred logic and should be contrasted with the circuit shown in Figure 13.32. This low-voltage-release circuit uses a normally closed stop switch coupled with an XIC instruction. If the stop switch loses continuity, the motor stops and cannot be restarted until the fault is repaired.

wired in series in options 2 and 3, so that actuation of any one switch breaks continuity to the shutdown control relays, causing the associated NO contacts in the permissive circuit for each boiler to return to their open positions. Option 3 is identical in function to option 2, but it has two shutdown control relays in parallel, where both relays need to be simultaneously deenergized to kill the boilers. In this arrangement, boiler operation continues upon failure of one relay, which actuates an alarm, alerting the operators to a controller problem.

In an FMEA analysis, the engineer would consider the system importance. For instance, in an attended heating plant, an unscheduled boiler shutdown event may be tolerable, and options 1 or 2 might be acceptable. However, in a hospital, unscheduled steam interruptions may not be permitted, in which case control-relay redundancy prevents failure of one relay from the drastic action of interrupting operation.

An FMEA engineer might examine the options and evaluate system response under the following faults:

1. On power loss to the controller:
   Option 1: The boilers continue to run without interruption; and worse, the safety controller is disabled.
   Options 2 and 3: The boilers are stopped in a failsafe manner.

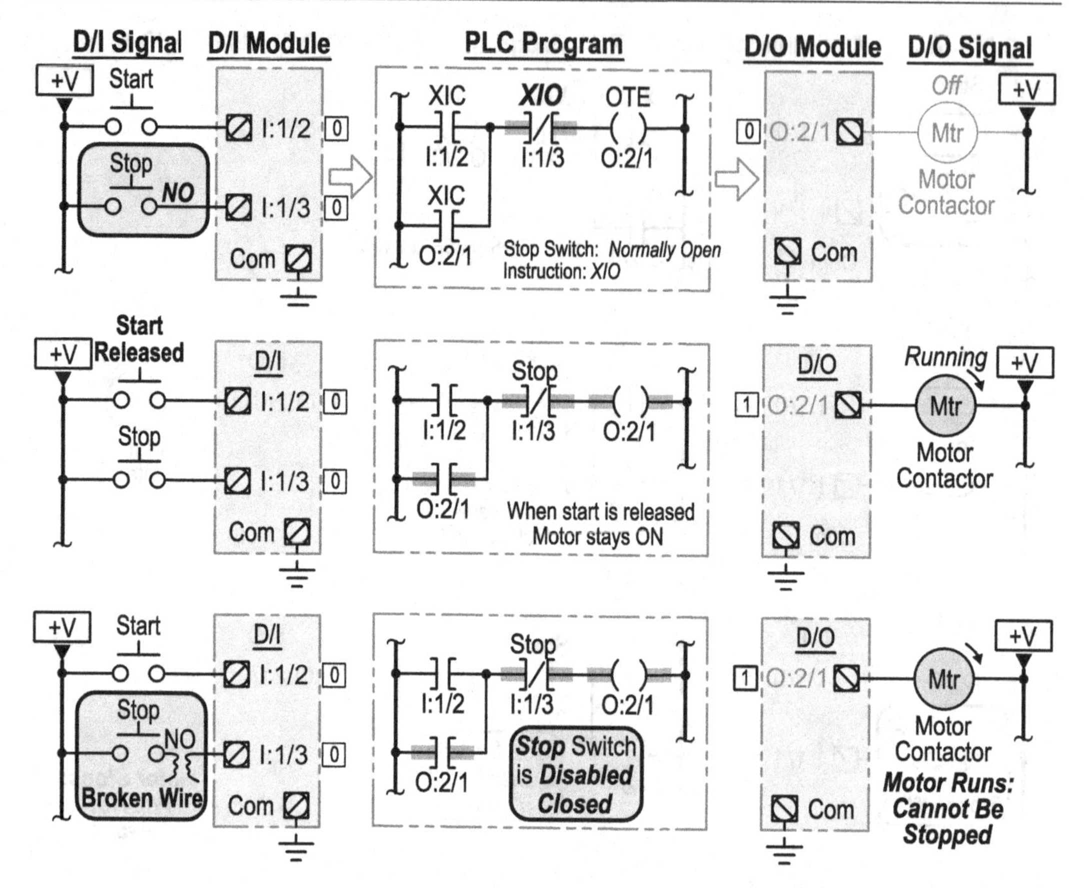

Figure 13.32 This diagram exemplifies faulty logic and should be contrasted with the circuit shown in Figure 13.31. This low-voltage-release circuit uses a normally open stop switch coupled with an XIO instruction. In this case, if the stop-switch wire breaks, the motor can still be started or the motor will continue running, however, the control circuit is unable to stop the motor. These examples show two ways to accomplish similar control, but one method has a logic flaw during an open-circuit fault.

2. If a wire to any one E-Stop switch were to come loose:
   Option 1: The control circuit is unknowingly deactivated for that one E-Stop switch, and the boilers could not be secured in an emergency from that location. All other E-Stops would continue to function.
   Options 2 and 3: The boilers are stopped in a failsafe manner and an E-Stop trip alarm is actuated.
3. If the wire to the E-Stop switch were to come loose in the controller:
   Option 1: The control circuit is unknowingly deactivated, and the boilers would never be secured in an emergency.
   Options 2 and 3: The boilers are stopped in a failsafe manner and an E-Stop trip alarm is actuated.
4. If one shutdown control relay coil burns out:
   Option 1: The boiler safety shutdown circuit would never work, and there would be no notification to the plant operators.
   Option 2: The boiler would safely shut down, and the operators would be alerted by the E-Stop trip alarm.
   Option 3. The boilers would continue to function without any interruption to service, and a faulty-relay alarm would be actuated, increasing boiler availability.

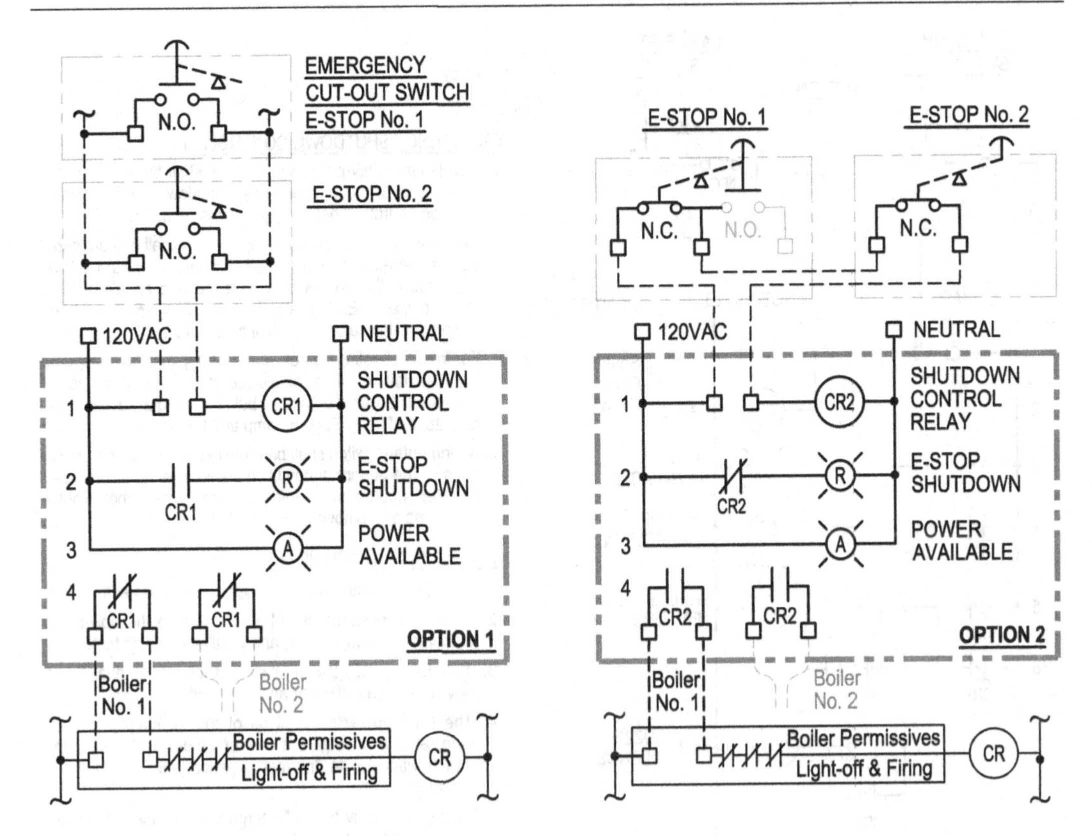

Figure 13.33 Two options for boiler emergency shutdown circuits, showing more than one way to achieve a safety-shutdown control result.

As a matter of academic interest, Figure 13.34 provides several good examples of Boolean statements being integrated into the control system. Note that during system operation, CR1 and CR2 are energized, and all associated contacts *are toggled* into the actuated position. That being the case, the Boolean functions are:

- CR1 and CR2 series NC contacts on line 4 form an *AND* function.
- CR1 and CR2 parallel NC contacts on lines 5 and 6 form an *OR* function.
- The combination of CR1 NO and CR2 NC contacts on line 5 with the CR1 NC and CR2 NO contacts on line 6 form and *XOR* function.
- The coil CR3 on line 7 with its associated NC contact on line 9 form an inverter or *NOT* function.
- The two parallel CR1 and CR2 contacts on lines 10 and 11 form CR1-NOT *OR* CR2-NOT. Using DeMorgan's Theorem, this reduces to a NAND function.

In making the final control-philosophy decision and despite the fact that option 1 consumes no power during operation, it has a serious design flaw, in that system operation and notification of trouble will not occur in the event of power loss, a signal-wire break, or a control-relay fault. If this system is installed at a site where continuous monitoring and periodic safety-system testing is mandatory, then there is a possibility that this option might be considered by some people. However, option 2 is energized continuously, and inherently

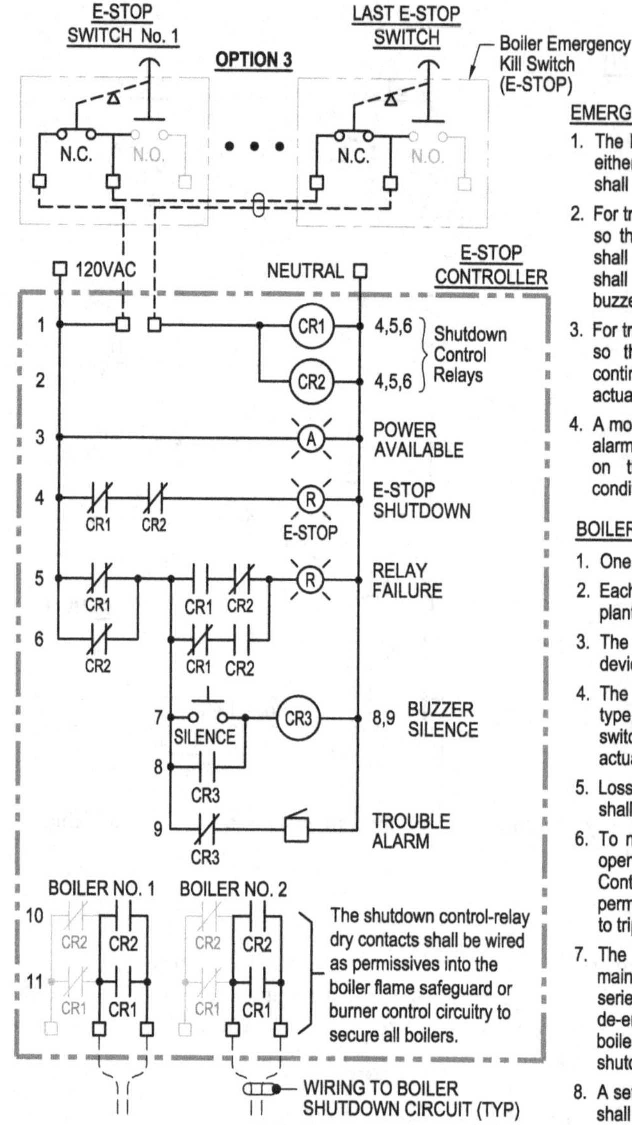

Figure 13.34 A boiler kill-switch wiring diagram is shown as an example. This circuit requires control-relay power to permit boiler operation, and will secure boilers upon loss of power. A pair of control relays ensures continued operation, even if one control relay fails. The circuit contains an alarm system and lights to indicate power available, one relay failure, and if the E-Stop has been actuated.

stops the boilers in the event of any E-Stop signal break leading to the shutdown control relay, whether intentional or accidental. Option 2 is more failsafe than option 1, and does not rely on close operator supervision. The E-Stop lamp distinguishes that service interruption occurred from the emergency hand-trip circuit, rather than the boiler safety devices. Option 3 provides the same failsafe benefits as option 2, however, the two shutdown control relays provide higher reliability, as failure of one relay does not interrupt service. Further, the alarm system distinguished between single-relay failures and the loss of power to both relays, which would be indicated by an E-Stop shutdown alarm. This circuit would be recommended in a high-importance system, where service interruptions need to be avoided.

QUESTIONS

1. List the purpose of the RSLinx, Boot-DHCP Tool, and RSLogix software programs.
2. Describe the XIC and XIO instructions. Describe the differences between the instructions and the similar, but different, normally open and normally closed contacts. Compare the OTE instruction to a relay-coil symbol and function.
3. How is sequential control achieved by a PLC?
4. How is feedback control achieved by a PLC?
5. Describe the difference between uploading and downloading a PLC program. Explain the benefit for "going online."
6. Explain the use of "forces," how to determine that forces are active, and the problem with running a program containing active forces.
7. Explain the data files, and how they interact with the programming effort.
8. Define rack, rung, instruction, bit, word, and hardwired I/O addresses. How does the B3 data file compare to the O0 and I1 data files?
9. List some instructions that are used in sequential logic.
10. Describe the process of receiving the sensor signal input, data manipulation, and producing the signal to control the output device. Include addressing and the PLC program scan steps in the description.
11. How are opto-isolators used with PLCs, and what are their benefits?
12. Describe flow charts and pseudocode and their purposes.
13. Explain the importance of documenting a PLC program.
14. Describe how the programmer needs to consider fail-safe conditions when putting together the program logic.
15. Modify the option 2 circuit in Figure 13.33 to create two alarm notifications, where one alarm indicates E-Stop shutdown occurred from any cause, while the second alarm shows that the shutdown was from a control-relay failure.
16. Sketch the switch arrangements in Figure 13.33 that produce AND, OR, NAND, Exclusive OR, and NOT Boolean functions. The sketches should show the switch states *after* the circuit is energized.

NOTES

1. Some instructions included in the RSLogix 5000 professional version software are not available in the RSLogix 500 version.
2. IEC-61131 is an international standard (International Electrotechnical Commission) for process-control software, where in theory, algorithms developed for one manufacturer's controller can be imported into another brand with minimal modifications. IEC-61131 standards address five PLC languages, Instruction List, Structured Text, Function Block, Ladder Diagram, and Sequential Function.
3. For discussion purposes, the terms internal devices or virtual devices shall be used interchangeably to mean the programming of instructions that exist only within PLC memory, but which emulate real-world hardware or devices.

4. 4–20mA is commonly used for analog I/O signals, where current has some transmission advantages over voltage in plant applications. However, although various levels of voltages can also be used, signal transmissions shall be referenced as using 4–20mA for simplicity.
5. The programmer should be careful to avoid confusing the letter “O” and numeral zero, and the uppercase letter “I” or lowercase “L” and the numeral 1. Additionally, PLC numbering typically begins at zero, so that device 00 is the local first device, which is the PLC, while the first connected I/O module is device 01. Similarly, I/O device terminal numbering begins at zero for the first wiring position, the second terminal position is 1, and so on.

Chapter 14

Electronic 4–20mA analog signals

Analog signals are time-varying inputs and outputs used by PLCs to continuously modulate systems and devices. An analog input is the signal representing the magnitude of a parameter to be controlled, and an analog output is the control signal whose strength is related to the amount of corrective action desired. Analog-input signals reflect measurements of temperature, pressure, flow, liquid level, and other parameters, and are typically used in conjunction with analog outputs used to continuously throttle valves, control positioners, vary motor speeds, and indicate meter readings, although analog values may be used to trigger digital outputs as well.

Analog signals are transmitted using either voltage or current. Common voltage signals are: 0–5Vdc, +/-5Vdc, 0–10Vdc, +/-10Vdc, and other voltages. The most common current signal is 4–20mA, with 0–20mA being less common. Although the 4–20mA analog signal was developed in the very early days of electronic control, it is still the most common signal used in plant control today. This chapter focuses on how 4–20mA signals are produced and used, given its prevalence and preference for plant control.

The preference for using current instead of voltage to transmit signals is its high reliability as a signal even when long or varying distances exist between the transmitter and receiver. The 4–20mA signal transmitter creates a constant-amperage signal even if the circuit resistance varies with conductor length, or if the supply voltage is not constant. When voltage is used as the signal, both the cable resistance and source voltage affect the signal. The longer the circuit, the higher the resistance, and the greater the signal voltage drop, thus potentially affecting the measurement.

Typically, voltage signals react faster than current, and consequently, voltage tends to be preferred in high-speed applications, such as some assembly-line and manufacturing processes, which can have very high-speed requirements. Relatively speaking, power plant mechanical systems tend to be low-speed applications as far as digital electronic control is concerned, and high-speed response is typically not needed. The speed limitation occurs because the current source generally has enough source impedance coupled with the cable's capacitive reactance to behave like a filter, which limits the high-frequency response and inhibits fast system changes.

In electronic control systems, the 4–20mA signal is analogous to the 3–15 psi pneumatic signal used to position regulating valves and other actuators. The 3-psi signal represents a "live-zero" signal and 15 psi represents the 100% signal. If a 3-psi pneumatic signal corresponds to a valve being completely closed, but on the verge of opening, 15 psi would correspond to the valve being fully opened. Similarly, 4mA represents a "live-zero" signal that corresponds to 0% level and 20mA corresponds to 100%, or vice versa in a reverse-acting controller (see Figure 14.1). If the pneumatic control system were to use a control actuation pressure of 0-30 psig, then a pressure signal of 0 psi would represent a "dead-zero." In electronic control, a signal between 3.2 and 4mA is interpreted as an abnormal or alarm condition, which is useful for identifying faults and troubleshooting, and where 0mA indicates

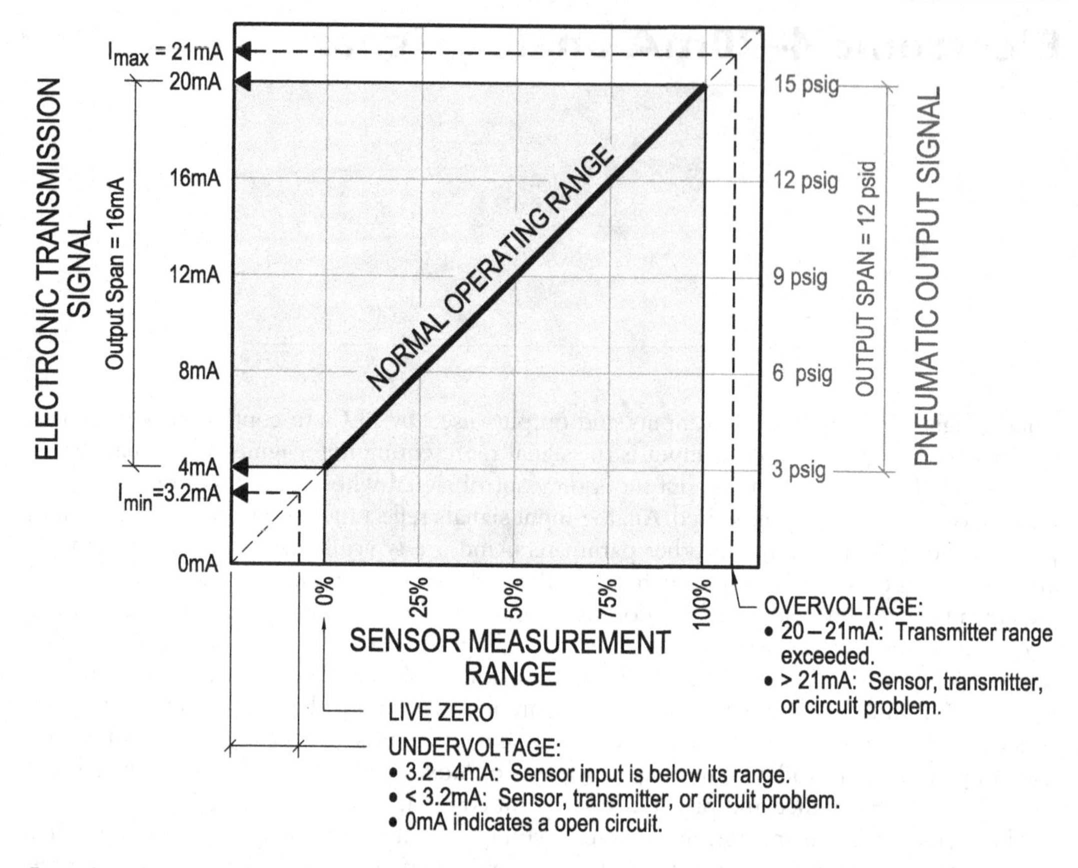

Figure 14.1 Electronic transmission signal vs. sensor measurement range.

loss of continuity. In cases where a 0–20mA signal is used, 0mA becomes a "dead zero" and the benefit of fault identification and troubleshooting is lost. In many plant control systems, the small 4–20mA analog-output signal is often converted into a more powerful 3–15 psi pneumatic-actuating signal to provide the large positioning forces needed by valves or linkages, using current-to-pneumatic or I/P converters.

The theory behind the 4–20mA current loop is that the sensor draws current in proportion to the mechanical property it measures. Surprisingly, the 4–20mA signal is not directly measured as amperage in the receiver. In the simplest case, the 4–20mA signal is measured indirectly as the voltage drop across a resistor in the receiver, and Ohm's Law is applied to translate voltage into the current measurement. For example, when a current between 4 and 20 milliamps is passed through a 250–ohm resistor, between 1 and 5Vdc is produced in accordance with Ohm's Law. A very precise voltage signal is generated to very accurately reflect the measured parameter, and the voltage is a very convenient signal for the receiver to process. For this reason, the receiver is often shown schematically as a resistor, even though voltage is the source and current is the transmitted signal.

Figure 14.2 shows a functional block diagram for following the signal processing from measurement, signal conversion, transmission, and finally for microprocessor data acquisition. The system starts with a measured analog parameter in a sensor, such as pressure or temperature. Often the sensor converts the measured parameter into a precise analog voltage that is measurable in a Wheatstone bridge. The voltage is then electronically converted into a precise 4–20mA analog current signal; usually through signal processing

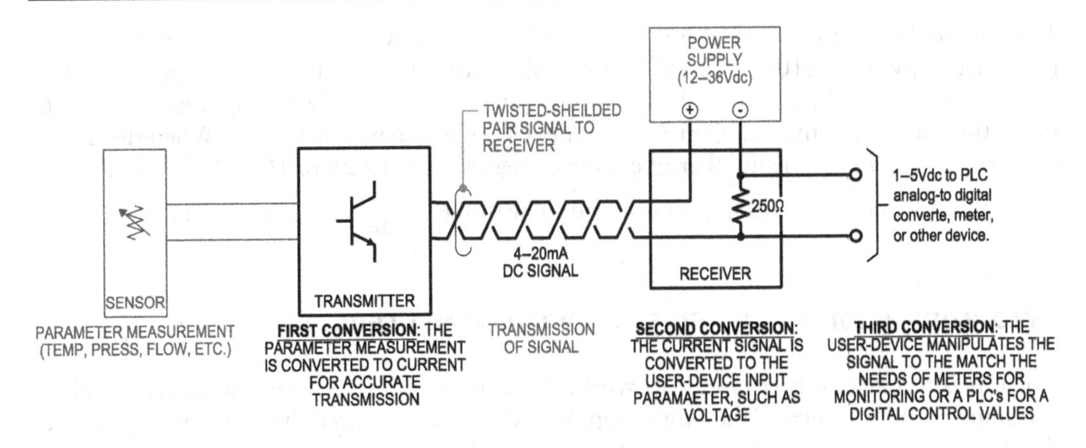

Figure 14.2 4–20mA loop functional block diagram.

and amplification (see Figure 14.5). The 4–20mA signal is robustly transmitted without signal losses through potentially long distances to the receiver, where the analog 4–20mA signal is converted back into a precise corresponding analog voltage. At this point, the analog voltage signal in the receiver is converted into a digital signal through an electronic analog-to-digital (A/D) converter, where the digital signal is required for manipulation in a microprocessor.

To provide an example of the signal processing, if a 4–20mA loop is used for a 0–1,000 psig pressure sensor, the 0 psig measurement would be converted into 4mA and the 1,000 psig measurement would correspond to 20mA. The minimum-to-maximum span of pressure readings is 1,000 psid, while the corresponding span of loop current is 16mA. The resolution of the loop device is $16mA / 1{,}000psid = 0.0160\,mA / psi$. At a mid-range pressure of 500 psi, the sensor produces 12mA, or half the current span.

Ultimately the 4–20mA transmitted signal is converted into voltage and then into digital "bits" or "counts" for use in a microprocessor controller. If a 4–20mA analog signal is applied into a receiver 250–ohm shunt resistor, the signal would be converted into voltage varying between 1 and 5Vdc. In this case, a 0–psig measurement produces a 4mA transmission signal, which, in turn, is converted into a 1–volt input into the receiver. A 1,000–psig measurement produces a 20mA transmission signal, which is converted into 5V in the receiver. The receiver full-range voltage spans a total of $4V_{\text{differential}}$. For the mid-range 500–psig measurement, the transmission signal would be 12mA, the receiver voltage would be 1V + 4V/2 = 3V.

To continue this example, assume that the receiver's 1–5V analog signal is brought into an analog-to-digital (A/D) module which is configured for an input range of ±10Vdc full scale. In this case, the A/D converter signal-input span is 20V, and consequently the signal processing within the A/D device only operates within 20% of its capability ($4V_{differential} / 20V_{span} = 20\%$). At the A/D module, the analog signal is further converted into a digital signal for use by a microprocessor. To continue this example, if a 10-bit data acquisition card was used, the full-scale resolution would be $2^{10} = 1{,}024$ bits, where 0 bits would correspond to -10V and 1024 bits to +10V. However, when using the 250–ohm shunt resistor, the signal is limited between +1V and +5V. The resolution of about 5 psi/bit can be determined by ratios:

$$\frac{4\text{V}}{20\text{V}} = \frac{\text{x}}{1{,}024\,\text{bits}} \qquad \text{x} = 204.8\,\text{bits F.S.} \qquad \Delta\text{p} = \frac{1{,}000\,\text{psid}}{204.8\,\text{bits}} = 4.88\,{}^{\text{psi}}\!/_{\text{bit}}$$

To improve the microprocessor's resolution, either the receiver's measurement span needs to be reduced, say from ±10Vdc to ±5Vdc full scale, and/or the number of microprocessor bits increased, say from 10-bits to 14 bits. Assuming the receiver's 1–5V analog signal is brought into a 0–5Vdc A/D converter, then 80% of the A/D signal span can be used. When the digital converter uses 14 bits, the number of counts increases from 1,024 to 16,384 (2^{14}). Following the procedure above, the resolution is found to be $0.076 \, {}^{psi}\!/_{bit}$, an improvement of 64 times.

TWO-WIRE 4–20MA CURRENT LOOP TRANSMITTERS

For a 4–20mA transmitter circuit to work, three devices are necessary; a power supply, a transmitter, and a receiver. The power supply is the voltage source that ultimately produces the 4–20mA current signal, and it often powers the measurement and signal-processing circuits. Power commonly originates from a nominal 24Vdc unregulated power supply, although other voltages are sometimes used. The transmitter is the device that upon being powered converts the measured parameter into a corresponding 4–20mA signal. Although the transmitter often does not contain the power source, it regulates the amount of circuit current, behaving as if it were an automatically modulated resistor. The receiver is the device at the plant-operator's end of the transmission line that "receives" and interprets the signal. Typically, the 4–20mA current signal is converted into voltage in the receiver for conversion into a computer signal and ultimately into a useful output.

The transmitter is typically constructed using integrated-circuit chips on a printed-circuit board (PCB), rather than the discrete components of the early-generation devices. It may contain an amplifier section along with signal filtering, and it may use a transistor to boost the signal into the 4–20mA range. Often the printed-circuit board will contain filtering to remove electromagnetic/radiofrequency interference (EMI/RFI noise), protection against overvoltage, protection to avoid damage from incorrect connections of power-supply polarity, temperature-correction circuitry, or even optical isolation. Optical isolation separates portions of circuits, so that a damaging electrical fault in the field device does not cascade into the sensitive electronic control system, providing robust control without the tendency to produce catastrophic controller failures.

Several arrangements of 4–20mA systems may be used, and are classified by ANSI/ISA standards as Type 2, Type 3, and Type 4, where the transmitter circuits for Type 2 uses two wires, Type 3 uses three wires, and Type 4 uses four wires (see Figure 14.3). The multi-conductor configurations are based on how the transmitter receives its power to operate.

The two-wire field device is "loop powered," i.e., it derives its power from the first 4mA of the transmitted analog signal. Some measuring devices require very little power and can operate in this simple arrangement, where the two conductors carry both power and signal. In some cases, measuring devices require more than 4mA available to operate, such as explosive-limit detectors as an example, or in other cases some devices need more than 24V, such as oxygen analyzers, which may require a separate 120V supply. In those cases, separate sources of signal and power are used. The 4–20mA loops are referred to as "active" or "passive" loops, depending on the source of instrument power. Passive loops are powered by the first 4mA obtained from the PLC analog-input cards, while active loops are powered by the transmitter/instrument and the PLC only senses the 0–21mA signal current. The terminations on the cards will be different for distinguishing between the two, as they cannot be interchanged, however, the devices often have built-in protection to prevent damage from misconnections.

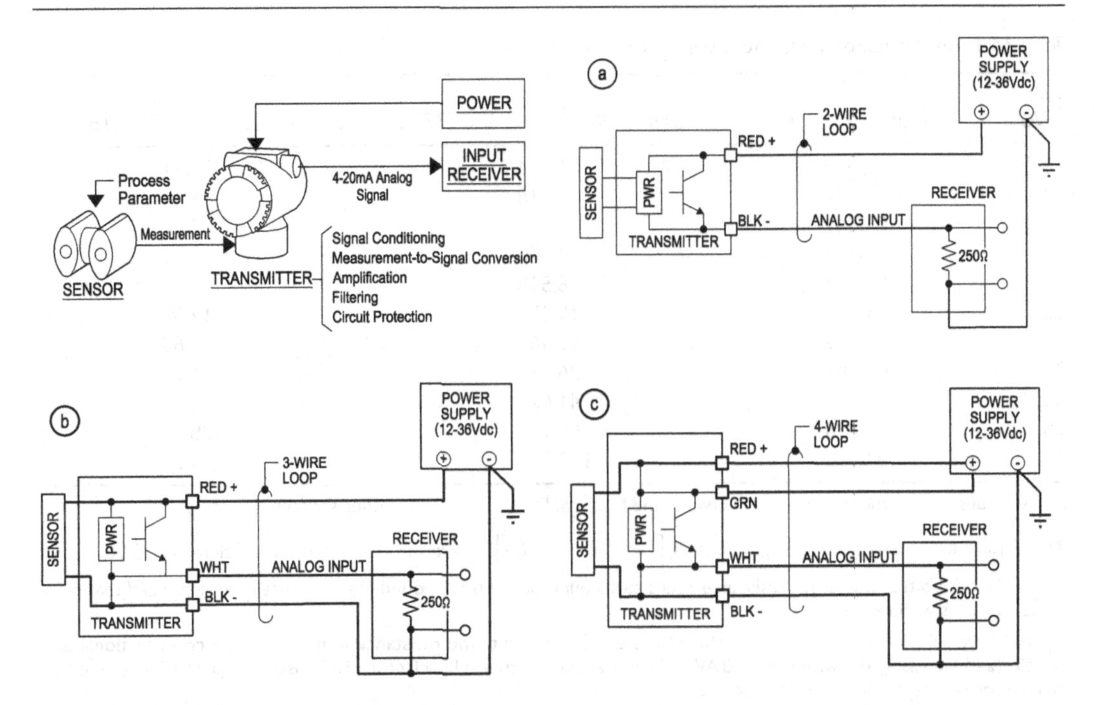

Figure 14.3 (Upper left) Block diagram for a loop controller. (a) Two-wire Type 2 loop controller. The transmitter/sensor is energized by the current loop. This device is passive; as it relies on the loop current to operate the sensor. The transmitter floats relative to ground. (b) Three-wire Type 3 loop controller. The transmitter/sensor assembly requires more than the 4mA minimum signal current to operate, so separate power sources are provided. The transmitter and sensor power share a common connection with the receiver return, and the 4–20mA analog signal is input to the receiver separately. (c) Four-wire Type 4 loop controller. The transmitter/sensor is powered independently from the 4–20mA loop, where both the transmitter and receiver float. The Type 3 and 4 devices are active devices, as they require a source of power to operate transmitter/sensor.

Two-wire transmitter connections are typically made using twisted-shielded pairs of conductors. The twisting is used to mitigate electromagnetic or radio-frequency interference (EMI or RFI noise). Noise is produced by inductive coupling of these fields into the signal wiring. By twisting the conductors, EMI/RFI tends to induce voltage in one direction in the supply conductor while simultaneously inducing opposite voltage in the return conductor, cancelling the effects. The shield absorbs some of the EMI/RFI fields before they reach the signal conductors. The shield is often grounded at one end, typically at the source, and it should never be grounded at both ends. If the shield is grounded at both ends, ground-loop currents can result, and noise will be readily induced into the signal conductors.

DETERMINING THE POWER SUPPLY VOLTAGE

The power supply may be located within the receiver or it may be a standalone device. The required voltage needs to produce the maximum 20mA for the total number of devices that are attached, plus it needs to overcome the conductor voltage drop, and then there needs to be enough voltage to drive the attached transmitter and sensor.

As an example, if the receiver's resistance is 250-ohms, which would convert 4–20mA into 1–5Vdc, and the total out-and-back conductor resistance is 500-ohms and the sensor/

Table 14.1 Resistance of uninsulated solid copper wire

Wire AWG	*Diameter (inches)*	*Ohms/1000-feet at 25°C (77°F)*	*Ohms/1000-feet at 75°C (167°F)*
10	0.1019	1.014	1.21
12	0.0808	1.618	1.93
14	0.0641	2.574	3.07
16	0.0508	4.100	4.89
18	0.0403	6.515	7.77
20	0.0320	10.35	12.3
22	0.0253	16.46	19.63
24	0.0201	26.17	31.21
26	0.0159	41.62	49.63
28	0.0126	66.17	78.91
30	0.0100	103.2	123.0

Notes: Values are in general agreement between the National Electric Code and other sources.

The formula for temperature change is $R_2 = R_1\left[1+\alpha\left(T_2 - 75°C\right)\right]$, where the temperature coefficient $\alpha = 0.00323$ for copper. Note also that the length/resistance needs to account for the total "round-trip" distance of wire run and that resistance increases with temperature.

Approximate rules of thumb: Increasing the wire every 3 AWG cuts the resistance in half while the cross-sectional area doubles; and increasing the wire every 10 AWG increases the resistance by a factor of 10. Also, changing the wire gage by 6 AWG, changes the diameter by a factor of 2.

transmitter assembly requires 12Vdc, the required power supply voltage would be the sum of the voltage drops, or:

$$V_{required} = [250\Omega \cdot 20mA] + [500\Omega \cdot 20mA] + 12Vdc$$

$$V_{required} = 5V + 10V + 12Vdc = 27Vdc$$

It is important to realize that the amount of excess applied voltage is not critical, as the transmitter circuit will only use the amount of voltage required to create signals in the range of 4-20mA. However, the voltage must be high enough to overcome the drop across the receiver's sensing resistance plus the conductor voltage drop, and then there needs to be enough voltage left over for the transmitter to function over its entire range without "signal roll-off" near 20mA. If the required circuit voltage exceeds the source rating, two options can be considered. The first option is to use the next-higher standard voltage level, or the second option is to increase the conductor wire gage so as to reduce the conductor voltage drop (Table 14.1).

In this control system, a 36V power supply could be used, as a 24V supply would be inadequate, or the conductor could be up-sized to bring the *IR* drop from 10V down to 7V. This trade-off decision would be based on the quantities of control circuits that require more than 24V as well as the length of the wire runs.

CREATING THE 4-20MA TRANSMITTER SIGNAL

4–20mA signals are generated from electronic circuits that are nowadays nearly always contained within integrated circuit (IC) chips. An illustration of a surface-mount 4–20mA converter IC chip is shown in Figure 14.4 and its schematic representation within a circuit having amplification, filtering, overvoltage protection, and polarity-miswiring protection is

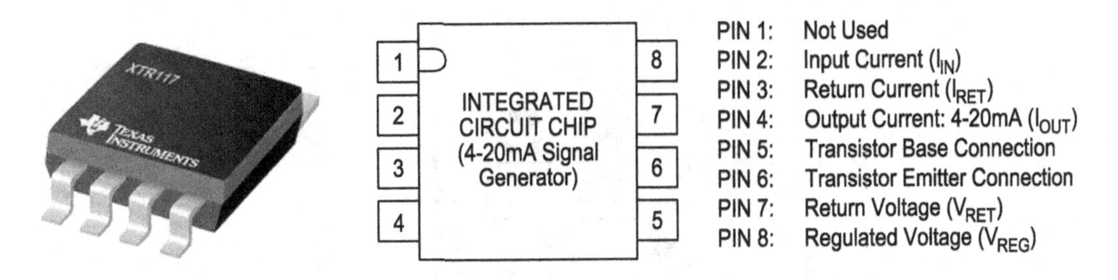

Figure 14.4 4–20mA current-loop transmitter chip XTR117. (Courtesy of Texas Instruments.)

shown in Figure 14.5. This chip is the fundamental building block for a precision-output 4–20mA analog signal generator designed to transmit over a current loop. The chip itself includes current scaling and output current-limiting functions. The chip shown has a built-in +5V voltage regulator that can be used to supply the low-power external circuitry found in Type 3 or Type 4 arrangements.

GROUND LOOPS: GROUNDING AND ISOLATED INSTRUMENTS

In transmitter circuits, grounds are common points of contact that are at the same electrical potential, which complete the circuit back to the source. Voltage measurements made between ground connections for various field sensors should read 0V for both DC and AC. In reality, many systems may measure several volts or more. A ground loop is illustrated in Figure 14.6. When grounds of different potentials are tied together, current loops are formed through the grounds. Ground-loop currents cause problems by adding or subtracting small currents or voltage levels to the process signal, corrupting the signal. Ground loops can result in noisy and inaccurate measurements, instrument damage, or instrument saturation where some measurement levels cannot be reached.

Two solutions are to eliminate all but one ground, or to isolate all grounds from each other. When local grounds cannot be isolated from each other, as with thermocouples or some analyzers that need a local ground for accuracy, signal isolators may be used.

To avoid ground-loop problems, some techniques are:

- use isolated instruments for 4–20mA measurements. If isolated instruments are used, then ground-loop problems are completely avoided, and all other grounding precautions are unnecessary.

If instruments cannot be isolated:

- isolate the control-loop 4–20mA dc sensor ground so that it is not connected to the ac line-input ground. If the ac and dc grounds are isolated from each other, all sensor 4–20mA grounds can be tied together without consequence;
- in self-powered applications, isolate the sensor-side of the loop from the power-source ground;
- where multiple power sources cannot be isolated, ensure that all devices, including power supplies, self-powered sensors, the instrument, and the PLC/controller, are all powered from the same physical location, so the grounding source is truly common (see Figure 14.7).

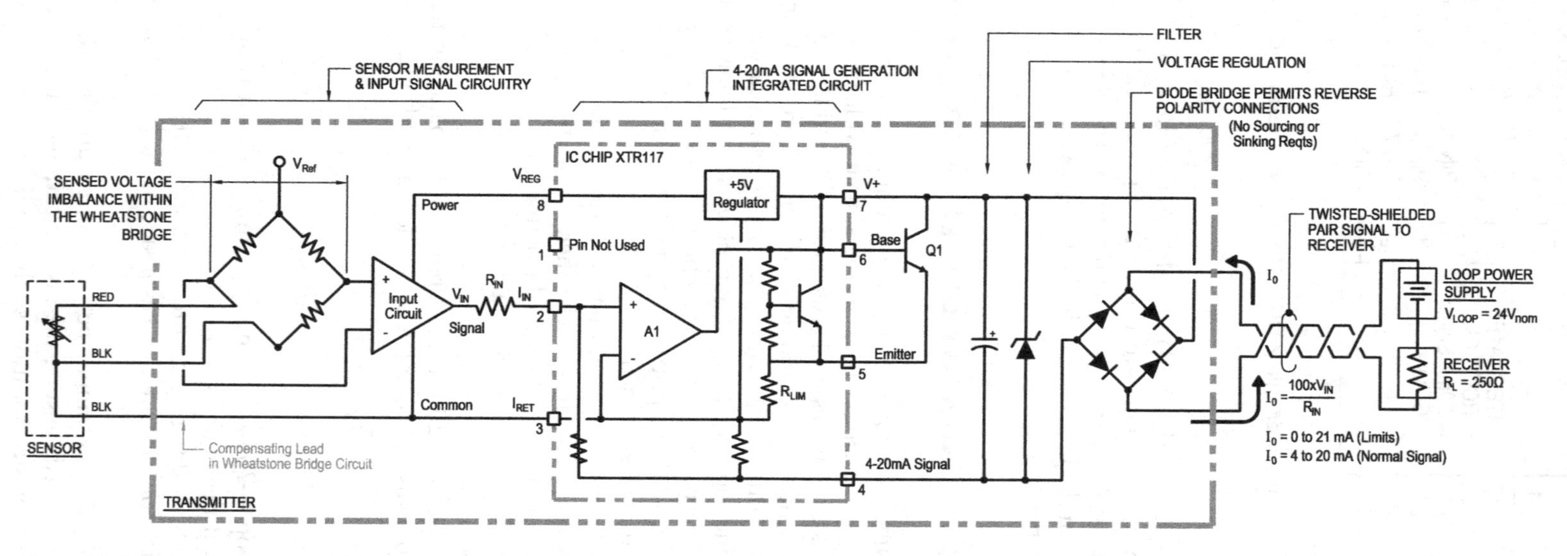

Figure 14.5 Basic circuit connections showing signal amplifier circuitry, figure adapted fomr the TI XTR117 4-20mA converter chip, filtering, overvoltage surge protection, and a diode wheel for reverse-polarity prevention. (Courtesy of Texas Instruments.)

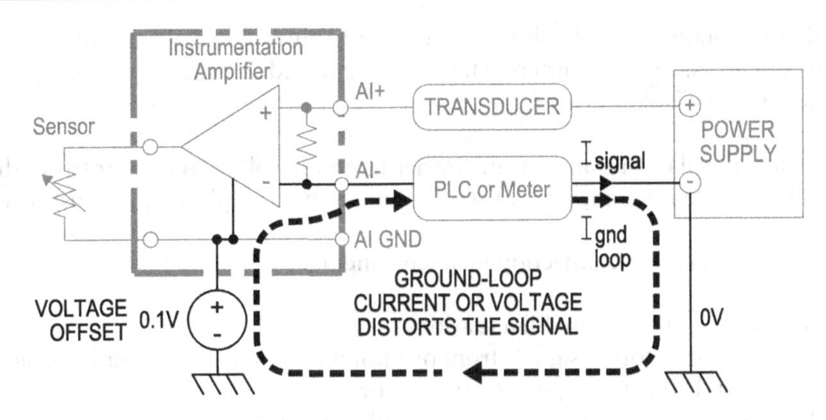

Figure 14.6 Ground-loop currents cause problems by adding or subtracting small corrupting currents or voltages to the process signal, resulting in error and control misbehavior.

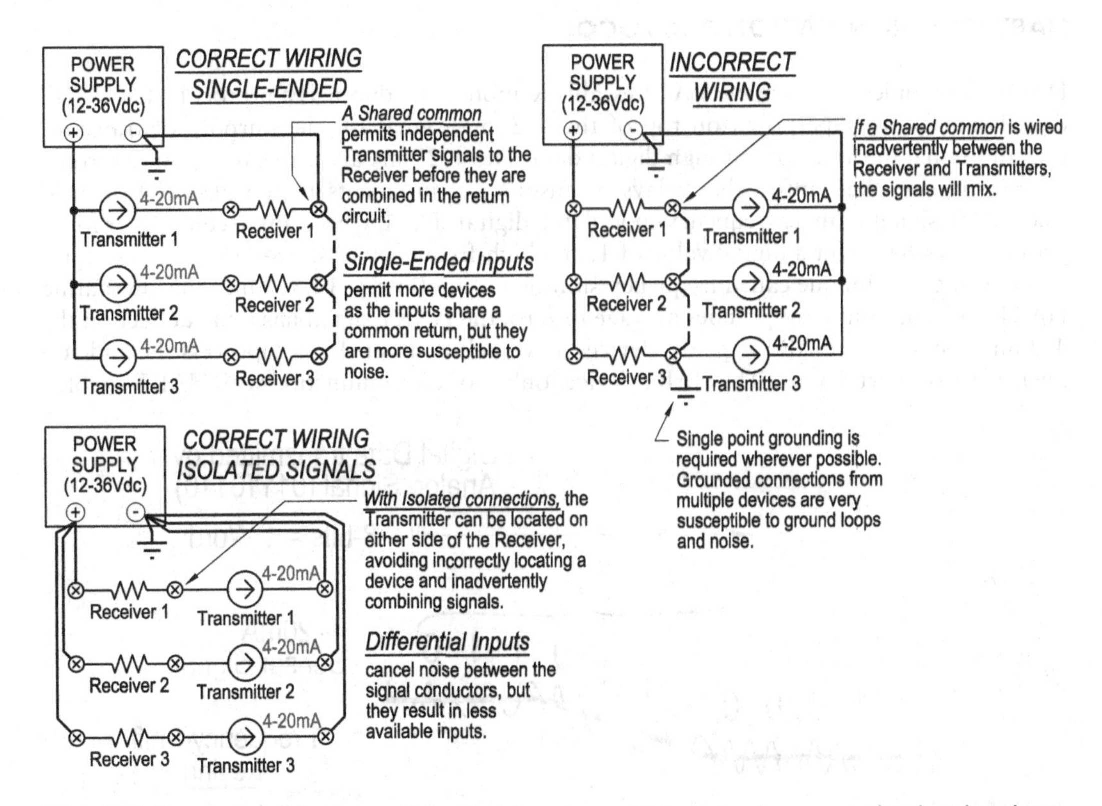

Figure 14.7 Care must be taken to avoid signal mixing, when multiple devices have non-isolated or shared commons. Signal isolation provides a clean alternative for avoiding ground loops and signal problems from shared commons.

Isolators work by galvanically decoupling all paths to ground while allowing the signal to pass unimpeded through the loop. This technique can also eliminate continuity that might couple the 4–20mA dc signal to ac noise, which is noise caused by the common-mode voltage. The signal isolators are selected based on the circuit application. Examples include resistance isolators used with RTDs, strain gages, and potentiometer transmitters, millivolt isolators used with thermocouples or millivolt transmitters, and current or voltage isolators used with alarm circuits or specialty transmitters. The isolators provide input, output, and power isolation. The two types of isolators are:

- analog signal isolation using a transformer to chop, isolate, and reconstruct the signal.
- discrete signal generation, which uses an opto-coupler in place of a transformer.

Symptoms of signal ground-connection problems include:

- unpredictable 4–20mA signal fluctuations.
- addition or subtraction of signals from one point in a loop to another causing measurement errors and sometimes out-of-range errors.
- signal averaging, causing same values on multiple devices.
- physical damage to components on rare occasions.

HART COMMUNICATION PROTOCOL

HART communication (Highway Addressable Remote Transducer) is a protocol where small digital signals are superposed on top of the 4–20mA signal for the purpose of transmitting additional information through digital data. The 4–20mA analog signal is an electronic dc transmission mechanism that relays a sensor's measurements to the PLC I/O devices. The HART signal is an ac frequency-modulated digital signal, where low-frequency or long-period waves represent a digital value of 1, and high-frequency or short-period waves represent 0, using a technique called frequency-shifting keying (Figure 14.8). Since the alternating HART plus-and-minus amplitudes average to zero, the HART signal has no net effect on the 4–20mA dc signal. HART-capable devices have self-contained microprocessors and are referred to as smart devices. The HART devices only work in conjunction with HART-enabled

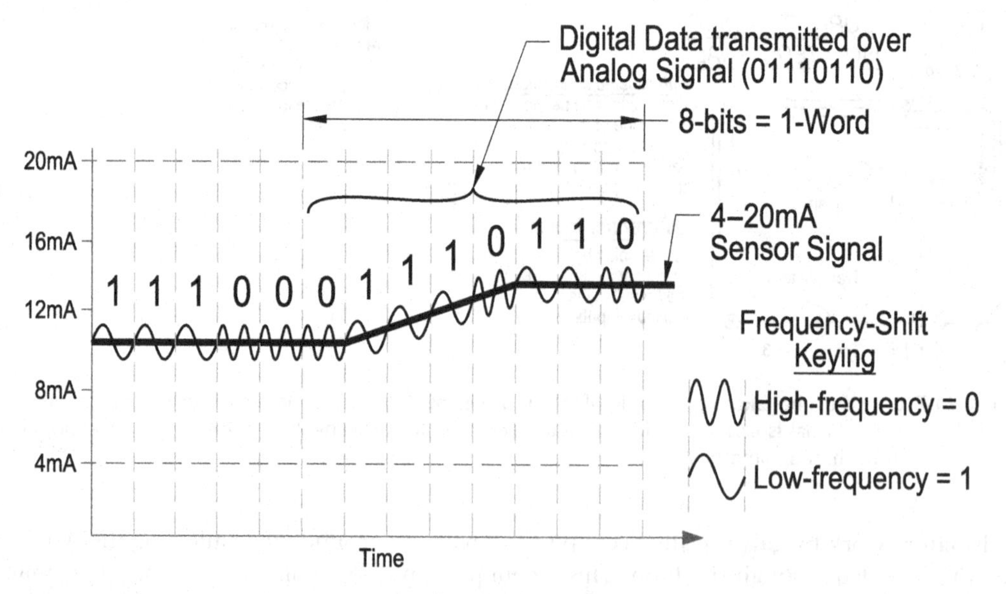

Figure 14.8 HART communication works by applying a digital signal, which is carried on top of the normal 4–20mA analog control signal. Frequency-shift keying of the low-level alternating-current HART signal is used to define binary 0 and 1 values for data transmission.

I/O devices. In this manner, the smart device can carry additional control information, where for example, a control valve's instantaneous stem position can be fed back on top of the 4–20mA signal that originally commanded the control valve adjustment. Additionally, the HART signal can convey other useful information, such as product make and model number, and incidental operating parameters.

SUMMARY

In summary, the 4–20mA signal is the dominant signal in plant control systems, but where high-speed is required, voltage signals tend to be used. The 4–20mA signal is a more-forgiving, robust signal, that is independent of circuit length without degradation up to a couple of thousand feet. Voltage drops in the circuit do not result in signal attenuation. Essentially, the circuit length is limited by the power supply voltage and the conductor resistance.

4mA forms a normal minimum value, called a "live zero," which can be used to observe signal integrity and for troubleshooting. Currents less than 3.2mA or between 20 and 21mA can be used to indicate a transducer problem, while 0mA indicates an open circuit and more than 21mA causes a PLC fault, if the error is not trapped. 4–20mA is less sensitive to background noise. Most installations require twisted-shielded pairs, or more commonly, twisted-shielded triplets, where the shielding tends to dissipate noise before it reaches the signal conductors, and the twisting tends to cancel any EMI and RFI noise that reaches the conductors. The source voltage must exceed the minimum voltage required by the device plus the voltage drop in the wiring, and care must be taken to avoid ground loops that corrupt signals, especially when a large number of process variables and control loops exist.

HART communication is a technique to transmit large amounts of digital data over the analog 4–20mA signal, but it requires both the I/O module and the field devices be HART capable.

QUESTIONS

1. List the typical analog input and output signals available to PLCs and their connected devices.
2. What is the advantage for using amperage as a signal in plant-control applications? What is the most common amperage signal used with PLCs?
3. Why might voltage signals be preferred in a manufacturing environment? What is the disadvantage of using voltage signals in a plant application?
4. What is an I/P device and an E/P device? How is a tiny 4–20mA signal stepped up to a strong actuating force in a PLC control application?
5. What is the common pneumatic-signal range? How is it related to the 4–20mA signal?
6. What does the term "live zero" mean? How are signals below 4mA and above 20mA interpreted? How is a signal of 0mA interpreted?
7. Describe the HART system of digital transmission of information over an analog signal. Provide a sketch and describe how the digital signal is achieved? How is it that the digital data transmission does not corrupt the analog signal? What are the hardware requirements for HART to work? What types of information can be carried on top of the analog signal?

Chapter 15

Analog functions using Allen-Bradley's RSLogix software

ANALOG INPUT AND OUTPUT

In addition to having a large amount of on-off logic-based control functions, plants and auxiliary equipment typically have large amounts of continuously varying modulated-control requirements. Continuous control uses analog-input and analog-output signals to regulate processes.

PLC programs can receive and manipulate continuously varying parameters that are scaled into analog values for real-time process monitoring, historical data archiving, and active control functions. The analog-input values are used to produce analog outputs, and they can even be used to produce digital outputs. Digital outputs originating from analog signals are used to trigger alarms, toggle solenoid valves, or start pumps when setpoint limits are exceeded. The analog-input data often produces continuously varying modulated analog-output signals, used for fine positioning of flow-regulating valves or adjusting motor speeds using variable-frequency drives. Commonly, the analog output is transmitted as either a low-voltage signal, such as 0–5Vdc, or a low-current signal, such as 4–20mA. By itself, the electronic signal is too weak to effect positioning of mechanical devices, so it is often transformed locally to a 3–15psig pneumatic signal using a current-to-pneumatic (I/P) converter or a voltage-to-pneumatic (E/P) converter. Pneumatic signals are robust and directly capable of producing high actuation forces to position final-control devices in proportion to the signal.

Referring to the block diagram in Figure 15.1, PLC-based analog controls fundamentally work by:

- Measure the parameter that best represents the process to be controlled, such as temperature, pressure, flow, level, etc.
- Convert the measurement into a proportional analog signal for transmission, such as 0–5Vdc or 4–20mA.
- Receive the transmitted analog signal into the PLC and convert it into a "digital" value.[1] The analog signal will then be in the form of a "digital" binary-number integer whose value represents the signal strength, and it will be stored in PLC input memory.
- Recall the "digital" binary-number input signal from memory and mathematically process the value in the PLC central processing unit to produce an appropriate output control signal. The output signal is also in the form of a binary-number integer and is placed in PLC output memory. The value of the output signal integer represents the amount of process adjustment that is required. Again, while the signal is being manipulated in the PLC as a binary word-based number, the signal is called "digital."
- Convert the digital output integer to an analog output signal for transmission, e.g., 0–5Vdc or 4–20mA.

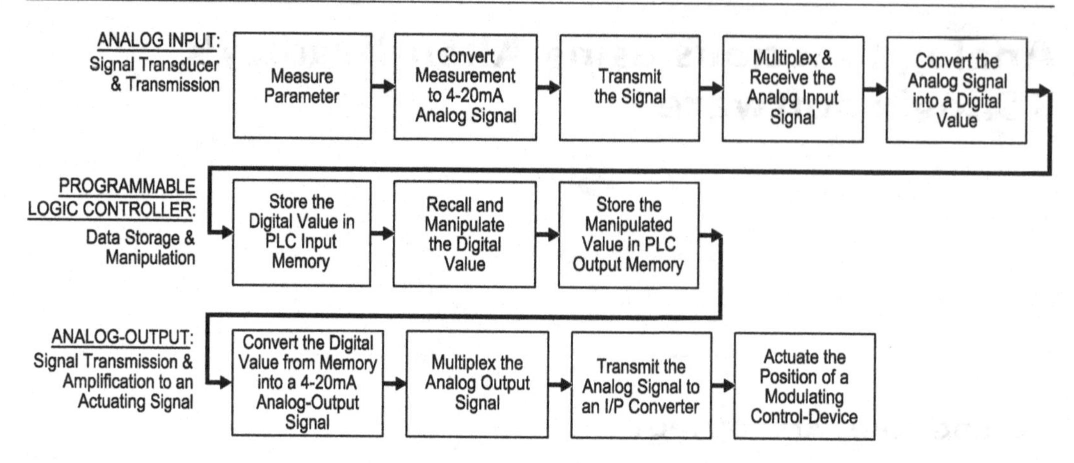

Figure 15.1 This block diagram represents all PLC analog-signal processing steps from input to output. Signal processing begins with the initial measurement, conversion into an analog signal, and transmission. The analog signal is multiplexed and converted into a binary number before being stored in input memory. The PLC manipulates the data to produce output control instructions, which are placed in output memory. The output undergoes digital-to-analog conversion, transmission as an analog signal, and amplification into a final-device actuation signal.

- Convert the analog output signal into a high-power actuation force, proportionate to the signal strength. As an example, a 4–20mA current signal may be converted into a 3–15psig pneumatic signal which can produce high actuating forces.
- Actuate the final control device using high-force pneumatic pressure.

Signal input and transmission

Analog systems begin with a sensor that measures an operating parameter, such as temperature, pressure, flow, level, etc. The sensor measurement is converted to either voltage or current for transmission. There are a variety of standard voltage levels that are used for signal transmission, such as 0–5Vdv, ±5Vdc, 0–10Vdv, ±10Vdc, etc.; however, the 4–20mA current signal is very common for PLC-based plant-control applications. See Chapter 14 for a discussion of the 4–20mA signal. For discussion purposes, 4–20mA will be used for many the following explanations, although it is understood that various voltages could be used. Table 15.1 shows the nominal- and full-range-signal options available for the Micrologix 1769 analog input and output modules.

Signal conversion to machine language

PLC analog-input modules use a built-in multiplexer (MUX) working in conjunction with an analog-to-digital (A/D) converter. The multiplexer sequentially polls each attached analog-input device for its signal value. When it is its turn during the MUX sequencing, the 4–20mA signal value is passed to an A/D converter. The A/D converter changes the milliamps into a binary-number integer, which is stored in memory. The binary number is now called a "digital signal." Binary numbers are capable of interpretation and mathematical manipulation by the PLC's microprocessor, and its digital value corresponds to the originating-signal strength. The integer is addressed by the PLC program into an input-storage location in the PLC's memory, using one or more computer 8-bit "words." When more "words" are used to define the signal, the signal resolution and fineness of control are improved. In summary, the

Table 15.1 The normal-operating analog-signal ranges and the full-module ranges are shown for the Micrologix 1769-IF4 four-channel analog module

Normal operating input range	*Full module range*
±10V dc	±10.5V dc
1–5V dc	0.5–5.25V dc
0–5V dc	-0.5–5.25V dc
0–10V dc	-0.5–10.5V dc
0–20 mA	0–21 mA
4–20 mA	3.2–21 mA

Table 15.2 Binary-number equivalents based on the 0–21 mA maximum span of analog-input signals for a 16-bit binary number that can be positive or negative ($2^{16}/2$ - 1 = 32,767). The full-module range of the Micrologix 1769-IF4 module spans 3.2–21 mA

Signal percentage:		*Min.*	*Live 0*	*25%*	*50%*	*75%*	*100%*	*Max*
Signal (mA):	0	3.2	4	8	12	16	20	21
Binary Number:	0	4,993	6,242	12,483	18,725	24,966	31,208	32,767

multiplexer incrementally "gates" the input signal to the analog-to-digital (A/D) converter, where it is changed into machine language.

The A-B MicroLogix PLC obtains the "digital" integer value as two 8-bit words for a total of 16-bits, to be stored as the analog-input-signal value. By default, the Allen-Bradley RSLogix 500 software stores the values in an "N7-integer data-file" address location. Sixteen-bit signals have the potential to be spread over a total of 2^{16} of 65,536 increments, for very fine resolution. However, to accept analog signals that may span between negative and positive values, the RSLogix software divides the 65,536-bit span between -32,767 and +32,767. By halving the total integer value, a negative-to-positive range can accommodate plus-or-minus input signals, such as ±10Vdc signals, as an example.[2] Conversely, the positive-only 4–20mA signal cannot use the negative integer values, so all values between 0 and negative 32,767 are unavailable. The 4–20mA normal span is from +6,242 to +31,208, so the signal resolution can be resolved over 31,208 - 6,242 = 24,966 increments. Table 15.2 shows the 0–21mA analog signal and its corresponding binary value.

Signal processing

The PLC program pulls the integer values from-memory, which were created from the analog-input signal. The raw integer value is manipulated by the program in the microprocessor to ultimately produce an output signal still in digital, i.e., binary-integer form. Typically, the program instructions retrieve the integer from memory and scale it into engineering units, such as psi or °F, that more intuitively represent the measured parameter to the programmer or user. The PLC program instructions could be as simple as comparing the scaled engineering-units value against a setpoint to create an output signal. The calculated output signal is subsequently reverse-scaled to become a digital integer, which is then placed into output memory.

Output signal and transmission

The process of creating and transmitting an analog-output control signal is similar but opposite to the analog-input signal transmission process. The digital value, which is stored as an integer, is retrieved from memory and passed to the digital-to-analog (D/A) converter. The D/A converter sequentially pulls each digital-output value from memory and converts it into a proportional voltage or current signal for analog transmission. The transmission signals by themselves are too weak to directly position the final control devices, so a signal converter/amplifier is installed locally, at or near the device. Voltage-to-pneumatic (E/P) or current-to-pneumatic (I/P) converters take voltage or current signals and change them into a corresponding pneumatic pressure. The pneumatic signal has the potential to create high actuation forces. In cases where pneumatic control-air pressure is not available, electrical actuators can be used. Electrical actuators use higher-voltage motors to create the strong device-positioning forces, where the positioner direction and excursion distances are controlled by the small output signals from the PLC.

GETTING THE SIGNAL INTO THE PLC

Electronic-signal transmission follows Ohm's Law (V = IR), where the analog signals are always in the form of either voltage or current. Resistance is a passive way of sensing a parameter and is not directly transmittable without conversion into voltage or current. Figure 15.2 shows a PLC rack having three expansion modules. In this case, the third expansion module is an Allen-Bradley four-channel analog I/O expansion module, having two inputs and two outputs. The input's analog signal originates from a transducer, which consists of a sensor element, power supply, and a transmitter. The analog signal is connected into the PLC I/O module's +/- input terminals corresponding to the channel being used, while maintaining correct polarity. The I/O module location in the rack and the connected input-channel terminal numbers combine to form the input-signal memory address for programming. Figure 15.2 (far right) and Figure 15.3 show the physical terminals in the I/O module, while Figure 15.4 shows a schematic wiring diagram for coordinating the wiring to the program's memory addresses. The PLC address is derived from these wiring connections. In this case, the address used by the PLC programming is I:3.0, where "I" specifies the signal is input, "3" indicates the fourth rack device—remembering that the device numbering begins with zero, and "0" is first or Channel 0 terminal connections. The PLC program assigns the value into the "I1 Input Data File" memory location, where the addressed data originates from the fourth device and occupies the first word, again remembering that the numbering starts at zero. Similarly, the channel 1 address would be assigned as I:3.1, where the ".1" specifies the second word in the input data file. Depending on the built-in features of any particular I/O device, various current or voltage input signals may be either hardware-selectable through DIP switches (dual in-line package) or software-selectable through menus (see Figure 15.5).

The analog-output signals originate as digital, word-based numbers during PLC program execution. The digital versions of the analog-output values may be real or integer numbers, and they are stored in an output memory address. The analog-output module consists or a multiplexer and digital-to-analog converter (D/A converter). The analog-output value is pulled from its memory address by the multiplexer and passed through the D/A converter, which transforms the number into a proportional voltage or current signal for transmission. Whether current or voltage is selected depends on the final control-device requirements, which itself, may have selectable signal options. Figure 15.2 and Figure 15.4 show dedicated voltage or current wiring terminals for each analog-output channel, and in either case, the output signal returns to common (COM). Both the input and output modules use opto-isolation technologies between the PLC and the field devices. Opto-isolation segregates and protects the sensitive low-voltage PLC circuits from damage if faults occur in the field devices or signal-transmission cabling.

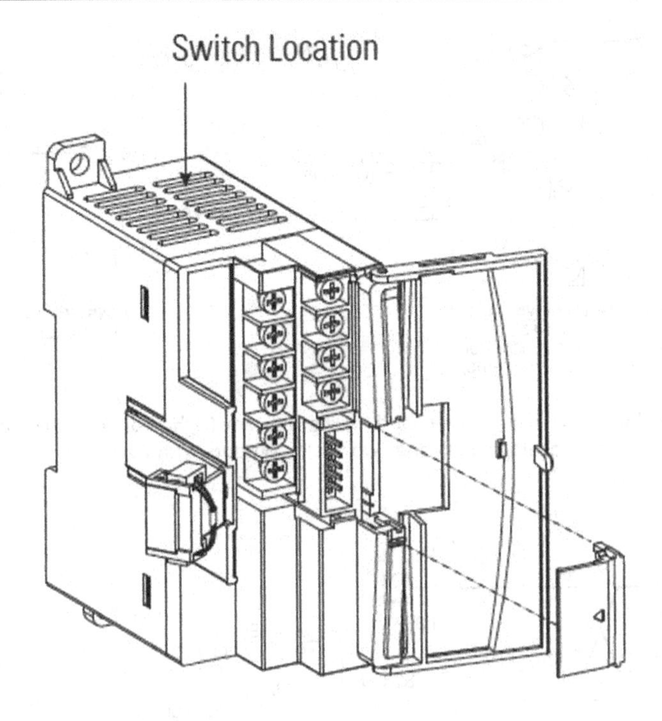

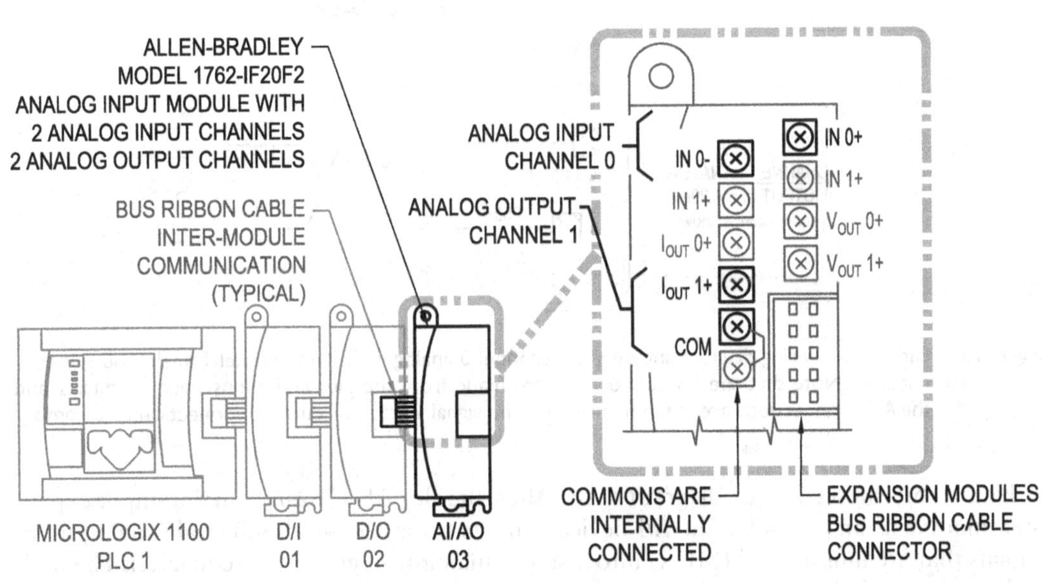

Figure 15.2 The Allen-Bradley 1762-IF2OF2 is shown on the left. This analog I/O expansion module is a four-channel device, having two inputs and two outputs. The location of the module in the rack defines its memory addresses, which are I:3.0, I:3.1, O:3.0, and O:3.1. The wiring terminals are shown at the right. The Channel-0-input device is wired between the IN 0- and IN 0+ terminals, shown in bold, regardless of whether the signal is voltage or current. The Channel-1- output device spans between the I_{OUT} 1+ and common terminals for 4–20mA, shown in bold. If the output signal were voltage, the Channel-1 connection would be made between the V_{OUT} 1+ and common terminals.

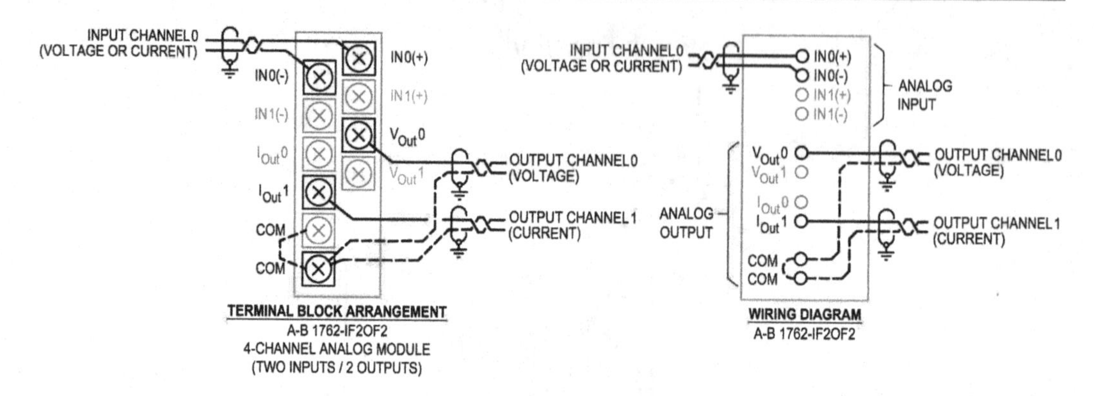

Figure 15.3 Terminal block arrangement (left) and wiring diagram (right) for four-channel analog expansion module having two analog inputs and two analog outputs with voltage or current options.

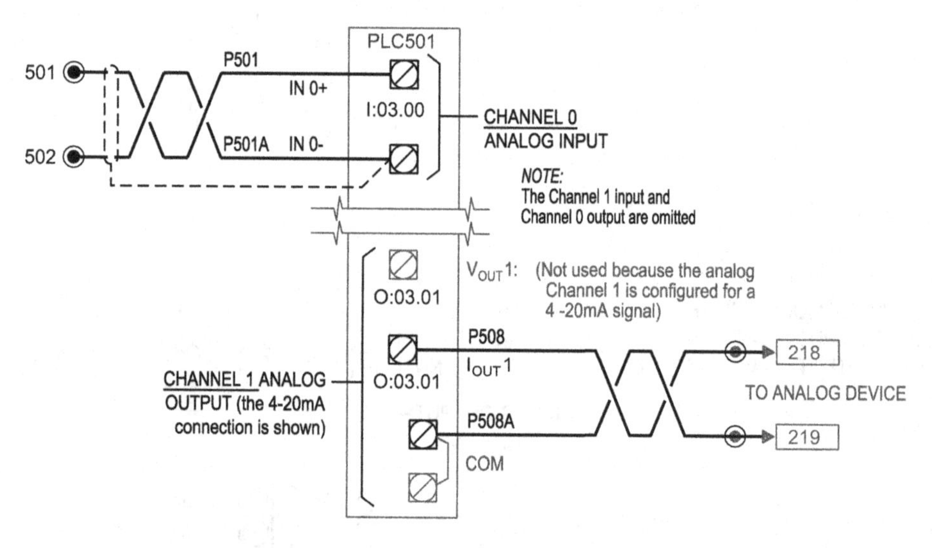

Figure 15.4 Partial PLC wiring diagram showing the channel 0 analog input and channel 1 analog output connections. Note that the A/I connections are made from the plus and minus signal terminals, and the A/O connections are made from either the signal voltage or current connection to common.

Figure 15.6 shows a block diagram of the Allen-Bradley AB 1762-*IF2* analog-input expansion module as an example. The A/I device consists of up to two differential[3] analog input signals that are multiplexed (MUX) into a single analog-to-digital (A/D) converter. The multiplexer sequentially switches through each input channel, temporarily directing the signal to the module's A/D converter. The A/D converter reads the MUX-selected signal and converts it to a digital signal in the form of a binary-number integer whose value 21 is proportional to the signal. The binary number is digital and is conducive to microprocessor machine language. The digital value of the signal is forwarded through a special-purpose Microcontroller Unit (MCU), which is an integrated-circuit chip. The MCU is a RISC-based processor (Reduced-Instruction-Set Controller) that is optimized for speed to perform very limited control functions. The MCU passes the digital value through opto-isolators to the Application-Specific Integrated Circuit (ASIC). The ASIC is a custom-designed processor, which is also optimized to perform its limited control functions. The digital signal is then passed to the PLC's data

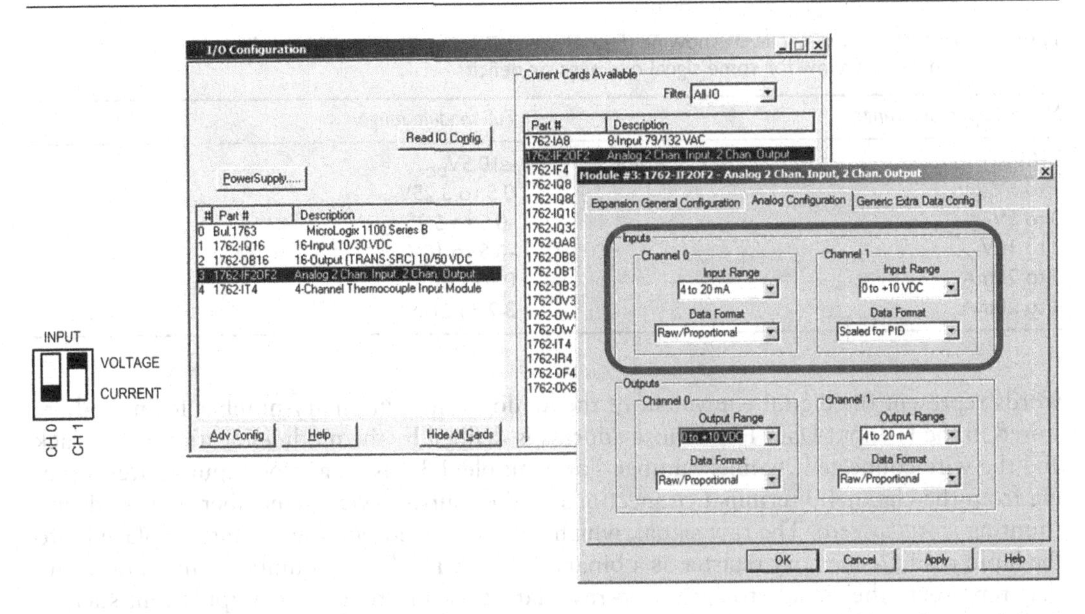

Figure 15.5 Analog input signals may be hardware-selected by small DIP switches (left) or from software-selections from an Analog Confinguration dialog box(right). Note that the voltage or current selection as an output signal requires that the wiring is made to the correct signal terminals *and* is coordinated with the coresponding software selection.

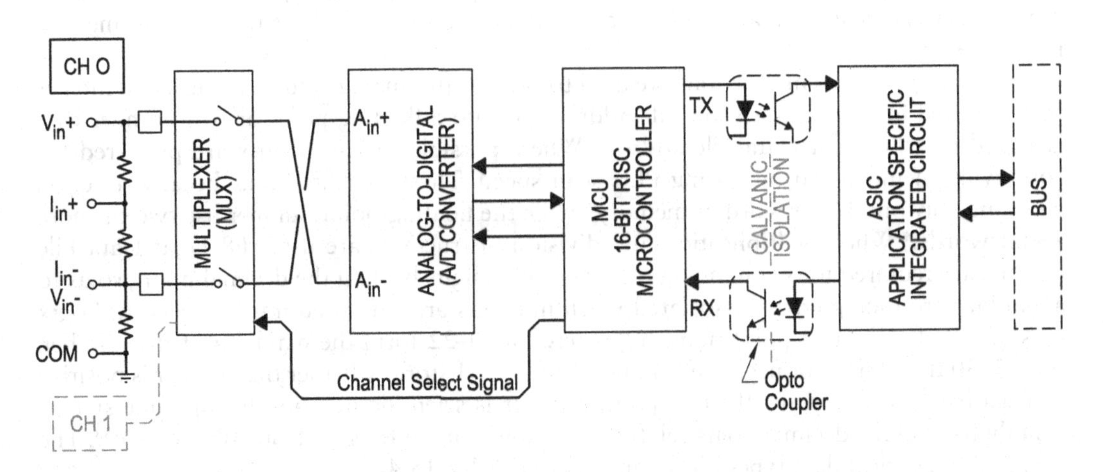

Figure 15.6 Block diagram for a two-channel analog input expansion module. Note that channel 0 is presently being poled by the multiplexer, and channel 1 will be poled during later scan cycles.

bus and routed to its memory location, where its value is stored as one or more words that are available for the RSLogix program.

USING THE PLC'S ANALOG SIGNAL

After the analog signal is processed through the A/I module, the signal is converted into one or more words of digital information corresponding to the strength of the input signal. The

Table 15.3 Analog input signal levels showing the normal signal spans and the full module ranges that allow for some signal overages or deficits

Normal operating range	*Full module range*
$\pm 10V_{DC}$	$\pm 10.5V_{DC}$
1 to $5V_{DC}$	0.5 to $5.25V_{DC}$
0 to $5V_{DC}$	-0.5 to $5.25V_{DC}$
0 to $10V_{DC}$	-0.5 to $10V_{DC}$
0 to 20mA	0 to 21mA
4 to 20mA	3.2 to 21mA

words represent the digital equivalent of the analog values in binary-number form, and are stored in the I1 Input Data File, whose address is defined by the module location in the rack and the wire-connected channel number. For example, I:3.0 is an analog input address coming from the channel 0 terminal connections of the fourth device, remembering that device counting starts at zero. The raw signal, which was converted into digital data, is placed into the input data file memory register as a binary number. The binary-number value is raw data that represents the signal strength. The raw data typically requires manipulation, such as scaling, to become meaningful as a real-world measurement parameter. From the input-data-file address, the binary-number raw data can be scaled into decimal values consistent with engineering units. The left-hand column in Table 15.3 shows normal-operating-ranges for analog input-signal, and the right-hand column shows the maximum signal span, including values above and below the normal signal range. Signal values that go beyond the normal ranges are indicative of an abnormal condition and are very useful for troubleshooting and isolating signal problems.

The RSLogix scaling instructions are used to convert the analog value into either an integer that is stored in a N7 Input Data File address, or into a floating-point or real number that is stored in a F8 Float Data File address. Where possible, integer values are preferred for conserving memory and increasing PLC scan speed. These benefits occur because integers consume only one 16-bit word of memory, while the floating-point number use two or more 16-bit words.[4] When multiplication and division instructions are used, F8 Float Data File outputs are required for maintaining the significant-digits beyond the decimal point, so as to avoid large rounding errors associated when numbers are truncated into integers. RSLogix uses 32-bits for floating-point numbers, where bits 0–22 form the mantissa of the number, bits 23–30 the positive or negative exponent, and bit 31 stores whether the number is positive or negative. Essentially, the floating-point number is scientific notation having nine significant digits, and the decimal floats relative to a power of ten ranging from 10^{-38} to 10^{+38}. The ranges of numerical data types are summarized in Table 15.4.

SCALING THE ANALOG SIGNAL

Raw analog-input data received from sensors is stored in the I1 Input Data File as a binary number. For manipulation and use, the data is usually scaled, however, in some cases the raw/proportional data is used directly. Programming data scaling instructions include:

- Raw/Proportional Data
- Engineering Units
- Scaled-for-PID
- Percent

Table 15.4 Data types, memory sizes, and numerical ranges

Data type	*Abbreviation*	*Memory bits*	*Range*
Boolean	BOOL	1	0–1
Short Integer	SINT	8	-128 to 127
Integer	INT	16	-32,768 to 32,767
Double Integer	DINT	32	-2,147,483,648 to +2,147,483,647
Real Number	REAL	32	$\pm 3.40282347 \times 10^{38}$ to $\pm 1.17549440 \pm 10^{-38}$

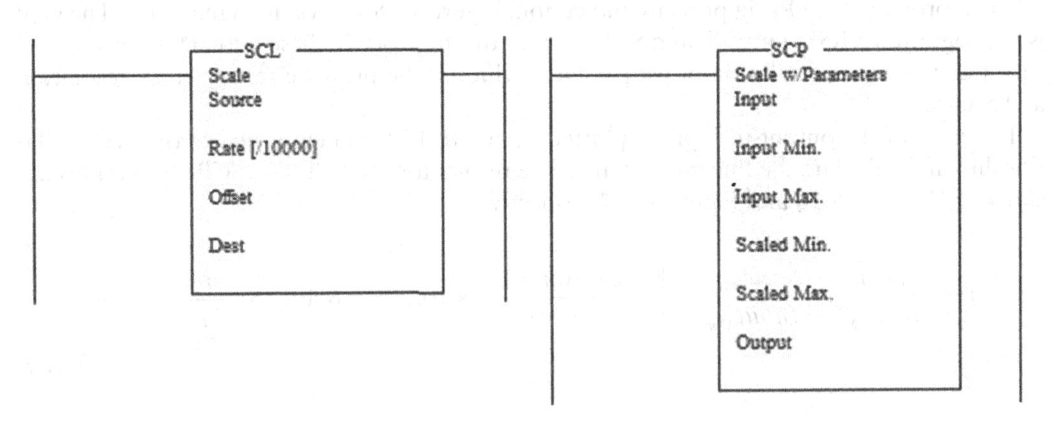

Figure 15.7 The RSLogix Scale (SCL) ladder instruction is shown on the left and Scale with Parameters (SCP) is shown on the right.

Raw analog input data from memory is scaled into engineering units using either the RSLogix Scale (SCL) or Scale-with-Parameters (SCP) instructions.[5] Both the SCL and SCP instructions are based on the equation of a straight line, $y = mx + b$, where x represents the source-input value, y the output value, m the slope or rate, and b is the *y-axis* intercept or offset above or below the $x=0$ axis. Both versions of the RSLogix ladder instructions are shown in Figure 15.7. In the SCL instruction, the rate and offset are calculated by the programmer from the relationships:

$$Rate(m) = \frac{Scaled_{Max} - Scaled_{Min}}{Input_{Max} - Input_{Min}} \quad \text{and} \quad Offset(b) = [Scaled_{Max} - Input_{Min}] \times m$$

The "Scale Source" variable in the SCL instruction is the *x value* pulled from the I1 Input-Data-File memory address. The scaled calculated result, "Dest," is the *y value* that is sent as the solution to either an N7 Integer Data File or an F8 Floating-Point Data File storage address. The "Rate" (m) and "Offset" (b) are values that are pre-calculated by the programmer and entered into the instruction. It should be noted that the "Rate" value calculated by the programmer is multiplied by 10,000 to shift the decimal point. Shifting the decimal point avoids difficult-to-read, very small fractional values having many leading zeros.

In contrast, the syntax for the SCP instruction uses "Input" and "Scaled" parameters, and the instruction, itself, calculates the rate and offset in a manner that tends to be simpler and more intuitive to the programmer. The "Input Min" and "Input-Max" values represent the smallest and largest possible digital values that can be pulled from the I:3.0 address, representing the binary-number digital values that were obtained from the original 4–20mA transmitter signal. The "Scaled" parameters represent smallest and largest user-defined engineering units, such as 0–500psig, 32–212°F, 4–20mA, 0–100%, etc. Several examples of scaling instructions follow.

PROBLEM 1: ANALOG-INPUT FOR A PRESSURE TRANSMITTER

A 4–20mA signal input is connected to the I:3.0 terminals. The output signal shall be sent to two meters on an HMI device, one to readout the 4–20mA transmission signal and the other to provide 0–100-psig pressure indication. Figure 15.8 shows the signal path. The goal is to program the RSLogix SCL and SCP instructions to produce these two HMI outputs and calculate the signal resolution in psi per bit, excluding the pressure transmitter's resolution and accuracy.

The ratio of the output-to-input is plotted in Figure 15.9 to help visualize the linear relationship and calculate the linear constants. The results for the SCL and SCP instructions are shown in Figure 15.10 and Figure 15.11 respectively.

$$Rate(m) = \frac{Scaled_{Max} - Scaled_{Min}}{Input_{Max} - Input_{Min}} = \left[\frac{21 - 0mA}{32{,}767 - 0bits}\right] \times 10{,}000 = \underline{\underline{6.408887 \frac{mA}{bit}}}$$

(4–20mA meter)

$$Peak\ Press(m) = \frac{100psig}{20mA} \times 21mA = 105psig \quad \text{(105psig corresponds to integer 32,767)}$$

$$Rate(m) = \frac{Scaled_{Max} - Scaled_{Min}}{Input_{Max} - Input_{Min}} = \left[\frac{100 - 0}{32{,}767 - 0}\right] \times 10{,}000 = \underline{\underline{30.51851}} \quad \text{(the 0–100psig meter)}$$

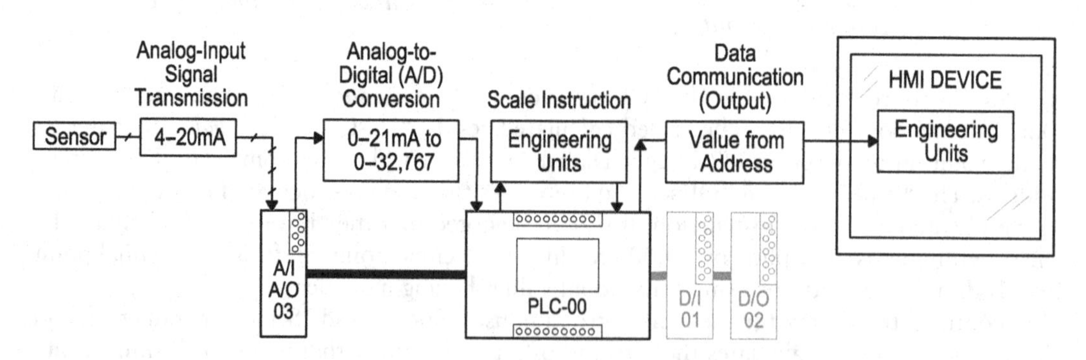

Figure 15.8 Block diagram for an analog-input signal and its output to an HMI display.

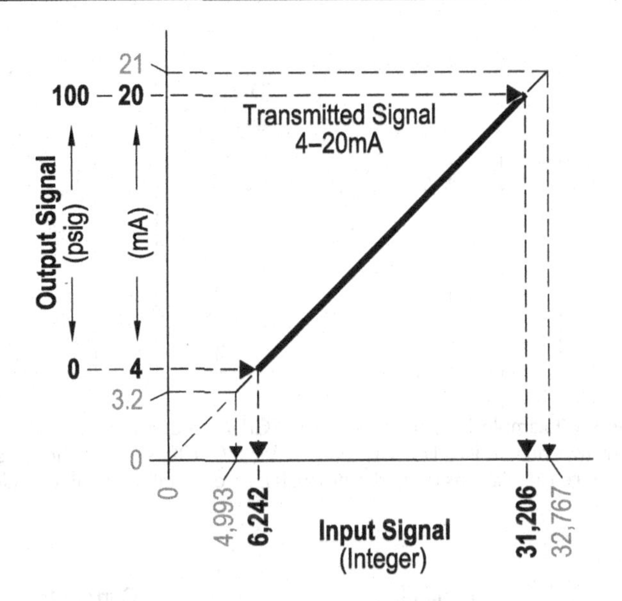

Figure 15.9 The scaled output, either in milliamps or psig, is plotted relative to its corresponding16-bit plus/minus Word, which has 32,767 increments. The analog-input values normally span the 4–20mA range. The maximum 3.2–21mA signal excursions are shown; however, the signal can go as low as zero. The SCL instruction is essentially the equation of a straight line, having a slope and y-intercept.

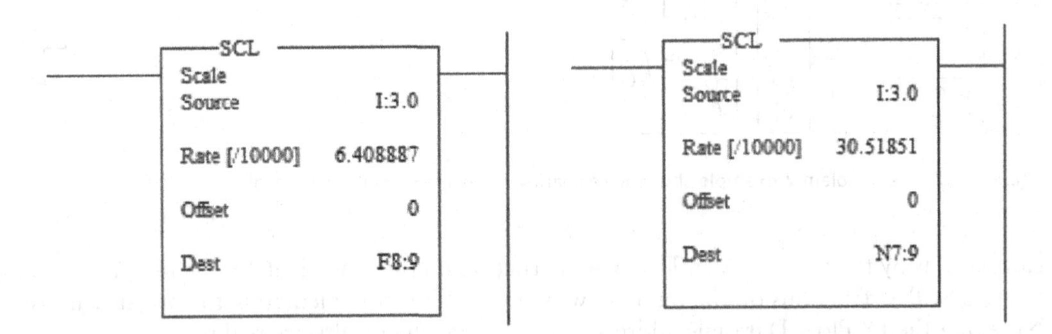

Figure 15.10 The two SCL instructions are used to manipulate the 4–20 mA signal (left) and the 0–100 psig pressure (right) into PLC memory. Note that for the fractional 4–20mA numbers, an F8 floating-point "Dest" is required to provide decimal resolution. For the 0–100psig pressure range, an N7 integer "Dest" would provide 1 psi resolution, which is acceptable for most plant operations.

$$Offset(b) = \underline{\underline{0}} \text{ for both the mA and psig meters}$$

$$Resolution = \frac{Scaled_{Max} - Scaled_{Min}}{InputSignd_{Max} - InputSignd_{Min}} = \left[\frac{100 - 0psi}{31{,}206 - 6{,}242bits}\right] = \underline{\underline{0.004\frac{psi}{bit}}}$$

Figure 15.10 shows the resulting SCL instructions for engineering units of mA on the left and psi on the right.

Figure 15.11 shows the same problem using the SCP instruction. This scaling method uses ratios of line endpoints to determine the straight-line equation and does not require

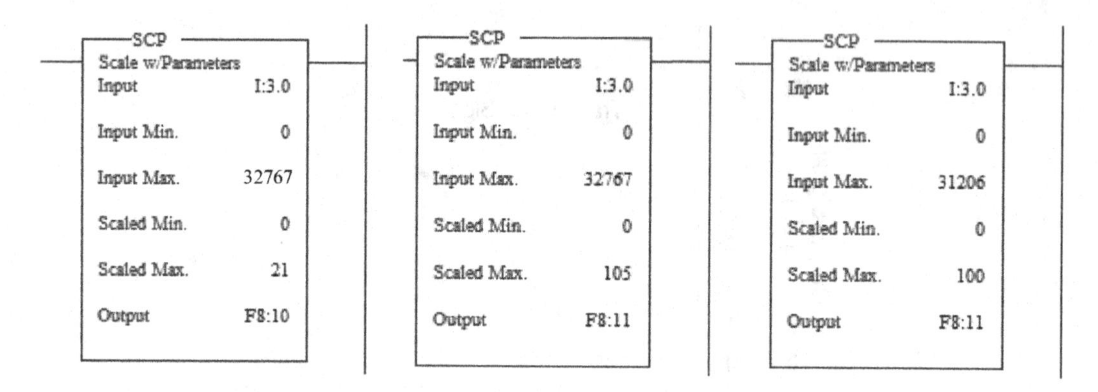

Figure 15.11 The solution to Example 1 is shown using the SCP instruction. The SCP instruction for a 4–20mA output is shown on the left. Two alternatives for a 0–100psig readout are shown on the center and right, where calculations using the normal-span or total-span values yield the same slope and intercept.

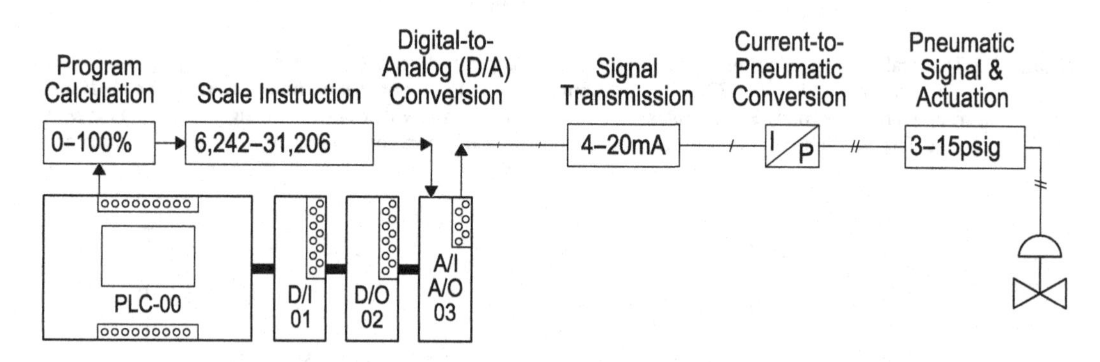

Figure 15.12 The problem 2 example showing regulating-valve analog-output control.

calculations by the programmer. The SCP instruction on the left calculates the mA signal and the two SCP instructions on the right show two variations for calculating the pressure in psi. Note that the F8 Float Data File address indicates that the results are real numbers.

PROBLEM 2: ANALOG-OUTPUT FOR A CONTROL VALVE

The result of a PLC output calculation is stored as a real number in the N7:10 address. The value represents the percent-open command to be transmitted to a control-valve positioner. The 4–20mA analog-output transmission signal is directed to a 3–15 psig I/P converter, which drives a pneumatic valve actuator. The block diagram in Figure 15.12 shows the signal path. The valve is a proportional, direct-acting type, so linear calculations are appropriate. The valve-disk excursion varies infinitely from 0% open (fully closed) to 100% open. The goal is to program the RSLogix SCL and SCP instructions to produce this output.

Figure 15.13 shows the linear plot of the valve-open percentage on the x-axis versus the digital-output signal on the y-axis. The input to the RSLogix instruction originates from a value calculated by the program and stored as a binary number in address N7:10. Both the SCL and SCP instructions serve to transform the valve-open percentage into a digital-integer value to be stored in the "O0" Output Data File, addressed here as "Dest" O:3.3. The integer

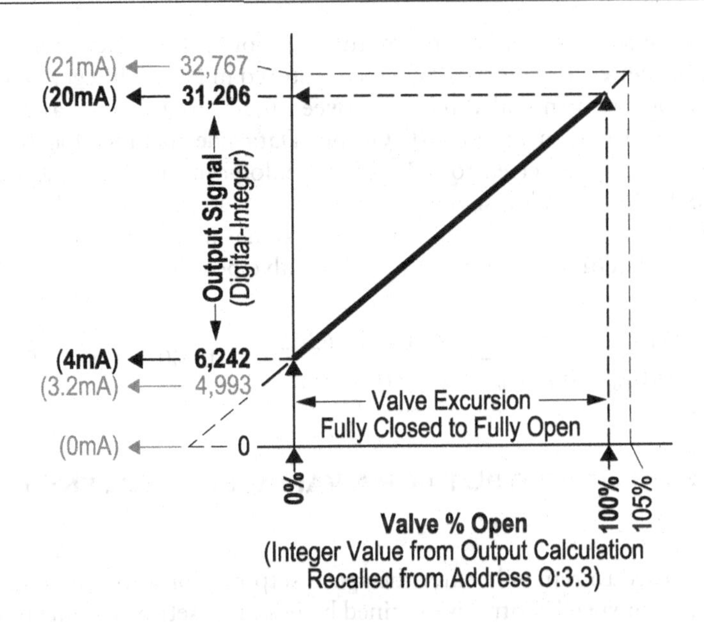

Figure 15.13 This plot shows the relationship of output to input. The digital form of the analog-output signal is shown before the PLC converts it to a 4–20mA signal, whereas the input is the 0–100% valve-open command signal that was stored in memory as an integer. Plotting these relationships is helpful in avoiding errors when calculating the SCL Rate and Offset values.

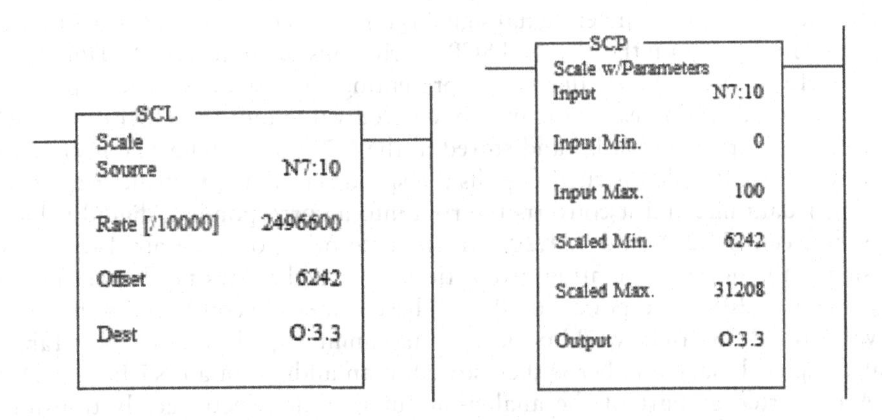

Figure 15.14 The RSLogix SCL (left) and SCP (right) instructions for setting a regulator-valve position, before the signal is converted to 4–20mA in the D/A converter. The SCP instruction is simpler.

digital signal can span between -31,208 and +31,208, but in the case of a positive-only 4–20mA signal, the normal range of proportionally scaled values is limited from 6,241 to 31,208, corresponding to 4–20mA or from 0 to 32,767, corresponding to 0–21mA. During scaling, it can be seen in Figure 15.2 that the maximum analog-output value is 21mA, which will be set to correspond to a maximum digital value of 32,767. Consequently, the normal maximum 20mA signal corresponds to 31,208. The digital-to-analog converter (D/A converter) as part of the analog-output module transforms the O0 Output-Data-File value into a 4–20mA analog control signal for transmission. The RSLogix SCL and SCP instructions are shown in Figure 15.14.

To summarize the analog-output process, the microprocessor calculates a desired valve-open setpoint value between 0 and 100%, which is stored in N7:10 Integer Data-File memory location. The setpoint is then scaled to fall between 6,242 and 31,208 as a binary-number digital value that is stored in the O:3.10 Output-Data-File memory location. The digital/integer value, in turn, is converted to a 4–20mA analog-output electronic signal by the D/A converter. The 4–20mA analog signal is transmitted over wire to an I/P converter. The I/P converter locally converts the very small electronic signal into a robust 3–15psig combined pneumatic signal and actuation pressure to set the valve positioner.

$$Rate(m) = \frac{Scaled_{Max} - Scaled_{Min}}{Input_{Max} - Input_{Min}} = \left[\frac{31208 - 6242}{100 - 0}\right] \times 10{,}000 = 2{,}496{,}600$$

PROBLEM 3: ANALOG-OUTPUT FOR A VARIABLE FREQUENCY DRIVE (VFD)

An HMI device is used to remotely enter the speed setpoint for a four-pole induction motor driving a pump. Motor speed control is obtained by infinitely setting the line frequency within a range of 25 to 60Hz[6] in a VFD. Accounting for motor slip, the pump is required to operate at any speed between 725 RPM and up to 1750 RPM when the motor reaches full load. Hydraulically, the system performance follows the pump affinity laws for speed versus flow, head, and power. The 4–20mA PLC output signal that is sent to the VFD originates from a manually entered speed setpoint entered by the user into the HMI device. The user-entered speed setpoint as a binary-number digital signal is directly sent to the PLC's N7 Integer Data File. The goal is to program the SCL and SCP instructions to control the VFD/motor speed.

Figure 15.15 shows a block diagram representing the signal path for speed control. The speed setpoint originates as a manually entered value into the HMI device, which is transmitted as computer "words" and stored in the N7 Integer Data File in the PLC. The RSLogix SCL or SCP scale instruction pulls the speed-setpoint rpm value from its address in the integer data file, and it converts the rpm into a corresponding digital value, which can vary between -31,208 and +31,208. In the case of a positive-only 4–20mA analog-output signal, the normal-operation proportionally scaled values fall between 6,242 and 31,208, where 31,208 corresponds to 20mA. These values are consistent with a maximum digital value of 32,767 that would occur at a maximum signal of 21mA (see Table 15.3). The scaled digital binary number is then stored in an address in an N7 Integer Data File. The D/A converter, as part of the analog-output module, electronically transforms the binary value of 6,242–31,208 into the 4–20mA transmission signal for VFD speed control. Figure 15.16 shows the linear input/output relationships between the 725–1750rpm speed setpoints and the 4–20mA command to be transmitted to the VFD. Figure 15.17 shows the RSLogix SCL and SCP ladder instructions.

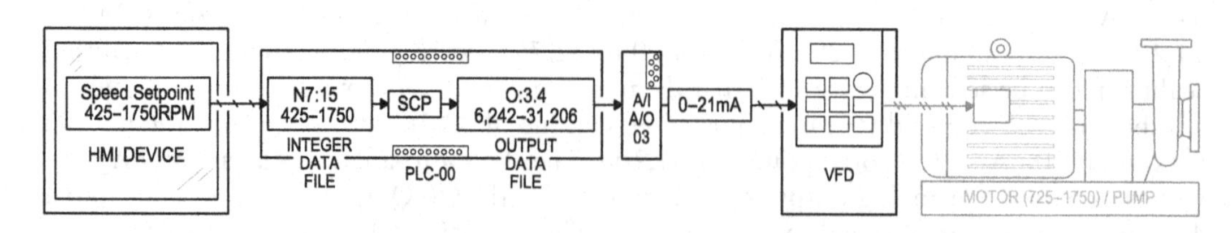

Figure 15.15 VFD analog-output example.

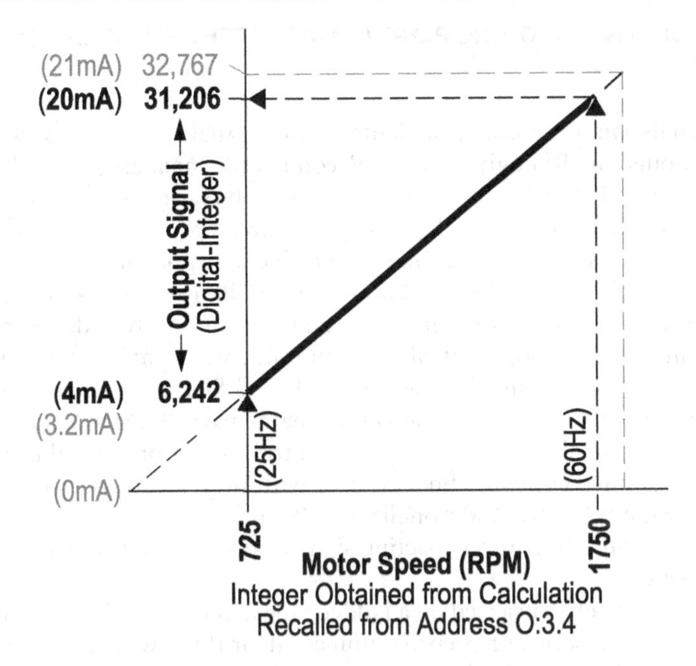

Figure 15.16 The relationship is shown between the 4–20mA control signal relative to the motor-speed- command for a VFD. The plotted relationship is useful for determining SCL and SCP parameters.

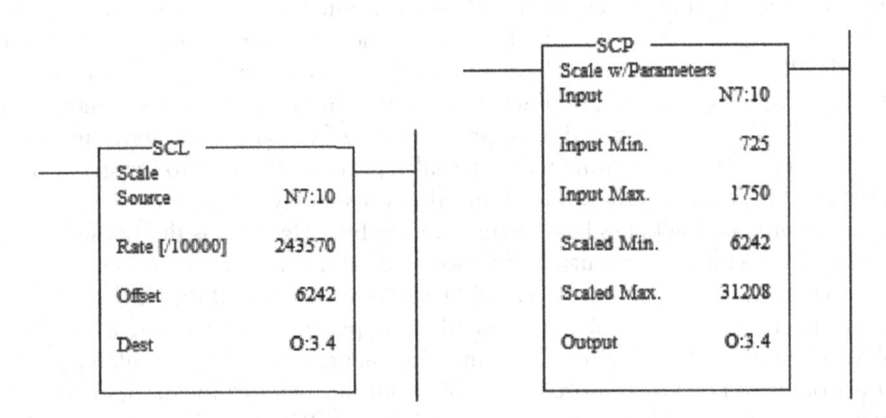

Figure 15.17 RSLogix ladder instructions are shown for VFD-speed control. The output is converted to 4–20mA by the D/A converter. The SCL instruction is on the left and the SCP instruction is on the right.

In summary, the speed setpoint is entered by the user in the HMI device. The setpoint is stored as an integer in the N7 Integer Data File memory location. The speed setpoint is scaled into a digital-output value between 6,242 and 31,208 value, which, in turn, is converted into a 4–20mA electronic analog-output transmission signal. The 4–20mA signal is used directly by the VFD to set the line frequency to the motor.

$$Rate(m) = \frac{Scaled_{Max} - Scaled_{Min}}{Input_{Max} - Input_{Min}} = \left[\frac{31208 - 6242}{1750 - 725}\right] \times 10{,}000 = \underline{\underline{243{,}570}}$$

PROGRAMMING ANALOG PROPORTIONAL-INTEGRAL-DERIVATIVE APPLICATIONS

Analog-input signals can produce logical digital outputs, analog outputs, or both. To produce logical digital outputs, the RSLogix "compare" commands, such as greater-than or less-than instructions, are useful for initiating alarms, opening valves, or starting pumps. The digital-output logic signal is initiated when the process variable (PV) exceeds a user-defined limit. Alternatively, analog-input signals can directly produce proportional analog-output signals using either the Scale (SCL) or Scale with Parameters (SCP) instructions in a relatively simple manner. Alternatively, the analog-output signal can be produced from the more sophisticated PID instruction in a closed-loop control configuration, where anticipatory responses having zero offset can be programmed. When the SCL and SCP proportional instructions are used in non-linear input-to-output relationships, mathematical instructions can be inserted to improve the final-control-device response. When truly superior control is needed, the PID instruction produces quicker-acting, finer control with little-to-no overshoot and zero control offset when properly tuned. Additionally, the PID instruction has other built-in features, such as signal deadband, direct/reverse-acting signals, auto/manual selection, and other useful controller functions.

The PID instruction itself is inserted on a ladder rung as shown on line 053 in Figure 15.18. The PID instruction is a series of preprogrammed algorithms whose variables are defined through a PID "Setup Screen" dialog box accessed from within the rung instruction. Essentially, the PID instruction behaves as a controller black box, where PID mathematical constants and real-time operating parameters are passed into the built-in subroutines that do the PID calculations. The Allen-Bradley MicroLogix 500 uses an integer-only PID program instruction, in lieu of floating-point numbers available in some other programs. Consequently, the integer version of the PID instruction generally requires additional scaling steps to change real numbers into integers, such as shown on lines 052 and 054 in Figure 15.18. The more-advanced Allen-Bradley PLC-5 and ControlLogix programming languages use more-intuitive, less-complex floating-point instructions that internally produce the real-to-integer conversions. The PID algorithms are pre-programmed into the software by the manufacturer; and consequently, some software packages have better, more robust algorithms than others.

Setpoint (input), control (output), and process variables (measurement/feedback) are the main variables used in PID control. The PID SetPoint variable (SP) can be a fixed value entered into memory, or it can be a calculated value placed into memory, while the output Control Variable (CV), which effects system adjustment, is the final calculated result from the PID instruction. The CV is stored in its own output memory location, exclusive of the PID-instruction variables. The feedback Process Variable (PV) is obtained from a scaled value originating from an "I1" Input Data File memory location, which contains the measured control parameter.

Figure 15.18 shows how the PID instruction fits into the RSLogix ladder program. A transducer's analog signal is received into an A/D converter where it becomes a digital value. The digital value is then routed into a PLC Input Data File address where it is available for program usage. That value in the Input-Data-File is scaled into engineering units and then stored into another address, where it is pulled into the PID instruction as its measurement/feedback Process Variable (PV). The PID instruction has a sizable quantity of function-related bit-based variables and word-based variables that are used in executing the PID algorithm, defining PID-instruction status, or triggering events. The PID algorithm computes the output Control Variable (CV), whose value ultimately produces the control signal and follow-up corrective actions. Using SCL or SCP, the Control Value is converted to a digital value representing the output transmission signal. For example, a digital value between 6,242 and

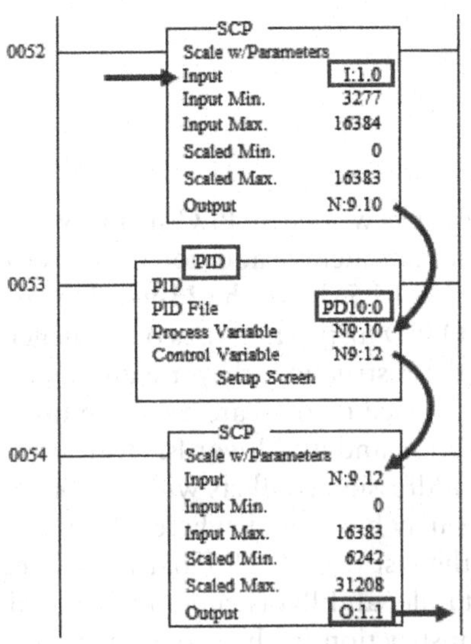

A transducer is connected to an A/D converter to produce a digital signal. The digital signal is stored in an Input Data File (I:1.0).

The I:1.0 input value is scaled into engineering units by the SCP instruction, which is then sent into a memory address (N9:10).

The N9:10 scaled-input value is passed into the PID instruction as its Process Variable (PV) by linking the N9:10 memory address.

PID-instruction functional variables are accessed through the "Setup Screen" and those values are stored in a user-defined Data File address (PD10:0). These variables are used in the PID calculations.

The N9:12 Control Variable (CV) address receives the PID solution in engineering units. The Control Variable is the PID-instruction result.

The N9:12 control variable is scaled into a digital value, which is directed to an Output Data File (O:1.1) address. This value will become the modulation signal for the final control device (e.g. 4-20mA).

The output value (O:1.1) is sent to the D/A converter, where the 4-20mA transmission signal is produced.

Figure 15.18 The signal-processing ladder logic for the PID instruction consists of the input scan and scaling function, the PID algorithm and control-signal production, and the output scaling function.

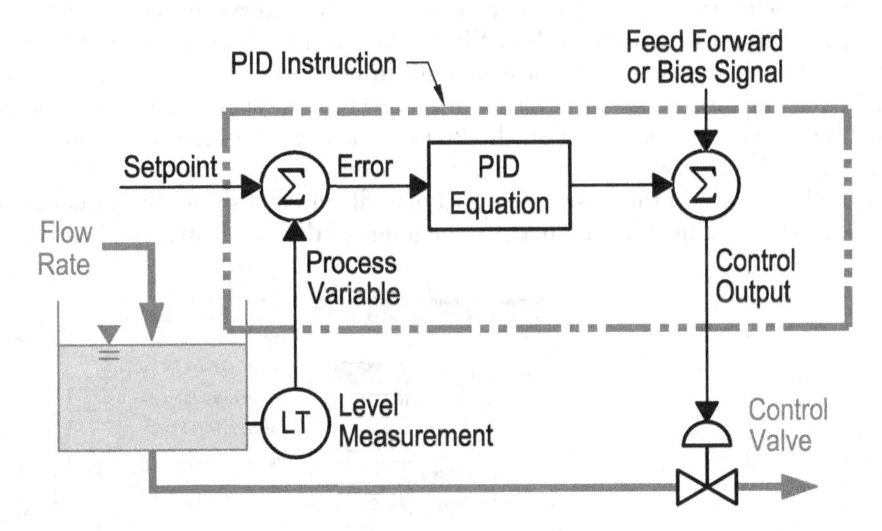

Figure 15.19 The PID closed-loop control algorithm keeps the process variable (PV) at a desired setpoint (SP) by modulating the control variable (CV). The PID equation is built into the PID instruction and uses variables obtained from memory addresses.

31,208 corresponds to an analog-output signal between 4–20mA. The PLC output address, such as O:1.1, corresponds to the device wiring connections on the D/A converter from where the digital value is electronically converted to a 4–20mA transmission signal.

$$Output = K \cdot \left[e + \frac{1}{T_i} \int e(t)dt + T_d \cdot \frac{d}{dt} e(t) \right] + FeedForward + Bias$$

Where e or e(t) = error or error as a function of time
K = Controller Gain
T_i= Integral Time Constant (minutes per repeat)
T_d= Derivative Time Constant (minutes)

The RSLogix 500 PID instruction consumes 23 *consecutive* words of PLC memory (to accomplish the functions shown in Figure 15.19), which does not include additional words to store the input data, output data, and scaling parameters. Table 15.5 shows the 23-word memory allocation used by the PID-instruction. The RSLogix 500 PID instruction uses only integer data types, and if any floating-point values are input into the instruction, they are automatically rounded to the nearest integer. Good programming practice is to create a unique user-defined Integer Data File dedicated solely to one PID instruction and its 23 words. Dedicating each PID instruction with its own Data File cleanly avoids addressing conflicts with any other data-file words or bits. As examples, N9 or PD10 Data Files may be created, where "9" would be the first available extended *user-defined* data file after the last default F8 Float Data File, and PD*10* represents the second available *user-defined* data file, and PD is chosen to remind programmers of the Data File purpose. Each PID ladder instruction in the program should have its own dedicated user-defined integer Data File, which is reserved by the programmer, and by doing so, any unintentional possibilities for conflicting addresses is easily avoided by exclusively associating each PID instruction with its own Data File.

PID programming starts by inserting the PID instruction onto the ladder rung. The PID instruction can be on its own rung so that the instruction is always executed, or it can have preceding permissive instructions, such as XIO or XIC, to limit its execution. If the program speed needs to be increased, the PID instruction can be bypassed during many scans, especially in slow-acting processes. The "PID Setup" dialog box shown in Figure 15.20 is accessed from the "Setup Screen" option within the PID instruction. The PID operational values as shown in Figure 15.21 are accessed in turn by viewing the associated user-defined Integer Data File, such as PD10 in this case. The PD10 data file directly shows the values contained in the 23 words/bits. Table 15.5 summarizes the usage of the PID-instruction's 23 words.

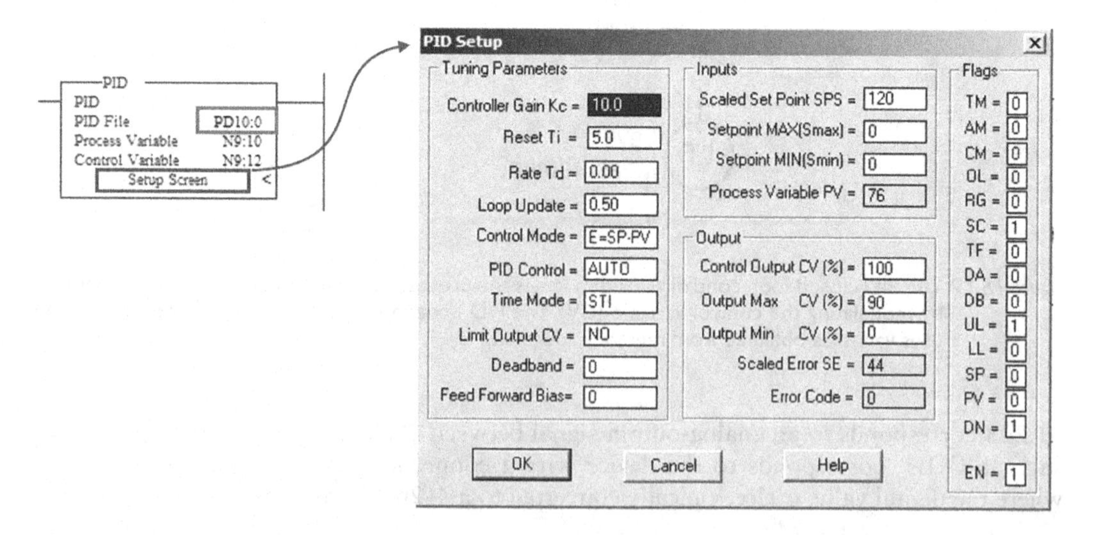

Figure 15.20 The PID ladder instruction (left) provides the PV-input and CV-output values to and from the PID algorithm. The algorithm variables on the right are accessed by clicking the "Setup Screen" text on the instruction symbol. All PID variables are stored in 23 words contained in the corresponding PD10 Data File (see Figure 15.21).

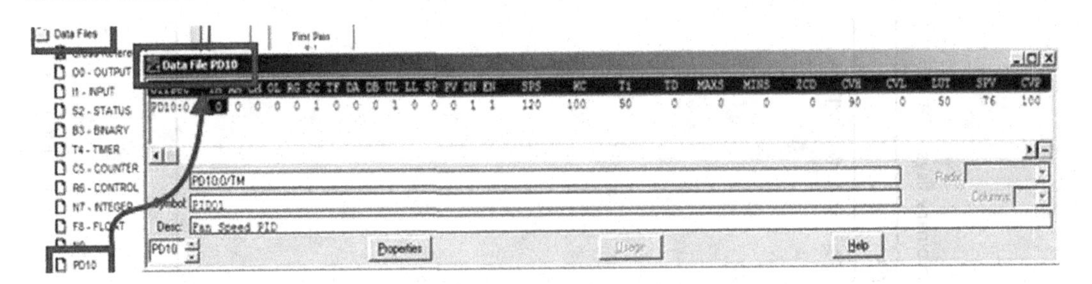

Figure 15.21 The PID Data File can be opened to show its values, which correspond to the 23 word locations shown in Table 15.5.

The PID "Setup Screen" dialog box shown in Figure 15.20 is organized into four main blocks; Inputs, Outputs, Tuning Parameters, and Flags. The following discussion briefly describes the PID variables stored as the 23 "words" listed in Table 15.5, as associated with the "PID Setup" dialog box shown in Figure 15.20. For discussion purposes, the variables are stored in the user-defined PD10:0 Integer Data File, where Figure 15.21 shows its interface screen. The following descriptions of the PID variables is referenced to Table 15.5.

- *Process Variable (PV)*: is the measured value used for feedback in the closed-loop control. The PV originates from a transducer measurement. The measurement is stored in an analog-input memory location, and subsequently converted into a 14-bit word to become the PV value. The PV memory-address value that is pulled into the PID ladder instruction. The 14-bit integer PV value comes from a 16-bit analog-input standalone "word" that is not part of the PID Setup's 23 words. The PV integer value ranges from 0 to 16,383 (2^{14}) and requires scaling to translate it from its original 16-bit raw-data source. The 16-bit raw-data value can come directly from the I1 Input Data File or from scaled engineering units, but in either case, the value needs to be translated into a 14-bit word by the SCL or SCP instructions. The 14-bit value is subsequently stored as the PV variable in its own dedicated Integer Data File address that is not part of the 23 PID words, such as its own "word" within the N7 Integer Data File, as an example. It should be stressed that the PV process value should not be confused with the "Scaled Process Variable" in *Word 14* location, which is PID instruction address PD10.14 in Table 15.5 of this example, and is a value used only for monitoring.
- *Setpoint (SP)*: is the desired output level in engineering units as stored in the *Word 2* location of the PD10 Data File. The SCL or SCP instructions are used to convert the raw data input to a 0-to-16,383 14-bit integer value from input units. For this example, the 14-bit setpoint word is stored in address PD10.2.
- *Control Variable (CV)*: is the output value. The CV is a 14-bit memory location selected by the programmer and containing a calculated output integer value (0 to 16,383). The output address is independent from the PID-Setup instruction. Note that the CV can be scaled to any units, but scaling may be as simple as multiplying by 2 if a 0 to 32,767 bits digital signal is applied directly to an analog-output device.
- *Controller Gain (K_c)*: is the proportional gain constant and is stored in *Word 3*. The K_c range of values depends on the *Word 0* RG flag reset-and-gain enhancement bit. When RG = 0, K_c varies from 0 to 3,276.7, and when RG = 1, K_c varies from 0 to 327.67. During tuning, K_c should be set to half the value that causes instability when the reset (T_i) and rate (T_d) terms are set to zero, resulting in proportional-only control. Gain is the inverse of the proportional band, which is the level variation where proportional control takes place. Beyond the proportional band the output is all or nothing. As an example, if a tank level is to maintain control within 5% of the level measurement span,

Table 15.5 The PID control block structure consists of 23 16-bit words programmed as an integer data file. Word 0 contains the two-state binary flags

15	*14*	*13*	*12*	*11*	*10*	*09*	*08*	*07*	*06*	*05*	*04*	*03*	*02*	*00*	
Word 0	EN	RA(1)	DN	PV	LL	UL	DB	DA(2)	TF	SC	RG(2)	OL(3)	CM(3)	AM(3)	TM(3)
Word 1	PID Sub Error Code (MSbyte)														
Word 2	Setpoint SP														
Word 3	Gain K_c														
Word 4	Reset T_i														
Word 5	Rate T_d														
Word 6	Feed Forward/Bias														
Word 7	Setpoint Max (SMax)														
Word 8	Setpoint Min (SMin)														
Word 9	Deadband														
Word 10	Internet Use Do Not Change														
Word 11	Output Max														
Word 12	Output Min														
Word 13	Loop Update														
Word 14	Scaled Process Variable														
Word 15	Scaled Error SE														
Word 16	Output CV% (0 to 100%)														
Word 17	MSW Integral Sun														
Word 18	LSW Integral Sun														
Word 19	Internet Use Do Not Change														
Word 20	Internet Use Do Not Change														
Word 21	Internet Use Do Not Change														
Word 22	Internet Use Do Not Change														

Table 15.6 Ranges and units of RSLogix 500 gain constants

Term	*Range*	*Reference*
Controller Gain (K_c)	0.1 to 25.5 (dimensionless)	Proportional
Reset Term (T_i)	25.5 to 0.1 (minutes per repeat)	Integral
Rate Term (T_d)	0.1 to 25.5 (minutes)	Derivative

it would have a proportional band of 5% of the 100% level measurement and the gain would be 1/0.05 = 20. Table 15.6 shows the units and ranges for the proportional-, integral-, and derivative-gain constants, which are adjusted during controller tuning.

- *Reset (T_i)*: is an integral-control variable stored in *Word 4*. Its value is the integral time constant having units of minutes per repeat. Integral gain T_i ranges from 0 to 3,276.7 when flag RG = 0, and from 0 to 327.67 when flag RG = 1. The Integral Gain (T_i) would be set to the period of controller-output oscillation occurring at critical instability, which is determined while obtaining the controller proportional gain K_c during the controller tuning process.
- *Rate (T_d)*: is a 16-bit derivative term stored in *Word 5*. Its value is the derivative time constant in minutes. The Derivative Gain T_d always varies from 0 to 327.67 and should be set to 1/8 of the integral gain (T_i) when doing proportional-plus-integral-plus-derivative control. Being a typically small value, the derivative gain is not affected by the RG flag, unlike the proportional and integral gains.
- *Feed Forward/Bias*: are signals that are summed to the PID Control-Variable (CV) output. The purpose of the feed-forward and bias signals is to shift the CV output. The CV integer is stored in *Word* 6 as a value between -16,383 and +16,383. The feed-forward signal may be automatically or manually generated. An example of automatic feed-forward occurs in the case of a two-element combustion-control system where immediate steam flowrate changes (disturbances) are used to immediately anticipate firing-rate changes before the steam pressure (the primary element) begins to drop. An example of manual biasing of the control signal is adjustment of the air-fuel ratio in a combustion control system, made to eliminate boiler smoking.
- *Timed Mode (TM)*: is a single-bit flag associated with the "Loop Update" word. The TM flag is stored in Word 0, Bit 0. TM is a toggle- between the timed and the sequenced-timed-interrupt modes (STI mode). When the TM bit is set to 1, the PID instruction is in timed mode, and when it is set to 0, the PID instruction is in STI mode.

In timed mode, the PID executes at the rate specified in the "Loop Update" parameter at *Word 13*, i.e., the PID instruction is ignored during each scan until the "Loop Update" time setting is reached. In timed mode, the processor scan should be at least ten times faster than the PID loop update, to avoid controller-induced disturbances and hunting.

When the PID block is in STI mode, the PID executes and updates the CV control variable each time the PID instruction is scanned from within the STI interrupt subroutine. Essentially, the main program runs continuously until the STI time is reached, at which point the main program is interrupted to run the PID routine. To work, the STI routine time interval needs to match the same value in the "Loop Update" parameter. The STI time interval is set within the Word S:30 located in the Status Data File, which is the value matched in the "Loop-Update" *Word 13* address.

Loop Update is the time interval between PID calculations and is stored in *Word 13*. The Loop Update is associated with the Timed Mode bit (TM) and defines the timed mode and STI-mode delay periods. The loop update is the time interval between PID calculations. The loop update time is in units of hundredths of a second, and it needs to be five-to-ten-times faster than the natural period of the controlled parameter. The natural period is obtained during PLC tuning to be the critical instability. The natural period is obtained as the first step in tuning where the reset and rate terms are eliminated, and the gain is increased until hunting just begins. The period of steady output oscillations is the loop-update value and its valid range is between 0.10 and 10.24 seconds.

Deadband is stored in *Word 9* and for the PID instruction, deadband represents a span above and below the setpoint when the controller takes no action. In essence, PID instruction deadband is analogous to differential in an automatic control switch, where the differential is the span of controller inactivity between cut-in and cut-out. The deadband feature commences only as the PV process variable passes through the setpoint. Usually the PID instruction deadband is set to zero, but like differential in a control switch, providing a deadband setting can sometimes be useful in avoiding short cycling.

Scaled Error (**SE**) is stored in *Word 15* and is the difference between the process variable (PV) and the setpoint (SP). The error is calculated as *E=SP-PV* or *E=PV-SP*, depending whether the controller is to be direct-acting or reverse-acting. When a calibrated signal source is applied, the scaled error may be used to adjust calibration.

Auto/Manual (**AM**) is a single bit stored in in the Word-0/Bit-1 location. This toggle is useful when switching from automatic to hand override. When AM is off (0), the program PID loop is toggled into automatic mode and the PID instruction sets the control variable (CV). When AM is on (1) the PID instruction is in manual operation, so that the PID block is essentially bypassed and the control variable (CV) must be set directly by the user. During tuning, the CV bit is set to manual.

Control Mode (**CM**) is a single bit sored in the Word-0/Bit-1 location and is used to set whether the PID instruction will be direct-acting or reverse-acting. Essentially, the CM bit changes the error calculation, so that when CM=0, *E=SP-PV* is used for a reverse-acting calculation and when CM=1, *E=PV-SP* is used for a direct-acting calculation.

Reset and Gain Enhancement Bit (**RG**) is a single bit stored in the Word-0/Bit-4 location and was discussed previously regarding the controller gain (K_c) and reset term (T_i). When the RG-bit is set (1), a multiplier of 0.01 is applied to both the controller gain and the reset term. When the RG-bit is clear (0), a multiplier of 0.1 is applied to both the controller gain and the reset term.

Setpoint Scale (**SC**) is a single read-only bit stored in the Word-0/Bit-5 location. SC is set to 0 when setpoint scaling values are specified and set to 1 when they are not.

Loop Update Time Too Fast (**TF**) is a single read-only bit stored in the Word-0/Bit-6 location. The TF bit is set by the PID algorithm and is normally 0. However, if the loop update time cannot be achieved by the controller due to scan-time limitations, TF is then set by the PID-instruction to 1. A set TF bit indicates a problem, which is corrected by slowing the PID update rate or by moving the PID instruction to an STI interrupt routine. The reset and rate gains values will be incorrect when this bit is set.

Derivative Action (**DA**) is a single read-only bit stored in the Word-0/Bit-7 location. When set (state 1), the DA-bit causes the derivative action (rate) calculation to be evaluated on the error instead of the process variable (PV). When cleared (state 0), this bit causes the derivative calculation to be based on the process variable.

Output Alarm Upper Limit (**UL**) and **Output Alarm Lower Limit** (**LL**) are limit bits stored in the Word-0/Bit-9 for the upper-limit and *Word 0/Bit 10* lower-limit addresses. When operating within normal ranges, these flags are clear. The flag is automatically toggled to state 1 when its calculated values of control variable (CV) falls above or below the normal range, indicating an error.

Setpoint Out of Range (**SP**) is a bit stored in the *Word-0/Bit-11* location. The SP bit is normally clear, but becomes set when either the maximum or minimum scaled value is exceeded, indicating an error.

Process Out of Range (PV) is a bit stored in the *Word-0/Bit-12* location. The PV bit becomes set when the unscaled digital-input process variable falls below zero or exceeds 16,383, indicating an error.

PID Done (DN) is a bit stored in the *Word-0/Bit-13* location that becomes set for one scan only when the PID algorithm computation is completed. It resets automatically after the scan is completed. This flag can be used to trigger an action when the PID instruction is done.

PID Rational Approximation Bit (RA) is a bit stored in the *Word-0/Bit-14* location. When this bit is set, the PID instruction uses a rational approximation method for its computation, resulting in a more accurate result, whether the calculation is user-defined or internal to the PID instruction.

PID Enable (EN) is a bit stored in the *Word-0/Bit-14* location. This bit becomes set when the PID rung is true and the PID instruction is initiated.

Integral Sum is a long-word or two-word value of 32 bits stored in the *Word 17* and *Word 18* locations. The integral sum is the result of the error integration over time $\frac{1}{T_i}\int E \cdot dt$.

SUBROUTINES: LADDER LOGIC RELATED TO COMPUTER PROGRAMMING

Although the PLC program looks like a motor-controller ladder-logic diagram, it is very much a computer program. As in all programming languages, subroutines, functions, or procedures can be user-defined to streamline computer code. The routines can be called only when needed and variables can be passed into and from the routine, so that the routine behaves as a black-box function producing a desired result. Common-usage subroutines can be shared by sections of code, when the programming has the same functions, resulting in shorter, easier to read, more efficient coding that runs faster.

Figure 15.22 shows three user-defined subroutines as ladder diagrams under the "Project\Program Files" menu in the left window. The subroutines are called by the main program using JSR "Jump to Subroutine" instructions in the ladder program in the right window. In this case, when the main program executes, the positions of mode-selection switches are examined using the XIO and XIC instructions. When its permissives are true, the corresponding subroutine is called to execute. In this example, the main program uses three switch positions to call the subroutine for "traffic," "flashing," or "lamp-test" mode. The subroutine's ladder-diagram code is contained within its individual functions, for each of the three modes.

Like classic computer program subroutines, the RSLogix subroutines can be labeled and called within the ladder, and then when the subroutine has reached a conclusion, program control can be sent back to the main program via a "Return" (RET) instruction.

SUMMARY

In summary, analog inputs and outputs require continuous measurements of time-varying parameters to accomplish modulating control. An electrical-signal measurement in a sensor is converted to either a voltage or current signal, such as 4–20mA, for transmission to the PLC analog-input module. During the input scan, the analog-input module examines each input signal one at a time via its multiplexer, converts the electronic signal into a binary number in an analog-to- digital converter, and then passes the digital form of the input value into

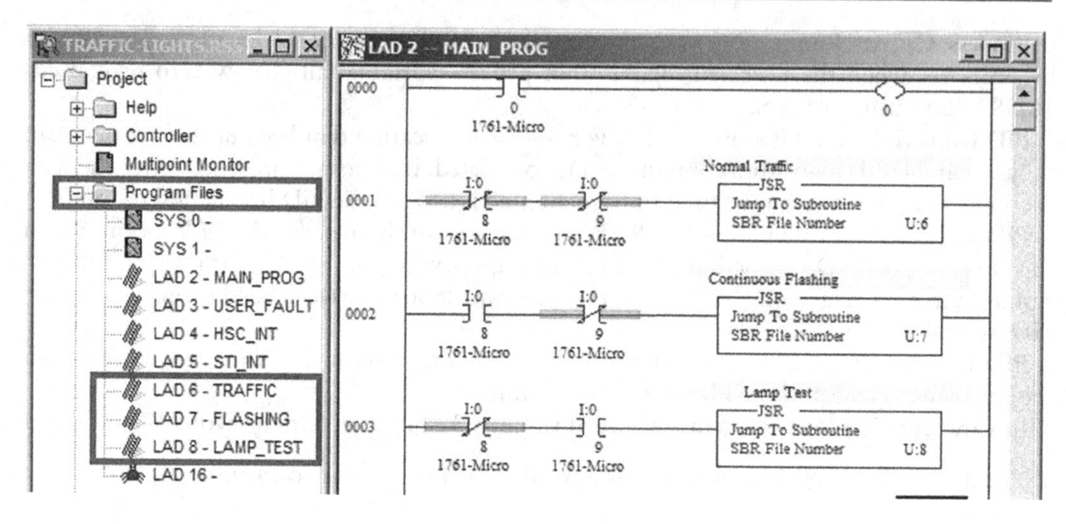

Figure 15.22 Subroutines are "Program Files" that are called by the active program using the "Jump to Subroutine" instruction (JSR).

memory. During the program-execution scan, the input values are often scaled to engineering units and saved to a different memory location. The analog-input values are pulled from memory by mathematical functions, which post their results in an output memory location. The output values are scaled into a digital-binary number and stored in output memory. During the update-outputs scan, the signal is directed to a digital-to-analog output device, where it is converted into an analog-output electronic voltage or current signal, such as 4–20mA. Often the analog output is converted later into a pneumatic signal via an I/P converter to produce high-actuation forces for mechanical devices, or it may be used directly by meters or VFD controllers.

The ladder-diagram programming language has built-in functions for producing analog outputs, including sophisticated PID instructions. The compare instructions, such as the greater or less than functions, can be used to trigger digital outputs, such as pump start or stop commands or alarms. Like any computer language, PLCs can make use of subroutines that can be called by a main program to streamline repetitive code and to speed the program operation.

QUESTIONS

1. Create a block diagram for an analog-input signal showing the steps from measuring a parameter to its storage in PLC memory. Briefly describe what happens during each step of the process.
2. Create a block diagram for the PLC manipulation of the analog signal to produce an output signal. Briefly describe what happens during each step of the process.
3. Create a block diagram for an analog-output signal showing the steps from reading the output signal from PLC memory to actuating a final device. Briefly describe what happens during each step of the process.
4. List some typical analog I/O signal types, including their nominal values and their full-range values.
5. How does a PLC I/O device manipulate an amperage input signal compared to an input voltage signal?

6. Describe in general terms what the PLC does with analog-input signals. What are some programming instructions used for input-data manipulation for determination of an output signal?
7. What is a multiplexer (MUX)? An analog-to-digital converter (A/D)? An MCU RISC-based microcontroller? An opto-coupler? And an application-specific integrated circuit (ASIC)?
8. What are the data types and ranges associated with analog instructions? What is the advantage of using an integer-data type, when possible, over a real-number data type?
9. Provide a brief overview how a PID-output control signal can be produced in a PLC.
10. What are program subroutines or functions? Describe the usefulness of programming using subroutines.

NOTES

1. Note the distinction between digital theory, where yes-and-no logic is used to start-and-stop devices, versus digital signals and digital data used with analog control. In analog systems, the term digital applies to continuously varying analog signals being converted into a "digital" or binary number consisting of a string of 0s and 1s. Binary numbers are conducive to manipulation by microprocessors.
2. In the case of a ±10VDC signal being converted to an integer spanning between -32,767 to +32,767, minus 10V would correspond to minus 32,767 and plus 10V to plus 32,767, for instance. The resolution would be [65,536 ÷ 20V] = 3,276 bits per volt. For 4–20mA positive-only signals, the usable integer span is limited between 0 and 32,767 for 0–20mA or 6,553-32,767 for 4–20mA. The 4–20mA signal resolution would be [(32,767 – 6,553) ÷ (20 – 4 mA)] = 1,638 bits per mA.
3. Single-ended versus differential signals: A differential analog signal uses two signal conductors that are isolated from all other signal conductors and are twisted for noise reduction. A single-ended analog signal uses one signal carrying conductor from a transmitter, and the return circuit is completed to a common analog ground connection that is shared between multiple sensors. The switches that are used for digital-inputs provide a good analogy for single-ended signals, where the voltage that is supplied to the switches are all returned to a common ground-potential connection.
4. As an example, the MicroLogix 1100 controller support 8K of memory, which is used for both program and data. The data memory is limited to less than 4K. This model of PLC supports up to 64K of battery-backed up memory for data logging. Program scan time is affected by the amount of used memory. For other PLC models, 4-byte/32-bit words may be standard and the word size for memory allocation is double.
5. The SCL instruction is found under the "File/Misc" tab and the SCP instruction is found under the "Advanced Math" tab within the RSLogix software.
6. Motor speed (n) is related to frequency (f), and number of poles (p) by the following relationship $f = \frac{pn}{120}$: or in terms of speed, $n = \frac{120 \cdot f}{p}$.

Chapter 16

IEC 61131-3 PLC programming languages (LD, FBD, SFC, ST, and IF)

Although ladder logic is the most common programming language in the United States for industrial applications, there are other programming languages that are used for commercial and industrial systems. Today, many manufacturers use programming languages that comply with Part 3 of the International Electrotechnical Commission standard IEC 61131. This standard is an attempt to create uniformity between PLC-based programming languages in terms of program architecture, syntax, and semantics, regardless of the manufacturer. This standard helps in providing various amounts of portability between control systems furnished by different PLC manufacturers, and programming is simplified by using automatic equipment that follows this standard, allowing system integrators to obtain debugged PLC code directly from a manufacturer when using their equipment. The standard also allows multiple PLC languages to be used within a single PLC.

IEC 61131-3 is organized into nine parts:

Part 1: General Overview
Part 2: Hardware
Part 3: Programming Languages
Part 4: User Guidelines
Part 5: Communication
Part 6: Functional Safety
Part 7: Fuzzy-Logic Control
Part 8: Guidelines for the Implementation of Languages
Part 9: Digital Communication Interface for Small Sensors and Actuators (IO Link)

The five IEC 61131-3 languages covered in the following subchapters include:

- *Graphical Languages*: Ladder Diagram (LD), Function Block Diagram (FBD), Sequential Function Chart (SFC)
- *Textual Languages*: Structured Text (ST), Instruction List (IF).

Figure 16.1 shows simplified examples using LD, FBD., IL, and ST syntaxes for a motor starter. There are some advantages and disadvantages to each language, which tend to make some languages more advantageous than others depending on the task. Generalizing, many U.S. engineers prefer Ladder Diagrams for industrial applications such as factories or plants, while in Europe the preference was originally for Instruction List or Structured Text, but has shifted to Function Block Diagrams, especially for manufacturing facilities. Engineers and scientists generally have no problem using Structured Text, but non-engineers might find Ladder Diagram programming challenging. Some languages are better suited for accomplishing particular functions than others. For example, memory management routines or complex

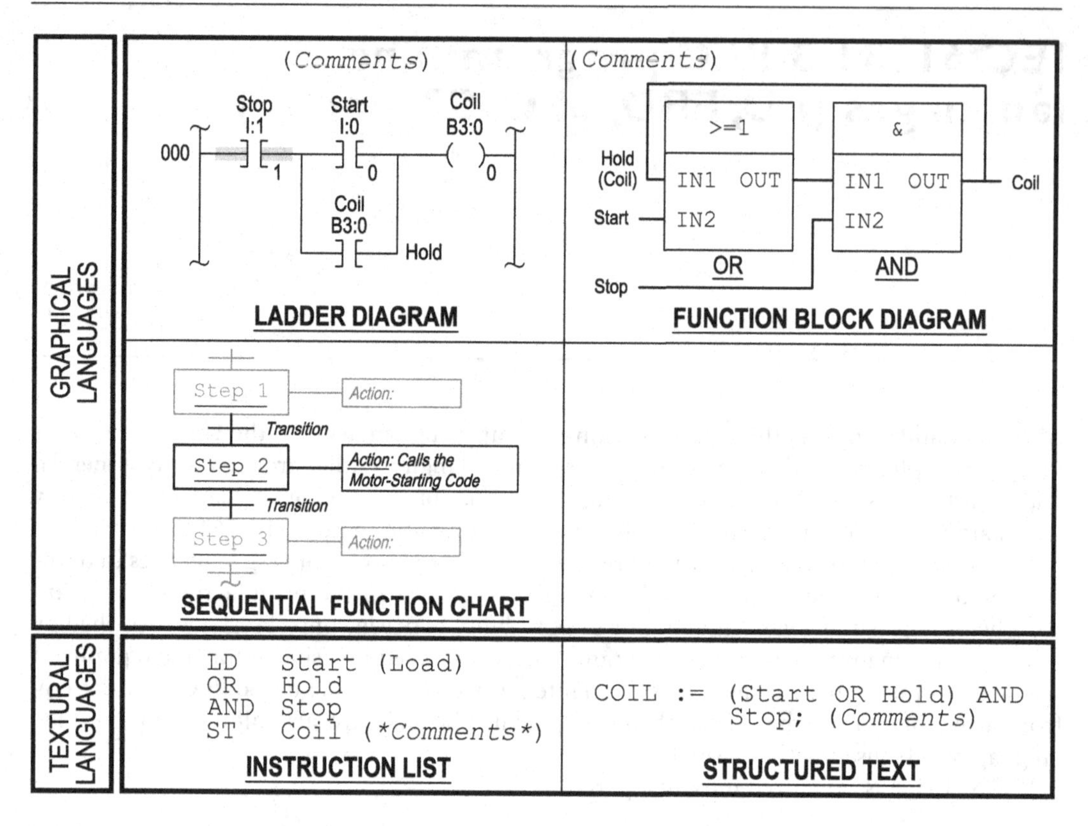

Figure 16.1 Comparison of the programming code for Ladder Diagram (LD), Function Block Diagram (FBD), Instruction List (IL), Structured Text (ST), and Sequential Flow Chart (SFC) languages for low-voltage-protection motor starting (LVP).

recursive mathematical calculations would be easier to follow in Structured Text than in Ladder Diagram. However, control of mechanical systems would be very clear and easy to follow in either Ladder Diagram or FBD. Sequential control functions might be programmed in a more straight-forward manner using the SFC language, especially when considering that any particular SFC sub-function can be programmed in any language that is more convenient, as long as it is supported by the PLC program, giving the programmer significant flexibility.

THE PROGRAM STRUCTURE

In general, programs are organized into blocks of routines or subroutines that perform needed tasks or functions. In many languages, customized subroutines are programmed once and then called by the program as predefined functions, where input values are passed into and results obtained from the function. Figure 16.2 shows a simplified program structure on the left, and the right side shows that each task can consist of a mix of any of the supported PLC programming languages. The language chosen for any task is based on the programmer's familiarity with the language and its convenience for accomplishing the task.

Regardless of the language chosen, controllers need to accomplish the same results; i.e., relay switching, timing, counting, calculating, comparing, and processing digital signals to perform logic or processing analog signals to modulate controlled devices. High-end PLC

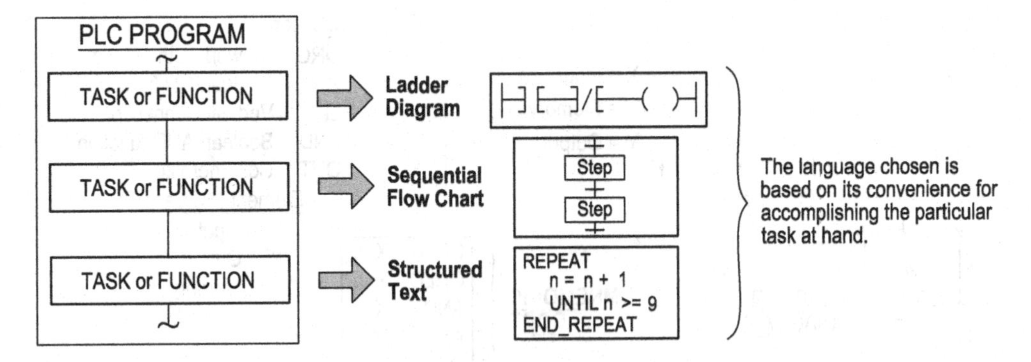

Figure 16.2 The generic program structure is shown of the left, with three examples on the right. Programs are often formed into blocks, which perform specific tasks or functions. In PLC programming, each sub-task can use any of the supported PLC languages, where the language is selected by the programmer based on convenience or efficiency in accomplishing the task.

programming languages, such as Allen-Bradley's RSLogix 5000 or Siemens Step 7 Simatic software, permit the user to choose from several of the IEC 61131-3 languages, and in fact, the programmer can mix different languages in a single program using block techniques called Program Organization Units (POUs). For instance, the Allen-Bradley RS Logix 5000 program is used for their Control Logix and Compact Logix PLCs, and the Siemens Step 7 Simatic program is used for the S7-300 and -400-level PLCs. These high-end programming languages can use Ladder Diagram, Function Block Diagram, and Structured Text programming, either as the sole programming language or combined, when it makes sense for simplicity, brevity, and readability of code or speed of the application.

LADDER DIAGRAM (LD)

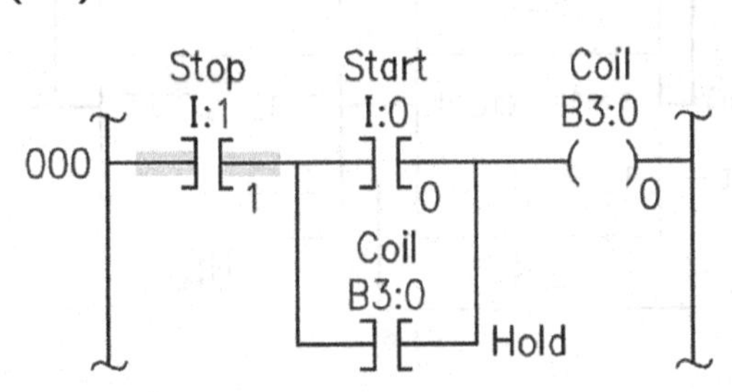

Ladder Diagram is often the main industrial programming language used for PLCs in the United States. The earliest generations of PLCs used a version of the ladder diagram language, which was based on Boolean or relay-logic as represented in a conventional motor-controller diagram. The logic diagram was subsequently coded into strings of text, where each line of code represented a text-version of a ladder rung in what was called mnemonic code. As an example, the GE Fanuc Series One mnemonic code is shown in the right side of Figure 16.3, based on a ladder rung shown on the left side. The lines of text are the code-equivalent instructions that correspond to relay coils, contacts, and PLC terminal connections and memory addresses.

Right-clicking on an Allen-Bradley RS Logix 500 ladder rung shows the textual code behind the graphical language, which is not too different from what programmers saw in the early days of PLCs, and is shown in Figure 12.5. Today's ladder-logic language uses

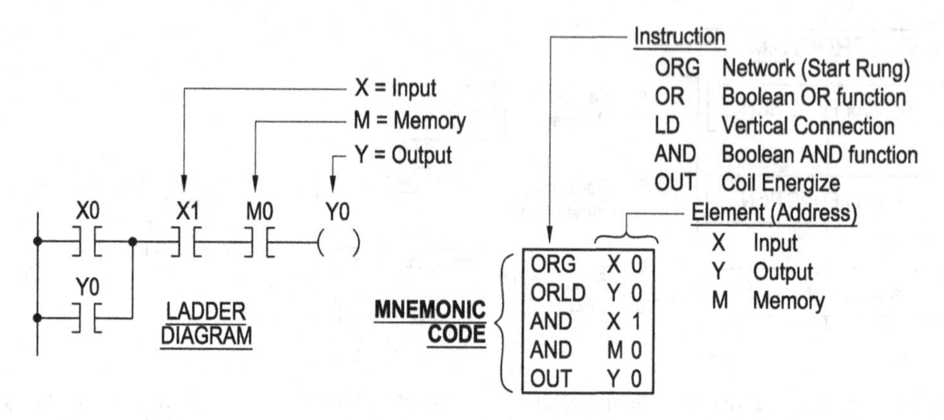

Figure 16.3 Ladder diagram or Boolean code as written using early generation mnemonic code.

drag-and-drop graphical symbols that directly emulate the relay-logic diagrams, while the textual equivalent resides behind the scenes. The LD language has a similar appearance to the motor-controller wiring diagrams that have been around for over a century, and the relay wiring diagram uses practical switching-theory logic, which is well understood by engineers, technicians, and electricians. The relay/ladder-logic is covered in much more detail in Chapter 5 on motor-controllers and the ladder diagram language is covered in Chapters 13 and 15 on RSLogix programming.

FUNCTION BLOCK DIAGRAM (FBD)

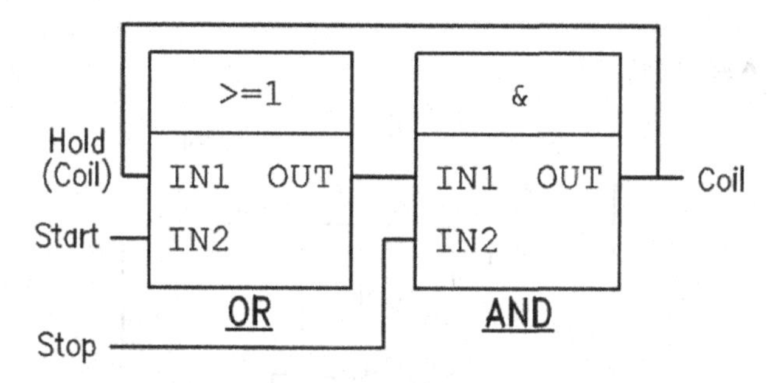

Function Block Diagram (FBD) is a graphical language, where the instructions are grouped into blocks that emulate Boolean algebra functions coupled with other mathematical and control-function blocks. Some functions are pre-defined and available from within the programming language, while others are user-defined, and can reach any level of complexity. Technically, all PLCs' built-in instructions, such as the Timer-On (TON), Limit Test (LIM), or the Scale with Parameters (SCP) instructions are pre-defined function blocks. The concept of using function blocks is to pass input data through a Boolean-like decision tree of logical, mathematical, and control instructions to produce outputs, using a visually based language. Some people believe FBD programming is more intuitive for technicians who do not have a good background in relay logic. Others believe that plant designs should be done by engineers, and the programming should also be done by the engineers who thoroughly understand the systems, rather than computer specialists, and that either LD or FBD programming is equally easy for an engineer. For that matter, ST routines to simplify mathematical computations should also be easy for an engineer, and when it is available, it should be used to simplify LD and FBD programs, and to perform complex mathematical operations.

As with LD programming, built-in pre-programmed functions and user-defined function blocks are available in FBD programming. Each use of a function or function block is called an instance. The distinction between a function and a function block is that each instance of a function block produces values that are retained in memory, whereas function outputs are not retained after the program moves to the next step. Both functions and function blocks are called as instructions by the program, and they themselves can also call other function blocks, as if the functions they are calling are subroutines. Using functions and function blocks provide an advantage, in that potentially large amounts of program code are placed into program memory only once, even if the function is called many times by the main program. In addition, some equipment manufacturers provide complicated pre-programmed function blocks that reduce programming and debugging time, if the engineer commits to that vendor's product.

FBD function-block symbols are shown below and include:

- Bit logic functions (Boolean logic) – Figure 16.5
- Bistable logic functions (flip-flops) – Figure 16.6
- Edge-detection functions (triggered events) – Figure 16.7
- Timer functions – Figure 16.8
- Counter functions – Figure 16.9
- Comparison functions – Figure 16.10
- Selection functions – Figure 16.11

Function blocks are snippets of code that are formed into a box-like symbol that are called by the main program. The box contains the function type located at the top of the box, often with an instance name entered by the programmer as a memory aid. The inputs are connected on the left side of the box and the outputs on the right as shown in Figure 16.4.

Other function blocks are pre-built into the program, such as the Timer On (TON), Limit Test (LIM), or the Scale with Parameters (SCP) instructions. However, a major advantage of function block programming is that the programmer can construct customized function blocks to be used as single user-defined instruction when input information is passed into and output results obtained from the block. These user-defined function blocks can be very complex, but after being programmed once, they are readily available as a straightforward instruction that can be used over and over.

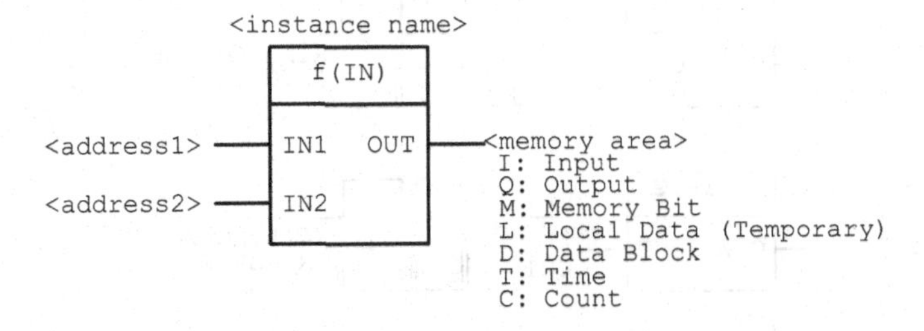

Figure 16.4 General arrangement of a Siemens Step 7 Simatic function block symbol. The function type is listed at the top of the box, the inputs on the left, and the outputs on the right.

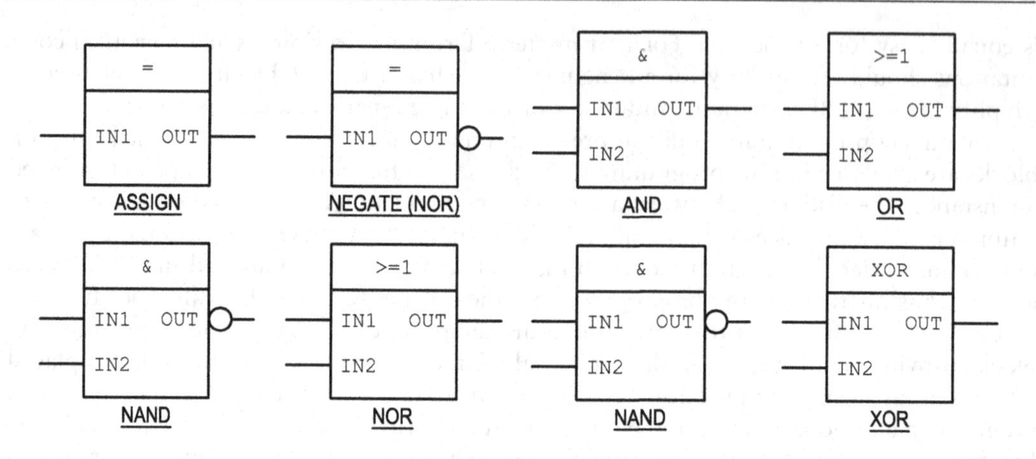

Figure 16.5 Boolean functions as they appear in IEC 61131-3 for function block programming.

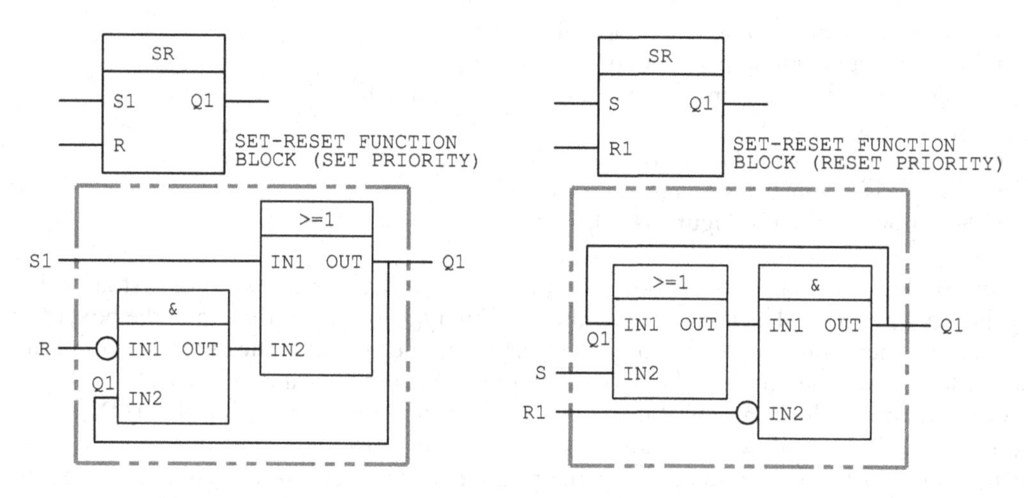

Figure 16.6 Bistable function block, like the S-R flip-flop, has Set priority on the left and Reset priority on the right.

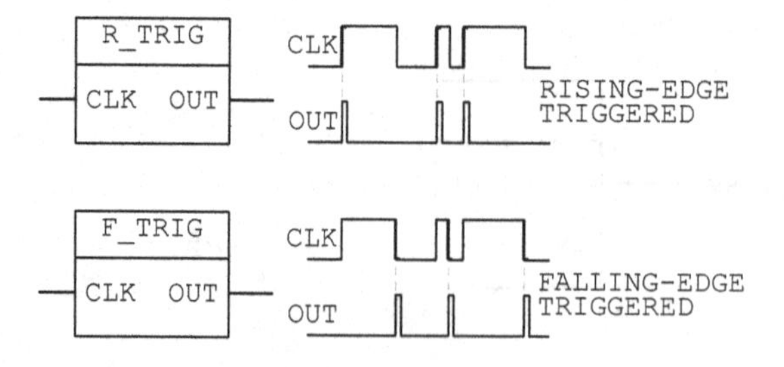

Figure 16.7 Edge-triggered block functions, rising-edge triggered (upper) and falling-edge-triggered (lower).

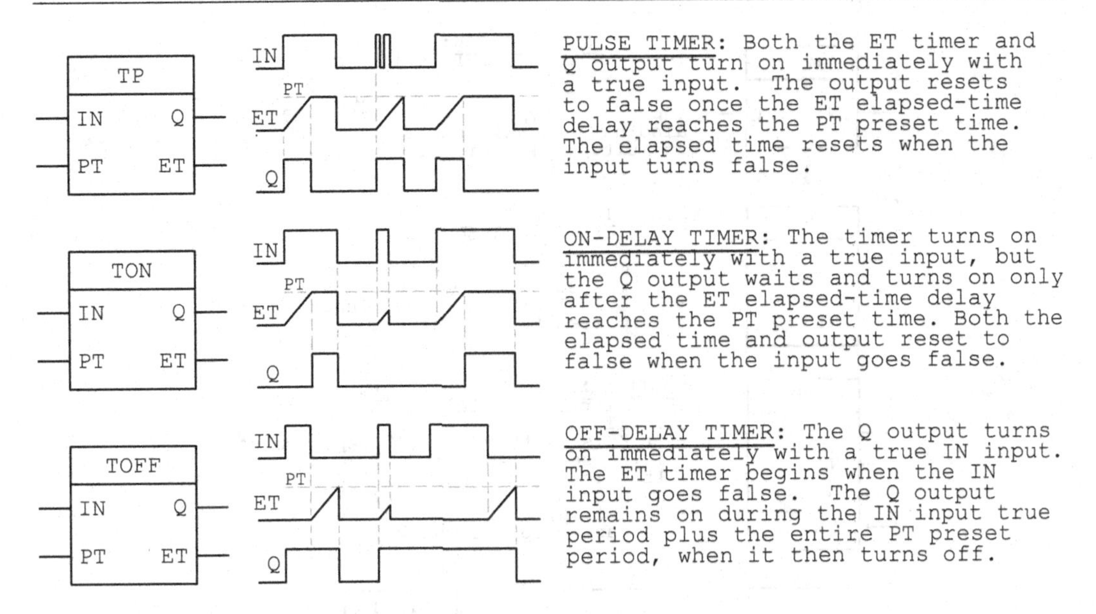

Figure 16.8 Timer function blocks, pulse timer (upper), on-delay (middle), and off-delay (lower).

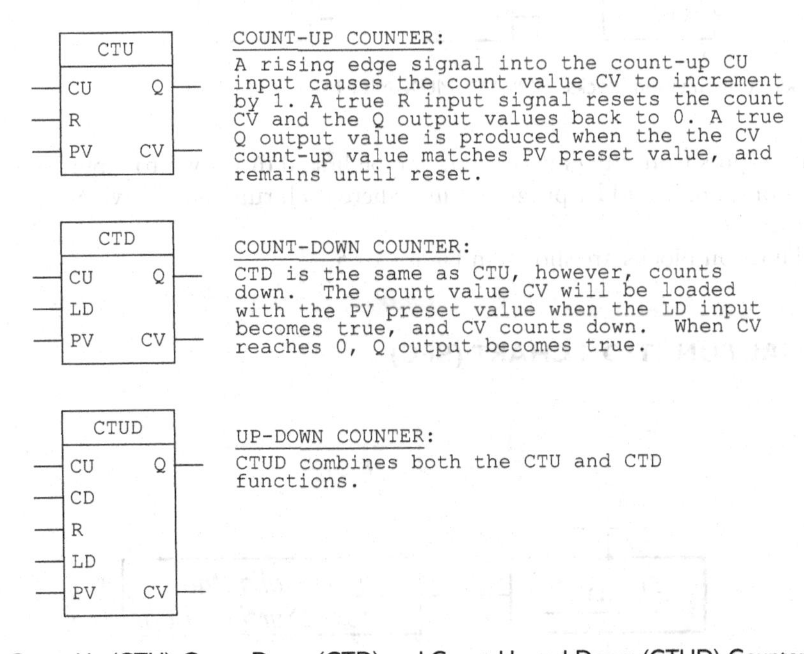

Figure 16.9 Count-Up (CTU), Count-Down (CTD), and Count-Up-and-Down (CTUD) Counters.

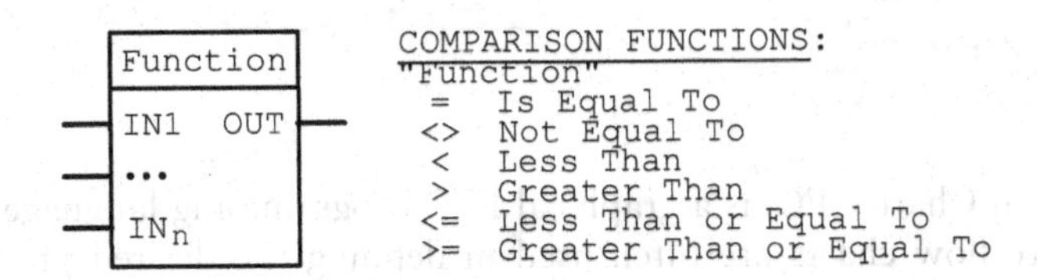

Figure 16.10 Comparison functions, such as greater-than or equal-to operations.

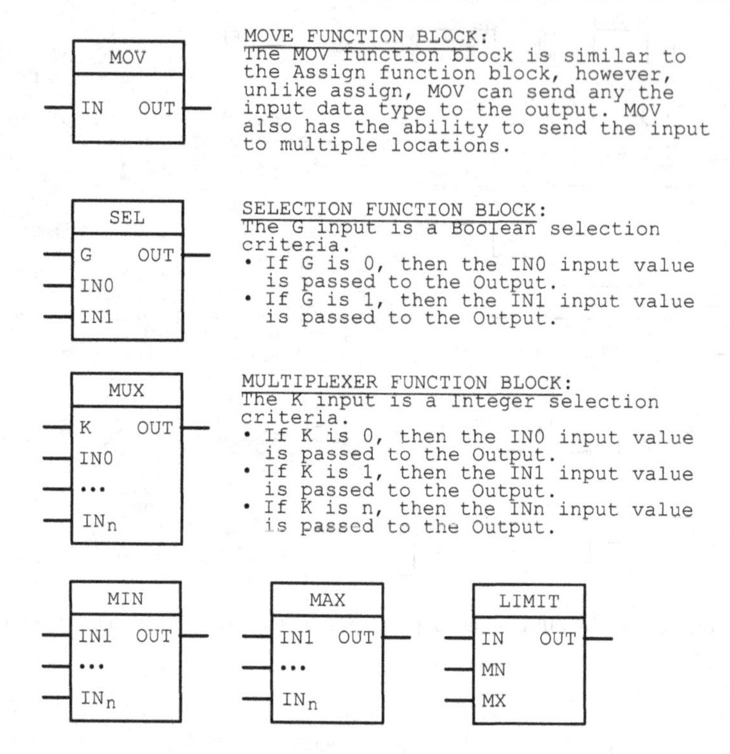

Figure 16.11 Selection function blocks: MOV, SEL, MUX, MIN, MAX, and LIMIT.

Note that outputs from FBD instructions can be linked directly into inputs of subsequent FBD instructions, unlike ladder programming where each rung must have one and only one output instruction.

Boolean Function Blocks are shown in Figure 16.5.

SEQUENTIAL FUNCTION CHART (SFC)

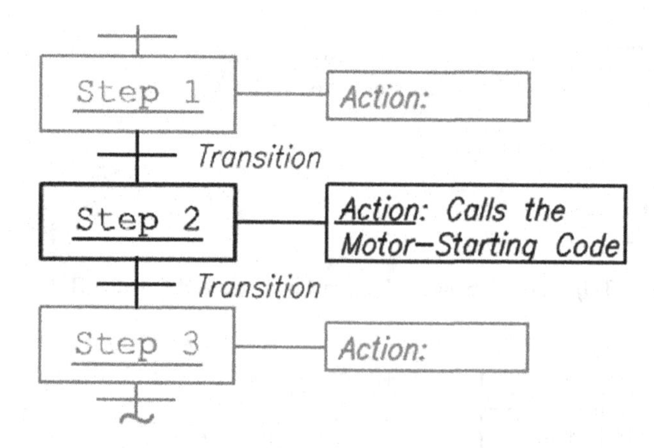

Sequential Function Chart (SFC) is a graphical PLC programming language. SFC resembles flow charts, where flow charts are often used in defining the desired program operation. Sequential function chart is very conducive for programs that follow many steps, particularly

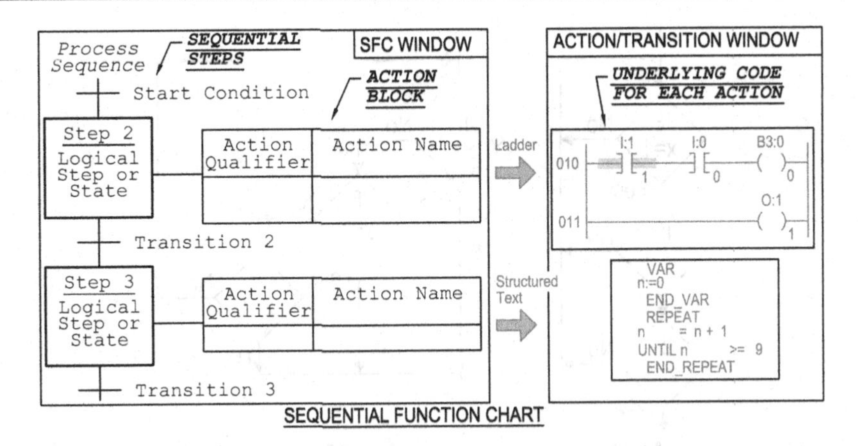

Figure 16.12 Sequential flow chart programming (SFC). The process sequence is defined by the logical steps or states. Each step has a file/action-name identifier, which in turn calls the underlying code that accomplishes the action.

in automated manufacturing processes, or in complex sequences used by boiler and burner management systems. An advantage of SFC is that the program can be easily adapted from simple flow charts produced by non-technical people. The sequences tend to be more straightforward and easier to follow, resulting in faster programming and troubleshooting. The main parts of the SFC program are shown in Figure 16.12 and consist of the following components:

- *Step* is a single process in the program. Contained within the step are a series of events or actions that are executed when the step becomes active. The first step defines the initial state of the system, and may not have an action, if there is nothing to process.
- *Action Block* contains the process sub-steps that comprise the overall single purpose of the step.
- *Action Qualifier* defines the execution time and memory-retention information for each action.
- *Action Name* is a link to the detailed sub-steps and logic functions that form the actions.
- *Transitions* are logic terminations, which cause the program to move to the next step. Essentially these are the decision instructions that cause a step to either loop back and repeat, or to proceed forward to the next step when conditions are satisfied. A transition is crossed if its Boolean conditions are TRUE, *and* all steps above the transition are deactivated *and* the steps after the transition become activated.

Typically, the program editor screen can be split to show the SFC steps with their associated actions in one window, and the detailed programing in the action/transition window. Simultaneous editing of the chart and the actions allows the programmer to keep an overview perspective while accomplishing the details of the tasks.

Program operation

Like flow charting, SFC steps are connected in series with each other. SFC is conducive to Boolean operations using AND *and* OR decisions. AND functions in flow charting are simple decisions that are placed in series with each other, while OR functions are decisions placed in parallel. Figure 16.13 shows a flow chart arrangement for AND *and* OR decisions. The SFC

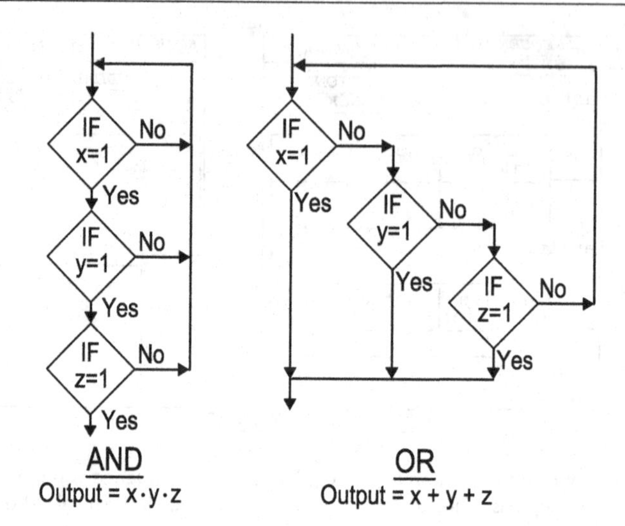

Figure 16.13 Flow chart equivalent of AND *and* OR functions.

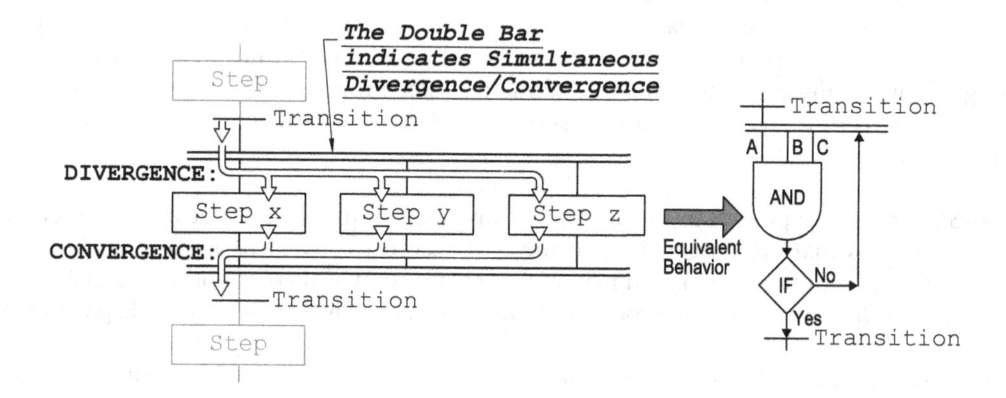

Figure 16.14 SFC syntax for simultaneous divergence-convergence, which behaves like an AND gate.

programming for the Boolean AND function is shown in Figure 16.14 using the double-bar syntax, and the OR function is shown in Figure 16.15 using the single bar syntax. SFC syntax uses the upper-and-lower bars to place the Boolean serial/parallel decisions into a more concise, easy-to-interpret flow path.

During program operation, the steps are followed in the order they occur. When a transition occurs, the preceding step becomes inactive and the following step becomes active. Once a step becomes active, the associated actions are called. The actions contain the code that accomplishes the step's tasks. The action code may contain any PLC language supported by the PLC program, including additional routines in the SFC language.

When parallel branches are used in the SFC program, all steps included within the parallel branch become active and are executed simultaneously. Parallel progression of each branch can take different amounts of time depending on the relative complexity of the branch code. The transition at the end of the branch takes place after the convergence is completed, also indicating synchronization of the branches has taken place. In some cases, an empty step is inserted into a branch representing a wait state for the other branches to reach completion.

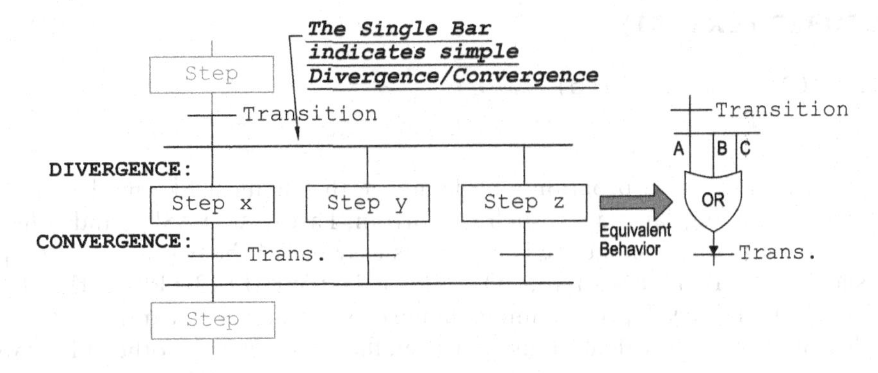

Figure 16.15 SFC syntax for simple divergence-convergence, which behaves like an OR gate.

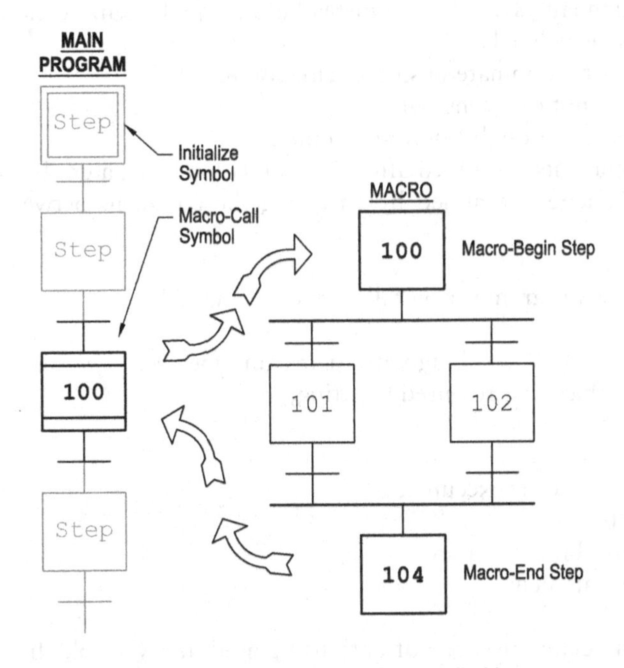

Figure 16.16 SFC macro. Note that a macro is a subroutine that is called by a program.

Like other PLC programming languages, sequential function charts are capable of calling macros, jumping to another SFC step, timeouts, parent-child routines, etc. The syntax for a macro is shown in Figure 16.16. A macro is a subroutine of steps that are pulled from the main program to improve readability. The macro can only be called one time in the program. A jump is a link from a transition to a step or steps that form an option or different chart path, such as normal versus emergency conditions. Timeouts are useful in troubleshooting, preventing locked-up programs, and in managing alarms. Parent-child programs are used in programming complex processes, where a parent action executes a series of detailed steps in the child, so that the parent program shows the big picture of the process. For example, a boiler light-off step in the main program may call a child program containing the detailed steps for pre-purge, i.e., start fan open dampers, increase fan speed, open burner register doors, begin pre-purge timer, etc.

STRUCTURED TEXT (ST)

```
COIL := (Start OR Hold) AND
        Stop; (Comments)
```

Structured Text is the PLC programming language that is most like modern high-level computer-programming languages, such as Fortran, Pascal, C, BASIC, and others. The strength of structured text language is it permits easy use of the decision and loop commands, such as the IF-THEN-ELSE, FOR-NEXT, DO-WHILE, REPEAT UNTIL statements found in high-level programming languages. Structured text permits iterative or complex mathematical calculations in a friendlier manner than other PLC language options.

Some structured text programming syntax rules include:

- := is an "assignment" symbol used instead of an equals sign. The := assignment can be interpreted as "is defined as."
- All statements are terminated using a semi-colon.
- The language is not case sensitive.
- Spaces have no function, beyond readability.
- Single-line comments are added after the semi-colon using a double slash // *comment*
- Multiline comments are added by enclosing the comment between slash-asterisk /* *comment* */

In structured text, the program is generally organized as follows:

- identification of variables along with their data types,
- storage of variables that required resetting,
- initialization of variables,
- reading inputs
- the program code to be executed,
- writing outputs
- resetting of variables, if required
- a termination statement.

Data types explicitly define the type of data to be used in a variable. Importantly, the data type defines the amount of memory in bits or bytes that must be set aside for storing the data, and each data type must be consistent with the operations it will undergo. In high-level programming languages, the data-type definitions are usually at the beginning of the program, but certainly before its variable is used. The data types correspond to the data used with the ladder-logic programming and consist of

- Integer numbers (Table 16.1)
- Floating point or real numbers (Table 16.2)
- Bit strings or logic information (Table 16.3)
- Strings or text (Table 16.4)
- Time values (Table 16.5)

Structured text programming also permits derived or custom data types. Derived data types are user-defined, formed by combining other data types. They can be structured, enumerated, sub-ranges, and array data types. Derived structured data types are formed by enclosing elements

Table 16.1 IEC Definitions for integers

IEC data type	*Description*	*Range*
SINT	Short Integer	-128 to +127
INT	Integer	-32768 to +32767
DINT	Double Integer	-2^{31} to $2^{31}-1$
LINT	Long Integer	-2^{63} to $2^{63}-1$
USINT	Unsigned Short Integer	0 to +255
UINT	Unsigned Integer	0 to $+2^{16}-1$
UDINT	Unsigned Double Integer	0 to $+2^{32}-1$
ULINT	Unsigned Long Integer	

Table 16.2 IEC definitions for floating-point numbers

IEC data type	*Description*	*Range*
REAL	Real Numbers	$\pm 10^{38}$
LREAL	Long Real Numbers	$\pm 10^{308}$

Table 16.3 IEC definitions for bit strings

IEC data type	*Description*
BOOL	1 bit
BYTE	8 bits
WORD	16 bits
DWORD	32 bits
LWORD	64 bits

Table 16.4 IEC definitions for strings

IEC data type	*Description*
STRING	Text Characters, numerals are text

Table 16.5 IEC definitions for time

IEC data type	*Description*	*Example*
TIME	Time Duration after an Event	TIME#<days>d<hours>h<minutes>m<sec>s<milliseconds>ms Ex.TIME#2d3h30m10ms500ms
DATE	Calendar Date	Ex. DATE#2018-02-28
TIME_OF_DAY	Time of Day	TIME_OF_DAY#Hour:Minutes:Seconds.Hundredths of Seconds Ex.TIME_OF_DAY#17:15:30.50
DATE_AND_TIME	Date and Time of Day	Ex. DATE_AND_TIME#2018-02-28-17:30.50

between the STRUCT and END-STRUCT commands. Each element in the data type includes its name, colon, and data type. Some examples of structured data types are defined as follows:

```
TYPE Circle: STRUCT // Comment – Circle is a structured data type
     Center: CenterPnt; // ( Note "CenterPnt" is a structured x & y data point.)
     Radius: INT; END_STRUCT;
END_TYPE

TYPE LightColor: (Green, Yellow, Red) // Structured Data Type
END_TYPE

VAR L1, L2 : LightColor := Red)

TYPE PumpState: (Power Available, Running, Stopped, Tripped, Error)
END_TYPE

VAR P1, P2: PumpState := Stopped)
END_VAR
```

Enumerated data type syntax:

```
TYPE <enum identifier>:
     (<enum_0>|:=<value>,<enum_1>|:=<value>, ...,<enum_n>|
       :=<value>)|<base data type>|:=<default value>;
END_TYPE
```

The ST operators can be divided into four groups, arithmetic, relational, logical, and bitwise, and the operations must be consistent with the data types. The standard programming operators as defined by IEC 61131-3 are summarized in Table 16.6. The operators are used to manipulate data, called the operands, as part of expressions, where expressions are designed to produce values or results. Like the rules of algebra, code operators follow rules of precedence. For example, expressions enclosed within parentheses are executed first, and multiplication and division operations take place before addition or subtraction.

INSTRUCTION LIST (IL)

```
LD   Start (Load)
OR   Hold
AND  Stop
ST   Coil (*Comments*)
```

Table 16.6 Standard structured text operators as described by IEC 61131-3

Operation	*Symbol*	*Operation*	*Symbol*
Parenthesizing	(expression)	Comparison	< > <= >=
Function Evaluation	MAX(A,B)	Equality Inequality	= <>
Negation Complement	– NOT	Boolean AND	& AND
Exponentiation	**	Boolean Exclusive OR	XOR
Multiply	*	Boolean OR	OR
Divide	/		
Modulo	MOD		
Add	+		
Subtract	-		

Instruction List (IL) is a low-level programming language, which is analogous to assembly language. Like assembly language, IL uses mnemonic commands that execute directly. Also, like assembly, IL is the lowest-level language where there is a direct correspondence between program statements and the processor's machine-code instructions. As a non-interpretive programming language[1], the Assembly program must be compiled into machine code to operate, which is the digital language understood by a computer. Beyond assembly language is machine code, itself. Machine code uses built-in instruction sets, which are patterns of bits that by the digital-circuit design can be executed directly by a CPU to perform very specific tasks. Machine code looks like a stream of 1s and 0s. Machine code might use 8-bit words, for instance, where the first four bits represent an instruction, called opcode, and the second four bits contain the operand, or in simpler terms, the data. Although it may be possible to write code in machine language, it would be very tedious and prone to errors, so an assembly-type language, like instruction list, would provide a more practical alternative to machine code.

Like assembly language, the instruction list syntax is specific to the PLC architecture. The IL language works by using mnemonic statements that correspond to the PLC's I/O components and their relationships to instructions and memory locations. IL, like assembly, is much closer to machine language than higher-level languages. Consequently, IL language takes less memory and runs faster. IL is an accumulator-oriented language, meaning that the instructions modify the present content in the accumulator register, which is fast-retrieving memory built into the CPU chip.

Instruction list programs achieve flow using jump instructions, and function calls. Although the IEC 61131-3 define the language, individual programs use their own calls and function blocks to suit their equipment. Siemens Simatic calls their IL software "Statement List," and they provide options where the instruction mnemonics may be in English or German and their function names vary from the IEC standard. As examples, Simatic uses "A" to represent the IEC AND function in English, and E in German.

Syntax

The IL program uses sequences of statements that are the equivalent of ladder diagram rungs. Each statement has an instruction, called the operator, and a variable, constant, or jump label, which is the operand, plus user comments to aid readability. Table 16.7 shows a ladder

Table 16.7 Example of instruction list programming compared to the equivalent ladder diagram code

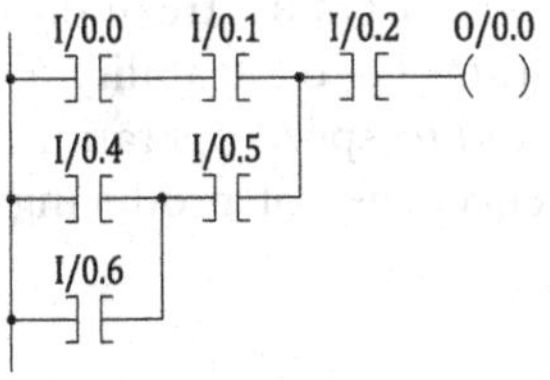

I/0.0 I/0.1 I/0.2 0/0.0
I/0.4 I/0.5
I/0.6

Function	*Step*	*Instruction & operand*	*Comment*
START:	00	LD I:0.0	/* LD: Load at Rung */
	01	AND I:0.1	/* I:0.0 And I:0.1 */
	02	LD I:0.4	/* I:0.6 Branch1 I:0.4 */
	03	OR I:0.6	/* Start I:0.6 Branch */
	04	AND I:0.5	/* I:0.5 And Branch1 */
	05	ORLD	/* End I:0.6 Branch */
	06	AND I:0.2	/* I:0.2 And */
		ST O:0.0	/* Output */

Table 16.8 Basic instruction list commands

Instruction	*Operand*	*Description*	*Instruction*	*Operand*	*Description*
OPERAND INSTRUCTIONS			**COMPARISON INSTRUCTIONS**		
LD / LDN	Var/Con	Load Operand	EQ	Var/Con	Equal to
ST / STN	Var/Con	Store Operand	NE	Var/Con	Not Equal to
S	Variable	Set Operand TRUE	LT	Var/Con	Less Than
R	Variable	Set Operand FALSE	LE	Var/Con	Less or Equal to
BOOLEAN INSTRUCTIONS			GT	Var/Con	Greater Than
NOT	Var/Con	Negation	GE	Var/Con	Greater or Equal to
AND / ANDN	Var/Con	AND *and* NAND	**SUBROUTINE INSTRUCTIONS**		
OR / ORN	Var/Con	OR and NOR	JMP/JMPN/JMPC	Label	Jump to Label
XOR / XORN	Var/Con	XNOR	RET		Return from Block
ARITHMETIC INSTRUCTIONS			Note: Var is short for Variables		
ADD	Var/Con	Addition	Con is short for Constants		
SUB	Var/Con	Subtraction	The N suffix means NOT		
MUL	Var/Con	Multiplication	The C suffix means execute when TRUE		
DIV	Var/Con	Division			

diagram program along with the equivalent instruction list code. Some instructions contain the modifier N or C, where N is the NOT condition and C indicates that the statement is only executed if the current register value is true. Some of the basic instruction list commands are listed in Table 16.8.

SUMMARY

There are five programming languages that are defined by Part 3 of the International Electrotechnical Commission standard IEC 61131, consisting of three graphical languages being Ladder Diagram, Function Block Diagram, and Sequential Function Chart, and two textual languages consisting of Structured Text and Instruction list. Many PLC manufacturers tend to use the IEC standard as a guide, with some nuances of syntax. The various programming languages have different strengths for different tasks, and many higher-end programming languages provide the ability to mix the languages within a single program. In some instances, program code can be obtained from equipment manufacturers to streamline the integration process between programs and devices. The PLC programs have the ability to use subroutines and jump statements to reduce the amount of code and to speed operation. Often, the choice of program language is a mix of personal preference, ease of producing desired certain functions, and speed requirements.

QUESTIONS

1. What is IEC 61131? What part pertains to PLC programming languages?
2. List the five IEC 61131-compliant languages.
3. Provide a representative sketch of each of the five IEC 61131-compliant languages.
4. Which language most closely approximates a motor-controller diagram? Which

language most closely approximates Boolean functions? Which language might be very useful for step-by-step procedures used by a burner management system? Which language might be most suitable for complex mathematical calculations associated with producing analog-output signals? Which language produces fast-acting, low-level code most like Assembly language?

5. What is the "grammar" associated with programming languages called, and why is it so important with PLC programming?

NOTE

1. Interpretive programming languages are executed directly, without the need to compile the program into machine language instructions. Compiled programs generally run faster but lack editability.

Chapter 17

Centralized control systems, DCS, and SCADA

Many plants have equipment and individual systems that operate autonomously, and often their measured parameters are transmitted to a central location. In this manner, individual controllers are distributed around the plant, where equipment is locally controlled, but operators can observe decision-making information at a central control station regarding the most important plant measurements. Oftentimes the individual devices, equipment, and systems communicate electronically with computer-based systems, which monitor the plant, but also have the ability to remotely operate the equipment and to change setpoints. Those systems are referred to as distributed control systems (DCS) or supervisory control and data acquisition systems (SCADA) when the SCADA computer performs real-time performance calculations, manages data, and creates historical archives for long-term efficiency assessments and troubleshooting. To complement this chapter, cabling, data transmission, and networking communications are covered in Chapter 18, and industrial control system security is covered in Chapter 19.

PLANT CONTROL ARCHITECTURES

Control systems are configured in a number of ways, some plants having very little automation while others are fully automated. Some architecture types include:

- Local and Remote Control
- Centralized Control and Monitoring
- Centralized Control
- Distributed Control and Supervisory Control and Data Acquisition

Local Control is where the machinery operation is managed from devices that are located within line-of-sight of the equipment.

Remote Control is where the operator can control machinery from a distance at a central location using electrical, electronic, pneumatic, or hydraulic actuators, or a combination of those techniques.

Centralized Control and Monitoring is where a central operating station is fitted with gages and meters for monitoring the status of important machinery and plant parameters. It uses hardwired switches and controls for remote actuation for start, stop, and adjustments. It includes alarms to indicate abnormal conditions. Although the purpose is to provide enough information at one location to emulate local supervision of machinery, it may contain limited automation-and-control ability.

Centralized Control uses an electronics-based automation system at a single location to receive inputs, perform logic, execute algorithms, and produce the output signals that are sent to remote equipment. Centralized control includes monitoring of status and alarm handling.

Distributed Control is a process-oriented automation system that links a central microprocessor or computer to a number of individual PLC-type controllers spread throughout the plant, located near the equipment. Each distributed PLC is dedicated to its own devices or system. With distributed control, data-acquisition and control signals are broadcast to and from the master to the local controller needing the signal, while that signal is ignored by all others. Distribution of control mitigates the susceptibility of a single-point failure of one controller to disrupt the entire plant. A characteristic of distributed control is that the master controller actively participates in the PLC scan cycle, and it uses real-time data.

Supervisory Control and Data Acquisition is a central automation system having many similarities to a DCS, but it is more data-oriented than process-oriented. The SCADA system is often spread over a geographical area. An example is a central building-energy-management system for a large campus that has SCADA engineering workstations and operating stations located in the physical-plant engineering office for equipment that is monitored or controlled in remote buildings. The SCADA system uses transmitted data that is retained in memory, rather than the real-time measurements from a PLC scan cycle. Like the DCS, SCADA systems may be programmed to permit remote control of equipment.

CENTRALIZED CONTROL AND MONITORING SYSTEM

Older plants used an engineer operating station where gages and indicators for the most important systems were piped to a panel or console in an arrangement referred to as central control and monitoring, shown schematically in Figure 17.1. The arrangement includes hardwired remote start-stop ability at the operator's station for the most important equipment, the most remote devices, or machinery that was operated often, in systems that were left aligned and ready-to-go, such as backup cooling-water pumps, fire pumps, ventilation systems, etc. Alarms were wired to a panel, where a watch engineer acknowledged the notifications. Generally, operators made periodic rounds of the plant to stay attuned to the plant status. Historical records were maintained using manual logbook entries and chart recorders, which kept a continuous record of the most important plant parameters, which were retained

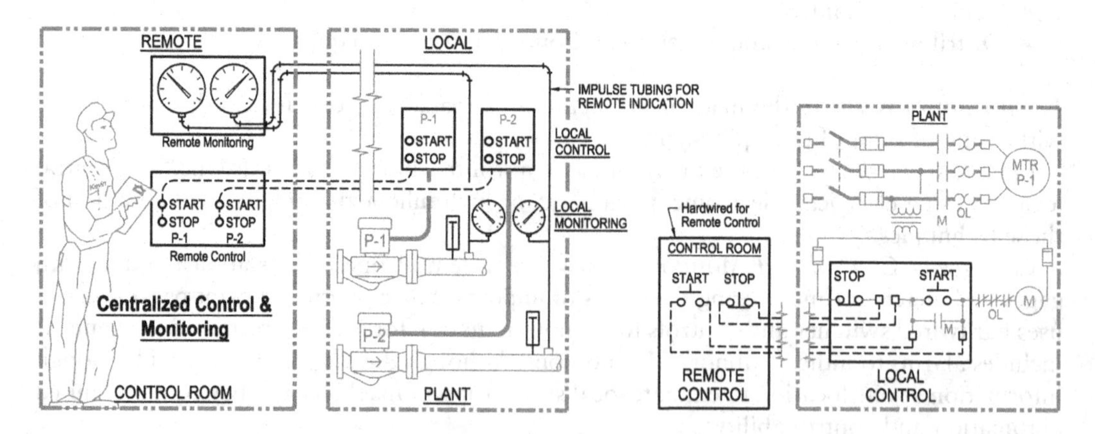

Figure 17.1 In the centralized control and monitoring system, gages are hard-piped and remote-control functions are hardwired.

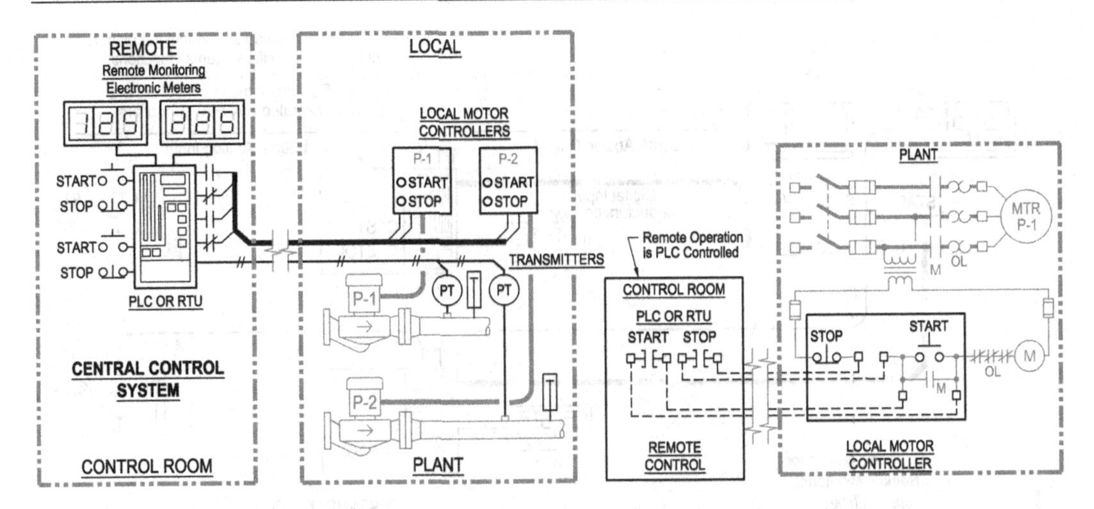

Figure 17.2 In the Central Control System, a microprocessor-based controller receives electronic signals for monitoring and control inputs, executes control routines, and conveys operating commands to local motor starters and controllers.

for hand trending or analysis. This control system is outdated, but it could be appropriate for small unsophisticated plants.

CENTRALIZED CONTROL SYSTEM

A centralized control system, shown in Figure 17.2, is similar to the central control and monitoring system, except it uses a microprocessor-based controller located at the operator's station to receive electronic signals, execute programs, and to remotely start-and-stop equipment, rather than hardwired switches to actuate remote devices. In this manner, the central controller delivers the operating commands to motor starters and valve controllers throughout the plant, where the automation originates from the central location. The microprocessors often consist of multiloop controllers, PLCs, or RTUs, which have the ability to run the logic and perform the algorithms to automate remote equipment. Many multiloop-type controllers use soft-touch switches to permit the operator to scroll through various displays or functions, where monitoring is through built-in bar graphs and digital readouts, which show operating parameters and permit setpoint changes. Some newer designs use touchscreen displays, however, the soft-touch scroll buttons may be preferred in plants where operators may have oily hands. Switches, gages, and meters are installed at the central panels to display important parameters in real time, provide remote actuation, and to simplify the system operation. In the past, these single-processor systems provided an economical option at a time when electronic controllers were very expensive. The downside of this arrangement is that the controller forms a potential single point of failure that could disrupt the plant. Today, PLCs are less expensive, and distributed-control systems have become the norm, however, these systems can be a cheaper option for small non-critical plants.

DCS AND SCADA SYSTEMS

Modern plant control systems integrate many measured parameters that are used for the control and coordination of plant functions to precision levels that were difficult to achieve

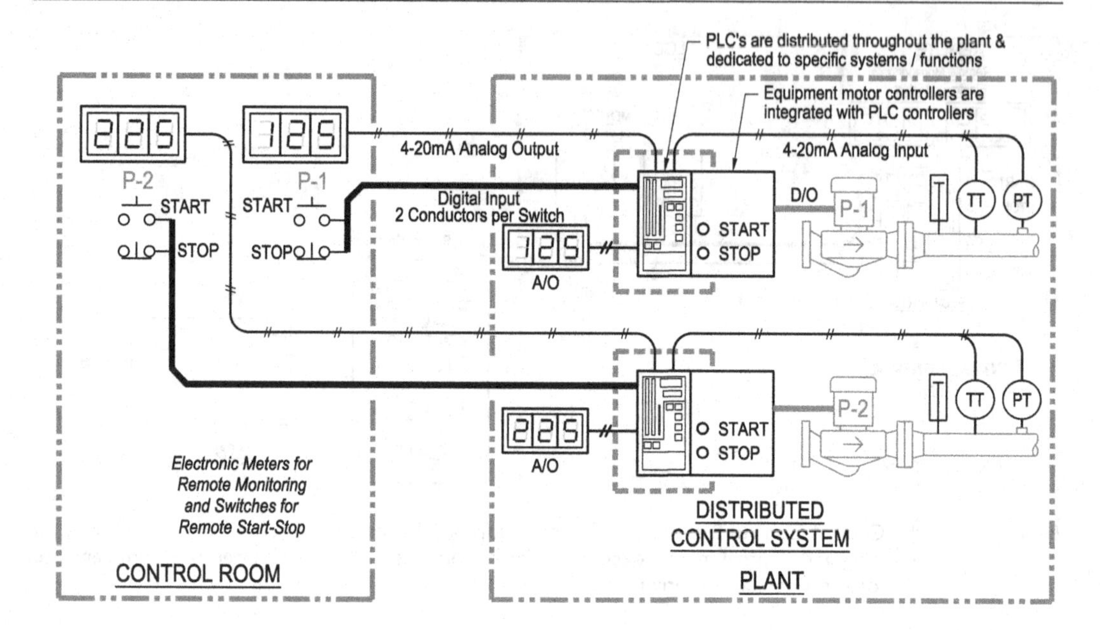

Figure 17.3 An early generation distributed control system (DCS) is shown, with individual controllers spread throughout the plant. Later generation DCS systems use a bus or networking system, typically in conjunction with computer workstations in the control room.

in the past. The electronic nature of the I/O signals lends itself to data collection in a central computer. These centralized systems are referred to as DCS systems or SCADA systems, where DCS stands for Distributed Control System and SCADA for Supervisory Control and Data Acquisition. The communication architecture consists of the control subsystems that are linked by a data-transfer bus, the synergistic integration of control strategies, and the organization of all controls into a single coordinated automation system capable of passing commands and information, and for sharing data.

There are many similarities between a modern DCS and SCADA system, to the point where the terms are sometimes used interchangeably. To generalize:

- The DCS is more process oriented, while a SCADA system is more data oriented. Although the modern DCS has a supervisory level to provide real-time data to the operator, as shown in Figure 17.3 and Figure 17.4, the DCS is more concerned with controlling the plant processes. SCADA systems, however, tend to manipulate transmitted data to provide more information to the operator beyond just mimicking the measured values, and the information is often superimposed onto detailed process diagrams.
- A DCS tends to be confined to a limited area, such as within a plant, and it tends to communicate over a local-area network (LAN), whereas a SCADA system tends to be spread over scattered areas, often using communication systems beyond the LAN, such as a wide-area network (WAN) or the internet.
- A DCS uses closed-loop control at the local PLCs and RTUs, but closed-loop control can also originate from the microprocessor located at the central-control station. The SCADA is more supervisory and typically leaves the closed-loop adjustments to the local microprocessor controllers.
- The DCS scans the process continually and updates the displayed information regularly, whereas the SCADA system tends to be event driven, waiting for signals to trigger

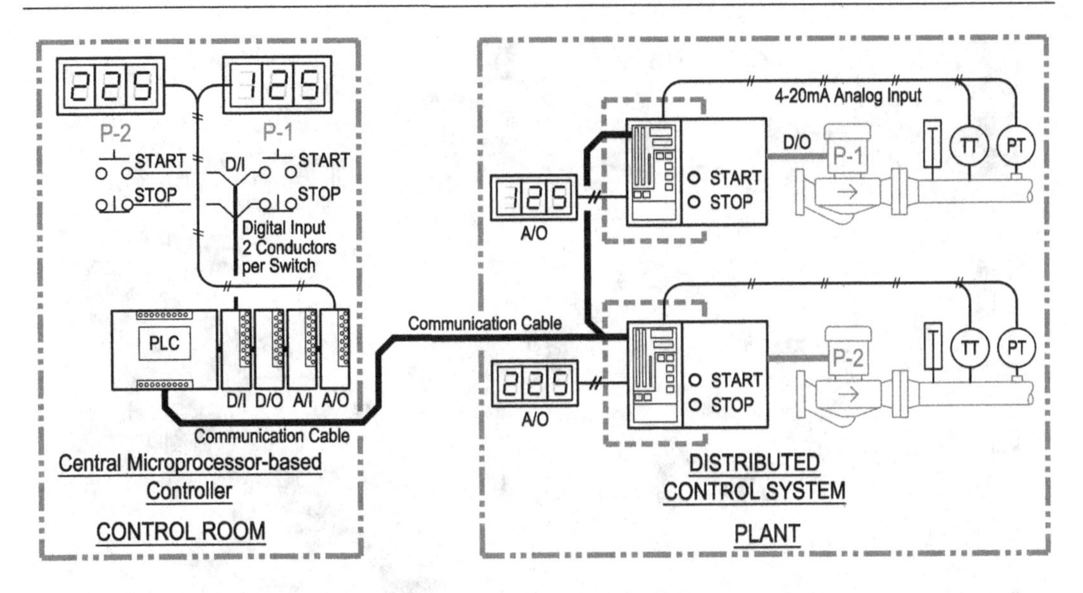

Figure 17.4 A modern DCS uses controllers that are distributed around the plant and coordinated with a central master microprocessor, so that many input and output functions are available at the control room.

actions. The DCS looks at present values of information, whereas the SCADA system looks at a database of recent plant information that is retrieved by the system or is archived for later analysis.

SCADA systems do the following:

- monitor I/O process variables in real time,
- pass along alarms,
- send notifications,
- record data periodically to automatically generate logbook functions,
- provide real-time plant performance calculations,
- provide recommendations for efficiency improvements,
- track equipment run-times and recommend scheduled maintenance or equipment changeover,
- create reports,
- troubleshoot based on received information and recommend operational changes.

In general, SCADA systems make use of a graphical user interface (GUI) that displays annotations and real-time operating parameters superimposed onto plant flow diagrams, such as pressures, temperatures, viscosities, tank levels, flow rates, plant load, valve/damper positions, stack-gas combustion products, salinity indications, equipment on-off status, and the list goes on. SCADA systems typically use page structures for navigating through different systems, and when programmed more thoroughly, shortcut links between pages are included for quick access. SCADA systems typically perform real-time calculations from the retrieved data for displaying plant performance, such as fuel-consumption rates, trending of loads, plant efficiency, etc. Real-time information permits operators to make fine-tune adjustments or configure plant systems to achieve best economy. See Figure 17.5. In addition to real-time performance indicators, SCADA systems archive data for historical purposes, to be used for

Figure 17.5 Central operating station of a ship's propulsion plant having a distributed control system. This system is a hybrid of central monitoring and control computers, console-mounted multiloop PLCs, start-stop switches for important equipment, and displays to indicate important plant parameters. Local microprocessor controllers are installed at the machinery throughout the plant.

trending, troubleshooting, load and economy analysis, future load predictions, proof of meeting contractual obligations, and so on.

In addition to showing annotated flow diagrams, SCADA objects can emulate conventional plant devices, such as analog pressure gages, bar-graphs, level indicators, and digital meters. Alarm mimic panels can be displayed in HMI arrangements that emulate the hardware systems, or via a database of information. The alarm panels can be color-coded to indicate alarm status and can display the measured values and the upper and lower "redline" setpoint limits and values. Screen pages can show emulations of the electrical motor-starter panels or PLCs, themselves, so that the operator would have the same visual image as if standing at the unit. When programmed with enough automation, plants can be controlled almost exclusively from a central SCADA location, or even from off-site.

The distributed control system is the predecessor to SCADA. Loosely, the terms DCS and SCADA are used interchangeably as there are many commonalties, but technically there are nuances. Distributed control more precisely means that individual pieces of equipment or systems use their own dedicated controllers, functioning autonomously from a central system, which does not need to be operating. With DCS, the individual distributed controllers can be networked to communicate with a group controller. The DCS is process-oriented, dealing directly with measured parameters, while the SCADA system is more event-driven and works with data. A DCS links devices through their local controllers, and its computer operating terminals provide high-end control, while a SCADA system is more Human Machine Interface oriented (HMI), interfacing through workstation screens that present a large amount of information in an organized manner. As a generalization, a DCS may be more appropriate to a factory having multiple independent computer/controllers that are dedicated to different production processes, while SCADA is more central-plant related, tending to network all subsystems. SCADA systems are considered to be more flexible than a DCS, as they use a graphical software environment to communicate directly with PLCs, and a single point workstation can remotely start and stop equipment or change setpoints. In today's systems, the distinction is blurred, but the term SCADA seems to be more widely used for plant control.

As an example for distinguishing DCS from SCADA: a boiler control system might use four loop controllers; one for firing rate and air-fuel ratio, another for burner management and flame safeguard, a third for controlling feed-pump discharge-pressure and feedwater-regulator-valve position, and a fourth for furnace/stack-draft regulation. Each of the four loop controllers may be connected to a common boiler master controller, and the arrangement by itself may be considered to be a distributed-control system. When multiple boilers are installed, each boiler may receive the same command signal from the one boiler master in a more complicated distributed-control scheme. The boiler master may then communicate to a central computer, which may also be networked to other plant controllers, whose functions are independent from the boilers. While the boiler master is performing distributed control as part of a localized DCS, the SCADA computer supervises multiple functions and passes the information. Depending on SCADA permissions and passwords, remote start-stop operations of equipment, machinery, or systems, or remote adjustments of the PLC-setpoints may be permitted or disabled, depending on the plant operating philosophy; such as, should an operator locally watch the equipment starting sequence for safety reasons, or should a person with access be permitted to start or readjust the equipment from another building.

SCADA system components

Supervisory computers form the core of the control system. The supervisory computer gathers, archives, and analyzes data obtained from PLCs, which are directly connected to field devices, other RTU microprocessors, or HMI devices. The supervisory computer contains the high-level SCADA software that presents information through an intuitive user interface. Often SCADA systems provide remote system operation and setpoint adjustment. The systems often use multi-monitor displays where several pages can be monitored simultaneously. Multiple supervisory computers may be networked, where some or all are given active control privileges while others are limited to monitoring only. The software programming package generally includes drawing objects, premade symbols, and easy methods for addressing data and devices.

Engineering workstations and *operating stations* are used as supervisory computers. The engineering station has control over the entire DCS and has the engineering software for programming and implementation. It also has the most unrestrictive supervisory capabilities. The operating stations monitor, operate, and control plant equipment and parameters. These computers generally lack the ability to modify programs, being limited to plant operations. In some cases, human-machine interfaces replace operating stations.

Remote Terminal Units are microprocessor controllers that interface the plant measuring devices with the host SCADA/DCS systems that supervise and control. The original RTUs were "dumb" devices that were used for connecting field instruments and relaying information to a host. The original RTUs essentially performed the I/O multiplexing and data-linking functions between the field devices and PLCs. Newer RTUs are now "smart" I/O devices having logic functions embedded in the unit.

Programmable-Logic Controllers (PLCs) are microprocessor controllers that are connected to sensor and switch inputs, or to output coils and actuators. PLCs have more sophisticated control capabilities than RTUs, and PLCs are often used in place of RTUs as they are more versatile, configurable, and nowadays not very expensive. PLCs may behave as RTU-like control black boxes, or they may permit accessibility via built-in touchpads and LCD displays, or they may be linked to HMI devices to provide a more robust user interface.

Human Machine Interface (HMI) is a graphical screen that gives the operator accessibility to the system-control functions via a visual interface. Older HMI devices use keypads or keyboards to navigate the controller, while newer models may have touchscreens for direct access. HMI devices often use mimic diagrams with graphics and symbols to represent the process, and they typically have pages to clearly show multiple systems or equipment, operating data, gages or instruments, trending charts, alarms/resets, etc. Photographic images and eye-catching animations are sometimes included to enhance the screen appearance.

Smart Field Devices are input or output devices that communicate directly with the SCADA system. Smart devices send or receive I/O signals as part of their normal operation but can also provide performance and diagnostic information on top of their control signals. As an example, a control valve may receive 4–20mA signals for positioning the valve, but it may also have the ability to transmit the supply-air pressure, the signal-air pressure, the valve-stem deviation from the setpoint position after a given time delay, as well as the make and model number of the device. This information can be poled or incorporated into an alert system to improve the plant reliability and troubleshooting. In addition to sending information to the SCADA, the intelligent instrumentation may have built-in LCD displays to locally show the measurements, bar charts, time-dependent graphs, fault messages and diagnostics.

Remote communications

Presently there are several options for DCS communications, from the original serial Modbus protocols to Modbus ASCII, Modbus RTU, and Modbus/Ethernet. Other communication protocols include Profibus, DeviceNet, HART, and Foundation Field Bus, which are used to directly connect smart devices to the controller. Ethernet TCP/IP is used for internet connectivity but can also be used for inter-device connections when gateway devices are provided. Nowadays, it is not necessary to use one protocol for the entire DCS, as different "layers" of the control system can use different communication protocols, interconnected through gateways to translate the information.

Newer generation communication may be expected to use XML and other web services, with the intention of simplifying the information technology support (IT). Rockwell Automation's Factory Talk is the web-based protocol used specifically for the Allen-Bradley PLC products. Some other examples of web-based protocols include Wonderware's SuiteLink, GE Fanuc's Proficy, I Gear's Data Transport Utility, OPC-UA, and there are others.

Designing the automation and controls systems

Before programming begins, several types of automation-and-controls drawings are developed. These drawings are used for fabrication, manufacturing, construction, assembly, and wiring. The design engineer should identify all plant equipment to be controlled, including options for local and remote starting, stopping, and operation, gage indications that are useful to the operator for visualizing plant status, electronic transmitters used in measurements, and the output devices that will be started, stopped, or modulated. In putting together the controls system, the design philosophy should identify critical items that affect personnel safety first, followed closely by equipment safety, and then reliable operation. It is important to adhere to the regulations, codes and standards governing the system design. The important functions that produce correct operation, sequencing, and modulation of plant equipment and sub-systems needs to be identified for programming the control scheme. The automation system needs to integrate control strategies that affect plant efficiency and economics, and any other items that are useful for troubleshooting, preventive or predictive maintenance, historical trending of data, as well as any other information that aids the operator in understanding the plant status. The engineer needs to consider alarming requirements, accounting for normal-operating variations, high-and-low-level notification points, and unsafe levels where automatic shutdown is required. The system importance needs to be factored into the control philosophy, as critical systems typically need safety devices that are independent from the control devices. In many cases, the automation-and-controls system is constrained by economics, and the determination of which I&C devices to be used needs to be prioritized. Where possible, the plant operators should be involved in the decision-making process, especially when control-and-monitoring items are to be omitted due to value engineering and budgetary tradeoffs.

The following drawing types define the panel or console arrangements, the quantity and types of devices, the PLC wiring, communication schemes, and the PLC programming.

- Piping and Instrumentation Drawings (P&IDs)
- Panel Layout Drawing
- I&C Plan and Elevation Drawings
- Instrumentation & Controls Wiring Diagrams
- Instrumentation & Controls (I&C) Schedules
- PLC Wiring Diagrams
- DCS and SCADA Communications Architecture

Piping and Instrumentation Drawings (P&IDs (P&IDs (Figure 17.6) are system flow diagrams that typically show all equipment, pumps, heat exchangers, valves, piping, line sizes, etc., but they also include all instrumentation-and-control devices, such as gages and indicators, transmitters, current-to-pneumatic converters, variable-frequency drives, control valves and actuators, etc. The P&IDs form the "road maps" for system design, equipment layout, pipe-routing details, etc., but they also aid in designing the control scheme and in visualizing the operation and control philosophies. The P&IDs are compiled into I&C schedules for purchasing and reordering, but they also aid in locating the instruments in the field as shown on plan and elevation drawings, and for producing the instrumentation-and-control wiring diagrams that show device interconnectivity.

Panel layout drawings

Many times, plant-control systems interface to the operating engineers through local cabinets, freestanding panels, or central-operating consoles. Nowadays, HMI devices are widely used for large and small applications, but even so, it is common practice to include panels that contain operating devices, such as start/stop switches, alarm-acknowledge switches, equipment selector switches, status-indicating lights, and even setpoint adjustment. A freestanding boiler control panel is shown in Figure 17.7. In some instances, control cabinets are eliminated in preference to SCADA workstations.

I&C plan & elevation drawings

Figure 17.8 shows an elevation-view instrumentation-and-control diagram, and Figure 17.9 shows a plan-view diagram. I&C plan and elevation drawings indicate the location of the control devices in the plant relative to the attached equipment and their connections to the controllers. In some cases, conduit runs are shown, and in other cases, home runs indicate device connectivity while reducing drawing clutter. In general, the cabling is field routed by the contractor with

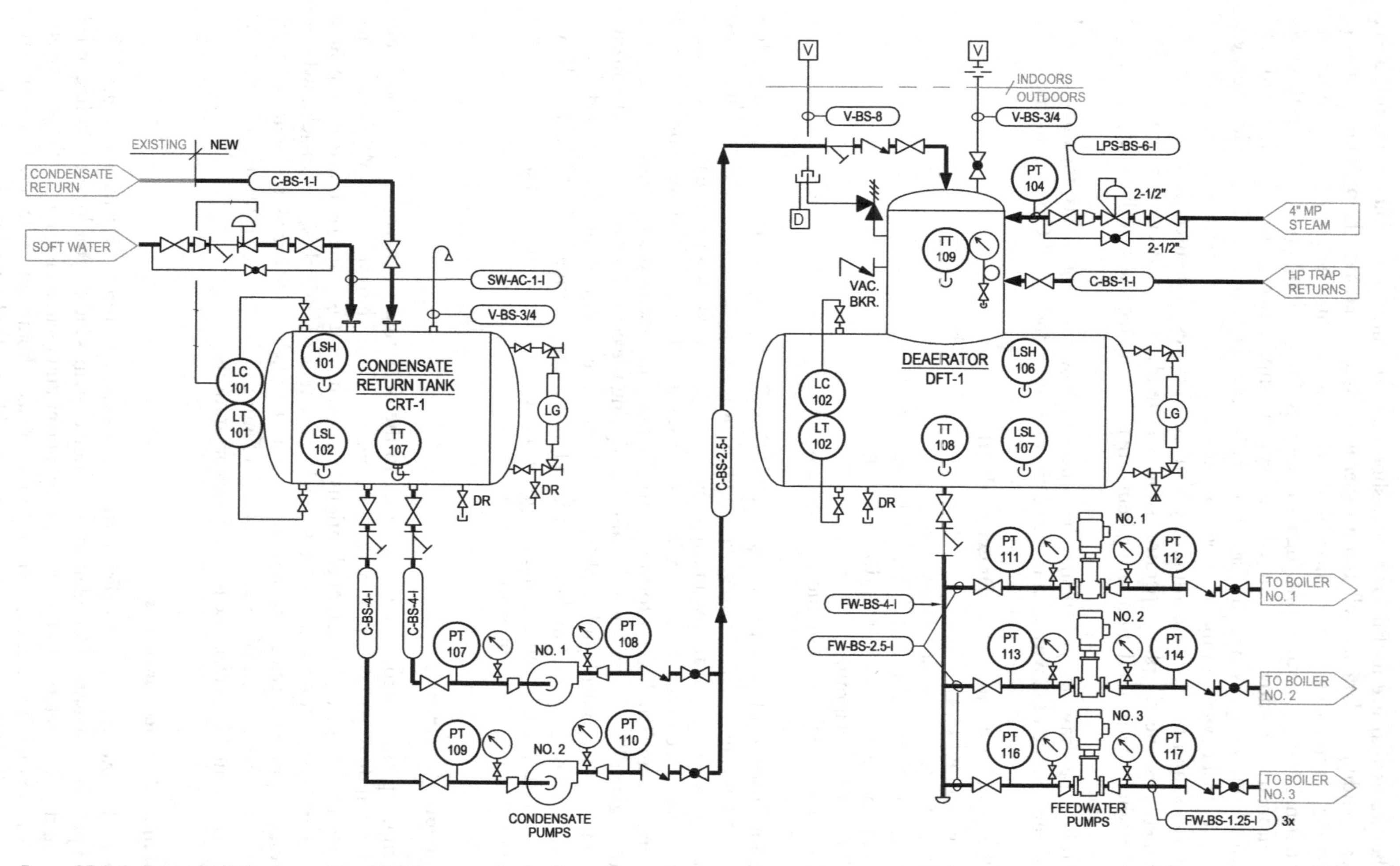

Figure 17.6 A partial P&ID is shown. The P&ID is generally the first step in designing an automation and controls system. The P&ID combines the system flow diagram with the instrumentation to provide an overview of how the system works.

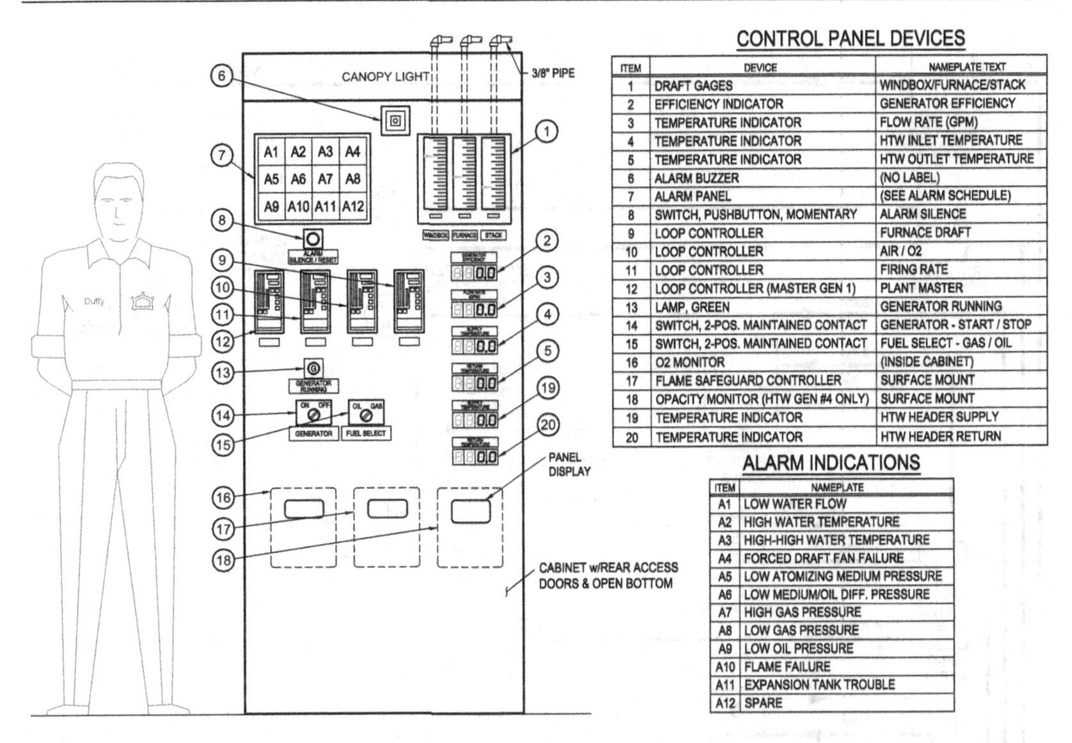

CONTROL PANEL DEVICES

ITEM	DEVICE	NAMEPLATE TEXT
1	DRAFT GAGES	WINDBOX/FURNACE/STACK
2	EFFICIENCY INDICATOR	GENERATOR EFFICIENCY
3	TEMPERATURE INDICATOR	FLOW RATE (GPM)
4	TEMPERATURE INDICATOR	HTW INLET TEMPERATURE
5	TEMPERATURE INDICATOR	HTW OUTLET TEMPERATURE
6	ALARM BUZZER	(NO LABEL)
7	ALARM PANEL	(SEE ALARM SCHEDULE)
8	SWITCH, PUSHBUTTON, MOMENTARY	ALARM SILENCE
9	LOOP CONTROLLER	FURNACE DRAFT
10	LOOP CONTROLLER	AIR / O2
11	LOOP CONTROLLER	FIRING RATE
12	LOOP CONTROLLER (MASTER GEN 1)	PLANT MASTER
13	LAMP, GREEN	GENERATOR RUNNING
14	SWITCH, 2-POS. MAINTAINED CONTACT	GENERATOR - START / STOP
15	SWITCH, 2-POS. MAINTAINED CONTACT	FUEL SELECT - GAS / OIL
16	O2 MONITOR	(INSIDE CABINET)
17	FLAME SAFEGUARD CONTROLLER	SURFACE MOUNT
18	OPACITY MONITOR (HTW GEN #4 ONLY)	SURFACE MOUNT
19	TEMPERATURE INDICATOR	HTW HEADER SUPPLY
20	TEMPERATURE INDICATOR	HTW HEADER RETURN

ALARM INDICATIONS

ITEM	NAMEPLATE
A1	LOW WATER FLOW
A2	HIGH WATER TEMPERATURE
A3	HIGH-HIGH WATER TEMPERATURE
A4	FORCED DRAFT FAN FAILURE
A5	LOW ATOMIZING MEDIUM PRESSURE
A6	LOW MEDIUM/OIL DIFF. PRESSURE
A7	HIGH GAS PRESSURE
A8	LOW GAS PRESSURE
A9	LOW OIL PRESSURE
A10	FLAME FAILURE
A11	EXPANSION TANK TROUBLE
A12	SPARE

Figure 17.7 Standalone boiler control panel using PLC-based multiloop controllers for a moderately sized high-temperature hot-water central heating plant. The panel provides local control and a SCADA-system interface for central monitoring.

general guidance provided by the drawings. The wiring and cabling requirements may be indicated on the plan or elevation drawings, included in an I&C schedule of devices, or contained in a set of job specifications. In rehabilitation work, it is common to distinguish new features using dark linework and existing equipment to be reused with faded-gray linework.

Instrumentation & controls wiring diagrams

Figure 17.10 shows an instrumentation-and-controls wiring diagram for the plant shown in Figures 17.8 and 17.9. Four multiloop controllers are used in a high-temperature water boiler and burner control system. The master controller receives the plant operating-parameter signals, compares the operation to the setpoints and produces a firing-rate control signal that is sent to the combustion-control system for each boiler in the plant. The master also performs lead-lag load-management to automatically cycle boilers to match the plant load. The combustion-control system uses the master signal to adjust the firing rate and to modulate the fuel and air for efficient, smokeless combustion. The signals are also used by the flame safeguard and burner-management systems. The I&C diagram shows the field devices that are identified using a tag number system which ties the devices to an equipment schedule, the device locations in the plant, and the communication cabling between controllers.

Instrumentation & controls schedules (I&C)

After the control requirements are addressed and the instrumentation-and-control diagrams are completed, the device details are summarized in an I&C schedule. The I&C schedule

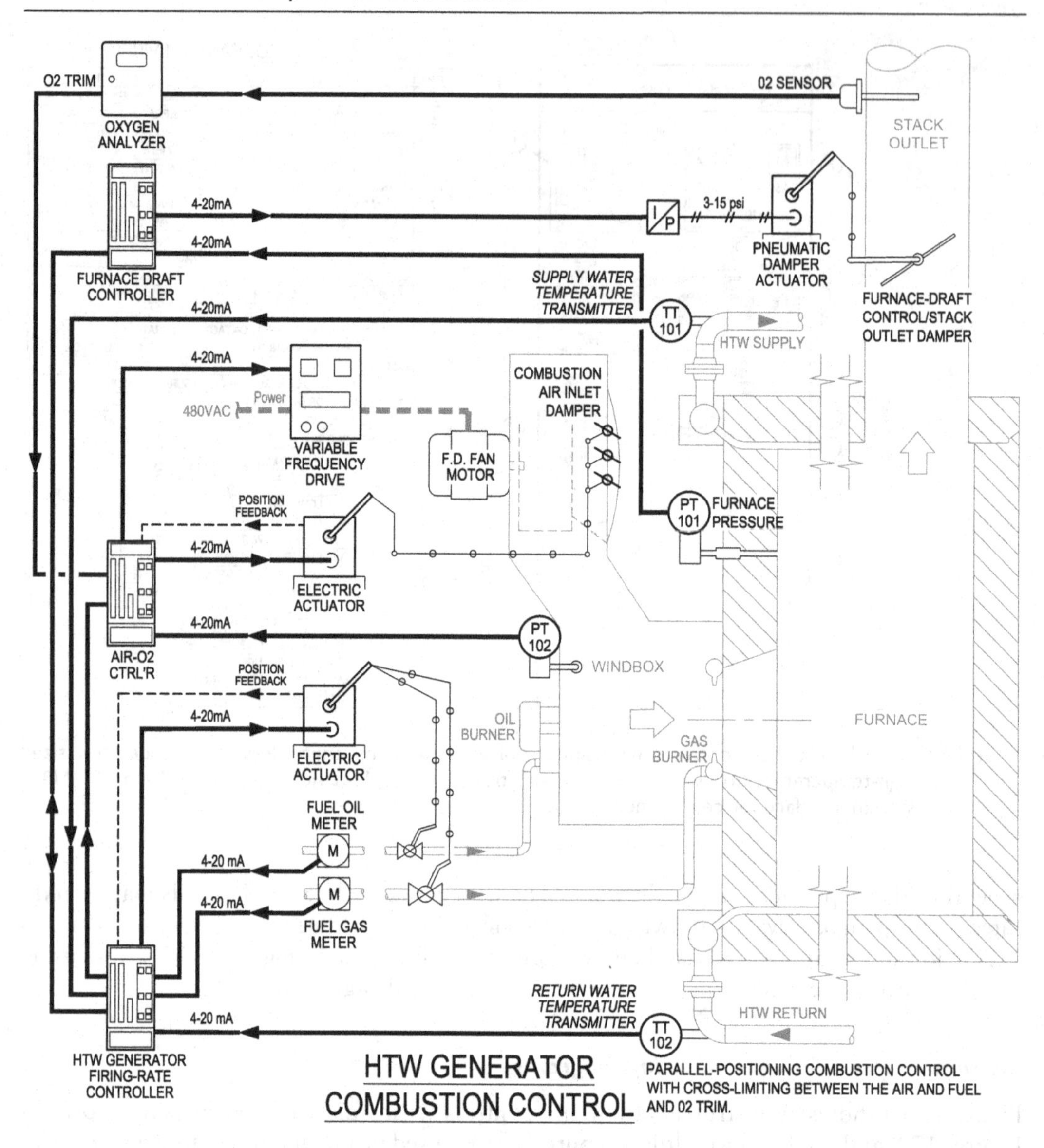

Figure 17.8 A combined I&C elevation drawing and signal-wiring diagram for a fully metered parallel-positioning combustion-control system, having cross-limiting and O2 trim. The diagram shows loop controllers and their final-device connections for a dual-fuel hot water boiler.

(Figure 17.11) identifies every instrument and control device in the system by its tag number, which is cross-referenced to all control drawings. The schedule includes the device function, a brief description, its location in the plant, the signal type and its output connectivity to its associated device. Along with specifications, the schedule provides the range of operation and other important information for purchasing.

DCS and SCADA communications architecture

See Figures 17.13 and 17.16 for Ethernet IP, and Figures 17.12 and 17.15 for Modbus RTU serial communication. Figure 17.13 shows an illustrative communications architecture using

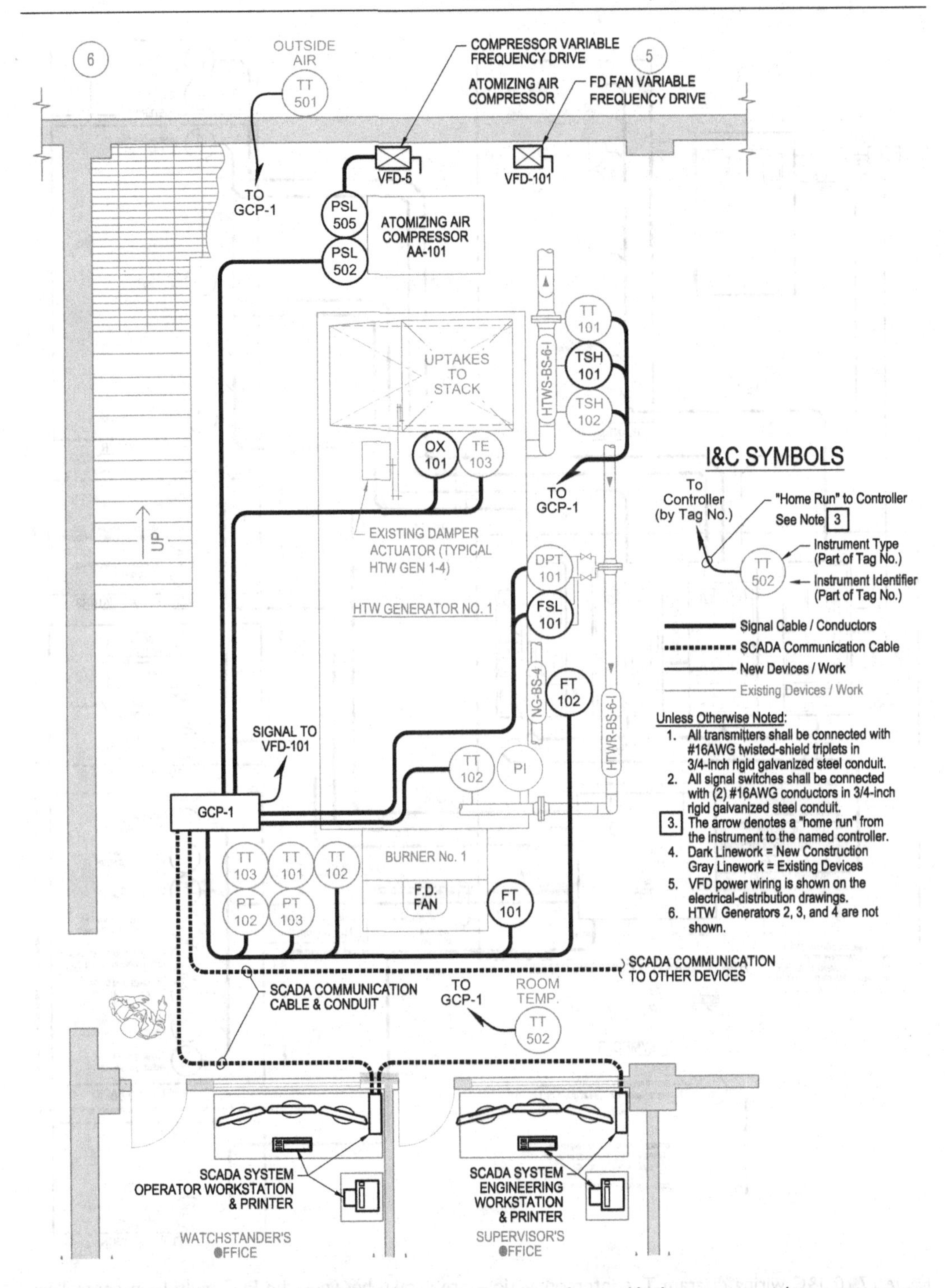

Figure 17.9 I&C plan drawing showing devices mounted on one of several high-temperature hot-water generators used in a large central-heating plant. The drawing shows the locations of all devices and their "home runs" back to their respective controllers, and it specifies the signal-wiring types.

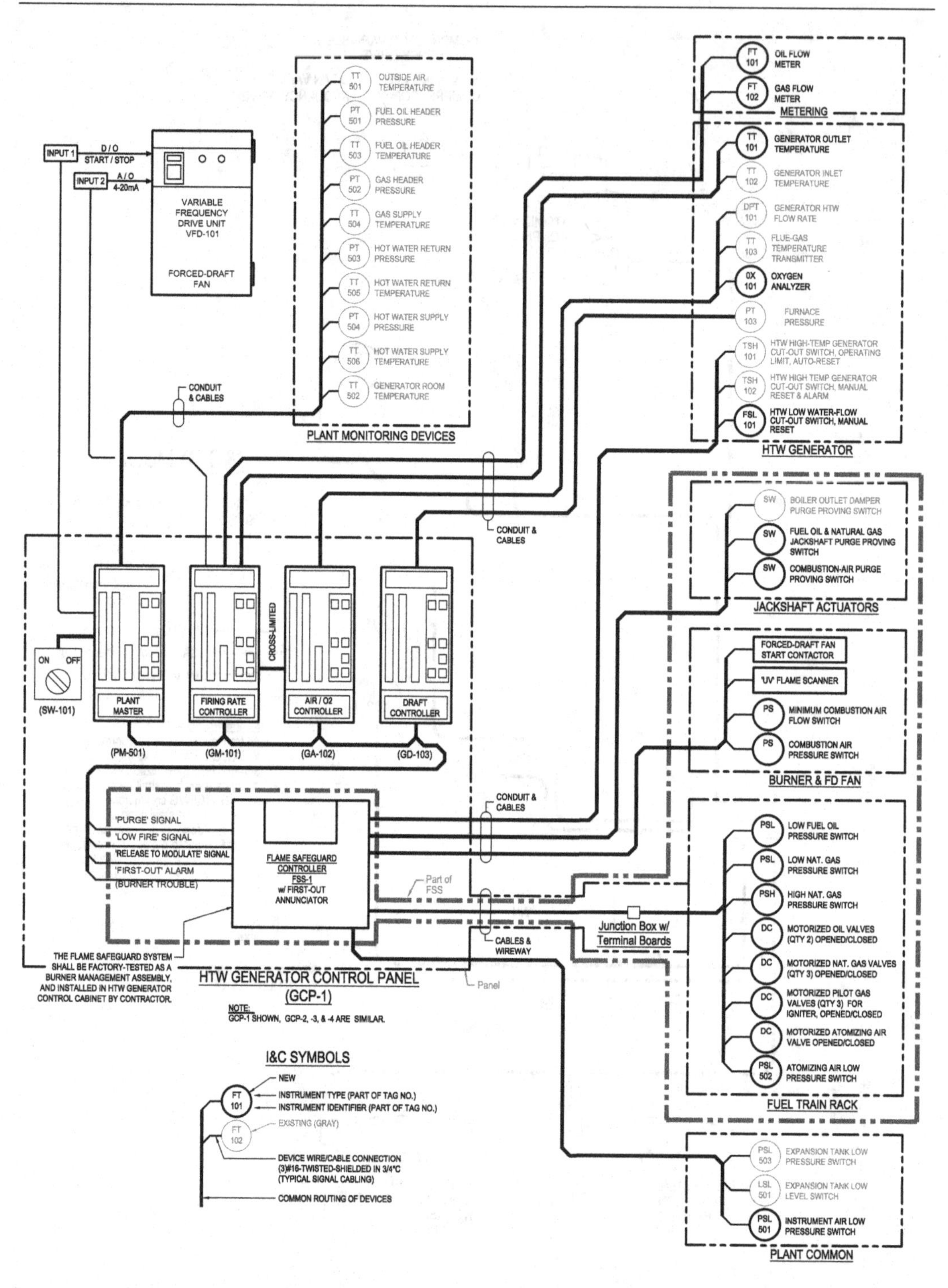

Figure 17.10 I&C wiring diagram. The interconnections are shown between the PLC multi-loop controllers, the attached input instruments, and the actuated output devices. The diagram works with plan and elevation drawings to define locations, and the equipment schedule for procurement.

INSTRUMENTATION & CONTROLS SCHEDULE

ITEM	INSTRUMENT TYPE	NEW TAG NO.	DESCRIPTION	LOCATION	SIGNAL to / from	RANGE	REMARKS
FLOW SWITCHES							
1	FLOW,LOW	FSL 101	SWITCH w/ LOCAL DISPLAY	GEN No.1	FSS-1	0-800gpm	EXTG MODEL 288A PRIME MEAS.
LEVEL SWITCHES							
2	LEVEL,LOW	LSL 501	EXPANSION TANK, LOW LEVEL	EXP TK	FSS-1-4	0-10'w.c.	MANUAL RESET
PRESSURE SWITCHES							
3	PRESSURE,LOW	PSL 501	INSTRUMENT AIR PRESS. LOW	PLANT	FSS-1-4	0-125psig	
4	PRESSURE,LOW	PSL 502	ATOMIZING AIR PRESS. LOW	PLANT	FSS-1	0-125psig	
5	PRESSURE,LOW	PSL 503	EXPANSION TANK, LOW PRESS.	EXP TK	FSS-1	0-200psig	MANUAL RESET
TEMPERATURE SWITCHES							
6	TEMP,HIGH	TSH 101	HIGH TEMP CUT-OUT,GEN 1	GEN No.1	FSS-1	0-500°F	AUTOMATIC RESET
7	TEMP,HIGH	TSH 102	HI-HI TEMP CUT-OUT,GEN 1	GEN No.1	FSS-1	0-500°F	MANUAL RESET
SWITCHES							
8	SWITCH	SW 101	ON-OFF, GEN 1	GEN No.1	FSS-1	0-500°F	2-POS, MAINT. CONTACT
9	DRY CONTACTS	DC 1XX	BURNER CONTROLS	VFD-1	FSS-1	0-100"w.c.	BURNER TRAIN DEVICES
FLOW TRANSMITTERS							
10	FLOW	FT 101	OIL TO BURNER,GEN 1	GEN No.1	GM-101	0-3gpm	POS.DISPL. w/PULSE OUTPUT
11	FLOW	FT 102	NAT GAS TO BURNER 1	GEN No.1	GM-101	0-25000cfh	VORTEX SHEDDING TYPE
LEVEL TRANSMITTER							
12	LEVEL	LT 501	EXPANSION TANK	EXP TK	PM-501	0-18'w.c.	MANUAL RESET
PRESSURE TRANSMITTERS							
13	PRESSURE	DPT 101	HTW FLOW, GEN 1	GEN No.1	FSS-1	0-110"w.c.	EXTG BAILEY MOD. PTDFD1212
14	PRESSURE	PT 102	BURNER WINDBOX, GEN 1	VFD-1	FSS-1	0-100"w.c.	EXTG BAILEY MOD. PTDFD1212
15	PRESSURE	PT 103	FURNACE PRESSURE	GEN No.1	FSS-1	0-30"w.c.	EXTG BAILEY MOD. PTDFD1212
16	PRESSURE	PT 501	FUEL-OIL HEADER PRESS.	PLANT	PM-501	0-150psig	
17	PRESSURE	PT 502	FUEL-GAS HEADER PRESS.	PLANT	PM-501	0-20psig	
18	PRESSURE	PT 503	HTW RETURN FROM CAMPUS	PLANT	PM-501	0-400psig	
19	PRESSURE	PT 504	HTW SUPPLY TO CAMPUS	PLANT	PM-501	0-400psig	
20	PRESSURE	PT 505	ATOMIZING AIR VFD SIGNAL	VFD-5	PM-501	0-125psig	FEEDBACK TO VFD CTRL'R
TEMPERATURE TRANSMITTERS							
21	TEMPERATURE	TT 101	HTW OUTLET TEMP,GEN 1	GEN No.1	FSS-1	0-500°F	EXTG BAILEY EQ10 Temp/mV
22	TEMPERATURE	TT 102	HTW INLET TEMP,GEN 1	GEN No.1	FSS-1	0-500°F	EXTG BAILEY EQ10 Temp/mV
23	TEMPERATURE	TT 103	FLUE GAS TEMP,GEN 1	GEN No.1	FSS-1	0-800°F	EXTG BAILEY EQ10 Temp/mV
24	TEMPERATURE	TT 501	OUTSIDE AIR TEMP	OUTSIDE	FSS-5	0-100°F	
25	TEMPERATURE	TT 502	GENERATOR ROOM TEMP	PLANT	FSS-5	0-100°F	
26	TEMPERATURE	TT 503	FUEL OIL SUPPLY TEMP	PLANT	FSS-5	0-100°F	
27	TEMPERATURE	TT 504	GAS SUPPLY TEMP	PLANT	FSS-5	0-1500°F	
28	TEMPERATURE	TT 505	HTW RETURN FROM CAMPUS	HTW RET	FSS-5	0-500°F	HTW PLANT RETURN PIPE
29	TEMPERATURE	TT 506	HTW SUPPLY TO CAMPUS	HTW SUP	FSS-5	0-500°F	HTW PLANT SUPPLY PIPE
CONTROLLERS							
30	LOOP CTRL'R	GM 101	FIRING RATE CTRL, GEN 1	GCP-1	GA-102		CALCULATE EFFICIENCY
31	LOOP CTRL'R	GA 102	AIR/O2 CTRL,GEN 1	GCP-1	GD-103		AIR/FUEL & O2 TRIM
32	LOOP CTRL'R	GD 103	FURNACE DRAFT CTRL, GEN 1	GCP-1	FSS-1		EXTG DAMPER ACTUATOR
33	LOOP CTRL'R	PM 501	PLANT MASTER	GCP-1	GM-101-4		FIRING-RATE SIGNAL
VARIABLE-FREQUENCY DRIVES							
34	VAR SPD DRIVE	VFD 101	FORCED-DRAFT FAN,GEN 1	GCP-1	GM-101		COMB CTRL, COORD w/ DAMPER
MISCELLANEOUS DEVICES							
35	OXYGEN ANALYZER	OX 101	OXYGEN ANALYZER,GEN 1	STACK	FSS-1		ROSEMOUNT WORLDCLASS 3000
36	OPACITY MONITOR	AT 401	OPACITY MONITOR,GEN 4	STACK	FSS-4		PREFERRED MOD. JC30F4C

Figure 17.11 I&C equipment schedule showing device types and their identifying tag numbers. The schedule defines the devices, locations, connecting points, and operating ranges. Switches are digital-input devices and transmitters are analog-input.

Ethernet IP as its backbone. This diagram shows various communication protocols at the lowest layers, with information linked to the SCADA system through an Ethernet IP system of switches and routers. In cases where the lowest layer communicates uses Ethernet IP, the connections are made directly through switches or routers. In cases where the lowest layer communicates using a series protocol such as Modbus RTU, a gateway device is used to

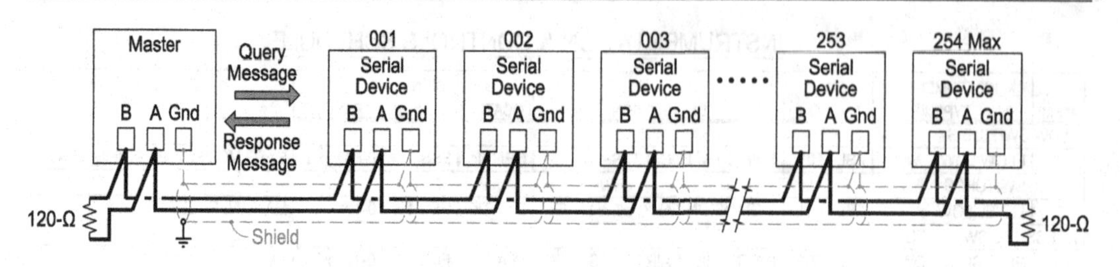

Figure 17.12 Modbus signal connection for half-duplex communication. (See also Figure 18.12.)

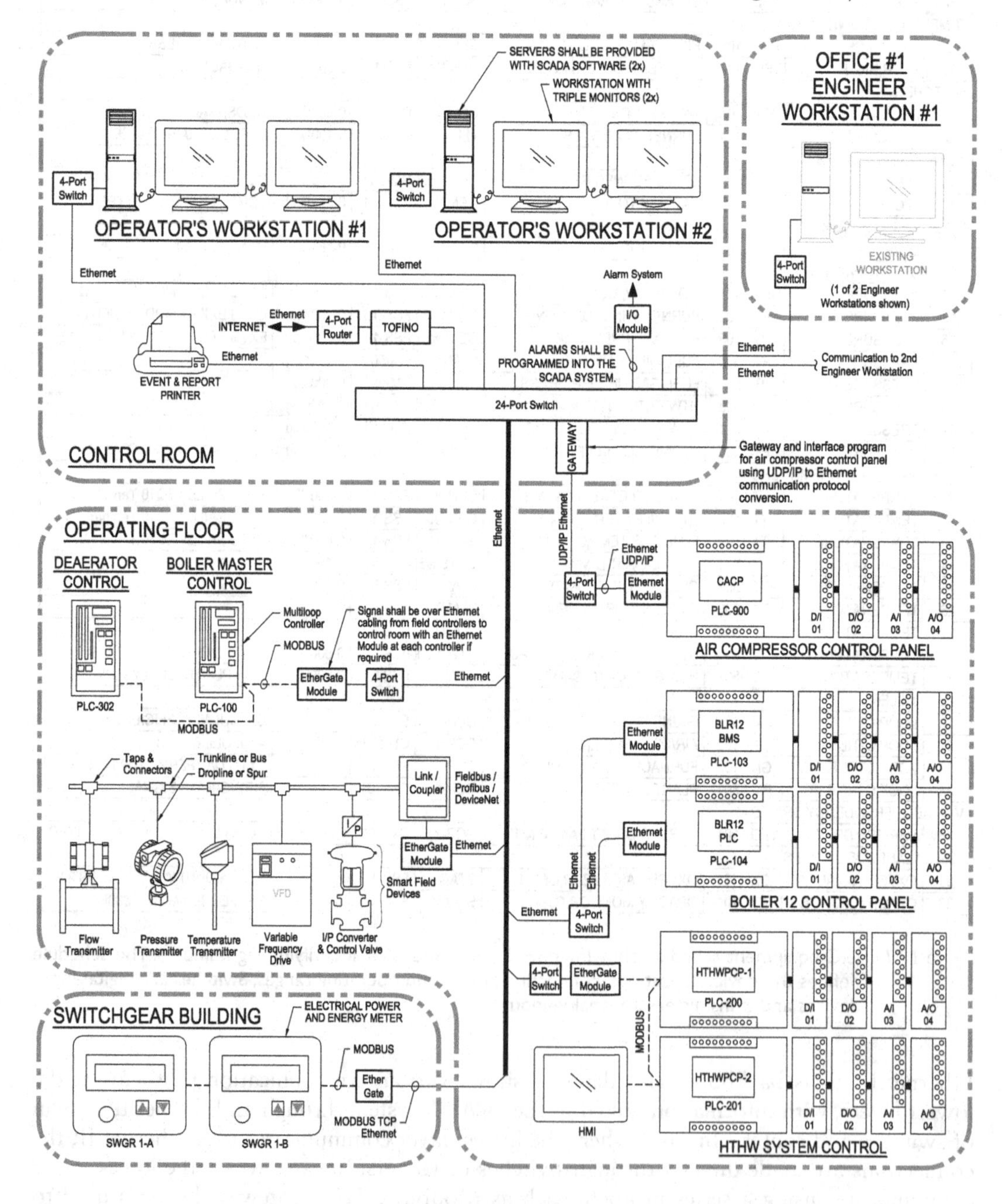

Figure 17.13 SCADA system using various communication protocols at different layers for different subsystems. This hybrid communication system is based on Ethernet IP connections to sub-system controllers using a mix of Ethernet UDP/IP, Modbus RTU, Modbus TCP, and field-bus protocols.

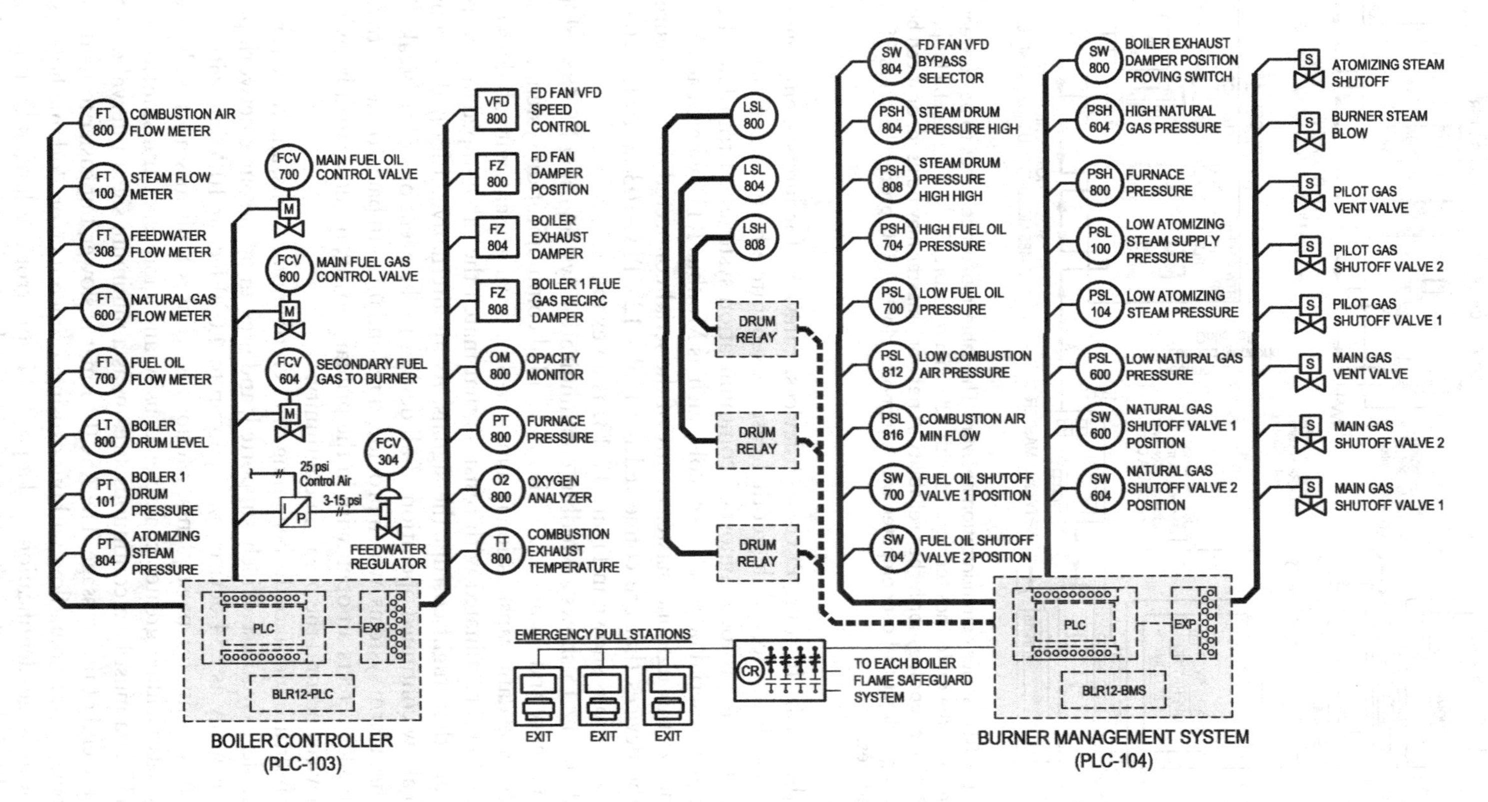

Figure 17.14 I&C wiring diagram for the boiler PLC controllers shown in Figure 17.13. This boiler PLC-based control system is an alternative to the multiloop controllers shown in Figure 17.10.

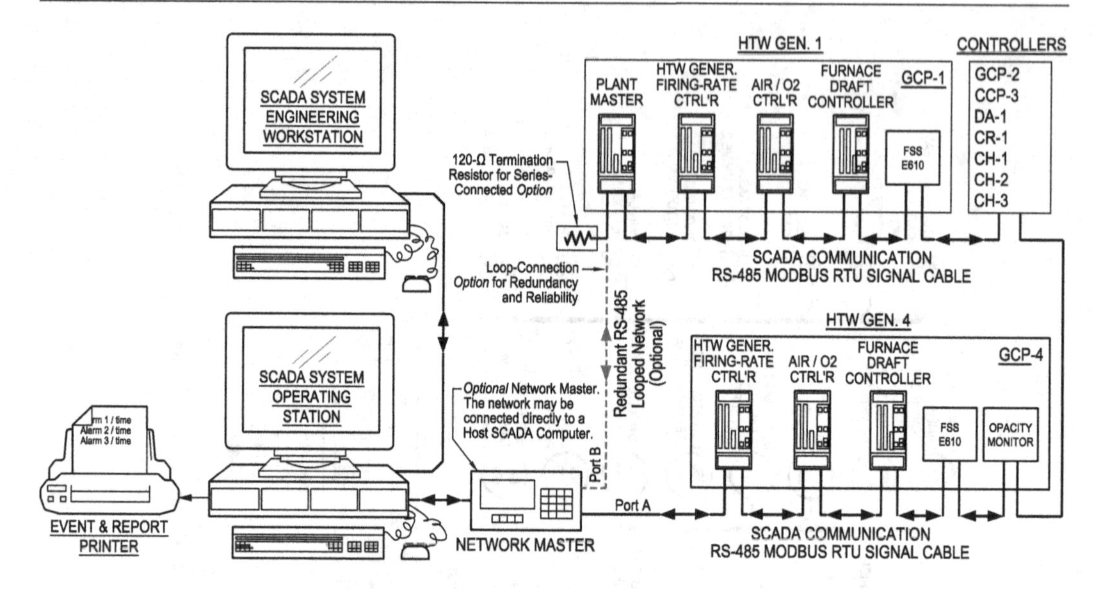

Figure 17.15 Modbus RTU series communication network. The cable impedance needs to be between 50–120 ohms, with the cable ends terminated with 120-ohm resistors. The cabling can be looped back to form a ring topology for higher reliability. A network master may be used, or it can be avoided by using a direct-to-host computer, which supports Modbus RTU protocol and has RS-485 serial interfaces.

translate to Ethernet IP. In some PLCs and devices, gateways for intersystem communications are built into the device, while in other cases, a separate ethernet gateway must be furnished. EtherGate is a "protocol-agnostic" communication system that translates TCP data to any of a number of serial network protocols, such as Modbus RTU, where each EtherGate layer communicates through one master to any of the serial-connected devices. Figure 17.14 relates all of the devices that are connected for Boiler-12's PLC-103 and PLC-104 to the SCADA communications shown in Figure 17.13, as an example.

Figure 17.12 and 17.15 shows a Modbus communication system using RS 485 cabling for inter-PLC connectivity and Figure 17.16 shows a radial communication system for Ethernet IP. Modbus uses serial data transmission in a master-slave arrangement, with all devices wired in series. The ends are terminated with resistors that match the cable impedance to prevent voltage reflections that might corrupt the signals. As an alternative, a looped network can be provided to allow continued operation of all devices in the event of a wire break, like the NFPA Class A fire-alarm systems, which provides a redundant path back to the control panel. When the controller detects an open wire in the primary string, it automatically switches to back feed all devices beyond the break in continuity.

Modbus has been around since the beginning of PLC communications, and there is no end in sight, as it is simple, easy to use, widely supported, and uses an open-source network protocol. It can connect as many as 254 devices and span up to 4,000 feet at 10 Mbps without signal repeaters, which is plenty for most plants. Modbus RTU uses 8-bit messages with two 4-bit hexadecimal words yielding greater character density and more data transmission per time period. The RTU data must be a continuous stream, as a minimum silent interval of 3.5 characters initiates the end of the message. The ASCII format is slower, but easier to read, and can tolerate short pauses in the data flow without causing a timeout error. Both methods use start and stop bits for message identification, and a parity bit for error checking. RTU is much more common for industrial-control communication. Although not shown in the figures, Modbus TCP can be used to transmit PLC data using an Ethernet IP client-server architecture, where the Modbus RTU data is essentially remapped into TCP/IP packets.

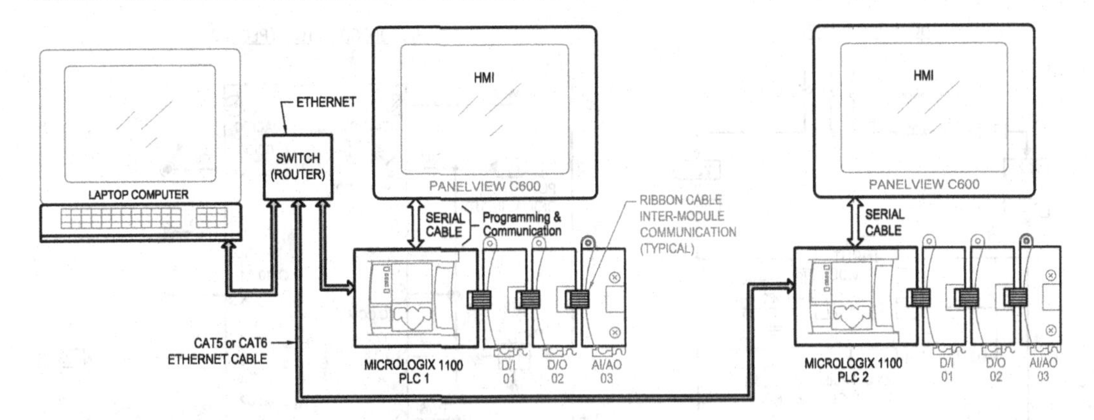

Figure 17.16 The I&C tag numbers are tied to the PLC wiring connections. The PLC program itself references the devices via the wire-terminal connections, which form the program address. The program is loaded from a laptop or engineering workstation through the communication system.

PLC wiring diagrams ***(Figures 17.17, 17.18, and 17.19)***

The PLC-wiring diagrams integrate the plant input and output devices into the controller itself. The wiring diagrams show the power connections that operate the PLC, as well as power connections to drive the field devices. More importantly, the wiring diagrams indicate the signal connections from the field devices into the PLC, where the PLC wiring terminals become the programming addresses. Figure 17.17 shows the dc power-supply wiring on the left ladder and the PLC connections on the right ladder. The PLC dc connections run the PLC itself and any connected I/O devices needing power. The PLC shown has on-board digital input-and-output-signal capability. Figure 17.18 shows two expansion modules, where the one on the left ladder adds more digital-input capability and the one on the right adds more digital outputs. Figure 17.19 shows a four-channel analog I/O module having two A/I channels and two A/O channels.

Line numbers are located to the left of each rung for cross referencing the wiring diagram to the devices. The system shown uses a technique called line-smart, page-smart, where the first digit indicates the sheet number, and devices are identified by their tag numbers relative to the sheet. Each line number identifies the attached conductor and device, which is incorporated into the tag number. As an example, line 401 in Figure 17.18 provides both a digital-input switch and it receives a digital-output signal for its light, and is identified by tag number PBLT401. The tag number directs the controls engineer to page 4 in the wiring diagrams, where it is identified as a lighted pushbutton. The switch digital-input signal is supplied with positive voltage by conductor P401, and is completed to module 01 through conductor P401A. Module 01 indicates that it is physically the first expansion module in the rack, and it is tagged as PLC401. P401A is attached to the "0" terminal and has programming address I:1/0. All four-hundred-tag devices are located on page 4 of the drawing set.

The "02" as part of "Module 02" on the right side of Figure 17.18 indicates that this device is physically installed in the second expansion location in the rack, and the wiring diagram indicates that it produces digital-output signals. Module 02 is tagged as PLC422, where the tag number was obtained from the line number located at its first terminal. It receives positive voltage from rail 401, shown via the continuation symbol numbered as 422. Line 423 connects the first terminal to lamp PBLT401 via conductor P423, and it completes the lamp signal to the negative-terminal rail, tagged as N402. Its PLC address is O:2/0, indicating that

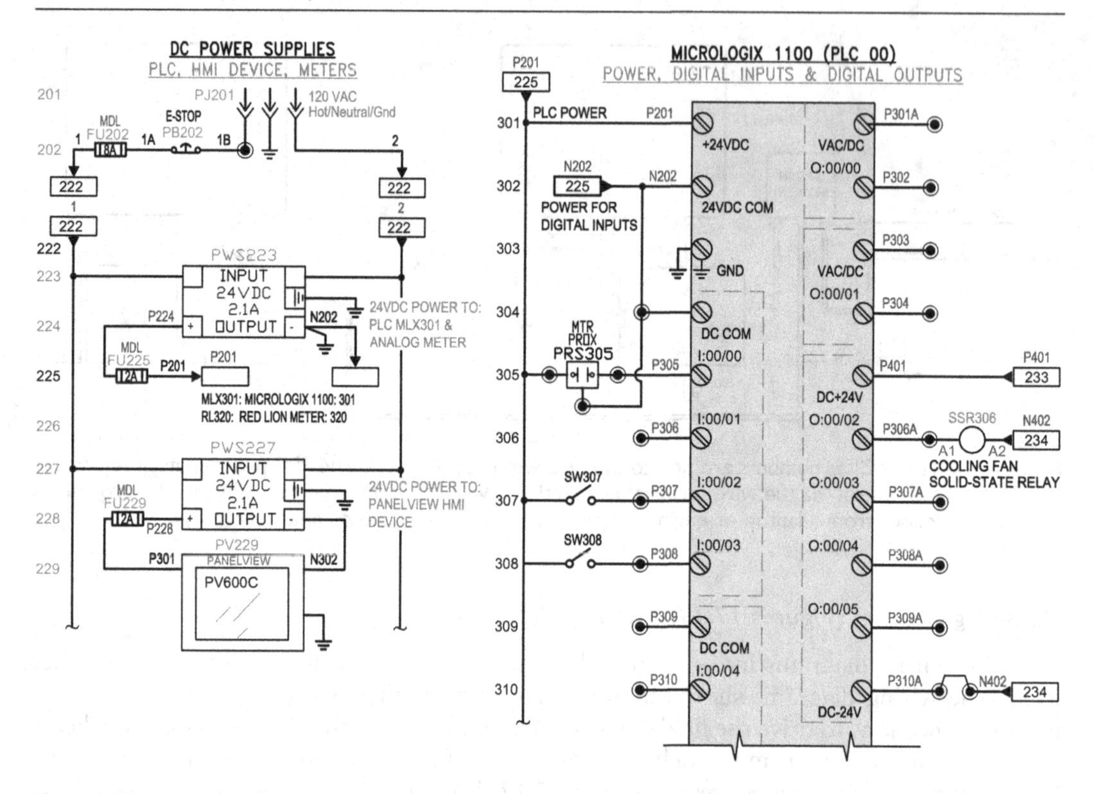

Figure 17.17 PLC wiring diagrams. The circuit for ac-to-dc power supplies is shown on the left. The PLC wiring is shown on the right.

its signal originates from the first terminal on the second expansion device. The number 401 below the lamp symbol cross references the light to the switch.

Figure 17.19 shows the wiring connections for a four-channel analog-input-and-output device, where two channels receive inputs and two channels send outputs. The five-hundred numbers indicate this drawing to be on page 5. The detail shows that both input signals are set to 4-20 mA by microswitches, and the input devices are wired through conductors surrounded by a grounded shield. The meter uses a voltage signal because the wire on line 501 is connected to V_{out}, whereas a current signal would be wired to line 502. The meter is separately energized from the 24 Vdc power supply shown on page 2, with connections made at line 225. In both cases, the voltage/current signals would be defined in the configuration data file within the PLC program.

SCADA system screens

A variety of SCADA-system screen captures are shown in Figure 17.20 through Figure 17.24, showing some presentation options. The SCADA screens include system diagrams that are similar to P&ID diagrams with indications of the active operational status, a simulated strip chart, and chiller equipment pages.

Programming the HMI device

Figure 17.25 through Figure 17.31 show various pages used in the construction of HMI screens for the Allen-Bradley PanelView 600. In general, SCADA systems and HMI devices are programmed by establishing communications, placing objects on the screen to create

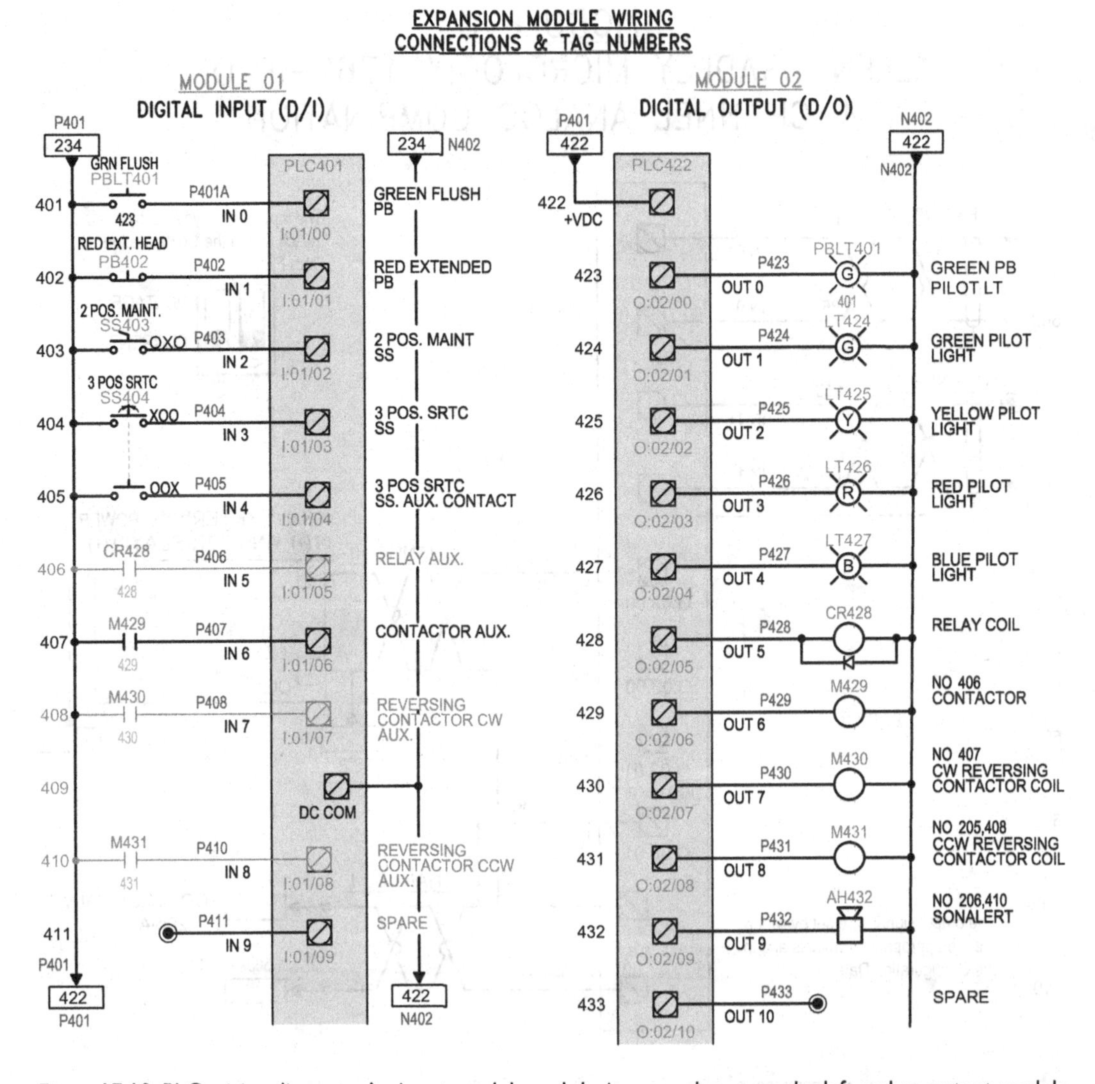

Figure 17.18 PLC wiring diagrams. An input module and devices are shown on the left and an output module and devices are shown on the right. This illustration provides an example of line-smart/page-smart annotations. The first digit on the rung line identifies that this detail is located on sheet 4 of the drawing set. Using rung line *401* as an example, the lighted pushbutton tag number is PBLT*401* and the wire label is P*401*. The 423 tag below the PBLT*401* switch symbol indicates that the output lamp is shown on the same page 4, having its conductor labeled as P423 and a lamp labeled as PBLT401.

system diagrams, dragging active virtual input and output devices onto the screen, applying color or other attributes to indicate status, creating tags that map back to the source PLC, the data type, and the PLC data addresses.

Security

Original automation and control systems were completely isolated from the world and communicated only within the plant. Nowadays, many, if not most SCADA systems are connected

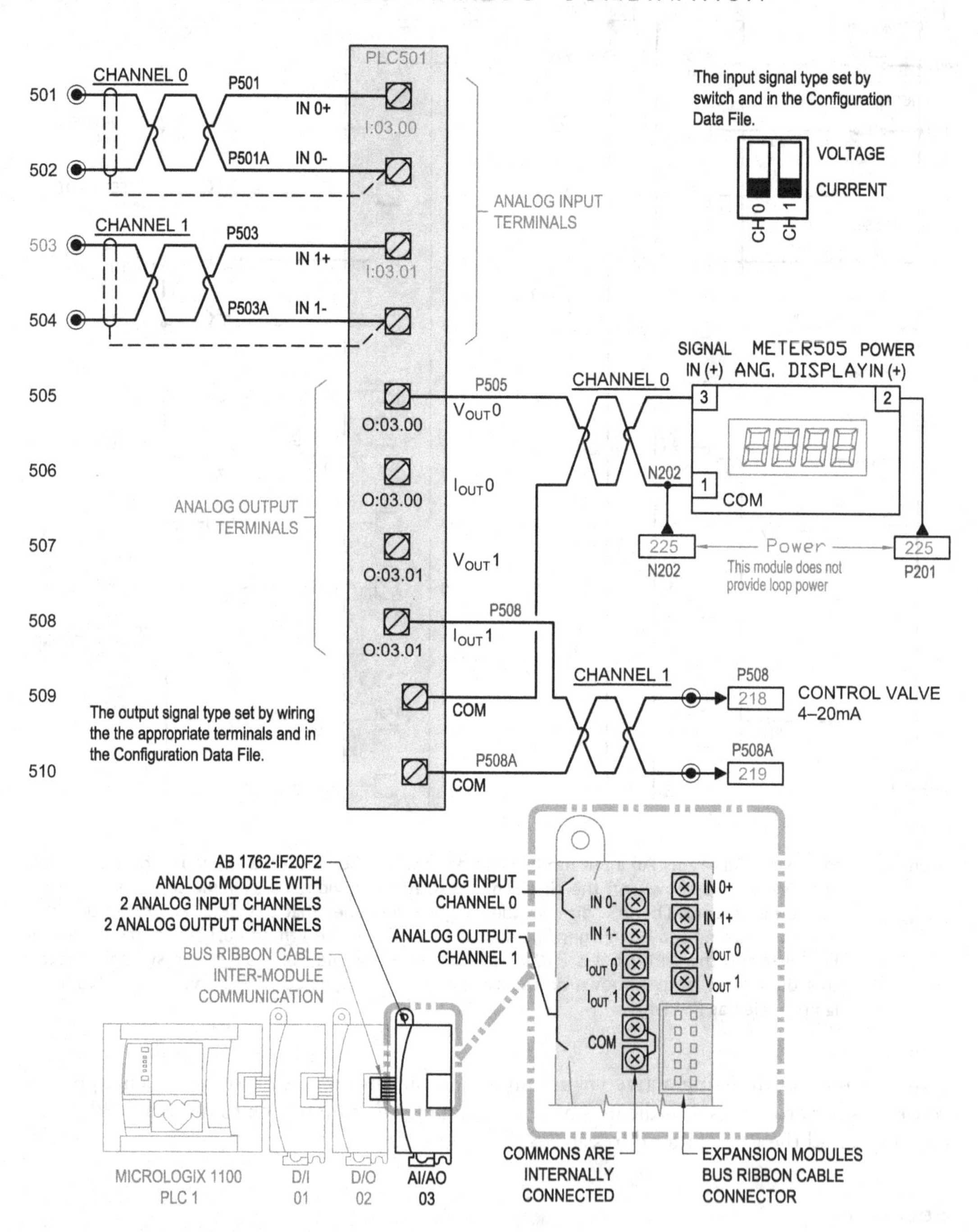

Figure 17.19 Wiring diagram for an analog I/O module.

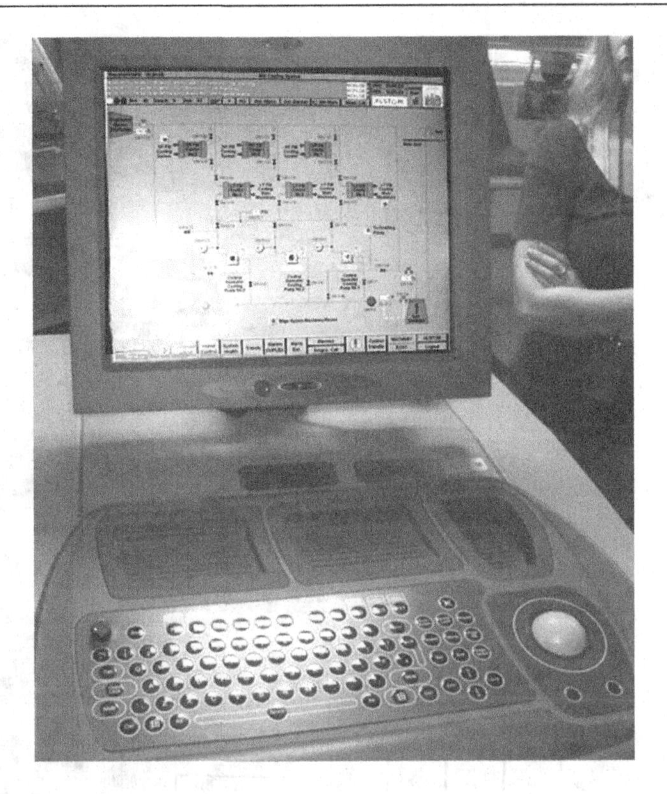

Figure 17.20 Shipboard SCADA system.

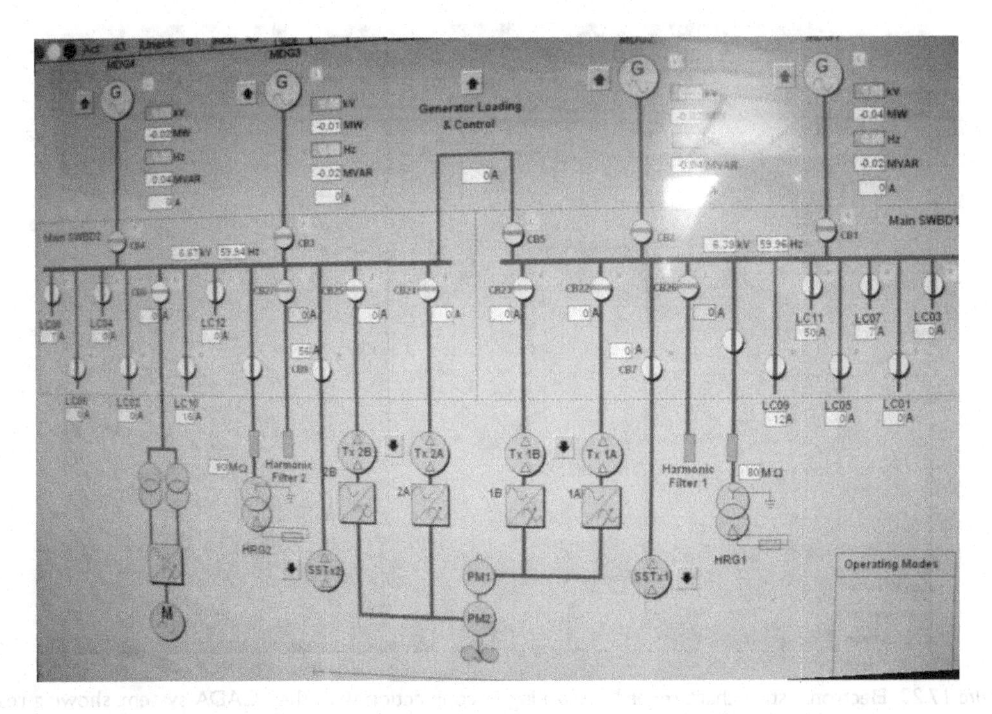

Figure 17.21 SCADA display of the electrical distribution system.

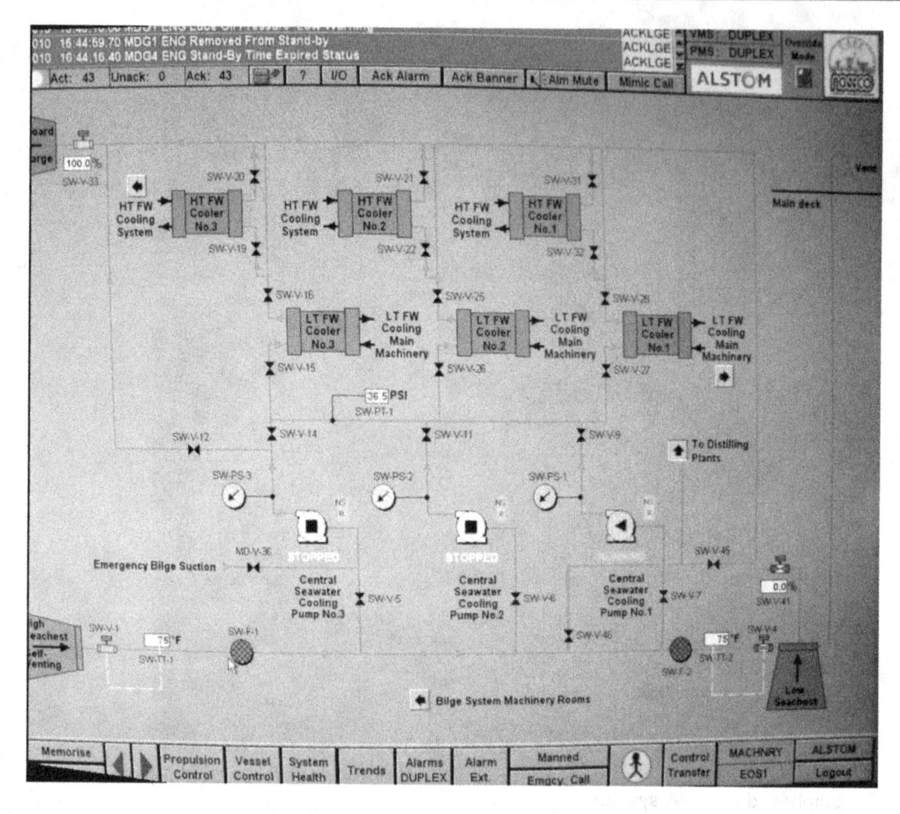

Figure 17.22 SCADA display of the ship's central cooling water system.

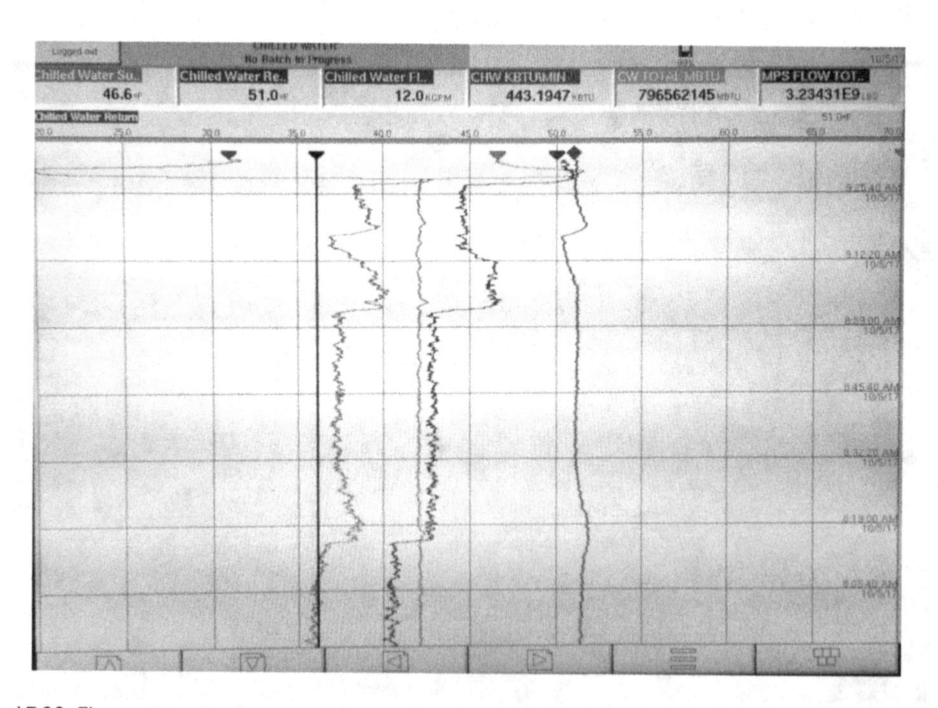

Figure 17.23 Electronic strip chart recorder, working in conjunction with the SCADA system, showing real-time parameters. Historical data is saved for efficiency calculations, trending, and troubleshooting.

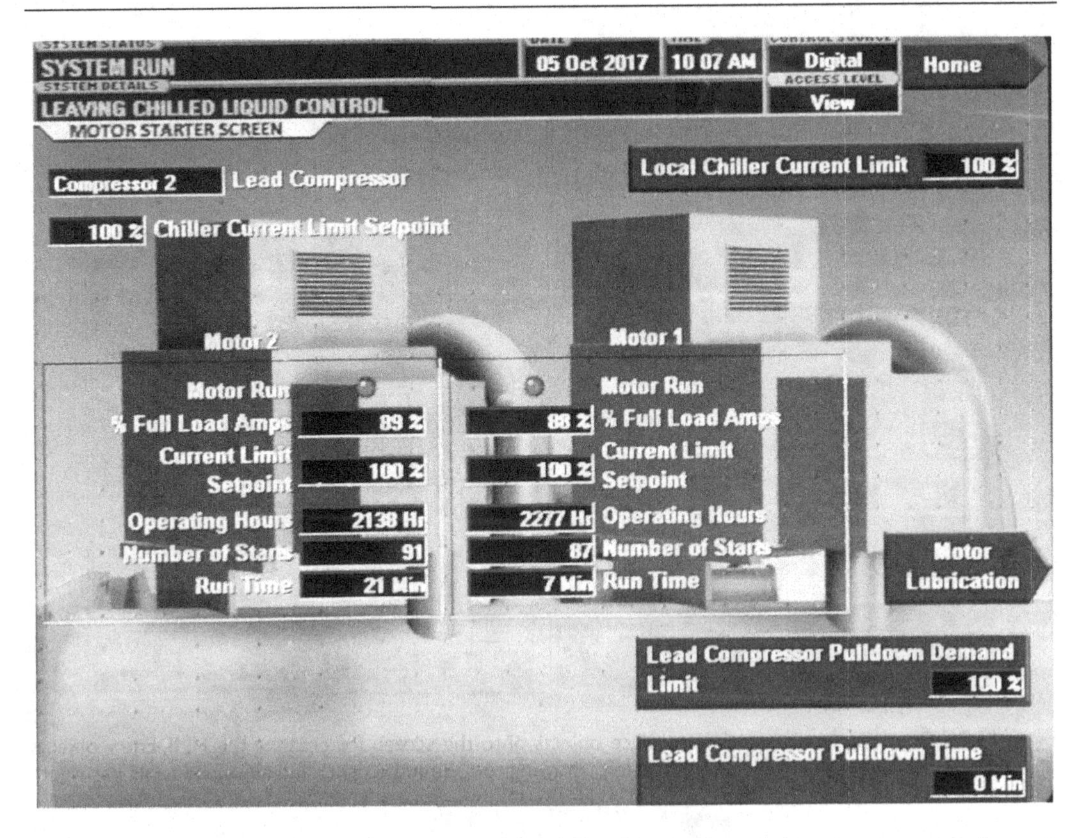

Figure 17.24 HMI display screen for a large centrifugal chiller showing the real-time motor status and operating parameters. Typically, the operator can change the setpoints.

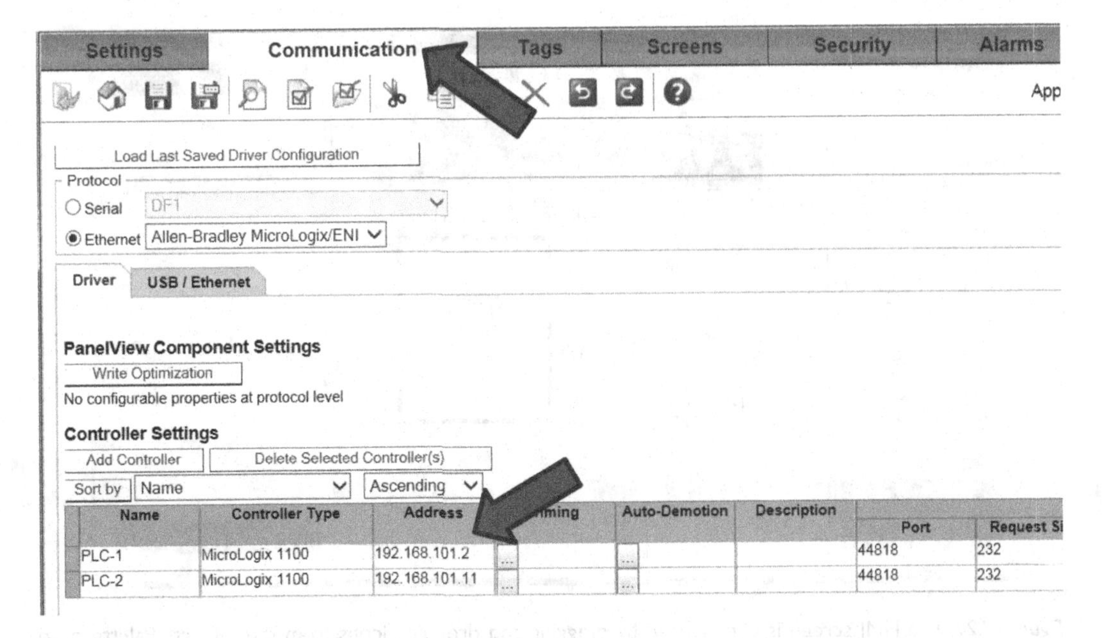

Figure 17.25 The communication tab is used to set the PLC Controller to its Model No. and IP Address.

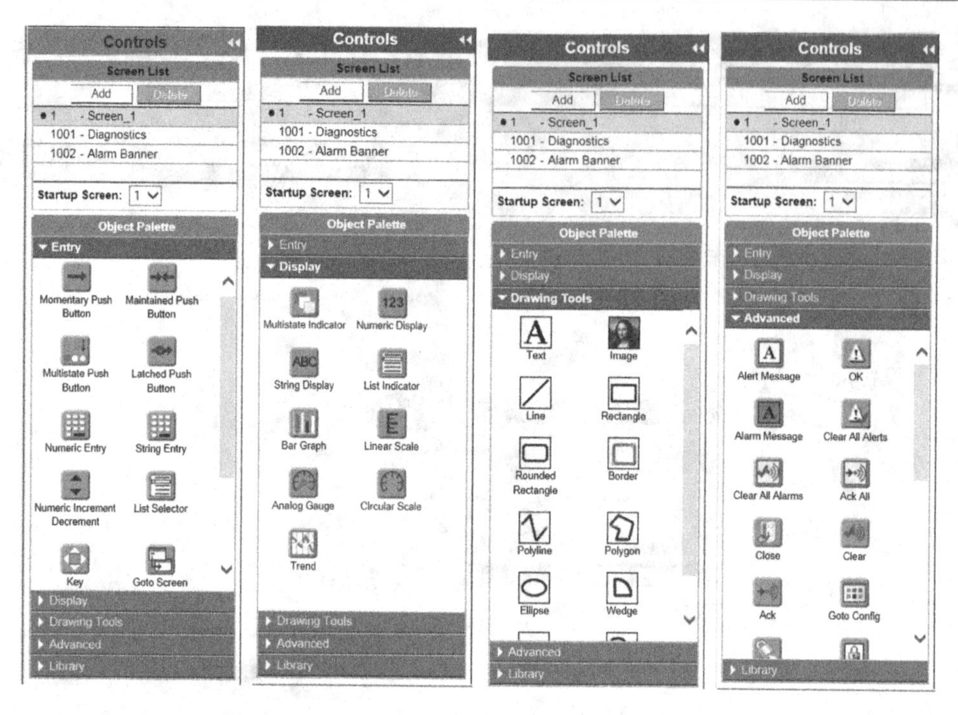

Figure 17.26 Controls menus are used to place objects onto the screen for creating the HMI. Entry objects (far left), display objects (middle left), drawing tools (middle right), and advanced tools (far right).

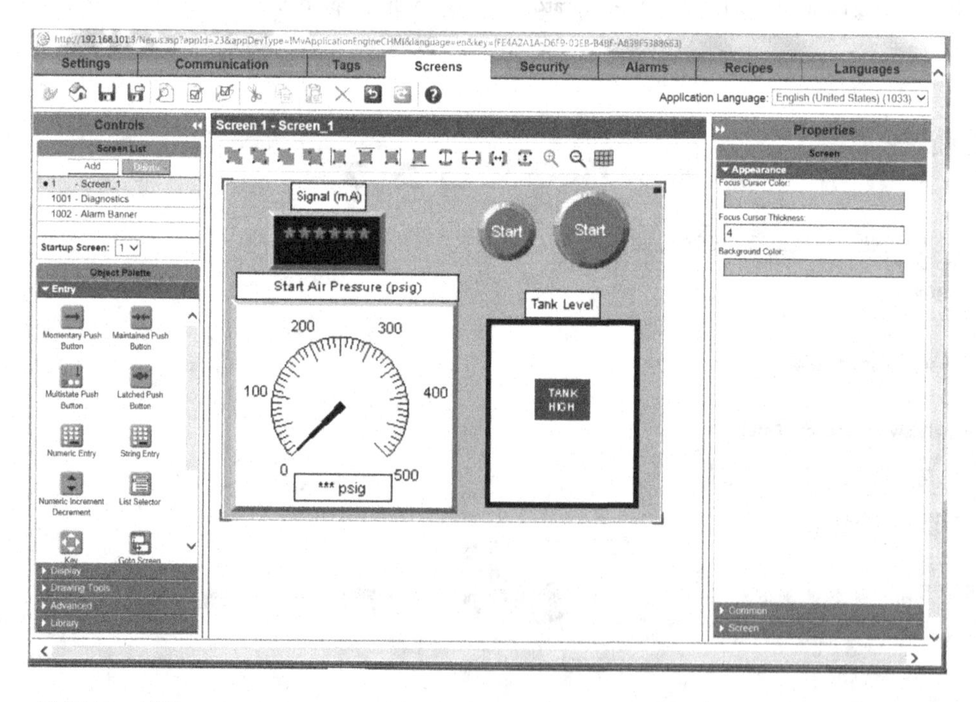

Figure 17.27 The HMI screen is constructed by dragging and dropping icons from the Objects Palette onto the screen.

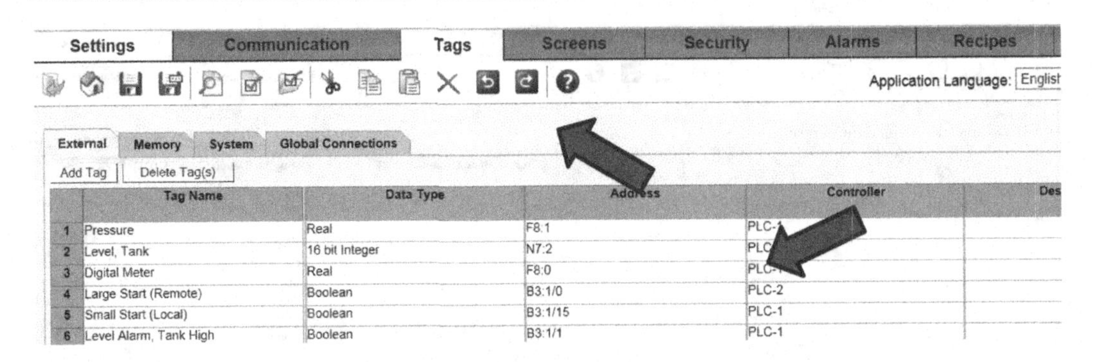

	Tag Name	Data Type	Address	Controller	Des
1	Pressure	Real	F8:1	PLC-1	
2	Level, Tank	16 bit Integer	N7:2	PLC	
3	Digital Meter	Real	F8:0	PLC	
4	Large Start (Remote)	Boolean	B3:1/0	PLC-2	
5	Small Start (Local)	Boolean	B3:1/15	PLC-1	
6	Level Alarm, Tank High	Boolean	B3:1/1	PLC-1	

Figure 17.28 "Tags" are defined to link the screen objects to the PLC program.

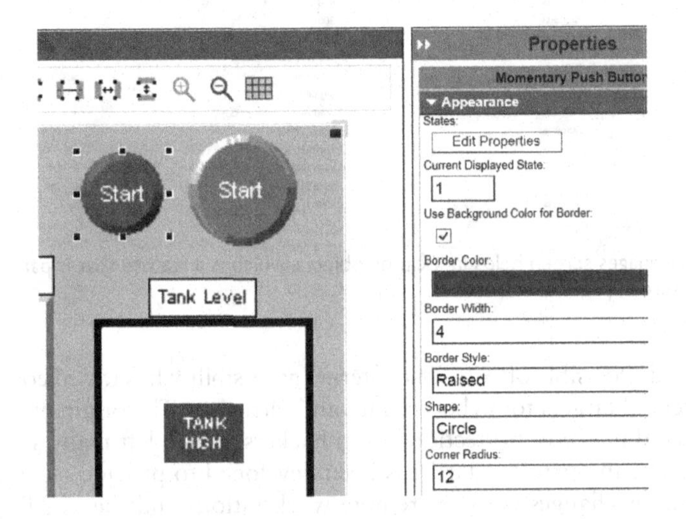

Figure 17.29 Screen objects are linked to the PLC program by attaching the "TAG."

States

Name: PushButton_1
Type: Momentary Push Button

Properties

	Value	Background Color	Background Fill Style	Fill Color	Caption Text	Text Color	Font Name	Font Size	Font Bold	Font Italic	Fo
1	0		Background Color		Start		Arial	14	☐	☐	
2	1		Background Color		Pressed		Arial	14	☐	☐	
3			Background Color		Error		Arial	14	☐	☐	

OK Cancel

Figure 17.30 The appearance of the objects is defined from a spreadsheet-like interface. For instance, a switch's color may be defined for the unpressed, pressed, or error state.

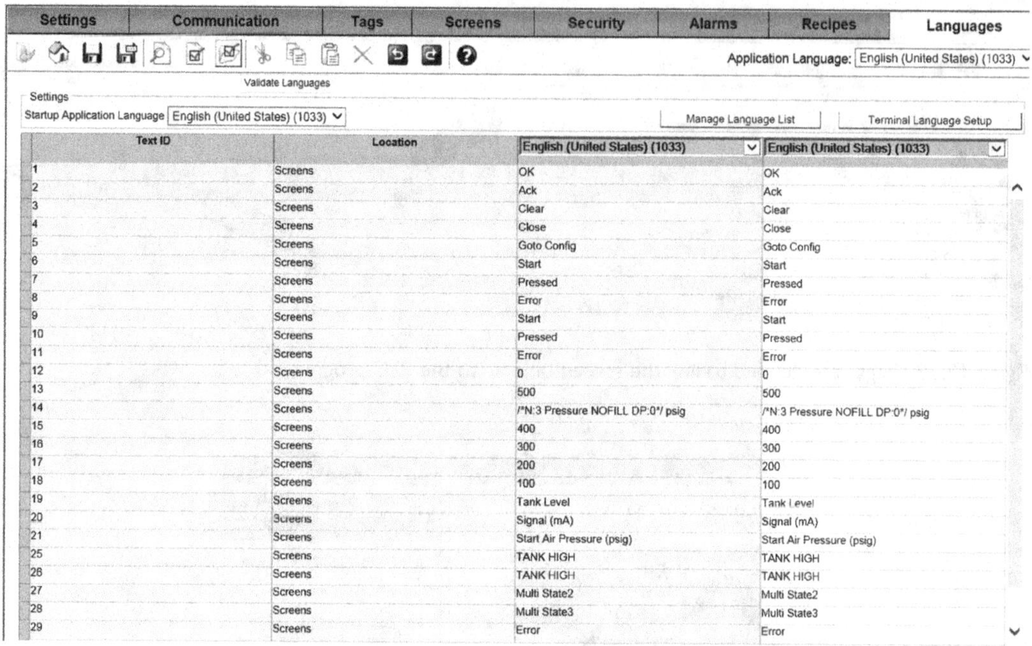

Settings | Communication | Tags | Screens | Security | Alarms | Recipes | Languages

Application Language: English (United States) (1033)

Validate Languages

Settings

Startup Application Language: English (United States) (1033)

Manage Language List | Terminal Language Setup

Text ID	Location	English (United States) (1033)	English (United States) (1033)
1	Screens	OK	OK
2	Screens	Ack	Ack
3	Screens	Clear	Clear
4	Screens	Close	Close
5	Screens	Goto Config	Goto Config
6	Screens	Start	Start
7	Screens	Pressed	Pressed
8	Screens	Error	Error
9	Screens	Start	Start
10	Screens	Pressed	Pressed
11	Screens	Error	Error
12	Screens	0	0
13	Screens	500	500
14	Screens	/*N:3 Pressure NOFILL DP:0*/ psig	/*N:3 Pressure NOFILL DP:0*/ psig
15	Screens	400	400
16	Screens	300	300
17	Screens	200	200
18	Screens	100	100
19	Screens	Tank Level	Tank Level
20	Screens	Signal (mA)	Signal (mA)
21	Screens	Start Air Pressure (psig)	Start Air Pressure (psig)
25	Screens	TANK HIGH	TANK HIGH
26	Screens	TANK HIGH	TANK HIGH
27	Screens	Multi State2	Multi State2
28	Screens	Multi State3	Multi State3
29	Screens	Error	Error

Figure 17.31 The Languages screen helps in locating objects, which is a feature that is particularly useful when many screen pages are used.

to the internet, and accessible off-site. The internet accessibility has raised concerns that plants can become potential targets for cyberwarfare and terrorism. The original internet-connected SCADA systems did not have protections from hackers, which left many systems vulnerable. Within recent years, an open standard has been developed to provide secure encryption and authenticated data exchanges between remote workstations and the SCADA host platform. Historically, most unauthorized accesses to SCADA and plant control systems came through backdoor accesses reserved for remote diagnostic or maintenance purposes. Thumb drive connections are another source of vulnerability, so that security issues can exist in systems that are isolated from the internet. Industrial Control System security is covered in Chapter 19.

DEFINE:

1. Local control versus remote control
2. Centralized control and monitoring
3. Centralized control
4. Distributed control
5. Supervisory control and data acquisition

QUESTIONS

1. Compare and contrast a DCS to a SCADA system.
2. List and describe the SCADA system components.
3. What is an HMI device? How are HMI devices helpful in operating a plant?
4. What are the benefits of using SCADA systems for plant operations and for long-term plant-performance analysis and evaluation?

5. What are smart devices? How do they enhance measurement and data acquisition for use by the PLC?
6. List seven types of drawings or drawing components that are used for defining a plant automation and control project.
7. What is an I&C diagram? What is an I&C schedule? How does the I&C schedule relate to the I&C diagram?
8. How are PLC wiring diagrams configured? What does line-smart/page-smart mean, and how is it related to PLC-to-device connections and troubleshooting?
9. List several alternative SCADA-system communication architectures and their communication protocols.
10. Describe the differences between a communications switch, router, and a modem.
11. List the advantages of the Modbus RTU series protocol, and explain why it is still popular in plant control systems, even after many decades and the availability of newer technologies.
12. Provide an overview description for programming an HMI device.

Chapter 18

PLC cabling, data transmission, and networking

Cables containing low-voltage, small-gage conductors are used to carry signals, data, and sometimes low amounts of power between PLCs and their externally connected field-mounted instrumentation devices, and between PLCs and the engineering workstations used for plant monitoring and control. On-off digital signal transmission to a field device is robust and is simply accomplished by a pair of conductors, while analog sensors carry very small signals with very tiny variations that must be accurately interpreted by the PLC. Analog-signal cables must also mitigate signal corruption from electromagnetic interference (EMI noise) and radio-frequency interference (RFI noise). Typically, analog sensors require cabling that carries both signal conductors and power conductors to drive the transducers.

For large-scale data sharing in a networking environment, information is moved serially using protocols such as Modbus, Profibus, Ethernet/IP, or other proprietary systems. These information systems are often referred to as Supervisory Control and Data Acquisition Systems (SCADA) or Distributed Control Systems (DCS). SCADA/DCS systems have multiple-conductor cables bound into a single-jacketed construction used as a data link in a Local Area Network (LAN). Wireless Local Area Networking (WLAN) using Wi-Fi routers is another option for moving data between SCADA system components without the hindrance of cabling, and in some cases, similar wireless technology is used to connect field devices to PLCs. Fiber-optic cables are used for long-distance transmission, with the added advantage of very high speeds and high bandwidths that can carry tremendous amounts of data. Where very long-distance transmission is needed, Ethernet/IP over the internet is often available, but if not available in very remote locations, control signals can be transmitted using radio transmitters and receivers.

SENSOR/TRANSDUCER CABLING—SIGNAL, POWER, EMI/RFI, SHIELDING, AND GROUNDING

Conductors are often bundled into jacketed cables where the individual wires are used to connect both digital and analog devices to PLCs. In general, digital devices simply require only two conductors with no other special provisions, as the signals are at higher voltages and are either completely on or completely off. Digital signals are robust, experience very low losses, and are impervious to EMI/RFI noise. Digital signals are sometimes just a pair of small-gage conductors routed directly through conduit[1], although twisted pairs bundled within a jacket are also used. The conductors are color-coded, such as black and white or black and red. On occasion, some engineers specify three-conductor signal cable, where the third wire is a spare.

Analog devices, on the other hand, transmit very small values of current or voltage where extremely small variations must be accurately discernable. Because the analog-signal variations are so small, signal conductors can be prone to induced noise superimposed by EMI/RFI. When very fine signal fidelity is required for accurate signal transmission and processing, EMI/RFI can alter the signal enough to cause signal corruption leading to erratic control. The distinction between EMI and RFI is that EMI generally refers to lower-frequency interference, while RFI to higher-frequency interference; however, either can cause problems. With analog-signal cable, foil or braided-wire shields mitigate interference from reaching the conductors (Figure 18.1), while twisting the wires tends to cancel the inductive effects of interference (Figure 18.2).

Analog transducers have two electrical needs, power and signal. For low-power transducers driven by less than 4mA, both power and signal can share the same conductors, however, for higher-power transducers, power and signal are transmitted on separate conductors. Transmitters come in two-, three-, and four wire versions, which are shown respectively as

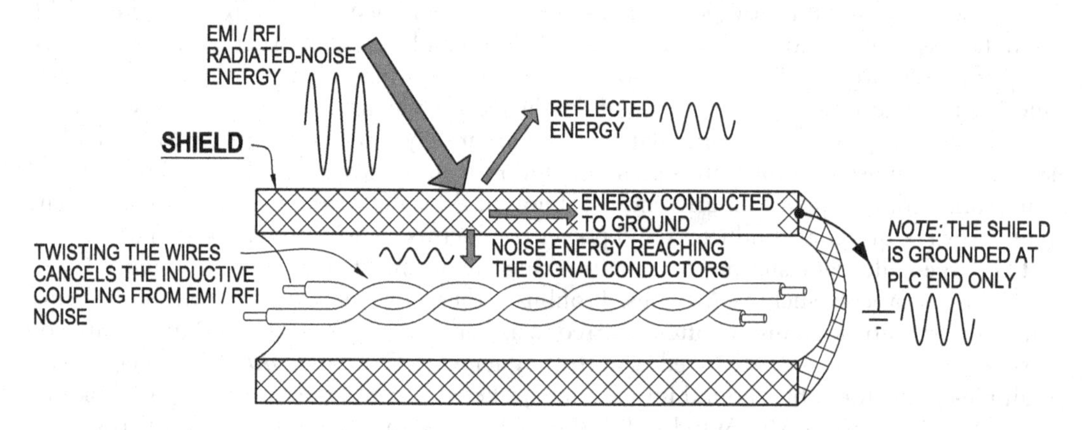

Figure 18.1 Shielding is used to reduce the amount of EMI/RFI reaching the signal conductors.

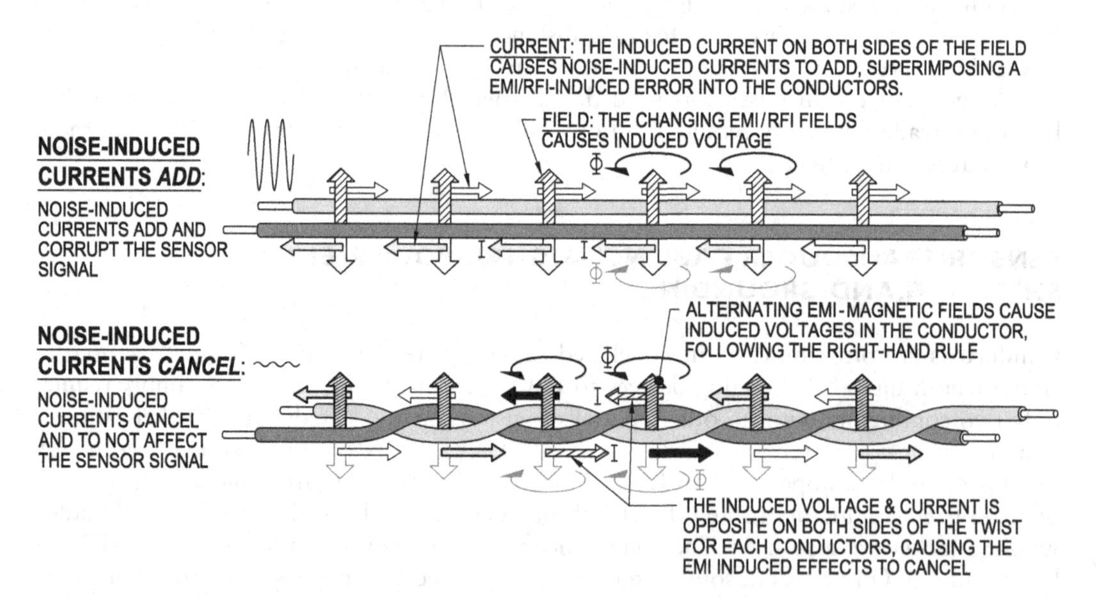

Figure 18.2 Changing magnetic fields between straight conductors cause noise-induced voltages and currents to be superposed onto the signal (upper), distorting the control signal. Twisting the conductors causes noise-induced currents to cancel (lower).

Type 2, Type 3, and Type 4 in Figure 18.4. Figure 18.3 shows the two- and three-conductor signal cable used with analog sensors.

Type 2 transmitters draw such miniscule amounts of power that they can be driven by the same 4–20mA current used to carry the analog-signal. Essentially, the sensor functions correctly using a current supply of less than 4 mA. Consequently, Type 2 low-power devices only require two conductors formed as twisted-shielded pairs (TSP) to mitigate noise interference. A Type 2 circuit is sometimes referred to as a loop system.

Most plant transducers are low power devices that consume more than the minimum of 4mA that is available from a 4–20mA signal loop. These devices are configured as Type 3 transmitters, using three wires formed into a twisted-shielded triplet (TST) to mitigate noise interference. In the three-wire arrangement, the signal is carried by one conductor, power by a second conductor, and the third conductor is used to complete both signal and power circuits through a common conductor.

As a generalization, most 4–20mA analog-signal transmitters are Type 3 devices, using three-conductor cables. A foil or wire-braid shield covering the conductors mitigates the EMI noise,

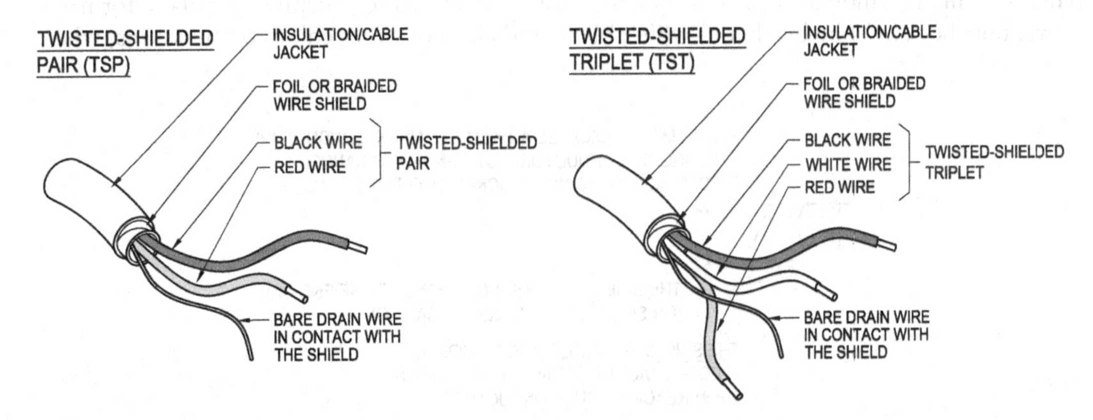

Figure 18.3 Signal cabling will be either a twisted-shielded pair (left) or a twisted-shielded triplet (right). The drain, which is in contact with the shield along the length, is terminated to ground at the PLC end only, and the far end is left floating, to avoid ground loops.

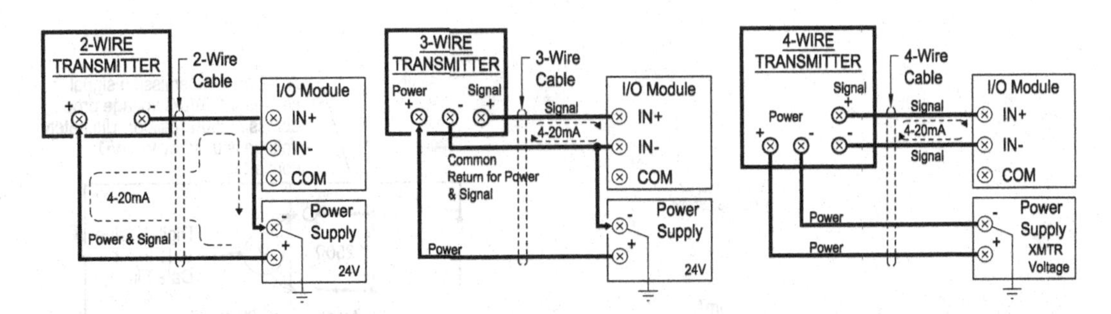

Figure 18.4 Type 2, Type 3, and Type 4 transmitters from left to right. The Type 2 two-wire transmitter is a low-power device that operates entirely on a single 4–20mA loop. The 3- and 4-wire transmitters require more power, which is provided separately from the signal. In the Type 3 three-wire transmitter, the signal and power wires are both returned on one common, shared conductor. The Type 4 transmitter uses four-wire cable for high power applications, such as when the electricity drives an actuator. In addition, separate sources may be required when the connected field device operates at high voltage or when it uses alternating current.

and twisting the wires cancels any noise transmitted past the shield (Figure 18.5). Twisted-shielded triplets are used for both analog input and output signals. Foil shielding is preferred for mitigating higher-frequency RFI, and wire-braid shielding is used for lower-frequency EMI. In installations where high EMI and RFI noise levels are anticipated, high quality cables are constructed having braid on top of foil. Good quality cables include a dedicated bare "drain" wire, which runs along the cable length in direct contact with the foil shield. The drain makes the shield termination easier and more reliable for proper grounding and noise dissipation.

When high-power devices are used, Type 4 transmitters are used with four-wire cables. Type 4 cables include two separate pairs of conductors; one pair is dedicated to carry the signal and the second pair the power. In some high-power devices, 120VAC line voltage may be furnished in a completely separate circuit from the signal cable.

In general, signal wires should be a minimum of 18AWG, stranded, plated copper, and preferably 16AWG for industrial applications. Long runs may require even larger conductors, if voltage drop is a concern. Likewise, the drain conductor should also be stranded, plated copper, typically the same gage as the signal conductors, but no more than one gage size smaller than the signal conductors. The cable outer protective insulation jacket should be high-quality thermoplastic. The conductors are typically red for positive and black for negative, but they could be black and white, black and clear, or other color combinations, where

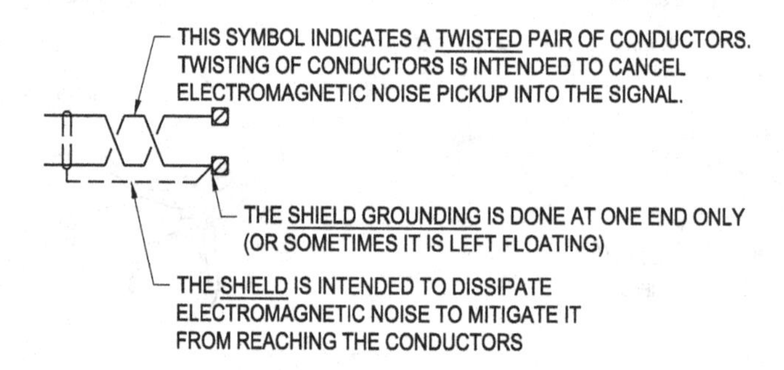

Figure 18.5 PLC wiring-diagram drawing convention for a twisted pair of conductors having a grounded shield. Twisting and shielding reduces EMI/RFI noise. The shield is attached to a common ground point at the PLC to dissipate the noise before it penetrates through to the signal conductors. A single grounding point avoids multiple ground paths and problematic ground-loop currents.

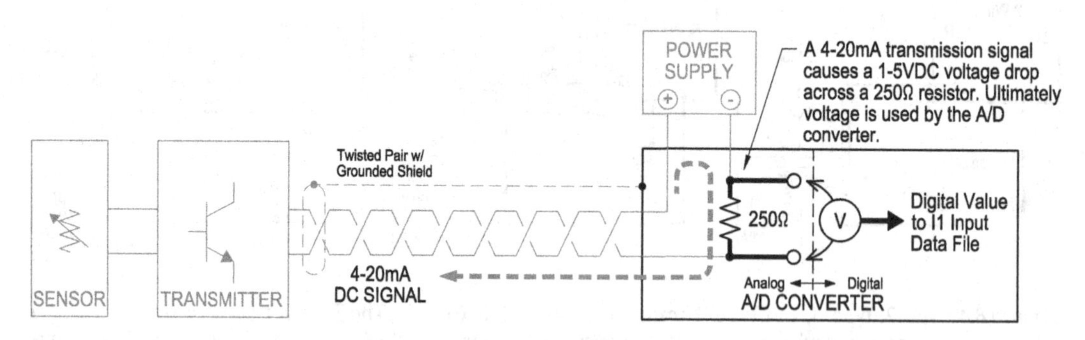

Figure 18.6 Although 4-20mA is often used for analog signal transmission, the signal is transformed into voltage drop by a precision resistor at the A/D converter before it is ultimately changed into its binary-digital form stored in the PLC's memory location. Twisted-shielded-pair cables avoid signal corruption.

the colors are used to coordinate correct connections at both ends and to maintain proper polarity. Sometimes the insulations of twisted conductors are fused together during the extrusion process while manufacturing, which keeps the conductors twisted tightly against each other, resulting in better noise reduction. Thermoplastics are sometimes used to fill the interstitial spaces between conductors, which produces an outer jacket that is more circular. During installation, cables should be labeled at both ends to match the wiring diagram. and individual conductors should be labeled at the PLC and terminal board connections.

The following are general recommendations for I&C signal cable:

- Always run signal cables and AC wiring in physically separated cable assemblies or conduits, to avoid inducing 60Hz noise into the signal. Note that most codes do not permit mixing high and low voltages in the same conduit.
- Physically, cross signal and AC wiring perpendicularly to avoid crosstalk noise and inductive coupling into the signal conductors.
- Avoid loops in signal cabling, where the loop can behave like an antenna and pick up noise. It is good practice to route cables in straight runs.
- Use twisted-shielded pairs or twisted-shielded triplets for signal cables. Twisting cancels the inductive effects of EMI, which are equal and opposite between both supply and return conductors, where tighter twists are even more effective. Better-quality cables are fabricated with the wire insulations fused together, which permanently keeps the twists tight.
- Ground the shield at one location only, i.e., at the PLC. The shield tends to spread out the EMI fields, and the ground connection provides a path to dissipate the EMI energy. Less EMI reaches conductors when a grounded shield is used, and the inductive effects of EMI can be essentially eliminated when the conductors are tightly twisted. By grounding the shield at a single-point, troublesome current loops formed by parallel-grounded circuits are avoided. Ground loops can be a serious source of noise and can produce significant operational problems.
- Use 4–20mA (or 0–20mA) current instead of voltage for signal transmission, where possible. Note that current is converted into voltage at the A/D converter, as shown in Figure 18.6.
  - 4–20mA signals are less susceptible to EMI noise than voltage signals
  - 4–20mA signals are independent of circuit length and will not degrade over rather long distances. Conversely, voltage signals experience voltage degradation in direct relation to cable length.
  - Amperages below 4mA or above 20mA can be used to indicate when a system or device has a problem and can be used to initiate an error notification. Voltage often lacks this ability.
  - Broken or disconnected 4–20mA conductors are clearly indicated as 0mA. Conversely when voltage is used for signal transmission, 0V can be a normal signal and does not necessarily indicate a broken wire. Moreover, the inductive effects of EMI can induce "phantom voltages" from the conductor capacitance and inductance, which could mask an incomplete circuit.
  - The A/D and D/A electronics can be electrically isolated by using opto-isolators with 4–20mA transmitters, allowing better EMI filtering and the avoidance of problematic ground loops.
- Minimize the lengths of unshielded signal conductors within control panels and use twisted supply-and-return conductors, as much as possible. Keep the signal conductors physically separated from AC conductors and other sources of EMI, such as transformers and fluorescent lamp ballasts.

SCADA AND DCS CABLING AND DATA TRANSMISSION

Methods for exchanging large amounts of data include Serial, Serial Packetized, and Transmission Control Protocol/Internet Protocol (TCP/IP) transmissions. Serial means that the information is sent over a single transmission line one bit at a time, encapsulated as computer words. The serial-data word includes start and stop bits as overhead used for synchronization, one or two parity[2] bits for error checking the information, and between five to eight bits of data itself. RS232 is the original method of serial data transmission and is asynchronous.[3] It uses an RS232 serial cable with either a 9- or 25-pin connector. Data bits are sent at a user-selected frequency, which is called the baud rate. The baud rates between the send device and receive device must be matched and the baud rate cannot exceed the speed of the slower device. RS232 transmission is notoriously slow, limited to short cable distances, and can be prone to picking up EMI-related noise. Serial communication can be half-duplex or full duplex. Half duplex means that the device cannot send and receive data simultaneously, while full duplex means it can. Half duplex requires a data conductor and signal ground, while full duplex requires separate send and receive data conductors and a common signal ground. Additional wires are needed for Request to Send (RTS), Data Set Ready (DSR), and Data Terminal Ready (DTR) signals required for transmission coordination, called handshaking. Serial communication occurs over twisted pairs within a cable (Figure 18.7). Today, RS232 is still used for connecting some PLCs to computers for programming and troubleshooting (see Figure 18.8 and Figure 18.9), and for connecting between some PLCs and HMI devices.

The Universal Serial Bus (USB) is the next generation of serial cables/connectors for transmission distances less than 15 feet (Figure 18.10). The cables can be "hot swapped" without powering down the system or causing computers to lock up; they are essentially "plug and play." They are fast; USB 2.0 is capable of up to 480Mbps, which is over 20,000 times faster than RS232 communication and USB 3.0 is capable of 5Gbps. USB cables carry power as well as signals. A USB 2.0 can carry up to 2.5W at 500mA, USB 3.0 up to 4.5W at 900 mA, and USB 3.1 Generation 2 with a Type-C connector up to 5A and 20V for as much as 100W. USB uses USAP (Unison Serial Access Protocol) transfer protocol for moving data between computers and devices.

Serial communications for PLCs are used for connecting peripheral equipment, such as printers and HMI devices. Serial interface standards that use RS232 cables include RS232, RS423, RS422, and RS485 protocols, where RS485 is most common in plants (see Figure 18.11 for the cable construction). Table 18.1 shows the major characteristics of the various serial interface standards. The most common options for central-plant data transmission

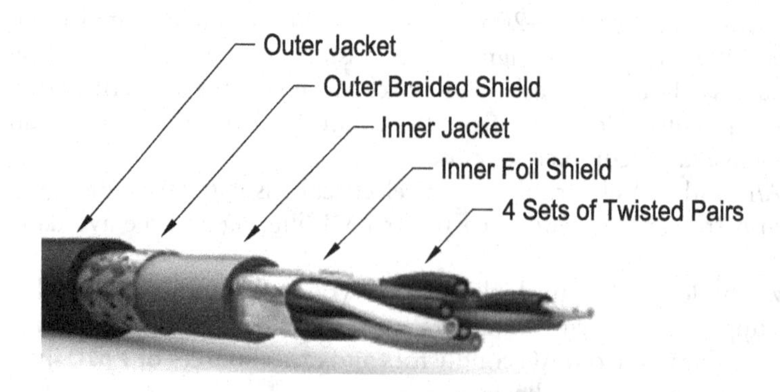

Figure 18.7 RS232 cable consists of 4-sets of twisted pairs and a shield/drain/ground.

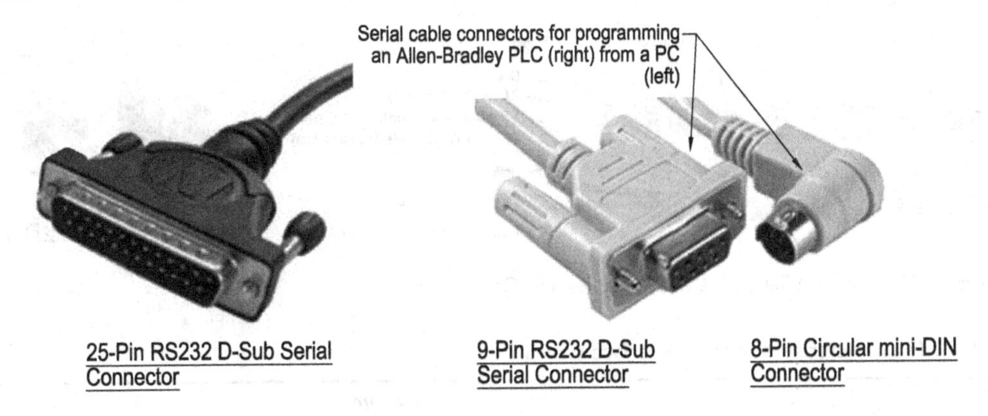

Figure 18.8 RS232 serial cable for connecting a laptop's DB-9 nine-pin connector (left) to an Allen-Bradley MicroLogix PLC's eight-pin circular Mini Din serial connector (right), for programming.

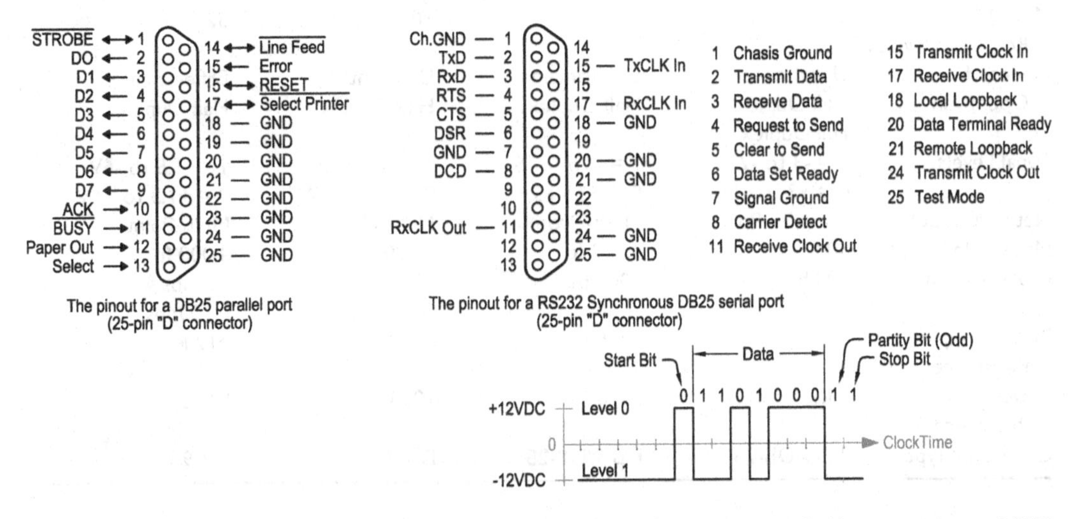

Figure 18.9 The DB25 is a 25-pin "D" type connector. Pinouts for a parallel port (left), pinouts for an RS232 serial connector (right), and synchronous data transmission signal (lower). Data transmission with odd parity is shown.

protocols include Modbus using RS485 cabling and protocol, Profibus fieldbus using Profibus DP or PA[4] cabling, or Ethernet/IP using Category 5 (CAT 5) or Category 6 (CAT 6) cables.

Modbus technology was developed during the first generations of PLCs and is so robust and reliable that it is still very commonly used in plant controls, even after more than five decades of technology advancements. Modbus is a versatile, easy-to-use, and hearty means of transmitting data over long or short distances. Modbus RS485 uses a serial data link, where the RS485-device cables are daisy-chained from one controller to the next (Figure 18.12). Failure of any device in the string does not interrupt the daisy chain wiring circuit, and all remaining devices continue to communicate. For relatively long runs, the signal conductors at each end of the RS485 daisy chain are connected through terminating resistors, in order to match the cable impedance to the impedance required by the hardware (Figure 18.13). Impedance matching is necessary to avoid signal reflections that would be very problematic at high data-transmission rates. Signal reflections can also distort the original signal, reducing reliability.

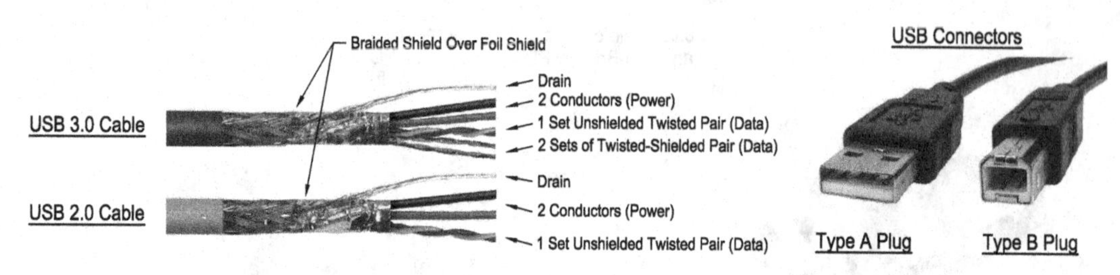

Figure 18.10 Universal Serial Bus (USB) Cables and Connectors.

Table 18.1 Characteristics of serial interface standards

	Standard			
	RS232	*RS423*	*RS422*	*RS485*
Max Qty of Send Devices	1	1	1	32
Max Qty of Receiver Devices	1	10	10	32
Modes of Operation	Unbalanced Half Duplex or Full Duplex	Unbalanced Half Duplex	Differential Half Duplex	Differential Half Duplex
Signal Levels	1 = -3 to -25V 0 = +3 to +25V	1 = -3.5 to -6V 0 = +3.5 to +6V	1 = -2 to -6V 0 = +2 to +6V	1 = -1.5 to -6V 0 = +1.5 to +6V
Network Topology	Point-To-Point	Multi-Drop	Multi-Drop	Multi-Point
Max Cable Length	50 feet	4,000 ft	4,000 ft	4,000 ft
Max Date Rate at 40 ft at 4,000 ft	20 kbs	100 kbs 1 kbs	10 Mbs 0.10 Mbs	35 Mbs 0.10 Mbs
Receiver Input Resistance	3 kΩ	≥4 kΩ	≥4 kΩ	≥12 kΩ
Driver Load Resistance	3 kΩ	≥450 kΩ	100 kΩ	54 kΩ
Connector Type	DB-9/DB-25	DB-9/DB-25	DB-9	DB-9

The RS485 interface has the following advantages:

- communications are made without using modems or switches
- connections are made in a network structure
- communications over long distances are possible
- communication rates are fast

CATEGORY 5 AND CATEGORY 6 ETHERNET TCP/IP CABLE

CAT 5 and CAT 6 cables are used for networking and internet connections. CAT 5E is the enhanced version replacement of now obsolete CAT 5 cable and it can handle data transmission rates of 10/100Mbps at a 100MHz bandwidth. CAT 6 and CAT 6A are newer versions that operate about 100 times faster than CAT 5 cables. CAT 6 can support a transmission rate of 1Gbps up to 164-feet at a bandwidth of 100MHz and CAT 6A up to 328 feet. The CAT 6 cables typically use shielded twisted wires, which are much less prone to crosstalk

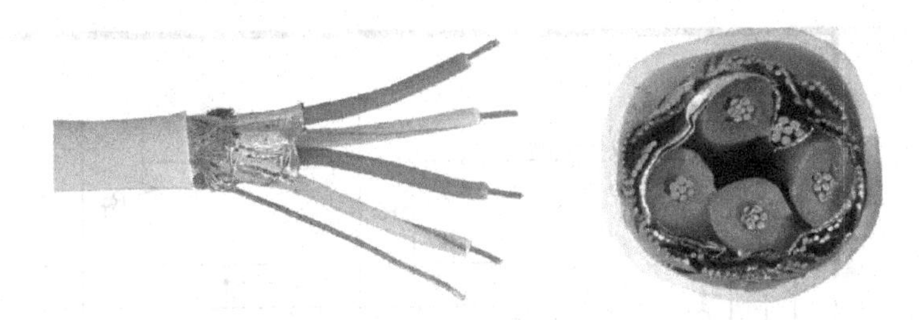

Figure 18.11 RS485 high-speed serial data transmission cable, consisting of two-twisted pairs of 7-stranded 24AWG conductors, where both pairs share a dual outer tinned-copper braided-wire shield over an inner foil shield and drain.

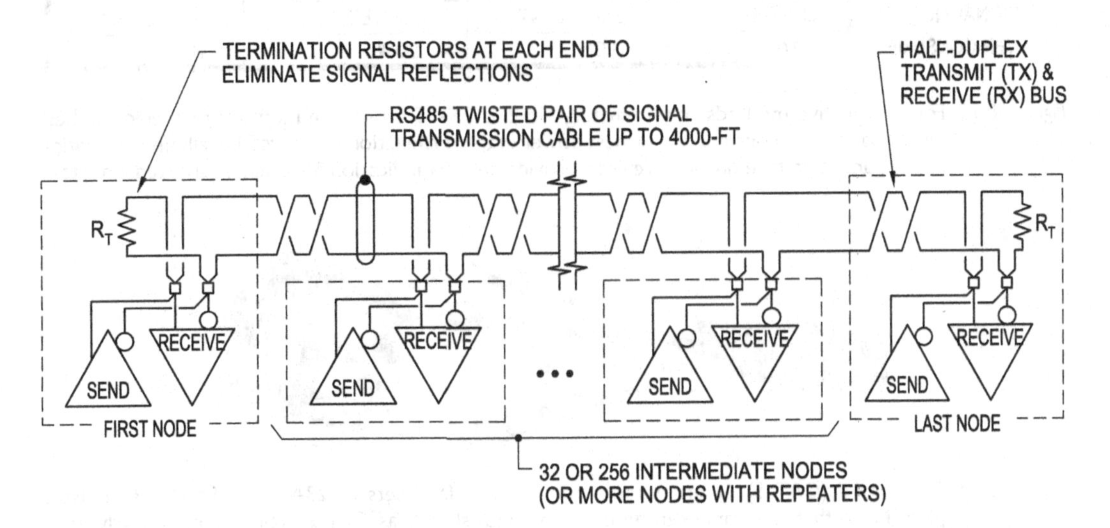

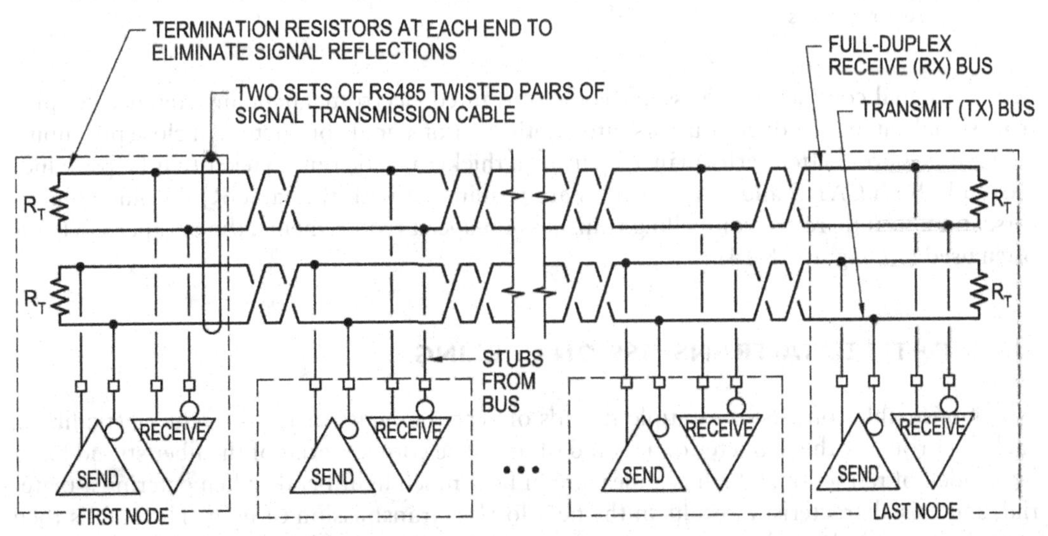

Figure 18.12 Upper: Modbus RS485 *two-wire half duplex transmission* using *daisy-chain* connections from a bus topology. Lower: Modbus RS485 *four-wire full duplex transmission* using *stub* connections from a bus topology. Half-duplex allows send or receive data transmissions in one direction at a time, whereas full-duplex permits simultaneous send and receive transmissions. Termination resistors at the first and last node mitigate signal reflections, distortion and corruption.

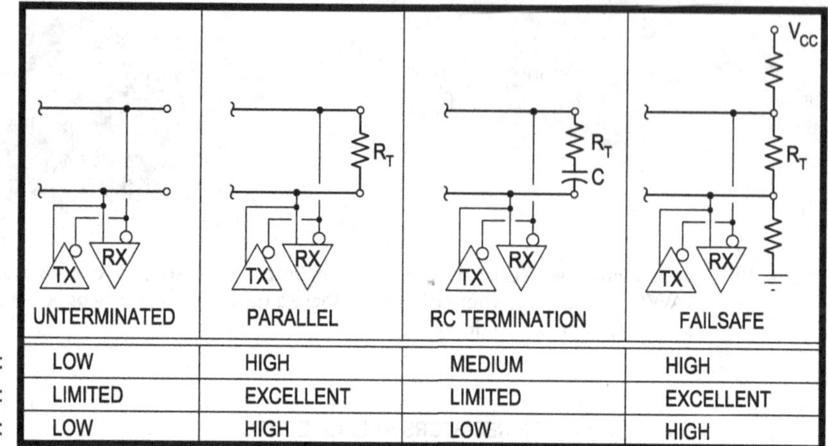

Figure 18.13 Four alternative methods of terminating Modbus RS485 buses. The methods that yield the best performance also consume the highest power in the termination resistors. The failsafe termination uses an active voltage-divider circuit to provide an alert indication for open or shorted circuits.

Figure 18.14 Category 5 (CAT 5) Ethernet cable consisting of four sets of 23AWG solid copper twisted pairs. Note that this cable can be procured with shields as CAT 6 cable for particularly noisy environments.

between signal conductors. Crosstalk is the unwanted electromagnetic interference coupled from signals in one set of conductors into another set of signal conductors in close proximity. CAT 6A achieves better performance by using a thicker plastic outer jacket that helps reduce crosstalk. Both CAT 5 and CAT 6 cables are terminated with the same RJ-45 plug connectors, and unless there are compelling transmission-speed requirements, the cheaper CAT 5 is often used (see Figure 18.14).

FIBER-OPTIC DATA-TRANSMISSION CABLING

Fiber-optic cable consists of multiple strands of very thin glass or transparent plastic fibers, each fiber roughly the diameter of a strand of hair. The core material of the fiber strand has a high index of refraction, which is a measure of how much light bends when entering a material, and this characteristic results in the near-lossless transmission of light. The core is then coated with a cladding having a low-refractive-index. The cladding reflects light back into the core, reducing transmission losses, like a mirror. The internal reflection turns the core into a waveguide for light. Signals are transmitted as pulses of light, representing 0s and 1s. Some advantages of fiber-optic cable include very high signal-transmission speeds at a very high bandwidth, which means it can simultaneously carry many different signals at many frequencies over very long transmission distances with low losses, fewer signal repeaters,

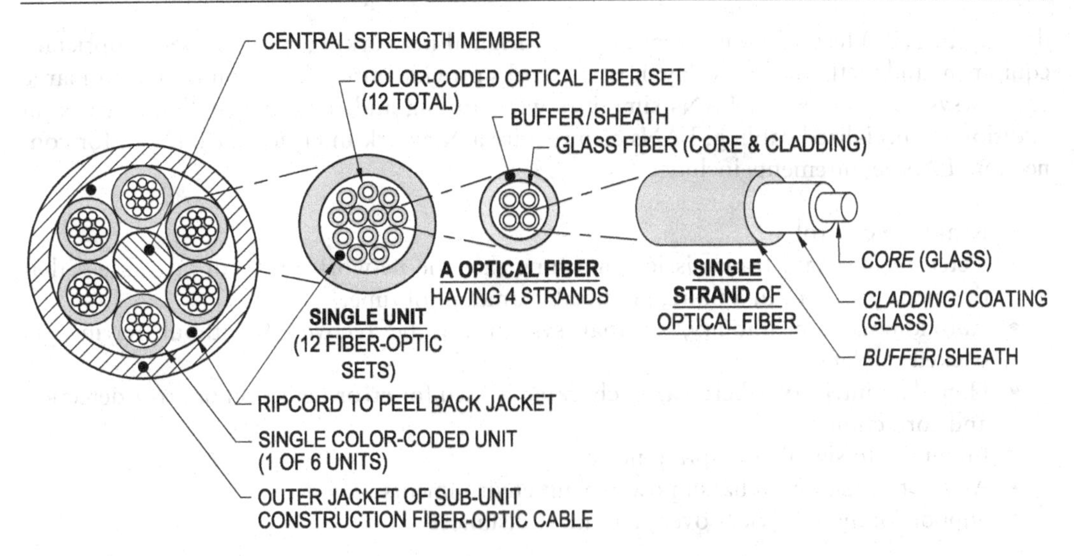

Figure 18.15 Complex fiber-optic cable using "sub-unit" construction for multiple-mode operation. This cable has six sub-units; where each sub-unit contains 12 optical-fiber-circuits, each having four strands of glass.

Figure 18.16 Gigabit Interface Converter (GBIC) optical-fiber-to-wire converter. Separate transmitter (TX) and receiver (RX) fiber-optic connections are provided for full-duplex data transmission, and an Ethernet IP connector links the optical signal wiring to the computer or PLC.

and complete immunity to EMI/RFI noise. Fibers may be bundled together to create multiple circuits and spares, as shown in Figure 18.15. Fiber-optic systems require transceivers that convert electronic pulses into light using LEDs or lasers for transmission, and photodetectors to convert the light back to electronic pulses at the receiver (Figure 18.16).

NETWORKING ARRANGEMENTS

Networking is the linking of multiple LAN (local area network) devices to each other and to a host device, such as a computer workstation, via a main cable, called a bus. The LAN is a high-speed communication system, capable of exchanging 10/100 Mbps or 1Gbps of data up to 1,500 feet without a repeater. The original networking systems used proprietary equipment and methods, but today most networks provide unrestricted connectivity. In large control systems, the use of LANs simplifies wire routing and reduces installation costs. In addition to specialized cabling, LANs also require a Network Interface Card (NIC) for connection. LAN requirements include:

- Real-time control
- Deterministic data transmission, meaning that the network protocol guarantees that the message is sent in a finite, predictable amount of time.
- Autonomous repeatability, so that system can be recovered without relying on backed-up data
- High data integrity, where parity checking is used for error detection or error detection and correction
- Immunity to signal-corrupting noise
- Ability to function in harsh power-plant environments
- Support of many devices over significant distances

Network topology is defined as the connection arrangement between devices, called nodes. Topology is subcategorized as physical and logical. Physical topology refers to the plantwide location of the devices, the cable routes, and the wiring interconnections. Figure 18.17 shows physical topologies for bus, star, linear, ring, and mesh networks. Logical topology refers to the communication protocol, the way that the signals act within the network, and the way that data is transferred reliably and without collisions.

Bus topologies are networks where each node is connected to a single cable through middle-of-the-run interface connectors. Bus topologies tend to be simple and reliable, and are easily expandable by connecting nodes anywhere onto the bus. If one node in a bus network fails, the remaining nodes still communicate with each other. To be reliable, the bus and nodes cable lengths must remain within prescribed electrical-loss limits.

Star topologies are networks where each node is connected to a central hub, so that each node is effectively joined to each other only through the hub. The hub is either a router or a switch. The router/switch connects the server and all other devices, which are peripherals or clients. All information passes through the router/switch, which acts as a signal repeater. The star topology is simple, reliable, also easily expandable. However, disadvantages of star topology are that the hub can form a single point of failure that brings down the entire system, and that the hub can become a bottleneck during periods of high data transmission rates.

Linear topologies are networks using serial daisy-chain connections of devices. Linear topologies are nearly as easy to implement as the star topology. In this arrangement, data transmissions that originate from one end are bounced through all nodes before reaching the far-end device. Linear topologies typically require termination resistors at each end to

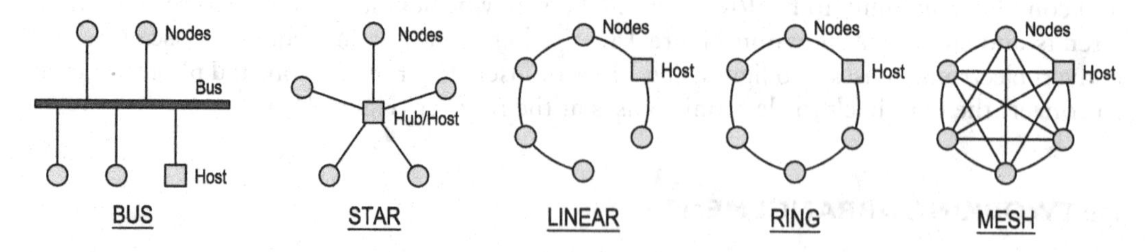

Figure 18.17 Network topologies. Note that one node in each topology would serve as the "host" or "master."

maintain the correct circuit impedance required by the devices. The disadvantage of linear topologies is that a middle-of-the-run cable failure disables the entire series of downstream nodes, becoming a reliability issue.

Ring topology is a network that is essentially linear topology where the two extreme ends are returned to form a loop. In the event of a single-cable failure, data transmission continues without interruption, as the signal is back-fed through the loop. Nowadays, the ring system can report the presence of a cable failure for troubleshooting a broken cable.

Fully connected Mesh topologies are networks where every node is physically connected to every other node. The fully connected mesh network is essentially a complete set of two-node devices, where packetized data and coordinated signal broadcasting is not required, since all nodes communicate directly. However, since the number of connections grows exponentially, this topology is limited to small networks, and typically is not found on industrial PLC control systems.

INDUSTRIAL NETWORK STANDARDS

Networking can be broken into serial-based Device Networks and ethernet-based Process Bus Networks. The networking consists of the physical wiring types and connections as well as the data-transmission protocols or rules. There are many network options. Some of the more prevalent industrial networking options are shown in Table 18.2. Presently about 48% of industrial control networks use serial-based networks, about 46% use ethernet-based networks, and the remaining use wireless networks, but the trend may be leaning toward ethernet-based networking. By geographic location, Modbus is used everywhere, while DeviceNet, Ethernet/IP, and EtherCAT predominates in North America, Profibus, Profinet, and EtherCAT in Europe, and CC Link in Asia.

Device Bus Networks are a control-device option where smart devices containing embedded product and diagnostic information are networked to a PLC rather than hardwired directly to a dedicated PLC I/O device. When used, Device Bus Networks are interfaced to low-level discrete devices, such as simple transmitters, using a common half-duplex data link, called a bus. The host or workstation communicates directly with many devices in a coordinated manner.

Process Bus replaces hardwired connections with communication lines. Process bus transmits and receives high-level information from "smart devices" in a networked environment, where the smart device may contain device information, such as make and model number, built-in self-diagnostic information, and sometimes control functionality that is independent from the PLC.

Table 18.2 Characteristics of the most common industrial networking protocols used worldwide

Serial-based network	*Relative usage*	*Max nodes*	*Speed Mbps*		*Controller Mfr*	*Ethernet-based network*	*Relative usage*	*Speed Mbps*	*Share wires*
Profibus	30%	32/126	12	→	Siemens	Profinet IO	24%	1,000	✓
Modbus RTU†	15%	32/254	0.1	→		Modbus/TCP†	9%	1,000	✓
DeviceNet	8%	64	0.5	→	Rockwell	Ethernet/IP†	24%	1,000	✓
CANopen†	10%	127	1	→		EtherCAT COE	15%	1,000	
Other	37%					Other	28		

Note: † There are many manufacturers whose controllers support Modbus RTU, CANopen, Modbus TCP, and Ethernet/IP.

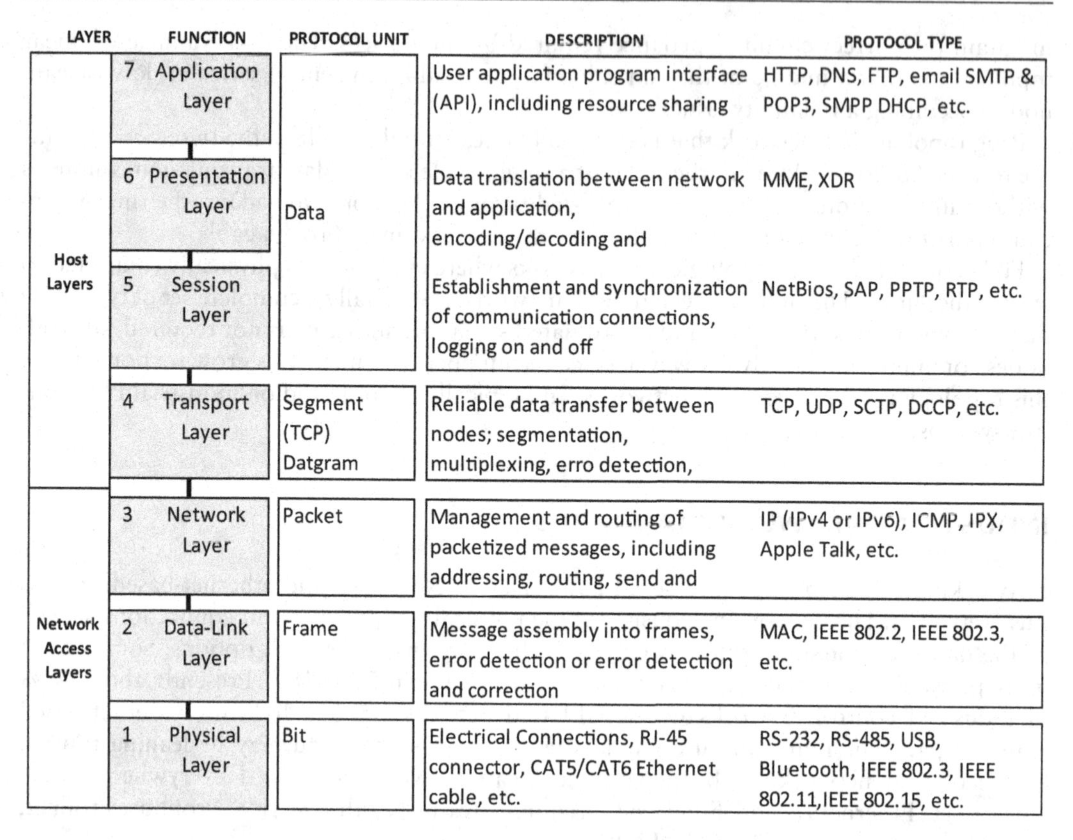

Figure 18.18 The standard OSI Reference Model stack is used for internet and network protocols and interoperability of devices. Some PLC controllers use manufacturer-specific, proprietary network-reference models, such as shown in Figure 18.20.

For devices to communicate on the same network, it can be inferred that they must use the same communication protocol. Historically, many manufacturers used their own proprietary protocols, and it was common to experience communication problems when hardware was obtained from various sources. To obtain interoperability, a translation system was developed which is called the Open System Interconnections Reference Model (OSI Network Model). The OSI model is an open-source seven-layer communications system using standardized data-transmission protocols (see Figure 18.18). OSI enables coordinated interoperability between different systems from different manufacturers. In the OSI Model, each layer is served by the layer below, and in turn, each layer serves the layer above it. In TCP/IP transmissions, Ethernet fundamentally handles the bottom two layers of the stack, which are the physical and data-link layers.

IEEE 802.3, 802.4, and 802.5 are the official standards used for networking. IEEE 802.3 governs the first two OSI layers when using a bus topology or Ethernet. IEEE 802.4 and IEEE 802.5 govern token-ring bus networks, using baseband and broadband digital-signal transmissions. Many PLCs use the IEEE 802.4 standards.

Networking protocols include Modbus, Profibus, Ethernet, ControlNet, DeviceNet, Foundation Fieldbus, Interbus, CAN, and others. Industrial networking with moderate levels of automation tend to use very robust serial-data transfer systems like Modbus RS 485 and Profibus, while applications with very large levels of automation and requirements for interconnectivity tend to use Ethernet/IP.

Modbus, Profibus, Ethernet/IP

Modbus

Modbus is the communication protocol has evolved from the original 1979 Modicon PLC systems as a link between PLCs and devices in a master-slave or a server-client arrangement. The Modbus messaging structure is the *application protocol* that defines the rules for organizing and interpreting data, regardless of the transmission mechanism/physical layer. Modbus' strength and continued popularity comes from its implementation simplicity, without the burden of royalty fees. Modbus runs serially on either the RS232 or RS485 physical layer, but it can be used with Wi-Fi or over phone lines. Modbus uses one master, and the individual slaves are queried on demand by the master. The slaves respond by either returning the requested data or by taking the actions demanded by the query. A single broadcasted command can be made to all slaves to produce a common action, but slaves will only respond with data when they are individually queried. The structure of the message frame is shown in Figure 18.19, where both the query and the response include an address; a function code that describes the actions to take—such as read or write; the data; and parity-bit error checking. If an error occurs in the query, the slave returns an "exception" message indicating that it could not perform the action. The error field originating from the slave back to the master is used to verify that the message is valid. The Modbus message frame is up to 252 bytes in length and supports up to 247 addresses.

Modbus ASCII is an old version, where data is encoded as bits. Modbus RTU is a later generation of Modbus, where the data is encoded using bytes instead of bits, greatly increasing the transmission throughput. RTU is more common in later installations. Modbus TCP/IP[5] was developed in 1999 and is the latest generation of Modbus protocol. Modbus TCP/IP is essentially Modbus RTU with a built-in TCP interface to run on the internet.

Profibus

Profibus gets its name from Process Field Bus. Fieldbus is an industrial network system for realtime distributed control, used in a variety of network topologies including daisy-chain, star, ring, and others, of which Profibus is one "type." Profibus is an open-source communication protocol that was developed in Germany and is used by Siemens for their fieldbus-networking technology used with their equipment and devices. Profibus is used in process automation as a smart technology, where devices are connected on a common bus/cable. The devices communicate information in an efficient manner, provide automation messages, and perform self-diagnosis of the device and connection.

Profibus uses only three of the OSI Reference Model network layers (see Figure 18.20). There are different versions of Profibus that handle different types of messaging at the layer 7 application level. Profibus supports both cyclic and acyclic data exchange, isochronous messaging, alarm handling, and diagnosis. Layers 3 through 6 are not used. The data-link Layer 2 is used through a Fieldbus Data Link (FDL), to combine both the master-slave and token-passing procedures. Profibus supports three types of media for the Layer 1 physical layer; the standard RS485 twisted-pair wiring, fiber-optic transmission, and their safety-enhanced system called Manchester Bus Power that they recommend for explosion-proof installations.

Address Field	Function Code	Data	Error Checking

Figure 18.19 Modbus message frame. Both the query and the response include an address.

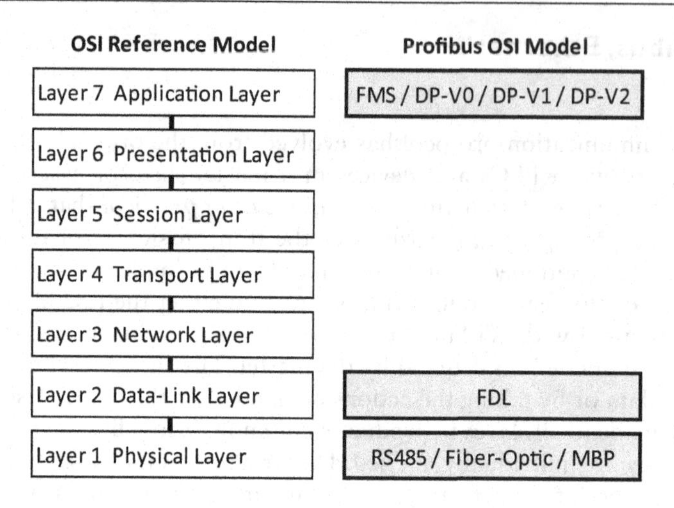

Figure 18.20 Profibus OSI Model.

The Profibus topology uses drop-type device take-offs attached directly to common cable conductors, called the bus. Using a bus avoids having to install a full-length wiring connection from the controller to each device. Older versions of Profibus used one central bus, but newer technologies now allow several buses to be connected into a ProfiNet Ethernet system for improved flexibility.

Profibus is a more complicated protocol than Modbus, having four versions tailored to the type of control. Versions of Profibus include Profibus FMS, DP, PA, and MBP. FMS is their first-generation Fieldbus Message Specification for communicating complex information between PLCs and PCs. Due to Profibus FMS complexity, many designers tend to use newer versions. Profibus DP is the Decentralized Periphery version, which has three subsets DP-V0, DP-V1, and DP-V2, where each version provides more features. Profibus DP is simpler and faster than the Profibus FMS version. Profibus PA is used for Process Automation and is a specialized type of Profibus DP that standardizes the transmission of measurements. Profibus PA is designed for specifically for hazardous environments. Most PA installations use RS485 twisted-pair cables to support both power and signal. When used in intrinsically safe explosion-proof applications, MBP cables are used.

Ethernet/IP is an application layer protocol for transferring data in a Transmission Control Protocol (TCP)/Internet Protocol (IP) packet as part of a Common Industrial Control (CIP). Ethernet/IP devices group their data into sets of attributes called objects., which must include its identity, TCP, and router address. The application objects include data for the specific device, such as a motor object. The two types of messages that are transferred are explicit messages, which may be asynchronous, and I/O messages, which are continuous. The Ethernet/IP scanner device opens the connections and initiates the data transfer, and the Ethernet/IP adapter provides the data to the scanner device.

QUESTIONS

1. In general, what kind of wiring is used for digital control I/O circuits? In general, what kind of cabling is used for analog I/O circuits? Why is there a wiring difference between the two signals?
2. What is EMI and RFI, and how do they affect analog signals?

3. What type of cabling is commonly used with analog signals? What is a shield, and what is its purpose?
4. Where are cable shields grounded? What is a ground loop and what is its effect on a system of analog inputs?
5. What is a drain wire? Why is a drain needed?
6. Describe the power versus signal wiring arrangement associated with two-wire, three-wire, and four-wire transmitters. In general, which cabling type is used most often with 4-20mA signals?
7. How is the 4-20mA signal manipulated before being changed into a computer-based digital signal in the A/D converter?
8. List some good practices for I&C signal cable wiring methods. Describe why the 4-20mA signal is popular in plant control systems.
9. What is parity, parity checking, checksum, and cyclic redundancy checking.
10. What is meant by a data packet, and how is information sent over a communication system?
11. What is the difference between RS485 cable and connectors, CAT 5 cable, and CAT 6 cable?
12. When and where is fiberoptic cable used? What PLC I/O devices are needed for fiberoptic systems to function?
13. List and sketch five network topologies.
14. Provide an overview and block diagram of layers used in the OSI model for data transmission.

NOTES

1. A conduit is a plastic or metallic pipe or tubing that encloses and protects insulated conductors. A cable is a group of conductors that are enclosed in a plastic or rubber jacketed covering.
2. Parity checking is either even or odd and is used to verify data integrity. Parity checking works by counting the number of high states or 1s in the data word before transmission. With "even parity," the parity bit value of 1 is tacked onto the end of the word if the quantity of high states or 1s in the word are even. The receiver performs the same bit-counting operation to verify that the parity bit is correct. "Odd parity" is similar, however, the parity bit is set to 1 when there are an odd number of 1s in the word. In some instances, additional parity bits can be added for both error checking *and* correction.

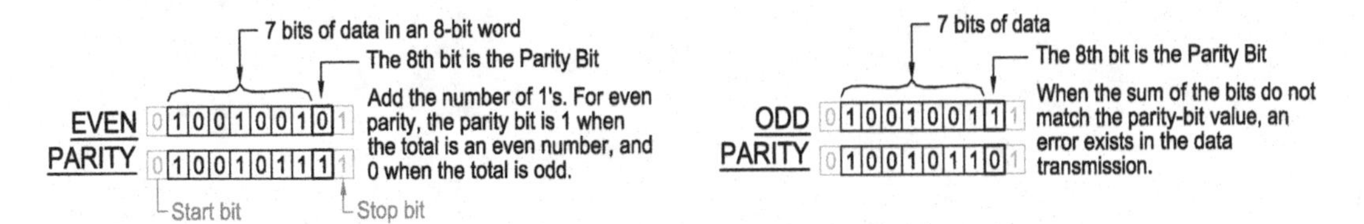

3. Asynchronous means that data transfer is independent time, and the receiver is tasked with detecting when data transfer starts and ends. Conversely, synchronous communication uses a clock signal to mark the beginning of data transfer.
4. For Profibus, DP is short for Decentralized Periphery and PA is short for Process Automation. Profibus was developed in Germany and is a smart "field-bus" technology that is used more with factory automation systems, while Modbus is common with power-plant automation systems. "Smart" sensors contain embedded information in addition to the measured signals, such as manufacturer model number and diagnostic data.
5. TCP/IP is Transmission Control Protocol/Internet Protocol.

Chapter 19

Industrial control system security

An Industrial Control System (ICS) is a network of SCADA systems, DCSs, and PLCs used to control plants and infrastructure, where the PLCs are a subset of the SCADA and DCS controllers. For reference, network and security-related definitions are included at the end of this chapter. Often, the ICS is connected to the internet for remote monitoring, management, and maintenance of plant software. Today there is a large amount of experience and effort in the information technology (IT) and network-security side of computers, however, security for automation-and-controls systems continues to develop and needs to catch up with IT security. Figure 19.1 shows general industrial control architectures and their relative risk levels from outside threats, like hackers. Isolated-area systems that are not network connected have the lowest security risks and are the most protected from outside threats. Greater user access and internet connectivity raises the possibility of system breaches.

Hackers have developed malware to compromise computer systems, such as using ransomware for illicit monetary gain, or for more sinister and damaging purposes. Some examples include the Dragonfly hacker group and the Havex malware, both known to target U.S. and European infrastructures, including electric grids, energy-related industries, and ICS automation-and-control equipment. The Havex malware uses a "backdoor approach," referred to as a "RAT," which provides a means for hackers to obtain remote computer access to PCs or servers. Malware from these hacker groups can destroy equipment, take down critical infrastructure, and jeopardize safety. Some hackers are state sanctioned groups and others are rogue individuals; and in either case, these hackers often have little risk of detection or retribution. Other sources of threats come from operators, criminal groups, industrial espionage, foreign intelligence agencies, terrorists, insiders, spyware, malware, and even untrained users. One of the best-known examples of breached ICS security is the 2010 Israeli Stuxnet malware, which successfully perpetrated a cyberattack on the Iranian Natanz nuclear-fuel enrichment plant. Stuxnet is a worm virus that is propagated through trusted networks, which was likely initiated through an "air gap" via a USB drive. This malware damaged roughly 1,000 centrifuges and disrupted the enrichment process during several years through overpressurization, overspeed-and-underspeed variations, all while the control-room indicated normal readings.

A few examples of malware known to specifically target Industrial Control Systems are: Stutnex, Havex, BlackEnergy, TRITON/TRISS, Industroyer, and others. Industroyer, for example, left one-fifth of the Ukrainian city of Kiev without electrical power for an hour, and that hack was conjectured to just be a large-scale test of the malware. The concerns go beyond simple inconvenience and are issues that affect life-safety or national security, where such outages can cause military vulnerability or long-term large-scale disruption of critical infrastructures.

An example of a possible hacker approach might be to:

- obtain an undocumented portal or backdoor, which is sometimes available for software upkeep or maintenance, to control network command and control components,

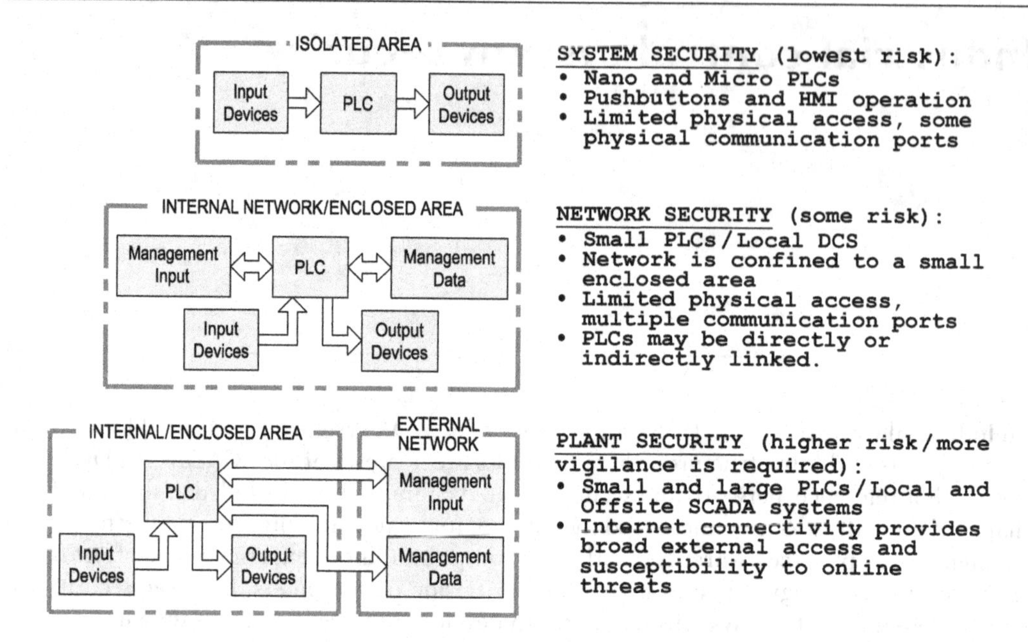

Figure 19.1 General control architectures and associated risk levels. Additional security risks come with the addition of internet connections and more extensive networking.

- create additional backdoors, using a remote access Trojan or RAT, to permit an easy means of regaining access if the initial backdoor portal is closed to the hacker,
- run executable files that launch the "payload" portion of the malware, such as to wipe data throughout all vulnerable nodes of the ICS, often at a coordinated date and time to simultaneously disrupt the working process of the ICS. For ICS systems, the payloads typically disrupt data transmission and communications systems associated with SCADA and HMI devices.
- wipe crucial system data, such as registry keys, in a manner that makes the system unbootable, requiring system recovery.

The National Institute of Standards and Technology publishes the NIST Special Publication 800-82 titled, *Guide to Industrial Control (ICS) Security*, and the International Society of Automation publishes a series of standards under ISA99 and ISA/IEC 62443 for cyber security of industrial automation and control systems. These documents provide guidance for securing ICS systems, including SCADA, DCS, and PLCs, and it addresses ICS-related performance, reliability, and safety requirements. These publications provide overviews of typical ICS topologies, types of threats and vulnerabilities, and recommendations for security countermeasures to mitigate the associated risks.

The basic operation of the control-loop portion of an ICS is shown in Figure 19.2, where the controller may include safety/shutdown systems that can be hacked through remote access via modems or the internet. Many ICS systems contain multiple control loops, human interfaces, local physical communication ports, SCADA remote connections, and internet-based diagnostics and maintenance tools, with various network layering and protocols. When networking, the control-system design and security decisions are based on the following factors,

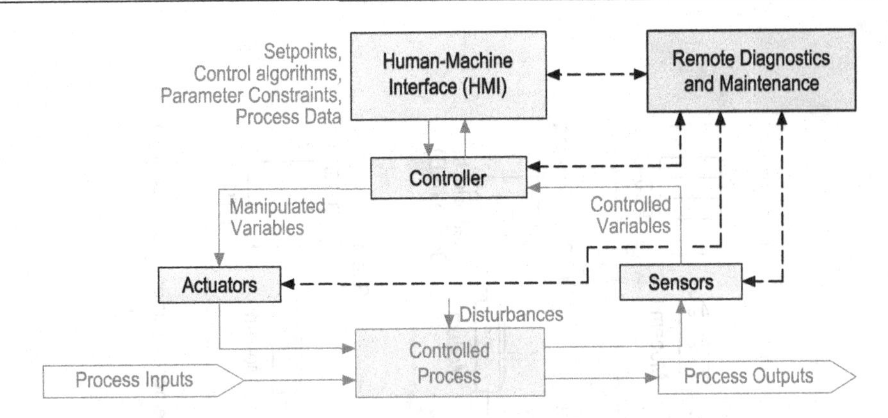

Figure 19.2 Electronic control system operation. The dashed lines show possible vulnerability points that may exist to accomplish remote diagnostics, or for routine software or firmware maintenance.

where higher security risks exist with more-complicated control architectures and larger numbers of external access points:

- *Control timing*/speed/synchronization requirements
- *Geographic distribution*, e.g., a small isolated plant compared to a large electric-utility network
- *Hierarchy*, supervisory control may include data from multiple locations to support a control decision, often by a human observing the entire system
- *Control complexity*, where control can be from a single pre-programmed algorithm or may be a composite of many complex systems
- *Availability/Reliability*, which is a function of the system importance and may require fail-safe designs, backup, and redundancy
- *Physical interaction* between plant operators and equipment
- *Managed support* between computers/control systems and software engineers
- *Communications* between devices, controllers, plant operators, and management personnel
- *Networking architecture security requirements*
- *Impact of failure*, where the seriousness of a failure is considered in determining the system redundancy and security
- *Safety*, where programming can detect unsafe conditions or limits and take actions to avoid damage or service interruptions

SCADA SYSTEM

SCADA systems are designed to spread out control assets around a central-control location. They integrate setpoint and control functions, coupled with data acquisition, transmission, and historical archiving. Generic hardware arrangements are shown in Figure 19.3, consisting of a central control center, a control server, communications equipment, all coupled to one or more remote sites containing local RTUs and PLCs, which obtain data and produce control signals. Some state-of-the-art Intelligent Electronic Devices (IED), such as smart sensors and smart actuators, can communicate directly with an RTU or SCADA system, thus providing more communication portals and increasing security exposure, especially if done wirelessly. Various SCADA connectivity methods include point-to-point, series, series-star, and multi-drop topologies, as shown in Figure 19.4. Other SCADA topologies are shown in the NIST Guide to Industrial Security (ICS) document. The topology helps to define how attacks can be propagated and the steps that may prevent and attack from being successful.

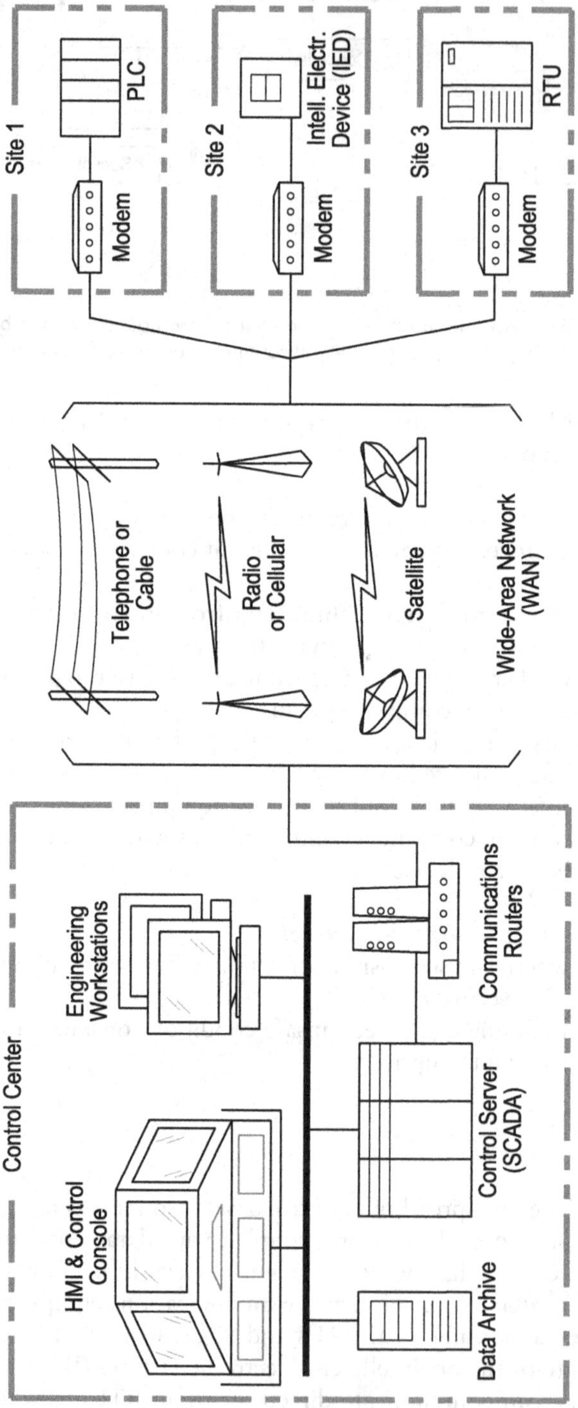

Figure 19.3 Central Control and SCADA system general arrangement using various modes of communications to remote locations, connecting to many diverse types of equipment. These links may provide security weaknesses.

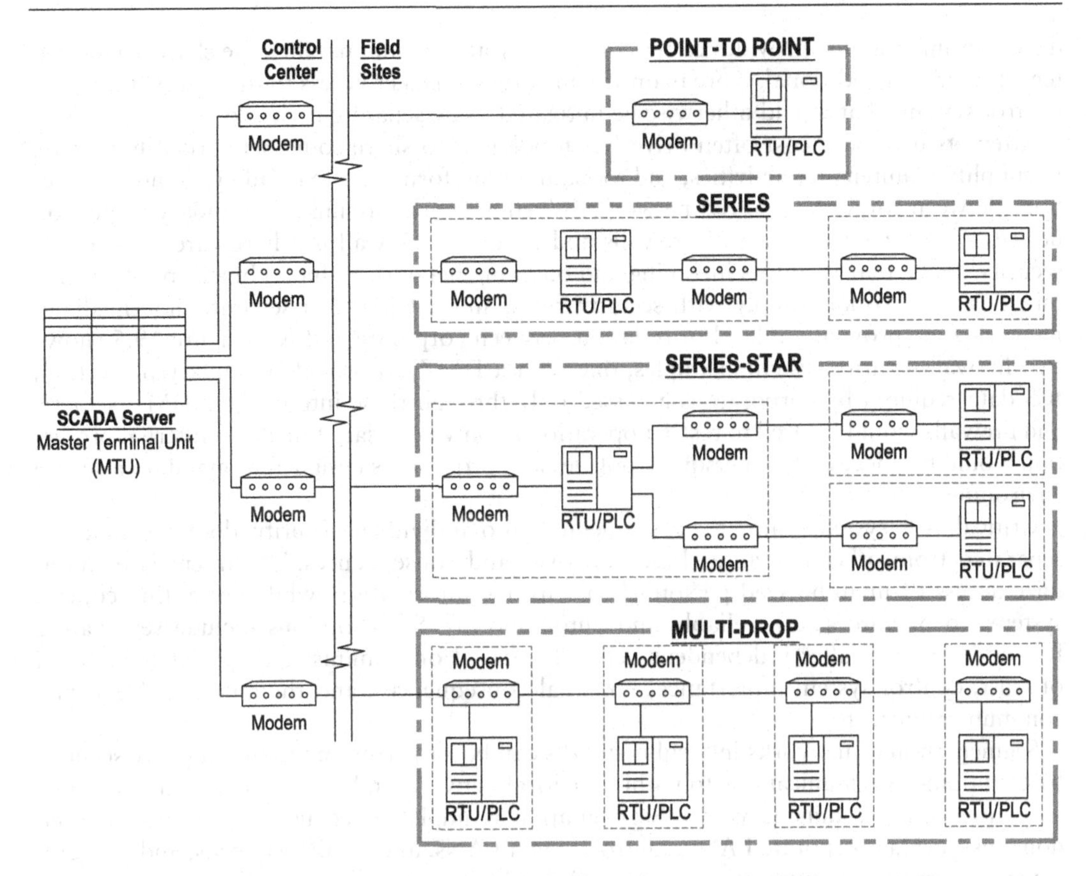

Figure 19.4 SCADA communication topologies; Point-to-Point, Series, Series-Start, and Multi-Drop.

Figure 19.3 shows wide area network (WAN) links that are not limited to internet connections, but can be hacked through the air from either radio, cellular, or satellite connections. Radio connections can be made when a system is controlled from a long distance and internet connections are not available. One example is signal and control transmissions made through radio broadcasts to and from a water-pumping station located at a remote reservoir designed to fill a level-controlled gravity storage tank many miles away; and another example is remote monitoring of a ship's propulsion plant through near-real-time satellite transmissions. The connected devices at the sites can be intelligent electronic devices, PLCs, or RTUs that are connected to more extensive computer-based control equipment.

ICS NETWORKING ARCHITECTURE SECURITY

For Industrial Control System hardware and software setups, also called deployments, it is good practice to separate the control ICS computer systems and corporate information-management networks, so IT security and performance problems on the corporate network do not affect controls on the ICS network. Generally, the networked information is different between the two systems. For example, corporate networking includes internet access, FTP, email, etc., functions which should be avoided on an ICS network, whereas ICS systems carry control signals and data for archiving. While ICS information carried on a shared corporate data transmission line provides easier connectivity and maintainability, the shared-data systems provide an additional, more mainstream path for hacker access to the automation-and-control systems that would otherwise be mitigated using separate networks.

Attempts to reduce cost often drive ICS topologies to share some networked hardware, to simplify maintenance activities, and to maintain uniformity, often leading to interconnections between corporate and ICS networks. When networks are linked, boundary protection devices should be used, such as firewalls and DMZs. A firewall is a hardware or software security device that monitors incoming and outgoing network traffic to block specific transmissions that do not comply with security rules, and a DMZ is a separate, intermediate-network zone providing a level of isolation between corporate and ICS. Figure 19.5 shows two firewall-security system examples, one with a DMZ, to provide boundary protection. ICS data required by corporate is accessed only through these intermediate DMZ servers, and firewalls would be used to restrict operation to only necessary functions. Internal protection would be addressed by personnel policies and procedures designed to maintain a secure environment.

Critical ICS systems should be segmented into individual security domains that are separated from other networks, based on risks and consequences. The intent is to minimize access of unauthorized personnel to sensitive information, while permitting control systems to operate safely, reliably and autonomously. Some options include segregating ICS nodes into smaller independent networks, based on administrator permissions, level of trust, control system importance, potential consequences, and the amount of external communications, etc.

Segmentation of networks into subnetworks can boost performance and improve security by containing communications to within the local network and hiding the internal network structure from the outside world. Segmentation is implemented using gateways between domains, such as corporate LANs, control-system LANs, and DMZ networks, and to segregate unnecessary communications from riskier internet connections. Segmentation is accomplished via a firewall's network-control rulesets that define IP addresses and the types of communicated information permitted past the boundary. The rules are typically established depending on the data source and destination addresses, and the type of data being transmitted. Some techniques that are used include:

- Physical separation of networks having no communications between domains.
- Logical separation of networks, where data encryption makes hacking essentially impossible.
  - Virtual Local Area Networks (VLANs)
  - Encrypted Virtual Private Networks (VPNs)
  - Unidirectional gateways, preventing reverse flow of data from being received into the ICS.
- Network traffic filtering that enforces security rules
  - Network layer filters that restrict communication to defined IP addresses
  - State-based filtering that limits data transmission based on the function or present state of operation
  - Port or protocol filters that limit the types of functions that can be executed
  - Filters that are built into applications, such as firewalls, proxies, and content.

The following suggestions provide depth of protection:

- segmentation and separation should be used at multiple levels, such as at the layers between data links and applications

- permissions should be limited on a need-to-use basis, and communications between networks should be prohibited, unless necessary
- systems should be segregated based on security requirements
- "whitelisting" of IP addresses that permit only a few known sites to communicate to an ICS, rather than trying to blacklist many sites in a deny-all, permit-by-exception manner.

Boundary protection is obtained through securing:

- gateways,
- routers,
- firewalls,
- routers having firewalls,
- unidirectional gateways,
- managed interfaces,
- encrypted tunnels,
- code that analyzes for malicious code,
- intrusion-detection systems,
- segregated systems using DMZ zones,
- examination of data packets before transfer,
- deny-all/permit by exception communication restrictions,
- use of proxy servers,
- passive monitoring of communication traffic for abnormal data,
- concealment of network and component addresses from automatic discovery,
- disabling plug-and-play,
- disabling of remote troubleshooting services,
- configurations of boundary devices so they fail in a safe state, and other methods.

Firewalls are system devices or software that control the flow of data between networks having different security schemes involving internet connectivity using User Datagram Protocol (UDP) or Internet Protocol (IP) communication formats. Firewalls are used in internet connections, but can also be applied to internal network connections.

The three classes of firewalls are packet-filtering firewalls, stateful-inspection firewalls, and application-proxy firewalls. These filters work on the seven-layer Open System Interconnection (OSI) model used for data transmission, as discussed in Chapter 18: "PLC Cabling, Data Transmission, and Networking." Packet-filtering is the most basic type of firewalling, and it is a routing device that adds access restrictions to some addresses using rulesets at the network layer (layer 3). The rulesets compare IP addresses to permission criteria, where the packet can be forwarded, dropped, or a message returned to the sender. Stateful-inspection firewalls filter packets and evaluate the transport layer (layer 4) to determine legitimacy of the packet contents. This firewall either forwards or drops the transmission. Application-proxy firewalls examine the packets at the application layer (layer 7) and filter the data based on browsers or protocols, such as FTP or TCP/IP. Application-proxy firewalls are very effective in preventing hacks.

In addition to firewalls, routers can be used for a more secure system. The router is installed before the firewall, and it also provides basic packet-filtering services. This arrangement is popular with internet-connected firewalls, as the faster router can remove the packet-filtering burden from the slower firewall. This arrangement is good for preventing denial-of-service attacks and it impedes hackers who then need to bypass two security

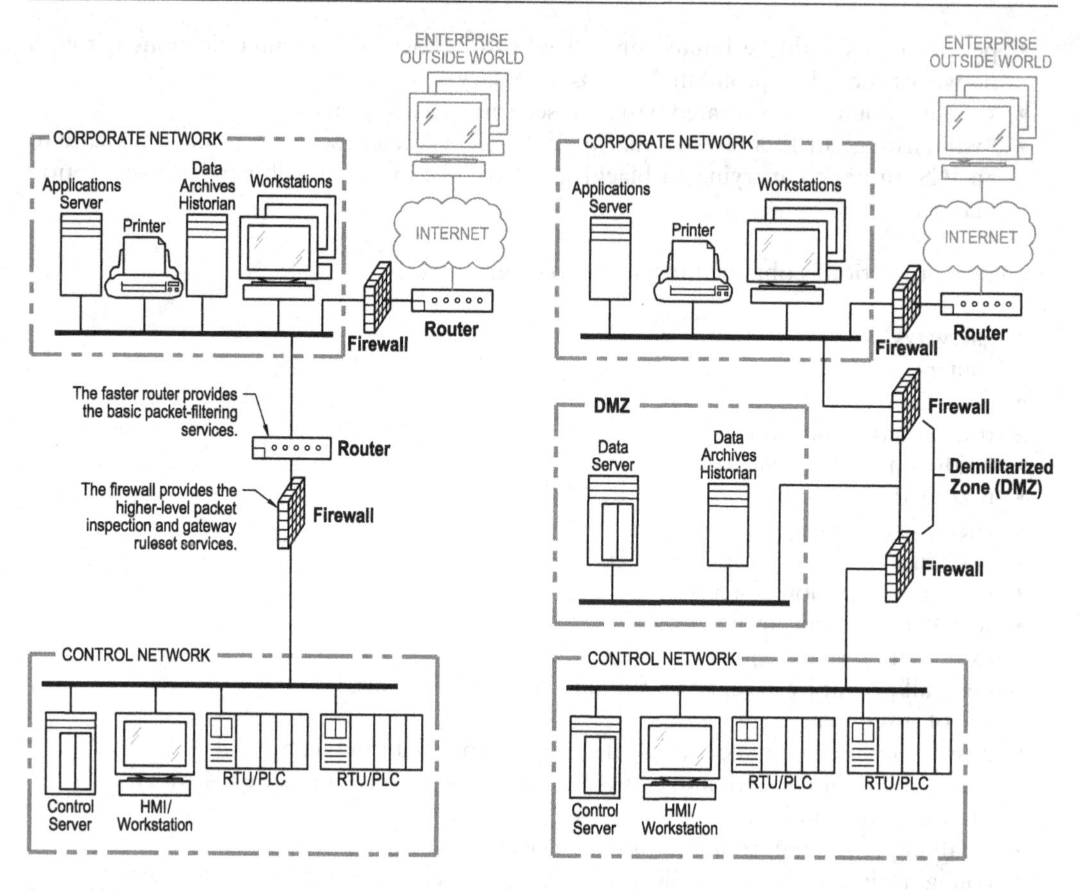

Figure 19.5 A router and firewall isolate the corporate and ICS networks (left). For better security, the corporate and ICS systems are isolated by an intermediate DMZ network, which prevents direct contact between the ICS and the outside world.

systems. Figure 19.5 shows two of many network architectures, designed to provide secure ICS systems, including one with an intermediate DMZ between a corporate network and an automaton network. The left illustration shows a corporate network that is isolated from the ICS simply using a router and firewall, while the right illustration includes an intermediate DMZ to better isolate the critical control system. One firewall protects the DMZ from incoming communications received through a router or modem linked to the corporate network, and the other firewall protects the control network from communications sent by the DMZ. To produce a more secure ICS network, the firewall ruleset should allow only the data communications to the ICS that are solicited by the control network devices. In this manner, the DMZ and corporate networks cannot initiate direct connection to the control network. Network isolation also allows each network to use its own anti-virus package, where a control network malware package may identify a virus that was missed by the corporate virus software.

THREAT TYPES

ICS threats can come from many sources, as shown in Table 19.1.

Table 19.1 Overview of threat types

Threat	*Description*
Human error and sabotage	Negligence and human error that compromises confidentiality or availability of resources. Human error can make it much easier for hackers. Internal or external deliberate sabotage can produce very severe consequences.
Unauthorized use of remote maintenance access	Maintenance access that provides deliberate openings that can be used as a backdoor entry into an ICS.
Online attacks via Office or Enterprise networks	The same vulnerabilities that affect IT computing can be used to provide a path for malicious activity into ICS using these networks.
Attacks against components used in ICS networks	Standard operating systems, application servers, and databases sometimes contain weaknesses or flaws that can be exploited, if they are part of the ICS.
Denial of service	Network connections and resources that can be made inoperable, causing ICS functionality problems and system crashes.
Loss of proprietary or confidential information	Information that can be used to impact competition or produce economic losses.
Malicious code via removable memory devices and code execution	Removable memory devices and mobile IT devices that provide a path for malware, which can then worm its way through a system and release a synchronized attack or provide an unauthorized back door.
Malicious code superimposed into ICS messages and code execution	Many control devices communicate using unprotected plain-text protocols, and it may be possible to insert malicious-code commands into the communications.
Attacks on network components	Network components can provide a path of attack from the middle or general access to the ICS. A good firewall helps.
Technical faults and force majeure	Some problems are beyond the control of the ICS, and a solid safety system can mitigate problems or damage.
Database corruption	Loss or change of information required for analysis or system operation.
Elevation of user privileges	Changing a user's privileges can provide a backdoor into a system at the least, or permit active malicious acts directly into the system.
Buffer overflows	A program anomaly where data being written into a memory buffer fills its allotted space and then overflows into other memory areas, where it can cause data corruption, erratic behavior, system crashes, or provide a link to unlimited access of resources.

THREAT-MITIGATION STRATEGIES

Typically, threat-mitigation strategies take a layered approach. As an example, if a DCS has only local control, there can be no security issues arising from the internet, and personnel may be the sole source of concern. In that case, the only security requirement may be to prevent unauthorized persons from accessing the ICS. Key locking of PLC equipment and the use of password protection will prevent modifications of the system by both negligent and malicious internal sources. ICS security is easier in these isolated systems.

However, when the ICS is connected to the outside, as through the internet, then the system is open to a myriad of threats. These can be mitigated by the physical protection of servers and computers against outsiders, and the addition of virus and malware software. The software needs to be updated often, and detection of attempted attacks can be used for eliminating weaknesses. The addition of strong firewalls avoids unauthorized access to computer resources, and data encryption can make hacking of information nearly impossible. Maintaining an offsite archive of programs and data, including a good backup and restoration system can mitigate downtime and damage.

Possibly the weakest link to network security is the social-engineering aspect, as it relies on human error rather than inherent vulnerabilities in software and operating systems. Psychological manipulation and trickery, such as phishing, spear phishing, scareware, pretexting, and baiting are some techniques for gaining a foothold into a computer or network. Good practices need to be followed by individuals, such as not leaving notes lying around that contain passwords, avoiding opening emails from sources who appear bogus, avoiding tempting too-good-to-be-true offers, not clicking on links attached to emails from unknown entities, not inserting thumb drives found lying around. Positive actions include multifactor login authentication, complex passwords, and keeping antivirus and antimalware software updated.

Physical access

- Place servers in locked areas
- Maintain a corporate security system, such as locked doors and badges
- Provide escorts to outsiders
- Educate employees to avoid unauthored personnel, such as individuals who piggyback physical entry into secure areas

Hardware

- Use key locks on PLCs to prevent tampering
- Use non-rewritable memory.
- Keep access to non-rewritable memory devices behind a lockable door, which is alarmed to indicate access was obtained by someone.

Computer, SCADA, and internet access practices

- *Most important*: Install and maintain the latest versions of virus software
- Do not permit computers to have local administrator privileges, especially user accounts that are set to different levels of access
- Set browsers at the highest security levels that can be tolerated
- Educate users to follow safe browsing habits and email usage, and to avoid social engineering and phishing scams
- Require strong passwords that are changed on a regular basis
- Run computer updates and patches regularly to fix bugs and eliminate vulnerabilities
- Install Intrusion Detection/Prevention Systems (IDS/IPS) to scan network packets for suspicious data, especially in looking for unrecognized signatures. The Intrusion Detection System provides alerts to system administrators, and the Intrusion Protection System adds rules to the network firewall to block unrecognized system access.
- Provide checksumming, such as parity checks, to verify that data transmission or data storage is error free, especially after software is installed. Checksumming and parity checks are techniques used for verifying data integrity during communications, which are used to re-initiate retransmission of data that contained errors. In some schemes, checksumming can be used for both data-transmission error-detection and correction. Checksum variations can also indicate that a hacker has altered a file, when retransmission cannot correct it.

General PLC programming and operation practices

- Keep programming as simple as possible, so that malicious code stands out.
- Always verify programs before downloading them to the PLC. Verification may be required by default by the software to ensure that errors are not loaded into the PLC, but verification may also identify malicious code through syntax errors.
- Provide program logic checks for unsafe states for equipment, PLC, or I/O before the program is permitted to continue
- Carefully consider scenarios in which some hacker-induced problems would be more damaging than others. It may be possible to add programming checks that examine for significant problems. For example, if a hacker were to disable high-pressure limits and alarms on a boiler and set its burners to the highest firing rate, the malicious action would eventually lift all safety-relief valves, and by design, the boiler would never reach destructive pressure levels. However, if a hacker's actions were to close a feedwater valve with all boiler tripping limits and alarms disabled, the boiler would eventually run dry and damage would be extensive. A workaround where damage could be catastrophic might be to use a safety system that is separate and independent from the SCADA/PLC system.
- Use comparator instructions on analog devices to determine and anticipate when critical safety routines should be initiated. The analog device can be used as a cross-check and backup for the primary safety device. As an example, the analog level transmitter can provide backup indication independently from the hardwired boiler-drum low-level alarm and a low-low level cut-out switch.
- Consider programming using pages, functions, or subroutines that are called by the main program. The pages should provide an intuitive function, and should be kept simple, so malicious code might stand out.
- Provide recovery-state conditions for reinitializing the program, error-checking, or user input that reinitializes the program for a fresh start.
- Provide recovery logic in the event of PLC crashes. This programming can be used for program restarting, bypassing bad code, and for reinstalling and rebooting a backup version of the program.
- For ladder diagrams, provide validation conditions to important rungs, so that only expected data types are accepted.
- Sparingly use latching instructions, which retain logic states.
- Program an always-true bit and an always-untrue bit in the program memory. These bits can be used to verify agreement between the program code and the proper output of a ladder rung. The always-true and always-untrue bits are useful as a troubleshooting tool, but can also be used to impede hackers from modifying code that can produce critical problems.
- Provide code that ensures that only data transfer protocols such as DNP3 and Modbus send signal information, and that transmissions do not contain commands that can alter code or directly modify another device's memory addresses.
- Keep PLCs locked in RUN mode to avoid sabotage from persons having insider access to the equipment. Key-lock the PLC to prevent program editing by unauthorized persons.
- When writing computer code, provide detailed comments to all lines of code, function blocks, or ladder rungs. Programmers use comments as memory aids describing the purpose of the code. Comments are good practice for describing program operation, troubleshooting, and for documenting edits, but the practice can be used to identify

Networking and security-related definitions

Access Control List (ACL)	A list of system entities that are permitted to access other resources.
Anti-Virus Tool	Software that detects and isolates malicious code to prevent it from infecting the system.
Application Server	A computer that hosts programs and applications used by workstations.
Attack	An attempt to gain unauthorized access to any part of an ICS.
Authentication	Verification of identity of a user or device before allowing access to other resources.
Authorization	Rights or permissions granted to a user or device.
Backdoor	An undocumented and unauthorized way of gaining access to a computer system. A backdoor is a security risk.
Broadcast	Transmission of information to all resources in a network without any acknowledgement from the receiving devices.
Buffer Overflow	A condition where information floods its allotted memory and then begins to write information into other memory addresses. A buffer overflow has the tendency to crash systems, but can be used to gain control of a system.
Certification	An assessment of the security controls to determine that the controls are implemented and operating correctly, and follow the security requirements.
Clear Text	Information that is not encrypted.
Communication Router	A communication device that transfers information between two networks, such as connecting a WANs to a LANs, or for connecting MTUs and RTUs for data transmission for a SCADA.
Configuration Control	System setup defining interconnections, locations, communication protocols, etc.
Control Center	The equipment from where processes are measured, monitored, and controlled.
Control Network	A network of connected equipment that controls physical processes. Control networks can be subdivided into zones, and there can be multiple control networks within an enterprise or site.
Control Server	A controller that also acts as a server to host control software that communicates over an ICS network with lower-level resources, such as RTUs or PLCs. A Control Server may also be called a SCADA, MTU, or supervisory controller.
Controller	A device that automatically regulates a controlled variable.
Cycle Time	The time to complete one control loop, from reading inputs, executing the control program, and updating the outputs.
Database	A repository for information.
Data Diode	A resource that allows unidirectional flow of data.
Data Historian / Archive	A centralized database that keeps information for trending or analysis.
Demilitarized Zone (DMZ)	A neutral interface location for a host computer/controller between two firewalls that separate both internal and external networks.
Denial of Service (DoS)	Prevention of authorized access to a system resource, or the delay of system operations.
Diagnostics	Information regarding a failure mode and its characteristics, used to troubleshoot the cause of a failure.
Disaster Recovery Plan (DRP)	A plan for processing applications in the event of a major hardware or software failure.

Networking and security-related definitions (Cont.)

Distributed Control System (DCS)	A system where physical control is accomplished by local, independent controllers that report to a central computer for data acquisition, archiving or setpoint adjustment.
Domain	An environment containing a set of resources and a set of system entities having the rights to access only the resources defined by the security policy.
Domain Controller	The server responsible for managing domain information, such as identification and passwords.
Encryption	Cryptographic transformation of data to conceal the original data. Decryption restores the data back to its plain-text state.
Enterprise	An organization that coordinates the operation of multiple processing sites.
Extensible Markup Language (XML)	A programming syntax to mark data with human-readable tags, enabling transmission, validation, and interpretation of data between the application and organizations.
Fault Tolerant	Capability to provide continued operation of a control function in the presence of a hardware or software problem.
Fieldbus	A digital, serial, multi-drop, bi-directional data bus or communication path between low-level field devices, such as sensors, actuators, and their controllers. Fieldbus eliminates the need for star-type point-to-point wiring between the controller and each device. The fieldbus protocol defines the addressing and messaging to the individual devices.
Field Device	Equipment, such as RTUs, PLCs, HMIs, sensors, actuators, etc., that are installed on the physical side of an ICS.
File Transfer Protocol (FTP)	An internet standard for transferring files over the internet. FTP protocols are used for uploading and downloading web pages, graphics, and files between local media and a remote server.
Firewall	An internetwork gateway that restricts data flow to and from one network inside the firewall, and another network outside the firewall. Typically, the firewall is configured with a series of rules that permit forwarding the rule-abiding data packets or the prohibition of suspicious data-packets.
Human-Machine Interface (HMI)	A controller screen that permits a plant operator to monitor, control, and adjust PLC-based machines.
Incident	An occurrence that jeopardizes information or the control process.
Industrial Control System (ICS)	General term that encompasses the SCADAs, DCSs, RTUs, PLCs, modems, routers, switches, and field devices in an automated control system.
Insider	A person within the security perimeter with authorized access that is used in an unapproved manner.
Intelligent Electronic Device (IED)	An electronic sensor or actuator having a processor that can communicate directly with a network, such as for sending status or troubleshooting information.
Intrusion Detection System (IDS)	A security service that monitors and analyzes network or system events to find unauthorized attempts to access the system.
Intrusion Prevention System (IPS)	A system that can detect and prevent unauthorized attempts to access the system before it reaches its target.
Jitter	A time or phase difference between a data signal and a clock for determining if extra data has been piggybacked.
Key Logger	A program that records key strokes to obtain passwords or encryption keys.

(continued)

Networking and security-related definitions (Cont.)

Local Area Network (LAN)	A group of computers or resources connected in a limited environment by a communications link that enables device interaction.
Malware	Software or firmware that performs an unauthorized process. Malware can range from a virus, worm, Trojan horse, to spyware, ransomware, and adware.
Master Terminal Unit (MTU)	A controller that also acts as a server to host control software that communicates over an ICS network with lower-level resources, such as RTUs or PLCs. An MTU may also be called a SCADA, Control Server, or supervisory controller.
Modem	A device used to convert serial data into a transmission signal over phone or cable lines, and then to reconvert it back into the original serial data at the receiver.
Network Interface Card (NIC)	A circuit card that creates transmission signals following a specified protocol and permits a computer to connect to a network.
Object Linking and Embedding (OLE or OPC for Process Control)	A set of standards to promote interoperability between disparate field devices and a computerized automation and control system.
Operational Controls	The security controls or safeguards in a system that are primarily implemented by people, rather than machines.
Phishing	Techniques to trick people into unwittingly disclosing sensitive information, usually by pretending to be a trustworthy entity.
Port	A connection point into or from a computer.
Protocol	A set of communication rules, formatting, and procedures that defines the transmission of data between sending and receiving devices.
Proxy Server	A server that acts as an intermediary between clients and other servers.
Ransomware	A virus that encrypts a user's data files with the promise of providing a decryption program upon the payment of a ransom, typically through untraceable bitcoins. This malware can require the complete reformatting of a hard drive and reinstallation of all software and user files from a backup.
Redundant Control Server	A backup to the control server that always maintains the current state of the control server, and is ready to take control upon failure of the primary server.
Remote Access	Access by authorized users from outside the security perimeter.
Remote Access Point	The a mobile or laptop device located outside the security perimeter used to access an ICS
Remote Maintenance	Maintenance done by individuals from outside the security perimeter.
Remote Terminal Unit (RTU)	RTUs are devices equipped with network capabilities to communicate between field devices and SCADA/DCS systems.
Resource Starvation	A condition where a computer process cannot be supported due to the unavailability of computer resources, which can be a hardware issue or when multiple processes are competing for the same computer resources.
Risk	The level of impact that a threat or attack might have on the organization and its operations.
Risk Management Framework (RMF)	A disciplined approach that combines information security and risk management into an ICS security system development.
Router	A computer that is a gateway between two networks at the OSI layer 3 level. A router relays and directs data packets through the LAN. The most common form of router data transmission is via internet protocol (IP)

Networking and security-related definitions (Cont.)

Safety Instrumented System (SIS)	A system of sensors, logic, and final control elements designed to take a process to a safe state when predetermined conditions are violated. The safety instrumented system is also called the emergency shutdown system (ESS), safety shutdown system (SSD), and safety interlock system (SIS).
SCADA Server	The server that serves as the master in a SCADA system.
Security Audit Security Controls Security Plan	The security plan provides documented security requirements and security controls, i.e., the safeguards and countermeasures to meet security requirements. A security audit is a review and examination by an independent entity to ensure that the controls meet the plan objectives.
Simple Network Management Protocol (SNMP)	A standard TCP/IP protocol for network management. The SNMP is used to monitor and map network availability, performance, and error rates. SNMP - network devices use distributed data storage called a Management Information Base (MIB), which provides static and dynamic attributes about the device.
Single-Loop Controller	A very simple and dedicated process controller.
Social Engineering	Attempts to obtain sensitive information by trickery.
Spyware	Malicious software that gathers and transmits information about individuals or organizations without their knowledge.
Supervisory Control and Data Acquisition (SCADA)	A central computer that gathers and processes data from remote PLC-based controllers. The SCADA can also be used to archive data, to change setpoints, and remotely start equipment.
Threat	An event that has the protentional to impact an ICS operation.
Transmission Control Protocol (TCP)	TCP is one of the main protocols used in internet-protocol data transmission (TCP/IP) TCP enables two hosts for the connection and exchange of data, while IP arranges the data packets that carry the data. TCP guarantees the delivery of data and guarantees that the packets are delivered in the same order in which they were sent.
Trojan Horse	A computer program that appears to have a worthwhile function, but it contains malicious code that is released upon its execution.
Unauthorized Access	A person or computer that gains access to a computer or network without permission, typically with malicious intent.
Unidirectional Gateway	Combination of hardware and software that permits the transmission of network data in one direction only, but cannot receive any information back from the receiver. The software is used to backup databases and to emulate protocol servers and devices.
Virtual Private Network (VPN)	A restricted-use, logical computer network that is uses a public network, such as the internet, to create a remote encrypted connection by "tunneling" a virtual network link through the real network.
Virus	A hidden, self-replicating section of computer software, typically having malicious intent that propagates by infecting other programs and other computers. A virus cannot run by itself, but must be activated through its host program.
Vulnerability	A weakness in the physical system, operating system, communication or programming that can permit a potential exploit by a threat source.
Wide Area Network (WAN)	A physical and logical network that provides data communication to a larger number of independent users, beyond the confines of a LAN. A WAN can make use of the internet for broad distribution of data.
Worm	A computer program that can run independently and propagates a copy of itself onto other computers or network hosts. This program may consume computer resources, or it may be programmed to run malicious software in a synchronized manner with other computers.

hacked programs by looking for a lack of comments, stylistic or idiomatic changes, or poorly written comments.

- Create jump subroutines to safe-shutdown program sequences, when the code reveals unsafe inputs, outputs, or bad data.
- Produce operation charts that can be used to verify the program logic based on I/O values. These charts can be used during first-article proof testing, commissioning, program verification after modifications have been made, and troubleshooting.
- To help make malicious code more obvious, use conventions for naming, abbreviations, tags, data typing, commenting, symbols, messaging, etc., and commenting nomenclature, acronyms, and variable names that are easily understood by the programmer and users, but do not disclose equipment details to outsiders.
- To make malicious code more obvious and if supported by the program editor, use standard colors and fonts between code, comments, etc., and keep the program style and appearance consistent. Formatting can be useful to help identify code that does not belong.
- Use a library of well-annotated and well-tested functions or subroutines during programming. This usage provides a means for checking that only vetted functions are contained in a program.
- Provide separately derived control, alarm, and safety-shutdown functions to make a hacker's damage more difficult to accomplish.
- Where possible, use the best-quality programming editor. For example, the A-B RSLogix 5000 provides higher-level functions than the RSLogix 500 version, plus it includes all the five IEC 61161–3 languages. By using multiple languages in a single program, the strengths of each can be used to leverage its power and efficiency to the task at hand.

QUESTIONS

1. What is meant by Industrial Control System Security (ICS)?
2. Describe the difference between system security, network security, and plant security.
3. What are some of the ways that security systems are breached?
4. What are some consequences that can result from security system breaches?
5. Describe two firewall-security systems.
6. Describe how routers can enhance security. What is a demilitarized zone (DMZ)?
7. List some methods for mitigation of ICS threats.
8. List some methods of mitigating human-error-caused ICS breaches.

Chapter 20

PLC terms and definitions

PLC-related definitions are separated into the following categories:

- Programmable Logic Controller
- PLC Input/Output
- PLC Programming
- PLC Data Transfer

PROGRAMMABLE LOGIC CONTROLLERS AND COMPUTERS

Address

A storage location in the PLC's memory to store I/O status, program mathematical variables, or calculated values. Addresses are accessed as numbered locations in the program formed by the concatenation of the type of data and bit or word location within the type.

Binary coded decimal (BCD)

A method used to express the numerals 0-thru-9 from the base 10 system as a machine-language binary value in the base 2 system.

Bit

A single binary digit of stored data. It can only take digital values of 0 or 1.

Byte

Eight single bits combined into a longer data-storage unit. Bytes may be combined into longer "words" for moving data more efficiently.

Central processing unit (CPU)

An integrated circuit chip (IC) that reads data, executes program instructions, and then returns data.

Electrically erasable programmable read-only memory (EEPROM)

Same as EPROM but can be erased electrically.

Erasable programmable read-only memory (EPROM)

Program-storage memory that can be erased by the programmer and reloaded with a new program many times as read-only memory (ROM). EPROM is nonvolatile memory and will not lose data during an electrical-power loss.

Image register/image table

A dedicated memory location reserved for I/O bit status. The image register columns hold values of 0 or 1 that are addressed/organized into words forming rows of data.

Index register

The effective address of any PLC information, consisting of a base, index, and relative address.

Millisecond

One thousandth of a second (1/1000 sec, 0.001 sec).

Operator interface

A device that allows a system operator to access a PLC and its I/O monitoring and control functions.

Random-access memory (RAM)

Memory where data can be accessed at any address without having to read a number of sequential addresses. Data can be read from and written to storage locations. RAM is volatile, meaning a loss of power will cause the contents in the RAM to be lost.

Read-only memory (ROM)

Memory from which data can be read but not written. ROMs are often used to keep programs or data from being destroyed due to user intervention. ROM is used to execute the bootup sequence, check for the operation of attached devices, and load the controller program into the CPU.

Register

A temporary holding place for data in the microprocessor's CPU. A register may hold instructions, addresses, or any functional data from I/O or calculated values. A register must be large enough to hold an instruction. As an example, a 64-bit machine has registers holding 64 bits or 8 words of length.

Shift register

A first-in/first-out (FIFO) memory circuit where bit enters the register from one end and pushes all other bits to shift in the other direction. The last bit on the end disappears during the shift operation. Shift registers are commonly used in serial-to-parallel data-transmission converters.

Word

A unit of bit length that is addressed and moved between memory storage and the CPU. There are eight bits in a byte, and a word consists of one or more byte, depending on the computer.

PLC INPUT/OUTPUT

Analog input (A/I)

A varying signal supplying real-time process-strength information from a sensor to the analog-input module.

Analog output (A/O)

A varying output-adjustment signal producing real-time process-change information from the analog output module, ultimately directed to a final-control device.

Analog-to-digital converter (A/D)

A module that transforms an analog signal into a digital word.

Digital-to-analog converter (D/A)

A module that converts a PLC-output digital word into a varying-strength analog signal for transmission to a final-control device.

Distributed control system (DCS)

A PLC arrangement were individual PLCs are installed to locally control individual systems or equipment, independently from a central control system. Loss of functionality of a central control system does not cause interruption of the individual PLCs that are distributed throughout the plant.

Input module

Receives and processes digital or analog signals from field sensors.

I/O points

Terminal connections on I/O modules that are wired between the field sensors/control devices and the PLC's input and output modules.

Output module

Produces digital or analog output signals directed to final-control devices.

SCADA (supervisory control and data acquisition)

A control system architecture that uses computers networked into a data communications arrangement with PLCs distributed throughout the plant. The SCADA provides a high-level supervisory computer interface for monitoring plant operation, typically having graphical-user interfaces (GUI) for easy visualization and user friendliness. Often, the SCADA is programmed to provide remote start/stop capability of equipment. It provides a central source for plantwide data acquisition, real-time performance assessments, and a central repository for historical data archiving for subsequent plant evaluations or troubleshooting.

Transducer

A device that measures an electrical parameter and then converts it into an electronic analog signal for transmission into an analog-input device.

PLC PROGRAMMING

Expression

A simple function that returns a value, such as *A=pi()*d^2/4* returns the area of a circle. Expressions and functions can include constants, such as *pi*; variables, such as *diameter*; and operators, such as multiply and divide.

Function

A part of a program that carries out a certain task, such as a complex mathematical operation. It is a type of routine, which returns a value. The terms function, procedure, routine, and subroutine are somewhat synonymous, and sometimes used interchangeably despite nuances. In general, a function can be called from a statement, as if it is a command.

Instruction

A command placed by a PLC program into the CPU's processor. The instruction is the basic building block for a control program, and may deal with input, output, comparisons and decision-making processes, calculations, etc.

Procedure

An ordered set of statements or instructions that executes commands or performs a specific task. A procedure can call functions to return values, but not vice versa.

Program

One or more computer-based instructions or statements that carry out a task.

Programming device

A unit, such as a laptop or hand-held device, that transfers programs to or from a PLC to create PLC-resident instructions.

Routine

A group of instructions that form functions or procedures.

Software

One or more programs that control a process.

Statement

This is the smallest standalone element of code for an action to be carried out to form the sequence of a program. Statements can be simple lines of code, or more complex actions, such as *if-then-else* or *do-while* statements. Typically, a statement does not return a value, or then it would be more accurately called a function.

Subroutine

Shorter groups of instructions that may be called by many routines, when needed, to avoid repetitive programming. Subroutines can be thought of as user-defined commands or instructions. Typically, variables are passed into and from subroutines.

Syntax

The rules to follow for correctly using computer commands or instructions.

PLC DATA TRANSFER

Bandwidth

The amount of data that can be simultaneously carried by a cable. Bandwidth is a function of the speed of transmission through the media, such as copper versus fiber optics, and the number of signal-carrying frequencies available for data transmission.

Baud rate

The number of symbols per second that is either transmitted or received in serial devices, where a symbol may have one or more bits. Baud is the speed of serial data transmission. The baud rate must be less than or equal to the rated speed of digital transmission that can be accepted by both the send and receive devices. Serial connections use RS-232 or RS-485 serial interfaces, cables, and connectors.

Collision detection

Used in multi-mode data transfer. Each node with information to send waits until there is no traffic on the network before it jumps in. Once a node is transmitting, the collision-detection circuitry looks for other nodes that may errantly begin to transmit. If a collision of data occurs, each transmission is stopped, and the nodes wait before making a new attempt. This detection method is sometimes referred to as CSMA/CD for Carrier Sense Multiple Access with Collision Detection.

Ethernet

A Local Area Network (LAN) communications linking protocol used with TCP/IP transmission layers and Category 5 (CAT 5) or Category 6 (CAT 6) cabling and connectors.

Ethernet uses packet-and-frame data transmission, where the packet affixes a device IP address and demarks where the frame begins and ends. The frame contains the "payload" of transmitted information and the local MAC address in the device. The TCP/IP protocol remaps a dynamic IP address to the device's static MAC address. Ethernet cable consists of four unshielded twisted pairs of conductors bundled in a single jacket, or fiber-optic cables and converter modules. Ethernet topology uses a star arrangement from an Ethernet hub, router, or switch; and communication is directed only to the device having the current IP address. An Ethernet device checks to see if other devices are transmitting, and if data collisions may occur, the Ethernet device will delay transmission before retrying. This protocol uses data parity checking for error detection and correction and is extremely reliable. The PLC data transmission alternative to Ethernet is Modbus/RS485 serial communication.

Master-slave protocol (client-server)

Method of data collisions avoidance where one device is the master, and all other nodes are the slaves. The master sends data requests or commands, while the slaves respond to queries or take actions based on commands. Conflicts are avoided, as no slave can speak unless granted permission by the master. Any exchange of information between nodes goes through the master, as contrasted with token-ring protocol.

Media-access control (MAC)

A unique identifier assigned to network devices for receiving communication data, used by Ethernet and Wi-Fi technologies. The MAC address is sometimes referred to as the Ethernet Hardware Address (EHA) or the physical address. MAC addresses are typically assigned by the manufacturer or programmer of the PLC system. Addresses can be universally or locally administered. A universally assigned MAC address is uniquely assigned and burned-in by the manufacturer, while a locally administered address is assigned by a network administrator or PLC programmer and overrides the burned-in address. For internet connection, a correspondence table translates the IP communication address to the device's physical address through the ARP (Address Resolution Protocol).

Modbus protocol

An open-ended, i.e., free from royalties, communication protocol developed by Modicon systems for transmitting PLC information over serial lines to up to 247 devices. Modbus uses RS485 transmission protocol and cables. The device requesting information is the master, and the device supplying the information is the slave. Information transmission is initiated by the master, and the information is broadcast serially through each device; however, only the addressed device will act on the information. Although the Modbus protocol was developed in 1979, developed versions of the communication method are so robust over long distances and easy to use that Modbus/RS485 is very often used in modern PLC communications. Some other alternatives are Profibus and Ethernet/IP.

Modem

Acronym for modulator/demodulator: a device that mixes signals (modulates) into an analog wave for transmission over telephone or higher bandwidth cable lines, and separates (demodulates) signals back into digital form for microprocessor use.

Network

Interconnection of PLCs as nodes to allow sharing of resources or information. Network connections can be made using cabling, such as CAT5 or CAT6 wiring, or using wireless connections.

Nodes

Devices that route data or form data termination locations, such as switches/modems, PLCs, HMIs, SCADAs, etc.

Packet

Sometimes called datagram: a unit of data containing "chunks" of information of originating information transmitted through a network, where each packet has the address for the destination node. After all packets are received by at the destination, they are reassembled into the original information.

Parallel communication

Data whose bytes or words are transmitted or received with all their bits present at the same time. Although the data is transmitted serially, parallel cables are used to send multiple amounts of data simultaneously.

Peer-to-peer network

Decentralized communications where each node has the same rights, and any node can initiate communication, as contrasted with master-slave. Peer-to-peer means that each node can act as a client or server at any time.

Polling/Query

Most common access method in master-slave protocols. The master sequentially polls each node (slave) to determine if it has data to transmit. The slave has a fixed time limit for responding before the master polls the next device.

Port

Local address at the transport layer, which identifies the source and destination of a packet. Ports allow TCP/IP to multiplex and demultiplex IP datagrams going simultaneously to different application processes, e.g., Modbus TCP/IP uses port 502 to receive Modbus messages over Ethernet. Well-known port numbers are defined between 0–1,024, user-registered port numbers between 1,024–49,151, and private/dynamic port numbers between 49,152–65,535.

Profibus DP

This is an open-ended networking protocol created in the 1990s for high-speed I/O factory automation applications, where DP refers to "Decentralized Periphery." Profibus uses a bus master to poll slave devices in a multi-drop fashion on an RS485 bus, as contrasted with Modbus.

Protocol

The rules/syntax that define how two or more machines communicate with each other. Protocol includes data-flow control, error checking, procedures to address detected errors and the message direction to a specific device. Protocols are used to avoid conflicts or collisions of data.

RS485 (or TIA-485)

A standard defining the electrical characteristics for drivers and receivers used in serial communications in industrial applications. This communication can be used over long distances, and in electrically noisy environments.

Serial communication

Data are transmitted sent bit by bit along a single cable in the form of words. The word has start and bits for synchronization, parity bits for error checking, and the data itself.

Socket

This is an application layer address formed by combining an IP address with a port number. Sockets are used by application protocols to keep track of the assigned port number when using TCP.

Supervisory control and data acquisition (SCADA)

See the section above titled "PLC INPUT/ OUTPUT."

Switch or switching hub

A multiport PLC networking device that connects multiple devices together for information sharing or for connection to a central supervisory and control computer. A switch receives, processes, and forwards data packets directly to only the addressed destination device(s) without broadcasting the information to all connected devices.

Token ring

This is a communication protocol for local area networks as an alternative to the master-slave arrangement. It uses a special tree-byte frame called a token that is passed around a ring of nodes. A node is only allowed to transmit or get serviced when it has possession of the token.

Transmission control protocol/Internet protocol (TCP/IP)

The computer language used to access the Internet for data connectivity and transmission, including addressing, mapping, and acknowledgement. TCP handles the message, breaking it into smaller units called packets, and IP handles the transmission. TCP/IP is a suite of communication protocols having four layers, where TCP and IP are the two main protocols. The other two layers include the Application and Datalink layers. The Application includes

HTTP (hypertext transfer protocol), SMTP, POP3, etc., for web or email communications. The Datalink layer includes Ethernet or Address-Resolution Protocol (ARP) to translate a "revolving" IP address to a "fixed" MAC or physical address at the node device. TCP/IP is designed to work independently, regardless of any differences in networking hardware, so that it can run using any connection media.

Index

Printed in the United States
by Baker & Taylor Publisher Services